Nanostructured and Advanced Materials for Fuel Cells

T0225421

Advances in Materials Science and Engineering

Series Editor

Sam Zhang

Aerospace Materials Handbook, *edited by Sam Zhang and Dongliang Zhao*

Biological and Biomedical Coatings Handbook: Applications, *edited by Sam Zhang*

Biological and Biomedical Coatings Handbook: Processing and Characterization, *edited by Sam Zhang*

Hydroxyapatite Coatings for Biomedical Applications, *edited by Sam Zhang*

Nanobiomaterials: Development and Applications, *edited by Dong Kee Yi and Georgia C. Papaefthymiou*

Nanostructured and Advanced Materials for Fuel Cells, *edited by San Ping Jiang and Pei Kang Shen*

Micro- and Macromechanical Properties of Materials, *Yichun Zhou, Li Yang, and Yongli Huang*

Nanostructured and Advanced Materials for Fuel Cells

Edited by
San Ping Jiang
Pei Kang Shen

CRC Press
Taylor & Francis Group
Boca Raton London New York

CRC Press is an imprint of the
Taylor & Francis Group, an **informa** business

CRC Press
Taylor & Francis Group
6000 Broken Sound Parkway NW, Suite 300
Boca Raton, FL 33487-2742

First issued in paperback 2017

© 2014 by Taylor & Francis Group, LLC
CRC Press is an imprint of Taylor & Francis Group, an Informa business

No claim to original U.S. Government works

ISBN-13: 978-1-4665-1250-4 (hbk)
ISBN-13: 978-1-138-07657-0 (pbk)

Visit the Taylor & Francis Web site at
http://www.taylorandfrancis.com

and the CRC Press Web site at
http://www.crcpress.com

Contents

Series Statement vii
Preface... ix
Authors xiii
Contributors...xv

1 **Introduction** 1
   San Ping Jiang and Pei Kang Shen

2 **Advanced Electrode Materials for Solid Oxide Fuel Cells** 15
   Ling Zhao and San Ping Jiang

3 **Seals for Planar Solid Oxide Fuel Cells: The State of the Art**............ 45
   Teng Zhang and Shaorong Wang

4 **Advances in Micro Solid Oxide Fuel Cells** 69
   Partha Sarkar and Saeid Amiri

5 **Advances in Direct Carbon Fuel Cells**.. 127
   Cairong Jiang, Andrew C. Chien, and John T. S. Irvine

6 **Advances in Electrode Material Development in Molten
   Carbonate Fuel Cells**.. 163
   Gang Chen

7 **Nanostructured Nitrogen–Carbon–Transition Metal
   Electrocatalysts for PEM Fuel Cell Oxygen Reduction Reaction** 195
   Gang Wu, Zhongwei Chen, and Jiujun Zhang

8 **Advances of the Nanostructured Carbon-Based Catalysts for
   Low-Temperature Fuel Cells** 223
   Fengping Hu and Pei Kang Shen

9 **Atomic Layer Deposition of Metals and Metal Oxides for Fuel
   Cell Applications** 249
   Gaixia Zhang, Shuhui Sun, and Xueliang Sun

10 **Noncarbon Material–Supported Electrocatalysts for Proton
   Exchange Membrane Fuel Cells** 291
   Vladimir Neburchilov and Jiujun Zhang

11 Understanding the Nanostructures in Nafion Membranes 315
    Hongwei Zhang and Pei Kang Shen

12 Advances in Proton Exchange Membranes for Direct Alcohol
    Fuel Cells 335
    Haekyoung Kim and Doo-Hwan Jung

13 Nanostructure Advances in Catalysts for Direct Alcohol Fuel
    Cells 375
    XiaoChun Zhou and Hui Yang

14 Carbon Nanotubes and Nanofibers and Their Supported
    Catalysts for Direct Methanol Fuel Cells 443
    Madhu Sudan Saha and Jiujun Zhang

15 Advanced Alkaline Polymer Electrolytes for Fuel Cell
    Applications 481
    Jing Pan and Lin Zhuang

16 High-Temperature Inorganic Proton Conductors for Proton
    Exchange Membrane Fuel Cells 505
    Haolin Tang, Junrui Li, and Mu Pan

17 Advances in Microbial Fuel Cells 523
    Chang Ming Li and Yan Qiao

Index 553

Series Statement

Materials form the foundation of technologies that govern our everyday life, from housing and household appliances to handheld phones, drug delivery systems, airplanes, and satellites. Development of new and increasingly tailored materials is key to further advancing important applications with the potential to dramatically enhance and enrich our experiences.

The *Advances in Materials Science and Engineering* series by CRC Press/Taylor & Francis is designed to help meet new and exciting challenges in Materials Science and Engineering disciplines. The books and monographs in the series are based on cutting-edge research and development, and thus are up-to-date with new discoveries, new understanding, and new insights in all aspects of materials development, including processing and characterization and applications in metallurgy, bulk or surface engineering, interfaces, thin films, coatings, and composites, just to name a few.

The series aims at delivering an authoritative information source to readers in academia, research institutes, and industry. The Publisher and its Series Editor are fully aware of the importance of Materials Science and Engineering as the foundation for many other disciplines of knowledge. As such, the team is committed to making this series the most comprehensive and accurate literary source to serve the whole materials world and the associated fields.

As Series Editor, I'd like to thank all authors and editors of the books in this series for their noble contributions to the advancement of Materials Science and Engineering and to the advancement of humankind.

Sam Zhang

Preface

A fuel cell is an electrochemical device to directly convert the chemical energy of fuels, such as hydrogen, methane, methanol, ethanol, hydrocarbons, gasified coal gas, carbon, etc., to electricity. A fuel cell is the most efficient and least polluting energy conversion technology and is a potential and viable candidate to moderate the fast-growing increase in power requirements and to minimize the impact of the increased power consumption on the environment. However, fuel cell technology is not new and was invented more than 160 years ago by Sir William Grove. In fact, the fuel cell is one of the oldest electricity-generation technologies. Surprisingly, the technological development of fuel cells has lagged far behind more-well-known energy-conversion technologies, such as steam and internal combustion engines. The reasons can be reduced primarily to materials problems and economic factors.

The last 20 years has seen a quantum leap in the fundamental understanding and development of advanced and nanostructured materials for fuel cells. The objective of this book is therefore to provide an updated and critical overview of the progress that has been made so far in the key material and catalyst development for fuel cells. All chapters were written by leading international experts. It was the intention of the editors and authors that the book be designed to help those involved in research and development of fuel cells and, at the same time, to serve as a valuable resource on materials-related issues for students, materials engineers, and researchers interested in fuel cell technology in general.

The book contains 17 chapters covering all aspects of synthesis, characterization, and performance of advanced materials and catalysts in fuel cells, starting with Chapter 1 with an overview on the principles, classifications, and types of fuels used in fuel cells. The next four chapters focus on high-temperature (500°C–1000°C) solid oxide fuel cells (SOFCs). Chapter 2 presents the development of high-performance and durable electrodes. As the operating temperatures of SOFCs are reduced to intermediate temperatures of 600°C to 800°C, the performance of SOFCs increasingly depends on the electrochemical activity and stability of both the anode and cathode. Chapter 3 discusses the critical properties, design, and advances made in various sealing materials, particularly for planar SOFCs. Micro-SOFCs or μ-SOFCs show a great potential as energy sources for portable electronic devices and applications, and Chapter 4 gives an extensive review on the design, configuration, fabrication, modeling, materials, and stack performance of μ-SOFC technology. Direct carbon fuel cells (DCFCs) use solid carbon as a fuel rather than gaseous hydrogen or hydrocarbon fuels and are a relatively new member of the SOFC family. Their classification, design, and critical material issues are discussed and reviewed in Chapter 5. Molten carbonate fuel cells (MCFCs) represent probably the most mature and advanced fuel cell technologies for stationary power generation applications. Chapter 6 places a particular emphasis on the development of advanced cathodes and anodes and electrode materials-related issues in MCFC.

The low-temperature proton exchange membrane fuel cell (PEMFC) is considered to be the most promising power source for applications ranging from portable electronic devices and vehicle transportation to station power because of the low operating temperature and quick start-up and shut-down cycles. The challenges in PEMFCs arise from the need for inexpensive and more active and durable proton exchange membranes (PEMs) and electrocatalysts. Chapters 7, 8, and 10 discuss, in detail, the advancement and challenges in the synthesis, characterization, and fundamental understanding of the catalytic activity of nitrogen-carbon, carbon, and noncarbon-based electrocatalysts for PEM fuel cells, respectively. Chapter 9 specifically chooses the topic of the atomic layer deposition (ALD) technique as ALD has emerged as one of the most versatile and powerful techniques in the deposition of high-performance components for various types of fuel cells. Another core component of PEMFCs is the PEM. Chapter 11 summarizes the advancements in the fundamental understanding of the most successful Nafion membranes, and Chapter 12 focuses on the development of alternative and composite membranes for direct alcohol fuel cells (DAFCs). The specific challenges and advances made in the development of nanostructured electrocatalysts for electrooxidation of liquid alcohol fuels, such as methanol and ethanol, are extensively reviewed and updated in Chapters 13 and 14.

The next three chapters discuss the development and critical materials in emerging fuel cells—specifically, alkaline polymer electrolytes (APEs), inorganic high-temperature proton conductors and microbial fuel cells (MFCs). The development of APEs with high anion exchange capability, mobility, and high stability under fuel cell operation conditions is the key in the advancement of APE fuel cells, which are expected to bring significant advantages over perfluorosulfonic acid–based PEM fuel cells. Operation of fuel cells at elevated high temperatures of 200°C or higher have the advantages of fast electrode kinetics, substantially reduced CO poisoning, and much better water and thermal management, but the development of an inorganic proton conducting materials–based PEM is still in the very early stages as shown in Chapter 16. Chapter 17 presents a concise description and critical review of the principles, classification, catalysis, nanostructure electrode materials, biocatalysts, and other components of MFCs. The main difference between a conventional fuel cell and an MFC is that the anode in the MFC uses microbes rather than noble metals as catalysts to convert the chemical energy of alcohol and hydrocarbons in aqueous solution into electricity. However, many factors that affect the overall performance of conventional fuel cells also affect the performance of MFCs, such as the nanostructure and materials of the anode and cathode, the electrode/electrolyte interface, the nature of the PEM, and the configuration of fuel cells.

The quality of the chapters reflects well the caliber of the contributing authors, and we would like to thank the authors for their tremendous effort and hard work. We would also like to thank the editors from the publisher CRC Press, especially Allison Shatkin and Kate Gallo for their professional assistance and strong support during this project.

San Ping Jiang
Perth, Australia

Pei Kang Shen
Guangzhou, China

Authors

 **Dr. San Ping Jiang** is a professor at the department of chemical engineering and deputy director of the Fuels and Energy Technology Institute, Curtin University, Australia, and adjunct professor of the University of Sunshine Coast University, Australia. He also holds visiting/guest professorships at Central South University, Harbin Institute of Technology, Guangzhou University, Huazhong University of Science and Technology, Wuhan University of Technology, University of Science and Technology of China, Sichung University, and Shandong University. After receiving his BEng from South China University of Technology in 1982 and his PhD from The City University, London in 1988, Dr. Jiang worked at Essex University in the U.K.; CSIRO Materials Science and Engineering Division; Ceramic Fuel Cells, Ltd. (CFCL) in Australia; and Nanyang Technological University in Singapore. His research interests encompass fuel cells and water electrolysis, solid state ionics, electrocatalysis, and nanostructured functional materials. With an h-index of 45, he has published ~250 journal papers, which have accrued ~7000 citations.

 **Dr. Pei Kang Shen** is a professor and director at the Advanced Materials Research Laboratory in Sun Yat-sen University, China. Dr. Shen obtained his BSc degree in electrochemistry at Xiamen University in 1982, and he continuously carried out his research and teaching at the same university for seven years before he became a visiting researcher in the United Kingdom. He received his PhD in chemistry at Essex University in 1992. From then on, he has been working at Essex University, Hong Kong University, the City University of Hong Kong, and the South China University of Technology. Since 2001, he has been a professor at the Sun Yat-sen University (Guangzhou, China). He is the author of over 200 publications in qualified journals or specialized books, of 40 patents, and of more than 100 meeting presentations. He has organized five international conferences on Electrochemical Energy Storage and Conversion as a conference chairman. His research interests include fuel cells and batteries, electrochemistry of nanomaterials and of nanocomposite functional materials, and electrochemical engineering.

Contributors

Saeid Amiri
Wave Control Systems, Inc.
Edmonton, Alberta, Canada

Gang Chen
State Key Laboratory of Electrical
    Insulation and Power Equipment
Xi'an Jiaotong University
Xi'an, China

Zhongwei Chen
Department of Chemical
    Engineering
Waterloo Institute for
    Nanotechnology
University of Waterloo
Waterloo, Ontario, Canada

Andrew C. Chien
School of Chemistry
University of St Andrews
St Andrews, Fife, United Kingdom

Fengping Hu
Advanced Energy Materials
    Laboratory
Sun Yat-sen University
Guangzhou, China

John T. S. Irvine
School of Chemistry
University of St Andrews
St Andrews, Fife, United Kingdom

Cairong Jiang
School of Chemistry
University of St Andrews
St Andrews, Fife, United Kingdom

San Ping Jiang
Fuels and Energy Technology
    Institute ang Department of
    Chemical Engineering
Curtin University
Perth, WA, Australia

Doo-Hwan Jung
New and Renewable Energy
    Division
Korea Institute of Energy Research
Daejeon, Korea

Haekyoung Kim
School of Materials Science and
    Engineering
Yeungnam University
Gyeungsan, Korea

Chang Ming Li
Institute for Clean Energy and
    Advanced Materials
Southwest University
Chongqing, China

Junrui Li
State Key Laboratory of Advanced
    Technology for Materials
    Synthesis and Processing
Wuhan University of Technology
Wuhan, China

Vladimir Neburchilov
Vancouver, British Columbia,
    Canada

Jing Pan
Department of Chemistry
Wuhan University
Wuhan, China

Mu Pan
State Key Laboratory of Advanced
    Technology for Materials
    Synthesis and Processing
Wuhan University of Technology
Wuhan, China

Yan Qiao
Institute for Clean Energy and
    Advanced Materials
Southwest University
Chongqing, China

Partha Sarkar
Environment and Carbon
    Management Division
Alberta Innovates-Technology
    Futures
Edmonton, Alberta, Canada

Pei Kang Shen
Advanced Energy Materials
    Laboratory
Sun Yat-sen University
Guangzhou, China

Madhu Sudan Saha
AFCC Automotive Fuel Cell
    Cooperation Corp.
Burnaby, British Columbia, Canada

Shuhui Sun
Center for Energie, Materiaux and
    Telecommunications
Institut National de la Recherche
    Scientifique
Varennes, Quebec, Canada

Xueliang Sun
Department of Mechanical and
    Materials Engineering
The University of Western Ontario
London, Ontario, Canada

Haolin Tang
State Key Laboratory of Advanced
    Technology for Materials
    Synthesis and Processing
Wuhan University of Technology
Wuhan, China

Gang Wu
Materials Physics and Applications
    Division
Los Alamos National Laboratory
Los Alamos, New Mexico,
    United States

Shaorong Wang
Shanghai Institute of Ceramics,
    Chinese Academic of Sciences
Shanghai, China

Hui Yang
Center for Energy Storage and
    Conversion
Shanghai Advanced Research
    Institute
Chinese Academy of Sciences
Shanghai, China

Gaixia Zhang
Center for Energie, Materiaux and
    Telecommunications
Institut National de la Recherche
    Scientifique
Varennes, Quebec, Canada

Hongwei Zhang
Advanced Energy Materials
    Laboratory
Sun Yat-sen University
Guangzhou, China

Jiujun Zhang
Energy, Mining and Environment
  Portfolio
National Research Council of
  Canada
Vancouver, British Columbia,
  Canada

Teng Zhang
College of Materials Science and
  Engineering
Fuzhou University
Fuzhou, Fujian, China

Ling Zhao
Fuels and Energy Technology
  Institute
Curtin University
Perth, WA, Australia

XiaoChun Zhou
Suzhou Institute of Nano-Tech and
  Nano-Bionics
Chinese Academy of Sciences
Suzhou, Jiangsu, China

Lin Zhuang
Department of Chemistry
Wuhan University
Wuhan, China

Contributors

1

Introduction

San Ping Jiang and Pei Kang Shen

CONTENTS

1.1   Introduction 1
1.2   Fundamentals of Fuel Cells 3
    1.2.1   Operation Principles 3
    1.2.2   Classification 6
1.3   Fuels for Fuel Cells 7
    1.3.1   Hydrogen, Methanol, Formic Acid, and Liquid Fuels 7
    1.3.2   Natural Gas, Hydrocarbons, and Coal- and Biomass-
        Derived Synthetic Fuels 9
1.4   Fuel Cell Applications 11
References 13

1.1 Introduction

Massive use of fossil fuels, especially coal, oil, and gas, represents a serious environmental threat and a global energy crisis, such as natural resource exhaustion, pollutant emission, and waste generation. Carbon dioxide emission from power generation by burning fossil fuels is considered to be a key contributor to climate change and related environmental problems. Consequently, there are urgent needs to increase electricity-generation efficiency and to develop clean, sustainable, and renewable energy sources. For example, coal is believed to be the bridging energy source with the diminishing of crude oil, and currently, half of electricity produced in the United States comes from coal-fired power plants. Coal's share of electricity production in many developing countries exceeds this amount, reaching more than 70% in China and India. However, the CO_2 and other pollutant emission can be reduced by integration of coal gasification and fuel cells (integrated gasification fuel cell or IGFC). The energy efficiency of the IGFC power plant can achieve 56%–60%, depending on gasification and fuel cell operating conditions [1]. With the increased efficiency, CO_2 emission will be dramatically decreased.

Fuel cells are an electrochemical energy conversion technology to directly convert the chemical energy of fuels, such as hydrogen, methanol, ethanol, and natural gas, to electrical energy, and hence, fuel cells inherently have a significantly higher efficiency than that of conventional energy conversion technologies, such as an internal combustion engine (ICE). In an ICE, the engine accepts heat from the combustion of the fuel, converts part of the energy into mechanical work, and rejects the remainder to a heat sink at a low temperature, the efficiency of which is limited by the Carnot cycle (see Figure 1.1). Fuel cells are considered to be the most efficient and least polluting power-generating technology and are a potential and viable candidate to moderate the fast increase in power requirements and to minimize the impact of the increased power consumption on the environment. Fuel cells are also versatile devices, ranging from room temperature fuel cells, such as proton exchange membrane fuel cells (PEMFCs), to intermediate temperature (600°C) molten carbonate fuel cells (MCFCs) and to high temperature (600°C–1000°C) solid oxide fuel cells (SOFCs). SOFCs allow internal reforming, promote rapid kinetics with nonprecious materials, and offer high flexibilities in fuel choice. The other advantages of fuel cells are quiet operation and flexible modulability.

Fuel cell technology is one of the oldest electrical energy conversion technologies and was invented more than 160 years ago by Sir William Grove [2]. Surprisingly, the technological development of fuel cells has lagged far behind the more well-known energy conversion technologies, such as steam engines and ICE. The reasons can be mainly related to economic factors, materials problems, and certain inadequacies in the operation of electrochemical devices. However, during the last 20–30 years, fuel cells have received enormous attention worldwide as alternative electrical energy conversion systems because of their huge potential for power generation in portable, transport, and stationary applications. One of the major factors that have influenced the development of fuel cell technologies in the last few decades is the increased public awareness of the finite reserves of fossil fuels and the environmental consequences of the increasing consumption of fossil fuels in electricity production and for vehicular transportation.

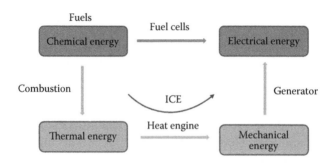

FIGURE 1.1
Graphic illusion of energy conversion differences between fuel cells and the conventional thermomechanical routes of ICEs.

This chapter presents a general introduction, classifications, and operation principles of fuel cell technologies. The general issues related to the various types of fuels used in fuel cells will also be briefly discussed. However specific challenges and advances in various fuel cells will be critically reviewed and discussed in relevant chapters in this book.

1.2 Fundamentals of Fuel Cells

1.2.1 Operation Principles

A fuel cell is an electrochemical cell, and thus the fuel cell reaction can be generally represented by two electrode reactions (anode and cathode) with an overall reaction described in the following equation:

$$A_{ox1} + B_{red1} \rightarrow C_{red2} + D_{ox2} \tag{1.1}$$

An ideal fuel cell can produce current while sustaining a steady voltage as long as the fuel is supplied. The theoretical voltage of the cell, E, can be thermodynamically predicted by the Nernst equation:

$$E = E^o - \frac{RT}{nF} \ln(\Pi) \tag{1.2}$$

where E^o is the cell voltage at standard conditions of 25°C and one atmosphere, R is the gas constant (8.314 J mol^{-1} K^{-1}), T is the temperature (K), n is the number of electrons involved in the reaction, F is the Faraday's constant (96,485 C mol^{-1}), and Π is the chemical activity of the products divided by those of the reactants.

Each type of fuel cell has a unique set of processes and reactions to describe its operation. For a H_2/O_2 fuel cell with a proton exchange membrane (PEM) electrolyte, hydrogen is catalytically oxidized at the anode to give electrons and protons.

$$H_2 \rightarrow 2H^+ + 2e^- \quad E^o = 0.0 \text{ V} \tag{1.3}$$

The protons enter the electrolyte and are transported to the cathode. Because the electrolyte is not an electronic conductor, electrons flow away from the anode in the external circuit to the cathode, where the supplied oxygen reacts with proton according to

$$\tfrac{1}{2} O_2 + 2H^+ + 2e^- \rightarrow H_2O \quad E^o = 1.229 \text{ V} \tag{1.4}$$

The overall reaction of the fuel cell is the sum of the anode reaction 1.3 and cathode reaction 1.4, that is, the combination of hydrogen and oxygen to produce water:

$$\frac{1}{2} O_2 + H_2 \rightarrow H_2O \quad E^o = 1.229 \text{ V} \tag{1.5}$$

The chemical energy present in the combination of hydrogen and oxygen is converted into electrical energy by controlled electrochemical reactions at the anode and cathode in the cell.

The theoretical efficiency of a fuel cell can be defined as a function of the energy ratio as

$$\mu_G = \frac{\Delta G}{\Delta H} = 1 - \frac{T \times \Delta S}{\Delta H} \tag{1.6}$$

where ΔH is the change in enthalpy, ΔG is the change in free energy, and ΔS is the change in entropy of the fuel cell system. ΔH is the total heat of the combustion of the reaction, and ΔG represents the net useful energy available (i.e., the free energy) from a given reaction. Thus, in electrical terms, the net available electrical energy from a reaction in a fuel cell is given by

$$\Delta G^o = -nFE^o \tag{1.7}$$

and

$$\Delta G = -nFE \tag{1.8}$$

The van't Hoff isotherm identifies the free energy relationship for the chemical reactions as

$$\Delta G = \Delta G^o + RT \ln(\Pi) \tag{1.9}$$

Combining Equations 1.7, 1.8, and 1.9 with the van't Hoff isotherm, the Nernst equation (Equation 1.2) for electrochemical reactions in a fuel cell is obtained.

The standard cell potential for a H_2/O_2 fuel cell is 1.23 V at 25°C. In practice, the actual voltage of a fuel cell is less than the predicted thermodynamic voltage due to irreversible losses when the current is higher than zero. The three major irreversible losses that affect fuel cell performance are activation losses or activation polarization (η_{act}), ohmic losses (η_Ω), and mass transport losses or concentration polarization (η_{conc}). In addition to the losses related to the fuel cell reactions, there are also parasitic losses ($E_{parasitic}$), which are mainly caused by a mixed potential at the electrode/electrolyte interface or by the internal shorting due to the mixed ionic and electronic conducting of the electrolyte membrane. For example, in the case of direct methanol fuel cells (DMFCs), $E_{parasitic}$ can be as high as 0.15–0.2 V due to the methanol crossover from the anode, through the Nafion membrane, to the cathode, producing a mixed potential at the cathode [3]. The extent of these losses varies from one system to another. The actual operational voltage output (E_{op}) of a fuel cell is determined by subtracting the voltage losses associated with each process from the E value as follows:

$$E_{op} = E - [(\eta_{act} + \eta_\Omega + \eta_{conc})_{cathode} + (\eta_{act} + \eta_\Omega + \eta_{conc})_{anode}] - E_{parasitic} \quad (1.10)$$

Figure 1.2 shows schematically the losses taking place in a fuel cell. Current generation in a fuel cell depends largely on the kinetics of the reaction that takes place at the anode and the cathode. The reaction kinetics is limited by an activation energy barrier, which impedes the reaction. When current is drawn from a fuel cell, a portion of the potential at the anode and cathode is then lost to overcome this activation barrier, that is, η_{act}, characterized by an exponential loss of potential on the current–voltage curve at low current densities. This energy barrier that the charge has to overcome to go from the electrodes to the reactants and vice versa highly depends on the catalytic properties of the electrode materials. Ohmic polarization, η_Ω, arises from the resistance of the cell components (electrolyte, electrodes, interconnect or bipolar plates, current collectors, etc.) to the electrons and ion flow. This loss generally follows Ohm's law. As the reactions proceed, the availability of active species at the electrode/electrolyte interface changes. Concentration polarization or overpotential, η_{conc}, arises from limited diffusion of active species or products to and from the electrode/electrolyte interface. η_{conc} usually occurs at high currents. In addition, there are parasitic losses, which are primarily related to the permeability properties and ionic transfer number of the electrolyte materials.

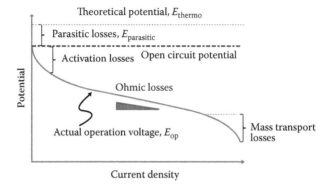

FIGURE 1.2
Schematic of the current–potential curve in a fuel cell.

Thus, the energy conversion efficiency for a fuel cell can be obtained by the cell voltage ratio:

$$\text{energy conversion efficiency} = E_{op}/E \qquad (1.11)$$

1.2.2 Classification

A fuel cell has three main components: the anode, cathode, and electrolyte. Assembly of these three components is often referred to as membrane-electrode assembly (MEA). Many individual cells are joined together through bipolar or interconnect plates to form a stack. A stack is the main component of the power section in a fuel cell power plant and consists of other auxiliary components, such as the interconnect, seal, current collector, etc. Generally, most fuel cells consume hydrogen (or hydrogen-rich fuels) and oxygen to produce electricity. In the case of the oxygen ion–conducting electrolyte, oxygen ions are produced at the cathode and transported through the electrolyte, producing water at the anode chamber. If a PEM is used, a proton produced at the anode will migrate through the electrolyte, forming water at the cathode chamber. Figure 1.3 shows the typical SOFCs with an oxygen ion–conducting electrolyte and PEMFCs with PEMs.

Fuel cells are usually classified according to their working temperature or to the electrolyte employed. There are, thus, low- and high-temperature fuel cells. Low-temperature fuel cells include alkaline fuel cells (AFCs), PEMFCs, DMFCs, and phosphoric acid fuel cells (PAFCs). The high-temperature fuel cells operate at 500°C–1000°C, including MCFCs and SOFCs. Due to their high operating temperature, MCFCs and SOFCs have a high tolerance to typical catalyst poisons, such as CO, produce high-quality heat for reforming of hydrocarbons, and offer the possibility of direct utilization of hydrocarbon fuels. The electrolytes in fuel cells can be solid (polymer or ceramic) or liquid (aqueous or molten) and must have high ionic (primarily O^{2-}, OH^-, H^+, or CO_3^{2-}) conductivity with negligible electronic conductivity. Areas of

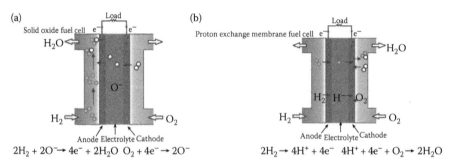

FIGURE 1.3
Fuel cells based on (a) an oxygen ion conductor and (b) a proton-conducting membrane.

TABLE 1.1

Types of Fuel Cells According to the Electrolyte Employed and Their Main Characteristics

	AFC	PEMFC	PAFC	MCFC	SOFC
Temperature (°C)	60–90	80	160–200	630–650	800–1000
Electrolyte	KOH	Nafion	H_3PO_4	$LiCO_3/K_2CO_3$	Y_2O_3–ZrO_2
Anode	Pt/C	Pt/C	Pt/C	Ni alloy	Ni/YSZ
Cathode	Pt/C	Pt/C	Pt/C	NiO	LSM
Primary fuel	H_2	H_2	H_2/ reformate	H_2/CO reformate	H_2/CO/CH_4 reformate
Oxidant	O_2	O_2/air	O_2/air	CO_2/O_2/air	O_2/air
Efficiency (%)	35–40	40–50	32–42[a]	47–57[a]	60–65[a]
Power range (W)	1 k	1–100 k	50 k–200 k	200 k–300 k	100–125 k
Main applications	Space transport	Portable transport	Stationary	Stationary	Auxiliary stationary

[a] The production of additional electric energy by means of thermal energy cogeneration is not considered.

application include battery replacement in portable electronic devices, transportation, prime movers and/or auxiliary power units (APUs) in vehicles, residential combined heat and power (CHP), and large-scale megawatt stationary power generation.

Table 1.1 lists the different types of fuel cells, classified according to the electrolyte employed, together with their main characteristics [4–7].

In addition to the clarification based on the electrolyte types, fuel cells can also be clarified based on fuels used, for example, DMFCs, direct ethanol fuel cells (DEFCs), direct formic acid fuel cells (DFAFCs), and direct carbon fuel cells (DCFCs). DMFCs, DEFCs, and DFAFCs can be considered as a subcategory of PEM fuel cell types, and DCFCs can be considered as a subcategory of SOFCs. Other emerging fuel cells include single chamber SOFCs (SC-SOFCs), micro SOFCs (μ-SOFCs), alkaline polymer electrolyte fuel cells (APEFCs), microbial fuel cells, and biofuel cells.

1.3 Fuels for Fuel Cells

1.3.1 Hydrogen, Methanol, Formic Acid, and Liquid Fuels

Hydrogen gas is the preferred fuel for fuel cells as the only product is water with no greenhouse gas emission. The main obstacle in the use of hydrogen as an energy carrier is that hydrogen is not a readily available fuel. Currently, steam reforming, partial oxidation, and autothermal reforming of hydrocarbons as well as biofeed stocks are the major processes for hydrogen generation, but all these methods produce a large amount of CO and/or CO_2 and certain levels of H_2S. Any trace amount of CO, CO_2, and H_2S in the hydrogen-rich reformat gas must be removed as low-temperature fuel cells, such as AFCs and PEMFCs, are extremely sensitive toward CO, CO_2, and H_2S impurities: PEMFCs are poisoned by very low (ppm) levels of CO, and AFCs are susceptible to CO_2 contamination of the electrolyte when air is used as well as the CO and H_2S poisoning of the Pt and Ni catalysts. Preferential oxidation of CO is one of the most effective methods for trace CO cleanup from the reformate stream [8]. High-purity hydrogen can be produced by water electrolysis; however, water electrolysis is often limited to small-scale and unique situations in which access to large-scale hydrogen production plants is not possible [9].

For transportation and portable applications of fuel cells, the challenge is the storage and transportation of hydrogen. There are several options in the development of storage technologies for hydrogen: liquid hydrogen, pressurized hydrogen, metal hydrides, borohydrides, carbon nanotubes, zeolites, and metal oxide frameworks. Despite the enormous effort toward the search for a material that could store an appropriate amount of hydrogen, the US Department of Energy (DoE) targets of 6% by mass appear to be very tough to meet [10]. State-of-the-art compressed hydrogen storage consists of lightweight tanks using polymer and carbon fibers containing hydrogen compressed to 700 bar. The infrastructure for a hydrogen fuel station is also very costly, and for safety reasons, a hydrogen filling station will be very different from a gasoline station. Also, from an energetic point of view, large-scale transportation of hydrogen and the necessary compression or liquefaction of hydrogen can be highly unattractive. Thus, effective and safe onsite or onboard generation of hydrogen could speed up the introduction of a fuel cell system without the presence of a widespread hydrogen infrastructure. In this aspect, formic acid is proposed as one of the alternative materials for hydrogen storage [11].

The direct use of liquid fuels in fuel cells is of significant importance due to potentially higher energy density and higher maximum thermodynamic efficiencies. Liquid fuels, such as methanol and ethanol, have several advantages with respect to hydrogen. They are relatively cheap; are easily handled, transported, and stored; and have a high theoretical energy density. However, apart from the energy density, the toxicological–ecological hazards of the liquid fuels and the environmental effects of the byproduct of the liquid fuel oxidation reactions should also be taken into account when selecting one particular fuel. For example, methanol, the most studied liquid fuel in fuel cells, is predominantly produced by steam reforming of natural gas and is also highly flammable and toxic. On the other hand, ethanol would be a preferred liquid fuel for portable fuel cells as ethanol can be easily produced by hydration of acetylene or can be derived by fermentation of sugar-rich raw materials and is less harmful as compared to methanol.

The electrochemical oxidation of the majority of liquid fuels in low-temperature fuel cells is hardly a complete reaction. Thus, the environmental effects of the byproducts need to be considered in the design of fuel cell systems. For example, the main byproducts of the electrooxidation of methanol are formaldehyde and formic acid [3]. Formaldehyde is highly irritant, corrosive, carcinogenic, and toxic. In the case of ethanol fuel, the main challenge is the breakage of the C–C bond. Final products of the ethanol oxidation are acetaldehyde and acetic acid [12], which are also highly flammable and carcinogenic. Dimethyl ether (DME) is also used as a liquid fuel for fuel cells as DME is the simplest ether with no C–C bonds and low toxicity as compared to methanol. However, as reported by Mizutani et al. [13], the main byproducts of direct DME fuel cells are methanol and methyl formate, and the formation of methanol does not depend on the current but increases with increasing temperature. Wang et al. [14] studied the oxidation of 1-propanol and 2-propanol in a PEMFC using online mass spectrometry. Propanol is the main product of the electrooxidation of 1-propanol, and the electrooxidation of 2-propanol mainly yields acetone. Thus, treatment processes must be incorporated into the fuel cell system or be created as the byproducts of the oxidation reactions of the liquid fuels cannot be directly discharged without proper cleaning treatment. Demirci [15] gave a detailed thermodynamic and environmental analysis of potential liquid fuels in fuel cells. Table 1.2 lists specific energy density, hazards, environmental effects, and main byproducts of the common liquid fuels used in fuel cells.

Direct liquid fuel cells using methanol, ethanol, propanol, formic acid, etc., suffer from an additional and more specific problem: CO from the indirect decomposition of liquid fuels significantly poisons the Pt-based electrocatalysts. Development of advanced nanoparticle synthesis methods and supporting materials is still needed to significantly enhance the electrocatalytic activity and durability of the Pt-based catalysts for liquid fuel–based fuel cell systems.

1.3.2 Natural Gas, Hydrocarbons, and Coal- and Biomass-Derived Synthetic Fuels

Given the lack of a supply infrastructure and difficulties in storing hydrogen, the ability of fuel cells to operate at high efficiency on hydrocarbon fuels is considered a major advantage of high temperature fuel cells, such as MCFCs and SOFCs. Natural gas (primarily methane) is seen as an ideal fuel due to its abundance, existing distribution infrastructure, and low cost.

The steam reforming of methane and hydrocarbons is an established process used on an industrial scale for the production of hydrogen. The first step of the steam reforming results in the formation of a mixture of CO and H_2 (syngas), and further reaction via a water gas shift (WGS) reaction converts CO and steam into more H_2. During internal reforming, these two reactions occur simultaneously; the equilibrium composition of the gas is dictated by

TABLE 1.2

Theoretical Specific Energy Density, Hazard, Environmental Effects, and Main Byproducts of the Most Common Liquid Fuels Used in Fuel Cells

Name	Chemical Formula	n_e	Specific Energy		Health Effects	Fire Hazard	Main Byproduct
			$kWh\ kg^{-1}$	$kWh\ L^{-1}$			
Hydrogen	H_2	2	39.42 (HHV)	0.53[a]		Highly flammable	Water
Methane	CH_4	4	15.29 (HHV)	0.011		Highly flammable	CO_2, water
Propane	$CH_3CH_2CH_3$	20	13.97 (HHV)	0.028		Highly flammable	CO_2, water
Ethanol	C_2H_5OH	12	8.297	6.546	Irritant	Highly flammable	Formic acid, acetaldehyde
Ethylene glycol	C_2H_6O	10	5.268	5.864	Irritant, harmful if swallowed	Flammable	Oxalic acid
DME	CH_3OCH_3	12	8.807	0.017		Highly flammable	Methanol, methyl formate
Methanol	CH_3OH	6	6.362	5.037	Toxic	Highly flammable	Formic acid, formaldehyde
Formic acid	$HCOOH$	2	1.63	1.989	Irritant, harmful	Flammable	CO
1-Propanol	$CH_3CH_2CH_2OH$	18	9.07	7.187	Irritant	Highly flammable	Propanol
2-Propanol	$CH_3CH(OH)CH_3$	18	8.99	7.066	Irritant	Highly flammable	Acetone

Source: Lamy, C. et al. *J Power Sources*, 105: 283–296, 2002. Wang, J.T. et al. *J Electrochem Soc*, 142: 4218–4224, 1995. Demirci, U.B., *J Power Sources*, 169: 239–246, 2007.

Note: n_e: number of electrons involved in the chemicals of the fuels; HHV: higher heating value.
[a] Compressed under 20 MPa.

the steam-to-carbon ratio, temperature, and pressure. Methane can be also directly or indirectly used as fuel in high-temperature MCFCs and SOFCs.

Renewable fuels derived from biomass (particularly from nonfood biomass) are attractive for fuel cells as the electricity generation from biomass-based renewable fuels through fuel cells is not only highly efficient but also carbon neutral. The typical product gas of the biomass gasification consists of H_2, CO, CH_4, CO_2, H_2O, and N_2 as well as some impurities, such as sulfur, chlorine, and alkali metals [16]. The fuel from the biomass gasifier can also contain considerable amounts of tars depending on the type of gasifier used. Tar is a complex of aromatics and can be represented by a mixture of toluene, naphthalene, phenol, and pyrene. Thus, certain pre-reforming and cleaning of biomass gasification fuels are needed for fuel cell applications. Alkalis are highly corrosive toward materials associated with syngas processing and utilization, including ceramic particulate filters and metallic components. The removal of alkali contaminants, consequently, is an important step in the cleaning of coal- and biomass-derived syngas [17].

1.4 Fuel Cell Applications

Fuel cells can be used in a wide range of applications meeting energy demands: electricity generation, land and marine transportation, and portable electronic devices. The range of fuel cell applications and the size of the potential markets for fuel cell–based energy devices are enormous.

PEMFCs have been the technology of choice for transportation because of their low-temperature operation and rapid start-up capability. The past decades have seen significant progress in power density and durability close to meeting the requirement for automotive and stationary applications. However, the cost of PEMFCs is still too high. In addition to the high cost of a hydrogen fuel infrastructure, the successful entry of hybrid and battery-powered vehicles suggests that they might achieve prominence before we might expect fuel cell vehicles to become widespread. Thus, it makes sense to consider the opportunities for PEMFCs in other markets. For example, fuel cells can compete favorably with lead-acid batteries for forklift applications.

Planar SOFCs offer a much high power density as compared to tubular SOFCs pioneered by Siemens-Westinghouse. One of the significantly growing application areas for planar SOFC systems in the 1 to 5 kW sizes is the residential CHP systems operating on natural gas developed by SOFC developers, such as Ceramic Fuel Cells, Ltd., (CFCL) in Australia [18] and Ceres Power in the United Kingdom [19]. For example, the BlueGen units with a power output of up to 2 kW and 60% electrical efficiency produced by CFCL can produce up to 17,000 kWh of electricity per annum, which is more than sufficient to power an average home. Surplus electricity can be sold back to the grid. Residential CHP systems could be the first commercial products on SOFCs. Lower operation temperature, faster start-up and shutdown time, and more rugged construction due to compliant seals and metallic interconnects make IT-SOFCs a viable technology for mobile applications. IT-SOFCs have been demonstrated for use as APUs and traction power in vehicles [20].

The power density requirement for portable power sources is ever increasing, and power consumption is forecast to pose long-term technical challenges for the portable electronics industries, which are working to find ways to extend the running time of mobile devices, such as portable computers, MP3 players, and mobile phones. The power range is 0.1–1 W for MP3 players, 2–5 W for mobile phones, and 15–30 W for laptop computers. In fact, the performance, mass and volumes, and lifetimes of power supplies presently limit most applications of microelectromechanical system (MEMS) technology [21]. Miniaturized fuel cells or microfuel cells are particularly attractive for powering portable electronic devices, such as laptops and mobile phones. The most appealing benefit over batteries is longer usage time between recharges, given the considerably greater energy density of liquid fuels than that of state-of-the-art batteries. DMFCs have been primarily the technology of interest for high-energy portable power sources. In this application, Pt catalysts supported on high surface carbon are the preferred catalysts for the cathode, and unsupported PtRu catalysts at the loading level of 2 to 4 mg cm^{-2} are needed to achieve practical power densities in DMFCs. There are various companies leading the development of small DMFC systems with aims to reduce the size and increase the power density for portable electronics applications.

There is significant and growing military interest in fuel cell technology for various applications, including submarine power, unmanned aerial and underwater vehicles, base camp power, and soldiers' personal power supplies. The power requirement for a future soldier is in the range of 25–150 W. The ability of high-temperature SOFCs and μ-SOFCs to operate directly on logistic fuels (e.g., JP-8) would be particularly attractive for military and defense applications. Fuel cell technologies offer the potential for extended mission length.

The integration of biofuels, such as bioethanol or biomass gasification fuel cells, is a promising and forthcoming technology for electricity and heat cogeneration along with profound environmental and socioeconomic benefits. A SOFC creates many synergies in the integrated system. The high-quality heat eluted from SOFCs can be used either in external thermal cycles, increasing the total yield, or to cover the thermal demands of the integrated process (reforming of the fuel, thermochemical treatment of biomass). Omosun et al. [22] showed that, theoretically, for a biomass-fueled SOFC system, the overall efficiency for the cogeneration of electricity and heat can be as high as 60%. Such integrated systems are well suited for the distributed power generation for remote areas due to the wide availability of biomass and flexible modulability of SOFC technology. One of the main focus areas of the Solid State Energy Conversion Alliance (SECA) program is to develop large (>100 MW) integrated coal gasification SOFC power systems [1].

In addition to the electricity and heat generation, fuel cells can also be used for chemicals and energy cogeneration [23]. In cogeneration mode, the fuel cells are characterized by their current efficiency, selectivity with respect to the product obtained, and/or current and power densities.

References

1. Surdoval, W.A. 2009. The status of SOFC programs in USA. *ECS Trans* 25 (2), 21–28.
2. Andujar, J.M. and Segura, F. 2009. Fuel cells: History and updating. A walk along two centuries. *Renew Sust Energ Rev* 13, 2309–2322.
3. Arico, A.S., Srinivasan, S. and Antonucci, V. 2001. DMFCs: From fundamental aspects to technology development. *Fuel Cells* 1, 133–161.
4. de Bruijn, F. 2005. The current status of fuel cell technology for mobile and stationary applications. *Green Chem* 7, 132–150.
5. Farooque, M. and Maru, H.C. 2006. Carbonate fuel cells: Milliwatts to megawatts. *J Power Sources* 160, 827–834.
6. Singhal, S.C. 2002. Solid oxide fuel cells for stationary, mobile, and military applications. *Solid State Ionics* 152, 405–410.
7. Haile, S.M. 2003. Fuel cell materials and components. *Acta Mater* 51, 5981–6000.
8. Choudhary, T.V. and Goodman, D.W. 2002. CO-free fuel processing for fuel cell applications. *Catal Today* 77, 65–78.
9. Zeng, K. and Zhang, D.K. 2010. Recent progress in alkaline water electrolysis for hydrogen production and applications. *Prog Energy Combust Sci* 36, 307–326.
10. Ross, D.K. 2006. Hydrogen storage: The major technological barrier to the development of hydrogen fuel cell cars. *Vacuum* 80, 1084–1089.
11. Grasemann, M. and Laurenczy, G. 2012. Formic acid as a hydrogen source: Recent developments and future trends. *Energy Environ Sci* 5, 8171–8181.
12. Lamy, C., Lima, A., LeRhun, V., Delime, F., Coutanceau, C. and Leger, J.M. 2002. Recent advances in the development of direct alcohol fuel cells (DAFC). *J Power Sources* 105, 283–296.
13. Mizutani, I., Liu, Y., Mitsushima, S., Ota, K.I. and Kamiya, N. 2006. Anode reaction mechanism and crossover in direct dimethyl ether fuel cell. *J Power Sources* 156, 183–189.
14. Wang, J.T., Wasmus, S. and Savinell, R.F. 1995. Evaluation of ethanol, 1-propanol, and 5-propanol in a direct oxidation polymer-electrolyte fuel cell: A real-time mass spectrometry study. *J Electrochem Soc* 142, 4218–4224.
15. Demirci, U.B. 2007. Direct liquid-feed fuel cells: Thermodynamic and environmental concerns. *J Power Sources* 169, 239–246.
16. Yu, J., Tian, F.J., McKenzie, L.J. and Li, C.Z. 2006. Char-supported nano iron catalyst for water-gas-shift reaction: Hydrogen production from coal/biomass gasification. *Process Saf Environ Protect* 84, 125–130.

17. Dolan, M.D., Ilyushechkin, A.Y., McLennan, K.G. and Sharma, S.D. 2012. Alkali, boron and phosphorous removal from coal-derived syngas: Review and thermodynamic considerations. *Asia-Pac J Chem Eng* 7, 317–327.
18. 2010. CFCL wins more European orders for BlueGen mCHP. *Fuel Cells Bulletin March* 4–5.
19. 2007. Ceres fuel cells set to provide back-up power for homes. *Prof Eng* 20, 54.
20. Mukerjee, S., Haltiner, K., Klotzbach, D., Vordonis, J. and Iyer, A. 2009. Solid oxide fuel cell stack for transportation and stationary applications. *ECS Trans* 25 (2), 59–63.
21. Cook-Chennault, K.A., Thambi, N. and Sastry, A.M. 2008. Powering MEMS portable devices: A review of non-regenerative and regenerative power supply systems with special emphasis on piezoelectric energy harvesting systems. *Smart Mater Struct* 17, 43001.
22. Omosun, A.O., Bauen, A., Brandon, N.P., Adjiman, C.S. and Hart, D. 2004. Modelling system efficiencies and costs of two biomass-fuelled SOFC systems. *J Power Sources* 131, 96–106.
23. Alcaide, F., Cabot, P.L. and Brillas, E. 2006. Fuel cells for chemicals and energy cogeneration. *J Power Sources* 153, 47–60.

2

Advanced Electrode Materials
for Solid Oxide Fuel Cells

Ling Zhao and San Ping Jiang

CONTENTS

2.1 Introduction 15
2.2 Development of Advanced Cathodes 16
    2.2.1 Lanthanum Manganite–Based Cathodes 17
    2.2.2 Lanthanum Strontium Cobalt–Based Cathodes 19
    2.2.3 Layered Perovskite Cathodes 22
    2.2.4 Modeling Approach to Cathode Materials Development 23
2.3 Development of Advanced Anode 26
    2.3.1 Anode Requirements for SOFCs 26
    2.3.2 Cermet Anode 27
    2.3.3 Perovskite Oxide Anode 29
    2.3.4 Other Anode Materials 33
2.4 Conclusions 34
References 35

2.1 Introduction

Solid oxide fuel cells (SOFCs) are high-temperature electrochemical devices to directly convert the chemical energy of fuels, such as hydrogen, natural gas, and other hydrocarbon fuels, to electric power with high efficiency and low greenhouse gas emissions. The major components of a single SOFC consist of a porous cathode, a porous anode, and a dense electrolyte sandwiching them. Yttria-stabilized zirconia (YSZ) and gadolinium-doped ceria are the most commonly used solid electrolyte. $(La,Sr)MnO_3$ (LSM) perovskite oxides and nickel-YSZ cermets often serve as the cathode and anode, respectively. Traditional SOFCs operate at high temperatures, 900°C–1000°C, because of the relatively low oxygen ion conductivity and high activation energy of oxide electrolytes, such as YSZ. However, a lowering of the operating temperature of SOFCs to an intermediate range of 500°C–800°C brings substantial technical and economic benefits. The cost of a SOFC system can be substantially reduced by using less-expensive components, and the durability of the system can be enhanced by reducing the role of the chemical interaction and thermomechanical incompatibility [1–4].

On the other hand, lowering the operation temperature results in a significant increase in the electrolyte resistivity and the polarization resistance of the electrode reactions. The high electrolyte resistance can be compensated by the use of a thin electrolyte layer on an anode- or cathode-supported cell structure. Thus, for the intermediate-temperature SOFCs (IT-SOFCs), the overall cell performance critically depends on the activity of the electrode materials. There are significant activities in the research and development in the electrode materials [5–8], electrolytes [9,10], interconnects [1,11], and seals [12]. The present chapter will be mainly on the review of the development of advanced cathode and anode materials for IT-SOFCs. Relevant reaction processes, microstructural aspects, potential limitations, and performance of the cells are also briefly covered.

2.2 Development of Advanced Cathodes

A cathode is the place where oxygen is reduced to oxygen ions as shown in the following Kröger-Vink notation:

$$O_2 + 4e^- \leftrightarrow 2O^{2-} \tag{2.1}$$

where O_2 represents oxygen molecular in air, e^- is the electron in the cathode material, and O^{2-} is an oxygen ion that will be transferred into an electrolyte. The cathode should possess not only high electronic conductivity for efficient current collection, high ionic conductivity for oxygen transfer, and catalytic activity toward oxygen reduction but also an appropriate porous structure to facilitate the diffusion and transportation of gas-phase reactants. Furthermore, the cathode should be thermally and chemically compatible with the electrolyte and other cell components under the operation and fabrication conditions, that is, exhibiting a minimum thermal or chemical expansion mismatch with the electrolyte and a negligible chemical reaction with the electrolyte during the high-temperature fabrication and operation conditions. In addition to the high activity and stability, a cathode also requires adequate tolerance toward poisoning of some contaminants in the systems, such as chromium from the Fe–Cr alloy–based interconnect and boron from the glass–ceramic sealants.

Among these various cathode materials, perovskite oxides, until now, are the most investigated cathode materials for SOFCs. A perovskite oxide has the general formula ABO_3, the lattice cell of which is shown in Figure 2.1. The ideal lattice of perovskite is a cubic structure. Usually, some distortions, such as cation displacements within the octahedral and tilting of the octahedral,

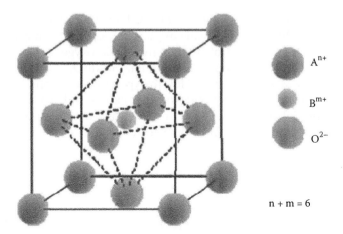

A^{n+}

B^{m+}

O^{2-}

n + m = 6

FIGURE 2.1
Unit cell of the ABO_3 perovskite lattice. (Reprinted by permission from Macmillan Publishers Ltd. *Nature Materials*, Boukamp, B.A., 2, 294–296, copyright 2003.)

appear in the perovskite lattice, which are related to the properties of the substituted atoms on A and B sites [13]. The distortion degree of perovskite can be described by the Goldschmidt tolerance factor (*t*):

$$r_A + r_O = t\sqrt{2}(r_B + r_O) \tag{2.2}$$

where r_A, r_B, and r_O represent the ionic radio of A site, B site, and O site, respectively. Due to the high tolerance factor ($0.7 < t < 1.1$) on the ion radius, different doping ions or different concentrations could be well accepted to create, as expected, electrochemical catalytic properties. For perovskite oxide cathode materials in SOFCs, the cation on A site is normally a rare and/or alkaline earth (such as La, and Ca, Sr, or Ba, etc.), and the B site is occupied by transition metals, such as Mn, Fe, Co, Ni, etc. or a mixture of them. The octahedral symmetry of oxygen around the transition metal often promotes a metallic or semiconducting band at high temperatures, resulting in high electronic conductivity. And oxygen ion vacancy could be introduced in perovskite lattice by a defect balance along with the doping [14].

2.2.1 Lanthanum Manganite–Based Cathodes

Perovskite-type manganites $La_{1-x}Sr_xMnO_{3-\delta}$ (LSM) with high chemical stability are typical electronic conducting cathode materials for the high-temperature SOFCs [7]. $La_{1-x}Sr_xMnO_{3-\delta}$ exhibits a rhombohedral phase when $0 < x < 0.5$, a tetragonal phase when $x = 0.5$, and a cubic phase when $x = 0.7$, depending on the Sr^{2+} content. Further, as compared to most perovskite oxides reflecting oxygen deficiency, LSM is somewhat unusual for the oxygen excess as well as the oxygen-deficient nonstoichiometries, which is generally written as $La_{1-x}Sr_xMnO_{3\pm\delta}$ ("+" means oxygen excess, and "−" means oxygen deficiency). The relationship of nonstoichiometric oxygen, oxygen partial pressure, temperature, and composition in $La_{1-x}Sr_xMnO_{3\pm\delta}$ has been studied systematically by Mizusaki et al. [15,16]. To explain the relationship between nonstoichiometric oxygen, including oxygen excess and oxygen deficient, and oxygen partial pressure in LSM, a model had been proposed by Van Roosmalen and Cordfunke [17,18] and Van Roosmalen et al. [19], shown as follows:

Oxygen excess:

$$\frac{3}{2}O_2 + 6Mn_{Mn}^X \leftrightarrow 3O_O^X + 6Mn_{Mn}^\bullet + V_{La}''' + V_{Mn}''' \tag{2.3}$$

Oxygen deficient:

$$2Mn_{Mn}^\bullet + O_O^X \leftrightarrow 2Mn_{Mn}^X + V_O^{\bullet\bullet} + \frac{1}{2}O_2 \tag{2.4}$$

$$2Mn_{Mn}^X + O_O^X \leftrightarrow 2Mn_{Mn}' + V_O^{\bullet\bullet} + \frac{1}{2}O_2 \tag{2.5}$$

where $Mn_{Mn}^\bullet$, Mn_{Mn}^X, and Mn_{Mn}' are Mn^{4+}, Mn^{3+}, and Mn^{2+} ions, respectively.

The electronic conductivity of LSM is expressed by a small polaron hopping, generally expressed by [20]

$$\sigma T = (\sigma T)^\circ \exp\left(-\frac{E_a}{kT}\right) = A\left(\frac{h\nu^\circ}{k}\right)c(1-c)\exp\left(-\frac{E_a}{kT}\right) \tag{2.6}$$

where $(\sigma T)^\circ$ is the pre-exponential constant, E_a is the activation energy, and c represents the ratio of carrier occupancy. As shown in Figure 2.2a [16], the electronic conductivity increases along with increasing Sr^{2+} contents and then decreases. The highest conductivity was obtained in $La_{0.5}Sr_{0.5}MnO_3$ with a value of 300 S cm^{-1} at 1000°C. Figure 2.2b [21] shows that the electronic conductivity is nearly constant irrespective of the P_{O_2} in the high P_{O_2} region (>10^{-8} Pa), but it decreased exponentially with P_{O_2} in the low region ($10^{-15} < P_{O_2} < 10^{-8}$ Pa), suggesting typical p-typed oxide semiconductors.

Compared with its electronic conductivity, the oxygen ionic conductivity of the LSM series is negligible [22]. Thus, the electrocatalytic activity of LSM for the O_2 reduction reaction decreases significantly with the temperature. One strategy is to introduce oxygen ionic phases, such as YSZ and

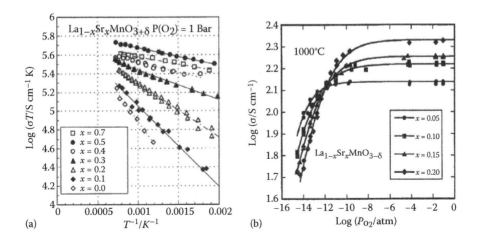

FIGURE 2.2
Electrical conductivity of $La_{1-x}Sr_xMnO_{3-\delta}$ as functions of (a) temperature. (Reprinted from *Solid State Ionics*, 132, Mizusaki, J. et al. Electronic conductivity, Seebeck coefficient, defect and electronic structure of nonstoichiometric $La_{1-x}Sr_xMnO_3$, 167–180, Copyright 2000, with permission from Elsevier.) (b) Oxygen partial pressure. (Reprinted from *Journal of Solid State Chemistry*, 123, Yasuda, I. and Hishinuma, M., Electrical conductivity and chemical diffusion coefficient of strontium-doped lanthanum manganites, 382–390, Copyright 1996, with permission from Elsevier.)

SDC to form composite cathodes. Ji et al. [23] reported that oxygen diffusion coefficient increases from 10^{-12} cm^2 s^{-1} with pure LSM to 10^{-8} cm^2 s^{-1} with LSM-YSZ. Wang et al. [24] showed that the surface exchange coefficient on LSM is increased three times after the introduction of YSZ. Therefore, the polarization resistance decreased from 1.5 Ω cm^2 with pure LSM to 0.17 Ω cm^2 with LSM-YSZ at 1000°C [25]. The electrochemical performance of an LSM cathode can also be enhanced substantially by introducing catalytically active nanoparticles, such as doped CeO$_2$ and Pd, into the LSM porous structure [26–28]. The use of composite cathodes can also significantly reduce the effect of the Cr poisoning [29].

2.2.2 Lanthanum Strontium Cobalt–Based Cathodes

Cobalt-containing perovskite oxides and their derivatives, such as $La_{1-x}Sr_xCoO_{3-\delta}$ (LSC), have been extensively investigated as the cathode materials for SOFCs due to their high level of electrochemical activity. LSC shows a complex behavior with regard to the dependence of oxygen nonstoichiometry, phase structure, electrical, and catalytic properties on the strontium content, temperature, and oxygen partial pressure. It has been found that $LaCoO_{3-\delta}$ ($x = 0$) crystallizes in a rhombohedral structure at room temperature. This could be changed to the cubic phase when the Sr content is up to 0.55 or the temperature is up to 1673 K [30]. Petrov et al. [31] proposed a defect model associated with the defect structure of LSC. $Co^{\cdot}_{Co}$ and oxygen vacancies are formed in the lattice to maintain electrical neutrality, which is caused by the substitution of Sr ions on La sites. The overall electrical neutrality condition is shown as follows:

$$Sr^x_{Sr} + Co^{\cdot}_{Co} + 2V^{\cdot\cdot}_O \leftrightarrow Sr'_{La} + Co^x_{Co} \tag{2.7}$$

$$2Co^{\cdot}_{Co} + O^x_O \leftrightarrow 2Co^x_{Co} + V^{\cdot\cdot}_O + \frac{1}{2}O_2 \tag{2.8}$$

$$\left[V^{\cdot\cdot}_O\right]\left[Co^x_{Co}\right]^2 \Leftrightarrow K_{V^{\cdot\cdot}_O}\left[Co^{\cdot}_{Co}\right]^2\left[O^x_O\right]P_{O_2}^{-1/2} \tag{2.9}$$

where $Co^{\cdot}_{Co}$ and Co^x_{Co} stand for the Co^{4+} and Co^{3+} ions at B site in perovskite lattice, respectively, and $K_{V^{\cdot\cdot}_O}$ is the defect equilibrium constant. There may be also some Co^{2+} ions in the lattice due to the delocalization of electrons [32]. The electronic conductivity of LSC samples with different Sr content as a function of temperature is shown in Figure 2.3. When $0.25 < x < 0.7$, LSC shows metallic behavior over the whole temperature range studied, whereas samples with $x < 0.20$ show semiconducting behavior.

LSC possesses a remarkable electrode activity due to fast oxygen diffusivity and the dissociation ability of oxygen molecules. An increase in the Sr^{2+}

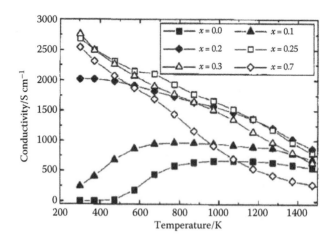

FIGURE 2.3
Temperature dependencies of electronic conductivity of $La_{1-x}Sr_xCoO_{3-\delta}$ samples measured in air. (Reprinted from *Journal of Solid State Chemistry*, 142, Mineshige, A. et al. Metal-insulator transition and crystal structure of $La_{1-x}Sr_xCoO_3$ as functions of Sr-content, temperature, and oxygen partial pressure, 374–381, Copyright 1999, with permission from Elsevier.)

content ($x \leq 0.5$) at the A site will increase the oxygen vacancy and thus the oxygen ion conductivity [33], but excess Sr substitution ($x > 0.5$) decreases the electronic and ionic transport because the defect balance is gradually governed by charge compensation of the B sites [3]. $Sm_{0.5}Sr_{0.5}CoO_{3-\delta}$ (SSC) is another potential cathode material for SOFCs with high electrochemical activity. Using SSC as a cathode on SDC electrolytes, the polarization resistance at 700°C is ~0.10 Ω cm² [34], two to three orders of magnitude lower than that of LSM. However, high cobalt content at the B sites in the perovskite structure results in an increased thermal expansion coefficient (TEC) mismatch, which may result in a delamination at the cathode-electrolyte interface or cracking of the electrolyte [35]. Baek et al. [36] reported that SSC on a $Ce_{0.8}Sm_{0.2}O_2$ (SDC) electrolyte experienced a fast degradation on thermal cycling and during operation.

The standard practice to suppress thermal and chemical incompatibility includes A site doping and extensive substitution of cobalt, which enable a decrease in TEC and an increase in the chemical stability, including compositions such as $La_{0.8}Sr_{0.2}Co_{1-x}Fe_xO_{3-\delta}$ [37], $Sm_{0.5}Sr_{0.5}Co_{1-x}Fe_xO_{3-\delta}$ [38], and $PrBaCo_{2-x}Fe_xO_{5-\delta}$ [39]. These strategies lead often to the decrease in the electronic and ionic conductivities of these oxides as the mobility of electrical holes for ferrite ($LaFeO_3$) is about three orders of magnitude lower than that for cobaltite ($LaCoO_3$) [40]. Among lanthanum strontium cobalt perovskite families, $La_{0.6}Sr_{0.4}Co_{0.2}Fe_{0.8}O_{3-\delta}$ is the most representative cathode in SOFCs due to its high electrical (340 S cm⁻¹ at 550°C [37]) and ionic conductivity (10^{-2} S cm⁻¹ at 800°C [41]). The polarization resistance of $La_{0.6}Sr_{0.4}Co_{0.2}Fe_{0.8}O_{3-\delta}$ on GDC electrolytes at 700°C is only 0.22 Ω cm² [42]. $SrCo_{1-x}Sb_xO_{3-\delta}$ with $x = 0.05$ on Nd-doped ceria (NDC) electrolytes was found to exhibit the highest conductivity and lowest polarization resistance, reaching 0.009 to 0.23 Ω cm² at 600°C–900°C [43]. However, the TEC is high and increases with Sb doping, reaching 29.3×10^{-6} for $x = 0.15$. Doping of $SrCoO_3$ with Fe stabilizes the structure but does not reduce the TEC substantially to the level compatible with common electrolytes, such as YSZ and doped ceria.

Replacing lanthanum with alkaline earth metals, such as barium, at the A site of LSCF substantially enhances its electrochemical activity for the O_2 reduction reaction at intermediate temperatures. Shao et al. [44] investigated $Ba_{0.5}Sr_{0.5}Co_{0.8}Fe_{0.2}O_{3-\delta}$ (BSCF) as a cathode material in SOFCs. The polarization resistances of BSCF at 600°C and 500°C are 0.055–0.071 and 0.51–0.61 Ω cm², respectively, which are considerably lower than that of other perovskite cathode materials under similar operation conditions. Although BSCF shows high electrochemical performance, the TEC of BSCF is very high, ~20×10^{-6} K⁻¹, and the conductivity is relatively low, ~25 S cm⁻¹ at 800°C [45,46]. Another issue is the interaction between the cathode materials containing alkaline-earth elements and CO_2 and H_2O. A BSCF cathode is susceptible to CO_2 poisoning at 450°C–750°C, and a decrease in the cell performance and an increase in the polarization resistance were observed when BSCF was exposed to CO_2-containing atmosphere due to the formation of carbonates of Sr and Ba [47,48]. Bucher et al. [49] observed that at 300°C–400°C, the surface of BSCF is passivated with respect to oxygen exchange at $4 \times 10^{-4} \leq p_{CO2}(bar) \leq 5 \times 10^{-2}$. A pronounced mass increase at 600°C–800°C has been observed in CO_2-rich atmospheres (20 vol% O_2 + 5 vol% CO_2 + Ar) due to the significant carbonate formation of BSCF powders. In order to improve the stability of BSCF material, different element doping on B site has been studied, such as $Ba_{0.6}Sr_{0.4}Co_{1-y}Ti_yO_{3-\delta}$ [50], $Ba_{0.6}Sr_{0.4}Co_{0.9}Nb_{0.1}O_{3-\delta}$ [51], and $BaCo_{0.7}Fe_{0.2}Nb_{0.1}O_{3-\delta}$ [52]. The stability of BSCF can be significantly enhanced by coating a CO_2 resistive shell on BSCF using an infiltration and microwave plasma technique [53].

However, the long-term activity and stability of a cobalt-based perovskite cathode is affected by the surface segregation. Simner et al. [54] reported segregation of SrO on the surface of LSCF after a long-term test in air at 750°C using x-ray photoelectron spectroscopy technology (XPS). Oh et al. [55] detected the segregation of SrO on the LSCF surface after heat treating in air at 600°C–900°C for 50 h using Auger electron spectroscopy (AES). Strontium and cobalt segregation is also reported for lanthanum cobaltite perovskites. Vovk et al. [56] studied the $La_{0.5}Sr_{0.5}CoO_3$ perovskite oxide surfaces under electrochemical polarization using an *in situ* XPS technique. Under cathodic polarization, the Sr/(La + Co) ratio at the oxide surface increased irreversibly by 5%, while the La/Co ratio remained constant, indicating the surface enrichment of strontium. Using *in situ* XRD during SOFC operation conditions (750°C in H_2/air), Hardy et al. [57] observed the continuous and gradual lattice expansion of LSCF cathodes, which has been caused by the segregation or diffusion of Sr and Co out of the cathode under polarization conditions of 0.8 V at 750°C. Our study also indicates that Ba and Sr segregation plays an important role in the chromium deposition and poisoning on LSCF and BSCF cathodes [58–60]. The surface segregation of barium may be related to the phase instability of BSCF perovskite [61].

2.2.3 Layered Perovskite Cathodes

Double-layered perovskites, such as $LnBaCo_2O_{5+\delta}$ (Ln = La, Pr, Nd, Sm, Eu, Gd, and Y), have been studied as cathode materials of SOFCs. Unlike the disordered A sites in LSC or SSC lattices, the ideal structure of these $LnBaCo_2O_{5+\delta}$ family oxides with ordered A sites can be represented by the stacking sequence, $LnO_\delta|CoO_2|BaO|CoO_2$, as shown in Figure 2.4. Transformation of a simple cubic perovskite with randomly occupied A sites into a layered crystal with alternating lanthanide and alkali earth planes reduces the strength of the oxygen binding and provides disorder-free channels for ionic motion, thereby theoretically increasing the oxygen diffusivity [62]. Zhang et al. [63] systematically studied the oxygen nonstoichiometric, electronic conductivity and cathode performance of these $LnBaCo_2O_{5+\delta}$. The polarization resistance of $PrBaCo_2O_{5+\delta}$ on SDC electrolytes at 700°C is as low as 0.08 Ω cm^2 [64]. The structure plays an important role in the electrocatalytic activity of

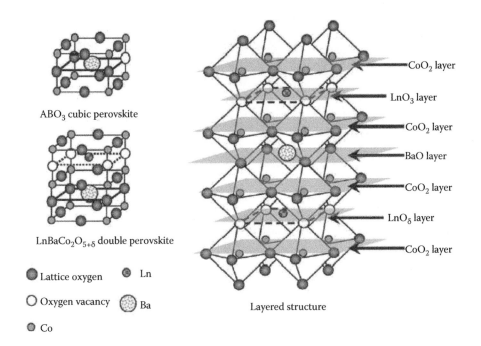

FIGURE 2.4

Schematic diagram of layered perovskite oxides $LaBaCo_2O_{5+\delta}$. (Reprinted from *Acta Materialia*, 56, Zhang, K. et al. Synthesis, characterization and evaluation of cation-ordered $LaBaCo_2O_{5+\delta}$ as materials of oxygen permeation membranes and cathodes of SOFCs, 4876–4889, Copyright 2008, with permission from Elsevier.)

cobalt-based oxide cathodes. For example, $SmBa_{0.5}Sr_{0.5}Co_2O_{5+\delta}$ with a double perovskite structure shows a low polarization resistance of 0.092 Ω cm^2 at 700°C [65] while $Ba_{1.2}Sr_{0.8}CoO_{4+\delta}$ with a K_2NiF_4 structure shows a much worse performance, ~0.5 Ω cm^2 at 750°C [66].

The layered cobaltites often have substantially lower TEC as compared to disordered perovskite analogs as the state of cobalt cations is more stable. For example, Nagasawa et al. [67] reported that $Ca_3Co_4O_{9-\delta}$ on a GDC electrolyte has a TEC of 10 × 10^{-6} K^{-1} and a polarization resistance of 0.68 Ω cm^{-2} at 800°C. $TbBaCo_3ZnO_{7+\delta}$ shows a high electrocatalytic performance for the O_2 reduction reaction, reaching a polarization resistance of 0.07 Ω cm^{-2} at 800°C with a low TEC of 9.45 × 10^{-6} K^{-1} [68].

2.2.4 Modeling Approach to Cathode Materials Development

Computational modeling is a powerful tool for examining the energy for the processes involved in oxygen reduction on cathodes for SOFCs, including adsorption, dissociation, incorporation, and diffusion to TPB [69,70]. Choi et al. [71] shows that the adsorption energies of oxygen species have a strong

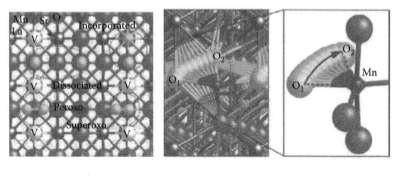

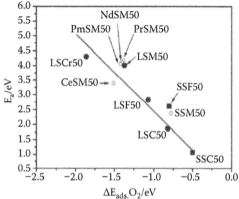

FIGURE 2.5
(See color insert.) Comparison of adsorption energies versus diffusion barriers of oxygen ions through the MIEC bulk phases. (Reprinted from *J Power Sources*, 195, Choi, Y. et al. Rational design of novel cathode materials in solid oxide fuel cells using first-principles simulations, 1441–1445, Copyright 2010, with permission from Elsevier.)

correlation with the diffusion barriers of oxygen ion transport as shown in Figure 2.5. A weaker binding between a surface and oxygen species also suggests a lower oxygen diffusion barrier, comparing $La_{0.5}Sr_{0.5}CoO_{3-\delta}$ (LSC), $La_{0.5}Sr_{0.5}FeO_{3-\delta}$ (LSF), $La_{0.5}Sr_{0.5}MnO_{3-\delta}$ (LSM), and $La_{0.5}Sr_{0.5}CrO_{3-\delta}$ (LSCr) oxides (adsorption energies and diffusion barrier Co < Fe < Mn < Cr). The calculation suggests that $La_{0.5}Sr_{0.5}CoO_{3-\delta}$ is a potential cathode for the oxygen reduction reaction.

Lee et al. [72] demonstrated that the experimentally measured area-specific resistance and oxygen surface exchange of the perovskites are strongly correlated with the DFT calculated oxygen p-band center and vacancy formation energy. As shown in Figure 2.6, the surface exchange coefficient, area-specific resistance, and oxygen p-band center correlate well with a range of energies relevant for the oxygen reduction reaction. The results suggested that the material properties for oxygen reduction are somewhat related to the oxygen

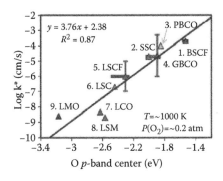
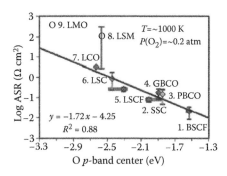

FIGURE 2.6
Experimental surface exchange coefficient and area-specific resistances at 1000 K versus the calculated bulk O p-band center of perovskites with simulated composition. (Lee, Y.-L. et al. 2011. Prediction of solid oxide fuel cell cathode activity with first-principles descriptors. *Energy & Environmental Science*, 4: 3966–3970. Reproduced by permission of The Royal Society of Chemistry.)

p-band center. However, the validity of the model is still needed to be established because the area-specific resistance is closely related with not only materials but also the microstructure of the cathode.

The electrochemical activity depends strongly on the microstructure and, in particular, the length of the TPB. Percolation theory has been used to predict the electrocatalytic properties of the composite electrodes of SOFCs. Percolation theory depends upon the concept of coordination numbers, representing the number of contacts that a certain particle makes with neighboring particles. The conceptual illustration of percolation theory is shown in Figure 2.7 [73].

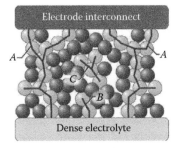
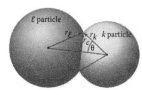

FIGURE 2.7
Conceptual illustration of a composite SOFC electrode. The dark particles represent the electrode particles, and the light particles represent the electrolyte. (Reprinted from *Journal of Power Sources*, 191, Chen, D. et al. Percolation theory to predict effective properties of solid oxide fuel-cell composite electrodes, 240–252, Copyright 2009, with permission from Elsevier.)

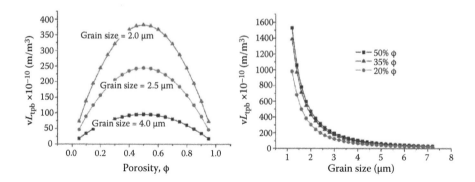

FIGURE 2.8

TPB length as a function of grain size and various porosities. (Reprinted from *Journal of Power Sources*, 178, Janardhanan, V.M. et al. Three-phase boundary length in solid-oxide fuel cells: A mathematical model, 368–372, Copyright 2008, with permission from Elsevier.)

Costamagna et al. [74,75] employed percolation models for composite electrodes. Janardhanan et al. [76] used percolation theory to predict three-phase boundary lengths as functions of particle dimensions and packing density in the case of monosized particles with equal volume distribution of ionic and electronic phases. As shown in Figure 2.8, the results suggest that a finer grain size leads to more TPB length.

Subsequently, to realize the TPB length of a traditional composite cathode and an impregnated composite cathode, Chen et al. [73] used the percolation theory to predict TPB lengths with different particle size of ionic and electronic phases. The results indicate that great improvement in TPB length in the impregnated electrode could be obtained. The calculation results provided strong evidence for the mechanism that the extension of TPB lengths by impregnation contributes to higher electrode performance.

2.3 Development of Advanced Anode

2.3.1 Anode Requirements for SOFCs

In SOFCs, an anode is an electrode in which the electrochemical oxidation of fuels, such as hydrogen and hydrocarbon, takes place. Thus, the anode should be catalytically active for the oxidation reaction, possess sufficient electronic and ionic conductivities, and be highly resistant to the contamination and poisoning of impurities, such as carbon deposition and sulfur poisoning. Similar to the cathode, the anode materials should also be chemically and thermally compatible with other cell components, such as the electrolyte, interconnect, and sealants, during the cell fabrication and operation conditions. Sufficient phase stability during reducing and oxidizing (redox) cycles is also essential for the anode of SOFCs.

2.3.2 Cermet Anode

The state-of-the-art anode materials are based on the Ni/YSZ cermets. The function of the oxide component is primarily to reduce the sintering of the Ni metal phase, to decrease the TEC, and to improve the electrochemical performance of the anode. The Ni phase in the cermet can offer not only catalytic activity for fuel oxidation and steam reforming of hydrocarbon fuels but also electronic conductivity. Due to a good compatibility between NiO and YSZ at high temperatures, NiO and YSZ can be sintered at high temperatures (1300°C–1400°C) to form a NiO/YSZ composite and then reduced to obtain porous Ni/YSZ cermets [77,78]. Typically 30 vol% of Ni is required to reach continuity of current collection [79]. Ni/YSZ or Ni/GDC cermets have been extensively investigated and optimized for the hydrogen oxidation reaction [80]. However, the conventional Ni/YSZ cermets are problematic in the case of direct utilization of hydrocarbon fuels, including low tolerance to carbon deposition (or coking) [81,82] and sulfur poisoning [83,84]. Carbon deposition involves carbon adsorption on the nickel surface, carbon dissolution into the bulk of the nickel, and precipitation of graphitic carbon on some facet of the nickel particle after it becomes supersaturated in carbon as shown in Figure 2.9 [85].

There have been extensive research efforts in enhancing the oxidation activity and stability of Ni-based cermet anodes. Murray et al. [86] reported

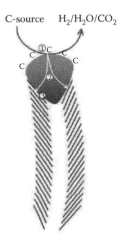
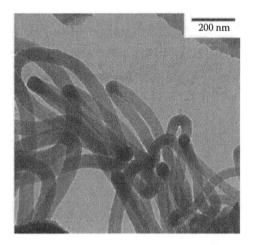

FIGURE 2.9
Diagram and microscopy image of carbon nanofiber formation over Ni catalysts. (Reprinted from *Catal Today*, 76, Toebes, M.L. et al. Impact of the structure and reactivity of nickel particles on the catalytic growth of carbon nanofibers, 33–42, Copyright 2002, with permission from Elsevier.)

the direct electrochemical oxidation of methane in a cell employed with a
0.5-μm-thick $(Y_2O_3)_{0.15}(CeO_2)_{0.85}$ (YDC) porous interlayer between YSZ electrolyte and Ni–YSZ cermet anodes, generating power densities up to 0.37 W
cm^{-2} at 650°C. Low operating temperatures were used to avoid carbon deposition. It has been reported that the nickel and doped ceria composites, such
as Ni–Ce$_x$Zr$_{1-x}$O$_2$ (x = 0, 0.25, 0.5, 0.75, and 1) catalysts, show excellent coking
resistance, and for the composition with x = 0.25, there is no coking after 20 h
of reactions at 700°C [87]. Yang et al. [88] developed a novel Ni-based cermet
anode using a mixed proton and ion conductor $(BaZr_{0.1}Ce_{0.7}Y_{0.1}Yb_{0.1}O_{3-\delta})$ that
allows rapid transport of both protons and oxygen ions, showing excellent
tolerance toward the carbon cracking and sulfur poisoning. As revealed by
Raman spectroscopy, the surface of the Ni/ZCYYb anode exposed to wet
propane (3 vol% H$_2$O) is free of carbon deposition. Further, using hydrogen
containing 10, 20, or 30 ppm H$_2$S as fuels, there is no change in the power
output of the cell with Ni-BZCYYb cermet anodes. The existence of the ionic
defects, such as proton and oxygen ion vacancy, may enhance the catalytic
activity for reforming and/or oxidation of hydrocarbons and for conversion
of H$_2$S to SO$_2$ [89]. The Ni/YSZ cermet anodes with the BaO interface demonstrated high power density and stability in C$_3$H$_8$, CO, and other gasified
carbon fuels at 750°C [90]. The DFT calculations indicate that the dissociated
OH from H$_2$O on BaO reacts with carbon on Ni near the BaO/Ni interface to
produce CO and H species, which are then electrochemically oxidized at the
triple phase boundaries of the anode (see Figure 2.10). The anodes with BaO
interface show excellent tolerance toward carbon deposition.

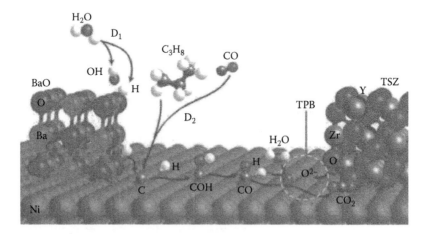

FIGURE 2.10
(See color insert.) Proposed mechanism for water-mediated carbon removal on the anode
with BaO/Ni interfaces. (Reprinted by permission from Macmillan Publishers Ltd. *Nature
Communications,* Yang, L., Choi, Y., Qin, W., Chen, H., Blinn, K., Liu, M., Liu, P., Bai, J., and
Tyson, T.A., Promotion of water-mediated carbon removal by nanostructured barium oxide/
nickel interfaces in solid oxide fuel cells, 2, 357–367, Copyright 2011.)

Gorte et al. [91–93] pioneered the development of copper-based cermet anodes for direct utilization in hydrocarbon fuels. Unlike Ni, Cu has no catalytic activity for carbon deposition and mainly functions as an effective current collector. Thus a catalytic active phase, such as ceria, needs to be added to provide the catalytic activity required for oxidation of hydrocarbons. The direct, electrochemical oxidation of various hydrocarbons (methane, ethane, 1-butene, n-butane, and toluene) at 700°C was realized [93]. To enhance the activity and stability of Cu-based anodes, Cu was alloyed with a second metal [94]. Kim et al. [95] examined activity of Cu/Ni alloys with a Ni composition of 0%, 10%, 20%, 50%, and 100% for the direct oxidation of methane at 800°C. A single cell with a Cu 80%–Ni 20% cermet was tested in dry methane for 500 h and showed a significant increase in power density with time due to the enhanced electronic conductivity in the anode caused by carbon deposits. Lee et al. [96] compared the performances of Cu–Ni or Cu–Co alloy anodes at 700°C and 800°C using n-butane as fuels. The results indicate that the Cu–Co appears to be more tolerant to carbon. Some noble metals, such as Pt, Rh, Pd, and Ru, have been found to be beneficial to the reforming reactions of hydrocarbon. Hibino et al. [97, 98] studied the catalytic activity of a Ru–Ni–GDC anode in hydrocarbon fuels, including methane, ethane, and propane, at 600°C. The results suggested that the function of the Ru catalyst in the anode reaction is to promote the reforming reaction of the unreacted hydrocarbons by the produced steam and CO_2. The cell with the Ru–Ni–GDC anode produced a peak power density of 750 mW cm^{-2} in dry methane. Zhan et al. [99] reported that a Ru–CeO_2 catalyst function layer combines with a conventional anode, achieving internal reforming of iso-octane without coking. A single cell with SDC as the electrolyte produced a maximum power density of 350 mW cm^{-2} at 570°C, comparable to those reported SDC electrolyte–based cells operated in H_2. Ni-Al_2O_3 [100–102] can also be used as the reforming function layer for hydrocarbon fuels.

2.3.3 Perovskite Oxide Anode

Electronic or mixed ionic and electronic conducting oxides based on perovskite structure have been extensively investigated as alternative anodes for SOFCs. Table 2.1 presents the reported conductivities of typical perovskite oxide anodes [103].

$SrTiO_3$ (STO)–based materials, showing a high sulfur tolerance and resistance to coking and having good chemical and redox stability, have been studied as potential anodes for SOFCs [122]. However, pure $SrTiO_3$ has low electronic conductivity in a reducing atmosphere [123]. A site doping of lanthanide series can significantly increase the electronic conductivity as well as maintain the stability of materials. The conductivity of $La_{0.3}Sr_{0.7}TiO_3$ [104] and $(Y_{0.08}Sr_{0.92})Sr_{0.95}TiO_3$ [107] is ~200 and 80 S cm^{-1}, respectively. However, $SrTiO_3$-based materials have very low catalytic activity for fuel oxidation reaction. Li et al. [104] studied $La_{0.3}Sr_{0.7}Ti_{0.93}Co_{0.07}O_3$ and found that doping

TABLE 2.1

Reported Conductivities of Typical Perovskite Anodes

Anode Composition	Anode Performance				
	Electronic Conductivity (S cm^{-1})	Ionic Conductivity (S cm^{-1})	Temperature (°C)	Fuel	Reference
$La_{0.3}Sr_{0.7}TiO_3$	~220	—	800	$H_2(<5.8\%)/N_2$	[104]
$Y_{0.08}Sr_{0.92}Ti_{0.92}Nb_{0.08}O_3$	~200	—	600	Wet H_2	[105]
$(La_{0.3}Sr_{0.7})_{0.93}TiO_3$	145	0.01	800	$H_2(<5.8\%)/N_2$	[106]
$(Y_{0.08}Sr_{0.92})_{0.95}TiO_3$	~80	~0.014	800	$H_2(5\%)/N_2$	[107]
$Y_{0.09}Sr_{0.91}TiO_3$	73.7	—	800	$H_2(<5.8\%)/N_2$	[108]
$La_{0.3}Sr_{0.7}Sc_{0.1}Ti_{0.9}O_3$	49	0.01	800	$H_2(<5.8\%)/N_2$	[109]
$La_{0.3}Sr_{0.7}Ti_{0.93}Co_{0.07}O_3$	~45	~0.011	800	$H_2(<5.8\%)/N_2$	[104]
$Y_{0.08}Sr_{0.92}Ti_{0.97}O_3$	~35	~0.015	800	$H_2(<5.8\%)/N_2$	[108]
$Sr_{0.86}Y_{0.08}TiO_3$	~23	—	800	—	[110]
$Y_{0.08}Sr_{0.92}Ti_{0.96}Co_{0.04}O_3$	20	0.0007	800	$H_2(<5.8\%)/N_2$	[111]
$La_{0.75}Sr_{0.25}Cr_{0.5}Mn_{0.5}O_3$	1.49	—	900	$H_2(5\%)/Ar$	[112]
$La_{0.75}Sr_{0.25}Cr_{0.5}Mn_{0.44}Ni_{0.06}O_3$	2.4	—	800	$H_2(5\%)/Ar$	[113]
$La_{0.65}Ce_{0.1}Sr_{0.25}Cr_{0.5}Mn_{0.5}O_3$	~0.4	—	700	Wet H_2	[114]
$Pr_{0.75}Sr_{0.25}Cr_{0.5}Mn_{0.5}O_3$	0.14	—	900	$H_2(5\%)/N_2$	[115]
$La_{0.75}Sr_{0.25}Cr_{0.5}Fe_{0.5}O_3$	~7.5	—	800	Wet $H_2(5\%)/Ar$	[116]
$Sr_2Mg_{1-x}Mn_xMoO_6$	~0.8	—	800	$H_2(5\%)/Ar$	[117]
Sr_2CoMoO_6	~0.5	—	800	$H_2(5\%)/Ar$	[118]
Sr_2NiMoO_6	~10	—	800	$H_2(5\%)/Ar$	[118]
$Sr_2FeMoO6$	~22	—	800	H_2	[119]
$Sr_2Fe_{4/3}Mo_{2/3}O_{6-\delta}$	16	—	800	Wet H_2	[120]
$Sr_2Fe_{1.5}Mo_{0.5}O_{6-\delta}$	32	0.00021	800	Wet H_2	[121]

Source: Tao, S.W. and Irvine, J.T.S.: Synthesis and characterization of $(La_{0.75}Sr_{0.25})(Cr_{0.5}Mn_{0.5})O_{3-\delta}$, a redox-stable, efficient perovskite anode for SOFCs. *Journal of the Electrochemical Society.* 2004. 151. A252–A259. Copyright Wiley-VCH Verlag GmbH & Co. KGaA. Reproduced with permission.

of cobalt at the B sites increased the oxygen ionic conductivity but decreased the electronic conductivity due to the increased oxygen vacancy concentration and the decreased Ti^{3+} concentration, according to defect equilibrium. The electronic and oxygen ionic conductivity is 45 and 0.011 S cm^{-1} at 800°C in H_2(<5.8%)/N_2, respectively. Marina et al. [124] systematically investigated (La,Sr)TiO$_3$ doped with several transition metals (Ni, Co, Cu, Cr, and Fe) and Ce at the B sites. It was found that the most effective dopant is cerium, which significantly decreases the polarization resistance. Using (La,Sr)(Ti,Ce)O$_3$ as the anode, the polarization resistances is 0.2 Ω cm^2 at 850°C and 1.3 Ω cm^2 at 700°C measured in wet hydrogen. A-site deficient $(La_{0.3}Sr_{0.7})_{0.93}TiO_3$ [106] and $(La_{0.3}Sr_{0.7})_{0.93}TiO_3$ [107] were found to have high electronic conductivity while gaining similar ionic conductivities as compared to the effect of B-site doping.

Adding a catalytic active phase is effective to enhance the catalytic activity of perovskite oxide–based anodes. By the use of nickel as a catalyst in a strontium titanate–based anode, the maximum power density of a LST-GDC anode cell was significantly increased by five times [125]. Palladium has been found to be a very effective anode catalyst. Lu et al. studied $Sr_{0.88}Y_{0.08}TiO_{3-\delta}$ (SYT)/$La_{0.4}Ce_{0.6}O_{1.8}$ (LDC) composite anodes impregnated with Pd as a sulfur-tolerant anode for the $La_{0.8}Sr_{0.2}Ga_{0.83}Mg_{0.17}O_{3-\delta}$ (LSGM) electrolyte cells. The impregnated Pd was found to significantly enhance the anode performance. The single cell with the Pd-impregnated SYT/LDC anode showed a maximum power density of 1006 and 577 mW cm^{-2} at 850°C and 800°C in H_2, respectively. The Pd-impregnated composite anode showed essentially no decay in performance in H_2 containing up to 50 ppm H_2S [126]. Recent study by Zheng et al. [127] showed that adding Pd nanoparticles slows performance degradation of Ni-GDC anodes in sulfur-containing hydrogen fuel. Impregnation of ceria nanoparticles also enhanced the stability of a Ni-YSZ anode in H_2 containing 40 ppm H_2S over 500 h at 700°C [128]. The ceria nanoparticles, which act as a sulfur sorbent to react with H_2S to form Ce_2O_2S, can avoid the formation of Ni_3S_3 at a H_2S concentration as high as 200 ppm. Ruthenium was also found to be effective to reduce anode polarization resistance for a YST–YSZ composite, achieving 0.8 Ω cm^2 at 800°C as compared to 2.6 Ω cm^2 for the anode without the Ru addition [129].

$La_{0.75}Sr_{0.25}Cr_{0.5}Mn_{0.5}O_3$ (LSCM) is a p-type semiconductor with an electronic conductivity of approximately 38 S cm^{-1} in air and 1.5 S cm^{-1} at 900°C in 5% H_2/Ar [130]. LSCM is stable in both oxidizing and reducing atmospheres and can be utilized as both an anode and cathode in a symmetrical SOFCs [116]. However, pure LSCM has very low activity for oxidation reaction of fuels, such as hydrogen and methane [131], and is not sufficiently stable with sulfur impurities in the fuel, forming MnS, La_2O_2S, and α-MnOS upon exposure to 10% H_2S [132]. Danilovic et al. [133] systematically studied the catalytic activity and conductivity of $La_{0.75}Sr_{0.25}Cr_{0.5}X_{0.5}O_{3-\delta}$ (X = Ti, Mn, Fe, Co). Among these oxides, cobalt substitution in LSCM showed the highest conductivity, and iron substitution exhibited the highest catalytic activity. However, the iron-substituted oxide was not sufficiently stable. When exposed to a highly reducing atmosphere at 900°C, partial degradation into the spinel phase was observed [134]. $La_{0.8}Sr_{0.2}Cr_{1-y}X_yO_3$, (X = Ni, Ru, LSCRu or LSCNi) was found to have high catalytic activity and low polarization resistance due to the formation of metal nanoparticles (Ru or Ni) on the surface during reduction [135,136]. Partial substitution of lanthanum with cerium, $La_{0.65}Ce_{0.1}Sr_{0.25}Cr_{0.5}Mn_{0.5}O_{3-\delta}$ (CeLSCM), significantly increased the catalytic activity of the anode, reducing the polarization resistance from 2.3 to 0.2 Ω cm^{-2} at 800°C [137].

Jiang et al. [131] studied the performance of LSCM–YSZ composites as the alternative anodes for the direct utilization of methane. The addition of the YSZ phase greatly improved the adhesion and reduced the electrode polarization resistance for the methane oxidation reaction in wet CH_4. The best electrode performance was achieved for the composite with LSCM contents of 50–60 wt% with polarization resistances of 2–3 Ω cm^{-2} in wet (3% H_2O) CH_4 at 850°C. Further studies show that impregnation of GDC substantially increases the electrocatalytic activity of the LSCM anode for the direct utilization of methane [138]. The electrode polarization resistance of a 4.0 mg cm^{-2} GDC-impregnated LSCM anode achieved 0.44 Ω cm^{-2} in wet (3% H_2O) CH_4 at 800°C, showing a superior polarization performance compared to the pure LSCM or LSCM/YSZ composite anodes. Liu et al. [139] investigated the performance of LSCM impregnated with GDC and Ni in H_2, CH_4, C_3H_8, and C_4H_{10} fuels and found that the anode performance is comparable to that of Ni–GDC anodes in hydrogen and methane fuels. The anodes also provided good performance with propane and butane. Unlike Ni–GDC, there was little or no coking. The results indicate that a small amount of Ni (e.g., 4%) can promote the electrocatalytic effect without detrimental effects on carbon deposition. Kim et al. [140] also showed that the addition of 0.5 to 1 wt% Pd, Rh, or Ni is effective to increase the power output of the LSCM anode, achieving 500 mW cm^{-2} in humidified H_2 at 700°C.

Double perovskite materials, such as $Sr_2FeNbO_{6-\delta}$ and $Sr_2MnNbO_{6-\delta}$, have been considered as alternative SOFC anodes, but their electronic conductivity is very low [141,142]. For example, $SrMn_{0.5}Nb_{0.5}O_{3-\delta}$ or $Sr_2MnNbO_{6-\delta}$ has a cubic double perovskite structure and is redox stable in a reducing atmosphere. The electronic conductivities of the oxide in air and 5% H_2 are 1.23 and 3.1 × 10^{-2} S cm^{-1}, respectively, at 900°C [141]. The apparent conduction activation energy is also high. Huang et al. [117] explored the use of double perovskite $Sr_2Mg_{1-x}Mn_xMoO_{6-\delta}$ (SMMO) as an anode operating in H_2 and CH_4 fuel and reported reasonable cell performance. Double perovskite oxides $Sr_2MMoO_{6-\delta}$ (M = Co, Ni) were synthesized and investigated as anode materials for SOFCs. $Sr_2CoMoO_{6-\delta}$ was shown to have better performance probably due to higher catalytic activity for fuel oxidation [118,143]. Vasala et al. [144] systematically studied a double perovskite oxide series $Sr_2MMoO_{6-\delta}$ (M = Mn, Fe, Co, Ni, Zn). Double perovskite phases with M = Co, Ni, and Zn appeared to be oxygen stoichiometric and were stable under oxidizing conditions, whereas those with M = Mn and Fe were oxygen deficient and stable under reducing conditions. For the M = Mg phase, it exhibited stability both under reducing and oxidizing conditions. The instability of the double phases, such as $Sr_2FeMoO_{6-\delta}$ in air, implies that it needs special care during SOFC fabrication in order to ensure the formation of the desirable phase. On the other hand, Xiao et al. [120] indicated that $Sr_2Fe_{4/3}Mo_{2/3}O_{6-\delta}$ exhibited good electrochemical activity and stability, and its conductivity at 800°C in wet (3% H_2O) H_2 is ~16 S cm^{-1}. At 800°C, the maximum power density was about 130 mW cm^{-2} in wet (3% H_2O) CH_4 and 472 mW cm^{-2} in H_2 with 100 ppm H_2S, suggesting a potential to utilize hydrocarbon and sulfur-containing fuels directly. $Sr_2Fe_{1.5}Mo_{0.5}O_{6-\delta}$ has been reported to be a potential anode for SOFC, achieving maximum power densities of 835 and 230 mW cm^{-2} at 900°C in wet H_2 and CH_4, respectively [145]. Zhang et al. [119] studied $A_2FeMoO_{6-\delta}$ (A = Ca, Sr, and Ba) series. Their study indicated that $Ca_2FeMoO_{6-\delta}$ showed high conductivity, but it was not stable in a reducing atmosphere, even in nitrogen. The best performance of single cells was observed with a $Sr_2FeMoO_{6-\delta}$ anode, 831 mW cm^{-2} in dry H_2 and 735 mW cm^{-2} in commercial city gas at 850°C, respectively. Adding SDC also substantially increased the electrode catalytic activity and reduced the interfacial polarization resistance of $Sr_2Fe_{1.5}Mo_{0.5}O_{6-\delta}$ anodes from 0.84 to 0.45 Ω cm^2 for 30 wt% SDC-$Sr_2Fe_{1.5}Mo_{0.5}O_{6-\delta}$ composites [121].

2.3.4 Other Anode Materials

Doped ceria exhibits MIEC behavior due to the partial oxidation of Ce^{4+} to Ce^{3+} upon reducing environments [146]. Although doped ceria has very low electronic conductivity, ceria anodes showed modest fuel cell performance. Ishihara et al. [147] studied the anodic performance of CeO_2 doped with Mn and Fe on a Co-doped LSGM electrolyte, and the cell using the $Ce_{0.6}(Mn_{0.3}Fe_{0.1})O_2$ anode produced a maximum power density of 620 mW cm^{-2} at 1000°C in wet H_2.

A pyrochlore-type oxide, $A_2B_2O_7$, can be derived from fluorite by removing one eighth of the oxygen ions and ordering the two cations and oxygen anions. The $A_2B_2O_7$ pyrochlore structure is formed if the cation radius of the two cations falls into a specific range [148]. The pyrochlore structure oxide $Gd_2Ti_2O_7$ was considered as a potential SOFC anode. Ca doping into the A site of $A_2B_2O_7$ increases the ionic conductivity, while Sr and Mg doping leads to an enhancement in electronic conductivity but a reduction in ionic conductivity due to a large dopant-host size mismatch. The conductivity of pyrochlore oxides is generally low. The highest total conductivity is about 1.20×10^{-2} S cm^{-1} at 900°C for the $GdSm_{0.9}Ca_{0.1}Zr_2O_{6.95}$ oxides [149]. The ionic conductivity of $(Gd_{0.98}Ca_{0.02})_2Ti_2O_7$ is about 10^{-2} S cm^{-1} at 1000°C [150,151]. Mo doping in the Ti sites was shown to increase the ionic and electronic conductivity under reducing conditions, implying that it may be suitable to be used as an anode material for SOFCs [152]. However, the performance of cells based on

$Gd_2Ti_{0.6}Mo_{1.2}Sc_{0.2}O_7$ anodes is very low; a maximum power density of 9.5 and 1.8 mW cm^{-2} in wet H_2 and wet CH_4, respectively, was reported at 932°C [153].

The tungsten bronze structure with the general formula $A_2BM_5O_{15}$ (with M = Nb, Ta, Mo, W, and A or B = Ba, Na, etc.) can be obtained from the perovskite structure by rotating some of the MO_6 octahedra. The distortion of the octahedra makes some B-O bonds stretched and the others shortened. The interconnection of short B-O bonds may supply a percolation path for charge transfer, possibly leading to high electronic conductivity [154]. Slater et al. [155,156] investigated the composition $(Ba/Sr/Ca/La)_{0.6}M_xNb_{1-x}O_{3-\delta}$ (M = Mg, Ni, Mn, Cr, Fe, In, Ti, Sn) as potential anode materials for SOFCs. The results suggested that the oxides with M = Cr, Mn, Fe, Ni, and Sn are not suitable for anode materials, possibly due to the low oxygen ionic conductivity and partial decomposition on prolonged heat treatment at 1000°C in reducing atmospheres. However, the oxides with M = Mg or In are stable under the reducing conditions but have relatively poor conductivity characteristics. The conductivity increased with the reducing atmosphere and reached 1–10 S cm^{-1} at $P_{O_2} < 10^{-17}$ bar. The fuel cell performance of tungsten bronze–based anodes is poor and limited by the low conductivity and poor catalytic activity. For example, A- and B-site doped Sm_2MO_6 (SMO) tungsten bronze structured oxides exhibit total conductivity of 0.12 S cm^{-1} at 550°C in wet H_2 [157], which is too low for the practical application as anodes of SOFCs.

2.4 Conclusions

A significant amount of work and progress have been made in the development of advanced cathodes and anodes of SOFCs with substantial improvement in the fundamental understanding and materials development. The increase in the cell performance has been made primarily through microstructure optimization and incorporation of a catalytic active phase of existing anode and cathode materials.

A wet impregnation or infiltration method is an effective technique to introduce nanosized catalytic active phases into pre-sintered, structurally stable, and compatible scaffolds, such as LSM, Ni/YSZ, YSZ, and doped ceria, forming a nanostructured electrode with substantially enhanced electrocatalytic activity. The advantage of the infiltration method is the separation of the formation temperature of the ionic and catalytic active phase from the high sintering temperature as required to establish the intimate electrode–electrolyte interface bonding. Because the catalytic active components are to be infiltrated in a postfiring step, the infiltrated phases can be formed at a much lower temperature. The infiltration method opens a new horizon in the electrode development as the technique expands the selection of variable electrode material combinations with the minimized TEC mismatch, reduced chemical reactions between electrode and electrolyte materials, and formation of nanosized ionic and catalytic active phases. The nanoscale engineering approach via the low-temperature infiltration process has shown promising potential as the most effective alternative technique in the development of nanostructured electrodes with high performance and advanced microstructure in a way that otherwise would not be possible with high-temperature processes for standard SOFC electrodes. There have been significant progresses in nanostructured electrode development via infiltration as indicated by detailed reviews available elsewhere [158–162].

References

1. Fergus, J.W. 2005. Metallic interconnects for solid oxide fuel cells. *Mater Sci Eng A-Struct Mater Prop Microstruct Process* 397, 271–283.
2. Sun, C.W., Hui, R. and Roller, J. 2010. Cathode materials for solid oxide fuel cells: A review. *J Solid State Electrochem* 14, 1125–1144.
3. Tsipis, E.V. and Kharton, V.V. 2008. Electrode materials and reaction mechanisms in solid oxide fuel cells: A brief review. *J Solid State Electrochem* 12, 1367–1391.
4. Tarancon, A., Burriel, M., Santiso, J., Skinner, S.J. and Kilner, J.A. 2010. Advances in layered oxide cathodes for intermediate temperature solid oxide fuel cells. *J Mater Chem* 20, 3799–3813.
5. Jiang, S.P. and Chan, S.H. 2004. A review of anode materials development in solid oxide fuel cells. *J Mater Sci* 39, 4405–4439.
6. Zhu, W.Z. and Deevi, S.C. 2003. A review on the status of anode materials for solid oxide fuel cells. *Mater Sci Eng A-Struct Mater Prop Microstruct Process* 362, 228–239.
7. Jiang, S.P. 2008. Development of lanthanum strontium manganite perovskite cathode materials of solid oxide fuel cells: A review. *J Mater Sci* 43, 6799–6833.
8. Tsipis, E.V. and Kharton, V.V. 2008. Electrode materials and reaction mechanisms in solid oxide fuel cells: A brief review. *J Solid State Electrochem* 12, 1039–1060.
9. Fergus, J.W. 2006. Electrolytes for solid oxide fuel cells. *J Power Sources* 162, 30–40.
10. Ishihara, T. 2006. Development of new fast oxide ion conductor and application for intermediate temperature solid oxide fuel cells. *Bull Chem Soc Jpn* 79, 1155–1166.
11. Zhu, W.Z. and Deevi, S.C. 2003. Development of interconnect materials for solid oxide fuel cells. *Mater Sci Eng A-Struct Mater Prop Microstruct Process* 348, 227–243.
12. Fergus, J.W. 2005. Sealants for solid oxide fuel cells. *J Power Sources* 147, 46–57.
13. Boukamp, B.A. 2003. Fuel cells: The amazing perovskite anode. *Nat Mater* 2, 294–296.
14. Adler, S.B. 2004. Factors governing oxygen reduction in solid oxide fuel cell cathodes. *Chem Rev* 104, 4791–4843.
15. Mizusaki, J., Mori, N., Takai, H., Yonemura, Y., Minamiue, H., Tagawa, H., Dokiya, M., Inaba, H., Naraya, K., Sasamoto, T. and Hashimoto, T. 2000. Oxygen nonstoichiometry and defect equilibrium in the perovskite-type oxides La1-xSrxMnO3+d. *Solid State Ionics* 129, 163–177.

16. Mizusaki, J., Yonemura, Y., Kamata, H., Ohyama, K., Mori, N., Takai, H., Tagawa, H., Dokiya, M., Naraya, K., Sasamoto, T., Inaba, H. and Hashimoto, T. 2000. Electronic conductivity, Seebeck coefficient, defect and electronic structure of nonstoichiometric La1-xSrxMnO3. *Solid State Ionics* 132, 167–180.

17. Van Roosmalen, J.A.M. and Cordfunke, E.H.P. 1994. The defect chemistry of LaMnO3+/-Delta.3. the density of (La,a)MnO3+Delta (a = Ca, Sr, Ba). *J Solid State Chem* 110, 106–108.

18. Van Roosmalen, J.A.M. and Cordfunke, E.H.P. 1994. The defect chemistry of LaMnO3+/-Delta.4. defect model for LaMnO3+Delta. *J Solid State Chem* 110, 109–112.

19. Van Roosmalen, J.A.M., Cordfunke, E.H.P., Helmholdt, R.B. and Zandbergen, H.W. 1994. The defect chemistry of LaMnO3+/-Delta.2 structural aspects of LaMnO3+Delta. *J Solid State Chem* 110, 100–105.

20. Kamata, H., Yonemura, Y., Mizusaki, J., Tagawa, H., Naraya, K. and Sasamoto, T. 1995. High-temperature electrical-properties of the perovskite-type oxide La1-XSrxMnO3-D. *J Phys Chem Solids* 56, 943–950.

21. Yasuda, I. and Hishinuma, M. 1996. Electrical conductivity and chemical diffusion coefficient of strontium-doped lanthanum manganites. *J Solid State Chem* 123, 382–390.

22. Carter, S., Selcuk, A., Chater, R.J., Kajda, J., Kilner, J.A. and Steele, B.C.H. 1992. Oxygen-transport in selected nonstoichiometric perovskite-structure oxides. *Solid State Ionics* 53–56, 597–605.

23. Ji, Y., Kilner, J. and Carolan, M. 2005. Electrical properties and oxygen diffusion in yttria-stabilised zirconia (YSZ)–LaSrMnO (LSM) composites. *Solid State Ionics* 176, 937–943.

24. Wang, Y., Zhang, L. and Xia, C.R. 2012. Enhancing oxygen surface exchange coefficients of strontium-doped lanthanum manganates with electrolytes. *Int J Hydrogen Energy* 37, 2182–2186.

25. Jørgensen, M.J. and Mogensen, M. 2001. Impedance of solid oxide fuel cell LSM/YSZ composite cathodes. *J Electrochem Soc* 148, A433–A442.

26. Jiang, S.P. and Wang, W. 2005. Fabrication and performance of GDC-impregnated (La,Sr)MnO3 cathodes for intermediate temperature solid oxide fuel cells. *J Electrochem Soc* 152, A1398–A1408.

27. Liang, F., Chen, J., Jiang, S.P., Chi, B., Pu, J. and Jian, L. 2009. High performance solid oxide fuel cells with electrocatalytically enhanced (La, Sr)MnO(3) cathodes. *Electrochem Commun* 11, 1048–1051.

28. Liang, F.L., Chen, J., Jiang, S.P., Chi, B., Pu, J. and Jian, L. 2008. Development of nanostructured and palladium promoted (La,Sr)MnO3-based cathodes for intermediate-temperature SOFCs. *Electrochem Solid-State Lett* 11, B213–B216.

29. Jiang, S.P., Zhen, Y.D. and Zhang, S. 2006. Interaction between Fe-Cr metallic interconnect and (La, Sr) MnO3/YSZ composite cathode of solid oxide fuel cells. *J Electrochem Soc* 153, A1511–A1517.

30. Mineshige, A., Kobune, M., Fujii, S., Ogumi, Z., Inaba, M., Yao, T. and Kikuchi, K. 1999. Metal-insulator transition and crystal structure of La1-xSrxCoO3 as functions of Sr-content, temperature, and oxygen partial pressure. *J Solid State Chem* 142, 374–381.

31. Petrov, A.N., Kononchuk, O.F., Andreev, A.V., Cherepanov, V.A. and Kofstad, P. 1995. Crystal structure, electrical and magnetic properties of La1-xSrxCoO3−y. *Solid State Ionics* 80, 189–199.

32. Yamamoto, O., Takeda, Y., Kanno, R. and Noda, M. 1987. Perovskite-type oxides as oxygen electrodes for high temperature oxide fuel cells. *Solid State Ionics* 22, 241–246.

33. Zipprich, W., Waschilewski, S., Rocholl, F. and Wiemhöfer, H.D. 1997. Improved preparation of La1-xMexCoO3-delta (Me = Sr, Ca) and analysis of oxide ion conductivity with ion conducting microcontacts. *Solid State Ionics* 101–103, 1015–1023.

34. Xia, C.R., Rauch, W., Chen, F.L. and Liu, M.L. 2002. Sm0.5Sr0.5CoO3 cathodes for low-temperature SOFCs. *Solid State Ionics* 149, 11–19.

35. Weber, A. and Ivers-Tiffée, E. 2004. Materials and concepts for solid oxide fuel cells (SOFCs) in stationary and mobile applications. *J Power Sources* 127, 273–283.

36. Baek, S.W., Bae, J. and Yoo, Y.S. 2009. Cathode reaction mechanism of porous-structured Sm0.5Sr0.5CoO3-delta and Sm0.5Sr0.5CoO3-delta/Sm0.2Ce0.8O1.9 for solid oxide fuel cells. *J Power Sources* 193, 431–440.

37. Tai, L.W., Nasrallah, M.M., Anderson, H.U., Sparlin, D.M. and Sehlin, S.R. 1995. Structure and electrical properties of LSCF part the system LSCF. *Solid State Ionics* 76, 259–271.

38. Lv, H., Wu, Y., Huang, B., Zhao, B. and Hu, K. 2006. Structure and electrochemical properties of Sm0.5Sr0.5Co1−xFexO3−δ cathodes for solid oxide fuel cells. *Solid State Ionics* 177, 901–906.

39. Zhao, L., Shen, J., He, B., Chen, F. and Xia, C. 2011. Synthesis, characterization and evaluation of PrBaCo2−xFexO5+δ as cathodes for intermediate-temperature solid oxide fuel cells. *Int J Hydrogen Energy* 36, 3658–3665.

40. Gaur, K., Verma, S.C. and Lal, H.B. 1988. Defects and electrical conduction in mixed lanthanum transition metal oxides. *J Mater Sci* 23, 1725–1728.

41. Stevenson, J.W., Armstrong, T.R., Carneim, R.D., Pederson, L.R. and Weber, W.J. 1996. Electrochemical properties of mixed conducting perovskites La(1-x)M(x)Co(1-y)Fe(y)O(3-delta) (M = Sr,Ba,Ca). *J Electrochem Soc* 143, 2722–2729.

42. Ralph, J.M., Rossignol, C.C. and Kumar, R. 2003. Cathode materials for reduced-temperature SOFCs. *J Electrochem Soc* 150, A1518–A1522.

43. Aguadero, A., Perez-Coll, D., de la Calle, C., Alonso, J.A., Escudero, M.J. and Daza, L. 2009. SrCo(1-x)Sb(x)O(3-delta) perovskite oxides as cathode materials in solid oxide fuel cells. *J Power Sources* 192, 132–137.

44. Shao, Z.P. and Haile, S.M. 2004. A high-performance cathode for the next generation of solid-oxide fuel cells. *Nature* 431, 170–173.

45. Wei, B., Lu, Z., Li, S.Y., Liu, Y.Q., Liu, K.Y. and Su, W.H. 2005. Thermal and electrical properties of new cathode material Ba0.5Sr0.5Co0.8Fe0.2O3-delta for solid oxide fuel cells. *Electrochem Solid-State Lett* 8, A428–A431.

46. Wei, B., Lu, Z., Huang, X., Miao, J., Sha, X., Xin, X. and Su, W. 2006. Crystal structure, thermal expansion and electrical conductivity of perovskite oxides BaxSr1−xCo0.8Fe0.2O3−δ (0.3≤x≤0.7). *J Eur Ceram Soc* 26, 2827–2832.

47. Karthikeyan, C.S., Nunes, S.P. and Schulte, K. 2006. Permeability and conductivity studies on ionomer-polysilsesquioxane hybrid materials. *Macromol Chem Phys* 207, 336–341.

48. Yan, A., Maragou, V., Arico, A., Cheng, M. and Tsiakaras, P. 2007. Investigation of a Ba0.5Sr0.5Co0.8Fe0.2O3-delta based cathode SOFCII. The effect of CO2 on the chemical stability. *Appl Catal B, Environ* 76, 320–327.

49. Bucher, E., Egger, A., Caraman, G.B. and Sitte, W. 2008. Stability of the SOFC cathode material (Ba,Sr)(Co,Fe)O3-delta in CO2-containing atmospheres. *J Electrochem Soc* 155, B1218–B1224.

50. Zhao, H., Teng, D., Zhang, X., Zhang, C. and Li, X. 2009. Structural and electrochemical studies of Ba0.6Sr0.4Co1−yTiyO3−δ as a new cathode material for IT-SOFCs. *J Power Sources* 186, 305–310.
51. Huang, C., Chen, D., Lin, Y., Ran, R. and Shao, Z. 2010. Evaluation of Ba0.6Sr0.4Co0.9Nb0.1O3−δ mixed conductor as a cathode for intermediate-temperature oxygen-ionic solid-oxide fuel cells. *J Power Sources* 195, 5176–5184.
52. Yang, Z., Yang, C., Xiong, B., Han, M. and Chen, F. 2011. BaCo0.7Fe0.2Nb0.1O3−δ as cathode material for intermediate temperature solid oxide fuel cells. *J Power Sources* 196, 9164–9168.
53. Zhou, W., Liang, F.L., Shao, Z.P. and Zhu, Z.H. 2012. Hierarchical CO2-protective shell for highly efficient oxygen reduction reaction. *Sci Rep* 2, 327.
54. Simner, S.P., Anderson, M.D., Engelhard, M.H. and Stevenson, J.W. 2006. Degradation mechanisms of La–Sr–Co–Fe–O3 SOFC cathodes. *Electrochem Solid-State Lett* 9, A478–A481.
55. Oh, D., Gostovic, D. and Wachsman, E.D. 2012. Mechanism of La0.6Sr0.4Co0.2Fe0.8O3 cathode degradation. *J Mater Res* 27, 1992–1999.
56. Vovk, G., Chen, X. and Mims, C.A. 2005. In situ XPS studies of perovskite oxide surfaces under electrochemical polarization. *J Phys Chem B* 109, 2445–2454.
57. Hardy, J.S., Templeton, J.W., Edwards, D.J., Lu, Z.G. and Stevenson, J.W. 2012. Lattice expansion of LSCF-6428 cathodes measured by in situ XRD during SOFC operation. *J Power Sources* 198, 76–82.
58. Kim, Y.M., Chen, X., Jiang, S.P. and Bae, J. 2011. Chromium deposition and poisoning at Ba0.5Sr0.5Co0.8Fe0.2O3−δ cathode of solid oxide fuel cells. *Electrochem Solid-State Lett* 14, B41–B45.
59. Kim, Y.-M., Chen, X., Jiang, S.P. and Bae, J. 2012. Effect of strontium content on chromium deposition and poisoning in Ba1−xSrxCo0.8Fe0.2O3−δ (0.3 ≤ x ≤ 0.7) cathodes of solid oxide fuel cells. *J Electrochem Soc* 159, B185–B194.
60. Jiang, S.P., Zhang, S. and Zhen, Y.D. 2006. Deposition of Cr species at (La,Sr) (Co,Fe)O3 cathodes of solid oxide fuel cells. *J Electrochem Soc* 153, A127–A134.
61. Fang, S.M., Yoo, C.Y. and Bouwmeester, H.J.M. 2011. Performance and stability of niobium-substituted Ba0.5Sr0.5Co0.8Fe0.2O3-delta membranes. *Solid State Ionics* 195, 1–6.
62. Taskin, A.A., Lavrov, A.N. and Ando, Y. 2005. Achieving fast oxygen diffusion in perovskites by cation ordering. *Appl Phys Lett* 86, 91910.
63. Zhang, K., Ge, L., Ran, R., Shao, Z. and Liu, S. 2008. Synthesis, characterization and evaluation of cation-ordered LnBaCo2O5+δ as materials of oxygen permeation membranes and cathodes of SOFCs. *Acta Mater* 56, 4876–4889.
64. Chen, D., Ran, R., Zhang, K., Wang, J. and Shao, Z. 2009. Intermediate-temperature electrochemical performance of a polycrystalline PrBaCo2O5+δ cathode on samarium-doped ceria electrolyte. *J Power Sources* 188, 96–105.
65. Kim, J.H., Cassidy, M., Irvine, J.T.S. and Bae, J. 2009. Advanced electrochemical properties of LnBa(0.5)Sr(0.5)Co(2)O(5+delta) (Ln = Pr, Sm, and Gd) as cathode materials for IT-SOFC. *J Electrochem Soc* 156, B682–B689.
66. Jin, C. and Liu, J. 2009. Preparation of Ba1.2Sr0.8CoO4+delta K2NiF4-type structure oxide and cathodic behavioral of Ba1.2Sr0.8CoO4+delta-GDC composite cathode for intermediate temperature solid oxide fuel cells. *J Alloy Compd* 474, 573–577.
67. Nagasawa, K., Daviero-Minaud, S., Preux, N., Rolle, A., Roussel, P., Nakatsugawa, H. and Mentre, O. 2009. Ca3Co4O9−δ: A thermoelectric material for SOFC cathode. *Chem Mat* 21, 4738–4745.

68. Vert, V.B., Serra, J.M. and Jorda, J.L. 2010. Electrochemical characterisation of MBaCo3ZnO7+delta (M = Y, Er, Tb) as SOFC cathode material with low thermal expansion coefficient. *Electrochem Commun* 12, 278–281.
69. Choi, Y., Lin, M.C. and Liu, M.L. 2007. Computational study on the catalytic mechanism of oxygen reduction on La(0.5)Sr(0.5)MnO(3) in solid oxide fuel cells. *Angew Chem Int Ed Engl* 46, 7214–7219.
70. Choi, Y.M., Mebane, D.S., Lin, M.C. and Liu, M.L. 2007. Oxygen reduction on LaMnO3-based cathode materials in solid. *Chem Mat* 19, 1690–1699.
71. Choi, Y., Lin, M.C. and Liu, M. 2010. Rational design of novel cathode materials in solid oxide fuel cells using first-principles simulations. *J Power Sources* 195, 1441–1445.
72. Lee, Y.-L., Kleis, J., Rossmeisl, J., Shao-Horn, Y. and Morgan, D. 2011. Prediction of solid oxide fuel cell cathode activity with first-principles descriptors. *Energy Environ Sci* 4, 3966–3970.
73. Chen, D., Lin, Z., Zhu, H. and Kee, R.J. 2009. Percolation theory to predict effective properties of solid oxide fuel-cell composite electrodes. *J Power Sources* 191, 240–252.
74. Costamagna, P., Costa, P. and Antonucci, V. 1997. Micro-modelling of solid oxide fuel cell electrodes. *Electrochim Acta* 43, 375–394.
75. Costamagna, P., Panizza, M., Giacomo, C. and Barbucci, A. 2002. Effect of composition on the performance of cermet electrodes: Experimental and theoretical approach. *Electrochim Acta* 47, 1079–1089.
76. Janardhanan, V.M., Heuveline, V. and Deutschmann, O. 2008. Three-phase boundary length in solid-oxide fuel cells: A mathematical model. *J Power Sources* 178, 368–372.
77. Iwai, H., Shikazono, N., Matsui, T., Teshima, H., Kishimoto, M., Kishida, R., Hayashi, D., Matsuzaki, K., Kanno, D., Saito, M., Muroyama, H., Eguchi, K., Kasagi, N. and Yoshida, H. 2010. Quantification of SOFC anode microstructure based on dual beam FIB-SEM technique. *J Power Sources* 195, 955–961.
78. Laosiripojana, N. and Assabumrungrat, S. 2007. Catalytic steam reforming of methane, methanol, and ethanol over Ni/YSZ: The possible use of these fuels in internal reforming SOFC. *J Power Sources* 163, 943–951.
79. Dees, D.W. 1987. Conductivity of porous Ni/ZrO2-Y2O3 cermets. *J Electrochem Soc* 134, 2141.
80. Jiang, S.P. and Chan, S.H. 2004. Development of Ni/Y2O3-ZrO2 cermet anodes for solid oxide fuel cells. *Mater Sci Technol* 20, 1109–1118.
81. Nikooyeh, K., Clemmer, R., Alzate-Restrepo, V. and Hill, J.M. 2008. Effect of hydrogen on carbon formation on Ni/YSZ composites exposed to methane. *Appl Catal A: Gen* 347, 106–111.
82. Alzate-Restrepo, V. and Hill, J.M. 2010. Carbon deposition on Ni/YSZ anodes exposed to CO/H2 feeds. *J Power Sources* 195, 1344–1351.
83. Rasmussen, J.F.B. and Hagen, A. 2009. The effect of H2S on the performance of Ni–YSZ anodes in solid oxide fuel cells. *J Power Sources* 191, 534–541.
84. Hagen, A., Rasmussen, J.F.B. and Thydén, K. 2011. Durability of solid oxide fuel cells using sulfur containing fuels. *J Power Sources* 196, 7271–7276.
85. Toebes, M.L., Bitter, J.H., van Dillen, A.J. and de Jong, K.P. 2002. Impact of the structure and reactivity of nickel particles on the catalytic growth of carbon nanofibers. *Catal Today* 76, 33–42.
86. Murray, E.P., Tsai, T. and Barnett, S.A. 1999. A direct-methane fuel cell with a ceria-based anode. *Nature* 400, 649–651.

87. Xu, S. and Wang, X. 2005. Highly active and coking resistant Ni/CeO2-ZrO2 catalyst for partial oxidation of methane. *Fuel* 84, 563–567.
88. Yang, L., Wang, S., Blinn, K., Liu, M., Liu, Z. and Cheng, Z. 2009. Enhanced sulfur and coking tolerance of a mixed ion conductor for SOFCs: BaZr(0.1)Ce(0.7)Y(0.2-x)Yb(x)O(3-delta). *Science* 326, 126–129.
89. Cheng, Z., Wang, J.-H., Choi, Y., Yang, L., Lin, M.C. and Liu, M. 2011. From Ni-YSZ to sulfur-tolerant anode materials for SOFCs: Electrochemical behavior, in situ characterization, modeling, and future perspectives. *Energy Environ Sci* 4, 4380–4409.
90. Yang, L., Choi, Y., Qin, W., Chen, H., Blinn, K., Liu, M., Liu, P., Bai, J. and Tyson, T.A. 2011. Promotion of water-mediated carbon removal by nanostructured barium oxide/nickel interfaces in solid oxide fuel cells. *Nat Commun* 2, 357–367.
91. Gorte, R.J., Kim, H. and Vohs, J.M. 2002. Novel SOFC anodes for the direct electrochemical oxidation of hydrocarbon. *J Power Sources* 106, 10–15.
92. McIntosh, S. and Gorte, R.J. 2004. Direct hydrocarbon solid oxide fuel cells. *Chem Rev* 104, 4845–4865.
93. Park, S.D., Vohs, J.M. and Gorte, R.J. 2000. Direct oxidation of hydrocarbons in a solid-oxide fuel cell. *Nature* 404, 265–267.
94. Lashtabeg, A. and Skinner, S.J. 2006. Solid oxide fuel cells-a challenge for materials chemists? *J Mater Chem* 16, 3161–3170.
95. Kim, H., Lu, C., Worrell, W.L., Vohs, J.M. and Gorte, R.J. 2002. Cu-Ni cermet anodes for direct oxidation of methane in solid-oxide fuel cells. *J Electrochem Soc* 149, A247–A250.
96. Lee, S.-I., Vohs, J.M. and Gorte, R.J. 2004. A study of SOFC anodes based on Cu-Ni and Cu-Co bimetallics in CeO2-YSZ. *J Electrochem Soc* 151, A1319–A1323.
97. Hibino, T., Hashimoto, A., Asano, K., Yano, M., Suzuki, M. and Sano, M. 2002. An intermediate-temperature solid oxide fuel cell providing higher performance with hydrocarbons than with hydrogen. *Electrochem Solid-State Lett* 5, A242–A244.
98. Hibino, T., Hashimoto, A., Yano, M., Suzuki, M. and Sano, M. 2003. Ru-catalyzed anode materials for direct hydrocarbon SOFCs. *Electrochim Acta* 48, 2531–2537.
99. Zhan, Z. and Barnett, S.A. 2005. An octane-fueled solid oxide fuel cell. *Science* 308, 844–847.
100. Wang, W., Zhou, W., Ran, R., Cai, R. and Shao, Z. 2009. Methane-fueled SOFC with traditional nickel-based anode by applying Ni/Al2O3 as a dual-functional layer. *Electrochem Commun* 11, 194–197.
101. Wang, W., Su, C., Wu, Y., Ran, R. and Shao, Z. 2010. A comprehensive evaluation of a Ni–Al2O3 catalyst as a functional layer of solid-oxide fuel cell anode. *J Power Sources* 195, 402–411.
102. Wang, W., Su, C., Ran, R. and Shao, Z. 2011. A new Gd-promoted nickel catalyst for methane conversion to syngas and as an anode functional layer in a solid oxide fuel cell. *J Power Sources* 196, 3855–3862.
103. Cowin, P.I., Petit, C.T.G., Lan, R., Irvine, J.T.S. and Tao, S. 2011. Recent progress in the development of anode materials for solid oxide fuel cells. *Adv Energy Mater* 1, 314–332.
104. Li, X., Zhao, H., Xu, N., Zhou, X., Zhang, C. and Chen, N. 2009. Electrical conduction behavior of La, Co co-doped SrTiO3 perovskite as anode material for solid oxide fuel cells. *Int J Hydrogen Energy* 34, 6407–6414.

105. Smith, B.H., Holler, W.C. and Gross, M.D. 2011. Electrical properties and redox stability of tantalum-doped strontium titanate for SOFC anodes. *Solid State Ionics* 192, 383–386.
106. Li, X., Zhao, H., Zhou, X., Xu, N., Xie, Z. and Chen, N. 2010. Electrical conductivity and structural stability of La-doped SrTiO3 with A-site deficiency as anode materials for solid oxide fuel cells. *Int J Hydrogen Energy* 35, 7913–7918.
107. Zhao, H., Gao, F., Li, X., Zhang, C. and Zhao, Y. 2009. Electrical properties of yttrium doped strontium titanate with A-site deficiency as potential anode materials for solid oxide fuel cells. *Solid State Ionics* 180, 193–197.
108. Gao, F., Zhao, H., Li, X., Cheng, Y., Zhou, X. and Cui, F. 2008. Preparation and electrical properties of yttrium-doped strontium titanate with B-site deficiency. *J Power Sources* 185, 26–31.
109. Li, X., Zhao, H., Shen, W., Gao, F., Huang, X., Li, Y. and Zhu, Z. 2007. Synthesis and properties of Y-doped SrTiO3 as an anode material for SOFCs. *J Power Sources* 166, 47–52.
110. Lu, X., Pine, T., Mumm, D. and Brouwer, J. 2007. Modified Pechini synthesis and characterization of Y-doped strontium titanate perovskite. *Solid State Ionics* 178, 1195–1199.
111. Li, X., Zhao, H., Gao, F., Zhu, Z., Chen, N. and Shen, W. 2008. Synthesis and electrical properties of Co-doped Y0.08Sr0.92TiO3−δ as a potential SOFC anode. *Solid State Ionics* 179, 1588–1592.
112. Tao, S.W. and Irvine, J.T. 2003. A redox-stable efficient anode for solid-oxide fuel cells. *Nat Mater* 2, 320–323.
113. Bao, W., Guan, H. and Cheng, J. 2008. A new anode material for intermediate solid oxide fuel cells. *J Power Sources* 175, 232–237.
114. Tao, S. and Irvine, J.T. 2004. Discovery and characterization of novel oxide anodes for solid oxide fuel cells. *Chem Rec* 4, 83–95.
115. Gu, H., Zheng, Y., Ran, R., Shao, Z., Jin, W., Xu, N. and Ahn, J. 2008. Synthesis and assessment of La0.8Sr0.2ScyMn1−yO3−δ as cathodes for solid-oxide fuel cells on scandium-stabilized zirconia electrolyte. *J Power Sources* 183, 471–478.
116. Bastidas, D.M., Tao, S. and Irvine, J.T.S. 2006. A symmetrical solid oxide fuel cell demonstrating redox stable perovskite electrodes. *J Mater Chem* 16, 1603–1605.
117. Huang, Y.H., Dass, R.I., Xing, Z.L. and Goodenough, J.B. 2006. Double perovskites as anode materials for solid-oxide fuel cells. *Science* 312, 254–257.
118. Huang, Y.H., Liang, G., Croft, M., Lehtimaki, M., Karppinen, M. and Goodenough, J.B. 2009. Double-perovskite anode materials Sr2MMoO6 (M = Co, Ni) for solid oxide fuel cells. *Chem Mater* 21, 2319–2326.
119. Zhang, L., Zhou, Q., He, Q. and He, T. 2010. Double-perovskites A2FeMoO6−δ (A = Ca, Sr, Ba) as anodes for solid oxide fuel cells. *J Power Sources* 195, 6356–6366.
120. Xiao, G., Liu, Q., Dong, X., Huang, K. and Chen, F. 2010. Sr2Fe4/3Mo2/3O6 as anodes for solid oxide fuel cells. *J Power Sources* 195, 8071–8074.
121. He, B., Zhao, L., Song, S., Liu, T., Chen, F. and Xia, C. 2012. Sr2Fe1.5Mo0.5O6−δ - Sm0.2Ce0.8O1.9 composite anodes for intermediate-temperature solid oxide fuel cells. *J Electrochem Soc* 159, B619–B626.
122. Mukundan, R., Brosha, E.L. and Garzon, F.H. 2004. Sulfur tolerant anodes for SOFCs. *Electrochem Solid-State Lett* 7, A5–A7.
123. Balachandran, U. and Eror, N.G. 1981. Electrical conductivity in strontium titanate. *J Solid State Chem* 39, 351–359.

124. Marina, O.A. and Pederson, L.R. 2002. Investigation of a Ba0.5Sr0.5Co0.8Fe0.2O3-delta based cathode SOFCII: The effect of CO2 on the chemical stability. In J. Huijsmans (Ed.), Fifth European Solid Oxide Fuel Cell Forum, European Fuel Cell Forum, Oberrohfdorf, Switzerland, pp. 481–489.

125. Yoo, K.B. and Choi, G.M. 2009. Performance of La-doped strontium titanate (LST) anode on LaGaO3-based SOFC. *Solid State Ionics* 180, 867–871.

126. Lu, X.C., Zhu, J.H., Yang, Z., Xia, G. and Stevenson, J.W. 2009. Pd-impregnated SYT/LDC composite as sulfur-tolerant anode for solid oxide fuel cells. *J Power Sources* 192, 381–384.

127. Zheng, L.L., Wang, X., Zhang, L., Wang, J.Y. and Jiang, S.P. 2012. Effect of Pd-impregnation on performance, sulfur poisoning and tolerance of Ni/GDC anode of solid oxide fuel cells. *Int J Hydrogen Energy* 37, 10299–10310.

128. Kurokawa, H., Sholklapper, T.Z., Jacobson, C.P., De Jonghe, L.C. and Visco, S.J. 2007. Ceria nanocoating for sulfur tolerant Ni-based anodes of solid oxide fuel cells. *Electrochem Solid-State Lett* 10, B135–B138.

129. Kurokawa, H., Yang, L., Jacobson, C.P., De Jonghe, L.C. and Visco, S.J. 2007. Y-doped SrTiO3 based sulfur tolerant anode for solid oxide fuel cells. *J Power Sources* 164, 510–518.

130. Tao, S.W. and Irvine, J.T.S. 2004. Synthesis and characterization of (La0.75Sr0.25)(Cr0.5Mn0.5)O3−δ, a redox-stable, efficient perovskite anode for SOFCs. *J Electrochem Soc* 151, A252–A259.

131. Jiang, S.P., Chen, X.J., Chan, S.H., Kwok, J.T. and Khor, K.A. 2006. (La0.75Sr0.25)(Cr0.5Mn0.5)O-3/YSZ composite anodes for methane oxidation reaction in solid oxide fuel cells. *Solid State Ionics* 177, 149–157.

132. Zha, S., Tsang, P., Cheng, Z. and Liu, M. 2005. Electrical properties and sulfur tolerance of La0.75Sr0.25Cr1−xMnxO3 under anodic conditions. *J Solid State Chem* 178, 1844–1850.

133. Danilovic, N., Vincent, A., Luo, J.-L., Chuang, K.T., Hui, R. and Sanger, A.R. 2010. Correlation of fuel cell anode electrocatalytic and ex situ catalytic activity of perovskites La0.75Sr0.25Cr0.5X0.5O3−δ(X = Ti, Mn, Fe, Co)†. *Chem Mater* 22, 957–965.

134. Haag, J.M., Barnett, S.A., Richardson, J.W. and Poeppelmeier, K.R. 2010. Structural and chemical evolution of the SOFC anode La0.30Sr0.70Fe0.70Cr0.30O3−δ upon reduction and oxidation: An in situ neutron diffraction study. *Chem Mater* 22, 3283–3289.

135. Kobsiriphat, W., Madsen, B.D., Wang, Y., Shah, M., Marks, L.D. and Barnett, S.A. 2010. Nickel- and ruthenium-doped lanthanum chromite anodes: Effects of nanoscale metal precipitation on solid oxide fuel cell performance. *J Electrochem Soc* 157, B279–B284.

136. Kobsiriphat, W., Madsen, B.D., Wang, Y., Marks, L.D. and Barnett, S.A. 2009. La0.8Sr0.2Cr1−xRuxO3−δ–Gd0.1Ce0.9O1.95 solid oxide fuel cell anodes: Ru precipitation and electrochemical performance. *Solid State Ionics* 180, 257–264.

137. Lay, E., Gauthier, G., Rosini, S., Savaniu, C. and Irvine, J.T.S. 2008. Ce-substituted LSCM as new anode material for SOFC operating in dry methane. *Solid State Ionics* 179, 1562–1566.

138. Jiang, S.P., Chen, X.J., Chan, S.H. and Kwok, J.T. 2006. GDC-impregnated, (La0.75Sr0.25)(Cr0.5Mn0.5)O-3 anodes for direct utilization of methane in solid oxide fuel cells. *J Electrochem Soc* 153, A850–A856.

139. Liu, J., Madsen, B.D., Ji, Z. and Barnett, S.A. 2002. A fuel-flexible ceramic-based anode for solid oxide fuel cells. *Electrochem Solid-State Lett* 5, A122–A124.

140. Kim, G., Lee, S., Shin, J.Y., Corre, G., Irvine, J.T.S., Vohs, J.M. and Gorte, R.J. 2009. Investigation of the structural and catalytic requirements for high-performance SOFC anodes formed by infiltration of LSCM. *Electrochem Solid-State Lett* 12, B48–B52.

141. Tao, S.W. and Irvine, J.T.S. 2002. Study on the structural and electrical properties of the double perovskite oxide SrMn0.5Nb0.5O3-delta. *J Mater Chem* 12, 2356–2360.

142. Tao, S., Canales-Vazquez, J. and Irvine, J.T.S. 2004. Structural and electrical properties of the perovskite oxide Sr2FeNbO6. *Chem Mat* 16, 2309–2316.

143. Zhang, P., Huang, Y.-H., Cheng, J.-G., Mao, Z.-Q. and Goodenough, J.B. 2011. Sr2CoMoO6 anode for solid oxide fuel cell running on H2 and CH4 fuels. *J Power Sources* 196, 1738–1743.

144. Vasala, S., Lehtimäki, M., Huang, Y.H., Yamauchi, H., Goodenough, J.B. and Karppinen, M. 2010. Degree of order and redox balance in B-site ordered double-perovskite oxides, Sr2MMoO6–δ (M = Mg, Mn, Fe, Co, Ni, Zn). *JoJ Solid State Chem* 183, 1007–1012.

145. Liu, Q., Dong, X., Xiao, G., Zhao, F. and Chen, F. 2010. A novel electrode material for symmetrical SOFCs. *Adv Mater* 22, 5478–5482.

146. Mogensen, M., Sammes, N.M. and Tompsett, G.A. 2000. Physical, chemical and electrochemical properties of pure and doped ceria. *Solid State Ionics* 129, 63–94.

147. Ishihara, T., Shin, T.H., Vanalabhpatana, P., Yonemoto, K. and Matsuka, M. 2010. Ce-0.6(Mn0.3Fe0.1)O-2 as an oxidation-tolerant ceramic anode for SOFCs using LaGaO3-based oxide electrolyte. *Electrochem Solid-State Lett* 13, B95–B97.

148. Sun, C. and Stimming, U. 2007. Recent anode advances in solid oxide fuel cells. *J Power Sources* 171, 247–260.

149. Liu, Z.G., Ouyang, J.H., Sun, K.N. and Zhou, Y. 2012. Effect of CaO addition on the structure and electrical conductivity of the pyrochlore-type GdSmZr2O7. *Ceram Int* 38, 2935–2941.

150. Kramer, S.A., Spears, M. and Tuller, H.L. 1994. Conduction in titanate pyrochlores role of dopants. *Solid State Ionics* 72, 59–66.

151. Kramer, S.A. and Tuller, H.L. 1995. A novel titanate-based oxygen ion conductor Gd2Ti2O7. *Solid State Ionics* 82, 15–23.

152. Porat, O., Heremans, C. and Tuller, H.L. 1997. Stability and mixed ionic electronic conduction in Gd-2(Ti1-xMox)(2)O-7 under anodic conditions. *Solid State Ionics* 94, 75–83.

153. Mailley, S.C., Kelaidopoulou, A., Siddle, A., Dicks, A.L., Holtappels, P., Hatchwell, C.E. and Mogensen, M. 2000. Electrocatalytic activity of a Gd(2) Ti(0.6)Mo(1.2)Sc(0.2)O(7-delta) anode towards hydrogen and methane electrooxidation in a solid oxide fuel cell. *Ionics* 6, 331–339.

154. Ge, X.-M., Chan, S.-H., Liu, Q.-L. and Sun, Q. 2012. Solid oxide fuel cell anode materials for direct hydrocarbon utilization. *Adv Energy Mater* 2, 1156–1181.

155. Slater, P.R. and Irvine, J.T.S. 1999. Synthesis and electrical characterisation of the tetragonal tungsten bronze type phases, (Ba/Sr/Ca/La)(0) 6MxNb1-xO3-delta (M = Mg, Ni, Mn, Cr, Fe, In, Sn): Evaluation as potential anode materials for solid oxide fuel cells. *Solid State Ionics* 124, 61–72.

156. Slater, P.R. and Irvine, J.T.S. 1999. Niobium based tetragonal tungsten bronzes as potential anodes for solid oxide fuel cells: Synthesis and electrical characterisation. *Solid State Ionics* 120, 125–134.
157. Li, Q. and Thangadurai, V. 2011. Novel Nd2WO6-type Sm(2-x)A(x)M(1-y)B(y) O(6-delta) (A = Ca, Sr; M = Mo, W; B = Ce, Ni) mixed conductors. *J Power Sources* 196, 169–178.
158. Jiang, S.P. 2006. A review of wet impregnation: An alternative method for the fabrication of high performance and nano-structured electrodes of solid oxide fuel cells. *Mater Sci Eng A-Struct Mater Prop Microstruct Process* 418, 199–210.
159. Jiang, S.P. 2012. Nanoscale and nano-structured electrodes of solid oxide fuel cells by infiltration: Advances and challenges. *Int J Hydrogen Energy* 37, 449–470.
160. Gorte, R.J. and Vohs, J.M. 2009. Nanostructured anodes for solid oxide fuel cells. *Curr Opin Colloid Interface Sci* 14, 236–244.
161. Vohs, J.M. and Gorte, R.J. 2009. High-performance SOFC cathodes prepared by infiltration. *Adv Mater* 21, 943–956.
162. Jiang, Z., Xia, C. and Chen, F. 2010. Nano-structured composite cathodes for intermediate-temperature solid oxide fuel cells via an infiltration/impregnation technique. *Electrochim Acta* 55, 3595–3605.

3

Seals for Planar Solid Oxide Fuel Cells: The State of the Art

Teng Zhang and Shaorong Wang

CONTENTS

3.1 Introduction 45
    3.1.1 Different Seals for Planar Cell Design 46
    3.1.2 Hermeticity of Different Seals 47
3.2 Critical Issues for the Design of Sealants 49
    3.2.1 Critical Properties 49
        3.2.1.1 Coefficient of Thermal Expansion 50
        3.2.1.2 Sealing Temperature 52
        3.2.1.3 Thermomechanical Stability 52
        3.2.1.4 Chemical Stability 55
        3.2.1.5 Glass Volatility 57
        3.2.1.6 Electrical Resistance 59
    3.2.2 Critical Process: Crystallization 59
3.3 Conclusions and Future Works 61
Acknowledgments 61
References 62

3.1 Introduction

The solid oxide fuel cell (SOFC) technology has attracted much attention in the last two decades. SOFC has many advantages over other power-producing technologies, including a high electric conversion efficiency that is not limited by Carnot cycle considerations because the chemical energy is directly converted into electricity together with lots of heat inside the SOFC chamber. A SOFC produces heat and power, which improves its efficiency up to about 80%. In addition, a SOFC has an excellent environmental performance because there is only water and CO_2 emission from a fuel cell with O_2 as the oxidant.[1]

The traditional SOFC operates at about 900°C–1000°C, which restricts the choice of materials for SOFC design.[2] Typical materials used in SOFCs are Ca- or Sr-doped $LaMnO_3$ as the cathode, Ni/zirconia cermets as the anode, and ~8%–10% Yttria-doped zirconia (YSZ) as the electrolyte. A high-temperature SOFC uses conductive oxides, such as $(La,Sr)CrO_3$, for intercon-nects. This material is difficult to sinter to a fully dense state and is expensive for manufacturing. Furthermore, the corrosion and volatilization of stack components and formation of an undesired phase at the interfaces between various cell components seriously affect the chemical stability of the high-temperature SOFC. Therefore, there is a strong drive to lower the operational temperature to around 700°C–800°C. The operational temperature can be decreased by the use of very thin film electrolytes, which are supported on the air or fuel electrode[2] by using new electrolytes with a higher oxide-ion conductivity, such as scandia-doped zirconia, gadolinia-doped ceria, and lanthanum gallate,[3,4] and by using alternative electrode materials.[5] If lower temperatures are practicable, inexpensive materials, including stainless steels, can be used as interconnects in SOFC design.

SOFCs can be classified into two groups by their unit cell designs: pla-nar and tubular.[6] The planar cell designs have a number of advantages over the tubular designs, including simpler manufacturing processes and a rela-tively short current path that results in higher power density and efficiency.[7,8] However, there are many edges that need to be sealed at high temperatures. Therefore, joining and sealing technologies are the most important research for the planar SOFC components.

3.1.1 Different Seals for Planar Cell Design

A variety of different joining and sealing techniques have been developed for planar SOFCs, including brazing, glass and glass-ceramic sealing, ther-mal spraying, and sol-gel deposition.[9] For the glass and glass-ceramic joining and sealing technologies, compressive sealing, rigid sealing, and compliant sealing comprise most of the design options.[10–12]

In compressive sealing, a compliant high-temperature material is captured between the two sealing surfaces and compressed, using a load frame out-side the stack, to form a dynamic seal. Because the sealing surfaces can slide past one another without destroying the hermeticity of the seal, a coefficient of thermal expansion (CTE) matching is not required between the ceramic cell and the metal separator. A number of materials, such as mica, nickel, and copper, have been considered, but each has been found deficient for different reasons, ranging from oxidation resistance in the case of the metals to poor hermeticity and through-seal leakage in the case of mica.[13] With regards to the poor hermeticity and leakage, a novel hybrid compressive mica seal was developed by Chou et al.[14–18] Compared with simple mica seals, the leak rate of this hybrid compressive mica was reduced by 4300 times at 800°C due to the special structure of two compliant glass layers sandwiching a mica layer.[16] However, an additional difficulty with the use of a compressive seal is in designing the load frame[10] as it must be capable of delivering loads in high temperatures, oxidizing environments over the entire period of operation.

Material oxidation and load relaxation due to creep as well as added expense and additional thermal mass that must be heated, cooled, and maintained at temperatures under equilibrium operation are all issues that need to be considered carefully.

Resilient sealing provides an opportunity to seal the glass with substrates without the required CTE match. Ideally, the glass should remain viscous at high temperatures, for example, 800°C, to release the thermal stress generated by any CTE mismatch during thermal cycles. The glass melt should also flow to heal any cracks in the seals that form during operation and still maintain the necessary mechanical strength to support the substrates. Raj[19] developed a composition that remains viscous after 500 h of exposure at 800°C. It takes about 5 min to heal the crack generated in the glass matrix at the temperature range of 550°C to 600°C. However, this resilient glass, with a viscosity that is too low to support other SOFC components, might also cause the leak in the seals at high temperatures due to the movement of glass at high temperature.[20] A thermal stable $SrO-La_2O_3-Al_2O_3-B_2O_3-SiO_2$ glass system was developed recently by suppressing the crystallization with the addition of Al_2O_3 (7–17 mol%).[21] However, the great content of La_2O_3 (20 mol%) and SrO (40 mol%) in this system still leads to the formation of a crystal phase of $LaBO_3$[21] under heat treatment. The great content of B_2O_3 (40 mol%) will be another problem due to its volatility under SOFC operational conditions.[22] An improved glass system of $La_2O_3-Al_2O_3-B_2O_3-SiO_2$ with the nominal composition of $35(SrO+La_2O_3)-5Al_2O_3-60SiO_2$ (mol%) was reported to remain amorphous after a heat treatment at 850°C for 200 h.[23,24] The thermal stability of this glass system was also explained in terms of the glass network connectivity, which decreases with the increase of B_2O_3 content.[23] The greater content of La_2O_3 in this composition tends to react with ferritic stainless steel interconnects such as Crofer 22 APU to form $La_2Cr_2O_9$ in SOFC operational environments.[25] In addition, the modifications on the species, such as polishing, accelerated the surface crystallization of such glass.[26] A recent work on a crystalline phase with a low melting point, such as CaB_4O_7 (~986°C), provides an additional approach for obtaining a desired self-healing property through the viscous flow of a ceramic phase at a temperature below its melting point (e.g., 840°C).[27] However, much work is needed to integrate such self-healing crystalline phases into a traditional sealing system.

3.1.2 Hermeticity of Different Seals

The primary function of a sealant is to prevent the mixing of fuel and oxidants within the stack and prevent leaking of fuel and oxidants from the stack. For planar SOFC stacks, a leak rate corresponding to ≤1% of total fuel flow for a 6 in. × 6 in. cell at a differential pressure of 0.147 kPa (~0.02 psi) was proposed by the Solid Energy Convergence Alliance as a criterion for sealant development. Compressive stress of 20.7 kPa (3 psi) and 689.5 kPa (100 psi)

TABLE 3.1

Helium Leak Rates of Different Seals

Compositions	Compressive Stress (kPa)	Pressure Gradient (kPa)	Temperature (°C)	Hold Time/ Thermal Cycle	Helium Leak Rate (sccm/cm)
Mica/borosilicate glass between YSZ and inconel[16]	689.5	13.8	800	0.5 h	1.6×10^{-4}
	344.7				2.4×10^{-4}
	172.4				3.6×10^{-4}
Ba–Al–Ca silicate glass/phlogopite mica flakes[18]	20.7	1.4	800	28 cycles	0.04–0.05
	82.7			36 cycles	0.011–0.015
	41.4			25 cycles	0.01
	41.4	13.8		25 cycles	0.04–0.05
H_3BO_3 infiltrated mica between YSZ and inconel[28]	689.5	13.8	800	15 cycles	5.0×10^{-4}
				36 cycles	2.5×10^{-3}
Hybrid phlogopite mica with silver interlayers[15]	82.7	1.4	800	28,366 h	0.01–0.02
Al_2O_3-based compressive seal[64]	140	10.5	750	96 cycles	0.04
Fumed silica-infiltrated alumina–silica fiber paper gaskets[63]	1000	10.3	850	N/A	0.04
	50.3	1.4			0.05
Gasket with compressive mica between Crofer[65]	689.5	20.7	800	400 h	1.5×10^{-3}
				2 cycles	2.1×10^{-3}
				34 cycles	2.4×10^{-3}
Aluminosilicate glass between YSZ and Crofer[66]	103.4	6.9	800	2500 h	No leak
				3 cycles	Leak and healed
Aluminosilicate glass mixed with ZrO_2 between YSZ and Crofer[32]	551.6	13.8	800	50 h	6.2×10^{-3}
SrO–CaO–NiO–B_2O_3 glass with 10 vol% Ni between YSZ and Crofer[35]	34.5	2.8	800	10 cycles	1.6×10^{-4}
Aluminosilicate glass between YSZ and Crofer[37–39]	10.3	150	800	100 h	10^{-3}–10^{-4}

were chosen as a lower and upper limit, respectively, to reduce the applied load for achieving such stress in SOFC stacks.[18]

The helium leak test provides a simple way to quantify the hermeticity of seals by monitoring the pressure change of helium through the interface between sealants and other SOFC components. The leak rate (in sccm, standard cubic centimeter per minute), calculating from the pressure drop across the closed system over time, makes it possible to compare the hermeticity in different systems.[16] The leak rate reported in Pa m²/s can also be converted into sccm. In addition, the *in situ* fixture allows the measurement to be made during the thermal cycles, which is more similar to the actual condition than other approaches. The leak rates of different seals are summarized in Table 3.1. Generally, the leak rate of seals decreases with increasing compressive stress and pressure gradient.

Because of problems with the mechanical stability and thermal stability of both the "resilient seals" and "compressive seals," much research has focused on developing SOFC seals using rigid glass and glass-ceramic sealants.[11,28–32] Various glass and glass-ceramic systems have been investigated in the last decade, including borosilicate, aluminosilicate, aluminoborosilicate, phosphate, and other systems.[6,29,33–54] In addition, many composite seals haven been developed for SOFC application by adding different oxides[23,55–61] as well as glass fibers[62,63] to a glass matrix. To accomplish the sealing target of planar SOFCs, some critical issues for the design of sealants need to be taken into account, which will be discussed in the following sections.

3.2 Critical Issues for the Design of Sealants

3.2.1 Critical Properties

The sealant within the SOFC stack is designed to prevent mixing of fuel and oxidants, to prevent leaking of fuel or oxidants from the stack, to electrically isolate cells within the stack, and also to provide mechanical bonding of components. Therefore, many requirements for the sealing design need to be fulfilled, including a thermal expansion match to the fuel cell components, electrically insulating properties, and thermochemical stability under the operational conditions of the stack. The seal also should be chemically stable with other cell components; should be stable under both the high-temperature oxidizing and reducing operational conditions, that is, 800°C for at least 40,000 h; should be created at a low enough temperature to avoid damaging other cell components (typically under 900°C for some materials); and should not migrate or flow from the designated sealing region during sealing or cell operation. In addition, the sealing system should be able to withstand thermal cycling between the operational temperature and room temperature.

3.2.1.1 Coefficient of Thermal Expansion

A model based on the classical beam bending theory and the fracture theory of ceramic materials was developed by Zhang et al.[67] to predict the crack extension in the seal. It was found that the resistance of the seal to cracking on cooling depends on the thickness of the seal as well as the CTE mismatch. Therefore, the CTE of the glass-ceramic sealants must be compatible with those of other SOFC components, such as the YSZ electrolyte ($\sim 10 \times 10^{-6}$/K)[7] and the stainless steel interconnects ($\sim 13 \times 10^{-6}$/K),[56] to minimize the thermal stress developed during the fabrication and operation of SOFC stacks.

Much work has been done to tailor the CTE by either changing the composition or adding second phase fillers, for example, Al_2O_3,[61] YSZ,[55] NiO,[23,56] ZrO_2,[58] and MgO[59] to a sealing glass. It is well established that additions of alkaline and alkaline earth oxide components to the glasses as the network modifiers can increase the overall CTE of sealants.[3,39,68,69] Our research on the CTE of more than 80 compositions based on the "invert glasses" system also reveals that the CTE of glass increases with the decreasing average field strength of the modifiers (RO)[70] as shown in Figure 3.1. The average field strength of some modifiers, for example, CaO, SrO, BaO, and ZnO, are also summarized in Table 3.2.

In addition, the CTE of glass also depends on the ratio of B_2O_3/SiO_2,[6,39,48] although the effect of boron oxide on the CTE of glass is complex. Greater CTE in the glasses with a higher ratio of B_2O_3/SiO_2 was reported for the BaO–Al_2O_3–La_2O_3–B_2O_3–SiO_2 system.[6]

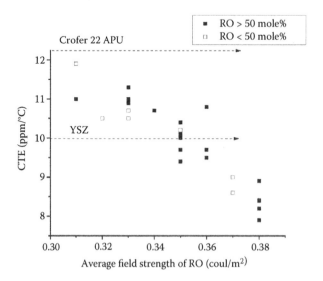

FIGURE 3.1
CTE values for "invert" silicate glasses as a function of average field strength of RO (R = Ca, Sr, Ba, and Zn). (From Zhang, T. 2008. "Glass-ceramics for solid oxide fuel cells seals." Ph.D. thesis. Missouri University of Science and Technology, Rolla.)

TABLE 3.2

Field Strength of RO

Oxides	Field Strength (Coul/m²)
MgO	0.44
CaO	0.35
SrO	0.31
BaO	0.26
ZnO	0.44

The change in composition often leads to the crystallization of specific phases, which therefore changes the overall CTE of glass-ceramics.[3,33,34,51–54,56,60,71–73] The CTEs of some common crystalline phases developed from the sealing glasses are summarized in Table 3.3. The formation of wollastonite ($CaSiO_3$) occurs at 950°C in a $CaO–Al_2O_3–SiO_2$ system and decreases the overall CTE of sealing materials due to its low CTE ($\sim 6 \times 10^{-6}$/K).[51] The decrease in the CTE of sealing materials leads to the decrease in the bond strength between sealing materials and substrates because of the thermal stress generated by the CTE mismatch. The CTE of barium-containing glass, such as $BaO–MgO–SiO_2$, increases with the increasing BaO content due to the formation of $BaSiO_3$, which has a much larger CTE than $MgSiO_3$ (9–13 × 10^{-6}/K vs. 7–9 × 10^{-6}/K).[54] The decrease in CTE for $MgO–B_2O_3–SiO_2$ glass was found to be caused by the formation of $Mg_2Al_4Si_5O_{18}$ (CTE $\sim 2 \times 10^{-6}$/K).[3]

TABLE 3.3

Coefficients of Thermal Expansion of Some Common Crystalline Phases Formed in Glass-Ceramics for SOFC Seals

Phase	CTE(×10⁻⁶/K)
$CaSiO_3$ (wollastonite)[51,74,75]	6–9
$BaSiO_3$[48,54,75]	9–13
$MgSiO_3$[54,74]	7–9
$BaAl_2Si_2O_8$ (hexacelsian)[54,76]	7–8
$Ca_2ZnSi_2O_7$ (hardystonite)[40]	11.2
$Mg_2Al_4Si_5O_{18}$ (cordierite)[44]	2
$Sr_2Al_2SiO_7$[41]	11.5
Sr_2SiO_4[56]	11–12
YBO_3[56]	6.4–7.5
Ca_2SiO_4[75]	11–14
$Ca(Mg_{0.85}Al_{0.15})(Si_{1.7}0Al_{0.30})O_6$ (Augite)[71,72]	9
Mg_2SiO_4[74,77]	9–10
$Ca_3Si_2O_7$[77]	10–12
Zn_2SiO_4[78]	5–7

The formation of celsian ($BaAl_2Si_2O_8$) with low CTE (~2.5 × 10^{-6}/K) was often found in Ba-aluminosilicates.[33,53] A patent composition of $CaO-SrO-ZnO-Al_2O_3-B_2O_3-SiO_2$ glass[34] has a good CTE match with that of 8YSZ because of the formation of hardystonite ($Ca_2ZnSi_2O_7$) (CTE~11.2 × 10^{-6}/K).[40] In addition, a glass composition of $Na_2O-MgO-CaO-Al_2O_3-ZnO-SiO_2$ was developed recently with a CTE of ~10 × 10^{-6}/K after the crystallization at 950°C for 30 min, in which $Al_{1.25}Ca_{1.94}O_7Si_{1.38}Zn_{0.43}$ and $Al_{1.2}Ca_{0.2}Na_{0.8}O_8Si_{2.8}$ were identified as the major and minor phases, respectively.[52]

Pacific Northwest National Laboratory (PNNL) developed glasses based on the $BaO-CaO-Al_2O_3-B_2O_3-SiO_2$ system.[49] The major crystalline phase developed from a glass matrix, such as $BaSiO_3$ (CTE 9–13 × 10^{-6}/K)[54] at 800°C yields an overall CTE of glass-ceramic about 11.6 × 10^{-6}/K. The formation of hexacelsian ($BaAl_2Si_2O_8$) at 850°C to 900°C causes the decrease in overall CTE of glass-ceramic due to its relative low CTE (7–8 × 10^{-6}/K).[54]

The formation of $BaZrO_3$ was found in the $BaO-CaO-MgO-B_2O_3-Al_2O_3$ glass doped with micrometer and nanometer YSZ.[60] Therefore, the overall CTE of composites decreases from 9.4 × 10^{-6}/K to 7.6 × 10^{-6}/K and 8.0 × 10^{-6}/K with nanoscaled (~27 nm) and micrometer size (~53 μm) YSZ additives, respectively.

The CTE of $SrO-CaO-B_2O_3$ glass increases with increasing content of doped NiO due to its high CTE (~14 × 10^{-6}/K) and good compatibility with the glass.[56] However, the CTE of similar composition with NiO in the melt decreases with the increasing NiO content due to the formation of YBO_3 with low CTE (6.4–7.5 × 10^{-6}/K).

3.2.1.2 Sealing Temperature

The sealing glass needs to initially flow sufficiently to get an adequate seal and then maintain enough rigidity for mechanical integrity. The dilatometric softening temperature corresponds to a viscosity of ≈10^{11} Pa·s and is often used as a direct guide for determining the sealing temperature of a glass. The desired operational temperature of many SOFC designs is in the range from 700°C to 800°C. This then fixes the desired sealing temperature in the range from 800°C to 900°C. The corresponding dilatometric softening temperature is in the range from 700°C to 800°C. The dilatometric softening temperatures are known to decrease with increasing B_2O_3 content[4,6,79] as well as ZnO content, for example, in the "invert glasses" system as shown in Figures 3.2 and 3.3, respectively.[70]

3.2.1.3 Thermomechanical Stability

The thermomechanical stability of sealants is another critical issue for the performance of the SOFC stack in operation. Some parameters, for example thermal stress and interface strength, have been applied to evaluate the thermomechanical stability of sealants.

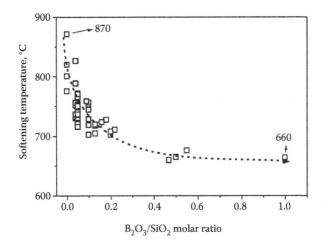

FIGURE 3.2
Dilatometric softening temperature versus the B_2O_3/SiO_2 molar ratio. The dashed line is a guide for eyes. (From Zhang, T. 2008. "Glass-ceramics for solid oxide fuel cells seals." Ph.D. thesis. Missouri University of Science and Technology, Rolla.)

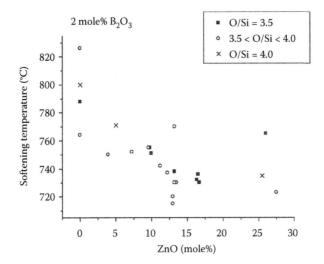

FIGURE 3.3
Dilatometric softening temperature versus the ZnO content in glasses. (From Zhang, T. 2008. "Glass-ceramics for solid oxide fuel cells seals." Ph.D. thesis. Missouri University of Science and Technology, Rolla.)

Nakata and Wakayama[80] simulated the thermal stress under thermal cycles by the finite element method (FEM). The cell was made from a LaSrCrCoO$_3$ (LSCC) separator, 8 mol% Y$_2$O$_3$ doped ZrO$_2$ (8YSZ) electrolyte, and glass sealant. It was found that the stress was generated by the thermal expansion mismatch between the sealant and the SOFC components during the thermal cycles, especially at the boundary of the constrained area during cooling. In addition, the viscous glass sealants reduce the stress compared with the rigid material. So a glass sealant with close CTE match to other SOFC components will be helpful to reduce the thermal stress during the thermal cycles.

The strain at fracture of a barium aluminosilicate glass was measured by Meinhardt et al.[48] to characterize the thermomechanical stability of the sealants. The greater strain at fracture was observed in the nascent specimen relative to the aged one due to the larger amount of residual glass in the former as well as the entanglement of the crystal phases in the latter. In addition, the mean interfacial tensile strength properties of G18 and the interconnect substrate, for example, Crofer 22 APU, decreases from 25 to 5.3 MPa with the temperature increases from 700°C to 800°C.[81]

Zheng et al.[4] investigated the thermal stability of $CaO–Al_2O_3–B_2O_3–SiO_2$ glass sealed with YSZ using an oxygen concentration cell and the electromotive force (EMF). The change of partial pressure of oxygen in the anode side (EMF) was recorded as a function of thermal cycles. Then, a constant partial pressure of oxygen indicates that the seal remains hermetic under certain thermal cycles.

Weil et al.[75] measured the strength of sealants using a modified rupture test. Alumina-forming ferritic steel substrates offer greater bond strength with the barium aluminosilicate-based glass under both thermal aging and cycling (Figure 3.4). The composition and thickness of the reaction zone that forms between the oxide scale of the metal and the bulk glass play important

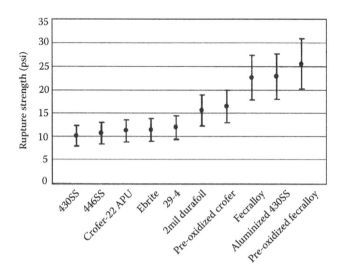

FIGURE 3.4
Rupture strength of glass-metal seals as a function of the metal substrate compositions. (From Weil, K.S. et al. *Journal of Materials Engineering and Performance*, 13(3), 316–26, 2004.)

roles for determining the interface strength. The barium chromate layer that develops on the chromia-forming steels exhibits poorer thermal expansion matching and thus lower strength than that of alumina-forming substrates. A similar result was also reported in a recent study on an improved barium aluminosilicate system, in which a fourfold difference in strength was observed between alumina-forming and chromia-forming steels.[48]

Therefore, the thermal stress still remains a critical problem for the rigid seal especially after the long-term thermal cycles. A sealant with minimum CTE mismatch with other SOFC components is necessary but not enough for a rigid seal design.

3.2.1.4 Chemical Stability

The reaction between the glass-ceramic sealants and other SOFC components often causes the failure of the SOFC seals. Therefore, the chemical stability of sealants in contact with other SOFC components has been investigated extensively as an important criterion for the design of SOFC seals.

The formation of a phase with low CTE, such as cordierite, was observed in $BaO–CaO–MgO–Al_2O_3–B_2O_3–SiO_2$ glasses in contact with anode couples. In addition, the presence of CaO in these glasses even leads to the formation of monoclinic ZrO_2 because yttrium diffuses from the electrolyte into the glasses.[44] The formation of Ni-phosphide is thermodynamically favorable in phosphate-based glasses contacting with a Ni/YSZ anode.[82] It was also found that the silica-based compositions show high stability under anodic conditions due to the high viscosity caused by the diffusion of silicon to the interface.

Compared with the reactions on the anode side, the interactions between glass-ceramic sealants and Cr-containing interconnects at the SOFC operational temperature were more often reported on the cathode side.[25,44,48,83–90] The reaction products, for example, $SrVSi_2O_7$ and AVO_4, were identified as the reaction products at the interface between the $30SrO–40SiO_2–20B_2O_3–10A_2O_3$ (A = La, Y, Al) glasses and the $Bi_4V_{1.8}Al_{0.2}O_{11}$ electrolyte (DBiV).[83] The phase with large volume change, such as cristobalite, was observed in the $BaO–CaO–MgO–Al_2O_3–B_2O_3–SiO_2$ glasses in contact with the interconnect, such as oxide dispersion strengthened (ODS) $Cr_5Fe_1Y_2O_3$ alloy.[44] The formation of $La_2Cr_2O_9$ was reported at the interface between a $La_2O_3–Al_2O_3–B_2O_3–SiO_2$ glass and the ferritic stainless steel interconnect, for example, Crofer 22 APU, in SOFC operational environments.[25] The formation of Na_2CrO_4 was observed at the interface between a glass sealant containing alkali metals and the metallic interconnect, for example, Crofer 22 APU, which reduces the joint quality due to the formation of cracks.[86]

Much attention has been focused on the notorious reactions between the alkaline earth oxide component of glass seals and the ferritic interconnects, which lead to the formation of some detrimental phases with a very high CTE, such as $BaCrO_4$[39,48,90,91] and $SrCrO_4$[87,92] (~18 to 20 × 10^{-6}/K), at the glass–metal interface. The formation of these phases causes the physical separation of sealing glass and stainless steels due to the large CTE mismatch between these phases and that of glass-ceramic sealants (~12.5 × 10^{-6}/K) or stainless steels (~13 × 10^{-6}/K).

The reaction between a barium–calcium–aluminosilicate (BCAS) based glass and ferritic stainless steels, for example, AISI446, depends on the exposure conditions.[88] At the edges of the joints, where oxygen or air is accessible, the interaction often leads to the formation of $BaCrO_4$ as shown in Figure 3.5. At the interior of the joints, where the oxygen or air access is blocked, chromium or chromia dissolves into the glass to form a thin layer of chromium-rich solid solution. In addition, the interaction of sealing glass and steel in the interior of the joints also generates pores lining up along interfaces, which may decrease the hermeticity of the seal. Moreover, the formation of barium chromate often leads to the physical separation of sealing glass and stainless steels due to its high thermal expansion and thus could lead to deterioration in seal quality under long-term operation.

Reactions between barium-containing glass-ceramic sealants and different types of interconnect alloys, for example, Nicrofer 6045 (austenitic chromia forming), 446 (ferritic chromia forming), and Fecralloy (alumina forming alloys), have also been investigated to further the knowledge of the interfacial reactions.[48,88,89] The thermodynamic calculations reveal that the extent and nature of the interactions and their products depend on the matrix alloy compositions and the exposure conditions. For chromia-forming alloys, such as 446 ferritic stainless steel, the edges of the seals where oxygen or air is accessible typically exhibit the $BaCrO_4$ formation; for a more oxidation-resistant alloy, for example, the Ni-based superalloy, the extent of the formation of $BaCrO_4$

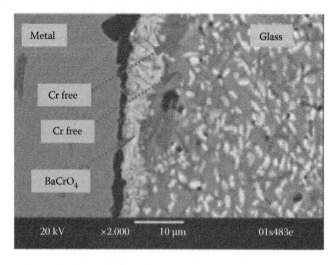

FIGURE 3.5
The presence of $BaCrO_4$ was detected at the interface between sealing glass and the ferritic SS 446 using EDS. (From Yang, Z. et al. *Journal of the Electrochemical Society*, 150(8), A1095–A101, 2003.)

can be limited; for alumina-forming alloys, the interaction between sealing glass and alloys still occurs but without the formation of $BaCrO_4$. However, the interaction typically causes large bubbles and small pores at the interface zone, which likely degrades the strength and hermeticity of interface. Similar reactions were also observed at the interface between glass-ceramic sealants and a newly developed ferritic stainless steel, Croferr22 APU, in spite of the growth of a unique scale on the alloy during high temperature exposures.[54] In addition, a rupture strength decrease of about 50% was observed (exposure 400 h at 750°C) in Ba–Ca–aluminosilicate glass joined with a ferritic interconnect steel (Crofer22 APU), which was caused by the mismatches in thermal expansion coefficients and the further growth of the interfacial oxide scale.[85]

Besides the pretreatment of metallic interconnects, for example, preoxidation,[93–96] deposition of the protective coatings,[97] as well as the aluminizing of the Cr-containing ferritic alloys,[98,99] the compositional design of glass-ceramic sealants has also been considered to reduce the interfacial reactions. For example, chromium was purposely added to glass-ceramic sealants to get the matrix saturated with Cr_2O_3 and subsequently terminate the further dissolution of Cr_2O_3 in the glass.[84] The CTE of the glass was stable at 9.3×10^{-6}/K after being saturated with Cr_2O_3, which indicates a compatibility with chromium-containing alloys.

A recent study showed that the formation of $SrCrO_4$ can be reduced by some additions, for example, ZnO, La_2O_3, and MnO_2, in the CaO–SrO–ZnO–B_2O_3–Al_2O_3–SiO_2 glasses contacted with SS 430.[100] For example, the percentage of Cr^{6+} decreased from 58 mol% for the base glass to 21 mol% for the glass containing 10 mol% MnO_2 after heat treatment at 900°C for 7 h. The significant reduction of the chromate phase by different additives is explained in terms of the possible competing reactions.

3.2.1.5 Glass Volatility

The volatility of glass components also restricts the design of candidate glass compositions for high-temperature sealing applications.[3,5,22,46,47,101,102]

At the SOFC operating temperature, the phosphate volatilizes and reacts with the anode to form nickel phosphide and zirconiumoxyphosphate.[46] In addition, these phosphate glasses usually crystallize to form metaphosphates or pyrophosphates, which exhibit low stability in a humidified fuel gas at the operating temperature.

Lahl et al.[3] reported that the presence of Na and K in aluminoborosilicate glass sealants significantly increased chromium volatility from a metallic interconnect at the operational temperature, thereby increasing the electrical degradation of other SOFC components. The poisoning of the cathode caused by the volatilization of chromium was also related to the alkaline metals in the glasses.[3]

Gunther et al.[5] found that glasses with B_2O_3 as the only glass former experienced up to a 20% weight loss and extensive interaction with SOFC components both in air and fuel gas atmospheres.

In addition, B_2O_3 forms volatile compounds with water vapor leading to the seal degradation at operational temperatures.[47] It was also found that B_2O_3 in the anhydrous state at 1000°C has a vapor pressure of 1.9×10^{-4} mm Hg (2.5×10^{-2} Pa), while it is quite volatile at 100°C in steam with $B(OH)_3$ in the vapor phase.[102] Similarly, the volatilization in sodium borate melts is not consistent with evaporation of boron as boric oxide but rather with the reaction of $B_2O_3(l) + H_2O(g) \rightarrow 2BHO_2(g)$.[101] In this study, the rate of boron loss from the melt was found to depend on the partial pressure of water vapor above the melt. In addition, reaction at the melt surface will generate more volatile HBO_2 species as the water molecules diffuse into the melt. So the volatility of boron in borate glass increases with the increasing partial pressure of water vapor.

A recent study showed that the volatilization of B_2O_3 after 7 days at 800°C in air and in forming gas with 30% water vapor was 0.3% and 1.5% of the total B_2O_3 content, respectively, from a glass with 20 mol% B_2O_3.[22] The vaporized species were predicted to be $BO_2(g)$ and $B_3H_3O_6(g)$ under oxidizing and wet reducing atmospheres, respectively. Obviously the volatilization of B_2O_3 from the glass depends on the partial pressure of oxygen and water vapor and the operating temperature and time as well. The calculated vaporized species and corresponding vapor pressure of some typical glass components, such as ZnO, CaO, SrO, BaO, and SiO_2, under SOFC operational conditions (e.g., at 800°C in dry air and wet forming gas [30% H_2O]), are summarized in Table 3.4.[70]

Therefore, phosphate and alkaline metals are generally avoided in the design of sealing glasses due to the undesired volatilization. B_2O_3 is still attractive for the glass-ceramic sealants due to its great effect on the decrease in the viscosity and crystallization tendency in spite of the high volatilization in glasses with B_2O_3 as the only glass former. Modified borosilicate glasses can be a promising system to yield the desired viscosity and glassy state without significant volatilization.

TABLE 3.4

Vaporized Species and Vapor Pressure for Typical Glass Components at 800°C in Dry Air and Wet Forming Gas (30% H_2O)

	Dry Air	Wet Forming Gas
B_2O_3	BO_2 (−4.82 log Pa)	$B_3H_3O_6$ (2.54 log Pa)
ZnO	ZnO (−5.45 log Pa)	Zn (2.20 log Pa)
SiO_2	SiO_2 (−14.09 log Pa)	SiO (−12.55 log Pa)
SrO	SrO (−14.35 log Pa)	$Sr(OH)_2$ (−3.97 log Pa)
BaO	BaO (−19.84 log Pa)	$Ba(OH)_2$ (−0.87 log Pa)
CaO	CaO (−18.94 log Pa)	$Ca(OH)$ (−15.79 log Pa)

Source: Zhang, T. 2008. "Glass-ceramics for solid oxide fuel cells seals." Ph.D. thesis. Missouri University of Science and Technology, Rolla.

3.2.1.6 Electrical Resistance

In addition, a high electrical resistance (>2 kΩ cm) is desired for sealing glasses to avoid a short circuit in the cell.[103] The alkaline metals are often excluded in the compositions of sealing glasses due to their high electrical conductivity. The reported resistances of alkaline earth–containing glasses fall in the range of 10^4–10^8 Ω cm at temperatures in the range of 700°C–800°C, which are suitable for SOFC seals.[37,39,69,72,104–106] For example, the addition of BaO (12 wt%) leads to a substantial decrease in the total conductivity from $10^{-6.4}$ to 10^{-7} S cm^{-1} and to an increase in the corresponding activation energy from 144 to 218 kJ/mol in the MgO–CaO–La$_2$O$_3$–Al$_2$O$_3$–SiO$_2$ glasses.[72]

3.2.2 Critical Process: Crystallization

Crystallization behavior is also important for the design of glass-ceramic sealants, especially in the glass-ceramic systems at the SOFC operational temperature range, that is, 800°C–1000°C. For example, a joining process in the Siemens-SOFC stack requires the glass seal to remain viscous at 950°C for 2 to 3 h to allow small displacements of the single stack elements after joining at 1000°C, which can be achieved by using only a slowly crystallizing glass.[107] In addition, an increase in the flexural strength with temperature, at a temperature below the glass transition temperature, was observed as the result of the crystallization of Ba$_3$La$_6$(SiO$_4$)$_6$ from the BaO–CaO–La$_2$O$_3$–ZrO$_2$–Al$_2$O$_3$–B$_2$O$_3$–SiO$_2$ glass matrix.[108,109] A similar trend was also reported in the famous G#18, in which creep increases with increasing temperature but decreases with further aging.[110]

Lara pointed out that the sintering stage should be completed before crystallization takes place to get a final dense and low porosity material for a SOFC sealing application.[45] An uncontrolled crystallization during the initial sintering process can lead to the formation of a porous sealing layer that can adversely affect the SOFC operation.

The effect of composition on crystallization has been investigated extensively.[2,4,36,44,50,68,111] A high B$_2$O$_3$-containing glass matrix with low viscosity in the softening range was also observed in borosilicate glass after crystallization because B$_2$O$_3$ tends to partition into the glassy phase.[78] Similarly, the addition of B$_2$O$_3$ was also found to decrease the viscosity and delay the crystallization in BaO–MgO–ZnO–B$_2$O$_3$–SiO$_2$ glasses.[50] The activation energy of crystallization was increased from 301 to 367 kJ/mol with the B$_2$O$_3$/Al$_2$O$_3$ ratio increase from 0.4 to 1 in B$_2$O$_3$–Al$_2$O$_3$–SiO$_2$ glasses.[112] The activation energy of crystallization in the AO–Al$_2$O$_3$–B$_2$O$_3$–SiO$_2$ (A = Ba, Ca and Mg) glasses increases significantly from 206 kJ/mol for BaO-containing glass to 420 kJ/mol for MgO-containing glass due to the increasing field strengths of the alkaline earth ions from Ba^{2+}(0.24) to Mg^{2+}(0.45).[113]

The nucleation agents, such as TiO$_2$ and ZrO$_2$, increase the crystallization rate in the BaO–CaO–Al$_2$O$_3$–SiO$_2$ glasses.[2] The increase in the crystallization rate with the presence of nucleation agents, for example, Ni and YSZ, was also observed in the CaO–SrO–ZnO–Al$_2$O$_3$–B$_2$O$_3$–SiO$_2$ glasses.[114] The decrease in the nucleation and crystallization temperatures was observed in the $AO-Al_2O_3-B_2O_3-SiO_2$ (A = Ba, Ca, and Mg) glasses with the addition of TiO_2, which contributes to the increase in nuclei of Mg_2SiO_4.[44]

Some other studies revealed that the crystallization can be restrained effectively in the $CaO-Al_2O_3-B_2O_3-SiO_2$ glasses (with the additions of ZnO, BaO, K_2O, and Na_2O) by controlling the Al_2O_3/CaO ratio in the range of 0.4–0.68.[4] In addition, the noncrystalline network in this glass was found after holding at 850°C for 100 h.

Al_2O_3 and Ta_2O_5 were also found to inhibit the crystallization in $BaO-SrO-B_2O_3-Al_2O_3-Ta_2O_5-SiO_2$ glasses as observed in hot stage microscopy (HSM).[36] So a critical sealing viscosity (10^5 Pa·s) can be achieved at 900°C in the glass with only 6 mol% of B_2O_3. In addition, an increase in the peak temperature of crystallization from 763°C to 838°C was observed in $BaO-Al_2O_3-B_2O_3-SiO_2$ glasses with the addition of Al_2O_3 (10 mol%).[115]

The crystallization onset temperature was also related to the glass-forming tendency (GFT) together with the glass transition temperature. The glasses that have large differences between their crystallization (T_c) and glass transition temperatures (T_g), $\Delta Ts = (T_c - T_g)$ can be sealed in a broader temperature window, which is critically important in sealing procedure designs.[111] In other words, the increase in onset crystallization temperature or the decrease in the glass transition temperature is favored in order to get a dense, hermetic seal. Our research on the "invert glasses" also indicates that the glass-forming tendency depends on the glass structure as well as the boron content as shown in Figure 3.6.[70] Glasses bases on pyrosilicate structures (O/Si = 3.5) have greater ΔTs and so

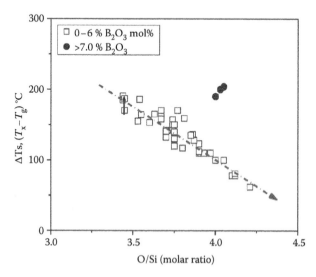

FIGURE 3.6
Glass-forming tendency, ΔTs, versus O/Si molar ratio. The dashed line is a guide for the eye. (From Zhang, T. 2008. "Glass-ceramics for solid oxide fuel cells seals." Ph.D. thesis. Missouri University of Science and Technology, Rolla.)

are generally easier to seal than glasses based on orthosilicate (O/Si) structures. Adding B_2O_3 to the compositions increases the ΔTs for a fixed O/Si ratio, indicating an increasing resistance to crystallization when the glass is heated above T_g.

Therefore, knowledge of the crystallization is important for the design of sealants. The controlled crystallization provides not only the crystal phases with desired CTE but also the desired viscosity to accomplish the joining process and self-healing during operation. In addition, a condensed seal can be obtained by selecting the appropriate sealing temperature range with the controlled crystallization.

3.3 Conclusions and Future Works

Besides the hybrid compressive seals, many glass-ceramic sealing materials have been investigated as either rigid seals or compliant seals for planar SOFC applications. Silicate-based glasses operate at too high temperatures; borate-based and phosphate-based glasses volatilize at the operational temperature. The silicate-based glass-ceramics are favored for the sealing materials of SOFC due to less volatility and better chemical stability with other SOFC components, compared with other glass-ceramics.

However, the thermomechanical stability of rigid seals still remains uncertain because the CTE mismatch between sealants and other SOFC components often leads to the generation of thermal stress and the mechanical failure of the sealed couples, especially under long-term operations. The resilient seals will be a promising approach to accomplish the sealing target. The thermomechanical challenge can be solved by the flow of viscous phases as demonstrated in the literature. However, some aspects, including the mechanical strength, the chemical stability, the glass volatility, as well as the electrical resistances of glasses, need to be addressed when designing the composition of resilient seals. In addition, some oxides for modifying the CTE of glass-ceramics should be excluded from the compositions to avoid the possible interfacial reaction between glasses and other SOFC components.

Acknowledgments

The authors gratefully acknowledge the financial support of the National Natural Science Foundation of China (No. 51102045), the Ph.D. Programs Foundation of Ministry of Education of China (No. 20103514120006), program for New Century Excellent Talents in Fujian Province University (No. JA12013), the Returned Overseas Chinese Scholars, State Education Ministry

(No. LXKQ0902) and funds for Distinguished Young Scientists from the Fujian Education Department (No. JA11007).

References

1. Singhal, S. 2003. Ceramic fuel cells for stationary and mobile applications. *American Ceramic Society Bulletin* 82 (11), 19–20.
2. Shwickert, T., Sievering, R., Geasee, P. and Conradt, R. 2002. Glass-ceramic materials as sealants for SOFC application. *Materialwissenschaften Werkstofftechnik* 33, 363–366.
3. Lahl, N., Singheiser, L. and Hilpert, K. 1999. Aluminosilicate glass ceramics as sealant in SOFC stacks. In 6th International Symposium on Solid Oxide Fuel Cells (SOFC), Vol. 99. Edited by S.C. Singhal and M. Dokiya, pp. 1057–1066.
4. Zheng, R., Wang, S.R., Nie, H.W. and Wen, T.L. 2004. SiO$_2$-CaO-B$_2$O$_3$-Al$_2$O$_3$ ceramic glaze as sealant for planar ITSOFC. *Journal of Power Sources* 128 (2), 165–172.
5. Gunther, C., Hofer, G. and Kleinlein, W. 1997. The stability of the sealing glass AF45 in H$_2$/H$_2$O and O$_2$/N$_2$ atmospheres. In 5th International Symposium on Solid Oxide Fuel Cells, Vol. 97, pp. 746–756.
6. Sohn, S.-B., Choi, S.-Y., Kim, G.-H., Song, H.-S. and Kim, G.-D. 2002. Stable sealing glass for planar solid oxide fuel cell. *Journal of Non-Crystalline Solids* 297 (2–3), 103–112.
7. Minh, N.Q. 1993. Ceramic fuel cells. *Journal of the American Ceramic Society* 76 (3), 563–588.
8. Taniguchi, S., Kadowaki, M., Yasuo, T., Akiyama, Y., Miyake, Y. and Nishio, K. 2000. Improvement of thermal cycle characteristics of a planar-type solid oxide fuel cell by using ceramic fiber as sealing material. *Journal of Power Sources* 90 (2), 163–169.
9. Simner, S.P. and Stevenson, J.W. 2001. Compressive mica seals for SOFC applications. *Journal of Power Sources* 102 (1–2), 310–316.
10. Badwal, S.P.S. 2001. Stability of solid oxide fuel cell components. *Solid State Ionics* 143 (1), 39–46.
11. Lessing, P.A. 2007. A review of sealing technologies applicable to solid oxide electrolysis cells. *Journal of Materials Science* 42 (10), 3465–3476.
12. Mahapatra, M.K. and Lu, K. 2010. Glass-based seals for solid oxide fuel and electrolyzer cells: A review. *Materials Science and Engineering: R: Reports* 67 (5–6), 65–85.
13. Ishihara, T., Matsuda, H. and Takita, Y. 1994. Doped LaGaO$_3$ perovskite type oxide as a new oxide ionic conductor. *Journal of the American Chemical Society* 116 (9), 3801–3803.
14. Chou, Y.-S. and Stevenson, J.W. 2005. Long-term thermal cycling of phlogopite mica-based compressive seals for solid oxide fuel cells. *Journal of Power Sources* 140 (2), 340–345.
15. Chou, Y.-S. and Stevenson, J.W. 2009. Long-term ageing and materials degradation of hybrid mica compressive seals for solid oxide fuel cells. *Journal of Power Sources* 191 (2), 384–389.
16. Chou, Y.-S., Stevenson, J.W. and Chick, L.A. 2002. Ultra-low leak rate of hybrid compressive mica seals for solid oxide fuel cells. *Journal of Power Sources* 112 (1), 130–136.

17. Chou, Y.-S., Stevenson, J.W. and Chick, L.A. 2003. Novel compressive mica seals with metallic interlayers for solid oxide fuel cell applications. *Journal of the American Ceramic Society* 86 (6), 1003–1007.

18. Chou, Y.-S., Stevenson, J.W. and Singh, P. 2005. Thermal cycle stability of a novel glass-mica composite seal for solid oxide fuel cells: Effect of glass volume fraction and stresses. *Journal of Power Sources* 152 (1), 168–174.

19. Raj, N.S. 2007. Sealing technology for solid oxide fuel cells (SOFC). *International Journal of Applied Ceramic Technology* 4 (2), 134–144.

20. Liu, W.N., Sun, X. and Khaleel, M.A. 2011. Study of geometric stability and structural integrity of self-healing glass seal system used in solid oxide fuel cells. *Journal of Power Sources* 196 (4), 1750–1761.

21. Mahapatra, M.K. and Lu, K. 2008. Effects of nickel on network structure and thermal properties of a new solid oxide cell seal glass. *Journal of Power Sources* 185 (2), 993–1000.

22. Zhang, T., Fahrenholtz, W.G., Reis, S.T. and Brow, R.K. 2008. Borate volatility from SOFC sealing glasses. *Journal of the American Ceramic Society* 91 (8), 2564–2569.

23. Lu, K. and Mahapatra, M.K. 2008. Network structure and thermal stability study of high temperature seal glass. *Journal of Applied Physics* 104 (7), 074910.

24. Mahapatra, M.K., Lu, K. and Bodnar, R.J. 2009. Network structure and thermal property of a novel high temperature seal glass. *Applied Physics A: Materials Science and Processing* 95 (2), 493–500.

25. Mahapatra, M.K. and Lu, K. 2009. Interfacial study of Crofer 22 APU interconnect-SABS-0 seal glass for solid oxide fuel/electrolyzer cells. *Journal of Materials Science* 44 (20), 5569–5578.

26. Jin, T. and Lu, K. 2010. Thermal stability of a new solid oxide fuel/electrolyzer cell seal glass. *Journal of Power Sources* 195 (1), 195–203.

27. Zhang, T., Tang, D. and Yang, H. 2011. Can crystalline phases be self-healing sealants for solid oxide fuel cells? *Journal of Power Sources* 196 (3), 1321–1323.

28. Chou, Y.-S. and Stevenson, J.W. 2004. Novel infiltrated phlogopite mica compressive seals for solid oxide fuel cells. *Journal of Power Sources* 135 (1–2), 72–78.

29. Fergus, J.W. 2005. Sealants for solid oxide fuel cells. *Journal of Power Sources* 147 (1–2), 46–57.

30. Huang, K., Lee, H.Y. and Goodenough, J.B. 1998. Sr- and Ni-doped LaCoO3 and LaFeO3 perovskites. New cathode materials for solid-oxide fuel cells. *Journal of the Electrochemical Society* 145 (9), 3220–3227.

31. Huang, K., Robin, S.T. and Goodenough, J.B. 1998. Superior perovskite oxide-ion conductor; strontium- and magnesium-doped $LaGaO_3$: I, phase relationships and electrical properties. *Journal of the American Ceramic Society* 81 (10), 2565–2575.

32. Seabaugh, M.M., Sabolsky, K., Arkenberg, G.B. and Jayjohn, J.L. 2008. Composite seal development and evaluation. In 30th International Conference on Advanced Ceramics and Composites, Vol. 27. *Ceramic Engineering and Science Proceedings*. Edited by A.W.E.L.-C. Narottam, and P. Bansal, pp. 265–272.

33. Bansal, N.P. 1997. Chemical vapor deposited SiC (SCS-0) fiber-reinforced strontium aluminosilicate glass-ceramic composites. *Journal of Materials Research* 12 (3), 745–753.

34. Brow, R.K., Reis, S.T. and Benson, G. 2006. Alkaline earth aluminosilicate glass and glass-ceramic sealants for solid oxide fuel cells. US Patent 2006044593 A2, USA.

35. Chou, Y.-S., Stevenson, J.W. and Gow, R.N. 2007. Novel alkaline earth silicate sealing glass for SOFC: Part II. Sealing and interfacial microstructure. *Journal of Power Sources* 170 (2), 395–400.

36. Flügel, A., Dolan, M.D., Varshneya, A.K., Zheng, Y., Coleman, N., Hall, M., Earl, D. and Misture, S.T. 2007. Development of an improved devitrifiable fuel cell sealing glass. *Journal of the Electrochemical Society* 154 (6), B601–B608.

37. Ghosh, S., Das, S.A., Kundu, P., Mahanty, S. and Basu, R.N. 2008. Development and characterizations of BaO-CaO-Al2O3-SiO2 glass-ceramic sealants for inter-mediate temperature solid oxide fuel cell application. *Journal of Non-Crystalline Solids* 354 (34), 4081–4088.

38. Ghosh, S., Sharma, A.D., Kundu, P. and Basu, R.N. 2008. Glass-ceramic sealants for planar IT-SOFC: A bilayered approach for joining electrolyte and metallic interconnect. *Journal of the Electrochemical Society* 155 (5), B473–B478.

39. Ghosh, S., Sharma, A.D., Mukhopadhyay, A.K., Kundu, P. and Basu, R.N. 2010. Effect of BaO addition on magnesium lanthanum alumino borosilicate-based glass-ceramic sealant for anode-supported solid oxide fuel cell. *International Journal of Hydrogen Energy* 35 (1), 272–283.

40. Haussühl, S. and Liebertz, J. 2004. Elastic and thermoelastic properties of syn-thetic Ca2MgSi2O7(åkermanite) and Ca2ZnSi2O7(hardystonite). *Physics and Chemistry of Minerals* 31 (8), 565–567.

41. Kobayashi, Y. and Inagaki, M. 2004. Preparation of reactive Sr-celsian powders by solid-state reaction and their sintering. *Journal of the European Ceramic Society* 24 (2), 399–404.

42. Kumar, V. 2006. Studies on some glass sealants for solid oxide fuel cells. Masters thesis. Thapar Institute of Engineering and Technology, Patiala.

43. Kumar, V., Arora, A., Pandey, O.P. and Singh, K. 2008. Studies on thermal and structural properties of glasses as sealants for solid oxide fuel cells. *International Journal of Hydrogen Energy* 33 (1), 434–438.

44. Lahl, N., Bahadur, D., Singh, K., Singheiser, L. and Hilpert, K. 2002. Chemical interactions between aluminosilicate base sealants and the components on the anode side of solid oxide fuel cells. *Journal of the Electrochemical Society* 149 (5), A607–A614.

45. Lara, C., Pascual, M.J., Prado, M.O. and Duran, A. 2004. Sintering of glasses in the system RO-Al2O3-BaO- SiO2 (R = Ca, Mg, Zn) studied by hot-stage micros-copy. *Solid State Ionics* 170 (3–4), 201–208.

46. Larsen, P.H., Bagger, C., Mogensen, M. and Larsen, J.G. Stacking of planar SOFCs. 1995. In 4th International Symposium on Solid Oxide Fuel Cells, Vol. 95. Edited by M. Dokiya, O. Yamamoto, H. Tagawa, and S.C. Singhal, pp. 69–78.

47. Ley, K.L., Krumpelt, M., Kumar, R., Meiser, J.H. and Bloom, I. 1996. Glass-ceramic sealants for solid oxide fuel cells: Part I. Physical properties. *Journal of Materials Research* 11 (6), 1489–1483.

48. Meinhardt, K.D., Kim, D.S., Chou, Y.S. and Weil, K.S. 2008. Synthesis and prop-erties of a barium aluminosilicate solid oxide fuel cell glass-ceramic sealant. *Journal of Power Sources* 182 (1), 188–196.

49. Meinhardt, K.D., Vienna, J.D., Armstrong, T.R. and Pederson, L.R. 2002. Glass-ceramic materials and method of making, USA. US Patent 6430966 B1.

50. Pascual, M.J., Guillet, A. and Duran, A. 2007. Optimization of glass-ceramic sealant compositions in the system MgO-BaO-SiO2 for solid oxide fuel cells (SOFC). *Journal of Power Sources* 169 (1), 40–46.

51. Sakaki, Y., Hattori, M., Esaki, Y., Ohara, S., Fukui, T., Kodera, K. and Kubo, Y. 1997. Glass-ceramics sealants in CaO-Al$_2$O$_3$-SiO$_2$ systems. In 5th International Symposium on Solid Oxide Fuel Cells, Vol. 97, pp. 652–660.
52. Smeacetto, F., Salvo, M., D'Herin Bytner, F.D., Leone, P. and Ferraris, M. 2010. New glass and glass-ceramic sealants for planar solid oxide fuel cells. *Journal of the European Ceramic Society* 30 (4), 933–940.
53. Sung, Y.M. and Kwak, W.C. 2002. Influence of various heating procedures on the sintered density of Sr-celsian glass-ceramic. *Journal of Materials Science Letters* 21 (11), 841–843.
54. Yang, Z., Xia, G., Meinhardt, K.D., Weil, K.S. and Stevenson, J.W. 2004. Chemical stability of glass seal interfaces in intermediate temperature solid oxide fuel cells. *Journal of Materials Engineering and Performance* 13 (3), 327–334.
55. Brochu, M., Gauntt, B.D., Shah, R. and Loehman, R.E. 2006. Comparison between micrometer- and nano-scale glass composites for sealing solid oxide fuel cells. *Journal of the American Ceramic Society* 89 (3), 810–816.
56. Chou, Y.-S., Stevenson, J.W. and Gow, R.N. 2007. Novel alkaline earth silicate sealing glass for SOFC: Part I. The effect of nickel oxide on the thermal and mechanical properties. *Journal of Power Sources* 168 (2), 426–433.
57. Gross, S.M., Koppitz, T., Remmel, J., Bouche, J.-B. and Reisgen, U. 2006. Joining properties of a composite glass-ceramic sealant. *Fuel Cells Bulletin* 2006 (9), 12–15.
58. Laorodphan, N., Namwong, P., Thiemsorn, W., Jaimasith, M., Wannagon, A. and Chairuangsri, T. 2009. A low silica, barium borate glass-ceramic for use as seals in planar SOFCs. *Journal of Non-Crystalline Solids* 355 (1), 38–44.
59. Wang, S.-F., Wang, Y.-R., Hsu, Y.-F. and Chuang, C.-C. 2009. Effect of additives on the thermal properties and sealing characteristic of BaO-Al$_2$O$_3$-B$_2$O$_3$-SiO$_2$ glass-ceramic for solid oxide fuel cell application. *International Journal of Hydrogen Energy* 34 (19), 8235–8244.
60. Xue, L., Yamanis, J., Piascik, J. and Ong, E.T. 2006. Alkali-free composite sealant materials for solid oxide fuel cells, USA. US Patent 7521387 B2.
61. Zhou, X., Sun, K., Yan, Y., Le, S., Zhang, N., Sun, W. and Wang, P. 2009. Investigation on silver electric adhesive doped with Al2O3 ceramic particles for sealing planar solid oxide fuel cell. *Journal of Power Sources* 192 (2), 408–413.
62. Ko, H., Lee, H., Lee, J., Lee, J., Song, H., Kim, J. and Noh, T. 2005. Method for producing solid oxide fuel cell sealant comprising glass matrix and ceramic fiber. US Patent 2005/0147866 A1.
63. Le, S., Sun, K., Zhang, N., An, M., Zhou, D., Zhang, J. and Li, D. 2006. Novel compressive seals for solid oxide fuel cells. *Journal of Power Sources* 161 (2), 901–906.
64. Dai, Z., Pu, J., Yan, D., Chi, B. and Jian, L. 2011. Thermal cycle stability of Al2O3-based compressive seals for planar intermediate temperature solid oxide fuel cells. *International Journal of Hydrogen Energy* 36 (4), 3131–3137.
65. Wiener, F., Bram, M., Buchkremer, H.P. and Sebold, D. 2008. Investigation of SOFC-gaskets containing compressive mica layers under dual atmosphere conditions. In 30th International Conference on Advanced Ceramics and Composites, Vol. 27. Edited by A.W.E.L.-C. Narottam, and P. Bansal, pp. 273–285.
66. Raj, N.S. and Shailendra, S.P. 2009. Performance of self-healing seals for solid oxide fuel cells (SOFC), 27. In 30th International Conference on Advanced Ceramics and Composites. Edited by A.W.E.L.-C. Narottam, and P. Bansal, pp. 287–295.
67. Zhang, T., Zhu, Q. and Xie, Z. 2009. Modeling of cracking of the glass-based seals for solid oxide fuel cell. *Journal of Power Sources* 188 (1), 177–183.

68. Ojha, P.K., Rath, S.K., Chongdar, T.K., Gokhale, N.M. and Kulkarni, A.R. 2011. Physical and thermal behaviour of Sr-La-Al-B-Si based SOFC glass sealants as function of SrO content and B2O3/SiO2 ratio in the matrix. *Journal of Power Sources* 196 (10), 4594–4598.

69. Sakuragi, S., Funahashi, Y., Suzuki, T., Fujishiro, Y. and Awano, M. 2008. Non-alkaline glass-MgO Composites for SOFC Sealant. *Journal of Power Sources* 185 (2), 1311–1314.

70. Zhang, T. 2008. Glass-ceramics for solid oxide fuel cells seals. Ph.D thesis. Missouri University of Science and Technology, Rolla.

71. Goel, A., Tulyaganov, D.U., Kharton, V.V., Yaremchenko, A.A., Eriksson, S. and Ferreira, J.M.F. 2009. Optimization of La_2O_3-Containing diopside based glass-ceramic sealants for fuel cell applications. *Journal of Power Sources* 189 (2), 1032–1043.

72. Goel, A., Tulyaganov, D.U., Kharton, V.V., Yaremchenko, A.A. and Ferreira, J.M.F. 2010. Electrical behavior of aluminosilicate glass-ceramic sealants and their interaction with metallic solid oxide fuel cell interconnects. *Journal of Power Sources* 195 (2), 522–526.

73. Kerstan, M. and Russel, C. 2011. Barium silicates as high thermal expansion seals for solid oxide fuel cells studied by high-temperature X-ray diffraction (HT-XRD). *Journal of Power Sources* 196 (18), 7578–7584.

74. Donald, I.W. 1993. Preparation, properties and chemistry of glass- and glass-ceramic-to-metal seals and coatings. *Journal of Materials Science* 28 (11), 2841–2886.

75. Weil, K.S., Deibler, J.E., Hardy, J.S., Kim, D.S., Xia, G.-G., Chick, L.A. and Coyle, C.A. 2004. Rupture testing as a tool for developing planar solid oxide fuel cell seals. *Journal of Materials Engineering and Performance* 13 (3), 316–326.

76. Bansal, N.P. and Hyatt, M.J. 1989. Crystallization kinetics of BaO-Al2O3-SiO2 glasses. *Journal of Materials Research* 4 (5), 1257–1265.

77. Skinner, B.J. 1996. *Handbook of Physical Constants*. The Geological Society of America: Connecticut.

78. Höland, W. and Beall, G.H. 1989. *Glass Ceramic Technology*. American Ceramic Society: Westerville.

79. Weil, K.S. 2006. The state-of-the-art in sealing technology for solid oxide fuel cells. *Journal of the Minerals, Metals and Materials Society* 58 (8), 37–44.

80. Nakata, M.A.K. and Wakayama, S. 1996. Study on durability for thermal cycle of planar SOFC. In Fuel Cell Seminar (Fuel Cell Information Series), pp. 143–146.

81. Stephens, E.V., Vetrano, J.S., Koeppel, B.J., Chou, Y., Sun, X. and Khaleel, M.A. 2009. Experimental characterization of glass-ceramic seal properties and their constitutive implementation in solid oxide fuel cell stack models. *Journal of Power Sources* 193 (2), 625–631.

82. Larsen, P.H., Primdahl, S. and Mogensen, M. 1996. Influence of sealing material on nickel/YSZ solid oxide fuel cell anodes. In 17th Risoe International Symposium on Materials Science: High Temperature Electrochemistry, Ceramics and Metals. Edited by F.W. Poulsen, N. Bonanos, S. Linderoth, M. Mogensen, and B. Zachau-Christiansen, pp. 331–338.

83. Kumar, V., Sharma, S., Pandey, O.P. and Singh, K. 2010. Thermal and physical properties of $30SrO-40SiO_2-20B_2O_3-10A_2O_3$ (A = La, Y, Al) glasses and their chemical reaction with bismuth vanadate for SOFC. *Solid State Ionics* 181 (1–2), 79–85.

84. Loehman, R. 2005. Engineered glass composites for sealing solid oxide fuel cells. In SECA Core Technology PeerReview Worshop.

85. Malzbender, J., Steinbrech, R.W., Singheiser, L. and Batfalsky, P. 2005. Fracture energies of brittle sealants for planar solid oxide fuel cells. In 29th International Conference on Advanced Ceramics and Composites, Vol. 26. Ceramic Engineering and Science Proceedings, Edited by Narottam P., Bansal, Dongming Zhu, and W.M. Kriven, pp. 285–291.

86. Ogasawara, K., Kameda, H., Matsuzaki, Y., Sakurai, T., Uehara, T., Toji, A., Sakai, N., Yamaji, K., Horita, T. and Yokokawa, H. 2007. Chemical stability of ferritic alloy interconnect for SOFCs. *Journal of the Electrochemical Society* 154 (7), B657–B663.

87. Stevenson, J.W., Chou, Y.S. and Sing, P. 2006. Rigid seals for SOFC. In SECA Core Technology Peer Review Worshop.

88. Yang, Z., Meinhardt, K.D. and Stevenson, J.W. 2003. Chemical compatibility of barium-calcium-aluminosilicate-based sealing glasses with the ferritic stainless steel interconnect in SOFCs. *Journal of the Electrochemical Society* 150 (8), A1095–A1101.

89. Yang, Z., Stevenson, J.W. and Meinhardt, K.D. 2003. Chemical interactions of barium-calcium-aluminosilicate-based sealing glasses with oxidation resistant alloys. *Solid State Ionics* 160 (3–4), 213–225.

90. Yang, Z., Xia, G., Meinhardt, K.D., Weil, K.S. and Stevenson, J.W. 2004. Glass sealing in planar SOFC stacks and chemical stability of seal interfaces. In 106th Annual Meeting of the American Ceramic Society, Vol. 158. Ceramic Transactions, pp. 135–146.

91. Peng, L. and Zhu, Q. 2009. Thermal cycle stability of BaO-B2O3-SiO2 sealing glass. *Journal of Power Sources* 194 (2), 880–885.

92. Zhang, T., Brow, R.K., Fahrenholtz, W.G. and Reis, S.T. 2012. Chromate formation at the interface between a solid oxide fuel cell sealing glass and interconnect alloy. *Journal of Power Sources* 205 (1), 301–306.

93. Chou, Y.-S., Stevenson, J.W. and Singh, P. 2008. Effect of pre-oxidation and environmental aging on the seal strength of a novel high-temperature solid oxide fuel cell (SOFC) sealing glass with metallic interconnect. *Journal of Power Sources* 184 (1), 238–244.

94. Smeacetto, F., Chrysanthou, A., Salvo, M., Zhang, Z. and Ferraris, M. 2009. Performance and testing of glass-ceramic sealant used to join anode-supported-electrolyte to Crofer22 APU in planar solid oxide fuel cells. *Journal of Power Sources* 190 (2), 402–407.

95. Smeacetto, F., Salvo, M., Ferraris, M., Casalegno, V., Asinari, P. and Chrysanthou, A. 2008. Characterization and performance of glass-ceramic sealant to join metallic interconnects to YSZ and anode-supported-electrolyte in planar SOFCs. *Journal of the European Ceramic Society* 28 (13), 2521–2527.

96. Smeacetto, F., Salvo, M., Ferraris, M., Cho, J. and Boccaccini, A.R. 2008. Glass-ceramic seal to join Crofer 22 APU alloy to YSZ ceramic in planar SOFCs. *Journal of the European Ceramic Society* 28 (1), 61–68.

97. Cabouro, G., Caboche, G., Chevalier, S. and Piccardo, P. 2006. Opportunity of metallic interconnects for ITSOFC: Reactivity and electrical property. *Journal of Power Sources* 156 (1), 39–44.

98. Choi, J.P., Weil, K.S., Chou, Y.M., Stevenson, J.W. and Yang, Z.G. 2011. Development of MnCoO coating with new aluminizing process for planar SOFC stacks. *International Journal of Hydrogen Energy* 36 (7), 4549–4556.

99. Chou, Y.-S., Stevenson, J.W. and Singh, P. 2008. Effect of aluminizing of Cr-containing ferritic alloys on the seal strength of a novel high-temperature solid oxide fuel cell sealing glass. *Journal of Power Sources* 185 (2), 1001–1008.

100. Zhang, H.Z. and Li, G. 2010. Reduction of chromate formation at the interface of solid oxide fuel cells by different additives. *Journal of Power Sources* 195 (19), 6795–6797.

101. Cable, M. 1978. In *Borate Glass: Structure, Properties, Applications*. Edited by L.D. Pye, V.D. Fréchette, and N.J. Kreidl. Plenum Press: New York, p. 399.

102. Günther, K.G. 1958. Measurement of the steam pressure and evaporation rate of the glass forming components. *Glastechnische Berichte-Glass Science and Technology* 31, 9–15.

103. Wang, R., Lu, Z., Liu, C., Zhu, R., Huang, X., Wei, B., Ai, N. and Su, W. 2007. Characteristics of a SiO_2-B_2O_3-Al_2O_3-$BaCO_3$-PbO_2-ZnO glass-ceramic sealant for SOFCs. *Journal of Alloys and Compounds* 432 (1–2), 189–193.

104. Ghosh, S., Kundu, P., Das Sharma, A., Basu, R.N. and Maiti, H.S. 2008. Microstructure and property evaluation of barium aluminosilicate glass-ceramic sealant for anode-supported solid oxide fuel cell. *Journal of the European Ceramic Society* 28 (1), 69–76.

105. Huang, S., Lu, Q. and Wang, C. 2011. Y2O3-BaO-SiO2-B2O3-Al2O3 glass sealant for solid oxide fuel cells. *Journal of Alloys and Compounds* 509 (11), 4348–4351.

106. Lara, C., Pascual, M.J., Keding, R. and Duran, A. 2006. Electrical behaviour of glass-ceramics in the systems RO-BaO-SiO2 (R = Mg, Zn) for sealing SOFCs. *Journal of Power Sources* 157 (1), 377–384.

107. Eichler, K., Solow, G., Otschik, P. and Schaffrath, W. 1999. BAS (BaO-Al_2O_3-SiO_2)-glasses for high temperature applications. *Journal of the European Ceramic Society* 19 (6–7), 1101–1104.

108. Chang, H.-T., Lin, C.-K. and Liu, C.-K. 2009. High-temperature mechanical properties of a glass sealant for solid oxide fuel cell. *Journal of Power Sources* 189 (2), 1093–1099.

109. Chang, H.-T., Lin, C.-K. and Liu, C.-K. 2010. Effects of crystallization on the high-temperature mechanical properties of a glass sealant for solid oxide fuel cell. *Journal of Power Sources* 195 (10), 3159–3165.

110. Milhans, J., Li, D.S., Khaleel, M., Sun, X., Al-Haik, M.S., Harris, A. and Garmestani, H. 2011. Mechanical properties of solid oxide fuel cell glass-ceramic seal at high temperatures. *Journal of Power Sources* 196 (13), 5599–5603.

111. Dietzel, A. 1968. Glass structure and glass properties. *Glastechnische Berichte-Glass Science and Technology* 22, 41–48.

112. Cheng, Y., Xiao, H., Shuguang, C. and Tang, B. 2009. Structure and crystallization of B2O3-Al2O3-SiO2 glasses. *Physica B: Condensed Matter* 404 (8–11), 1230–1234.

113. Lahl, N., Singh, K., Singheiser, L., Hilpert, K. and Bahadur, D. 2000. Crystallisation kinetics in AO-Al2O3-SiO2-B2O3 glasses (A = Ba, Ca, Mg). *Journal of Materials Science* 35 (12), 3089–3096.

114. Zhang, T., Brow, R.K., Reis, S.T. and Ray, C.S. 2008. Isothermal crystallization of a solid oxide fuel cell sealing glass by differential thermal analysis. *Journal of the American Ceramic Society* 91 (10), 3235–3239.

115. Sun, T., Xiao, H., Guo, W. and Hong, X. 2010. Effect of Al_2O_3 content on BaO-Al_2O_3-B_2O_3-SiO_2 glass sealant for solid oxide fuel cell. *Ceramics International* 36 (2), 821–826.

4

Advances in Micro Solid Oxide Fuel Cells

Partha Sarkar and Saeid Amiri

CONTENTS

4.1 Introduction 69
4.2 Single Chamber Micro SOFC 73
    4.2.1 Advancement of Single Chamber Solid Oxide Fuel Cell 74
4.3 Thin Film Based Planar Micro Solid Oxide Fuel Cell 78
    4.3.1 Thin Film Deposition Techniques 80
    4.3.2 Fabrication Method of Forming a Free-Standing
        Electrolyte on Silicon Wafer 82
    4.3.3 Advancements of Planar Micro-SOFC Technology 84
4.4 Microtubular Solid Oxide Fuel Cell 97
    4.4.1 Electrolyte-Supported First-Generation Microtubular SOFC ... 98
    4.4.2 Anode-Supported Second-Generation Microtubular SOFC 99
    4.4.3 Porous Electrolyte-Supported Third-Generation
        Microtubular SOFC 107
    4.4.4 Microtubular SOFC Stack 110
4.5 Modeling of Microtubular Solid Oxide Fuel Cell 114
4.6 Conclusion 118
References 118

4.1 Introduction

This chapter will cover mini-solid oxide fuel cells (mSOFCs) and micro-solid oxide fuel cells (μSOFCs). One of the fascinating things about fuel cells is that one can design and build a power-generating system capable of producing from microwatts to the megawatts level. Not only that, it can also be designed and produced in a modular format to achieve the desired electrical output capacity. Figure 4.1 shows the SOFC potential electrical power range from milliwatts to multi-megawatts and its applications spectrum [1]. Ten kilowatts and above is strictly for stationary applications; the 1 to 5 kW range is for transportable and stationary (mainly residential) applications. This chapter discussion will be limited to the 50 mW to 250 W power range, which corresponds to a portable power spectrum. The chapter will cover systems

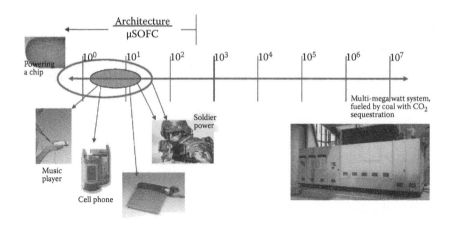

FIGURE 4.1

Illustration of potential power-generation range and application spectrum of SOFC. The typical power range of μSOFC architecture and some of its applications are given. (Modified and adopted from Richard, C., "Navy fuel cell programs," 1999. Available at http://www.netl.doe.gov/publications/proceedings/99/99fuelcell/fc2-1.pdf.)

ranging from single chamber μSOFC to planar thin film (<1 μm) μSOFC and microtubular tubular SOFC (μT-SOFC) as well as the modeling of μT-SOFC. Single chamber and thin film μSOFC are situated at the left-hand corner of the power spectrum (≤5 W), whereas submillimeter μT-SOFCs are more suited for ≤20 W power sources. Two to three millimeter diameter μT-SOFCs are suitable for ≤250 W power generation. It is practically less problematic to build a stack with a lesser number of single cells with higher power generation capacity than a large number of single cells with smaller power. In a series connection of single cells, one bad cell can reduce the series performance, and because SOFCs operate at high temperatures, it is generally not easy to detect the faulty cell and replace it. A modular design approach of the stack can increase the power-generation limit in any of above-mentioned cases.

The worldwide significant development activity in advanced technology areas, such as nanofabrication and microfabrication, has resulted in electronic device miniaturization and the creation of more and more efficient multifunctionalization of micromechanical, portable, communications, imaging, sensing, chemical, analytical, and biomedical devices. With the rapid growth of portable consumer electronics, such as cellular phones, digital cameras, and notebook computers, the market request for small power sources continues to grow. A large chunk of these devices' power needs falls within a few milliwatts to ≤20 W. Currently, batteries are mainly used to provide the power for these segments. Among all the batteries, rechargeable lithium batteries offer the highest density, no memory effect, and relatively low self-discharge and are accordingly the most commonly used. Unfortunately, the energy density of existing batteries is low: advanced lithium ion batteries only have an energy density of about 0.20 kWh/kg, and their development is not keeping pace with portable electronic devices. On the other hand, hydrocarbon fuel energy densities are considerably higher. Therefore, hydrocarbon fuel–based energy conversion devices, such as micro fuel cells, are potential candidates to provide an alternative to the rechargeable lithium-ion battery.

Typically a battery is composed of only a few components: anode, cathode, and electrolyte (semipermeable membrane). An advanced battery pack system also has a control circuitry to protect it from overheating. But a fuel cell power-generation system is much more complex as is shown in Figure 4.2. It is basically a chemical plant. A SOFC power-generating system has several components, called the balance of plant (BOP). Some of the important BOP components are the fuel reformer, thermal recuperator, air blower, catalytic combustor, power conditioning unit/or DC to DC converter, etc. To translate the high chemical energy density of a hydrocarbon fuel to high electrical energy density requires an efficient and compact SOFC system. To be successful, the SOFC system must have high electrochemical performance, high fuel utilization, and low parasitic loss. Large power-generation SOFC systems have demonstrated electrical efficiency >50%, but it is yet to be seen if smaller systems, mainly <5 W systems, can provide reasonable efficiency levels. For integrated electronic applications, the size, form factor, and SOFC device's outside surface temperature are important.

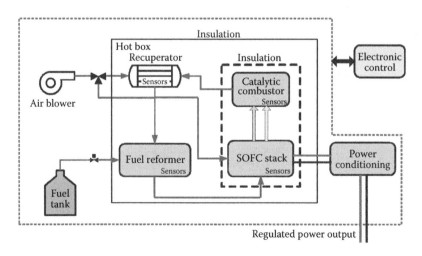

FIGURE 4.2
Schematic diagram of µSOFC power-generation system showing all the major components.

There are various types of micro fuel cells. Micro fuel cells based on polymer electrolyte membrane (PEM) are as follows:

- Hydrogen-fed micro fuel cell: pure hydrogen is directly fed into the anode of the PEM cell. This is the simplest in operation and provides high efficiency. However, it needs compressed hydrogen or metal hydride to run it.
- Micro-reformed hydrogen fuel cell: on-board reformer unit provides the hydrogen for the PEM cell. This design overcomes the on-board hydrogen storage problem.
- Direct methanol fuel cell (DMFC): here, liquid methanol is directly used in the anode to produce electricity; there is no need for a reformer.

One of the great advantages of the PEM–based micro fuel cell is its low operating temperature (60°C–90°C). This lower operating temperature makes it easy to thermally integrate with electronic devices, and that is why most of the development activities are concentrated on PEM micro fuel cells. This chapter will only cover micro fuel cells based on SOFCs. The typical operating temperature of a large conventional SOFC is 750°C–950°C. Steele and Heinzel [2] in 2001 detailed some of the basic design parameters of the SOFC. Using electrolyte conductivity data as a function of temperature and assigning an area specific resistance (ASR) 0.15 Ω cm^2 of electrolyte, they calculated the operating temperature of different types of 8 m/o yttria-stabilized zirconia (YSZ)–based SOFC cells. Figure 4.3 is an adaptation from Steele and Heinzel [2]. For electronic device applications, the µSOFC's operating temperature needs to be ≤500°C, preferably ≤400°C; otherwise thermal

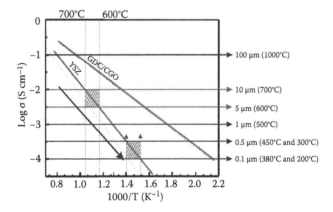

FIGURE 4.3
Logarithm of electrolyte conductivity plotted as a function of reciprocal of absolute temperature. Electrolyte thicknesses and estimated operating temperature of the SOFCs are leveled on the basis of electrolyte ASR 0.1 Ω cm^2. (Adapted from Steele, B.C.H. and Heinzel, A., *Nature*, 414, 345–352, 2001.)

integration will be difficult. Using a 0.1 Ω cm² ASR criterion in Figure 4.3, it is shown that an electrolyte thickness of 1 μm corresponded to cell operating temperatures of 500°C for YSZ and 300°C for ceria-based electrolytes. Therefore, it can be said that an electrolyte thickness in the range of ≤1 μm is required for μSOFC applications in which it needs to be integrated with electronic devices. It is important to remember that this criterion assumes that the electrode polarization is not affected by the cell operating temperature.

4.2 Single Chamber Micro SOFC

As the name suggests, in single chamber fuel cells (SCFCs), the anode and cathode are situated in the same gas chamber. In a conventional fuel cell, the oxidant and the fuel are separated by a gas-tight electrolyte, that is, the cell has two separate gas chambers. An SCFC has no separate stream of fuel and oxidant; instead it has single gas stream of a mixture of fuel, oxidant, and diluent. The operation of a single chamber SOFC (SC-SOFC) is based on the selectivity of the cathode and the anode because both are exposed to the same gas mixture of fuel and air. Figure 4.4 depicts the single chamber SOFC and conventional SOFCs side by side to highlight the differences. In conventional SOFCs, the cathode comes in contact with the oxidant but not with the fuel. Therefore, only the reduction of oxygen takes place at the cathode. Similarly, the oxidation of the fuel occurs at the anode. Therefore, conventional SOFC electrodes do not need to be selective. In a single chamber

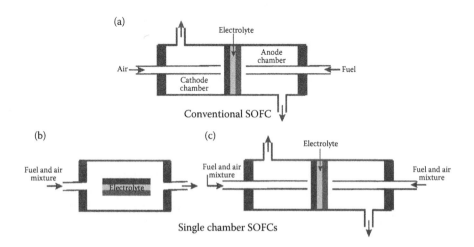

FIGURE 4.4
(a) Schematic of conventional SOFC and (b and c) schematics of single chamber SOFCs in two different configurations.

SOFC, the operating principle is based on the catalytic selectivity difference between electrodes. The anode catalyst preferentially adsorbs fuel and promotes an anodic reaction. Similarly, the cathode adsorbs oxygen and promotes a cathodic reaction. Although anode and cathode are exposed to the same gas mixture, due to the electrodes' relative selectivity differences, there is a different oxygen partial pressure set up in each electrode–electrolyte interface, and as a result, a voltage is generated between the anode and cathode. This type of fuel cell is also called "one chamber," "single compartment," "mixed-gas," "mixed-fuel," "mixed-reactant," and "separator free" in the literature, but the most accepted name in the scientific community is single chamber fuel cell. Normally, a cell with two electrodes made of the same materials will not generate any voltage/power. But it may be possible to generate voltage between electrodes made of the same materials but with a different microstructure; however, will it be able to produce enough power with practical significance?

4.2.1 Advancement of Single Chamber Solid Oxide Fuel Cell

The basic concept of the SCFC was proposed in the early 1960s [3,4]. van Gool in 1965 [4] from the Philips Research Lab described SCFCs in a paper titled "The Possible Use of Surface Migration in Fuel Cells and Heterogeneous Catalysis." Dyer [5] in a 1990 article in *Nature* reported the study results of SCFC-type electrochemical devices using hydrated aluminum oxide, pseudoboehmite, and Nafion membranes with submicron thicknesses at room temperature. The authors used different combinations of sputtered Pt, Ni, and Pd as anodes and cathodes and observed cell voltages close to 1 V in a hydrogen air mixture.

Japanese researchers from Nagoya University pioneered the high-temperature SC-SOFC development [6–13]. Hibino and Iwahara [6], in 1993, operated an $Ni\text{-}YSZ|YSZ|Au$ SC-SOFC system using a CH_4–air mixture and obtained only 2.36 mW/cm^2 at 950°C. In 1995, the same research group reported [7] a SC-SOFC using a ceramic proton conductor; the cell consisted of $Pt|BaCe_{0.8}Y_{0.2}O_{3-\delta}|Au$. This cell was operated between 750°C and 950°C, generating an open circuit voltage of 0.7–0.8 V and producing a maximum of 170 mW/cm^2 at 950°C. In 1996 [8], this group fabricated a coplanar SC-SOFC by positioning the anode (Pd) and cathode (Au) in the same face of the ceramic proton conductor. Hibino et al. [9,10] in 1999 studied YSZ electrolyte-based SC-SOFC and reported that the addition of MnO_2 in the electrodes enhanced the cell performance. This group [11], using a $La_{0.9}Sr_{0.1}Ga_{0.8}Mg_{0.2}O_3$ electrolyte, operated an SC-SOFC with hydrocarbon fuels and found that a $Ni\text{–}Ce_{0.8}Sm_{0.2}O_{1.9}$ cermet anode and $Sm_{0.5}Sr_{0.5}CoO_3$ oxide cathode provided the best performance. An article published in *Science* by Hibino et al. in 2000 [12] reported an SC-SOFC study using 10 wt% samaria-doped ceria (SDC) electrolyte and hydrocarbon fuel–air mixtures at temperature ≤500°C. In this study, YSZ and LSGM electrolyte-based SC-SOFCs were included for performance comparison. OCV of ~0.9 V was observed in a mixture of ethane or propane and air; the cells provided ~400 mW/cm² at 500°C. Yano et al. [13] in 2007 published a review article on SC-SOFC covering design, materials, and performances and is a recommended consult for additional information.

In 1981, Louis et al. [14] from United Technologies Corp. of the United States patented a coplanar design. This patent included various design configurations of series and parallel connections of the cells to enhance output voltage and power. In the coplanar design concepts of SC-SOFC, the anode and cathode are positioned on the same side of the electrolyte surface. Because all the active components of the cell are in the same plane, it is called a coplanar SCFC. This coplanar design is especially suitable for microfabrication techniques compared to conventional positive-electrolyte-negative (PEN) design. In this design, the electrolyte can be deposited on a dense substrate, some amount of porosity in the electrolyte is acceptable, and the thickness of the electrolyte does not determine the ohmic resistance of the electrolyte of the cell. Electrolyte ohmic resistance is dependent on the lateral distance between the anode and cathode.

Hibino et al. [8,15] studied coplanar SC-SOFC. This group used several types of electrolytes: proton-conducting $BaCe_{0.8}Gd_{0.2}O_{3-\delta}$, oxide ion–conducting 8 mol% YSZ, $La_{0.9}Sr_{0.1}Ga_{0.8}Mg_{0.2}O_3$ (LSGM), and $Ce_{0.8}Sm_{0.2}O_{1.9}$ (SDC). They used a conventional technique, brush painting, to apply the electrodes on thick (~2 mm) electrolyte surfaces, keeping the gap between electrodes between 0.5 and 3 mm. The best performing cell consisted of Ni-SDC-Pd (anode) $|Ce_{0.8}Sm_{0.2}O_{1.9}$ (SDC)$|$ $Sm_{0.5}Sr_{0.5}CoO_3$ (cathode). The effect of electrolyte surface roughness was studied using an impedance analyzer, and it was found that the internal ohmic resistance of the cell reduced with decreasing surface roughness. Figure 4.5 highlights schematic cell configuration with I-V characteristics and power densities for different gaps between the electrodes. These results show that at 600°C, 0.5 mm gap between the elec-

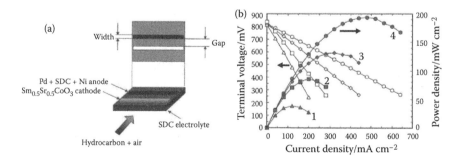

FIGURE 4.5
(a) Schematic of the co-planar cell showing the electrode configuration and (b) I-V and power characteristics of SC-SOFCs with two electrodes separated from each other by different gaps (1: 3 mm, 2: 1.5 mm, 3: 1 mm, and 4: 0.5 mm) on the same surface of electrolyte at 600°C. (Hibino, T. et al. *J. Electrochem. Soc.*, 149, A195–A200, 2002.)

trodes has a maximum power density of 193 mW/cm^2, and same cell at 500°C achieved 122 mW/cm^2.

Fleiga et al. [16] considered miniaturization of SOFC and concluded co-planar microfabricated SC-µSOFC, in principle, can be integrated onto a single chip with other electronic circuits in a device. Their analysis indicated that to realize high-performance SC-SOFC with co-planar electrodes on a thin film electrolyte, a comb-like interdigitated electrode structure is required with electrode stripe distances as small as a few micrometers and stripe widths ~10 µm. Rotureau et al. [17,18] employed screen printing for fabrication of a co-planar cell. They studied two types of configurations: (1) electrodes were screen-printed on the YSZ electrolyte substrate surface, and (2) the electrolyte, anode, and cathode were screen printed on an alumina substrate anode. These cells had a very poor performance at 800°C in a CH$_4$–air mixture.

Ahn et al. [19,20] fabricated a YSZ electrolyte–based SC-SOFC using two different techniques. In the first instance [19], the YSZ surface was micropatterned with electrodes by a microfluidic lithography technique. The interdigitated micropatterned electrodes had an area of 5 × 5 mm, in which individual electrodes had 100 µm line width with anode-to-cathode distances of 50 µm. *I-V* characteristics of the SC-SOFC were analyzed using a dc source meter. The OCV of the resulting cell in the CH$_4$–air mixture was ~350 mV, and the power density was 73.5 mW/cm^2 at 910°C. In the second case [20], a co-planar SC-SOFC was fabricated by applying the electrodes on the YSZ electrolyte surface via a robo-dispensing method. The influence of electrode geometry and spacing on the performance was investigated. Cells with a single electrode pair exhibit high and uniform OCVs in the range of 782–804 mV where OCV had shown a tendency of reduction in value with decreasing anode-to-cathode spacing. However, the power density increases with a decrease of electrode spacing due to a decrease in ohmic resistance of the cell. For cells with multiple pairs of electrodes, the OCV was drastically decreased with an increasing number of closely spaced electrode pairs potentially due to intermixing of product gases.

Yoon et al. [21] fabricated a comb-like electrode configuration by screen printing on a SDC electrolyte surface and observed 40 mW/cm^2 at 700°C in the CH$_4$–air gas mixture. Buergler et al. [22] studied a co-planar cell of Ni-Ce$_{0.8}$Gd$_{0.2}$O$_{1.9}$ (anode) |GDC| Sm$_{0.5}$Sr$_{0.5}$CoO$_{3-\delta}$ (cathode). Comb-type electrodes were fabricated either by screen printing or by micro-molding in capillaries. The cells were operated in CH$_4$–air mixtures, an open circuit voltage of 0.65–0.75 V was observed. Maximum cell power output was 17 mW/cm^2. Kuhn et al. [23] from Ecole Polytechnique, Canada, reported an experimental study on a current collection of SC-SOFC using a comb-like electrode configuration. This Canadian research group used direct-write microfabrication techniques [24,25] to apply the comb-like co-planar electrodes as shown in Figure 4.6a. It is a pressure-driven extrusion of ink through a micro-nozzle where the nozzle position is controlled by a XYZ-scanner to create an intricate electrode design on a substrate, as schematically illustrated in Figure

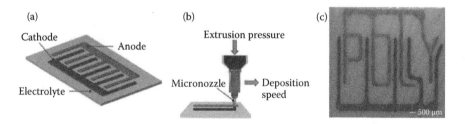

FIGURE 4.6

(a) Schematic illustration of co-planar single chamber µSOFC with interdigitated electrodes [24], (b) schematic illustration of the computer controlled direct-write microfabrication of electrode lines on an electrolyte plate [24], and (c) optical micrographs of sintered co-planar SC-µSOFC. (From Kuhn, M. et al. *J. Micromech. Microeng.* 18, 015005, 2008; Kuhn, M. et al. *J. Power Sources*, 177, 148–153, 2008.)

4.6b. Sintered co-planar SC-µSOFC is fabricated by direct-writing as shown in Figure 4.6c.

Kim et al. [26] in 2010 reported co-planar SC-SOFC fabrication using a novel microfabrication method, called photoresist molding with a thermosetting polymer (PRM-TP). This technique is very efficient for fabricating multiple ceramic structures with a sub-micron to a several millimeter scaled dimension. This is a hybrid process combining lithographic tools and ceramic slurry process. An electrochemical performance test provided low OCV and power density of 67 mW/cm² at 500°C.

An article published in *Nature* by Shao et al. [27] in 2005 reported the demonstration of a thermally self-sustaining SC-µSOFC two-cell stack operation on propane. The µSOFC consisted of $Ni + SDC|SDC|BSCF + SDC$ (BSCF: $Ba_{0.5}Sr_{0.5}Co0.8Fe_{0.2}O_{3-\delta}$). They used catalytic partial oxidation of the gas mixture to generate the heat for maintaining the fuel cells at 500°C–600°C. A schematic of the single cell and experimental setup for SC-µSOFC is shown in Figure 4.7. The anode was coated with a porous layer of $Ru-CeO_2$ to enhance the partial oxidation of the fuel at the anode (Figure 4.7a). The presence of a $Ru-CO_2$ layer lowered the partial oxidation initiation temperature. Here, a µSOC stack generated peak power densities in the range of ~180–250 mW/cm² between 500°C and 600°C depending on experimental conditions, such as fuel flow rate and a startup time of less than 1 min. Shao et al. [28] from the same, research reported in 2006 that at a 790°C cell temperature $Ni + SDC|SDC|BSCF + SDC$–based SC-µSOFC produced 760 mW/cm² in a gas mixture of methane, oxygen, and helium flow rates of 87, 80, and 320 mL/min [STP], respectively.

Hao et al. [29] performed a systematic simulation study backed by experimental validation. They investigated cell efficiency and fuel utilization of a methane-powered SC-SOFC as a function of operating parameters, including gas flow rate, balance gas, fuel-to-oxygen ratio, and fuel cell configuration. Their model predicted that the maximum achievable efficiency of a single-cell SC-SOFC is >10%, and the efficiency at typical operating conditions is >5%. These efficiency

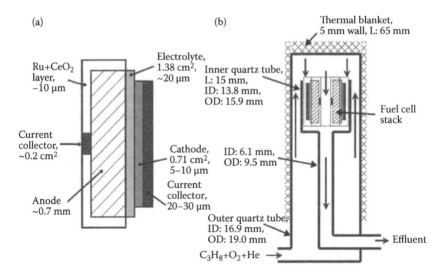

FIGURE 4.7
(a) Schematic diagram of the cell architecture and (b) experimental setup for the thermally self-sustaining two-cell μSOFC stack. (Shao, Z. et al. *Nature*, 435, 795–798, 2005.)

values are orders of magnitude more than experimentally observed results. The authors noted that the high efficiencies could be achieved by the optimization of flow rate and the adoption of efficient cell and stack design.

SC-SOFC design is considerably simpler than conventional SOFC because it does not need gas-tight seals, does not need a reformer for hydrocarbon fuel, is not susceptible to coking problems, has a quick startup, and is thermally self-sustainable. The electrolyte can have a connected porosity (which reduces the stringent manufacturing requirement); this does not adversely affect the cell performance as long as the electrodes do not get shorted through the pores. The electrolyte ASR increase due to porosity will reduce cell performance, so a highly porous electrolyte is not desirable. Although cell power density can reach high values, consumption or burning of fuel in a gas phase is a major issue. As a result, low fuel utilization, typically <1%, is the biggest challenge for SCFCs toward practical application. Kuhn et al.'s review article [30] covered most of the aspects of SC-SOFC extensively and is recommended for further consultation.

4.3 Thin Film Based Planar Micro Solid Oxide Fuel Cell

This section covers thin film planar μSOFCs, which are targeted for mobile electronics applications to replace rechargeable lithium ion batteries. A conventional fuel cells' typical minimum operating temperature is ~750°C. Integrated µSOFC power sources for electronic devices must work at temperatures as low as possible. For integration with mobile electronics, the µSOFC system must have a small form factor, must be similar in size to rechargeable batteries, and the outer surface temperature must be in the 30°C–40°C range. From a thermal integration point, the lower the operating temperature, the easier the integration. The operation temperature of a SOFC can be lowered by reducing the ASR and the polarization losses of the cell. Figure 4.3 shows that the ASR can be lowered by reducing the electrolyte thickness. Planar µSOFC operating temperatures must be ≤600°C with the desirable operating temperature of 300°C–500°C. Ultrathin film (electrolyte thickness <1 µm) planar µSOFC are potential candidates for mobile electronics, and there is considerable worldwide R&D activity to realize this goal. In the SOFC fuel cell spectrum diagram (Figure 4.1), these planar µSOFC power sources are situated in the extreme left, i.e., a fraction of a watt to the 20 W level or, more precisely, the ~0.1 W to 5 W level. Like conventional SOFCs, these thin film µSOFCs also comprise three active layers: two porous electrodes, i.e., an anode and cathode, separated by a dense oxygen-ion–conducting electrolyte layer. The combined thickness of these three layers is typically ≤1 µm. This three-layer fuel cell membrane structure (membrane electrode assembly) is sometimes referred to as PEN, i.e., the positive electrode–electrolyte–negative electrode element. To produce a few hundred mW of power typically requires a membrane electrode assembly active surface area in the order of 1 cm² (i.e., 10,000 µm × 10,000 µm). This thin membrane assembly is very fragile; therefore, to fabricate a practical device, the membrane electrode assembly has to be supported by a porous layer like conventional SOFCs or it has to be suspended on a support substrate. The suspended film configuration is commonly called a free-standing membrane in the literature. These two types of planar µSOFC configurations are illustrated in Figure 4.8. In conventional SOFCs, different types of thick (typically 200 µm–1000 µm) porous supports have been used to provide the mechanical support to the cell, such as a thick, dense electrolyte (electrolyte-supported); a thick, porous anode (called an anode-supported SOFC); a thick, porous cathode (called a cathode-supported SOFC); or a thick, porous metallic current collector (called current collector supported or metal-supported SOFC) support cells.

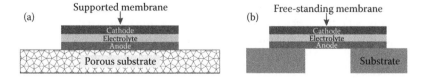

FIGURE 4.8
Schematic drawing of a planar µSOFC cell with the micrometer-thick PEN membrane assembly (a) supported by a porous substrate and (b) suspended on a substrate called a free-standing PEN membrane.

The current major R&D effort is to develop a planar μSOFC on a silicon wafer using microfabrication technology so that eventually this micro power source can be batch produced at a commercially acceptable cost. Another motivation is to improve the performance using micro- or nanostructures. In this section, a brief description of typical planar μSOFC fabrication will be covered followed by discussion of some of the major R&D efforts around the world.

4.3.1 Thin Film Deposition Techniques

Thin film techniques are the key for developing successful planar μSOFC power generation devices. In 2007 Beckel et al. [31] discussed different thin film techniques for SOFC. Pederson et al. [32] in 2006 reviewed the applicability of vacuum deposition techniques, such as magnetron sputtering, vacuum plasma deposition, laser ablation, and electrochemical vapor deposition (EVD), to SOFC technology.

As the name suggests, vacuum deposition technique is conducted under vacuum, where desire materials atoms from the source are transported to the substrate via a gas phase. On the substrate, a thin film will form via a nucleation and growth process. There are several types of vacuum deposition techniques in physical vapor deposition (PVD). In PVD techniques, the desired atoms are brought to the gas phase from a solid or molten target (source) by evaporation, sputtering, laser ablation (pulsed laser deposition, PLD) and hybrid methods. Evaporation techniques work well for metal thin films, such as gold and Ag deposition, but this technique is not suitable for complex SOFC materials, like YSZ, LSM, CGO, and others. Sputtering and PLD are the two most popular techniques to deposit the electrolyte membrane for μSOFC.

Sputtering is a process whereby atoms are ejected from a solid target material (source) due to bombardment by high-speed particles. The target and the substrate are placed in a vacuum chamber containing a noble gas; a high voltage is applied between them so that the target is the cathode, and the substrate is attached to the anode. Application of high voltage ionizes the low-pressure noble gas creating plasma, and positive ions from the plasma bombard the target ejecting atoms. These ejected atoms go toward the anode and deposit on the substrate. This is not a line-of-sight process and is able to coat a large surface with uniform thickness.

Although *radio frequency (RF) sputtering* using an oxide (dielectric materials) target and DC reactive sputtering using metallic targets both have been utilized to produce YSZ thin films, RF is the most popular one because it can sputter both metallic and dielectric oxides. In the case of YSZ, during RF sputtering, the target ejects a mixture of oxygen, metal, and metal oxides that deposit on the substrate. Often oxygen is added to the sputtering gas mixture to control the metal-to-oxygen ratio. Typically, the sputtering rate is less than 1 μm/h.

In pulsed laser deposition (PLD), the ablation of the target material upon laser irradiation and the creation of plasma are very complex processes. The removal of atoms from the bulk material is done by vaporization of the bulk at the surface region in a state of non-equilibrium. The laser beam usually has a wavelength in the UV range and applies ns (e.g., 10 ns) pulses. It can be seen from Figure 4.9b that the laser source is situated outside the vacuum chamber, i.e., the target and laser are not coupled. The vacuum chamber has a UV-transparent window for the laser beam. Typically, targets are dense, sintered, polycrystalline, fine-grain ceramics in a disk shape. To avoid surface roughening, the target moves with respect to the laser beam. The fine-grain, dense ceramics have a high absorption efficiency and also a lower thermal conductivity due to the high grain boundary density. Both properties are important in enhancing the material–radiation interaction. The focused laser beam strikes the target surface and forms a plume (atoms, molecules, electrons, ions, clusters), which is congruent with the target. This plume deposits as a thin film on a substrate, such as a silicon wafer.

Atomic layer deposition (ALD) is used for fabrication of ultrathin and conformal thin film structures for semiconductor and thin film device applications. The unique attribute of the ALD process is its thickness control ability through sequential self-limiting surface reactions to achieve control of film growth in the monolayer or sub-monolayer thickness regime. ALD has a close similarity in chemistry to chemical vapor deposition (CVD) except that the ALD reaction breaks the CVD reaction into two distinct half-reaction steps. Figure 4.10 [33] is an illustration of typical ADL cycles. The majority of ALD reactions use two chemicals. These precursors react with a surface one at a time in a sequential, self-limiting, manner. By exposing the precursors

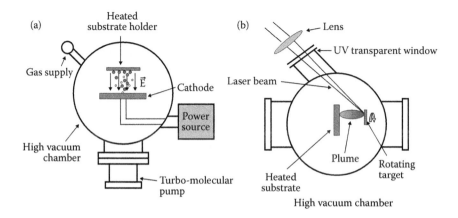

FIGURE 4.9
Schematics of generic simplified (a) sputtering system and (b) pulsed laser deposition system. (From Beckel, D. et al. *J Power Sources*, 173, 325–345, 2007.)

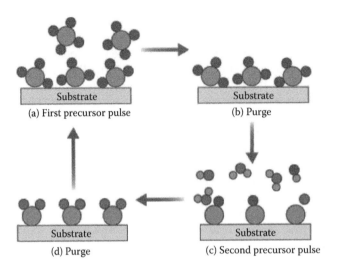

FIGURE 4.10
Schematic of atomic layer deposition (ALD) process cycle. (From Irvine, G., "High-k semiconductor materials from a chemical manufacturer perspective," http://www.electroiq.com/articles/sst/2011/02/high-k-semiconductor-materials-from-a-chemical-manufacturer-pers.html 2011.)

to the growth surface repeatedly, a thin film is deposited. By keeping the precursors separate throughout the coating process, atomic layer control of film growth can be obtained on a scale as fine as a sub-angstrom per cycle. Typically, the ALD technique provides uniform conformal, adherent, and homogeneous nano-scaled films, which are often crystalline as deposited without the need of high-temperature annealing treatments. Atomic-level control ensures that extremely thin films and complex nanostructures can be processed. ALD is also used for doping and modification of interfaces. Cassir et al.'s [34] 2010 publication provided extensive coverage and a thorough analysis of ADL literature dealing with SOFCs and is recommended for further information on ALD.

4.3.2 Fabrication Method of Forming a Free- Standing Electrolyte on Silicon Wafer

Bruschi et al. [35] in their 1999 article described the fabrication of free-standing YSZ on a silicon wafer. They used two different processes for the fabrication of a suspended YSZ membrane by means of micromachining technology. These processes are compatible with standard integrated circuits fabrication. The schematics of these two processes are shown in Figure 4.11.

Initially, in both the processes, the wafer surface was oxidized. In Process I, a photoresist was applied in side 1 and side 2 of the wafer except for a square hole in side 1. By chemical etching, the silica was removed from the square

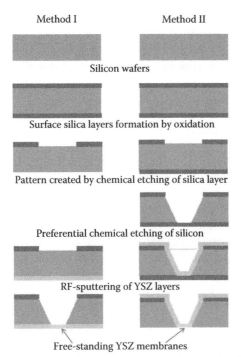

Method I   Method II

Silicon wafers

Surface silica layers formation by oxidation

Pattern created by chemical etching of silica layer

Preferential chemical etching of silicon

RF-sputtering of YSZ layers

Free-standing YSZ membranes

Preferential etching of silicon to form free-standing YSZ membranes

FIGURE 4.11
Schematics of typical microfabrication steps to fabricate free-standing electrolyte membranes.

area of side 1. In the next step in Process I, side 2 of the wafer was coated with a 300 nm YSZ film by RF-sputtering followed by annealing at 400°C. In the last step, by using a preferential chemical etching technique for silicon, a pyramid-shaped hole was created on the silicon substrate to expose the other side of the YSZ membrane, thereby creating a free-standing YSZ membrane. In Process II, a square window was created by silicon oxide removal on only side 1 of the wafer by a photolithography-etching technique. In the next step, silicon from the square was removed by preferential etching, leaving the silica layer at the bottom. The YSZ layer was deposited on the side 1 of the wafer. After annealing of the YSZ layer, the silica layer was removed from side 2 of the wafer to produce a free-standing YSZ membrane. By these processes, they were able to produce square membranes with a side as large as 170 μm. To complete the fabrication of planar μSOFC, both sides of the electrolytes are deposited with porous electrodes to form a membrane electrode assembly.

Currently, instead of forming oxide films by thermal oxidation processes on the wafer surface, silicon nitride (SiNx) is deposited by CVD techniques on both sides of a polished (100) silicon wafer. One side the silicon nitride is patterned using standard photolithographic methods to define a silicon etch mask followed by anisotropically etching through the wafer thickness to the topside silicon nitride. A thin electrolyte oxide film is deposited on the top side. The final step is the release to form free-standing electrolyte oxide membranes by removing the supporting SiNx layer with plasma etching.

4.3.3 Advancements of Planar Micro-SOFC Technology

In 1992, Kueper et al. [36] from Lawrence Berkeley Laboratory reported fabrication of crack-free YSZ films deposited on nonporous quartz substrates at thicknesses of ~0.2 to 0.5 μm. Their effort indicated that crack-free submicron electrolyte films can be deposited on porous substrates. Deposited films were sintered to full density at temperatures typically in the range of 500°C to 600°C. Fleiga et al. [16] in a 2004 publication used numerical calculations to examine the correlation between decreasing electrolyte film thickness and ohmic polarization for the planar cell. They concluded from the electrical point of view that electrolyte film thicknesses below 300 nm were not reasonable to achieve low electrolyte resistance due to the current constriction on electrode particles. Baertsch et al. [37] reported the fabrication process for self-supporting electrolyte thin films and methods for complex oxide electrolyte deposition and incorporation into a silicon processing platform. They studied two electrolytes, YSZ and gadolinium doped ceria (GDC). YSZ were deposited using two different techniques: electron beam (e-beam) evaporation and RF sputtering. All GDC thin films were deposited by e-beam evaporation. They achieved high yields of free-standing membranes ranging from 65 to 1025 μm for YSZ RF sputtered films, but GDC membranes were less resistant to fracture, resulting in poorer process yields. They observed compressive residual stresses in RF sputtered YSZ films leading to membrane buckling (Figure 4.12a) upon release. But annealing at 773 K removed the YSZ film buckling and the residual compressive stresses as shown in

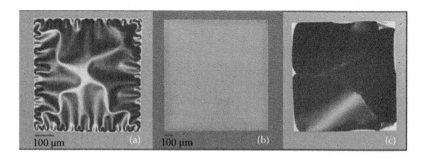

FIGURE 4.12
Optical images of free-standing 100 nm sputtered YSZ membranes (a) after release which results in buckling, (b) after annealing in air at 773 K which removes compressive stresses and flattens the membrane, and (c) after annealing in air at 923 K resulting in tensile fracture. (From Baertsch, C.D. et al. *J. Mater. Res.*, 19, 2604–2615, 2004.)

Figure 4.12b. Higher temperature annealing in air evolved tensile stresses in the sputtered YSZ membranes resulting in tensile fracture as shown in Figure 4.12c.

Srikar et al. [38] in 2004 analyzed the structural design of planar electrolyte-supported μSOFCs. An analytical model was used to determine the effect of electrolyte thickness on fuel cell electrochemical performance. Design diagrams were developed to identify geometries that are resistant to fracture and buckling. Their analysis found thermal losses are significantly higher than the electrochemical power produced. They recommended that more research effort is needed for developing appropriate schemes for thermal isolation of planar μSOFC. Tang et al. [39] in a 2005 paper presented analytical models for thermal stress and heat transfer for planar μSOFC. Thermal–mechanical reliability is vital due to intrinsic stresses in microfabrication and thermal stresses under operating conditions. They analyzed critical stresses in compressive and tensile states for flat and corrugated circular films and found that the corrugated films were more reliable. They proposed the cell architecture of a flat, thin YSZ film with a corrugated supported edge to decrease thermal stress in YSZ films. Thermal analysis found that air convection and radiation, which increase with the film radius, are the main source of heat loss in large circular thin films with radii of several millimeters.

Prinz's research group at Stanford University have been actively developing planar μSOFC since the mid 2000s [40–54]. The Stanford group has done considerable work, currently focused on silicon wafer thin film μSOFC but initially they had considered nanoporous aluminium oxide (AAO) and nanoporous nickel substrates. Although most of their work was focused on YSZ-based μSOFC, they also investigated proton conducting yttria-doped barium zirconate (BYZ) membranes [40,41] and achieved a power density of 136 mW/cm^2 at 400°C. For electrochemical performance enhancement the Stanford group is not only focusing on reduction of the electrolyte thickness but also on improving the cathode electrochemical reaction kinetics. They have been studying electrolyte surface modification [42] and the use of an yttria-doped ceria (YDC) interlayer between the electrolyte and cathode [43–46]. Their studies have shown that a few nanometer interlayer of YDC on the cathode side of the low temperature μSOFC can effectively reduce the interfacial activation energy and enhance the current exchange density.

Huang et al. [47] from the Stanford University group reported in 2007 the fabrication of free standing 50–150 nm YSZ thick films on silicon wafers by a combination of RF sputtering, lithography, and etching. Fuel cells were formed by the DC sputter deposition of an 80 nm porous Pt cathode and anode on the YSZ surface. In a 4-in. wafer, they fabricated 832 individual fuel cells of different sizes. The maximum size of the cell was 240 × 240 μm^2. The cells were tested using air–3% humid hydrogen; at 350°C the OCVs were 1.05 V, which is close to the theoretical value. Ultrathin 50 nm YSZ film cells at 350°C produced a maximum power of 130 mW/cm^2. To improve the cathode electrochemical charge-transfer reaction, they deposited nano-thin GDC as an active interlayer between the YSZ electrolyte and cathode. This cell generated 200 mW/cm^2 and 400 mW/cm^2 of power at 350°C and 400°C, respectively. Although the ASR of the electrolyte was increased by 1%–5% as a result of adding the 50 nm thick GDC interlayer, it is proposed that the maximum power density of the cell was increased considerably due to enhancement of the cathode oxygen reduction reaction. In a publication of Shim et al. [48], the Stanford University group in 2007 used an alternate microfabrication technique, atomic layer deposition (ADL), to fabricate ultrathin (60 nm) YSZ with the stoichiometry of ZrO2/Y2O3 = 0.92:0.08 (8 m/o Yttria-doped zirconia) on a silicon wafer. Sputtered porous Pt layers were deposited as an anode and cathode to form the μSOFC. A maximum power of 270 mW/cm^2 at 350°C was obtained. This ALD YSZ fuel cell performed considerably better than the Stanford group's 50 nm RF sputtered YSZ fuel cell [47]. It was concluded that the performance enhancement is due to an increase in exchange current density at the electrode-electrolyte interface.

Su et al. [49] of the Stanford group in 2008 used a corrugated electrolyte structure in their effort to boost the power density of the cell on the projected area basis, i.e., the power density is calculated using the cell active area, the same as an equivalent flat cell. With the electrolyte layer thickness already at the ~50 nm level, there was no room to enhance cell performance, i.e., power density, by reducing the ASR of the electrolyte. Therefore, for a given flat area, the cell power can be boosted by increasing the cell active surface area by introducing corrugation on the electrolyte layer. To fabricate the desired electrolyte structure, first a corrugated silicon mold was fabricated using photolithography. This mold was subsequently coated with ALD YSZ films, and the mold was removed, leaving a free-standing corrugated fuel cell architecture as shown in the images of Figure 4.13. The white square in Figure 4.13a is the top view of the free-standing YSZ membrane released after KOH etching, and the corresponding bottom view is captured in Figure 4.13d. The black dots represent cup-shaped (diameter 15 μm and depth 20 μm) corrugation in the YSZ film.

The corrugated cell increased the electrolyte surface area fivefold but provided cell power density boosting little less than twofold. The free-standing active cell projected area in the study was considerably higher (≥36 times), 0.0036 cm^2 (600 μm × 600 μm) as compare to most other cases ≤0.001 cm^2 (100 μm × 100 μm). This μSOFC provided a peak power density of 677 mW/cm^2 (i.e., total power of the cell is 2.4 mW) and 861 mW/cm^2 (i.e., total power of the cell is 3.1 mW) at 400°C and 450°C, respectively.

Chao et al. [50] from the Stanford group in a 2011 article reported the fabrication of corrugated thin film electrodes-electrolyte by using nano-sphere lithography, in which self-assembled silica nano particles were deposited on a silicon wafer in a hexagonal close-packed structure and used as an etching mask to pattern a silicon mold with a corrugated geometry. Schematics of the free-standing corrugated *Pt-anode|YSZ|Pt-cathode* μSOFC design together with SEM micrographs are shown in Figure 4.14a through 4.14e. A maximum

FIGURE 4.13
Optical microscopy images of corrugated free-standing YSZ membrane of μSOFC: (a through c) are the top view and (d through f) are the bottom. The white square in the middle is representing the free-standing YSZ membrane released after KOH etching. (From Su, P.-C. et al. *Nano Lett.*, 8, 2289–2292, 2008.)

power density of 820 mW/cm^2 and 1.34 W/cm^2 was measured for fuel cells based on this process at 450°C and 500°C, respectively. Although the cell achieved very high power density, there was considerable polarization loss which was evident from the I-V curve as well as the power curve. A maximum power corresponding to around 0.3 V was considerably lower than the expected value of 0.5 V for a low polarized cell.

Su et al. [51] from the Stanford group in a 2012 article described a modified simple silicon-based MEM fabrication process for a low-temperature planar μSOFC, featuring a large number of free-standing nano-thin film electrolyte membranes. This modified method provided an array-structured free-standing cell with a high surface area utilization. One of the common methods to produce a free-standing nanoscale electrolyte membrane is to deposit thin film electrolytes on (100) silicon substrate and perform through-wafer etching in alkaline solutions (Figure 4.15a). Typically this method creates wrinkles and easily induces cracks due to stress. One of the major practical disadvantages of this method is that the undersized free-standing membrane (usually less than 100 μm in lateral dimension), along with very inefficient wafer surface area utilization (<3%), gives an insignificant total power output. Using the modified fabrication method the spacing between each free-standing membrane is reduced to 10 to 20 μm, fabricated by silicon surface micromachining and KOH etching. The membranes are circular for even stress distribution to alleviate wrinkles and are arranged in a

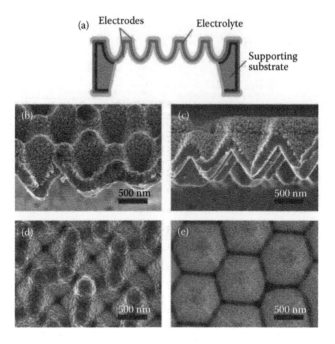

FIGURE 4.14
(a) Schematic of the free-standing corrugated Pt-anode|YSZ|Pt-cathode μSOFC design. (b) SEM image of the nanostructured MEA with porous platinum electrodes and an ALD YSZ electrolyte obtained after electrochemical characterizations. (c) Side view of the MEA. (d) Top view from the cathode side. (e) Bottom view from the anode side. The thickness of the MEA was approximately 200 nm. (From Chao, C.-C. et al. *ACS Nano*, 5, 5692–5696, 2011.)

close-packed fashion to achieve a high density of membranes (Figure 4.13 is illustrating a similar structure from the 2008 article [49]).

Figure 4.15c illustrates the modified method of fabrication of the planar μSOFC electrodes-electrolyte membrane assembly array. On the front side of the wafer, circular trenches were etched by photolithography and deep reactive ion etching. Low-stress silicon nitride with a thickness of 100 nm was deposited by low-pressure chemical vapor deposition on both sides of the wafer. On the reverse side of the wafer, square openings in the silicon nitride layer were patterned by photolithography and plasma etching. YSZ electrolyte with a thickness of 70 nm was deposited on the trenched surface by ALD. After electrolyte deposition, the silicon substrate was etched in KOH solution, and the etching was timed to stop at the vicinity of the trench bottom. Next, a silicon nitride layer underneath the YSZ electrolyte was etched by plasma etching in SF_6 and oxygen plasma. In the following step, porous Pt layers were deposited by DC sputtering as the cathode (top side) and anode to complete the planar μSOFC structure. Figure 4.15d shows a 6 mm by 6 mm and 70 nm thick YSZ membrane electrolyte array on a silicon chip. A 2×2 mm^2 μSOFC array was tested at 350°C, 400°C, and 450°C, and maximum power outputs of 1.78 mW, 2.43 mW,

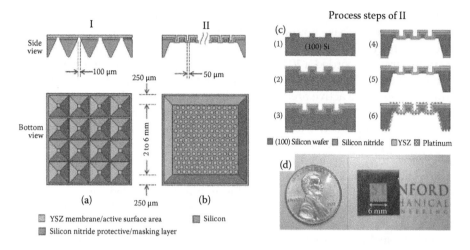

FIGURE 4.15
(See color insert.) (a and b) Illustration of schematic comparison of µSOFC free-standing membrane active areas (yellow) of two different fabrication processes. (a) Membrane array structure fabricated by common KOH through-wafer etching, (b) membrane array structure with high surface utilization by the modified method. The upper pictures in (a and b) are cross-section views of a membrane array on a (100) silicon substrate, and the bottom pictures present the bottom side view of the array. (c) Modified process steps of fabrication of planar µSOFC array on silicon substrate; (d) A 6 mm by 6 mm YSZ membrane electrolyte array on a silicon chip. (From Su, P.-C., and Prinz, F.B., *Electrochem. Comm.*, 16, 77–79, 2012.)

and 2.98 mW were obtained, respectively. This is a significant achievement; the planar µSOFC power level was boosted from the µW to mW level.

Ramanathan's research group at Harvard University is actively involved in planar µSOFC R&D [55–60]. Lai et al. [55] in 2010 reported the finding of dense $La_{0.6}Sr_{0.4}Co_{0.8}Fe_{0.2}O_3$ (LSCF) thin film cathode-based YSZ µSOFC. LSCF is a mixed conductor; therefore, dense LSCF can transport both oxide ions and electrons for the cathode's electrochemical reaction. They used RF sputtering to deposit the oxide layers, the LSCF cathode, and the YSZ electrolyte and DC sputtering to deposit the porous Pt anode. A thin film dense cathode microstructure and thermo-mechanical behavior had been investigated to identify the conditions for increasing the active area for µSOFC. Through careful engineering, the Harvard group was able to produce a failure-resistant planar square (250×250 µm²) µSOFCs of *LSCF (dense)-cathode|YSZ|Porous-Pt anode*. A peak power density of ~120 mWcm² and open circuit voltage of 0.6 V at 560°C was achieved on a µSOFC array chip containing a 10-membrane assembly. Lai et al. [56] in their 2011 paper presented the finding of the *LSCF-cathode|YSZ|LSCF-anode* symmetric planar µSOFC. The LSCF anodes had shown good stability and did not react with the YSZ electrolyte. Tests at 500°C for more than 100 h did not produce any cracking in the cell. But this symmetric cell had a very low peak power density in the sub-mW level. Kerman et al. [57] in 2011 using *Porous Pt-anode|YSZ|Porous*

Pt-Cathode symmetric μSOFC had shown that the electrode microstructure, mainly the triple phase boundaries (TPB) play a very important role on cell electrochemical performance. By careful optimization of anode TPB, the Harvard group recorded a 1037 mW/cm^2 peak power density at 500°C.

Johnson et al. [58] of the Harvard group has devoted attention to scale up and increase of the active area of the cell by mechanical reinforcement of the μSOFC's membrane assembly. Two design aspects to reinforce membrane assemblies for large active area were studied. In one approach, it was shown that a nickel plated micron-sized grid structure on the top surface of the cathode successfully supported a square of 5 × 5 mm^2 area of a ~200 nm thick *Dense-LSCF Cathode|YSZ|Porous-Pt Anode* membrane assembly while covering less than 20% of the membrane area. This architecture yielded very low maximum power density, 1 mWcm^2. In a second approach, a highly porous layer of aerogel/carbon fiber composite was employed to reinforce the membrane assembly. Tsuchiya et al. [59] from Harvard in 2011 reported further advancing the micron-sized metal grid reinforcement μSOFC scale up to the centimeter level and boosting the peak power density >100 mW. Here a Pt grid was deposited on the top surface of dense LSCF cathode instead of a Ni-grid [58]. One of the achievements here is to generate >20 mW of power from a single fuel-cell chip with the area utilization of ~54%. Microfabricated 5 mm × 5 mm grid-supported structures on a silicon wafer are shown Figure 4.16a. Square holes, 5 mm × 5 mm, are produced by KOH etching. The gap between holes is 1 mm. Using the same fabrication method, a membrane assembly with a 4 cm × 4

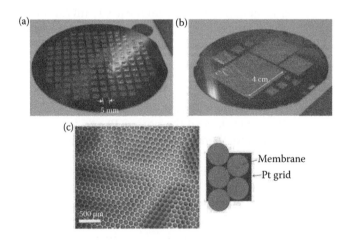

FIGURE 4.16
Image of a 4-in. wafer with grid-reinforced μSOFCs. (a) Array of 5 mm × 5 mm etched holes viewed from the top (LSCF cathode side). (b) Various sizes of active μSOFCs fabricated on a 4-in. wafer. The largest etch hole shown here is 4 cm × 4 cm. (c) Optical micrograph of free-standing LSCF/YSZ membranes and platinum grids of 100 μm in diameter, and the width is 10 μm. (From Tsuchiya, M. et al. *Nature Nanotech.*, 6, 282–286, 2011.)

cm active area was produced as shown in Figure 4.16b. Figure 4.16c shows the 100 μm diameter Pt grid. This Pt grid also is used as a cathode current collector.

The ETH Zurich research group is developing planar μSOFC using alternate non-silicon-based substrates [61,62]. Muecke et al. [61] in 2008 reported the successful demonstration of the feasibility of fabricating planar μSOFC on a Foturan® substrate. Foturan is a photosensitive glass manufactured by Schott Glass Corp. It is used as a MEMS substrate. It combines the unique glass properties (transparency, hardness, chemical resistance, etc.) and the opportunity to achieve very fine structures with tight tolerances. Microfabrication in Foturan is achieved through patterning by a pulsed UV laser, a follow-up heat treatment step, and chemical etching. Exposure to UV radiation precipitates silver atoms in the illuminated areas. Followed by heat treatment between 500°C and 600°C, the glass crystallizes as a crystalline lithium silicate phase around these silver atoms. The crystalline regions, when etched with a 10% solution of HF acid at room temperature, have an etching rate up to 20 times higher than that of the glassy regions. The authors noted that a silicon substrate is conducting, but the Foturan substrate is electrically insulating; therefore, unlike a silicon wafer, it does not need insulating SiO_2 or Si_3N_4 coating. The fabrication steps of μSOFC are schematically shown in Figure 4.17(i). To provide gas access to the anode, the substrate underneath the fuel cell membrane is removed by chemical etching. It is worthwhile to point out that except for the cell electrolyte layer all other components are porous to allow the fuel and oxidant to reach the electrode-electrolyte interfaces. DC sputtered porous Pt was used as an anode; single- or bilayer YSZ as an electrolyte; and Pt paste, porous sputtered Pt, or LSCF as a cathode. Figure 4.17(ii) [62] shows the schematic of a typical μSOFC on a Foturan substrate. The total thickness of all layers is less than 1 μm, and the cell is a free-standing membrane with a diameter up to 200 μm. The OCVs of the single-layer electrolyte cells range from 0.91 to 0.56 V at 550°C. No electronic leakage was detected; the low OCV was attributed to gas leakage through pinholes in the columnar microstructure of the electrolyte. By using a bilayer electrolyte of PLD–YSZ and spray pyrolysis YSZ, an OCV of 1.06 V was obtained, and the maximum power density reached 152 mW/cm^2 at 550°C. One major advantage of this glass substrate is that its thermal expansion coefficient is very close to that of the YSZ film; therefore, thermal stress management is easier. All fuel cell membranes were crack free after electrochemical testing at 600°C.

The fabrication process of μSOFCs involves dry- or wet-chemical etching for opening the substrate below the free-standing fuel cell membranes. Rupp et al. [62] from the ETH Zurich group reviewed dry- and wet-chemical etchants for structuring of ceria-based electrolyte materials and compared them to the etch-rates of common μSOFCs substrates. Wet-chemical etchants (e.g., HF) allow homogeneous etching of ceria-based electrolyte thin films contrary to common dry etching methods. In the case of the Foturan substrate, HF acid is the only choice. The authors noted that, in general, the etch-rates of GDC electrolyte thin films, Si or Si_3N_4 with HF acid, are extremely close,

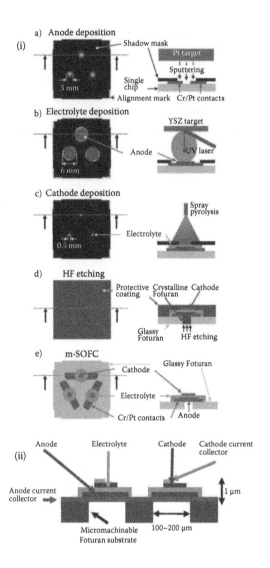

FIGURE 4.17
(See color insert.) (i) Schematic showing the fabrication steps of a planar μSOFC on Foturan substrate. Top views on the left and cross sections on the right. (ii) Schematic view of a μSOFC cell on Foturan substrate showing all the components. (From Muecke, U.P. et al. *Adv. Funct. Mater.*, 18, 3158–3168, 2008; Rupp, J.L.M. et al. *J. Power Sources*, 195, 2669–2676, 2010.)

thus presenting a challenge for silicon wafer–based microfabricated devices. But a strong difference exists in the case of Foturan glass-ceramic wafers and GDC electrolyte. They found that, in the case of GDC spray pyrolysis thin films and HF etchant, the etch-resistance shows a strong dependency on post deposition annealing of GDC, i.e., the film microstructure. This article provided guidance for thermal annealing and etching of GDC thin films for the fabrication of Foturan-based µSOFCs.

Park et al. [52] from the Stanford group in a 2005 publication reported the study of the formation of gas-tight alumina films on nanoporous anodic aluminium oxide (AAO) (Anodisc®, Whatman) substrates through oxidation. In this investigation, sputtered submicron aluminum was oxidized to a hydrogen impermeable dense oxide film. Authors noted that this study was a preliminary step for ultrathin electrolyte film deposition for low-temperature SOFCs. As suggested in their earlier paper, Park et al. [53] extended the fabrication process of gas-tight alumina for the fabrication of submicron YSZ films on AAO substrates for µSOFC. In a 2006 paper, [53] the Stanford group reported deposition of 40–230 nm thick Y-Zr alloys on AAO nanoporous substrates by DC magnetron sputtering. The deposited Y-Zr alloy thin film oxidation was studied as a function of temperature, and it was found that heat treatment at 700°C for 2 h in air was required to form the cubic zirconia phase. Figure 4.18 is a schematic presentation of dense oxide film formation on a columner nanoporous substrate from submicron thick sputtered metal film. Oxidation was associated with high volumetric change and caused severe deformation of the substrate. A fuel cell constructed for testing consisted of *Pt (50 nm-sputtered) anode|AAO-oxidized YSZ film|Pt coating (using Pt-paste) cathode.* Electrochemical performance tests resulted in low OCV,

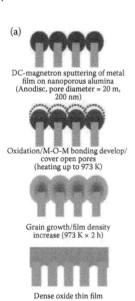

(a)

DC-magnetron sputtering of metal film on nanoporous alumina (Anodisc, pore diameter = 20 m, 200 nm)

Oxidation/M-O-M bonding develop/ cover open pores (heating up to 973 K)

Grain growth/film density increase (973 K × 2 h)

Dense oxide thin film

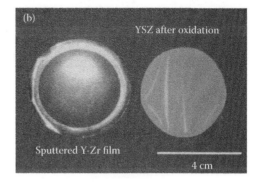

FIGURE 4.18
(a) Schematic presentation of dense oxide film formation on a columnar nanoporous substrate from submicron thick sputtered metal film and (b) photograph of the samples with the sputtered Y-Zr film, as-sputtered and after oxidation at 973 K for 2 h. (From Park, Y.-I. et al. *J. Electrochem. Soc.*, 153, A431–A436, 2006.)

0.8 V, and extremely low power density, 0.037 mW/cm² at 340°C. Cells operated at higher temperatures developed cracks on the thin film–AAO substrate assembly. This result shows that the test cell had leaks, and the power density was very low. This study concluded that AAO–supported µSOFCs have serious issues of thermomechanical and electrochemical performances, which need to be solved.

Joo et al. [63] from Pohang University of Korea in 2007 reported fabrication and study of an AAO–supported planar µSOFC of *AAO-support||Pt-anode|GDC|Pt-cathode*. Porous Pt electrodes were deposited using DC sputtering, and 2 µm GDC was deposited by the PLD technique. At 450°C, the cell's peak power was just a fraction of a milliwatt. In a joint publication from Seoul National University and the Korea Institute of Science & Technology, Kwon et al. [64] reported a thorough study to find out the origin of the low OCV of the AAO–supported YSZ fuel cell. By correlating the electrochemical properties with the microstructure of the YSZ layer, they identified pinholes in the YSZ layer that were responsible for low performance. They proposed and demonstrated a new µSOFC fabrication scheme to produce high-performance cells. In the new scheme, first a 600 nm AAO free-standing nanoporous membrane was integrated with a silicon substrate (Figure 4.19), followed by sputter-deposition of dense Pt on the AAO surface without blocking the AAO pores, then deposition of YSZ electrolyte by the pulsed laser deposition (PDL) technique.

This type of cell provided OCVs in the range of 0.05 V to 0.78 V depending on the average pore size of the AAO–supported layer and the thickness of the YSZ electrolyte. To repair, i.e., to block these pinholes in YSZ, a 20 nm thick Al_2O_3 layer was deposited by ALD on top of the 300 nm thick YSZ layer and another 600 nm thick YSZ layer was deposited by PLD after removing

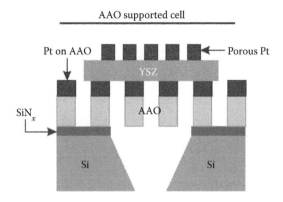

FIGURE 4.19
Schematic diagrams of AAO–supported planar µSOFC. (From Kwon, C.-W. et al. *Adv. Funct. Mater.*, 21, 1154–1159, 2011.)

the top intermittent Al_2O_3 layer. This new cell consisted of the following layers: *AAO membrane (600 nm)||Pt-anode|YSZ electrolyte (900 nm)|porous Pt-cathode*. Planar μSOFCs fabricated by this new scheme provided OCVs of 1.02 V and a maximum power density of 350 mW/cm^2 at 500°C. Kwon et al. [65] in 2012 reported the study of the thermomechanical stability of AAO–supported μSOFCs. The authors concluded that the use of a 600 nm thick AAO membrane significantly improved the thermo-mechanical stability due to its well-known honeycomb-shaped nanopore structure. Moreover, the Pt anode layer deposited between the AAO membrane and the YSZ electrolyte preserved its integrity in terms of maintaining the TPB and electrical conductivity during high-temperature operation. As a result the longevity of the AAO–supported μSOFC is considerably improved when compared with that of common, flat, freestanding membrane μSOFCs.

Kang et al. [54] from the Stanford group produced a unique nanoporous Ni-supported thin film μSOFC assembly. They recognized that pinhole-free submicron thin films can only be formed successfully on nanoporous supports because the film thickness is several times larger than the pore size. These nanoporous Ni-supports were prepared in the following steps: First, a Pt seed layer 20 nm thick was sputtered on the AAO disks' 20 nm porous side, followed by filling the AAO porous structure with PMMA. After PMMA polymerization by UV radiation, the AAO mother structure was dissolved. This process provided a negative geometry of the mother structure with the Pt seed layer. Using the Pt seed layer as an electrode, the final metal structure was electroplated. Finally, the PMMA negative structure was dissolved in acetone, and a nanoporous Ni-support was formed. The Ni-support structure had columnar pores with the diameter varying along the thickness. One surface of the Ni nanoporous support had an average pore size of ~20 nm, and the opposite side had ~200 nm diameter pores. YSZ films of ~200 nm thickness were formed on the ~20 nm porous surface. At the top surface, a Pt-porous layer was sputtered to form the cathode. Here, nanoporous nickel provided the structural support to the cell and acted as an anode as well as an anode current collector. The Pt seed layer acts as an anode catalyst during cell operation. This Ni-supported cell's OCV was 0.87 V and generated low, 7 mW/cm^2, power at 400°C.

Overmeere et al. [60] from the Harvard group in 2012 used an innovative concept to use μSOFC as energy storage. To implement an energy storage strategy, the fuel cell needs electrode materials that reversibly transfer charges like batteries. Vanadium's propensity to change its oxidation states makes vanadium oxide a candidate electrode material for rechargeable batteries. Another interesting property of vanadium oxide is that hydrogen can be inserted with H/V ratios in the range of 0.3–1.9 depending on the oxide. Knowing these properties, the Harvard group designed and fabricated planar YSZ–based μSOFCs with anodes consisting either of a porous Pt film, a VO_x film, or a bilayer VO_x/porous Pt film (VO_x/Pt). All the thin film electrodes and electrolyte were deposited by sputtering on a silicon wafer, and free-standing membrane assemblies were made using MEMS technology. As was shown schematically in Figure 4.20a, fuel cells were fabricated with anodes consisting either of a porous Pt film, a VO_x film, or a bilayer VO_x/porous Pt film (further referred to as VO_x/Pt). As was later shown, the VO_x/Pt anode had a higher performance than the VOx anode due to lower anodic polarization losses. A schematic cross-sectional view of all three thin film µSOFCs is shown in Figure 4.20a.

The results of testing the three cells during galvanostatic fuel cell operation at 0.2 mA/cm² at 300°C by interrupting the hydrogen fuel supply to the anode and monitoring the cell potential evolution are shown in Figure 4.20b. In the case of the porous Pt anode, for all current densities and temperatures tested, the cell potential decreased to 0 V in ~15 s. But the potential decay rates were slower for the VO_x and VO_x/Pt anode cells, 0 V being reached in 32–210 s depending on the vanadium oxide film thickness and current density. A portion of the charge delivered was attributed to the reversible oxidation of the anode to V_2O_5 and hydrogen storage in the electrode. The key point demonstrated through this work is that the vanadium oxide–containing anode µSOFC can generate electricity in the absence of fuel flow for short time periods. This result opens up the potential for future development of a single hybrid device with a combination of batteries and fuel cell characteristics.

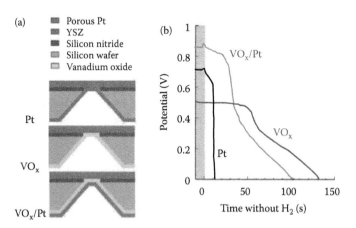

FIGURE 4.20
(a) Schematic cross-sectional view of the µSOFC with the three different anodes. (b) Evolution of the potential difference during galvanostatic operation at 0.2 mA cm² at 300°C when the hydrogen fuel supply is switched off. The fuel cells with vanadium oxide anodes (dark gray curve, 25 nm VO_x; light gray curve, 25 nm VO_x/porous Pt bilayer) continue producing electricity longer than fuel cells with porous Pt anodes (black curve). (From Overmeere, Q.V. et al. *Nano Lett.*, 12, 3756–3760, 2012.)

4.4 Microtubular Solid Oxide Fuel Cell

Microtubular SOFCs typically mean that the outer diameter of the single cell tube is less than 5 mm, and the multilayer tubes' thickest layer has a thickness of less than 500 μm, typically less than 250 μm. The thicker layer provides the mechanical support to the multilayer SOFC structures. The three key benefits of this design are high volumetric power density, high thermal shock resistance, and low thermal mass. High volumetric power density is essential to develop miniaturized compact stacks for portable power systems. Larger single-cell diameters do not provide a significant advantage from the viewpoint of volumetric power density. The tubular cell surface area (πDL; tube dia. = D and length = L) per unit volume is inversely proportional to the tube diameter, i.e., a decrease of tubular cell diameter increases the cell's surface area per unit volume. If the tubular SOFC's area specific power density is constant, i.e., it does not vary with tubular cell diameter; then the cell volumetric power density (VPD) is inversely proportional to the cell diameter. Sarkar et al. [66] estimated that a tubular SOFC stack consisting of 22-mm-diameter single-cell tubes has an electrolyte surface area of approximately 0.1 m²/L (estimated using square configuration of the single-cells arrangement in a stack).

If the diameter of the tube is reduced to 2 mm, then this surface area is ~0.8 m²/L (calculated using close-packed configuration of the single cells in a stack), which is eight times the surface area compared with the large 22-mm-diameter tube (in this calculation, volume of the stack insulation, manifold, etc., were ignored). If the area surface-specific power density (PD) of the cell is 0.250 W/cm² and is not affected by single cell diameter, then a stack containing a 2-mm-diameter tubular single cell will produce ~2000 W/L, i.e., its VPD is ~2 kW/L (see Figure 4.21). In Figure 4.21, the "surface area ratio" is a

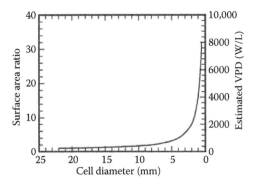

FIGURE 4.21
Electrolyte surface area ratio and corresponding estimated volumetric power density (VPD) in a stack as a function of stack single-cell diameter. (From Sarkar, P. et al. *Int. J. Appl. Ceram. Technol.*, 4, 103–108, 2007.)

normalized number to indicate the electrolyte surface area enhancement per unit volume of a stack for a particular single cell diameter and compared with respect to a stack containing 22-mm-diameter single cells. Figure 4.21 indicates that as the tube diameter gets smaller, the stack's predicted electrolyte surface area gets very high. It is yet to be verified that this extremely high volumetric surface can translate to high volumetric power density in a microtubular stack consisting of a single cell with diameter in the micrometer range. High thermal shock resistance is achieved due to the thin wall of the tubular cell. High thermal shock resistance together with low thermal mass are appealing for rapid startup and shutdown of the SOFC system.

4.4.1 Electrolyte-Supported First-Generation Microtubular SOFC

The development of a first-generation novel microtubular μSOFC design, based on ~2 mm diameter thin-walled (~200 μm) dense YSZ-electrolyte tubes, was pioneered by the Kendall research group from the UK in the early 1990s [67–70]. In this instance, the YSZ tube is self-supported. Anode, cathode and current collector layers are subsequently applied to the dense YSZ tubes' outer and inner surfaces to complete the SOFC single cell. Typical electrodes and current collectors layers are thin and require structural support. This type of microtubular SOFC cell is called an electrolyte-supported cell. The UK group–developed dense YSZ-electrolyte-supported microtubular SOFC single cell's schematic design is shown in Figure 4.22a [69]. YSZ electrolyte tubes were extruded followed by drying and sintering at 1450°C.

The inside of the YSZ tube was then coated with two layers of Ni-based anode using a slurry; this was followed by sintering at 1300°C. The outside of the tube surface was then coated with two layers of LSM-based cathode and sintered at 1100°C. For both electrodes, two layers were applied to improve the electronic conductivity. Typically these cells were based on 5 cm long, extruded, thin walled (~200–250 μm) YSZ tubes with an active surface area of 0.5 cm². The active surface area means the area of the electrolyte outer surface

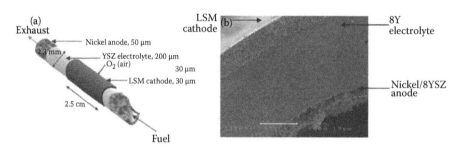

FIGURE 4.22
(a) Schematic of dense electrolyte (YSZ)–supported microtubular SOFC design and (b) cross-sectional SEM micrograph of dense electrolyte, Ni-based anode, and LSM cathode. (From Mallon, C. et al. *J. Power Sources*, 145, 154–160, 2005.)

coated with cathode. The overall cross-sectional SEM micrograph in Figure 4.22b shows the dense YSZ electrolyte with two thin porous electrodes.

Kendall was awarded a patent [70] in 1994 for the sealing and thermal shock resistance–related design improvement in an electrolyte-supported microtubular SOFC system. Alston et al. in 1998 [71] reported a major achievement in the building of a 1000 cell stack for cogeneration of heat and electricity. They used a modular approach to build the stack; 40 modules of 25 cells were used for the stack. This stack was capable of withstanding a 200°C/min heating rate. This microtubular stack power density was 82 mW/ cm^2 at 850°C. Kendall et al. [72] in 1998 reported the design and construction of a small μT-SOFC three-cell system design. This system was operated using a butane-air mixture for a microelectronic type of applications. This SOFC system provided between 0.1 and 1 W of power at 700°C. They validated that the small-diameter electrolyte-supported cells were mechanically robust and had high thermal shock resistance. Kendall and Sammes [73,74] studied various alloys as coatings for current collectors for their μT-SOFC cell and stack as a means to enhance the electrochemical performance. They discussed [75] the application of μT-SOFCs in small systems utilizing propane/ butane. Here, they investigated the leisure CHP systems and micro-hybrid vehicles. This system consisted of a co-generating gas burner with catalytic pre-reforming catalyst capability, μT-SOFC stack, and an afterburner. It is mentioned in the Introduction section, referring to Figure 4.3, that YSZ with thickness ≥100 μm needs an operating temperature of ~1000°C due to its high ASR. Therefore, this first-generation YSZ electrolyte-supported (electrolyte thickness ~200 μm) μT-SOFC cell exhibited a relatively low power density at ≤800°C and was not well suited for intermediate temperature application.

4.4.2 Anode-Supported Second-Generation Microtubular SOFC

The second-generation μT-SOFC, i.e., anode-supported μT-SOFC, was developed to reduce the operating temperature below 800°C and enhance the electrochemical performance [76,77]. The Kendall research group's co-extrusion of a nickel-YSZ cermet anode with a 30 μm thick YSZ electrolyte demonstrated the potential of fabricating anode-supported microtubular cells [77]. In the early 2000s, a number of activities on anode-supported μT-SOFC were initiated to develop fabrication processes that could reduce electrolyte thickness to ≤15 μm [66,78–85]. Kendall [83] noted that Acumentrics Corporation, USA, acquired Kendall's μT-SOFC patents [70,84] to start its R&D activities on μT-SOFC. Later on, Acumentrics moved to an anode-supported μT-SOFC architecture. In early 2000, Adaptive Materials, Inc., (presently known as Ultra Electronics AMI) of the United States started developing propane-powered portable μT-SOFC for military applications. For single-cell manufacturing, they utilized a proprietary [85,86] thermoplastic-based co-extrusion. Their micro co-extrusion technique is capable of manufacturing a small-diameter anode-supported cell with a ~10 μm thick electrolyte. Sammes et al. [82] studied the long term (~100 h) electrochemical performance of anode-supported μT-SOFC (dia. ~1.8 mm) consisting of *NiO-GDC-Anode|GDC| La$_{0.6}$Sr$_{0.4}$Co$_{0.2}$Fe$_{0.8}$O$_{3-\delta}$ (LSCF)–GDC-cathode* at 500°C–550°C. These cells produced 150 mW/cm^2 and 340 mW/cm^2 at 500°C and 550°C, respectively, at 0.7 V in humidified H$_2$ fuel flow. They found that the cell degradation rate was reasonably low, 0.25%/100 h under operation conditions of 200 mA and 0.75 V.

In the Advanced Ceramic Reactor (ACR) project, Japanese research groups, which included government research institutes, universities, and the industrial R&D sector, collaborated to advance anode-supported μT-SOFC technology. Japan's New Energy and Industrial Technology Organization (NEDO) initiated this μT-SOFC development project. This collaborative effort has contributed a tremendous amount of technical progress on anode-supported μT-SOFC [87–90]. Suzuki et al. [87,88] reported the fabrication method of an anode-supported μT-SOFC consisting of the following layers: an Ni-GDC cermet acts as an anode-cum–support layer, GDC-electrolyte layer, and LSCF-GDC composite cathode layer. The tubular anode–support layer was extruded using a plastic mass of NiO-GDC mixture followed by drying. Dried tubes were cut to size and dip coated with the GDC electrolyte layer. The electrolyte layer was dried, followed by sintering at 1450°C for 6 h in air. At the top of the dense sintered electrolyte layer, the LSCF cathode layer was applied by dip coating followed by drying and sintering at 1050°C for 1 h. The final cell diameter was 1.6 mm and produced ~0.85 W/cm^2 at 550°C [87]. The Japanese team produced sub-millimeter cells, as small as 0.8 mm diameter, as shown in Figure 4.23 [88]. Figure 4.23a shows all three layers of single cells. Figure 4.23b is the cross-sectional fracture surface viewed in an SEM. This cell GDC electrolyte thickness is ~10 μm, the cathode is ~15 μm, and the anode-support layer is ~175 μm. Large spherical voids in the electrodes are created by a spherical polymeric poreformer. Figure 4.23c shows the cells I-V characteristics and power density at 450°C, 500°C, and 550°C. This cell produced ~1 W/cm^2 at 550°C in wet hydrogen. Suzuki et al. [89,90] had studied

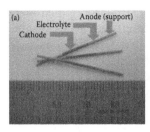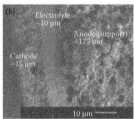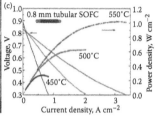

FIGURE 4.23
(a) Photograph of anode-supported 0.8 mm diameter μT-SOFC cells in which all the layers are marked. (b) Cross-sectional SEM image of the μT-SOFC, consisting of three layers: cathode, electrolyte, and anode-supported tube. (c) Cell electrochemical performance as a function of temperature. (From Suzuki, T. et al. *Electrochem. Solid-State Lett.*, 10, A177–A179, 2007.)

the effect of the anode functional layer (AFL) on cell electrochemical performance. The cell AFL consisted of NiO-Sc–stabilized zirconia (ScSZ) between
the electrolyte (ScSZ) and the anode (NiO-YSZ). Figure 4.24 [89] is the cross-
sectional SEM micrograph of a tested (reduced) single cell on which we can
see the AFL between the electrolyte and the anode. The thickness of the electrolyte is ~5 μm, and the AFL is ~4 μm. The thickness of the cathode is ~20
μm. The porosity of the anode was ~33% after reduction. Figure 4.24b shows
the performances (voltage and cell efficiency as a function of fuel utilization)
of the microtubular SOFCs with the AFL and non-AFL samples at 700°C
at the fuel flow rate of 6 mL/min. The energy efficiency of the lower heat
value (LHV) was calculated from the fuel flow rate and I-V results. The positive impact of AFL on cell performance is clear from Figure 4.24b; the peak
power fuel utilization increased from ~63% to ~85%, and energy efficiency
increased from ~36% to ~47% in the presence of the AFL layer.

In early 2000, Alberta Innovates - Technology Futures (AITF) (formerly the
Alberta Research Council, Canada) established a research team to develop
microtubular SOFC using an electrophoretic deposition (EPD) technique
[91–98]. Sarkar et al. [92] from the AITF SOFC research group developed
and patented a sequential electrophoretic deposition technique to fabricate
μT-SOFC single cells with successful single cell testing carried out in 2002.
A sequential EPD technique, developed by Sarkar and Nicholson [99–101] to
fabricate planar laminated ceramic was adapted to fabricate tubular multilayered μSOFC single cells. EPD is a colloidal-forming technique in which
charged, colloidal particles from a stable suspension are deposited onto an
oppositely charged substrate by application of a DC electric field. It is a facile technique to fabricate high-quality ceramics that can have a complicated
geometry. The sequential EPD process for fabricating a small diameter (≤5
mm) microtubular SOFC cell is schematically shown in Figure 4.25.

 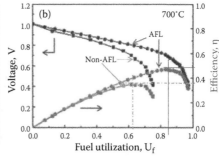

FIGURE 4.24
(a) The cross-sectional fracture SEM image of the cell with the AFL after the cell test (reduced).
(b) The voltage and the energy efficiency as a function of fuel utilization for the cells with the
AFL at 700°C furnace temperature along with the results of the non-AFL cell [18] at the fuel
flow rate of 6 mL min⁻¹ cm⁻². (From Suzuki, T. et al. *Electrochem. Comm.*, 13, 959–962, 2011.)

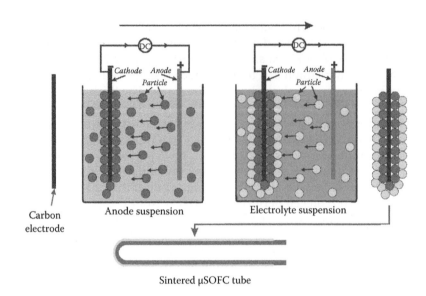

FIGURE 4.25
Schematic drawing to illustrate the EPD-based microtubular SOFC single cell fabrication.

A typical AITF anode-supported cell with an outer diameter of a little less than 3 mm, the innermost layer being the anode-support layer (ASL), is shown in Figure 4.25. Graphite rods have been used as an electrically conducting EPD substrate, that is, on the outer surface of the graphite rod, three layers of material are deposited by a sequential EPD method (the schematic in Figure 4.25 shows a two-layer deposition) using a semi-automated EPD setup. The first layer deposited by EPD is the ASL with a thickness of ~200–250 μm from a suspension containing a pore former and a ~55/45 weight fraction coarse NiO (average particle size ~0.65 μm) and Tosoh's 8 m/o YSZ (Tosoh, Japan). This layer is followed by a <10-μm-thick anode functional layer (AFL) from a suspension containing a 55/45 fraction of fine NiO (average particle size ~0.3 μm) and YSZ. The final and last layer is a ~5 μm YSZ electrolyte layer. It is important to note that sequential EPD can deposit the cathode layer in a single process step, but it was not because the cathode cannot be co-fired with the other layers at ~1400°C. The cathode was applied on the EPD sintered sample either by dip coating or by painting, followed by drying and sintering.

Figure 4.26a displays the large number of microtubular cells fabricated by the EPD technique, and Figure 4.26b shows a typical fracture surface cross-sectional SEM of the single cell. The micrograph clearly shows a ~6 μm dense YSZ electrolyte layer and a highly porous anode support layer; generally, graphite powder is used as an anode poreformer. The SEM micrograph clearly shows that the AFL has a fine microstructure and not large pores like the ASL. To demonstrate the EPD technique's ability in manufacturing micron

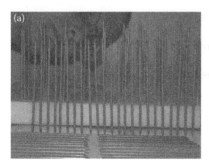

FIGURE 4.26
(a) Sintered microtubular SOFC without cathode, fabricated using sequential EPD technique;
(b) typical SEM fracture surface cross-sectional micrograph showing all the layers of the
μT-SOFC.

diameter tubes, a ~5 μm diameter graphite fiber was used as a depositing
electrode. Figure 4.27a shows an YSZ electrolyte tube with an outer diameter
in the range of 10 μm, and Figure 4.27b shows the smallest microtubular cell
(without cathode) fabricated using the AITF developed EDP technique.

Usually, an LSM/YSZ composite cathode was used in AITF's microtubular single cell. This cathode has two layers (Figure 4.28b) with a total thickness of 25–30 μm. As shown in Figure 4.28b, the cathode functional layer is
a composite of $La_{0.8}Sr_{0.2}MnO_3$ (LSM)-YSZ with a typical thickness of ~10 μm.
Typical polarization and power density curves of SOFC cell at 800°C in wet

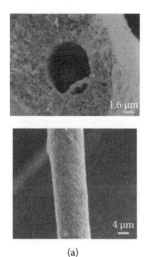
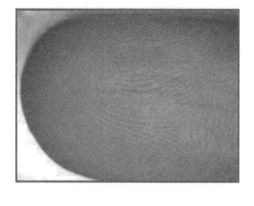

(a)   (b)

FIGURE 4.27
(a) YSZ-electrolyte microtube fabricated by EPD and (b) micron-sized microtubular SOFC
(without cathode layer) fabricated using sequential EPD technique.

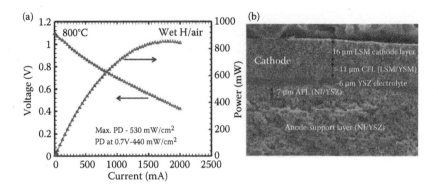

FIGURE 4.28
(a) Polarization and power curve of anode-supported microtubular single cell with LSM-YSZ cathode and (b) cross-sectional SEM micrograph showing double-layer cathode structure.

hydrogen are shown in Figure 4.28a. This cell had a peak power density of 0.53 W/cm^2.

To improve the cells' electrochemical performance, Sarkar's SOFC research group at AITF developed a new cathode design [102]. In a sintered cell, the YSZ-electrolyte outer surface was dip coated with a thin layer of YSZ followed by soft sintering (segment C1 in Figure 4.29a) to form a porous layer. This YSZ porous layer was named the reticulated electrolyte matrix (REM). Figure 4.29b shows the cross-sectional view of the YSZ-REM layer. Soft sintering bonded the REM layer with the dense electrolyte layer. This

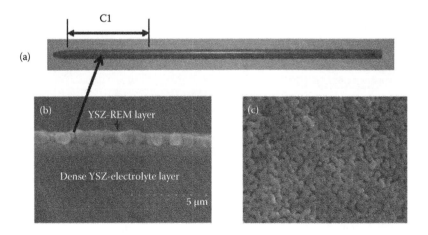

FIGURE 4.29
(a) Tubular single cell indicating the location of porous thin YSZ layer; (b) cross-sectional SEM micrograph of the sub-micrometer YSZ porous layer, and (c) SEM micrograph showing the top view of the YSZ-REM layer.

soft-sintering temperature and time is a function of starting powder characteristics, such as particle size, surface area, etc. Sintering temperature should
be high enough to provide sufficient strength for handling but not too high
to keep the grain size small and high surface area.

This micrograph also shows that this REM is a single grain in thickness.
The top view of the SEM micrograph is shown in Figure 4.29c. In the next
step, a samarium-doped strontium cobaltite (SSC:$Sm_{0.6}Sr_{0.4}CoO_{3-\delta}$) cathode
was formed by solution impregnation of the REM layer. The electrochemical test result for this single cell is shown in Figure 4.30. The ~2.5-cm-long
segment of the tubular cell generated ~1.9 W at 700°C and ~1.5 W at 600°C
in wet hydrogen. In power density terms, the SSC cathode cell has produced
peak powers of 870 mW/cm², 760 mW/cm², and 670 mW/cm² at 700°C, 650°C,
and 600°C, respectively. An 800°C processing temperature solution impregnation of SSC resulted in a nanostructured electrochemically highly active
cathode in the void structure of the REM. The thin sub-micrometer YSZ-
REM layer allows the mechanical stability of the cathode, enhances the triple
phase boundary, and the <1 μm thickness drastically reduced the ohmic voltage drop across the REM-based cathode layer. The thin REM layer made the
impregnation process simple and practical.

The AIFT research group demonstrated [66] that the anode-supported
cell has extremely high thermal shock resistance. Tubular cells were given
rapid thermal shock in localized regions using a micro burner as shown
in Figure 4.31a. Microscopic examination found no micro cracks or thermo
mechanical damage in the cell.

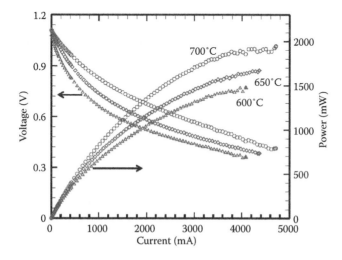

FIGURE 4.30
Polarization and power density curves of anode-supported YSZ cell with $Sm_{0.6}Sr_{0.4}CoO_{3-\delta}$
(SSC)–based cathode.

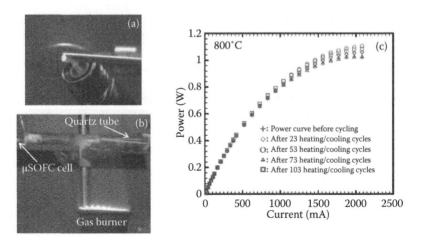

FIGURE 4.31
(a) Thermal shock resistance test of µT-SOFC using micro-burner; (b) rapid start-up and turn-off ability of µT-SOFC demonstrated by operating the cell using a gas burner, and (c) single cell power output as a function of cell current during thermal cycling.

Figure 4.31b shows the operating and testing of a single cell using a gas burner as a heater. The cell was in a quartz tube with a continuous airflow inside the quartz tube, and wet hydrogen was introduced in the anode side of the cell. The power output was monitored. From time to time, the burner was turned off to cool down the cell and then turned back on. It was observed that within seconds the cell started producing power. This thermal cycling test using a burner was not well controlled, but it indicated that the cell survived fast heating/cooling cycles and produced power, but it did not show if there was any power degradation. To find out the effect of thermal cycling on cell electrochemical performance, a microtubular cell was thermally cycled more than 100 times in a cell-testing furnace. After every few cycles, cell electrochemical performance was measured, and the power curves as a function of cell current are shown in Figure 4.31c. This thermal cycling test shows there is no degradation of cell performance.

Polymer hollow fiber (HF) is made using a spinning technique followed by a phase inversion process. A similar technique is employed with some modification to fabricate ceramic HF membrane. A typical diameter of this type of membrane manufactured by this spinning technique is less than 3 mm. This is an interesting way to fabricate small-diameter ceramic membrane tubes. Consequently, considerable research activity has recently been reported on the fabrication of microtubular SOFC based on this technique [103–107]. This small-diameter tubular membrane fabrication starts with preparation of polymer-based ceramic powder spinning suspension. In the case of SOFC porous support, this spinning suspension may contain a pore-former. In the next step, the HF of polymer-ceramic composite is made using the spinning technique. This polymer-ceramic composite HF is called HF

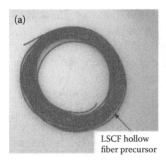
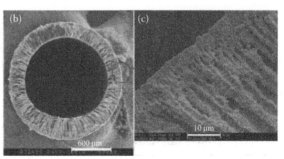

LSCF hollow
fiber precursor

FIGURE 4.32
(a) Photograph of polymer-ceramic (LSCF) hollow fiber precursor for fabrication of LSCF hollow fiber membrane [103], (b and c) SEM micrographs are showing the ceramic hollow fiber wall's asymmetric microstructure. (Reprinted from *Vol. 1: Basic Aspects of Membrane Science and Engineering of Comprehensive Membrane Science & Engineering*, Li, K., "Ceramic hollow fibre membranes and their applications," 253–274, Copyright 2010, with permission from Elsevier; Liu, Y. et al. *Sci. China Chem.*, 54, 850–855, 2011.)

precursor. Figure 4.32a [103] shows the LSCF (cathode) HF precursor. The high polymer content makes this precursor flexible.

The next processing step is the polymer burnout and sintering to form the HF ceramic membrane. The resultant sintered ceramic membrane is brittle and no longer flexible. Like the polymer HF, the ceramic HF also has an asymmetric structure. The inner side of the HF wall microstructure is dominated by finger-like voids followed by a sponge-like microstructure in the outer layer. The SEM micrographs of these microstructural features can be found in Figure 4.32b [104]. By manipulating the sintering process, one can make the outer screen of the HF membrane dense or porous. Yang et al. [105] presented a µT-SOFC fabrication process in which polyethersulfone was initially dissolved into N-methyl-2-pyrrolidone (NMP) and mixed with NiO–YSZ powders to form a viscous slurry for spinning. They used a spinneret with an orifice dimension/inner diameter of 3.0/2.0 mm to obtain a HF anode-support precursor. This tubular precursor was pre-sintered at 1200°C to obtained porous anode-support. The electrolyte and cathode are applied by a typical dip coating method to form the single cell. The peak power densities reported were 0.54 W/cm^2 and 0.78 W/cm^2 at 700°C and 800°C, respectively. Hafiz et al. [107] prepared the anode-support (Ni-CGO) layer and the electrolyte (CGO) layer of the µT-SOFC in a single fabrication step for a triple-orifice spinneret. They reported achieving very high power density, 2.32 W cm^2 at 600°C.

4.4.3 Porous Electrolyte-Supported Third-Generation Microtubular SOFC

The Ni/YSZ anode–supported µT-SOFC cells exhibited excellent electrochemical performance and thermal shock resistance. However, a major practical drawback of the anode-supported cell is anode oxidation-reduction related to microcracking. Any accidental oxidation of the anode can terminally damage an anode-supported SOFC device. To minimize redox cycling–related problems and also to enhance the mechanical robustness including thermal shock resistance, the AITF group started developing a third-generation microtubular SOFC in 2005 [108], i.e., a porous electrolyte-supported (PES) µT-SOFC.

A schematic of the porous electrolyte-support cell without any electrodes (partial cell) are shown in Figure 4.33a through 4.33b, in which the thickness of the dense electrolyte layer is ≤ 10 µm. The thickness of the porous electrolyte-support layer will depend on the diameter of the single cell. Figure 4.33e is the photograph of an YSZ-based porous electrolyte-supported microtubular cell without any electrodes. Figure 4.33c through 4.33d are the SEM micrograph of the cell showing the layers. Because this partially fabricated cell does not initially contain any electrodes, it can be adapted to a variety of electrodes in subsequent steps. In the later stages of processing, the porous electrolyte-support layer will be transformed to an electrode (in the present case, an anode) by infiltration techniques. Photographs of a sequential EPD–fabricated porous electrolyte-supported partial µT-SOFC cell (#ii) and a NiO-impregnated PES layer (dark gray (#i)) are shown in Figure 4.34a. SEM micrographs in Figures 4.34c through 4.33d are showing the nano-sized NiO in the interconnected void network of the porous electrolyte-support layer. An important feature of this cell design is that the gas-impermeable dense electrolyte layer and porous support structure are made of the same material. Therefore, there is no sharp interface or material discontinuity between the cell support structure and the electrolyte. If the anode goes through an oxidation-reduction cycle during service, anode volume changes associated with this cycle will occur inside the voids of the porous electrolyte-support, so the resulting volume change will not result in high stresses to the porous electrolyte support structure or the dense electrolyte layer. The porous electrolyte support and dense electrolyte layer will therefore not suffer from microcracking or mechanical damage. This important feature of the PES µT-SOFC has been confirmed by AITF using porous YSZ-supported cells.

Electrochemical performances of the cells were evaluated, and one set of typical results is shown in Figure 4.35. A photograph of tested PES cell is also shown in Figure 4.35. This cell has peak power densities of 110 mW/cm^2, 155 mW/cm^2, and 200 mW/cm^2 at 700°C, 750°C, and 800°C, respectively. These power densities are relatively low due to high ASR of the Ni-impregnated PES layer. Success of the PES architecture depends on developing a practical fabrication technique where porous electrolyte-support layers can be converted to a highly electro-catalytically active electrode with a low ASR. Presently Canadian research groups are developing porous electrolyte-supported (PES) tubular SOFC cells [109–113].

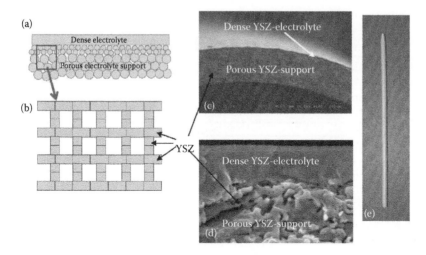

FIGURE 4.33
(a) Schematic of a porous electrolyte-supported thin (≤10 μm) dense electrolyte layer; (b) schematic of porous electrolyte support layer; (c and d) fracture surface cross-sectional micrographs of porous YSZ electrolyte-supported dense YSZ electrolyte layer, and (e) photograph of porous electrolyte-supported microtubular cell without electrodes.

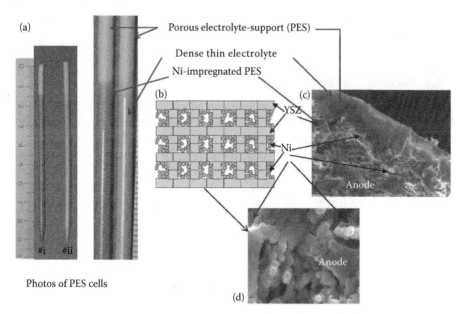

FIGURE 4.34
(a) Photographs of EPD–fabricated porous electrolyte-supported cell without electrodes (#ii) and cell with NiO impregnated PES layer (dark gray [#i]); (b) schematic of Ni-impregnated porous YSZ-support layer, and (c) SEM micrograph is showing both dense thin electrolyte and the NiO impregnated PES; (d) magnified SEM micrograph of PES layer showing the nano-sized NiO in the interconnected void network.

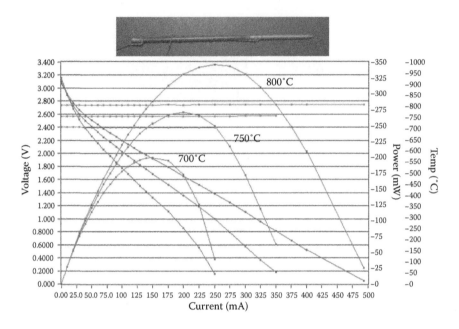

FIGURE 4.35
Polarization and power density curves of electrolyte-supported YSZ cell in 700°C, 750°C, and 800°C. Photograph of tested PES cell is at the top.

4.4.4 Microtubular SOFC Stack

The Japanese team developed a new concept for making microtubular cell bundles [114–118] as shown in Figure 4.36. Here, tubular single cells are embedded in a porous cathode matrix. Figure 4.36b is the photograph of actual anode-supported (Ni-GDC cermet) GDC–based cells embedded in a cube (with 1 cm sides) of porous LSCF-cathode. A patent was awarded to Sarkar et al. [119] of AITF of Canada in 2004 for a similar concept where a stack is embedded in a porous matrix. The Japanese research team has developed an extrusion–dip coating–based fabrication technology for microtubular SOFCs and bundles. They achieved high performance at 550°C, 0.46 W/cm² at 0.7 V in wet H_2 fuel and a volumetric power density of the bundle of 2.5 W/cc. They used numerical simulation [117] to optimize the cell and bundle/stack design.

Early in 2013, AIST in Japan announced the release of a prototype of a portable μT-SOFC system, which can operate on a range of hydrocarbon fuels, including liquefied petroleum gas (LPG) [120].

Figure 4.37a [120] is a 3-D sketch of the system with a listing of the eight main components. The schematic diagram of the SOFC system in Figure 4.2 of the Introduction section has a separate reformer for hydrocarbon fuel, but AIST avoided the need for this reformer by the internal reforming of fuel.

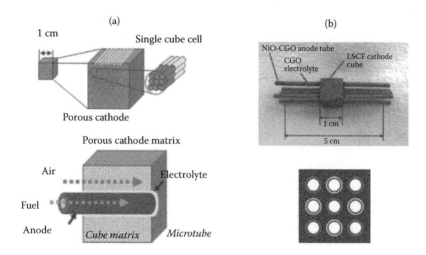

FIGURE 4.36
(a) Schematics of microtubular single cell inside porous cathode matrix [116]; (b) top: photograph of anode-supported single cell assembly in LSCF porous matrix; bottom: schematic cross-sectional view of the top tubular cell bundle. (From Suzuki, T. et al. *J. Fuel Cell Sci. Tech.*, 7, 031005, 2010; Hashimoto, S. et al. *J. Fuel Cell Sci. Tech.*, 8, 021010, 2011.)

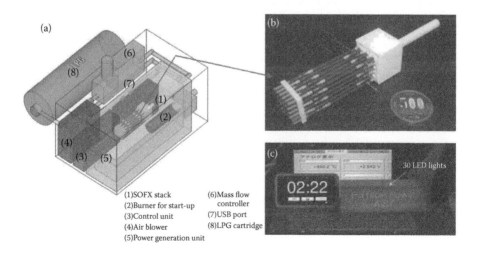

FIGURE 4.37
(a) 3-D schematic diagram of the AIST's microtubular SOFC system, (b) photograph of 36-cell microtubular SOFC stack, and (c) photograph is showing the prototype μT-SOFC system running the DC 5 volt USB devices. (From "Development of Handy Fuel Cell System Advanced at Industrial Science and Technology (AIST)," Japan, http://www.aist.go.jp/aist_e/latest_research/2013/20130321/20130321.html, 2013.)

Nano structure ceria is used in the anode for internal reforming. This system has a LPG burner whose exhaust is used for rapid startup. The microtubular stack for the system is shown in Figure 4.37b, it consists of 36 single cells in a square array. Testing demonstrated that in two minutes, the unit was able to drive a DC 5-volt USB devices (Figure 4.37c).

As mention earlier, Sarkar et al. [119] from the AITF SOFC research team was awarded a patent in 2004 on a porous matrix–embedded μT-SOFC stack. The patent application was filed early 2002. Figure 4.38a [119] shows a schematic of a porous matrix–embedded microtubular SOFC. This concept was developed considering that small-diameter ceramic tubes are brittle; therefore, the porous matrix will provide the mechanical robustness to the microtubular bundle. If the porous matrix is electrically conductive, then it will act as a current collector, e.g., cathode materials, metals, or alloys. Metal may be able to absorb mechanical shock or stresses by deforming itself without damaging the cell or its electrical properties. Ceramic porous materials may form cracks, and as a result, its mechanical and electrical performance will deteriorate. High thermal conductivity porous matrices will help in thermal management by aiding the achievement of temperature uniformity and will behave as a heat exchanger. Figure 4.38b is a photograph showing anode-supported cells embedded in a metallic form. This was fabricated using the following steps:

1. An open cell polyurethane form was cut to desired size.
2. Appropriately sized tubular holes were introduced along the length direction of the foam where the hole size is similar to the outer diameter of the microtubular cell.

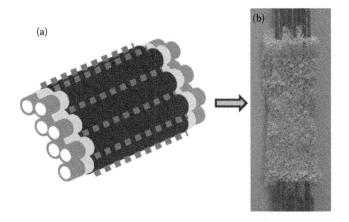

FIGURE 4.38
(a) Schematic of porous matrix–embedded microtubular stack and (b) photograph of metal foam embedded anode-supported microtubular SOFC. (From Sarkar, P. et al. US Patent#6,824,907, 2004, Nov. 30.)

3. The form was impregnated using particulate slurry (in the present case, metal particles).

4. Microtubular cells were inserted in these holes.

5. The assembly was sintered to obtained a metallic foam–embedded µT-SOFC bundle.

Within the scope of this program, the AITF team has designed, fabricated, and tested several stacks. It designed and operated thermally sustainable first-generation prototype stacks using wet hydrogen in early 2005 (Figure 4.39a). The bottom of the stack includes the fuel distribution and exhaust manifold. Small-diameter Inconel tubes were used for injecting wet hydrogen near the closed end of the microtubular cell. Figure 4.39b is a photograph of a 60-cell stack. A machined glass-ceramic plate was used for holding the cells for the stack. A mica gasket was used to place the glass-ceramic holding plate at the top of the fuel exhaust chamber for gas-tight sealing. No furnace was required to operate this prototype µSOFC system. The middle of the tubular stack arrangement in Figure 4.39a has a cylindrical space, in which a spe-cially designed and in-house–fabricated catalytic burner was embedded to start up the stack. With hydrogen fuel, the catalytic burner automatically started producing the required heat and did not require starter. In this first-generation prototype, gold current collectors were used as shown in Figure 4.39c. To improve temperature uniformity in the stack, the outer surfaces of

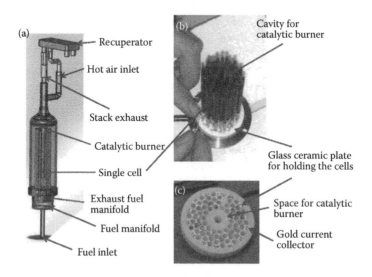

FIGURE 4.39
(See color insert.) (a) CAD drawing of thermally sustainable 60-cell anode-supported microtu-bular SOFC system, (b) photograph of AITF–built 60-cell stack, and (c) machined glass-ceramic holding plate showing gold current collector, space for catalytic burner.

the stack's cells, i.e., the cathode current collector surfaces of the single cells, were covered with a high emissivity coating. Also, the inside of the stack was surrounded by a high-emissivity foil to achieve temperature uniformity. The catalytic burner heated the stack to operating temperature in a few minutes. This prototype system was operated over 1000 h with several on/ off stages. Its power output was 20 W maximum, lower than expected. One of the reasons was that the thin-wall small-diameter Inconel fuel-injection tubes sagged and touched the inner anode surface. As a result, the stack was shorted and, therefore, generated lower power than expected. AITF was also awarded other patents on stack and system design [121–123]. If more information on microtubular SOFC stacks is required, a review article by Lawlor et al. [124] offers a good overview of μT-SOFC stack design issues and R&D activities.

4.5 Modeling of Microtubular Solid Oxide Fuel Cell

Mathematical modeling and computer simulation has been extensively used to model SOFCs to study and/or predict various aspects of their performance, some of which are difficult if not impossible to assess experimentally. These models can be used to investigate the effect of design parameters, operating conditions, and geometric designs; hence they are suitable for optimizing different aspects of SOFCs, such as electrochemical, thermal, and mechanical performance. Modeling μSOFCs in particular, especially μT-SOFC, has gained increasing traction since 2005, which indicates the increased attention to its development.

Several phenomena occur in an operating SOFC: charge transfer, mass transfer, momentum transfer, heat transfer, and electrochemical reactions. The set of mathematics modeling these phenomena form a set of highly interdependent nonlinear differential equations that have to be numerically solved. Some groups have developed their own solvers, which offers more flexibility and customizability, while most efforts take advantage of commercially available computational fluid dynamic (CFD) packages, such as COMSOL, FLUENT, and START-CD. Some or all of the SOFC phenomena are included in these models, depending on their objectives. The more recent efforts use more complete models, partly due to increasing availability of affordable high-performance computing as well as more capable CFD packages. While simpler models are easier to develop and solve, more complete models present more detailed information and are more suitable for comprehensive performance optimization and diagnosis. Lawlor et al. [124] and Howe et al. [125] provide useful reviews of μT-SOFC research activities, including the ongoing modeling efforts.

Mathematical modeling can provide a superb tool for studying the effect of various design parameters and operating conditions. Hotz et al. [126] developed one of the earlier comprehensive μT-SOFC models. They studied the effect of operational and geometric conditions on system efficiency and concluded that the energy efficiency of the system can be significantly improved by proper choice of parameters. Serincan et al. [127] developed a comprehensive μT-SOFC model to study the effect of various operating conditions, including temperature and fuel composition as well as anode/cathode flow rates and pressures on cell performance and fuel utilization. Mirahmadi et al. [128] modeled a μT-SOFC and compared μT-SOFC performance based on hydrogen, direct internal reforming (DIR) of methane, and indirect gradual reforming (IGR) of methane. They concluded that IGR provides a higher performance at the same fuel utilization, and DIR results in lower cell temperatures due to the cooling effect caused by reforming. Sammes et al. [129] developed a model to study the performance of a GDC–based μT-SOFC. Amiri et al. [130] presented a detailed μT-SOFC model, assessing the contribution of each performance loss mechanism associated with every cell component.

Modeling charge transfer enables ohmic loss estimation and facilitates designing optimal current collectors. Sammes et al. [131] modeled current flow in a stack of μT-SOFC to assess electric current flows and associated ohmic losses. Nehter [132] developed a model for a methane μT-SOFC, focusing on current collector design. They concluded that a cascade current collector design provides a higher average cell voltage under steady state and lower temporary voltage drop under load change. Suzuki et al. [133] modeled current collection losses in an anode-supported μT-SOFC as a function of geometric design and operating temperature. They suggested that selection of the length and thickness of the anode is critical in minimizing ohmic losses. Cui et al. [134] studied the performance of μT-SOFC current collection modes and found that if current collection is done on both ends of the cell, losses can be reduced by several-fold. Similarly, the model developed by Suzuki et al. [135] to study geometrical design for maximum current collection efficiency in an experimental 25-cell 1 cm^3 μT-SOFC stack concluded that double terminal current collection can reduce losses significantly.

Mass transfer modeling enables a concentration loss estimation in SOFCs. A detailed enough model that includes mass transfer can inherently predict the limiting current density. The μT-SOFC model developed by Serincan et al. [136] compared a current collector design that coats the entire electrode with a four-wire current collector in terms of mass transfer and performance. They showed that utilizing the entire electrode surface for current collection reduces ohmic losses, decreases thermal gradients along the electrodes, and introduces negligible mass transfer limitation for a large range of parameters. They found that the fully covering current collector results in a more uniform current density distribution compared to the four-wire method.

They also commented that the non-uniform current density generation in the four-wire method results in a non-uniform temperature distribution severe enough to cause mechanical failures, such as warping, distortion of cathode, and separation of wires. Serincan et al. [137] investigated the effect of co- and counter-flow of reactants on the cathode and anode sides in a μT-SOFC. Izzo et al. [138] developed a μT-SOFC model to study gas transport in the anode and to optimize anode design. They found that diffusion in the anode has a key role in cell performance and can cause significant loss in poorly designed anodes. Their model was capable of predicting the limiting current density as well.

Thermal management and temperature distribution is one of the significant aspects of SOFC operation. Too low temperatures reduce the electrochemical performance due to lower ionic conductivity of electrolyte material and lower electro-catalytic activity of electrodes. On the other hand, too high temperatures speed up agglomeration of metal-based electrodes and reduce their performance. Also, hot spots can cause thermal stress–induced mechanical failure. Hence maintaining a uniform temperature distribution is highly desirable. Lockett et al. [139] developed one of the first μT-SOFC models for heat transfer and fluid flow. They studied a single cell as well as a 20-cell stack and investigated the effect of inlet gas flow-rates and temperatures to obtain the optimal temperature distribution. The detailed μT-SOFC model developed by Amiri et al. [140] showed that local heating within cell layers is negligible. Modeling results by Serincan et al. [137] for μT-SOFC studied cell performance under various operating conditions and concluded that smaller cells result in smaller thermal gradients. They also suggested rules for μT-SOFC furnace design. Serincan et al. [141] developed a model for a μT-SOFC and studied the effect of temperature distribution on current density generation and cell performance. García-Camprubí et al. [142] studied the effect of gas flow rates and temperature distribution on cell performance and concluded that large flow rates of un-preheated feed gas can reduce cell performance due to cooling.

Reliability is one of the major hurdles to SOFCs becoming commercially viable. Thermal stresses can jeopardize μT-SOFC's micro- and macro-scale mechanical integrity and has become a focus of modeling work in recent years. Once a thermal model is developed, the addition of solid mechanics equations to the model can readily enable it for thermal stress analysis. Cui et al. [143] developed a detailed model to analyze thermal stresses and their effect on μT-SOFC mechanical integrity. They concluded that μT-SOFCs can resist high thermal shock, and the stresses are mainly caused by the components' thermal expansion mismatch. Hence, they suggested that matching cell component's thermal expansion is the most effective strategy for increasing μT-SOFC reliability. Serincan et al. [144] used modeling to study thermal stresses in an operating GDC–based μT-SOFC. Their detailed model accounted for residual stresses in each cell component caused by the high-temperature fabrication process. Their model also accounted for interactions between the cell, the sealant, and the connection tube. They found that stresses change significantly near the cell/sealant interface. They also analyzed the effect of temperature gradients on stress distribution and found that it has minimal effect for typical cells operating at mid-range current densities. They also found that oxygen vacancies caused by GDC reduction alleviate stresses due to reduction in Young's modulus mismatch.

Li et al. [145] developed one of the most comprehensive μT-SOFC models to study the effect of electrode composition on cell performance and its mechanical stability. They applied the Weibull approach for failure analysis to the thermal and residual stresses calculated in their model and assessed the effect of the electrode's composition and microstructure on the μT-SOFC's mechanical stability. They found that low LSM/Ni content in the cathode/anode are detrimental to cell performance, and high Ni content significantly undermines the cell's mechanical stability due to severe thermal expansion mismatch. Based on mechanical failure analysis, they concluded that the anode's Ni volume fraction at room temperature and 800°C should be no more than 43% and 33%, respectively. Overall, they suggested that the anode's Ni content should be as low as possible to ensure mechanical stability of the cell. On the other hand, they suggested that LSM content of the cathode has limited consequences for mechanical integrity of the cell, and therefore, the focus should be on enhancing the cathode's electrochemical performance.

Some models go beyond the cell and include other components of a μT-SOFC–based power-generation system. The model developed by Hotz et al. [126] for energy analysis of a butane based "micro-powerplant" included a detailed μT-SOFC model as well as models for the vaporizer, reformer, preheaters, and post-combustor. Raberger et al. [146] developed a model for analyzing a post-combustor for a butane-based μSOFC system to treat the toxic and flammable exhaust gases.

While most of the fuel cell models are developed to analyze steady-state conditions, a real cell experiences major transience, including start-up/shut-down cycles, load changes, and thermal dynamics. In addition, abnormal conditions, such as fuel loss, air loss, cell mechanical failure, and thermal management issues, can introduce significant dynamics. Nehter [132] developed one of the earlier transient models of a methane μT-SOFC to study the performance of various current collector designs under load change. Serincan et al. [147] developed one of the most complete μT-SOFC models, aiming at analyzing its transient characteristics. Their simulations showed that μT-SOFC transients are governed by heat transfer on the order of 20 seconds. They also noticed an overshoot in cell voltage when the current load is increased. They also simulated a loss of fuel scenario, which showed that it takes about 4 s for hydrogen to be completely depleted while the cell stops generating current much earlier than that, only after about a single second.

4.6 Conclusion

It is known that SOFCs have the highest energy conversion efficiency among various power generators and that the hydrocarbon fuel energy densities are considerably higher. But a fuel cell power generation system is much more complex than a rechargeable battery. Therefore, in a SOFC system's small form factor, it is a considerable challenge to translate the fuel's high chemical energy density to high electrical energy density while keeping the high-energy conversion efficiency. It is apparent from this chapter that a large amount of R&D effort has been deployed worldwide since the early 2000s in the micro-SOFC area and that a tremendous amount of progress has been made. Key developments include the appropriate implementation of microfabrication technology to reduce the electrolyte thickness to a few hundred nanometers and highly electro-catalytically active nanostructured anodes and cathodes that allow reducing the operating temperature of thin film planar SOFC well below 500°C. Its simpler design, rapid startup, and thermal self-sustainability are the major attributes of single chamber micro-SOFC, but low efficiency is the major drawback. A micro-tubular SOFC has demonstrated its high volumetric power density and rapid start-up ability. Porous matrix-embedded micro-tubular SOFCs are a relatively new concept, and results show that it is beneficial for small diameter (≤ 2 mm) tubular bundles. Appropriate implementation of the porous matrix-embedded micro-tubular SOFC concept will allow the extension of the development of longer length ceramic hollow fiber (CHF) membrane modules for other membrane applications. CHF membranes are brittle; therefore, it is difficult to build membrane modules using long CHF. Although significant progress has been made in single cell fabrication and performance enhancement, future R&D must put more emphasis on cell reproducibility and long-term durability. These are key to developing reliable micro-SOFC stacks and systems. More R&D effort is also needed in system design to develop simple architecture, efficient thermal management, and reliable operation.

References

1. Richard, C. 1999. Navy fuel cell programs, http://www.netl.doe.gov/publications/proceedings/99/99fuelcell/fc2-1.pdf.
2. Steele, B.C.H. and Heinzel, A. 2001. Materials for fuel cell technologies. *Nature* 414, 345–352.
3. Eyraund, C., Lenoir, J. and Gery, M. 1961. Piles a combustibles utilisant les proprietes electro-chimiques des adsorbats. *C.R. Acad. Sci. Paris* 252, 1599–1600.

4. van Gool, W. 1965. The possible use of surface migration in fuel cells and hetero-geneous catalysis. *Philips Res. Reports* 20, 81–93.
5. Dyer, C.K. 1990. A novel thin-film electrochemical device for energy conversion. *Nature* 343, 547–548.
6. Hibino, T. and Iwahara, H. 1993. Simplification of solid oxide fuel cell system using partial oxidation of methane. *Chem. Lett.* 1131–1134.
7. Asano, K., Hibino, T. and Iwahara, H. 1995. A novel solid oxide fuel cell system using the partial oxidation of methane. *J. Electrochem. Soc.* 142, 3241–3245.
8. Hibino, T., Ushiki, K. and Kuwahara, Y. 1996. New concept for simplifying SOFC system. *Solid State Ionics* 91, 69–74.
9. Hibino, T., Wang, S., Kakimoto, S. and Sano, M. 1999. Single chamber solid oxide fuel cell constructed from an yttria-stabilized zirconia electrolyte. *Electrochem. Solid-State Lett.* 2, 317–319.
10. Hibino, T., Kuwahara, Y. and Wang, S. 1999. Effect of electrode and electro-lyte modification on the performance of one-chamber solid oxide fuel cell. *J. Electrochem. Soc.* 146, 2821–2826.
11. Hibino, T., Hashimoto, A., Inoue, T., Tokuno, J., Yoshida, S. and Sano, M. 2000. Single-chamber solid oxide fuel cells at intermediate temperatures with various hydrocarbon-air mixtures. *J. Electrochem. Soc.* 147, 2888–2892.
12. Hibino, T., Hashimoto, A., Inoue, T., Tokuno, J., Yoshida, S. and Sano, M. 2000. A low-operating-temperature solid oxide fuel cell in hydrocarbon-air mixtures. *Science* 288, 2031–2033.
13. Yano, M., Tomita, A., Sano, M. and Hibino, T. 2007. Recent advances in single-chamber solid oxide fuel cells: A review. *Solid State Ionics* 177, 3351–3359.
14. Louis, G.A., Lee, J.M., Maricle, D.L. and Trocciola, J.C. 1981. Solid electrolyte electrochemical cell. U.S. Pat. 4248941.
15. Hibino, T., Hashimoto, A., Suzuki, M., Yano, M., Yoshida, S. and Sano, M. 2002. A solid oxide fuel cell with a novel geometry that eliminates the need for pre-paring a thin electrolyte film. *J. Electrochem. Soc.* 149, A195–A200.
16. Fleiga, J., Tuller, H.L. and Maier, J. 2004. Electrodes and electrolytes in micro-SOFCs: A discussion of geometrical constraints. *Solid State Ionics* 174, 261–270.
17. Rotureau, D., Viricelle, J.-P., Pijolat, C., Caillol, N. and Pijolat, M. 2005. Development of a planar SOFC device using screen-printing technology. *J. Euro. Ceram. Soc.* 25, 2633–2636.
18. Viricelle, J.-P., Pijolat, C., Riviere, B., Rotureau, D., Briand, D. and de Rooij, N.F. 2006. Compatibility of screen-printing technology with micro-hotplate for gas-sensor and solid oxide micro fuel cell development. *Sensors and Actuators B: Chem.* 118, 263–268.
19. Ahn, S.-J., Lee, J.-H., Kim, J. and Moon, J. 2006. Single-chamber solid oxide fuel cell with micropatterned interdigitated electrodes. *Electrochem. Solid-State Lett.* 9, A228–A231.
20. Ahn, S.-J., Kima, Y.-B., Moona, J., Lee, J.-H. and Kimb, J. 2007. Influence of pat-terned electrode geometry on performance of co-planar, single-chamber, solid oxide fuel cell. *J. Power Sources* 171, 511–516.
21. Yoon, S.P., Kim, H.J., Park, B.-T., Nam, S.W., Han, J., Lim, T.-H. and Hong, S.-A. 2006. Mixed-fuels fuel cell running on methane-air mixture. *J. Fuel Cell Sci. Tech.* 3, 83–86.
22. Buergler, B.E., Ochsner, M., Vuillemin, S. and Gauckler, L.J. 2007. From macro-to micro-single chamber solid oxide fuel cells. *J. Power Sources* 171, 310–320.

23. Kuhn, M., Napporn, T.W., Meunier, M. and Therriaulta, D. 2008. Experimental study of current collection in single-chamber micro solid oxide fuel cells with comb-like electrodes. *J. Electrochem. Soc.* 155, B994–B1000.

24. Kuhn, M., Napporn, T., Meunier, M., Vengallatore, S. and Therriault, D. 2008. Direct-write microfabrication of single-chamber micro solid oxide fuel cells. *J. Micromech. Microeng.* 18, 015005.

25. Kuhn, M., Napporn, T., Meunier, M., Therriault, D. and Vengallatore, S. 2008. Fabrication and testing of coplanar single-chamber micro solid oxide fuel cells with geometrically complex electrodes. *J. Power Sources* 177, 148–153.

26. Kim, H., Choia, S.-H., Kim, J., Leea, H.-W., Song, H. and Lee, J.-H. 2010. Microfabrication of single chamber SOFC with co-planar electrodes via multi-step photoresist molding with thermosetting polymer. *J. Mater. Process. Tech.* 210, 1243–1248.

27. Shao, Z., Haile, S.M., Ahn, J., Ronney, P.D., Zhan, Z. and Barnett, S.A. 2005. A thermally self-sustained micro solid-oxide fuel-cell stack with high power density. *Nature* 435, 795–798.

28. Shao, Z., Mederos, J., Chueh, W.C. and Haile, S.M. 2006. High power-density single-chamber fuel cells operated on methane. *J. Power Sources* 162, 589–596.

29. Hao, Y. and Goodwin, D.G. 2008. Efficiency and fuel utilization of methane-powered single-chamber solid oxide fuel cells. *J. Power Sources* 183, 157–163.

30. Kuhn, M. and Napporn, T.W. 2010. Single-chamber solid oxide fuel cell technology: From its origins to today's state of the art. *Energies* 57, 57–134.

31. Beckel, D., Bieberle-Hütter, A., Harvey, A., Infortuna, A., Muecke, U.P., Prestat, M., Rupp, J.L.M. and Gauckler, L.J. 2007. Thin films for micro solid oxide fuel cells. *J. Power Sources* 173, 325–345.

32. Pederson, L.R., Singh, P. and Zhou, X.D. 2006. Application of vacuum deposition methods to solid oxide fuel cells. *Vacuum* 80, 1066–1083.

33. Irvine, G. 2011 High-K semiconductor materials from a chemical manufacturer perspective, http://www.electroiq.com/articles/sst/2011/02/high-k-semiconductor-materials-from-a-chemical-manufacturer-pers.html, accessed 30/03/2013.

34. Cassir, M., Ringued, A. and Niinisto, L. 2010. Input of atomic layer deposition for solid oxide fuel cell applications. *J. Mater. Chem.* 20, 8987–8993.

35. Bruschi, P., Diligenti, A., Nannini, A. and Piotto, M. 1999. Technology of inte-grable free-standing yttria-stabilized zirconia membranes. *Thin Solid Films* 346, 151–254.

36. Kueper, T.W., Visco, S.J. and De Jonghe, L.C. 1992. Thin-film ceramic electrolytes deposited on porous and non-porous substrates by sol-gel techniques. *Solid State Ionics* 52, 251–259.

37. Baertsch, C.D., Jensen, K.F., Hertz, J.L., Tuller, H.L., Vengallatore, S.T., Spearing, S.M. and Schmidt, M.A. 2004. Fabrication and structural characterization of self-supporting electrolyte membranes for a micro solid-oxide fuel cell. *J. Mater. Res.* 19, 2604–2615.

38. Srikar, V.T., Turner, K.T., Ie, T.Y.A. and Spearing, S.M. 2004. Spearing, structural considerations for micromachined solid oxide fuel cells. *J. Power Sources* 125, 62–69.

39. Tang, Y., Stanley, K., Wu, J., Ghosh, D. and Zhang, J. 2005. Design consid-eration of micro thin film solid-oxide fuel cells. *J. Micromech. Microeng.* 15, S185–S192.

40. Shim, J.H., Park, J.S., An, J., Gür, T.M., Kang, S. and Prinz, F.B. 2009. Intermediate-temperature ceramic fuel cells with thin film yttrium-doped barium zirconate electrolytes. *Chem. Mater.* 21, 3290–3296.
41. Shim, J.H., Gür, T.M. and Prinz, F.B. 2008. Proton conduction in thin film yttrium-doped barium zirconate. *Appl. Phys. Letts.* 92, 253115.
42. Chao, C.-C., Kim, Y.B. and Prinz, F.B. 2009. Surface modification of yttria-stabilized zirconia electrolyte by atomic layer deposition. *Nano Lett.* 9, 3626–3628.
43. Kim, Y.B., Holme, T.P., Gür, T.M. and Prinz, F.B. 2011. Surface-modified low-temperature solid oxide fuel cell. *Adv. Funct. Mater.* 21, 4684–4690.
44. Fan, Z. and Prinz, F.B. 2011. Enhancing oxide ion incorporation kinetics by nanoscale yttria-doped ceria interlayers. *Nano Lett.* 11, 2202–2205.
45. Fan, Z., Chao, C.-C., Babaeia, F.H. and Prinz, F.B. 2011. Improving solid oxide fuel cells with yttria-doped ceria interlayers by atomic layer deposition. *J. Mater. Chem.* 21, 10903–10906.
46. An, J., Kim, Y.-B., Gür, T.M. and Prinz, F.B. 2012. Enhancing charge transfer kinetics by nanoscale catalytic cermet interlayer. *ACS Appl. Mater. Interfaces* 4, 6790–6795.
47. Huang, H., Nakamura, M., Su, P., Fasching, R., Saito, Y. and Prinz, F.B. 2007. High-performance ultrathin solid oxide fuel cells for low-temperature operation. *J. Electrochem. Soc.* 154, B20–B24.
48. Shim, J.H., Chao, C.-C., Huang, H. and Prinz, F.B. 2007. Atomic layer deposition of yttria-stabilized zirconia for solid oxide fuel cells. *Chem. Mater.* 19, 3850–3854.
49. Su, P.-C., Chao, C.-C., Shim, J.H., Fasching, R. and Prinz, F.B. 2008. Solid oxide fuel cell with corrugated thin film electrolyte. *Nano Lett.* 8, 2289–2292.
50. Chao, C.-C., Hsu, C.-M., Cui, Y. and Prinz, F.B. 2011. Improved solid oxide fuel cell performance with nanostructured electrolytes. *ACS Nano* 5, 5692–5696.
51. Su, P.-C. and Prinz, F.B. 2012. Nanoscale membrane electrolyte array for solid oxide fuel cells. *Electrochem. Comm.* 16, 77–79.
52. Park, Y.-I., Chab, S.-W., Saitob, Y. and Prinz, F.B. 2005. Gas-tight alumina films on nanoporous substrates through oxidation of sputtered metal films. *Thin Solid Films* 476, 168–173.
53. Park, Y.-I., Su, P.C., Cha, S.W., Saito, Y. and Prinz, F.B. 2006. Thin-film SOFCs using gastight YSZ thin films on nanoporous substrates. *J. Electrochem. Soc.* 153, A431–A436.
54. Kang, S., Sua, P.C., Park, Y.I., Saito, Y. and Prinz, F.B. 2006. Thin-film solid oxide fuel cells on porous nickel substrates with multistage nanohole array. *J. Electrochem. Soc.* 153, A554–A559.
55. Lai, B.K., Kerman, K. and Ramanathan, S. 2010. On the role of ultra-thin oxide cathode synthesis on the functionality of micro-solid oxide fuel cells: Structure, stress engineering and in situ observation of fuel cell membranes during operation. *J. Power Sources* 195, 5185–5196.
56. Lai, B.K., Kerman, K. and Ramanathan, S. 2011. Nanostructured $La_{0.6}Sr_{0.4}Co_{0.8}Fe_{0.2}O_3/Y_{0.08}Zr_{0.92}O_{1.96}/La_{0.6}Sr_{0.4}Co_{0.8}Fe_{0.2}O_3$ (LSCF/YSZ/LSCF) symmetric thin film solid oxide fuel cells. *J. Power Sources* 196, 1826–1832.
57. Kerman, K., Lai, B.K. and Ramanathan, S. 2011. $Pt/Y_{0.16}Zr_{0.84}O_{1.92}/Pt$ thin film solid oxide fuel cells: Electrode microstructure and stability considerations. *J. Power Sources* 196, 2608–2614.
58. Johnson, A.C., Baclig, A., Harburg, D.V., Lai, B.-K. and Ramanathan, S. 2010. Fabrication and electrochemical performance of thin-film solid oxide fuel cells with large area nanostructured membranes. *J. Power Sources* 195, 1149–1155.

59. Tsuchiya, M., Lai, B.-K. and Ramanathan, S. 2011. Scalable nanostructured membranes for solid-oxide fuel cells. *Nature Nanotech.* 6, 282–286.
60. Overmeere, Q.V., Kerman, K. and Ramanathan, S. 2012. Energy storage in ultra-thin solid oxide fuel cells. *Nano Lett.* 12, 3756–3760.
61. Muecke, U.P., Beckel, D., Bernard, A., Bieberle-Hütter, A., Graf, S., Infortuna, A., Müller, P., Rupp, J.L.M., Schneider, J. and Gauckler, L.J. 2008. Micro solid oxide fuel cells on glass ceramic substrates. *Adv. Funct. Mater.* 18, 3158–3168.
62. Rupp, J.L.M., Muecke, U.P., Nalam, P.C. and Gauckler, L.J. 2010. Wet-etching of precipitation-based thin film microstructures for micro-solid oxide fuel cells. *J. Power Sources* 195, 2669–2676.
63. Joo, J.H. and Choi, G.M. 2007. Open-circuit voltage of ceria-based thin film SOFC supported on nano-porous alumina. *Solid State Ionics* 178, 1602–1607.
64. Kwon, C.-W., Son, J.-W., Lee, J.-H., Kim, H.-M., Lee, H.-W. and Kim, K.-B. 2011. High-performance micro-solid oxide fuel cells fabricated on nanoporous anodic aluminum oxide templates. *Adv. Funct. Mater.* 21, 1154–1159.
65. Kwon, C.-W., Leea, J.-I., Kim, K.-B., Leeb, H.-W., Leeb, J.-H. and Son, J.-W. 2012. The thermomechanical stability of micro-solid oxide fuel cells fabricated on anodized aluminum oxide membranes. *J. Power Sources* 210, 178–183.
66. Sarkar, P., Yamarte, L., Rho, H. and Johanson, L. 2007. Anode-supported tubular micro-solid oxide fuel cell. *Int. J. Appl. Ceram. Technol.* 4, 103–108.
67. Kendall, K. 1992. Development of microtubular solid oxide fuel cell. Proceedings of the International Forum on Fine Ceramics, Japan Fine Ceramics Center, Nagoya, pp. 143–148.
68. Kendall, M. 1993. A novel SOFC design. Final year report. Middlesex University, Hendon, North London, England.
69. Mallon, C. and Kendall, K. 2005. Sensitivity of nickel cermet anodes to reduction conditions. *J. Power Sources* 145, 154–160.
70. Kendall, K. 1994. Solid oxide fuel cell structures. International Patent# WO 94/22178.
71. Alston, T., Kendall, K., Palin, M., Prica, M. and Windibank, P. 1998. A 1000-cell SOFC reactor for domestic cogeneration. *J. Power Sources* 71, 271–274.
72. Kendall, K. and Palin, M. 1998. A small solid oxide fuel cell demonstrator for microelectronic applications. *J. Power Sources* 71, 268–270.
73. Hatchwell, C.E., Sammes, N.M. and Kendall, K. 1998. Cathode current-collectors for a novel tubular SOFC design. *J. Power Sources* 71, 85–90.
74. Hatchwell, C.E., Sammes, N.M., Brown, I.W.M. and Kendall, K. 1999. Current collectors for a novel tubular design of solid oxide fuel cell. *J. Power Sources* 77, 64–68.
75. Tompsett, G.A., Finnerty, C., Kendall, K., Alsta, T. and Sammes, N.M. 2000. Novel applications for micro-SOFCs. *J. Power Sources* 86, 376–382.
76. Singhal, S.C. and Kendall, K. 2003. *High Temperature Solid Oxide Fuel Cells*, (Chapter 8.4). Elsevier, Oxford, U.K., p. 221.
77. Kendall, K. and Prica, M. 1994. Integrated SOFC tubular system for small scale cogeneration. Proceedings of the 1st Euro. SOFC Forum, Ed. U. Bossel, Switzerland, pp. 163–170.
78. Du, Y., Sammes, N.M. and Eberly, B. 2004. Anode-supported tubular SOFC systems: Fabrication & properties. Proc. of 6th Euro. SOFC Forum, Ed. M. Mogensen, pp. 125–134.

79. Sammes, N.M., Bove, R. and Du, Y. 2006. Assembling single cells to create a stack: The case of a 100 W microtubular anode-supported solid oxide fuel cell stack. *J. Mater. Eng. Perf.* 15, 463–467.
80. Sammes, N.M. and Du, Y. 2007. Fabrication and characterization of tubular solid oxide fuel cells. *Int. J. Appl. Ceram. Technol.* 4, 89–102.
81. Calise, F., Restuccciaa, G. and Sammes, N.M. 2010. Experimental analysis of micro-tubular solid oxide fuel cell fed by hydrogen. *J. Power Sources* 195, 1163–1170.
82. Yamaguchi, T., Galloway, K.V., Yoon, J. and Sammes, N.M. 2011. Electrochemical characterizations of microtubular solid oxide fuel cells under a long-term testing at intermediate temperature operation. *J. Power Sources* 196, 2627–2630.
83. Kendall, K. 2010. Progress in microtubular solid oxide fuel cells. *Int. J. Appl. Ceram. Technol.* 7, 1–9.
84. Kendall, K. and Kilbride, I. 2004. Fuel cell power generating system. U.S. Patent #6696187.
85. Crumm, A.T. and Halloran, J.W. 2004. Method for preparation of solid state electrochemical device. U.S. Patent #6,747,799.
86. Crumm, A.T. and Halloran, J.W. 1998. Fabrication of microcon figured multi-component ceramics. *J. Am. Ceram. Soc.* 81, 1053–1057.
87. Suzuki, T., Yamaguchi, T., Fujishiro, Y. and Awano, M. 2006. Improvement of SOFC performance using a microtubular, anode-supported SOFC. *J. Electrochem. Soc.* 153, A925–A928.
88. Suzuki, T., Funahashi, Y., Yamaguchi, T., Fujishiro, Y. and Awanoa, M. 2007. Design and fabrication of lightweight, submillimeter tubular solid oxide fuel cells. *Electrochem. Solid-State Lett.* 10, A177–A179.
89. Suzuki, T., Sugihara, S., Yamaguchi, T., Sumi, H., Hamamoto, K. and Fujishiro, Y. 2011. Effect of anode functional layer on energy efficiency of solid oxide fuel cells. *Electrochem. Comm.* 13, 959–962.
90. Suzuki, T., Liang, B., Yamaguchi, T., Sumi, H., Hamamoto, K. and Fujishiro, Y. 2012. One-step sintering process of gadolinia-doped ceria interlayer–scandia-stabilized zirconia electrolyte for anode supported microtubular solid oxide fuel cells. *J. Power Sources* 199, 170–173.
91. Sarkar, P. 2003. Production of hollow ceramic membranes by electrophoretic deposition. US Patent# 6,607,645, Aug. 19.
92. Sarkar, P. and Rho, H. 2003. Micro solid oxide fuel cell. The Electrochemical Society Proceedings Series, SOFC VIII, Eds. S.C. Singhal and M. Dokiya, Pennington, NJ, pp. 135–138.
93. Sarkar, P., De, D. and Rho, H. 2004. Synthesis and microstructural manipulation of ceramics by electrophoretic deposition. *J. Mat. Sci.* 39, 819–823.
94. Sarkar, P. and Kovacik, G. 2005. Micro-SOFC for electronic application. 7th Annual Small Fuel Cell for Portable Power Applications, Knowledge Foundation Inc., USA.
95. Sarkar, P., Rho, H., Liu, M., Yamarte, L. and Johanson, L. 2005. High power density tubular SOFC for portable application. The Electrochemical Society Proceedings Series, SOFC IX, Eds. S.C. Singhal and J. Mizusaki, Pennington, NJ, pp. 411–418.
96. Sarkar, P., Yamarte, L. and Johanson, L. 2007. Tubular micro-SOFC for remote power applications. *ECS Trans.* 7 (1), 603–608.

97. Soderberg, J.N., Sun, L., Sarkar, P. and Birss, V.I. 2009. Oxygen reduction at LSM-YSZ cathode deposited on anode-supported microtubular solid oxide fuel cells. *J. Electrochem. Soc.* 156 (9), B721–B728.
98. Soderberg, J.N., Sarkar, P. and Birss, V.I. 2010. Evaluation of Ni-YSZ anode performance in microtubular solid oxide fuel cells. *J. Electrochem. Soc.* 157, B607–B613.
99. Sarkar, P. and Nicholson, P.S. 1996. Electrophoretic deposition (EPD): Mechanisms, kinetics and applications to ceramics. *J. Am. Ceram. Soc.* 79, 1987–2002.
100. Sarkar, P., Huang, X. and Nicholson, P.S. 1992. Structural ceramic microlaminates by electrophoretic deposition. *J. Am. Ceram. Soc.* 75, 2907–2909.
101. Nicholson, P.S., Sarkar, P. and Huang, X. 1993. Electrophoretic deposition and its use to synthesize ZrO_2/Al_2O_3 micro-laminate ceramic/ceramic composites. *J. Mats. Sci.* 28, 6274–6278.
102. Sarkar, P., Richardson, M., Yamarte, L. and Johanson, L. 2013. Solid state electrochemical cell having reticulated electrode matrix and method of manufacturing same. US Patent#8,349,510, Jan. 8.
103. Li, K. 2010. Ceramic hollow fibre membranes and their applications. *Basic Aspects of Membrane Science and Engineering of Comprehensive Membrane Science & Engineering*, Vol. 1, Eds. E. Drioli and L. Giorno, Elsevier, pp. 253–274.
104. Liu, Y., Liu, N. and Tan, X.Y. 2011. Preparation of microtubular solid oxide fuel cells based on highly asymmetric structured electrolyte hollow fibers. *Sci. China Chem.* 54, 850–855.
105. Yang, C., Jin, C. and Chen, F. 2010. Micro-tubular solid oxide fuel cells fabricated by phase-inversion method. *Electrochem. Comm.* 12, 657–660.
106. Droushiotis, N., Torabi, A., Othman, M.H.D., Etsell, T.H. and Kelsall, G.H. 2012. Effects of lanthanum strontium cobalt ferrite (LSCF) cathode properties on hollow fibre micro-tubular SOFC performances. *J. Appl. Electrochem* 42, 517–526.
107. Hafiz, M., Othman, D., Droushiotis, N., Wu, Z., Kelsall, G. and Ang Li, K. 2011. High-performance, anode-supported, microtubular SOFC prepared from single-step-fabricated, dual-layer hollow fibers. *Adv. Mater.* 23, 2480–2483.
108. Sarkar, P., Yamarte, L. and Amow, G. 2009. Advanced materials for soldier power applications through waste-heat recovery: Development of porous electrolyte-supported of microtubular fuel cell. Proc. CF/DRDC International Defence Materials Meeting, Victoria, BC, Canada, p. 55–63.
109. Hanifi, A.R., Torabi, A., Etsell, T.H., Yamarte, L. and Sarkar, P. 2011. Porous electrolyte-supported tubular micro-SOFC design. *Solid State Ionics* 192, 368–371.
110. Hanifi, A.R., Shinbine, A., Etsell, T.H. and Sarkar, P. 2011. Fabrication of thin porous electrolyte-supported tubular fuel cell using slip casting. *J. Ceram. Process. Res.* 12, 336–342.
111. Torabi, A., Hanifi, A.R., Etsell, T.H. and Sarkar, P. 2012. Effect of porous support microstructure on performance of infiltrated electrodes in solid oxide fuel cells. *J. Electrochem. Soc.* 159, B201–B21.
112. Hanifi, A.R., Shinbine, A., Etsell, T.H. and Sarkar, P. 2012. Development of monolithic porous and dense YSZ layers through multiple slip casting for ceramic fuel cell applications. *Int. J. Appl. Ceram. Technol.* 9 (6), 1011–1021.
113. Vincent, A.L., Hanifi, A.R., Luo, J.-L., Chuang, K.T., Sanger, A.R., Etsell, T.H. and Sarkar, P. 2012. Porous YSZ impregnated with La0.4Sr0.5Ba0.1TiO3 as a possible composite anode for SOFCs fueled with sour feeds. *J. Power Sources* 215, 301–306.

114. Funahashi, Y., Suzuki, T., Fujishiro, Y., Shimamori, T. and Awano, M. 2007. Optimization of configuration for cube-shaped SOFC bundles. *ECS Trans.* 7, 643–649.

115. Suzuki, T., Funahashi, Y., Yamaguchi, T., Fujishiro, Y. and Awano, M. 2008. Development of microtubular SOFCs. *J. Fuel Cell Sci. Tech.* 5, 031201.

116. Suzuki, T., Yamaguchi, T., Fujishiro, Y., Awano, M. and Funahashi, Y. 2010. Recent development of microceramic reactors for advanced ceramic reactor system. *J. Fuel Cell Sci. Tech.* 7, 031005.

117. Funahashi, Y., Shimamori, T., Suzuki, T., Fujishiro, Y. and Awano, M. 2010. Simulation study for the series connected bundles of microtubular SOFCs. *J. Fuel Cell Sci. Tech.* 7, 051012.

118. Hashimoto, S., Liu, Y., Asano, K., Yoshiba, F., Mori, M., Funahashi, Y. and Fujishiro, Y. 2011. Power generation properties of microtubular solid oxide fuel cell bundle under pressurized conditions. *J. Fuel Cell Sci. Tech.* 8, 021010.

119. Sarkar, P. and Rho, H. 2004. Tubular solid oxide fuel cell stack. US Patent# 6,824,907, Nov. 30.

120. Development of handy fuel cell system advanced at industrial science and technology (AIST), Japan, 2013. http://www.aist.go.jp/aist_e/latest_research/2013/20130321/20130321.html, accessed 27/03/2013.

121. Sarkar, P., Rho, H. and Johanson, L. 2005. Solid oxide fuel cell system. US Patent# 6,936,367, Aug. 30.

122. Zheng, R., Kovacik, G., Rho, H., Sarkar, P., Yamarte, L. and Richardson, M. 2010. Heating solid oxide for fuel cell. US Patent# 7,732,076, Jan. 8.

123. Zheng, R., Rho, H., Yamarte, L., Kovacik, G. and Sarkar, P. 2007. Controlling solid oxide fuel cell operation. US Patent# 7,258,936, Aug. 21.

124. Lawlor, V., Griessera, S., Buchingerd, G., Olabib, A.G., Cordinere, S. and Meissner, D. 2009. Review of the micro-tubular solid oxide fuel cell part I. Stack design issues and research activities. *J. Power Sources* 193, 387–399.

125. Howe, K.S., Thompson, G.J. and Kendall, K. 2011. Micro-tubular solid oxide fuel cells and stacks. *J. Power Sources* 196, 1677–1686.

126. Hotz, N., Senn, S.M. and Poulikakos, D. 2006. Exergy analysis of a solid oxide fuel cell micropowerplant. *J. Power Sources* 158, 333–347.

127. Serincan, M.F., Pasaogullari, U. and Sammes, N.M. 2009. Effects of operating conditions on the performance of a micro-tubular solid oxide fuel cell (SOFC). *J. Power Sources* 192, 414–422.

128. Mirahmadi, A. and Valefi, K. 2011. Study of thermal effects on the performance of micro-tubular solid-oxide fuel cells. *Ionics* 17, 767–783.

129. Sammes, N.M., Song, J., Roy, B., Galloway, K., Suzuki, T., Awano, M. and Serincan, A.M.F. 2009. A study of GDC-based micro tubular SOFC. *Thermec 2009*, Pts 1–4, Eds. T. Chandra, N. Wanderka, W. Reimers and M. Ionescu, Trans Tech Publications, Switzerland, pp. 1152–1157.

130. Amiri, S., Hayes, R.E., Nandakumar, K. and Sarkar, P. 2010. Mathematical modeling of a novel tubular micro-solid oxide fuel cell and experimental validation. *Chem. Eng. Sci.* 65, 6001–6013.

131. Sammes, N.M., Du, Y. and Bove, R. 2005. Design and fabrication of a 100 W anode supported micro-tubular SOFC stack. *J. Power Sources* 145, 428–434.

132. Nehter, P. 2006. Two-dimensional transient model of a cascaded micro-tubular solid oxide fuel cell fed with methane. *J. Power Sources* 157, 325–334.

133. Suzuki, T., Funahashi, Y., Yamaguchi, T., Fujishiro, Y. and Awano, M. 2007. Anode-supported micro tubular SOFCs for advanced ceramic reactor system. *J. Power Sources* 171, 92–95.

134. Cui, D., Liu, L., Dong, Y.L. and Cheng, M.J. 2007. Comparison of different current collecting modes of anode supported micro-tubular SOFC through mathematical modeling. *J. Power Sources* 174, 246–254.

135. Suzuki, T., Yamaguchi, T., Fujishiro, Y. and Awano, M. 2007. Current collecting efficiency of micro tubular SOFCs. *J. Power Sources* 163, 737–742.

136. Serincan, M.F., McPhee, W.A.G. and Sammes, N.M. 2007. Mass transport effects of a current collector coated on the entire cathode surface of a micro tubular solid oxide fuel cell. *ECS Trans.* 3, 35–43.

137. Serincan, M.F., Smirnova, A. and Sammes, N. 2007. Modeling and analysis of a micro-tubular solid oxide fuel cell operating at intermediate temperatures. *ECS Trans.* 7 (1), 1955–1965.

138. Izzo, J.R., Peracchio, A.A. and Chiu, W.K.S. 2008. Modeling of gas transport through a tubular solid oxide fuel cell and the porous anode layer. *J. Power Sources* 176, 200–206.

139. Lockett, M., Simmons, M.J.H. and Kendall, K. 2004. CFD to predict temperature profile for scale up of micro-tubular SOFC stacks. *J. Power Sources* 131, 243–246.

140. Amiri, S., Hayes, R.E., Nandakumar, K. and Sarkar, P. 2013. Modelling heat transfer for a tubular micro-solid oxide fuel cell with experimental validation. *J. Power Sources* 233, 190–201.

141. Serincan, M.F., Pasaogullari, U. and Sammes, N.M. 2008. A computational analysis to identify the current density characteristics of a micro-tubular solid oxide fuel cell. Proceedings of the 6th International Conference on Fuel Cell Science, Engineering, and Technology – 2008, pp. 347–351.

142. Garcia-Camprubi, M., Jasak, H. and Fueyo, N. 2011. CFD analysis of cooling effects in H-2-fed solid oxide fuel cells. *J. Power Sources* 196, 7290–7301.

143. Cui, D.A. and Cheng, M.J. 2009. Thermal stress modeling of anode supported micro-tubular solid oxide fuel cell. *J. Power Sources* 192, 400–407.

144. Serincan, M.F., Pasaogullari, U. and Sammes, N.M. 2010. Thermal stresses in an operating micro-tubular solid oxide fuel cell. *J. Power Sources* 195, 4905–4914.

145. Li, J.Y. and Lin, Z.J. 2012. Effects of electrode composition on the electrochemical performance and mechanical property of micro-tubular solid oxide fuel cell. *Int. J. Hydrogen Energy* 37, 12925–12940.

146. Raberger, N.B., Stutz, M.J., Hotz, N. and Poulikakos, D. 2009. Simulation of the postcombustor for the treatment of toxic and flammable exhaust gases of a micro-solid oxide fuel cell. *J. Fuel Cell Sci. Tech.* 6, 041002.

147. Serincan, M.F., Pasaogullari, U. and Sammes, N.M. 2009. A transient analysis of a micro-tubular solid oxide fuel cell (SOFC). *J. Power Sources* 194, 864–872.

5

Advances in Direct Carbon Fuel Cells

Cairong Jiang, Andrew C. Chien, and John T. S. Irvine

CONTENTS

5.1 Introduction 127
5.2 Fuel Cells 128
5.3 Direct Carbon Fuel Cells 129
    5.3.1 Molten Salt DCFCs 134
        5.3.1.1 Molten Hydroxide 134
        5.3.1.2 Molten Carbonate 136
    5.3.2 Solid Oxide DCFCs 139
        5.3.2.1 Physical Contact 140
        5.3.2.2 Fluidized Bed 142
        5.3.2.3 In Situ–*External Gasification* 142
        5.3.2.4 Solid Carbon Anode 145
        5.3.2.5 Molten Metal 147
        5.3.2.6 Deposited Carbon 149
        5.3.2.7 Other Contact 150
5.4 Hybrid Direct Carbon Fuel Cells 150
    5.4.1 Influence of Carbonate Content 151
    5.4.2 Effect of SOFC Geometry 154
    5.4.3 Optimization of the HDCFC 155
5.5 Summary and Prospects 156
References 157

5.1 Introduction

Coal is almost certainly going to be the dominant fuel in coming decades due to its extensive reserves in important energy economies (see Figure 5.1). Indeed, the global reserve/production rate ratio for coal is around three times that of the next most abundant fossil fuel. Far from reduction of global CO_2 emissions, there is likely to be a very significant increase, undermining our climate change efforts, unless more efficient means of carbon conversion are achieved. That is why new clean carbon conversion technologies must be at the forefront of energy technology development.

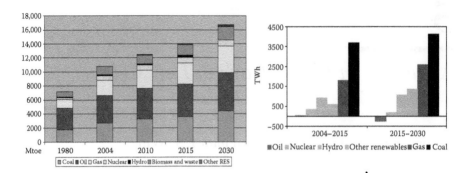

FIGURE 5.1

(See color insert.) Primary world energy consumption is set to march on with the majority from fossil sources (left) contributing to CO_2 emission. Coal is expected to dominate future electricity production (right). (Reproduced from Malzbender, J. et al. *Solid State Ionics*, 176, 2201–2203, 2005. With permission.)

The global abundance of coal, and the present low cost per unit of electricity generated using coal compared with other fossil fuels and uranium, ensures that its usage for electricity generation will continue for many decades [1]. Coal is not a "clean" fuel, owing to its CO_2 and SO_x/NO_x emissions upon burning; however, emerging technologies may reduce its environmental footprint. For example, *in situ* gasification of coal into CO produces a more clean-burning fuel. Sequestration of CO_2 emissions from coal combustion (carbon capture and storage, CCS) is another option, and as CCS technology develops, existing coal-powered stations could be retrofitted with the appropriate technologies. A very promising alternative method is to produce electricity by oxidizing coal electrochemically with oxygen in a direct carbon fuel cell (DCFC).

5.2 Fuel Cells

A fuel cell is an energy conversion device that produces electricity and heat by electrochemical combination of a fuel with an oxidant, which can be viewed as a battery with an external fuel supply. A fuel cell consists of four components: two electrodes, the anode and the cathode, separated by an electrolyte and connected by an external circuit or interconnect. Fuel is fed to the anode where it is oxidized, releasing electrons to the external circuit. The oxidant is fed to the cathode where it is reduced using the electrons delivered by the external circuit. The electrons flow through the interconnect from the anode to the cathode producing electricity.

In theory, any combination of chemicals capable of electrochemical oxidation and reduction can be used as the fuel and oxidant in a fuel cell. Oxygen is the most common oxidant for fuel cells because it is readily and economically available from air. Hydrogen, which offers high electrochemical reactivity, is the most common fuel. However, high-temperature fuel cells can be developed to work with alternative fuels to hydrogen.

The key feature of a fuel cell is its high energy-conversion efficiency. Because fuel cells convert the chemical energy of the fuel directly to electrical energy without the intermediate of thermal energy (unlike indirect conversion in conventional systems), their conversion efficiency is not subject to the Carnot limitation. Further energy gains can be achieved when the produced heat is used in combined heat and power or gas turbine applications. The improved efficiency as compared to conventional energy-conversion devices is the main reason that fuel cells are receiving considerable attention. Besides their high energy-conversion efficiency, fuel cells offer several additional advantages over conventional methods of power generation. They offer a much lower production of pollutants. A fuel cell fuelled with H_2 and air only produces water. Other significant advantages offered are modular construction and size flexibility, which makes them well suited for decentralized applications; high efficiency at different loads; fuel flexibility; and remote/unattended operation. Moreover, their vibration-free operation eliminates the noise usually associated with conventional power generation systems. More details concerning general features of fuel cells can be found in the literature [2,3].

5.3 Direct Carbon Fuel Cells

A DCFC is an electrochemical system in which the chemical energy of solid carbon fuel is directly converted into electrical energy while releasing carbon dioxide.

The concept and configuration of a DCFC are similar to those of other high-temperature fuel cells with the key difference that the fuel is solid as opposed to gaseous as in traditional molten carbonate fuel cells (MCFCs) or solid oxide fuel cells (SOFCs). A single unit cell comprises a cathode, an electrolyte, and an anode compartment where the carbon is electrochemically converted.

The main advantage of the DCFC is its high theoretical electrical efficiency, which results from the high electrochemical conversion efficiency, slightly in excess of 100% due to a positive entropy change for the cell reaction, combined with a high fuel utilization that can reach 100%. Practical systems feature an overall efficiency as high as 80%, which is a major improvement from the current 40% efficient traditional coal-fired power plants. Moreover,

DCFC systems are scalable, hence suitable for decentralized electricity production, and quiet due to the absence of rotating parts. Opportunities for DCFC applications are further strengthened by the abundance of fuels, which include both fossil fuels, such as coal (which is the most abundant fossil fuel on earth) or petroleum coke, and renewable fuels, such as biomass (e.g., wood or nut shells) or wastes. The DCFC therefore offers a cleaner and quiet means to convert the high chemical energy of a solid carbon fuel to electricity.

Although a prototype DCFC based on molten sodium hydroxide was invented in 1896 [4,5], fuel cells utilizing other fuels, such as hydrogen, have since received wider attention. This is despite the fact that the energy released per unit volume upon the oxidation of carbon with oxygen (23.95 kW h/L) substantially exceeds that of many other fuels, for example, that of liquid methane (5.9 kW h/L), liquid hydrogen (2.6 kW h/L), or diesel (10.6 kW h/L) [6].

The intrinsic difficulty in implementing carbon fuels is that most fuel cells are based on a solid membrane and with solid fuel there is very little interaction between the solid fuel and the solid electrode–electrolyte membrane. Therefore, most research to date has focused on high-temperature liquid electrolyte concepts. An alternative way to solve this problem uses a solid electrolyte with a fuel electrode–electrolyte based on a slurry of liquid electrolyte and solid electrode [7–9]. This liquid interface removes solid–solid contact problems and seems to provide a very attractive way forward. By employing a solid oxide electrolyte to separate the cathode and anode compartments while utilizing a molten carbonate electrolyte on the anode side, it is possible to avoid the need for CO_2 circulation. A similar idea utilizing a liquid metal–oxide redox shuttle has recently been published by Tao et al. in the United States [10].

In DCFCs, carbon is the fuel that is oxidized electrochemically at the anodes. The type of DCFCs can be categorized by the electrolytes: molten salts, solid membranes, and a combination of both in present DCFC research communities as shown in Table 5.1. In molten salts–based DCFCs, carbonates (CO_3^{2-}) or hydroxides (OH^-) are the major oxidant ions; in solid electrolyte–based DCFCs, oxygen anions (O^{2-}) are a major oxidant [11]. Figure 5.2 shows the schematic of a DCFC and its anode and cathode reactions. The anode and cathode reactions on the carbonate-based and solid electrolyte–based DCFCs are listed, respectively, as follows:

Molten carbonate–based

$$\text{Anode: } C + 2CO_3^{2-} \rightarrow 3CO_2 + 4\,e^- \tag{5.1}$$

$$\text{Cathode: } O_2 + 2CO_2 \rightarrow 4\,e^- \rightarrow 2CO_3^{2-} \tag{5.2}$$

TABLE 5.1

Some Present/Recent DCFC Research Activities

Countries	Center and Lab	Electrolytes	Development	Reference
United States	Hawaii Natural Energy Institute, University of Hawaii at Manoa	Molten hydroxide electrolyte	Aqueous-alkaline/carbonate carbon fuel cell	[12,13]
	West Virginia University	Molten hydroxide electrolyte	Char production from biomass Evaluation of carbon materials for use in a DCFC	[14]
	Lawrence Livermore National Laboratory, Livermore	Molten carbonate electrolyte	Direct conversion of carbon fuels in a MCFC	[15]
	Stanford University; Direct Carbon Technologies, LLC	Solid electrolyte membrane (YSZ)	Fluidized bed DCFCs Gasification-driven DCFCs	[16–18]
	University of Utah; Materials and Systems Research, Inc.	Solid electrolyte membrane (YSZ)	Dry gasified coal SOFCS CO-fuelled SOFCs	[19,20]
China	Tsinghua University	Solid electrolyte membrane (ScSZ)	Catalytic gasification process Pyrolysis of hydrocarbons as carbon fuels	[21,22]
	Tianjin University	Composite electrolyte; samarium-doped ceria and a ternary carbonate	Modeling and simulation of a single DCFC	[23–25]
	Nanjing University of Technology	Solid electrolyte membrane (ScSZ)	Carbon fuel cell integrated with *in situ* catalytic reverse Boudouard reaction	[26]
	Shanghai Inorganic Energy Materials and Power Source Engineering Center	Solid electrolyte membrane (ScSZ)	DCFC with tubular SOFCs	[27]
	Harbin Eng. University; Tianjin University	Molten carbonate electrolyte	Electrochemical oxidation activity of pretreated carbon	[28,29]

(continued)

TABLE 5.1 (Continued)

Some Present/Recent DCFC Research Activities

Countries	Center and Lab	Electrolytes	Development	Reference
Europe	School of Chemistry, University of St. Andrews, UK	Composite electrolyte (solid electrolyte + molten carbonate)	Hybrid electrolyte carbon fuel cells Carbon fuels from pyrolyzed biomass	[30–37]
	ZAE Bayern, Garching, Germany	Solid electrolyte membrane (YSZ)	Direct carbon conversion in a SOFC system with a nonporous anode	[38]
	Chemical Engineering, Imperial College, London, UK	SOFC with separate combustion reactor, oxygen shuttle	Indirect carbon oxidation with molten metal	[39]
Australia	School of Chemical Engineering, University of Queensland	Molten carbonate electrolyte	Carbon fuel evaluation DCFCs	[40,41]
	CSIRO	High-temperature SOFC	Electrode materials for coal oxidation, Tubular cells	[42–44]
Japan	Tokyo Institute of Technology	Solid electrolyte membrane	Rechargeable carbon SOFCs by pyrolysis of hydrocarbons	[45–48]

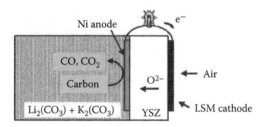

FIGURE 5.2
Schematic of a solid oxide DCFC with molten carbonate extended anode inside the DCFC chamber.

Solid electrolyte–based

$$\text{Anode: } C + 2O^{2-} \rightarrow CO_2 + 4\,e^- \tag{5.3}$$

$$\text{Cathode: } O_2 + 4\,e^- \rightarrow 2O^{2-} \tag{5.4}$$

$$\text{Overall: } C + O_2 \rightarrow CO_2 \tag{5.5}$$

Although carbon can be converted to CO_2 via direct electrochemical oxidation, partial oxidation of carbon to CO may occur, and CO can get oxidized further to CO_2. On the other hand, the oxidation product CO_2 can react with carbon to produce CO via a Boudouard reaction in DCFC testing conditions, that is, >700°C. These side reactions mean that it is difficult to obtain an accurate estimate for the efficiency of energy conversion, and their effects on the performance of DCFCs are still under investigation [20,49].

On the other hand, both molten carbonate–based and solid electrolyte–based DCFCs face different challenges and difficulties for improving power densities and achieving stable power generation. For molten carbonate–based DCFCs, concentration polarization (i.e., fuel or oxidant diffusion) and current collection are major issues. Therefore, ultrafine carbon particles and an effective-contact metal mesh current collector are needed [30]. For solid oxide DCFCs, the contact of carbon fuel with the electrode and interface of electrode–electrolyte, the so-called three-phase boundary (TPB), are limited, and the diffusion of oxidant ions in solid ceramic electrolyte is temperature dependent. Thus, most of research is directed to *in situ* carbon gasification to generate electricity in a direct DCFC [26] (here, "direct" does not mean one elementary step reaction, but instead indicates a direct conversion in one process, which can include gasification and fuel cell reactions in one chamber).

The fundamentals governing reaction mechanisms in both DCFC types are not well understood. Such limited information can be attributed to the complexities of the three-phase reaction (solid: carbon, gas: carbon dioxide, liquid: molten salt) in DCFCs compared with the two-phase reaction (gas and solid) in hydrogen fuel cells and nonstandardized fuels from different carbon sources.

There are some reviews on DCFCs summarized by other researchers: a general review on DCFC technology and status [42], the utility of coal in a DCFC [43], or the carbon oxidation mechanism at the anode in a molten carbon fuel cell [50], or different delivery modes for carbon conversion in DCFC systems [11]. There are still some parts that have not been covered in these review papers, and this review is going to bridge the gap. In this chapter, recent developments on DCFCs are reviewed based on present extensive research on three types of electrolyte: molten salts, solid oxide, and a combination of both. Discussion focuses on the contact mode design of solid oxide DCFCs. The advantages and disadvantages are presented on each type of DCFCs. DCFCs using molten hydroxide or molten carbonate operate at relatively low temperatures, while solid oxide DCFCs have no corrosion issues and enhance reaction rate at higher operation temperatures. The challenge for the solid oxide DCFC is the little interaction between the solid fuel and the solid electrode–electrolyte membrane due to contact problems. To increase the contact, researchers are making efforts toward designing various systems, for example, a fluidized bed to increase the contact chance between the solid fuel and solid electrode–electrolyte, *in situ*–external gasification to get gas fuels, deposited carbon to increase the TPB length, and utility of molten-metal anodes to increase the contact between the solid fuel and solid oxide [10,51]. Emphasis is given to a hybrid DCFC (HDCFC), a combination of a SOFC and a MCFC, which provides a possible route for carbon oxidation in order to benefit from both molten carbonate DCFC and solid oxide DCFC [24,34].

5.3.1 Molten Salt DCFCs

5.3.1.1 Molten Hydroxide

The main attractive features of molten hydroxides as an electrolyte are the high ionic conductivity, the high activity for the carbon electrochemical oxidation, and the large liquid temperature range offered, which spans 200°C to 800°C. This allows cells to be efficiently operated at relatively low temperatures (e.g., around 400°C) where less expensive materials can be used as compared to higher operating temperatures, hence bringing manufacturing costs down. In 1896, Jacques [4,5] assembled the first single carbon fuel cell and stacks using sodium hydroxide as the electrolyte. Despite the advantages over other types of electrolyte and being at the origins of the DCFC, molten hydroxides have been mostly ignored as a viable option due to their instability in a CO_2-containing atmosphere, which is the product of the carbon oxidation. Indeed, chemical and electrochemical reactions with CO_2 lead to carbonate formation, as shown in Equations 5.6 and 5.7, respectively:

$$2OH^- + CO_2 \leftrightarrow CO_3^{2-} + H_2O \text{ chemical} \tag{5.6}$$

$$C + 6OH^- \leftrightarrow CO_3^{2-} + 3H_2O + 4e^- \text{ electrochemical} \tag{5.7}$$

The chemical carbonate formation process, Equation 5.6, occurs in the two steps shown in the following equations:

$$6OH^- \leftrightarrow 3O^{2-} + 3H_2O \quad \text{chemical} \tag{5.8}$$

$$C + 3O^{2-} \leftrightarrow CO_3^{2-} + 4e^- \quad \text{electrochemical} \tag{5.9}$$

The adverse effect on the cell operation has led to many attempts to minimize the carbonate formation, which have focused on the addition of basic oxides, such as MgO, ZnO, SiO, and CaO, or oxyanions, such as pyrophosphate and persulfate. While the mechanism through which basic oxide addition reduces the carbonate formation is still unclear, magnesium oxide had been recognized by Jacques as an efficient additive in his patent, based on the obtained prolonged cell life. Jacques suggested that magnesium oxide served as a carrier to convey the carbonic acid through the hydroxide electrolyte.

It can be seen from Equations 5.6 and 5.8 that the carbonate formation is related to oxygen ion and water concentration. Therefore, an alternative effective way to reduce the formation of carbonate could be achieved by increasing the water content in an attempt to shift the equilibria in Equations 5.6 and 5.8 to the left [6,52]. A successful configuration using this concept is designed by Scientific Applications and Research Associates, Inc. (SARA). Using a humidified atmosphere, carbon is introduced above the sodium hydroxide melted electrolyte. The carbonate formation rate was reported to be significantly reduced, while the corrosion rate of various metals, such as Ni, Fe, or Cr, was nearly zero. Through the optimization of various parameters, including gas bubbling, electrolyte composition, electrode spatial arrangement,

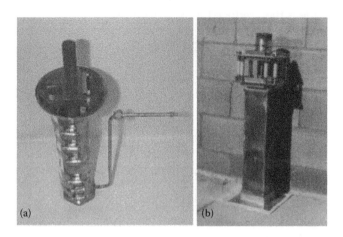

FIGURE 5.3
DCFC devices using hydroxide as the electrolyte (a) single cell and the cell's area is 26 cm²; (b) a stack and the area is 300 cm². (Reproduced from Zecevic, S. et al. *Chem Eng Commun*, 192, 1655–70, 2005. With permission.)

and cathode material, a maximum current density of 250 mA cm^{-2} was achieved using a coal-derived anode with a reported 540 h of stable operation at a current density of 140 mA cm^{-2} corresponding to a 40 mW cm^{-2} power output. The practical device, shown in Figure 5.3 [53], produced 8 A current when the cell's area was 26 cm^2 and 40 A current when the cell's area was 300 cm^2.

Although developments have been made on the use of molten hydroxides as an electrolyte for DCFCs, the effect of carbonate formation on the cell durability remains a serious problem hindering the development of practical durable systems.

5.3.1.2 Molten Carbonate

Molten carbonates, which are commonly used in conventional MCFCs, have received attention for direct carbon-oxidation applications. Molten carbonates offer a high conductivity, a suitable melting temperature range, and a good stability in CO_2-containing atmospheres. The investigation of molten carbonates dates to the 1970s when Anbar and Weaver fed carbonaceous fuel into a molten carbonate–lead bath to generate power under a combined flow of CO_2 and air [54]. The various reactions occurring in this variant can be described as follows: Lead is first oxidized to lead oxide:

$$Pb(l) + 1/2O_2 \rightarrow PbO \tag{5.10}$$

and carbon is oxidized into carbon monoxide:

$$C + 1/2O_2 \rightarrow CO \tag{5.11}$$

The lead oxide is then reduced back to lead:

$$PbO + CO \rightarrow Pb + CO_2 \tag{5.12}$$

The lead oxidation, as shown in Equation 5.10, can be decomposed into the following electrochemical reactions:

$$CO_2 + 1/2O_2 + 2e^- \rightarrow CO_3^{2-} \tag{5.13}$$

$$Pb + CO_3^{2-} \rightarrow PbO + CO_2 + 2e^- \tag{5.14}$$

Application of molten carbonate in DCFCs was proposed by Vutetakis et al. in 1987 [55], before being further developed by Peelen et al. in 2000 [56] and Cooper et al. [57,58] through the introduction of the tilted design as shown in Figure 5.4 [60]. This concept involves the conventional MCFC configuration by using lithiated nickel oxide as a cathode and a porous separator saturated with molten salt. A maximum power density of 100 mW cm^{-2} has been

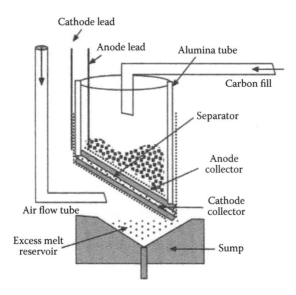

FIGURE 5.4
Schematic of DCFC with a molten lithium–potassium carbonate electrolyte, investigated by Cooper et al. (With kind permission from Springer Science+Business Media: *Recent Trends in Fuel Cell Science and Technology*, 2007, 248–66, Cooper, J.F.)

obtained in Cooper's design [59], and a stable voltage was observed under a 27 mA cm^{-2} current load for 30 h. The main problem with this design is that the porous separator must be thick enough to avoid the short circuit of the DCFC.

The composition of the carbonate electrolyte is an important factor. Table 5.2, which provides the melting points for various carbonate electrolyte compositions, shows that carbonate mixtures allow a decrease in the operating temperature compared to single-carbonate compositions. Hence, binary or ternary carbonate mixtures appear to be suitable options. Among the potential compositions, 62% Li_2CO_3–38% K_2CO_3 and 38% Li_2CO_3–62% K_2CO_3 are the common electrolytes for DCFCs as their melting temperature is lower

TABLE 5.2

Melting Temperatures of Carbonate Eutectic Mixtures

Salt	Melting Point(°C)
Li_2CO_3	723
Na_2CO_3	851
K_2CO_3	891
$52Li_2CO_3$–$48Na_2CO_3$	500
$59Na_2CO_3$–$41K_2CO_3$	710
$62Li_2CO_3$–$38K_2CO_3$	488
$38Li_2CO_3$–$62K_2CO_3$	510
$43.5Li_2CO_3$–$31.50Na_2CO_3$–$25K_2CO_3$	390

than 550°C. Peelen et al. [56] and Cooper et al. [57,58] applied either a binary carbonate 62/38 mol% lithium/potassium mixture or a 32/68 mol% lithium/potassium mixture. Vutetakis et al. employed a ternary carbonate of $32.1Li_2CO_3/34.5K_2CO_3/33.4Na_2CO_3$ for coal electrochemical oxidation [55] and could run the cell down to 500°C.

The possible reactions through which carbon is consumed by the carbonates can be described as in Equation 5.1. Also CO can be produced as follows:

$$C + CO_3^{2-} \rightarrow CO + CO_2 \rightarrow 2e^- \tag{5.15}$$

$$2C + CO_3^{2-} \rightarrow 3CO + 2e^- \tag{5.16}$$

Carbonate ions are regenerated by the combination of CO_2 with oxygen as shown in Equation 5.2.

At 600°C–700°C, the dominating product from the electrochemical reaction is CO_2 with a yield of 0.75. This indicates that the Boudouard reaction is not the governing reaction for this temperature range. The yield of CO_2 and CO increases at 800°C, which shows that the Boudouard reaction and some other reactions, Equations 5.15 and 5.16, may be responsible for the CO. Lee et al. [61] proposed that carbon oxidation did not result from direct oxidation (Figure 5.5, as shown in Equation 5.1). Instead, carbon was oxidized by carbonate into carbon monoxide as shown in Equations 5.15 and 5.16 and into carbon dioxide according to Equation 5.17. This result was confirmed by the step chronopotentiometric studies at 850°C (Figure 5.6).

$$CO + CO_3^{2-} \rightarrow 2CO_2 + 2e^- \tag{5.17}$$

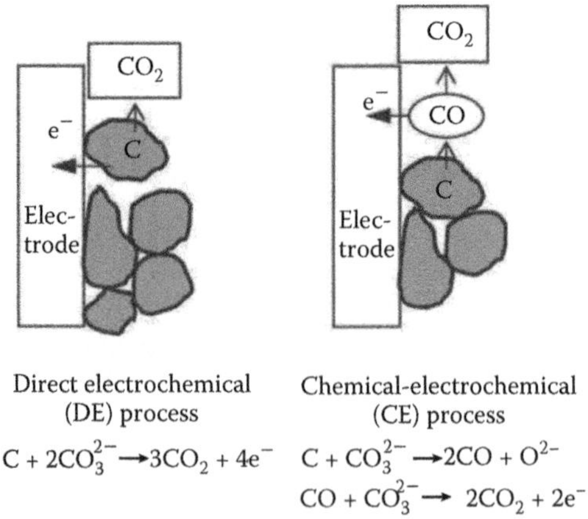

Direct electrochemical
(DE) process
$$C + 2CO_3^{2-} \rightarrow 3CO_2 + 4e^-$$

Chemical-electrochemical
(CE) process
$$C + CO_3^{2-} \rightarrow 2CO + O^{2-}$$
$$CO + CO_3^{2-} \rightarrow 2CO_2 + 2e^-$$

FIGURE 5.5
Carbon oxidation mechanism in molten carbonate DCFC, proposed by Lee et al. (Reproduced from Lee, C.G. et al. *J Electrochem Soc* 2011; 158: B410–B5. Copyright 2011 The Electrochemical Society. With permission.)

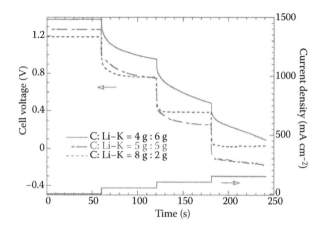

FIGURE 5.6
Step chronopotentiometric results with various carbon fuels (C:Li–K = 4:6 g, 5:5 g, and 8:2 g) (Li$_2$CO$_3$:K$_2$CO$_3$ = 62 mol:38 mol) at 850°C in a molten carbonate-DCFC. (Reproduced from Lee, C.G. et al. *J Electrochem Soc* 2011; 158: B410–B5. Copyright 2011 The Electrochemical Society. With permission.)

5.3.2 Solid Oxide DCFCs

Although the operation of DCFCs is possible at low temperatures and has advantages in terms of materials and costs, a higher operating temperature would yield a higher charge transfer and mass transport rates. Higher temperatures can be problematic for molten electrolytes due to salt decomposition and corrosion occurring over prolonged operation. Hence, solid oxygen ion conductors, commonly used in SOFCs, have been proposed for direct carbon conversion with a view to achieving enhanced reaction rates with a higher operating temperature. The first attempt to directly convert solid carbon at high temperatures using a solid electrolyte was reported by Baur and Preis [62]. The system was run at 1100°C with carbon powder and a carbon rod as the anode and anode current collector, respectively. In Baur and Preis's design, oxygen molecules are reduced into oxide ions by the magnetite oxygen electrode and transported through the solid electrolyte to react with solid carbon to form carbon dioxide. Other researchers [63–65] have investigated carbon direct oxidation in solid oxide electrolyte systems. In this type of design, SOFC technology was adopted for carbon oxidation, and carbon is oxidized by oxygen ions from the cathode at relatively high temperatures.

The requirements for a solid electrolyte are to be mechanically and chemically stable at high temperatures, high anionic conductivity, and low electronic conductivity at the operating temperature under reducing atmosphere and oxidizing atmosphere, as well as low ohmic resistance. Electrolytes that can be used include a mixture of CeO$_2$, WO$_3$, and clay; however, only modest current densities were achieved, and the cell has a very short life. YSZ (e.g., $Y_{0.15}ZrO_{0.85}O_{1.92}$) and CGO (e.g., $Ce_{0.8}Gd_{0.2}O_{1.9}$) are the conventional electrolytes for the SOFC and have been adopted for the DCFC. The oxygen molecules are oxidized into oxide ions and transferred into the anode to the reaction with carbon. Nickel and platinum are the common anodes for direct carbon oxidation. Carbides have been investigated as well. It was found that vanadium carbide is a good catalyst and shows stability in carbon compared with other carbides, such as ZrC, WC, and TiC [65,66].

The main drawback with solid electrolytes is the poor contact area between the electrolyte and the solid fuel, leading to slow fuel oxidation rates. Some direct methods have been used to increase the contact area by pressing the porous carbon pellets against the YSZ electrolyte [38]. The carbon used in this mode should have good conductivity, and therefore, carbon black and graphite were suitable for this system. Other indirect methods were applied to increase the chance of carbon contact with electrode. Two physical approaches, including fluidized bed and *in situ–external* gasification, were applied. Some other methods, such as the assistance by the mediator of molten metal, will be discussed in this section. A HDCFC merging molten carbonate into solid oxide electrolyte, which is supposed to have all the advantages for these two types of DCFCs, will be discussed in Section 5.4.

5.3.2.1 Physical Contact

The placement of solid carbon on the anode and the method for direct oxidation in a physical contact is straightforward (see Figure 5.7b). In a contact scheme, the electrochemical oxidation of solid carbon is expected to be

$$C + O_O^X \rightarrow CO + V_O^{\cdot\cdot} + 2e^- \qquad (5.18)$$

Because the contact area between the carbon and the electrode is limited, solid carbon is less able to extract oxygen from TPB for direct oxidation. The anodic reaction would largely be similar to that in the non-contact scheme (see Figure 5.7a). In fact, there is pratical difficulty to using a non-contact scheme for carbon oxidation. Here, Figure 5.7a is only used for comparison. The cell performance would depend on anodic oxygen activity, which is in equilibrium with the oxidation of carbon to CO and subsequent CO electrochemical oxidation (Equation 5.19).

$$CO + O_O^X \rightarrow CO_2 + V_O^{\cdot\cdot} + 2e^- \qquad (5.19)$$

The effect of anodic oxygen activity and its associated oxidation reactions on the cell performance is obvious when the carbon of low activity is used as fuel [67]. For example, nickel (Ni), the current state-of-the-art anode in SOFCs,

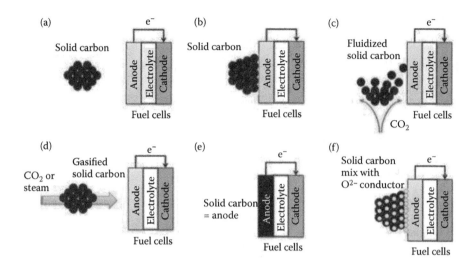

FIGURE 5.7
Different contact modes of solid carbon in DC-SOFCs.

tends to be oxidized to nickel oxide (NiO) (Equation 5.20) in the atmosphere of high oxygen activities due to sluggish oxidation reactions.

$$Ni + O_O^X \rightarrow NiO + V_O^{\cdot\cdot} + 2e^- \tag{5.20}$$

The oxidation of nickel causes volume expansion, which is responsible for cell degradation and failure of operation. On the other hand, nickel oxide can be reduced back to nickel by reducing gases, such as CO produced in the carbon-containing environment.

$$CO + NiO \rightarrow CO_2 + Ni \tag{5.21}$$

When the reducing gas is sufficient to reduce nickel oxide fast, the governing reaction in the DC-SOFC would be the electrochemical oxidation of CO (Note: Equation 5.19 is the sum of Equations 5.20 and 5.21).

This is similar to that in H_2-fueled SOFCs, in which the kinetics of NiO reduction are fast, and H_2 oxidation is the governing reaction. In contrast, the Ni-NiO equilibrium (Equation 5.20) would determine the performance of DC-SOFCs once the reduction of nickel oxide is limited. The OCV of nickel-anode based DCFCs, therefore, would follow the Ni-NiO equilibrium potential, ~0.7 (volt) [68], which agrees well with most of the reported values in the literature as shown in Table 5.3.

TABLE 5.3

Open Circuit Voltage Values Reported in Literature

Fuel Supply	Fuel	Anode/Electrolyte/Cathode	OCV (V)	T (°C)	Reference
Non-contact	Graphitic carbon	Pt/YSZ/Pt	0.9–1.1	725–955	[64]
Physical contact	Amorphous and graphitic carbon	Carbon/YSZ/LSM	0.7–0.8	800–1000	[38]
Physical contact	Coconut carbon	NiO/YSZ/LSM	0.7–0.9	800	[69,70]
Physical contact	Carbon black	NiO/YSZ/LSM	0.4–0.8 (0.71)	550–900 (800)	[35,37]
Physical contact	Carbon black, coal	Pt/YSZ or ScSZ or GDC/pt	1.0	600–800	[71]
Physical contact	Carbon black	NiO-YSZ/ScSZ/LSM- ScSZ	0.7–0.8	800	[72]
Fluidization	Synthetic carbon	Ni-Ce/YSZ/LSM	0.6–0.7	900	[16]
Paste	Carbon black, graphite	LSCF/CGO/LSCF-Ag	0.7–0.8	700	[73]
Hydrocarbon deposition	Pyrolyzed carbon	NiO-GDC/YSZ/LSM	0.8–1.0	900	[45,48]

5.3.2.2 Fluidized Bed

The fluidized bed concept has been applied to the DCFC to reduce mass transportation limitations and improve the contact between the solid fuel and the solid electrolyte (schematic is shown in Figure 5.7c). The FB-SOFC fluidizes solid carbon in the anode chamber with the advantages of easy fueling and separation of ash left from carbon fuels. In addition, the integrated process provides the advantage that waste heat from SOFCs can be recycled directly for a endothermic gasification reaction, thus improving the overall efficiency.

In a typical fluidized bed system developed by Gur et al. (see Figure 5.8) [16,49], the fuel is kept suspended through the action of a bubbling gas. The turbulence induced by the bubbling gas increases the contact opportunities between the solid fuel and the solid electrolyte. As CO_2 is generally used, the reaction with solid carbon can produce carbon monoxide, which will participate in electrochemical reactions to generate electricity. Moreover, the turbulent nature of the bed increases the heat transfer rate allowing for the fast reaction rate. It was found that the highest power density was 0.45 W cm^{-2} at 0.64 V and an electrical conversion efficiency of 35.7% based on CO utilization [19].

5.3.2.3 In Situ–External Gasification

Although the performance of a DC-SOFC in the contact scheme is subject to Ni-NiO equilibrium, a higher OCV can be obtained in a concentrated fuel

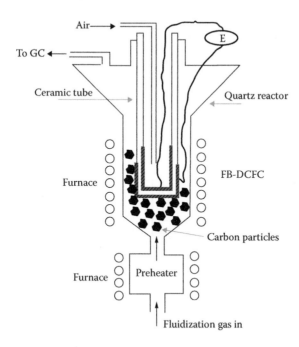

FIGURE 5.8
Fluidized bed DCFC developed by Gur et al. [16]. (Reproduced with permission from Li, S. et al. *Solid State Ionics, 79,* 1549–52, 2008.)

environment, i.e., CO. Such an atmosphere could easily form by gasification and localize at the anode surface enclosed by carbonaceous fuels. The external gasification device is shown in Figure 5.9. The external or *in situ* gasification (Figure 5.7d) occurs via reverse Boudouard reaction:

$$CO_2 + C \rightarrow 2CO \qquad (5.22)$$

The kinetics of the reverse Boudouard reaction is fast with a conversion to CO up to 90% at 900°C. The produced CO then diffuses to TPB for electrochemical oxidation. Thus, the CO and CO_2 proceed in a reaction cycle close to the anode, as illustrated in Figure 5.10, thereby delivering electric power.

The effect of the reverse Boudouard reaction and its contribution to the cell performance has been investigated at different carrier gas flow rates by transient techniques [69,70]. The results indicated that low carrier gas flow rates enhanced the cell performance by increasing the extent of the reverse Boudouard reaction and electrochemical oxidation of CO due to a long residence time. At lower flow rates (i.e., 50 ml/min^{-1}) and long residence time, the OCVs increased with higher CO concentrations as seen in Figure 5.11; at a high flow rate (i.e., 1000 ml/min^{-1}), the OCV is 0.71 V corresponding to Ni-NiO equilibrium.

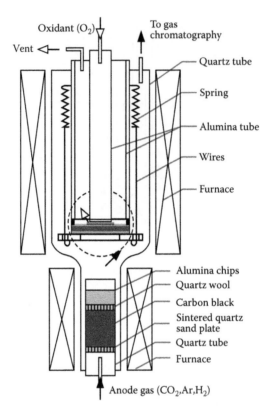

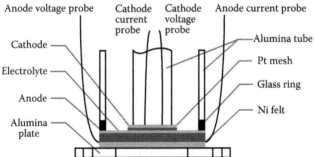

FIGURE 5.9
DCFC with the gasification of carbon. (Reproduced with permission from Li, C. et al. *J Power Sources*, 195, 4660–6, 2010.)

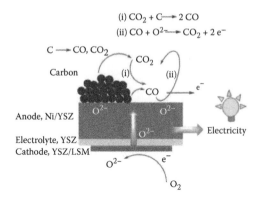

FIGURE 5.10
Reaction mechanisms on the anode of DC-SOFCs [70]. (Reproduced with permission from Chien, A.C. et al. *ECS Trans,* 33, 75–85, Copyright 2011 The Electrochemical Society.)

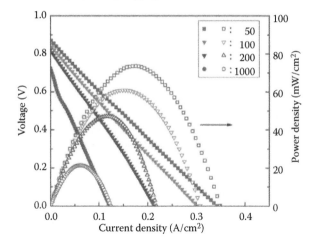

FIGURE 5.11
Effect of carrier gas flow rate on the performance of a DC-SOFC. (Reproduced with permission from Chien, A.C., and Chuang, S.S.C., *J Power Sources,* 196, 4719–23, 2011.)

5.3.2.4 Solid Carbon Anode

As discussed previously, the performance of a solid-electrolyte based DCFC is driven mainly by the electrochemical oxidation of CO produced from the gasification reaction at the anode. Recently, direct oxidation of carbon in contact mode (Figure 5.7e) was reported by pressing a non-porous carbon anode on an oxygen anion conductor (Figure 5.12) [38,74]. The direct oxidation was inferred from no noticeable electrical contribution of gasification products (i.e., CO) in different carrier gas flow rates. Although this result is interesting, how the oxidation occurs via a physical contact was not discussed. In addition, the reported OCV is lower than the value of carbon oxidation, making the direct oxidation more questionable.

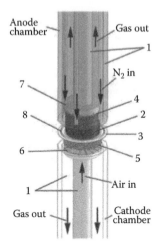

FIGURE 5.12
Schematic of the cell (1) alumina tubes (2) carbon pellet (3) YSZ electrolyte (4) current collector mesh (Ni) (5) current collector mesh (Pt) (6) cathode flow field (7) anode flow field (8) gold seal. (Reproduced with permission from Nurnberger, S. et al. *Energy Environ Sci*, 3, 150–3, 2010.)

Hence, we speculate that the direct oxidation did not occur according to the low OCV and propose a more plausible mechanism, as pictured in Figure 5.13. The reaction mechanism resembles that in non-contact mode. Solid carbon is oxidized by gaseous oxygen and consumed without direct electrochemical contribution. The generation of electricity occurs when oxygen is liberated from the ionic conductor and releases electrons.

$$O_O^X \rightarrow \frac{1}{2}O_2 + V_O^{\bullet\bullet} + 2e^- \tag{5.23}$$

The released electrons are then conducted by the carbon anode, which is in intimate contact with the ionic conductor to the current collector and external circuit. The liberated oxygen would subsequently oxidize the adjacent carbon to CO or CO_2.

$$C + O_2 \rightarrow CO \text{ or } CO_2 \tag{5.24}$$

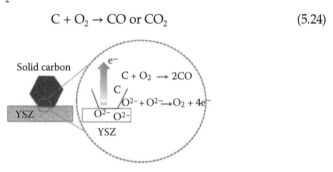

FIGURE 5.13
A proposed mechanism for a DCFC with solid carbon anode.

Because there are limited active sites (no TPB) using a non-porous carbon anode, the electric contribution from gaseous species, such as CO is low. On the other hand, the liberated oxygen is quickly depleted for carbon oxidation because of the short diffusion distance and fast oxidation kinetics. These explain little effect on the electric performance by varying carrier gas flow rate. Consequently, the electric performance depends mainly on the carbon anode, which functions both as an electronic conductor and an oxygen scavenger. However, these two functions are not equal. The solid carbon with higher oxidation activity, i.e., activated carbon, gave better performance than the one with a higher conductivity, i.e., graphite. Thus, the activity of a carbon fuel is a major factor in the cell performance because a lower oxygen activity can be kept at the anode for continuous oxygen diffusion and electric power generation.

5.3.2.5 Molten Metal

A molten metal has initially been used to promote carbon gasification rather than carbon direct oxidation. The earliest theoretical investigation on utilizing molten metal electrodes for coal gasification was done by Yentekakis et al. in 1989 [75]. The requirement of the molten metal is that the molten metal is a good electronic conductor, which is either stable in oxygen environments or, if it is oxidized, the resulting oxide is a good ion conductor for oxygen ion transportation from the electrolyte to the molten anode layer. Another important parameter is the melting temperature, which is shown in Table 5.4. The melting points of the metals are in the temperature range of 200°C–1000°C. It is favorable to use those metals having low melting points, such as indium, tin, or bismuth (Figure 5.14) [51,76]. Bismuth was used in DCFCs due to its low melting point, and also its oxides have good ionic conductivity (3 S/cm at 825°C) [77]. This allows oxygen ion transfer from the electrolyte to bismuth metal even if bismuth oxide is formed during the electrochemical reaction process. The oxygen diffusion coefficient is also important when molten

TABLE 5.4

Melting and Boiling Points of Metals

Metal	Melting Point (°C)	Boiling Point (°C)	Melting Point of the Oxide (°C)
In	137	2072	1910
Sn	232	2602	1630
Bi	271	1564	817
Pb	327	1749	888
Sb	631	1587	656
Ag	962	2162	–
Cu	1085	2562	1201
Fe	1538	2862	1566

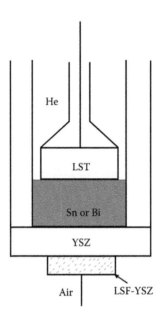

FIGURE 5.14
The DCFC with Sn or Bi molten metal as an anode. (Reproduced with permission from Jayakumar, A. et al. *J Electrochem Soc*, 157, B365–B9, Copyright 2010 The Electrochemical Society.)

metal is chosen for DCFCs. Metals, such as bismuth, silver, and lead tha have a good oxygen diffusion coefficient, which are 5.3×10^{-5} cm^2/s, 1.5×10^{-5} cm^2/s [78] and 1.29×10^{-5} cm^2/s [79], have been used in DCFC.

Recently, a molten Sb anode was reported as the anode for direct utilization of carbon fuels at 700°C [80]. Although the open circuit voltage value of 0.75 V is ot so high as the value in other designed systems, the power density of this DCFC is promising, which is more than 300 mW cm^{-2} at 0.5 V potential. The reactions for molten metal DCFC are Equations 5.25 and 5.26.

$$M + nO^{2-} \rightarrow MO_n + 2ne^- \tag{5.25}$$

$$(n/2)C(M) + MO_n \rightarrow (n/2)CO_2 + M \tag{5.26}$$

Taking Sn as an example, the reactions for carbon oxidation are as follows:

$$Sn(l) + O_2(g) \rightarrow SnO_2 \tag{5.27}$$

$$SnO_2 + 2C \rightarrow Sn(l) + 2CO(g) \tag{5.28}$$

$$SnO_2 + C \rightarrow Sn(l) + CO_2(g) \tag{5.29}$$

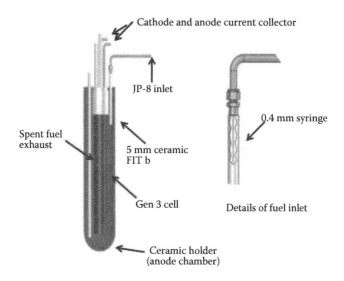

Cathode and anode current collector

JP-8 inlet

0.4 mm syringe

Spent fuel exhaust

5 mm ceramic FIT b

Gen 3 cell

Details of fuel inlet

Ceramic holder
(anode chamber)

FIGURE 5.15
A single cell set up for direct fuel conversion in the liquid tin Anode (LTA)-SOFC. (Reproduced with permission from Tao, T. et al. *Ecs Trans*, 5, 463–72, Copyright 2007 The Electrochemical Society.)

For the molten tin anode, the electromotive force is generated as a result of the metal oxidation to tin oxide. Typically the tin oxide is reduced back to metal by supplied carbonaceous fuel. Under this system, a low open circuit voltage is generally obtained: 0.7 V at 600°C.

The electrochemical potential of the metal oxidation and capability of the melt to be oxidized into corresponding oxides have major impacts on the cell performance through its contribution to polarization resistance [81]. Figure 5.15 shows one design with molten tin oxide for JP-8 fuel oxidation. The benefit of the molten tin oxide was demonstrated by using a high sulfur concentration, 1350 ppm, and no significant cell performance degradation was observed after 200 h operation; sulfur was oxidized into SO_2 and escaped as a waste gas [82].

5.3.2.6 Deposited Carbon

The pyrolysis of hydrocarbon seems to be effective in depositing carbon near TPB, thus making direct carbon oxidation feasible. The popular research was done by Ishara [45–48,83–86]. In 2004, Ihara developed a DCFC based on pyrolytic methane or propane on the nickel anode using various electrolytes, such as YSZ, gadolinia doped-ceria, or scandia-stabilized zirconia [45,46,83,85,86]. This concept has been proved by Kellogg [87]. Recently, the maximum power density of 55 mW cm^{-2} has been obtained at the charging time for 6 h. In order to obtain higher power density, parameters, such as the argon gas flow rate or the anode thickness, composition, porosity, etc., that could affect the cell power output need to be studied [46,87]. The issue in this system is that this type of DCFC only operates in a battery mode and needs recharge, the so-called rechargeable DCFC (RDCFCs). Due to limited carbon available in the matrix of the anode, the RDCFCs may find application in restricted areas, but not in medium- to large-scale power generation.

5.3.2.7 Other Contact

Several other types of contact between solid carbon and anodes were reported for DC-SOFC, such as the contact by pasting [73] or by mixing solid carbon and O^{2-} conductive oxides [71]. Figure 5.7f depicts the contact type of mixing solid carbon and O^{2-} conductive oxides. The concept of this design is to extend the contact zone between carbon and oxygen anions that can be additionally supplied from dispersed nanoparticles of oxide electrolytes. The result showed that the OCV at 880°C reached the theoretical value of carbon oxidation when a mixture of carbon and 8YSZ electrolyte powders was used. Nevertheless, the OCVs were still less than expected below 880°C with 8YSZ or other electrolytes of GDC and ScSZ. The mechanism of direct oxidation and long-term performance was also missed.

5.4 Hybrid Direct Carbon Fuel Cells

A HDCFC is to merge SOFC and MCFC technologies, using a solid oxide electrolyte to separate the cathode and anode compartments while molten carbonate electrolyte is utilized to extend the anode-electrolyte region (Figure 5.2). Oxygen is reduced to O^{2-} ions at the cathode and transported across the solid electrolyte membrane to the anode compartment, where carbon is oxidized to CO_2.

Molten carbonate could enhance the carbon oxidation in two ways: as a fuel carrier or as an electrochemical mediator. The ideal anode reaction is the direct electrochemical oxidation of carbon to CO_2 (Equation 5.3).

However, the actual anode reactions are more complicated. A partial oxidation of carbon to CO is also possible (Equations 5.15 and 5.16). These two reactions require solid–solid interaction as O^{2-} ions are supplied from solid electrolyte; however, the fluidity of the molten carbonate would enhance the transport of the carbon fuel to the anode. Enhancement of the anode reaction by the molten carbonate as an electrochemical mediator is also expected as shown in Equations 5.1, 5.15, and 5.16. These reactions should be followed by the regeneration of CO_3^{2-} ions shown in Equation 5.2 in order that the electric charge of the molten carbonate can be kept neutral.

The carbon can be converted via a non-electrochemical reaction known as the reverse Boudouard reaction (Equation 5.22). Although considerable amounts of CO can be produced via Equations 5.15 and 5.16, the chemical energy of CO can be converted to electric power via the electrochemical oxidation of CO at the anode in the DCFC (Equation 5.19).

The maximum energy density can be achieved by fully oxidizing carbon to CO_2. This concept avoids the need for CO_2 circulation and the protection of the cathode from molten carbonate and allows the use of materials already developed for SOFC applications. In ongoing studies at St Andrews, this concept has been demonstrated using a wide range of carbons and carbon-rich fuels, such as carbon black, and pyrolyzed medium density fiberboard (PMDF) [30,33,35], in tubular cells. The electrochemistry of the direct oxidation of solid carbon in the HDCFC with solid oxide and molten carbonate binary electrolytes was also recently investigated using a planar test cell [30,35,37]. The carbon-carbonate slurry increased the active reaction zone from a two-dimensional Ni/YSZ anode to a three-dimensional slurry and significantly enhanced carbon oxidation [35]. The electrochemical oxidation in the anode compartment is quite selective to CO_2 formation, but the final distribution of products is dominated by the equilibrium of the Boudouard reaction, which increasingly yields CO as temperature increases. Gur circumvented the difficulty of attaining solid fuel-to-anode contact by fluidizing the carbon in flowing CO_2, which generates CO *in situ* via the Boudouard reaction [49].

In this design, oxygen ions are produced at an SOFC cathode, such as lanthanum-doped strontium manganite $((La_{0.8}Sr_{0.2})_{0.95}MnO_{3-\delta}$, LSM), and transported by the YSZ electrolyte to the anode chamber, where a lithium-potassium carbonate having a melting temperature of 500°C, which is the lowest in the binary carbonate systems, is used as the second electrolyte.

Intensive investigations have been conducted to understand the fundamentals of the HDCFC concept, and optimize their performance. A typical single-cell test unit is represented in Figure 5.16. The advantage of this cell design is that carbon is oxidized in the alumina chamber, which is sealed using the SOFC. The atmosphere of the anode chamber can be controlled by purging with various gases. Inert gas, such as argon or nitrogen, not only prevents the direct oxidation of carbon, but also purges the reaction products from the cell.

Many parameters influence cell performance, such as the carbonate content, the solid oxide cell design, electrode or electrolyte materials, the nature of the purge gas, or the flow rates of gases employed at the anode and the cathode during operation. This section hereafter reviews the influence of these parameters on the HDCFC performance.

5.4.1 Influence of Carbonate Content

Investigation of the influence of the carbonate content on cell performances has been performed using cells featuring a 1 mm thick YSZ electrolyte, and

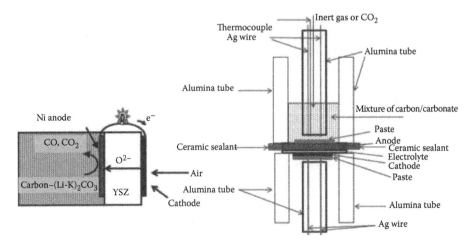

FIGURE 5.16
Schematic of the HDCFC, above left, and the used test geometry, right. The HDCFC is the combination of a SOFC and MCFC; the SOFC is composed of a nickel-based cermet, a YSZ electrolyte, and a $(La_{0.8}Sr_{0.2})_{0.95}MnO_{3-\delta}$/YSZ composite cathode or a $La_{0.6}Sr_{0.4}CoO_{3-\delta}$ (LSC) cathode; a mixture of carbon and lithium-potassium carbonate (62 mol% Li_2CO_3: 38 mol% K_2CO_3) in a weight ratio of $C:CO_3^{2-} = 4:1$ is placed in the anode chamber, which is sealed to the SOFC; the reaction at the cathode is $O_2 + 4e^- = 2O^{2-}$, and the total reaction of the anode is that carbon is oxidized to CO_2. (Reproduced with permission from Jiang, C. et al. *Energy Enviro Sci*, 5, 6973–80, 2012.)

the testing device is shown in Figure 5.16. Electrolytes with 20 mm diameters were formed by dry pressing, before the anode, a composite of 60 wt% nickel oxide and 40 wt% YSZ, and the cathode, a 50 wt% composite of lanthanum-doped strontium manganite $((La_{0.8}Sr_{0.2})_{0.95}MnO_{3-\delta}$, LSM) and YSZ were screen printed on each side, hence forming a conventional solid oxide cell. To perform the electrochemical investigations, a mixture of 62 mol% lithium carbonate and 38 mol% potassium carbonate was used with carbon black (XC-72R), or PMDF was used as a fuel.

The impedance spectra recorded for various carbon-to-carbonate ratios are shown in Figure 5.17. The results indicate that among the various carbon-to-carbonate ratios, the best performance is obtained at 25 wt% carbonate content (ratio of carbon to carbonate is 4:1 by weight). The cell resistance decreases from 30 Ω cm^2 without the carbonate addition to 6.35 Ω cm^2 when 25 wt% carbonate is added into carbon. The carbonate addition decreases the cell ohmic resistance, indicating that carbonates improve the connectivity in the anode chamber and possess a good conductivity. When the carbonate content is increased above 25 wt%, the overall cell performance degrades with the carbonate content. However if more carbonate is added, the polarization resistance of the hybrid direct carbon is much bigger, possibly due to a slower electrochemical reaction rate because of the lower carbon fuel content. Based on this optimal result on carbon black fuel, PMDF fuel was

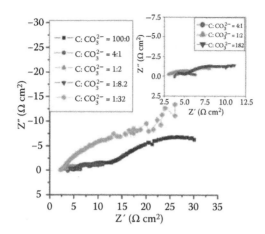

FIGURE 5.17
AC impedance spectra of the HDCFCs operating at 750°C in 20 ml/min of N₂ flow using carbon
black (Vulcan XC72R) fuel with various carbon to carbonate (62 mol% Li₂CO₃: 38 mol% K₂CO₃)
ratios. (Reproduced with permission from Jiang, C.R., and Irvine, J.T.S., *J Power Sources*, 196,
7318–22, 2011.)

added with 25 wt% carbonate, and it showed the electrochemical reactivity
was improved (Figure 5.18) as well even if it is not that significant as it shows
when carbon black added with the same amount of carbonate. As carbon
black has fine particle size, it is likely to need more carbonate to wet the
surface.

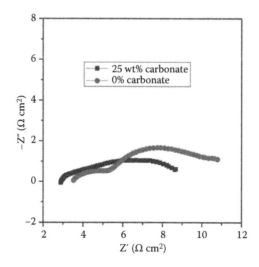

FIGURE 5.18
Effect of carbonate content (62 mol% Li₂CO₃: 38 mol% K₂CO₃) on AC impedance spectra of the
HDCFC using PMDF as the fuel operating at 750°C under 20 ml/min N₂ of flow, tested under
open circuit conditions.

5.4.2 Effect of SOFC Geometry

To optimize cell performance, it is important to reduce the thickness of the solid electrolyte. The contribution from the solid electrolyte to the cell ohmic resistance can indeed limit the cell performance at modest temperatures. Results in Figure 5.19 indicate that with a 1 mm thick electrolyte, the ohmic resistance of 2.89 Ω cm^2 accounts for 33% of the total cell resistance, which is 8.69 Ω cm^2. To decrease the cell's ohmic resistance and improve the overall cell performance, the 1 mm thick YSZ supported design was replaced by an anode-supported design, and the same test using PMDF as a fuel was repeated. Figure 5.19 shows the schematic of both designs with the associated impedance spectra for both designs. By using a thin electrolyte in the anode-supported design, the cell total resistance decreases to 0.95 Ω cm^2 while the cell ohmic resistance decreases to 0.14 Ω cm^2.

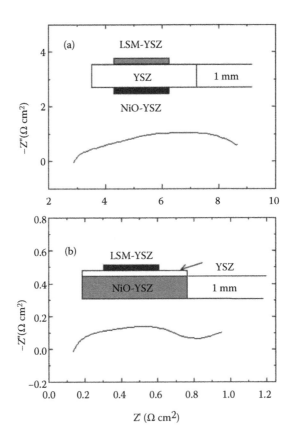

FIGURE 5.19

AC-impedance spectra of the HDCFC operating on two types of SOFC geometries (a) thick electrolyte-supported SOFC (b) anode-supported SOFC. Note: temperature: 750°C, open circuit condition, N$_2$ flow rate: 20 ml/min, PMDF/carbonate: 4:1 by weight.

5.4.3 Optimization of the HDCFC

Initial studies on the HDCFC have been conducted using standard LSM cathodes over the temperature range of 600°C–800°C. However, other cathode materials may be more suitable for this system. Further testing has been conducted using lanthanum-doped strontium cobalt (LSC, $La_{0.6}Sr_{0.4}CoO_{3-\delta}$), this is expected to provide enhanced catalytic activity. Figure 5.20 compares the impedance spectra recorded at OCV for cells with either a (a) LSM or a (b) LSC cathode [36]. The spectra indicate that the cell performance is further improved by the replacement of LSM by LSC, resulting in total cell resistance being reduced to ca. 0.44 Ω cm^2. Two arcs can be seen on the impedance spectra. The high-frequency arc is likely to be related to the charge transfer process, thus showing that the charge transfer process is faster for the LSC cathode.

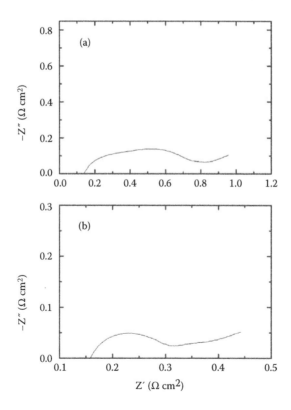

FIGURE 5.20
AC-impedance spectra of the HDCFC operating at 750°C in 20 ml/min of N_2 flow using PMDF as a fuel with (a) $(La_{0.8}Sr_{0.2})_{0.95}MnO_{3-\delta}$ (LSM) cathode; (b) $La_{0.6}Sr_{0.4}CoO_{3-\delta}$ (LSC) cathode, ratio of PMDF to carbonate is 4:1 by weight.

5.5 Summary and Prospects

In summary, present technology and status in DCFCs are reviewed. Followed by a brief introduction on molten salt DCFCs, the research activities on solid oxide DCFCs are discussed. Many efforts are made to increase the contact chance between the solid fuel and the solid electrolyte, for example, fluidized bed, *in situ*–external gasification, or utility of molten metal. A HDCFC, which is the combination of molten salts DCFCs and solid oxide DCFCs, increases the solid–solid contact chance. The cell performance is discussed based on carbonate content and solid oxide cell geometry and the selection of the cathode. It shows promising performance in this system, but there are some issues that should be further studied, such as the oxidation mechanism in this complex system, the dominant process for carbon oxidation, and the

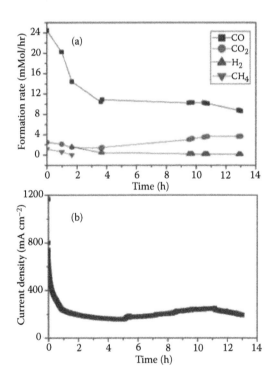

FIGURE 5.21
Stability test of the HDCFC under 0.7 V potential tested at 750°C under ~7 ml/min of N_2 flow. The cell is supplied with a 2 g mixture of PMDF and Li_2CO_3-K_2CO_3 (62:38 mol) (PMDF:carbonate = 4:1 by weight); the cell is composed of NiO-YSZ support/5 μm YSZ thin film/CGO interlayer/LSC cathode. (a) Gas production evolution with time, measured by online gas chromatography; and (b) current evolution with time. (Reproduced with permission from Jiang, C. et al. *Energy Environ Sci*, 5, 6973–80, 2012.)

electrode and electrolyte materials choice to reduce the corrosion due to the carbonate to allow long-term operation.

Although there are so many attractive features to develop DCFCs, some issues need to be solved before this technology can be developed on an industrial scale. One of them is the long-term durability, especially when molten salts are used in the system. CGO is supposed to be stable under molten carbonate conditions, but no experimental results prove that this is a viable DCFC electrolyte due to possible electronic leakage. For solid oxide-DCFCs, in order to improve the cell efficiency, effective catalysts should be found. Except for nickel, other metal catalysts, palladium, iron metal, or even a ceramic catalyst could be worthy of study, which is especially useful for the conversion of high-sulfur coal. At present, prolonged running under current load is limited by the discontinuous feed of the fuel, for example, 2 g of PMDF was used in the investigated system, and the cell has run under 0.7 V voltage for only 13 h (Figure 5.21) [36]. Although the short-term performance for the DCFC is promising, a system with a continuous fuel feed is necessary to generate electricity for sufficient durability.

References

1. Malzbender, J., Wessel, E. and Steinbrech, R.W. 2005. Reduction and re-oxidation of anodes for solid oxide fuel cells. *Solid State Ionics* 176, 2201–2203.
2. EG&G Technical Services I. 2004. *Fuel Cell Handbook*, 7th edition. US Department of Energy, West Virginia.
3. Singhal, S.C. 2002. Solid oxide fuel cells for stationary, mobile, and military applications. *Solid State Ionics* 152, 405–410.
4. Jacques, W.W. 1896. An improved electric battery, and method or process of converting the potential energy of carbon or carbonaceous materials into electrical energy. US Patent 555,551.
5. Jacques, W.W. 1896–1897. Electricity direct from coal. *Harper's Magazine* 94, 144.
6. Zecevic, S., Patton, E.M. and Parhami, P. 2004. Carbon-air fuel cell without a reforming process. *Carbon* 42, 1983–1993.
7. Pointon, K., Lakeman, B., Irvine, J., Bradley, J. and Jain, S. 2006. The development of a carbon-air semi fuel cell. *J Power Sources* 162, 750–756.
8. Lipilin, A.S., Balachov, I.I., Dubois, L.H., Sanjurjo, A., Mckubre, M.C., Crouch-Baker, S., Hornbostel, M.D. and Tanzclla, F.L. 2006. Liquid anode electrochemical cell. US Patent 20060019132.
9. Lakeman, J.B., Pointon, K.D., Irvine, J.T.S. and Badley, J. 2004. The development of a carbon-air semi-fuel cell. *Distillation* 7, 401–407.
10. Tao, T.T., Bai, W., Rackey, S. and Wang, G. 2003. Electrode layer arrangements in an electrochemical device. PCT Patent Applicant No. WO 2003/001617 A2.
11. Gur, T.M. 2010. Mechanistic modes for solid carbon conversion in high temperature fuel cells. *J Electrochem Soc* 157, B751–B759.

12. Antal, M.J. and Nihous, G.C. 2008. Thermodynamics of an aqueous-alkaline/carbonate carbon fuel cell. *Ind Eng Chem Res* 47, 2442–2448.
13. Nunoura, T., Dowaki, K., Fushimi, C., Allen, S., Meszaros, E. and Antal, M.J. 2007. Performance of a first-generation, aqueous-alkaline biocarbon fuel cell. *Ind Eng Chem Res* 46, 734–744.
14. Hackett, G.A., Zondlo, J.W. and Svensson, R. 2007. Evaluation of carbon materials for use in a direct carbon fuel cell. *J Power Sources* 168, 111–118.
15. Cherepy, N.J., Krueger, R., Fiet, K.J., Jankowski, A.F. and Cooper, J.F. 2005. Direct conversion of carbon fuels in a molten carbonate fuel cell. *J Electrochem Soc* 152, A80–A87.
16. Li, S., Lee, A.C., Mitchell, R.E. and Gur, T.M. 2008. Direct carbon conversion in a helium fluidized bed fuel cell. *Solid State Ionics* 179, 1549–1552.
17. Lee, A.C., Mitchell, R.E. and Gur, T.M. 2009. Thermodynamic analysis of gasification-driven direct carbon fuel cells. *J Power Sources* 194, 774–785.
18. Lee, A.C., Mitchell, R.E. and Gur, T.M. 2009. Modeling of CO_2 gasification of carbon for integration with solid oxide fuel cells. *Aiche J* 55, 983–992.
19. Gur, T.M., Homel, M. and Virkar, A.V. 2010. High performance solid oxide fuel cell operating on dry gasified coal. *J Power Sources* 195, 1085–1090.
20. Homel, M., Gur, T.M., Koh, J.H. and Virkar, A.V. 2010. Carbon monoxide-fueled solid oxide fuel cell. *J Power Sources* 195, 6367–6372.
21. Li, C., Shi, Y.X. and Cai, N.S. 2010. Performance improvement of direct carbon fuel cell by introducing catalytic gasification process. *J Power Sources* 195, 4660–4666.
22. Zhao, X.Y., Yao, Q., Li, S.Q. and Cai, N.S. 2008. Studies on the carbon reactions in the anode of deposited carbon fuel cells. *J Power Sources* 185, 104–111.
23. Liu, Q.H., Tian, Y., Xia, C., Thompson, L.T., Liang, B. and Li, Y.D. 2008. Modeling and simulation of a single direct carbon fuel cell. *J Power Sources* 185, 1022–1029.
24. Xia, C., Li, L., Tian, Y., Liu, Q.H., Zhao, Y.C., Jia, L.J. and Li, Y.D. 2009. A high performance composite ionic conducting electrolyte for intermediate temperature fuel cell and evidence for ternary ionic conduction. *J Power Sources* 188, 156–162.
25. Huang, J.B., Yang, L.Z., Gao, R.F., Mao, Z.Q. and Wang, C. 2006. A high-performance ceramic fuel cell with samarium doped ceria-carbonate composite electrolyte at low temperatures. *Electrochem Commun* 8, 785–789.
26. Wu, Y.Z., Su, C., Zhang, C.M., Ran, R. and Shao, Z.P. 2009. A new carbon fuel cell with high power output by integrating with in situ catalytic reverse Boudouard reaction. *Electrochem Commun* 11, 1265–1268.
27. Liu, R.Z., Zhao, C.H., Li, J.L., Zeng, F.R., Wang, S.R., Wen, T.L. and Wen, Z.Y. 2010. A novel direct carbon fuel cell by approach of tubular solid oxide fuel cells. *J Power Sources* 195, 480–482.
28. Cao, D.X., Wang, G.L., Wang, C.Q., Wang, J. and Lu, T.H. 2010. Enhancement of electrooxidation activity of activated carbon for direct carbon fuel cell. *Int J Hydrogen Energy* 35, 1778–1782.
29. Chen, M.M., Wang, C.Y., Niu, X.M., Zhao, S., Tang, J. and Zhu, B. 2010. Carbon anode in direct carbon fuel cell. *Int J Hydrogen Energy* 35, 2732–2736.
30. Nabae, Y., Pointon, K.D. and Irvine, J.T.S. 2009. Ni/C slurries based on molten carbonates as a fuel for hybrid direct carbon fuel cells. *J Electrochem Soc* 156, B716–B720.
31. Jain, S.L., Lakeman, J.B., Pointon, K.D. and Irvine, J.T.S. 2007. A novel direct carbon fuel cell concept. *J Fuel Cell Sci Technol* 4, 280–282.

32. Jain, S.L., Lakeman, J.B., Pointon, K.D. and Irvine, J.T.S. 2007. Carbon content in a direct carbon fuel cell. *ECS Trans* 7, 829–836.

33. Jain, S.L., Lakeman, J.B., Pointon, K.D., Marshall, R. and Irvine, J.T.S. 2009. Electrochemical performance of a hybrid direct carbon fuel cell powered by pyrolysed MDF. *Energy Environ Sci* 2, 687–693.

34. Jain, S.L., Nabae, Y., Lakeman, B.J., Pointon, K.D. and Irvine, J.T.S. 2008. Solid state electrochemistry of direct carbon/air fuel cells. *Solid State Ionics* 179, 1417–1421.

35. Nabae, Y., Pointon, K.D. and Irvine, J.T.S. 2008. Electrochemical oxidation of solid carbon in hybrid DCFC with solid oxide and molten carbonate binary electrolyte. *Energy Environ Sci* 1, 148–155.

36. Jiang, C., Ma, J., Bonacorroso, A.D. and Irvine, J. 2012. Demonstration of high power, direct conversion of waste-derived carbon in a hybrid direct carbon fuel cell. *Energy Environ Sci* 5, 6973–6980.

37. Jiang, C.R. and Irvine, J.T.S. 2011. Catalysis and oxidation of carbon in a hybrid direct carbon fuel cell. *J Power Sources* 196, 7318–7322.

38. Nurnberger, S., Bussar, R., Desclaux, P., Franke, B., Rzepka, M. and Stimming, U. 2010. Direct carbon conversion in a SOFC-system with a non-porous anode. *Energy Environ Sci* 3, 150–153.

39. Agbede, O., Hellgardt, K. and Kelsall, G. 2012. Indirect carbon-air fuel cell for reforming carbonaceous fuels. http://conftikvivbe/iscre22/programme/10-ABSTRACT-552-INDIRECT%20CARBON-AIR%20FUEL%20CELL%20FOR%20REFORMING_OOAgbede_KHellgardt_GHKelsallpdf.

40. Li, X., Zhu, Z.H., De Marco, R., Dicks, A., Bradley, J., Liu, S.M. and Lu, G.Q. 2008. Factors that determine the performance of carbon fuels in the direct carbon fuel cell. *Ind Eng Chem Res* 47, 9670–9677.

41. Li, X., Zhu, Z.H., De Marco, R., Bradley, J. and Dicks, A. 2010. Evaluation of raw coals as fuels for direct carbon fuel cells. *J Power Sources* 195, 4051–4058.

42. Giddey, S., Badwal, S.P.S., Kulkarni, A. and Munnings, C. 2012. A comprehensive review of direct carbon fuel cell technology. *Prog Energy Combust Sci* 38, 360–399.

43. Rady, A.C., Giddey, S., Badwal, S.P.S., Ladewig, B.P. and Bhattacharya, S. 2012. Review of fuels for direct carbon fuel cells. *Energy Fuels* 26, 1471–1488.

44. Kulkarni, A., Ciacchi, F.T., Giddey, S., Munnings, C., Badwal, S.P.S., Kimpton, J.A. and Fini, D. 2012. Mixed ionic electronic conducting perovskite anode for direct carbon fuel cells. *Int J Hydrogen Energy* 37, 19092–19102.

45. Ihara, M. and Hasegawa, S. 2006. Quickly rechargeable direct carbon solid oxide fuel cell with propane for recharging. *J Electrochem Soc* 153, A1544–A1546.

46. Hasegawa, S. and Iharaz, M. 2008. Reaction mechanism of solid carbon fuel in rechargeable direct carbon SOFCs with methane for charging. *J Electrochem Soc* 155, B58–B63.

47. Ihara, M., Matsuda, K., Sato, H. and Yokoyama, C. 2004. Solid state fuel storage and utilization through reversible carbon deposition on an SOFC anode. *Solid State Ionics* 175, 51–54.

48. Saito, H., Hasegawa, S. and Ihara, M. 2008. Effective anode thickness in rechargeable direct carbon fuel cells using fuel charged by methane. *J Electrochem Soc* 155, B443–B447.

49. Lee, A.C., Li, S., Mitchell, R.E. and Gur, T.M. 2008. Conversion of solid carbonaceous fuels in a fluidized bed fuel cell. *Electrochem Solid St* 11, B20–B23.

50. Cooper, J.F. and Selman, R. 2009. Electrochemical oxidation of carbon for electric power generation: A review. *ECS Trans* 19, 15–25.

51. Jayakumar, A., Lee, S., Hornes, A., Vohs, J.M. and Gorte, R.J. 2010. A comparison of molten Sn and Bi for solid oxide fuel cell anodes. *J Electrochem Soc* 157, B365–B369.

52. Zecevic, S., Patton, E.M. and Parhami, P. 2003. Electrochemistry of direct carbon fuel cell based metal hydroxide electrolyte. Direct Carbon Fuel Cell Workshop, http://www.netl.doe.gov/publications/proceedings/03/dcfcw/Zecevic.pdf.

53. Zecevic, S., Patton, E.M. and Parhami, P. 2005. Direct electrochemical power generation from carbon in fuel cells with molten hydroxide electrolyte. *Chem Eng Commun* 192, 1655–1670.

54. Anbar, M., McMillen, D.F., Weaver, R.D., Jorgensen, P.J. and Weaver, A. 1976. Method and apparatus for electrochemical generation of power from carbonaceous fuels. US Patent No. 3970474.

55. Vutetakis, D.G., Skidmore, D.R. and Byker, H.J. 1987. Electrochemical oxidation of molten carbonate-coal slurries. *J Electrochem Soc* 134, 3027–3035.

56. Peelen, W.H.A., Olivry, M., Au, S.F., Fehribach, J.D. and Hemmes, K. 2000. Electrochemical oxidation of carbon in a 62/38 mol% Li/K carbonate melt. *J Appl Electrochem* 30, 1389–1395.

57. Cooper, J.L., Cherepy, N. and Krueger, R.L. 2002. Fuel cell apparatus and method thereof. US Patent Application No. 2002/0106549 A1.

58. Cooper, J.L., Cherepy, N. and Krueger, R.L. 2003. Tilted fuel cell apparatus. US Patent Application No. 2003/0017380 A1.

59. Cooper, J.F. 2003. Reactions of the carbon anode in molten carbonate electrolyte. Direct Carbon Fuel Cell Workshop, http://www.netl.doe.gov/publications/proceedings/03/dcfcw/Cooper%202.pdf.

60. Cooper, J.F. 2007. Direct conversion of coal derived carbon in fuel cells. In Basu S., ed. *Recent Trends in Fuel Cell Science and Technology*, Springer, New York, pp. 248–266.

61. Lee, C.G., Hur, H. and Song, M.B. 2011. Oxidation behavior of carbon in a coin-type direct carbon fuel cell. *J Electrochem Soc* 158, B410–B415.

62. Baur, E. and Preis, H. 1937. Uber brennstoff-ketten mit festleitern. *Zeitschrift für Elektrochemie und angewandte physikalische Chemie* 43, 727.

63. Nakagawa, N. and Ishida, M. 1988. Performance of an internal direct-oxidation carbon fuel-cell and its evaluation by graphic exergy analysis. *Ind Eng Chem Res* 27, 1181–1185.

64. Gur, T.M. and Huggins, R.A. 1992. Direct electrochemical conversion of carbon to electrical energy in a high-temperature fuel-cell. *J Electrochem Soc* 139, L95–L97.

65. Horita, T., Sakai, N., Kawada, T., Yokokawa, H. and Dokiya, M. 1995. An investigation of anodes for direct-oxidation of carbon in solid oxide fuel-cells. *J Electrochem Soc* 142, 2621–2624.

66. Horita, T., Sakai, N., Kawada, T., Yokoyama, H. and Dokiya, M. 2001. Solid oxide fuel cell and a carbon direct-oxidizing-type electrode for the fuel cell. US Patent 6183896 B1.

67. Chien, A.C. 2011. *Methane and Solid Carbon Based Solid Oxide Fuel Cells*. The University of Akron, Akron.

68. Horita, T.S., Natsuko Kawada, T., Yokokawa, H. and Dokiya, M. 1996. Oxidation and steam reforming of CH4 on Ni and Fe anodes under low humidity conditions in solid oxide fuel cells. *J Electrochem Soc* 143, 1161–1168.

69. Chien, A.C. and Chuang, S.S.C. 2011. Effect of gas flow rates and Boudouard reactions on the performance of Ni/YSZ anode supported solid oxide fuel cells with solid carbon fuels. *J Power Sources* 196, 4719–4723.

70. Chien, A.C., Siengchum, T. and Chuang, S. 2011. Investigation of Boudouard reactions on carbon-based solid oxide fuel cells by transient techniques. *ECS Trans* 33, 75–85.
71. Dudek, M. and Tomczyk, P. 2011. Composite fuel for direct carbon fuel cell. *Catal Today* 176, 388–392.
72. Li, C., Shi, Y.X. and Cai, N.S. 2011. Effect of contact type between anode and carbonaceous fuels on direct carbon fuel cell reaction characteristics. *J Power Sources* 196, 4588–4593.
73. Kulkarni, A., Giddey, S. and Badwal, S.P.S. 2011. Electrochemical performance of ceria-gadolinia electrolyte based direct carbon fuel cells. *Solid State Ionics* 194, 46–52.
74. Desclaux, P., Nurnberger, S., Rzepka, M. and Stimming, U. 2011. Investigation of direct carbon conversion at the surface of a YSZ electrolyte in a SOFC. *Int J Hydrogen Energy* 36, 10278–10281.
75. Yentekakis, I.V., Debenedetti, P.G. and Costa, B. 1989. A novel fused metal anode solid electrolyte fuel-cell for direct coal-gasification: A steady-state model. *Ind Eng Chem Res* 28, 1414–1424.
76. Gur, T.M. 2006. Direct carbon fuel cell with molten anode. US patent US20060234098 A1.
77. Yarlagadda, V.R. and Nguyen, T.V. 2011. Conductivity measurements of molten Bi_2O_3. *ECS Trans* 33, 119–125.
78. Sunde, S., Nisancioglu, K. and Gur, T.M. 1996. Critical analysis of potentiostatic step data for oxygen transport in electronically conducting perovskites. *J Electrochem Soc* 143, 3497–3504.
79. Bandyopadhyay, G. and Ray, H. 1971. Kinetics of oxygen dissolution in molten lead. *Metall Mater Trans B* 2, 3055–3061.
80. Jayakumar, A., Ungas, R.K., Roy, S., Javadekar, A., Buttrey, D.J., Vohs, J.M. and Gorte, R.J. 2011. A direct carbon fuel cell with a molten antimony anode. *Energy Environ Sci* 4, 4133–4137.
81. Jayakumar, A., Vohs, J.M. and Gorte, R.J. 2010. Molten-metal electrodes for solid oxide fuel cells. *Ind Eng Chem Res* 49, 10237–10241.
82. Tao, T., Bateman, L., Bentley, J. and Slaney, M. 2007. Liquid tin anode solid oxide fuel cell for direct carbonaceous fuel conversion. *ECS Trans* 5, 463–472.
83. Ihara, M., Hasegawa, S., Saito, H. and Jin, Y. 2007. Power generations of rechargeable direct carbon fuel cells with Ni/GDC anodes and Ni/ScSZ anodes. *ECS Trans* 7, 1733–1740.
84. Ihara, M., Kusano, T. and Yokoyama, C. 2001. Competitive adsorption reaction mechanism of Ni/yttria-stabilized zirconia cermet anodes in H-2-H2O solid oxide fuel cells. *J Electrochem Soc* 148, A209–A219.
85. Tagawa, Y., Ohba, F., Takei, C. and Ihara, M. 2009. Carbon deposition and power density in rechargeable direct carbon fuel cells with gadolinium-doped ceria and scandium-stabilized zirconia. *ECS Trans* 25, 1133–1142.
86. Tagawa, Y., Saito, H. and Ihara, M. 2009. Rechargeable direct carbon solid oxide fuel cells with a gadolinium-doped ceria (GDC) anode and a La[sub 0.15]Sr[sub 0.85]MnO[sub 3]/GDC anode. *ECS Trans* 16, 287–298.
87. Kellogg, I.D., Koylu, U.O. and Dogan, F. 2010. Solid oxide fuel cell bi-layer anode with gadolinia-doped ceria for utilization of solid carbon fuel. *J Power Sources* 195, 7238–7242.

6

Advances in Electrode Material Development in Molten Carbonate Fuel Cells

Gang Chen

CONTENTS

6.1 Introduction.. 163
6.2 Advanced Cathode Materials.. 164
6.3 Development of Anode in DIR-MCFC... 179
6.4 Conclusions.. 186
References.. 187

6.1 Introduction

The molten carbonate fuel cell (MCFC) is another type of high-temperature fuel cells. Using carbonate salt as the electrolyte, the system can be operated around 650°C, above the melting point of carbonate salt. At such a high temperature, different from the low-temperature fuel cells, it is evident that the heat produced in an MCFC system can be utilized in fuel reforming and heat-generation systems. Generally, considering the heat generation together with the power consumption, an MCFC can provide higher efficiency, approaching 90% [1]. A high operation temperature also results in the application of cheaper materials rather than the precious noble metals. Therefore, MCFCs are considered to be the most promising fuel cell system for the near future.

The state-of-the-art MCFC consists of a porous nickel anode, a porous lithium-doped nickel-oxide cathode, and a lithium aluminate matrix filled with lithium and potassium/sodium carbonates as the electrolyte. Hence, in a certain sense, the power generation of an MCFC could be interpreted simply by the production or consumption of carbonate ions (CO_3^{2-}). The electrochemical equations on the anode and cathode sides can be described as follows:

$$\text{cathode:} \quad CO_2 + 1/2O_2 + 2e^- \rightarrow CO_3^{2-} \tag{6.1}$$

$$\text{anode:} \quad H_2 + CO_3^{2-} \rightarrow CO_2 + H_2O + 2e^- \tag{6.2}$$

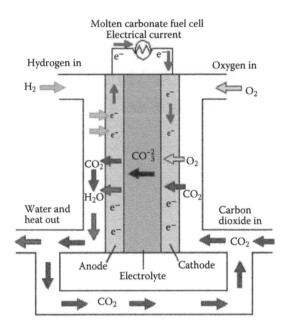

FIGURE 6.1
Schematic diagram of an MCFC system.

As seen in Figure 6.1, on the cathode side, the air or oxygen was fed with carbon dioxide to produce the carbonate ion. When the ions transfer from the cathode through the electrolyte layer to the anode side, the carbonate ion will react with a reductive gas, such as hydrogen or hydrocarbon, by the catalyst effect of anode.

On the other side, the development of MCFCs has to solve a number of problems to compete with current commercially available power-generation devices, such as diesel engines and gas turbines. One of the most serious problems is guaranteeing the long-term lifetime of MCFC stacks [2]. For a commercial MCFC system, a lifetime of more than 40,000 h and a voltage decay rate of less than 10% are required. In order to achieve such goals, the chemical and physicochemical stability should be improved.

6.2 Advanced Cathode Materials

Among these problems, the instability of the cathode during cell operation is now considered as one of the major technical difficulties facing MCFC development [3,4]. Generally, the prepared cathode is the presintered porous Ni plate. During cell operation, the plate will transfer to the NiO by *in situ* oxidation. Normally, NiO is a p-type impurity semiconductor material. However, in a molten carbonate environment, the lithium from Li_2CO_3 can dope into the crystal lattice of NiO. The conductivity of the cathode is enhanced because such a doping process provides more electron holes by replacing Ni^{2+} with Ni^{3+}. It was found that lithiation takes place at a potential around -0.4 V versus (1:2) $O_2:CO_2$ reference electrode according to the parabolic rate law, involving the oxidation of Ni^{2+} to Ni^{3+} and the diffusion of Li^+ into the solid phase [5].

$$Li_2CO_3 \rightarrow Li_2O + CO_2 \tag{6.3}$$

$$\frac{x}{2}Li_2O + (1-x)NiO + \frac{x}{4}O_2 \rightarrow Li_x^+Ni_{1-2x}^{2+}Ni_x^{3+}O \tag{6.4}$$

According to the calculation and simulation, the adequate cation fraction of Li^+ in NiO is known to be in the range of 0.02–0.06 mol [4]. However, as further study and practice show, the nickel oxide will dissolute into the molten carbonate in the operation environment. Theoretically, nickel oxide has a small degree of solubility in the carbonate electrolyte in the fuel cell cathode environment (about 10–15 ppm). But the solubility value strongly depends on CO_2 partial pressure and the amount of Li present in the NiO lattice [6]. NiO is an amphoteric oxide and can dissolve as a base or as an acid. Under MCFC operating conditions, the following acid mechanism is the main dissolution process [7]:

$$NiO(s) + CO_2(g) \leftrightarrow Ni^{2+}(1) + CO_3^{2-}(1)$$

or

$$NiO(s) \leftrightarrow Ni^{2+}(1) + O^{2-}(1)$$

$$CO_2(g) + O^{2-}(1) \leftrightarrow CO_3^{2-}(1) \tag{6.5}$$

Based on the above mechanism, the typical NiO solubility in Li–K_2CO_3 (62–38 mol%) under the cathode condition ($P_{O2}/P_{CO2} = 33/67$) at 650°C is about 30 ppm.

Dissolution of NiO may cause the following problems: (1) loss of cathode material, (2) internal short circuit of the cell, and (3) low total efficiency [8]. Most testing to date has shown that the common failure mode is cell shorting [5]. In 2001, Morita et al. [9] found that high pressure can accelerate the dissolution of cathode nickel oxide. They also examined the mechanics of nickel short-circuiting in Li/Na cells at ambient pressure. Due to a concentration gradient, the dissolved Ni ions can diffuse or transfer from the cathode through the electrolyte toward the anode. Under the influence of reducing conditions on the anode side, the dissolved Ni will precipitate into the electrolyte as Ni metal according to the following equation:

$$Ni^{2+} + CO_3^{2-} + H_2 \rightarrow Ni + CO_2 + H_2O \tag{6.6}$$

Additionally, the precipitation of Ni creates a sink for the Ni^{2+} cations, which facilitates further NiO dissolution [5].

As we know, the traditional thickness of electrolyte is less than 1 mm. So the existence of Ni metal between electrodes will give rise to the short circuit and then cause degradation or even the failure of performance in the short term. The voltage degradation of an MCFC is drawn as Figure 6.2. Generally, the curve could be divided into two phases [10]. The first phase with a linear relationship represents a degradation caused by the ohmic resistance and the electrode polarization due to the carbonate electrolyte loss. A rapid degradation of the second phase may be caused by Ni short circuit or the gas cross leakage. Because the transfer from the first phase to the second phase determines the operating period of an MCFC power unit, the dissolution of NiO and the reduction will be a vital factor to limit the development of MCFCs.

Many efforts have been made to improve the stability of the cathode materials. Current state of the art on solving the Ni dissolution problem is focused on varying the molten salt constituents or using alternate cathode materials. The basicity of the molten carbonate can be modified by a change in the carbonate melt composition. More basic molten carbonate melts, such as Li/Na carbonate eutectic, have been used to decrease the Ni dissolution rate in the melt. It was proved that a Li/Na (60:40) cell has a lower internal resistance and a lower cathode carbon dioxide, compared with Li/K carbonate. So its performance is higher than the average Li/K cell performance at any pressure. Fujita et al. [11] also found that by the use of Li/Na, the Ni deposited within the electrolyte was reduced, and particle growth of NiO in the

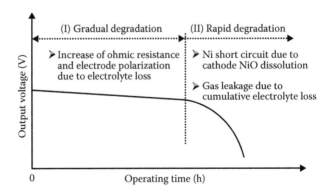

FIGURE 6.2
Schematic diagram of MCFC degradation upon operating time (at a constant current). (Reproduced from Morita, H. et al. *J. Power Sources*, 195, 6988–6996, 2010. With permission.)

cathodes was suppressed. Based on the Li/Na electrolyte system, Yoshiba et al. [12] demonstrated the stability in commercial-sized 10 kW class MCFC stacks with a high current density operating test and a 10,000 h endurance test. However, the advantage of using Li/Na in terms of reduced cathode dissolution is offset by the lower solubility of oxygen in Li/Na compared with Li/K. Therefore, when decreasing the operation temperature, the voltage drop of the Li/Na cells was larger than that of the average Li/K cell. In other words, an MCFC cathode suffers from a very high diffusion polarization when using the alternative Li/Na.

In early works, it was shown that alkaline earth (AE) or rare earth (RE) metal oxides/carbonates added to the molten carbonates have also been used to increase the basicity of the melt [13,14]. Reported by Doyon et al. [13], the addition of small amounts of SrO, MgO, or BaO is very effective in reducing the solubility of cathode. For example, an addition of 1 wt% SrO resulted in a 15-fold decrease in the dissolved NiO concentration [14]. In recent decades, the RE metal oxides were proved to have much greater effect on improving the stability of NiO. And the La oxide was considered as the more stable additive because the stable phase $La_2O_2CO_3$ was founded in molten carbonate. However, the addition of the additives has not been appreciable to the cathode modification because these additives cause the degradation of the electrical conductivity during long-term operation [15].

Besides, in searching for alternative cathode materials, it is a natural consideration to develop the cathode by coating or planting a stable layer on the NiO surface. It is also beneficial to decrease the complication and lower the cost. Here, in this chapter, the stabilization of NiO cathode coated by an electrochemically active layer with lower solubility was investigated and discussed.

At the original stage, it was shown that an addition of coexisting oxides had a positive effect on restraining the NiO solubility in molten carbonates [16]. And various methods were chosen to obtain an effective homogenous structure, including the sol-gel, electroplating, vacuum suction, and so on [5]. Actually, the morphology of the coating layer has a strong dependence on the synthesis process. By electrochemical potentiostatic deposition, Mendoza et al. [17] produced a cobalt oxide coating on a Ni substrate, and the X-ray characterizations indicated that the cobalt oxide has a cubic Co_3O_4 form. It is likely that the oxidation treatment allows a better adherence of the cobalt oxide film because of a better compatibility between the oxidized nickel foil and the cobalt oxide deposit [18]. By an electroless deposition method, Hong et al. [19] also confirmed the presence of Co_3O_4 after the treatment at 300°C. In the research of Fukui et al. [20], the CoO–NiO particle with a core-shell structure was fabricated by the mechanofusion method. With different treatments, different morphologies were observed. It showed that the preparation process has a great influence on the morphology and, further, on the NiO dissolution.

Usually, the NiO solubility of Co–NiO composites could be decreased to a range of 40%–70% than that of pure NiO. Lee et al. [21] ascribed the improvement of NiO stability to the formation of the stable $LiCo_{1-y}Ni_yO_2$ phase on the surface of NiO in molten carbonate. Ryu et al. [22] prepared the Co-coated Ni electrode using the galvanic pulse plating method. Less than 0.1% of Co can be found in the Co_3O_4 form, and most Co turns to the solid solution with NiO. Kim et al. [23] modified the NiO using Co_3O_4. They also found that the diffraction features of Co_3O_4 completely vanish after oxidation. And the lattice parameter of CoO–NiO is 4.1844 Å determined by XRD. According to Vegard's law, the lattice parameter of a solid solution varies linearly as a function of its composition. The lattice parameters of NiO and CoO are 4.177 and 4.261 Å, respectively. The theoretical calculation of Li–Co–Ni solid solution is 4.1854 Å, quite good agreement with the experimental result. Therefore, it provides strong evidence that the premixed powder forms a (Co, Ni)O solid solution after the oxidation process at 650°C. Escudero et al. [24] analyzed the XPS spectra. After exposure in the molten carbonate, the Co 2p spectrum confirmed the loss of cobalt and showed a poor resolution for the $2p_{3/2}$ peak at 781.4 eV and the $2p_{1/2}$ peak at 796.0 eV. But coupled with the O 1s spectra analysis, XPS could not confirm the formation of $LiCo_{1-y}Ni_yO_2$ layer. In the research of Hong et al. [19], two kinds of cathode—pure Co powder and Co–NiO powder—were compared together by Raman spectroscopy after the lithiation in Li–K_2CO_3 at 650°C under cathode gas condition.

In Figure 6.3, for the lithiated pure Co Raman spectrum data, two sharp peaks (485 and 597 cm^{-1}) of $LiCoO_2$ appeared, which are attributed to the $A1g$ and Eg vibrational modes, respectively [25], while only one band near 510 cm^{-1} exists for the lithiated Co–NiO powder. They proposed that the

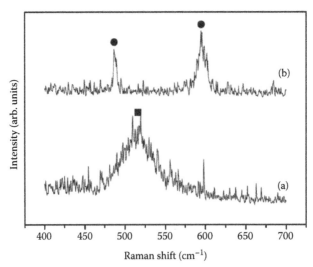

FIGURE 6.3
Raman data of (a) Co-coated Ni powder and (b) pure Co powder lithiated in $(Li_{0.62}K_{0.38})_2CO_3$ at 650°C under $CO_2{:}O_2 = 66.7{:}33.3\%$ for 120 h. (●) $LiCoO_2$, (■) $LiCo_{1-y}Ni_yO_2$. (Reproduced from Hong, M.Z. et al. *Electrochim. Acta*, 48, 4213–4221, 2003. With permission.)

stable phase of $LiCo_{1-y}Ni_yO_2$ seems attributable to the solid solution of $LiCoO_2$ and $LiNiO_2$ [26] formed from Co_3O_4 and Ni during lithiation.

This raises a new question: how does the Co or $LiCo_{1-y}Ni_yO_2$ affect the NiO solubility? During the oxidation process, nickel oxide was rapidly formed with molten carbonate salt in the presence of cathode gas, while the cobalt-coated Ni shows much slower oxidation [27]. Theoretical weight increase for the conversion of Ni or Co to their oxides (NiO/CoO) is around 27% based on stochiometric calculation as in Figure 6.4a. But in the presence of molten carbonate, the percentage increase in weight for cobalt-coated Ni was about 22% (Figure 6.4b). Actually, the actual weight increase is much lower in the

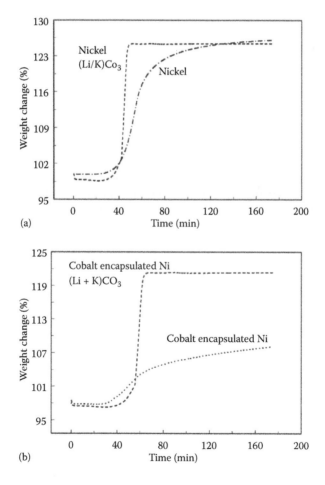

FIGURE 6.4
TGA analysis of sintered nickel (a) and cobalt-encapsulated nickel (b) tapes under cathode gas conditions in the presence and absence of molten carbonate melt. (Reproduced from Durairajan, A. et al. *J. Power Sources*, 104, 157–168, 2002. With permission.)

absence of the carbonate melt (near 9%). It is obviously due to the presence of cobalt, which is thermodynamically more stable.

Moreover, Ryu et al. [22] investigated the cubic lattice constants of Li–Ni–Co–O with different Co contents. It was found that the crystal shrinks much more in the samples with higher Co contents. Generally, Li^+ incorporates into cubic NiO by filling the cation vacancy [28]. Due to the valence difference between Ni^{2+} and Co^{3+}, the vacancy concentration will be increased in the formation of Ni–Co solid solution. Consequently, with higher Co content in the NiO–CoO solid solution, larger amounts of Li^+ could be incorporated and thus the cubic lattice constant was reduced [22]. Sato et al. [29] also found the similar rule that the lattice constants of Li–Ni–O or Li–Co–O decrease with the increase of Li contents.

On the other hand, the samples with higher Co content show the lower NiO solubility, as seen in Figure 6.5a [22]. For the Co mole fraction measurement in Figure 6.5b, the higher the Co content is, the larger the Co solubility observed. When the Co content is 12 wt%, the NiO solubility reaches its lowest value near 13 ppm, while the Co solubility is about 20 ppm after the dissolution experiment. Also, they found that the constant Ni mole fraction can be obtained at an earlier period. It inferred that the Co on the surface of the NiO acts as a barrier to reduce the NiO solubility. Furthermore, Durairajan et al. [27] analyzed the reaction process of Ni and Co in carbonate, respectively. After treatment in molten carbonate under cathode gas conditions for a period of 300 h, the XRD pattern revealed that pure Ni tape was characterized by peaks corresponding to NiO, while that of Co–NiO showed a different pattern due to the presence of mixed oxides, which was in agreement with the above analysis by Ryu et al.

However, Fukui et al. gave a different result for mechanofusion-treated samples. It has been reported that a Ni and Co oxide binary system, whose cobalt content is less than 15 mol%, forms a solid solution at 650°C in air [30].

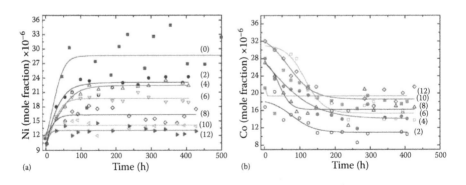

FIGURE 6.5
Time dependency of the ions' mole fraction; Ni (a) and cobalt (b) in molten $Li_{0.62}/K_{0.38}$ carbonate eutectics at 650°C under 1 atm; the numbers in brackets are the amount of Co in NiO–CoO electrodes. (Reproduced from Ryu, B.H. et al. *J. Power Sources*, 137, 62–70, 2004. With permission.)

Clearly, only 5 mass% CoO–NiO exhibits two thirds of Ni solubility of the pure NiO cathode. Although this result remained unclear, they proposed an explanation that if the content of Co in these composites becomes higher, the outer layer of (Co,Ni)O was separated from the NiO core when it was lithiated. The corrosion and electrochemical behaviors of $Li_{1-x}(Ni_yCo_{1-y})_{1+x}O_2$ (LNCO) oxides were performed by Pérez et al. [31]. They ascribed the change of NiO stability to the lithium content. Observed by SEM in Figure 6.6, a significant degradation occurred on the surface of the cathode after 4 days of immersion. For the LNCO oxide with lower Li-content ($0.5 > x > 0.2$), it exhibited the best behavior with a mainly $LiCoO_2$ structure, although a slight transformation of its surface showed up after 7 days immersion. In Figure 6.6b, the external zone was composed of small equiaxial grains ($LiCoO_2$–like), and the internal zone of homogeneous platelet-like grains corresponded to the LiNiCoO structure. So the writers concluded that Co addition coupled with lithium content lower than the metal content is the best option to improve the NiO properties.

Totally different, Kim et al. [23] argued that maybe the high lithium content is more stable for the LiCoNiO structure in their experiment. Recently, they employed neutron diffraction to investigate the structural changes of the modified cathode materials in Figure 6.7; the presence of two different phases is clearly observed in the lithiated sample. The outer phase has an edge shape, while the surface is a spherical shape. Generally, the lithium content in a NiO cathode *in situ* oxidized and lithiated at 650°C is about 2 to 4 mol% [20]. However, neutron diffraction data demonstrated a Li concentration of 1.6% and 6.7% for bulk and surface phases, respectively. When Li is doped into a NiO cathode, the Ni^{2+} will turn into Ni^{3+}, according to Equation 6.4. A high concentration of Li indicates a high content of Ni^{3+}. Thus, because the interaction between Ni^{3+} and O^{2-} is stronger than that between Ni^{2+} and O^{2-}, the chemical binding of Ni and O can be strengthened. Because the dissolution

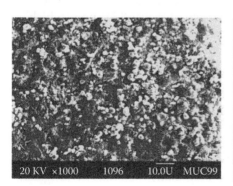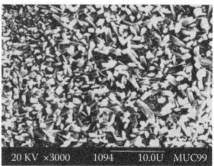

FIGURE 6.6
SEM micrograph of $Li_{1-x}(Ni_yCo_{1-y})_{1+x}O_2$ surface for (a) higher Li-content after 4 days of immersion and (b) lower Li-content after 7 days of immersion. (Reproduced from Pérez, F.J. et al. *J. Power Sources*, 86, 309–315, 2000. With permission.)

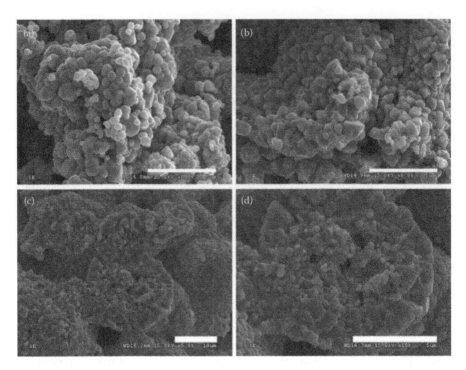

FIGURE 6.7
SEM images of (a) NiO and (b) $Ni_{0.9}Co_{0.1}O$ after lithiation under cathodic conditions for 100 h. SEM images of a cross section of a $Ni_{0.9}Co_{0.1}O$ cathode after lithiation for 300 h. (c) SEM image and (d) enlarged view of (c). (Reproduced from Kim, Y.-S. et al. *J. Power Sources*, 196, 1886–1893, 2011. With permission.)

of the cathode material and the electrochemical reactions occur at the surface, the high concentration of Li on the surface phase can act as a barrier for the bulk phase. In their experiment, the Ni solubility was 8.311 ppm, which is lower than that of the Co-coated NiO cathode (about 12 ppm [32]). As a result, they concluded that the presence of cobalt in the lithiation process results in the formation of a highly lithiated phase on the surface that is quite stable in the molten carbonate mixture. Anyway, the discussion about the action mechanism in Co-coated NiO needs more and deeper investigation.

To reduce the NiO dissolution, developing alternative materials that are more stable than nickel oxide is another choice. They should be corrosion resistant in the carbonate melt under MCFC cathode operation conditions. Moreover, the candidate material should possess an adequate electrical conductivity and electrocatalytic activity. Among many materials investigated, $LiFeO_2$ and $LiCoO_2$ were supposed to be the most promising candidates due to their lower solubility in molten carbonate and comparable electrochemical performance. For example, by ICP-AES, Veldhuis et al. [33] referred to the dissolution mechanism of $LiCoO_2$ as

$$LiCoO_2 + 1/2\ CO_2 \rightarrow Co(II)O + 1/4\ O_2 + 1/2\ Li_2CO_3$$

According to transport models for the cathode materials in an MCFC, this would result in a more uniform dissolution/precipitation behavior in the matrix and would thus reduce the chance of short-circuiting. Another advantage is that $LiCoO_2$ shows a minor dependency of $p^{1/4}$ on the system's total pressure P while NiO solubility increases linearly. The data showed that from 1000, 1500, and 3500 h tests at 160 mA cm^{-2}, the dissolution rate of $LiCoO_2$ and NiO cathodes were calculated to be around 0.5 and 4 µg cm^{-2} h [34], respectively.

However, the application of such new cathode candidates is limited from their intrinsic properties [5,35]. $LiFeO_2$ shows very low performance due to poor oxygen reduction kinetics and electrical conductivity. Porous $LiCoO_2$ cathodes have the problem of mechanical brittleness and the high cost of the raw materials. Especially, it was difficult to fabricate alternative cathodes sized over 1 m^2 because of the difficulties in the sintering process. Therefore, the direct substitution of lithiated NiO by $LiFeO_2$ or $LiCoO_2$ was obstructed [36].

Naturally, a number of attempts were made to cover the core surface of NiO cathodes with $LiCoO_2$ or $LiFeO_2$ layers to alleviate NiO dissolution [33,37,38]. And the dissolution rate of NiO in the modified cathode can be reduced to 60% than that of pure NiO. $LiFeO_2$ is cheaper, but its electrical conductivity is very low and sensitive to gas composition even if doped with other elements [39]. Many researchers have developed cathodes in $LiCoO_2$. Han et al. [37] investigated the relationship between the amount of $LiCoO_2$ incorporated into a cathode and Ni content in a matrix at different pressures. When the $LiCoO_2$-coated cathodes were used, the amount of Ni precipitates in the matrix was reduced to 50%. From the Ni distribution in the matrices after the 1000 h operation, it was also clear that the Ni precipitation was significantly reduced by using a $LiCoO_2$ coating on the NiO cathode [40].

The improvement can be proved by impedance spectroscopy. During the experiment by a sol-impregnation technique [40], although the cathode polarization increased with the addition of $LiCoO_2$, the impedance arcs decreased significantly with the time, as seen in Figure 6.8. As a result, the cell performance could be maintained with a much smaller degradation during the 1000 h.

Based on the work on $LiFeO_2$- or $LiCoO_2$-coated NiO, a ternary composition of $LiFeO_2$–$LiCoO_2$–NiO has been investigated and tested in MCFCs [41]. Containing low $LiCoO_2$ contents, $LiFeO_2$–$LiCoO_2$–NiO solid solutions of the *Fm3m* cubic rock salt structure rich in $LiFeO_2$ and NiO was formed. A high concentration of $LiFeO_2$ permits a decrease in the Ni dissolution and also affects the cobalt solubility. Kudo et al. [42] confirmed that the Ni^{2+} solubility decreases with the increase in the Fe content for Ni–Fe–Co ternary oxide addition. $LiCoO_2$ was expected to obtain better electrical and electrochemical

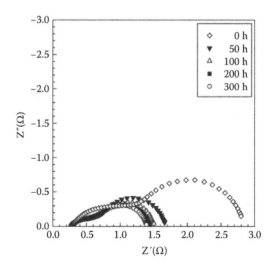

FIGURE 6.8
Evolution of impedance spectra with operation time for the cell with 3 mol% LiCoO₂-coated NiO cathode. (Reproduced from Kim, S.-G. et al. *J. Power Sources*, 112, 109–115, 2002. With permission.)

performances. Park et al. [43] concluded that the sensitivity of the solid solution to CO_2 partial pressure is dramatically reduced.

Otherwise, some work reported the NiO cathode coated by Mg-doped $LiCoO_2$ [39,43]. Because the conductivity of $LiCoO_2$ can be increased by partial substitution of Co^{3+} with Mg^{2+} [44], and MgO could make a solid solution with Co_3O_4 at any ratio [45], Co/Mg solid solution–coated Ni powder was studied to improve electrochemical performance. Characterized by XRD, significant $LiCoO_2$ peaks were detected above 650°C. It was confirmed by Raman that Mg diffused into $LiCoO_2$ to form $Li(Co_{1-x}Mg_x)O_2$. Then, the NiO solubility could be decreased up to 6.6 ppm, which was about half of the Co-coated Ni cathode and one fifth of the NiO cathode. The results suggested that a Co_3O_4/MgO solid solution deposited in NiO cathode might be more stable than the Co-coated NiO. Therefore, the structures, particle sizes, and pore sizes could be maintained after the solubility test.

Besides, some research works are focusing on the modification of NiO coated by other oxides, such as ceria and titanium dioxide. Cerium oxide is already being used as a catalyst to produce a good electrical contact between oxides. Daza et al. [46] measured the NiO dissolution rate in the molten carbonate mixture, showing a minimum corrosion for cathodes with 0.3–1 wt% CeO_2 content. Here, Kim et al. [47] synthesized a new Co/Ce–coated nickel powder based on the Pechini method. By the EPMA characterization, it can be confirmed that the fine Co/Ce particles are coated homogeneously on the surface of Ni particles. Comparatively, the solubility of the NiO and the Co-coated NiO cathodes were about 30.4×10^{-6} and 12.71×10^{-6} mole fractions, respectively [21]. Clearly, for the new Co/Ce (9.5/0.5 mol%)–coated NiO, the solubility was reduced to 5.5×10^{-6} mole fraction, almost one sixth of that of NiO and half of that of Co-coated NiO cathode. Concerning of the role of ceria, Albin et al. [48] investigated the CeO_2-coated dense Ni film. They found that after 48 h of exposure to the molten carbonate, the amount of remaining cerium is about 60% of the initial amount. It was inferred that cerium could fall in the molten carbonate first to prevent the NiO dissolution.

Also, as observed from Figure 6.9, the dissolution rate has a quite big difference that depends on the ceria content. Except for the Co/Ce (9.5/0.5 mol%) sample, other Co/Ce–coated NiO exhibited higher solubility. Although there were no explanations in the articles, we could deduce the effect of CeO_2 as follows: When a small content of CeO_2 was incorporated on the surface of the cathode, the NiO dissolution will be retarded significantly due to the stable $CeOCO_3$ presence in molten carbonate [49]. It was also noted that cerium oxide favors the lithiation process, resulting in a mixed oxide material with a higher Li^+ content that is more conductive and stable [50]. On the contrary, with the increase in CeO_2 content, it will make the NiO dissolution faster. CeO_2, unlike other lanthanide oxides, is also an oxygen ion conductor. As mentioned in Equation 6.5, NiO is easy to convert to Ni^{2+} and O^{2-} in the molten carbonate. Because the fine ceria is coated on the surface of NiO, it will transfer the produced O^{2-} from the interface of NiO–carbonate to the electrolyte system, which accelerates the reaction toward positive direction. As a result, the Ni solubility became larger at a rapid rate. Such assumption still needs to be verified further.

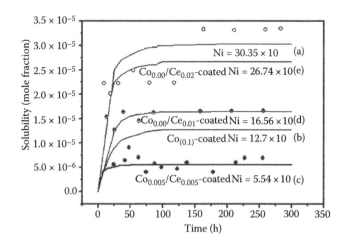

FIGURE 6.9
Time dependence of Ni solubility in $(Li_{0.62}K_{0.38})_2CO_3$ at 650°C under $CO_2:O_2$ (66.7:33.3%): (a) the pure Ni cathode, (b) the 10 mol% Co-coated Ni cathode, and the Co/Ce–coated NiO cathode at the molar ratio of Co/Ce (c) 9.5/0.5, (d) 9.0/1.0, and (e) 8.0/2.0 (mol%). (Reproduced from Kim, M.H. et al. *Electrochim. Acta,* 51, 6145–6151, 2006. With permission.)

Compared with CeO_2, the effect of TiO_2 on the NiO dissolution is quite different. TiO_2 deposits have already been applied as protective coatings of the stainless steels for MCFC bipolar plates [51–53]. TiO_2 seems to be stable in a molten carbonate environment. According to the OCP measurement, the starting potential of titanium dioxide–coated Ni in carbonate was $-0.9V/Ag|Ag^+$, probably corresponding to the oxidation of TiO_2 into Li_2TiO_3:

$$TiO_2 + 2Li^+ + CO_3^{2-} \rightarrow Li_2TiO_3 + CO_2(g) \tag{6.7}$$

The Li content has an increasing tendency as the TiO_2/Ni ratio increases. Thermodynamically, the Li_2TiO_3 is stable and should not be oxidized or reduced, but the dissolution of TiO_2 was observed. Figure 6.10 shows the SEM micrographs of TiO_2 coatings on dense nickel substrates with thicknesses of 200 nm. After 8 h of exposure to the carbonate, the surface was etched by the Li/Na carbonate. And only 20% of the titanium remains on the surface of a Ni foil [48]. Chauvaut and Cassir [54] suggested that Li^+ can be intercalated within the Li_2TiO_3 monoclinic structure, which possess at least two octahedral and two tetrahedral sites. Therefore, due to the different intercalation–deintercalation processes, Li_2TiO_3 actually exists in the presence of $Li_{2+x}TiO_3$ compounds as

$$Li_2Ti^{IV}O_3 + xLi^+ + xe^- \rightarrow Li_{2+x}Ti^{IV}_{1-x}Ti^{III}_xO_3 \tag{6.8}$$

The solubility of the TiO_2 (2.5, 5.0 mol%)–coated Ni cathodes decreases by about 48% and 77%, compared with that of a pure NiO cathode. The results suggest that the Li_2TiO_3 and $LiTi_{1-x}Ni_xO_2$ phase formed on the NiO prevents dissolution of Ni^{2+} ions into the electrolyte during MCFC operation.

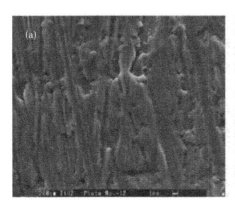

FIGURE 6.10
SEM micrographs of 200 nm TiO_2 deposited on Ni. Immersion condition: 8 h in Li_2CO_3–Na_2CO_3 (52–48 mol%) at 650°C under air/CO_2 (70/30). (a) Before immersion and (b) after immersion. (Reproduced from Albin, V. et al. *J. Power Sources*, 160, 821–826, 2006. With permission.)

The other approach to counter the nickel dissolution is to modify NiO by perovskite ABO_3 compounds. Initially, the ABO_3 compounds were used as a cathode for solid oxide fuel cells. Such perovskite structure compounds are also mixed ionic and electronic conductors (MIECs). They have received much attention because of their high ionic and electronic conductivities and their high catalytic activity for electrochemical reactions [55–58]. Because the MCFC was operated at intermediate temperatures, they were also expected to enhance cathode performance for MCFCs.

Song et al. [59] investigated the $Gd_xSr_{1-x}CoO_3$ (GSC)–coated NiO. It is known that GSC has a high conductivity and electrochemical catalytic activity at low temperatures [60], which is due to many oxygen vacancies on its surface. On the one side, the oxygen vacancies are the carriers to transport O^{2-} to the triple-phase boundaries (TPBs). Additionally, they provide the active sites for the adsorption and dissociation of oxygen sources in carbonate melts [59]. Generally, the ORR mechanisms in MCFCs are explained by three pathways: the peroxide, superoxide, and percarbonate pathways, listed in Table 6.1 [31,61]. For the uncoated cathode, at 600°C–700°C, the cathode reaction mechanism shifts from the superoxide to the peroxide. In the case of the GSC-coated cathode, the reaction order does not change corresponding to the peroxide reaction mechanism over the temperature range of 600°C–700°C. Meanwhile, the GSC-coated cathode has a lower dependence on CO_2 concentration. It was suggested that two O^{2-} ions are produced from (O_2^-) and three electrons, and CO_2 does not participate in the reaction at the electrode surface. The reaction of CO_2 with oxide ion in the melt can offset the negative dependence on the CO_2 concentration.

Similarly, in the experiment of $La_{0.8}Sr_{0.2}CoO_3$ (LSC)–coated NiO by Ganesan et al. [62], the partial pressure of O_2 and CO_2 has different effects for the polarization resistance of new cathode. The same negative order was observed

TABLE 6.1

Theoretical Mechanisms of Oxygen Reduction in Carbonate Melts [61]

Peroxide pathway	$O_2 + 2CO_3^{2-} \leftrightarrow 2O_2^{2-} + 2CO_2$
	$O_2^{2-} + 2e^- \leftrightarrow 2O^{2-}$ (rate-determing step [rds])
	$2O^{2-} + 2CO_2 \leftrightarrow 2CO_3^{2-}$
	$i_o \sim (O_2)^{0.375}(CO_2)^{-1.25}$
Superoxide pathway	$3O_2 + 2CO_3^{2-} \leftrightarrow 4O_2^- + 2CO_2$
	$O_2^- + e^- \leftrightarrow O_2^{2-}$ (rds)
	$O_2^{2-} + 2e^- \leftrightarrow 2O^{2-}$
	$2O^2 + 2CO_2 \leftrightarrow 2CO_3^{2-}$
	$i_o \sim (O_2)^{0.625}(CO_2)^{-0.75}$
Percarbonate pathway	$0.5O_2 + CO_3^{2-} \leftrightarrow CO_4^{2-}$
	$CO_4^{2-} + e^- \leftrightarrow CO_3^{2-} + (O^-)$ (rds)
	$O^- + CO_2 + e^- \leftrightarrow CO_3^{2-}$
	$i_o \sim (O_2)^{0.375}(CO_2)^{-0.25}$

by Yuh and Selman with 2% CO_2 [63]. Furthermore, the two distinct loops for NiO occur at high and low frequencies, while they seem to be merged with each other in the case of LSC–NiO. It also can be inferred that LSC or GSC improves the electrochemical kinetics during the oxygen diffusion and reduction processes. As a result, the cell using the GSC-coated cathode showed a stable performance for approximately 2200 h and no rapid drop-off [59]. However, the short-term stability test indicated that the NiO solubility was only reduced to 21 ppm mole fraction with the LSC-coated cathode. Maybe the excess oxygen vacancies in perovskite compounds impair the Ni stability as discussed with CeO_2-coated NiO. The effect and long-term stability for ABO_3-coated NiO should be investigated in detail.

Besides, except for ABO_3 compounds, the materials having AB_2O_4 structure, such as $MgFe_2O_4$, were also studied for the coating on the surface of NiO by mechanical coating technique. As shown in Figure 6.11, in the $MgFe_2O_4$/NiO composite cathode, the stability of the cathode is improved four and two times higher than that of the NiO and CoO (5 mol%)–coated NiO cathodes, respectively [65]. At our best knowledge, the 5×10^{-6} mole fraction in $MgFe_2O_4$/NiO is the smallest reported value of Ni solubility for various alternative cathodes in an MCFC. It suggested that a highly stable outer layer containing $LiFeO_2$ was formed on the surface of NiO grains using $MgFe_2O_4$ [64]. Then, a single cell with a $MgFe_2O_4$/NiO composite cathode was tested at 150 mA cm^{-2} for about 5000 h. The lower NiO solubility makes the new material maintain a cell voltage of 0.81–0.77 V from 500 to about 5000 h.

Summarizing, various methods have been experimented with to increase the stability of the cathode in MCFC conditions. Regarding the coating layer,

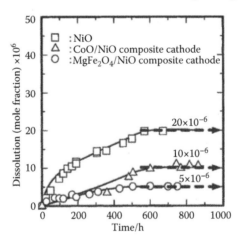

FIGURE 6.11

Dissolution of the composite cathodes into $(Li_{0.52}K_{0.48})_2CO_3$ melt at 650°C in 10%CO_2–7%O_2–83%N_2 at 1.2 MPa. □: Ni dissolution of normal NiO cathode; △: Ni dissolution of the CoO/NiO composite cathode; and ○: Ni dissolution of the $MgFe_2O_4$/NiO composite cathode. (Reproduced from Fukui, T. et al. *J. Eur. Ceramic Soc.*, 23, 2835–2840, 2003. With permission.)

the presence of solid solutions with other oxides is effective to reduce NiO dissolution into the electrolyte. The experimental results imply that the chemical stability of the coated NiO cathode could be enhanced not only by the protection of the deposited layer but also by the formation of new phases with good chemical stability against molten carbonate. With the development of a cathode in an MCFC, new preparation methods and candidate materials will be expected to improve the stability in the near future. However, the specific efforts should be concentrated on the mechanical stress and chemical reaction between cathode substrate and coating layer.

6.3 Development of Anode in DIR-MCFC

In a prototype of an MCFC, hydrogen is applied as fuel. Actually, for a practical stack, steam reforming is a mature large-scale technology characterizations for hydrogen production [66]. Various hydrocarbons, including methane and ethanol from biomass [67–70], are used to produce the hydrogen by a steam reforming reaction. The research on their compatibility in the MCFC environment has been undertaken and became a new interest since the last decade.

According to the location of the reformer, an MCFC can be classified into three types: (1) external reforming, (2) indirect internal reforming, and (3) direct internal reforming (DIR) [71–73]. For DIR-MCFC, the reforming chamber and anode channels are merged together, as shown in Figure 6.12 [74]. Therefore, this configuration has the advantages of simplicity and thermal benefit, which has attracted particular attention.

In the case of methane and ethanol fuel, during the reforming process, they would be transformed to CO_2 and H_2 by the following routes:

$$CH_4 + H_2O \rightarrow CO + 3H_2 \tag{6.9}$$

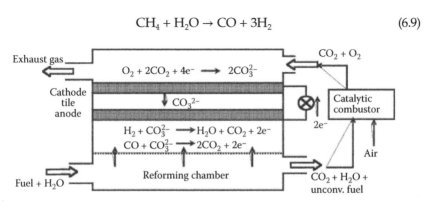

FIGURE 6.12
MCFC system: DIR configuration. (Reproduced from Frusteri, F. and Freni, S., *J. Power Sources*, 173, 200–209, 2007. With permission.)

$$C_2H_5OH + H_2O \rightarrow 2CO + 4H_2 \qquad (6.10)$$

Then,

$$CO + H_2O \rightarrow CO_2 + H_2 \qquad (6.11)$$

All the above reactions are of endothermic characterization. In DIR-MCFC, it has been demonstrated that the heat generated in an MCFC anodic reaction can sustain directly the reforming reaction [75]. Simultaneously, the reforming reaction will produce hydrogen, which is immediately consumed by an electrocatalytic cell reaction [71], which makes an improved conversion of methane with a simplified stack size. In particular, the DIR system has the potential to produce a uniform fuel cell temperature distribution [76]. As a result, 12% improvement in electrical efficiency was reported for DIR-MCFC units [77].

Furthermore, once these reactants enter the reforming chamber, the reforming reaction should take place through the dissociation of methane adsorbed onto the metallic Ni sites and the dissociation of water adsorbed at the metal–support interface to produce CO and H_2 [78]. Among different catalysts, nickel-based catalysts supported on oxide carriers are the most favorable candidates for their high activity and strong capability of breaking the C–C bond with a lower cost [79]. Even though Co-based catalysts show a better catalytic activity [80], they are usually affected by deactivation mainly due to sintering and surface Co oxidation [81].

However, Matsumura and Hirai [82] reported the severe deterioration of support stability by alkali metals. The deterioration of catalytic activity is supposed due to both the liquid-phase creep and the vapor phase pollution [83–85]. Liquid-phase status means that catalytic activity is hindered by molten salts adhering to the catalyst [86,87]. Some researchers applied the ceramic membranes as porous SiC or dense copper foils as a protective barrier against the alkali poisons [88] while nonuniform catalysts were also investigated to improved reaction performance [89]. Conversely, the solution of vapor phase pollution is quite difficult and critical. It has been reported that one reason is primarily due to alkali poisoning originating from the molten carbonate electrolyte [83–85].

Recently, there was progress in elucidating transport mechanism of alkali species from the electrolyte to catalyst [90,91]. Sugiura et al. [92] employed a noncontact image measurement technique on observation of the electrolyte volatile phenomenon. They found that the volatile matter flows in a belt-like manner from the inlet side toward the outlet during the operation. It is understood that the volatile matter, hydroxide, is generated from the carbonate and water generated by the cellular reaction. Once the alkali hydroxide is present on the catalyst surface, it can be retransformed to the corresponding alkali carbonate or be converted into other alkali compounds [5].

As well as for the alkali poisoning effect, there are two other problems that should be paid attention to. One is the sintering of the Ni particle under the reforming condition; the sintering behavior was demonstrated in Reference 93,94. The high temperatures and high water content will make Ni particles aggregate to be six times larger in size [95], which reduces directly the catalytic surface area and activity. The other critical factor is coke formation on the surface of the nickel catalyst, mainly due to the very high steam reforming reaction temperature [73]. It was convinced that ethanol steam reforming would take place based on the following mechanism [96]: (1) ethanol dehydrogenation to acetaldehyde; (2) decomposition of acetaldehyde into CH_4 and CO; and (3) steam reforming of CH_4 and water gas shift (WGS) reactions. There exists another route or possibility that the ethanol may be dehydrated to ethylene (in Figure 6.13), which is well known as a coke formation promoter.

In Figure 6.14 [74], a comparison of literature data in terms of coke formation rate over different catalysts is reported. It can be observed that catalysts based on the use of acidic carriers like alumina and silica are affected by the formation of a huge amount of coke. The effect of alkali is stronger on less acidic supports because the less acidic support has a looser bonding of alkali ions, resulting in easier adsorption on the metal [76,83]. Meanwhile, upon given alkali partial pressure in the gas phase, the acidic supports will absorb more alkali than the basic supports [83,97]. Therefore, the benefits of using a basic carrier are identified in preventing the ethanol dehydration to ethylene. In addition, alkali could induce an electronic enrichment of the active phase. It was considered as a reflection of both in depressing the Boudouard reaction and hydrocarbon decomposition activity, which is also the main cause of coke formation in reforming reactions [98].

Based on the above analysis, nickel catalysts supported on magnesia, alumina, and La_2O_3 are commonly employed [76]. Although Ni/La_2O_3 shows the higher activity and stability attributed to formation of a lanthanum

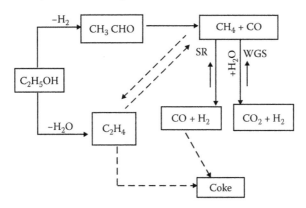

FIGURE 6.13
Steam reforming of ethanol: reaction mechanism scheme. (Reproduced from Frusteri, F. et al. *Catalysis Comm.*, 5, 611–615, 2004. With permission.)

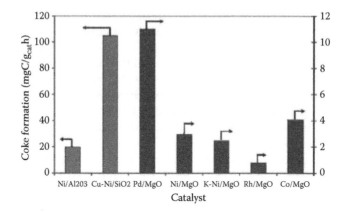

FIGURE 6.14
Coke formation over different supported catalysts. (Reproduced from Frusteri, F. and Freni, S., *J. Power Sources*, 173, 200–209, 2007. With permission.)

oxycarbonate species ($La_2O_2CO_3$), which prevents the surface carbon deposition [99], its hydrogen selectivity is lower than 77% due to the formation of appreciable amounts of CH_3CHO and CH_4 [74]. It was observed that coke was formed on Ni/MgO with almost orders of magnitude lower rate than other carriers [100–103]. An endurance test of 650 h performed at 10,000 h^{-1} with Ni/MgO shows a good performance with H_2 selectivity higher than 96% at total ethanol conversion [101]. Furthermore, the NiO–MgO system is easy to make an ideal solid solution in any atomic fraction ranges [104], implying a strong metal–support interaction between nickel and MgO [105].

However, the catalytic activity of Ni/MgO is known to decrease drastically due to severe poisoning by electrolytes [76]. There is still controversy about understanding the role of alkalis in the physicochemical changes of catalysts supported [106,107]. Rostrup-Nielsen and Christiansen [83] insisted that the presence of alkali does not stimulate the sintering of nickel crystals in the catalyst, but the deactivation might be caused by the faceting of the active nickel crystal planes to the less active (111) plane. Parmaliana et al. [108] found that Li decreases the catalytic activity; then Wee [109] analyzed the compositions of alkali metal–poisoned catalysts after 200 h of operation, and a higher Li content than K content was observed. As shown in Figure 6.15, as the amount of lithium increased, the catalytic activity decreased. For the 20 wt% Li_2CO_3–Ni/MgO sample, the catalyst deactivated completely. Even though a composition carbonate is used in a real stack of an MCFC rather than single Li_2CO_3, it also implies the large effect of lithium [107].

Park et al. [72] compared the morphology with different alkali. With the Li-added catalyst, many lithium aluminates ($LiAlO_2$) shaped of quill were generated on the Ni surface. Moon et al. [102] investigated the deactivation of the Ni/MgO catalyst for 72 h. They attributed the catalyst deactivation to the chemical poisoning by alkali species, particularly Li. Lithium was supposed

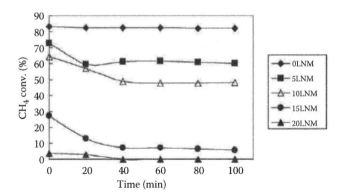

FIGURE 6.15
Activity of deactivated catalysts in methane steam reforming by a different amount of lithium.
(Reproduced from Choi, J.-S. et al. *J. Power Sources*, 145, 652–658, 2005. With permission.)

to be the most harmful reason for the formation of ternary $Li_yNi_xMg_{1-x-y}O$ solid solution, which resulted in nickel sintering. Arena et al. [110] also observed a great change of the morphology and dispersion of Ni/MgO by Li, and they concluded that the lithium poisoning effect is via the change in NiO–MgO interaction. Antolini [111] studied the influence of thermal treatment on Li-doped Ni/MgO and reported that lithium facilitated the dissolution of nickel into the solid solution, resulting in losing their activities. The explanation was confirmed by Choi et al. [107].

Conversely, some researchers ascribed the deterioration of the catalyst to potassium. The effect is most pronounced for K and less pronounced for Li and Na [83,85,92,112]. Berger et al. [90] proposed that not only Li_2CO_3 was harmful to the catalyst, and K_2CO_3 should be the more poisonous species. By visual observation, Sugiura et al. [85] elucidate that the volatile matter might be potassium, such as KOH at the DIR-MCFC anode site. Cavallaro et al. [96] observed that alkali species deactivated the catalyst by a pore-blocking mechanism, that is, the KOH glassy layer blocks the active sites by covering the whole external surface of the catalyst. In the report of Frusteri et al. [99], a Li-doped catalyst does a slow deactivation and will be stabilized at the higher conversion level (about 83%–85% for ethanol). However, in the test of coke formation, the large amount of filamentous and condensed carbon was observed compared with the bare, Li- or Na-doped Ni/MgO, as shown in Figure 6.16. That is a strong evidence that the presence of K is more harmful to the reforming reaction.

Because the solid solution is believed to be the major factor of deactivation, the support materials need to be substituted by the materials with more resistance to Li-poisoning [107].

Choi et al. [104] investigated the Ni/MgO–TiO_2 catalyst with a different Mg/Ti fraction. Usually, there are three different crystalline structures for the MgO–TiO_2 composite: karooite ($MgTi_2O_5$), geikielite ($MgTiO_3$), and qandilite

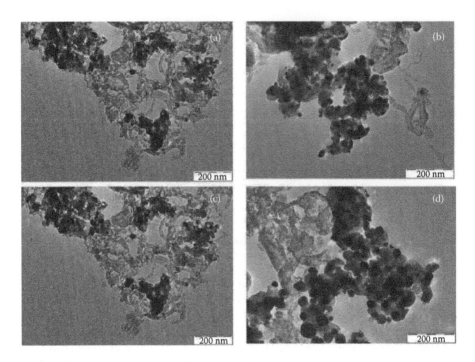

FIGURE 6.16
TEM images of bare and alkali-doped catalysts: bare Ni/MgO (a); Li-, Na-, and K-doped Ni/MgO (b, c, and d, respectively). (Reproduced from Frusteri, F. et al. *Appl. Catal. A: General*, 270, 1–7, 2004. With permission.)

(Mg_2TiO_4). So different crystals show the different behavior in the role of deterioration. After Li poisoning, the Ti-rich and Mg-rich catalysts lost almost all of their activities with the dramatic decrease in the BET surface area. However, the intermediate composition catalysts maintained their activities at more than half of their original activities [104]. Adding some content of Ti to the NiO/MgO system is suggested to be effective in obstructing $Ni_xMg_{1-x}O$ solid solution, which could weaken the Li poisoning on catalytic activity.

Li et al. [113] added the SiO_2 into the Ni/MgO to obtain a $Ni/MgSiO_3$ catalyst. In a DIR-MCFC environment, NiO, Ni_2O_3, and Ni were found coexisting in the $Ni/MgSiO_3$ catalyst due to strong interaction between nickel and $MgSiO_3$. By changing the content of SiO_2, the Ni/Mg_2SiO_4 was also prepared. Comparing the above ratios with catalyst activities and XPS data, the activities seem to be sensitive to the amount of Ni^{3+} present [71].

The equilibrium CH_4 conversion progresses due to consumption of H_2 by the electrochemical reaction in an MCFC [83,84]. Then time-programmed reduction (TPR) experiments were carried out to understand the interaction between the nickel oxide and the support. In Figure 6.17, the first peak at 200°C–250°C corresponds to the $Li(OH)_2$ reduction [114]. The decreased peak temperature in

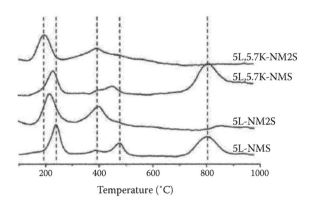

FIGURE 6.17
TPR profiles of electrolyte-added catalysts. (Reproduced from Li, Z. et al. *Int. J. Hydrogen Energy*, 35, 13041–13047, 2010. With permission.)

the Ni/Mg_2SiO_4 is attributed to the worse stability against the electrolyte than $MgSiO_3$. The peaks around 400°C and 450°C are reflected to be the reduction peaks of Ni^{2+} and Ni^{3+}, respectively. They proposed that the reduction of Ni^{3+} to Ni^{2+} supplies active oxygen for the conversion of CH_4 to CHO. Therefore, the Ni/$MgSiO_3$ catalyst showed a high activity of about 80% methane conversion. In the case of Ni/Mg_2SiO_4, however, the catalyst was poisoned and activity decreased.

Other work is focusing on the Al-doped Ni/MgO. Al was used as a sintering barrier by forming a Ni–Al solid solution [115]. Some researchers have found that Ni_3Al also acts as an active site for steam reforming even with a smaller level of activity than Ni [116]. Park et al. [73] employed a layered double hydroxide (LDH) method to prepare the Ni/(Al-MgO) material. When the Al^{3+} cation was incorporated into the framework of Mg–O, it inhibited the formation of the Ni–Mg–O solid solution [117]. If the Al/Mg molar ratio increased to 1.5, the number of Ni reduction peaks in the TPR spectra decreased from 3 to 1. This means that the Ni became more stable and homogeneous in the mixed oxides. On the other hand, $MgAl_2O_4$ spinel produced by Al and MgO interacts more strongly with the active metal Ni than MgO. Gadalla et al. [118] also reported that there exists a NiO phase after the heat treatment of $Ni/MgAl_2O_4$ in H_2, confirming a strong interaction between $MgAl_2O_4$ and Ni. Therefore, during their experiment, once Al was introduced, high activity was maintained [95]. Even over a 2000 h test, there was no performance degradation, and carbon deposition was not detected [95]. These results confirmed the durability of the catalyst in a DIR-MCFC system.

Based on the results, they proposed a model to explain the good catalytic activity, especially in the case of Al-rich catalysts (Al/Mg ≥ 1) [73]. In Figure 6.18, the aluminum oxide absorbs the Li species to form lithium aluminate, which hinders the coverage of Li species on Ni particles. For the Mg-rich catalyst, because the Li species could not be introduced to the MgO

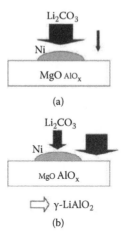

FIGURE 6.18
Mechanism of poisoning by Li_2CO_3 over Mg-rich (a) and Al-rich (b) catalysts. (Reproduced from Park, K. et al. *Fuel Cells*, 7, 211–217, 2007. With permission.)

surface, most Li species will agglomerate around Ni sites. Therefore, the catalyst lost its activity.

In summary, from the viewpoints of solid-state chemistry and of metal-to-support interaction, understanding the effects of alkalis on support material and Ni must be essential for further development of the reforming catalysts in DIR-MCFC. Furthermore, modifying or controlling the compositions and/or other metal oxides addition will stimulate the development of the DIR-MCFC. By the way, the practical operation in a certain scale of an MCFC stack should be carried out to evaluate the effect on cell lifetime.

6.4 Conclusions

For the commercialization of fuel cells, the stability of the electrodes is still one of the most important factors to affect the cell performance. Various methods and techniques have been investigated to improve the electrochemical activity with better stability for the MCFC electrodes.

Concerning the MCFC cathode, NiO will inevitably dissolve into the molten carbonate electrolyte and transfer from the cathode site to the anode where it should be reduced to Ni, which impacts the performance. By far, more attention has been focused on the protection of a stable oxide layer on the Ni surface. Actually, in most cases, the methods have a positive effect on reducing NiO dissolution in MCFC conditions. Generally, the stable layer can be divided into two classes: metal oxide–doped NiO and mixed-oxide catalysts, such as perovskite. For the former, there may be an inter-facial reaction, which increases the NiO stability. Another explanation is the different Li content on the interface than the bulk, while in the case of perovskite protection, the main contribution is supposed to be the oxygen transfer ability.

In a DIR-MCFC, the main problems affecting the stability of the Ni anode include the alkali poisoning effect, coke formation, and sintering of metal particles with the presence of a high temperature steam. SiO_2, MgO, and Al_2O_3 were discussed to improve the catalyst activity. Among the modifica-tion, MgO shows a positive effect for coke formation reduction with a forma-tion of $Li_yNi_xMg_{1-x-y}O$ solid solution. Whereafter, the adding of TiO_2 or Al_2O_3 was investigated widely to improve the alkali resistance.

Researchers are still devoted to enhancing the stability of electrode by var-ious attempts. But nonetheless, the long-time tests and the protection mecha-nism should be carried out to evaluate the realistic effect on MCFCs.

References

1. McPhail, S.J., Aarva, A., Devianto, H., Bove, R. and Moreno, A. 2011. SOFC and MCFC: Commonalities and opportunities for integrated research. *Int. J. Hydrogen Energy* 36, 10337–10345.
2. O'Hayre, R.P., Cha, S.-W., Colella, W. and Prinz, F.B. 2006. *Fuel Cell Fundamentals.* New York: John Wiley & Sons, pp. 290–292.
3. Brandon, N.P. and Brett, D.J. 2006. Engineering porous materials for fuel cell applications. *Phil. Trans. R. Soc. A* 364, 147–159.
4. Park, H.-H., Jang, C.-L., Shin, H.-S. and Lee, K.-T. 1996. Fabrication and charac-teristics of porous Ni and NiO(Li) cathodes for MCFC. *Korean J. Chem. Eng.* 13 (1), 35–39.
5. Antolini, E. 2011. The stability of molten carbonate fuel cell electrodes: A review of recent improvements. *Appl. Energy* 88, 4274–4293.
6. Antolini, E. 2000. Behaviour of Ni, NiO and $Li_xNi_{1-x}O$ in molten alkali carbon-ates. *J. Mater. Sci.* 35, 1501–1505.
7. Orfield, M.L. and Shores, D.A. 1989. The solubility of NiO in binary mixtures of molten carbonates. *J. Electrochem. Soc.* 136, 2862–2866.
8. Iacovangelo, C.D. 1986. Metal plated ceramic: A novel electrode material. *J. Electrochem. Soc.* 133, 1359–1364.
9. Morita, H., Komoda, M., Mugikura, Y., Izaki, Y., Watanabe, T., Masuda, Y. et al. 2002. Performance analysis of molten carbonate fuel cell using a Li/Na electro-lyte. *J. Power Sources* 112, 509–518.
10. Morita, H., Kawase, M., Mugikura, Y. and Asano, K. 2010. Degradation mecha-nism of molten carbonate fuel cell based on long-term performance: Long-term operation by using bench-scale cell and post-test analysis of the cell. *J. Power Sources* 195, 6988–6996.

11. Fujita, Y., Nishimura, T., Yagi, T. and Matsumura, M. 2003. Degradation of the components in molten carbonate fuel cells with Li/Na electrolyte. *Electrochem.* 71, 7–13.

12. Yoshiba, F., Morita, H., Yoshikawa, M., Mugikura, Y., Izaki, Y., Watanabe, T. et al. 2004. Improvement of electricity generating performance and life expectancy of MCFC stack by applying Li/Na carbonate electrolyte: Test results and analysis of 0.44 m²/10 kW and 1.03 m²/10 kW-class stack. *J. Power Sources* 128, 152–164.

13. Doyon, J.D., Gilbert, T. and Davies, G. 1987. NiO solubility in mixed alkali/alkaline earth carbonates. *J. Electrochem. Soc.* 134, 3035–3038.

14. Tanimoto, K., Miyazaki, Y., Yanagida, M., Kojima, T., Ohtori, N. and Kodama, T. 1995. Effect of addition of alkaline earth carbonate on solubility of NiO in molten Li_2CO_3–Na_2CO_3 eutectic. *Electrochem.* 63, 316–318.

15. Mitsushima, S., Matsuzawa, K., Kamiya, N. and Ota, K. 2002. Improvement of MCFC cathode stability by additives. *Electrochim. Acta* 47, 3823–3830.

16. Yang, B.Y. and Kim, K.Y. 1997. Molten Carbonate Fuel Cell Technology, the Electrochemical Society Proceeding Series PV97-4, in: D. Shores, H. Maru, I. Uchida, and J.R. Selman, Eds., Pennington, NJ: pp. 286–295.

17. Mendoza, L., Albin, V., Cassir, M. and Galtayries, A. 2003. Electrochemical deposition of Co_3O_4 thin layers in order to protect the nickel-based molten carbonate fuel cell cathode. *J. Electroanal. Chem.* 548, 95–107.

18. Kuk, S.T., Song, Y.S., Suh, S.I., Kim, J.Y. and Kim, K. 2001. The formation of $LiCoO_2$ on a NiO cathode for a molten carbonate fuel cell using electroplating. *J. Mater. Chem.* 11, 630–635.

19. Hong, M.Z., Bae, S.C., Lee, H.S., Lee, H.C., Kim, Y.-M. and Kim, K. 2003. A study of the Co-coated Ni cathode prepared by electroless deposition for MCFCs. *Electrochim. Acta* 48, 4213–4221.

20. Fukui, T., Ohara, S., Okawa, H., Hotta, T. and Naito, M. 2000. Properties of NiO cathode coated with lithiated Co and Ni solid solution oxide for MCFCs. *J. Power Sources* 86, 340–346.

21. Lee, H., Hong, M., Bae, S., Lee, H., Park, E. and Kim, K. 2003. A novel approach to preparing nano-size Co3O4-coated Ni powder by the Pechini method for MCFC cathodes. *J. Mater. Chem.* 13, 2626–2632.

22. Ryu, B.H., Han, J., Yoon, S.P., Nam, S.W., Lim, T.-H., Hong, S.-A. and Kim, K.B. 2004. Dissolution behavior of Co-coated NiO cathode in molten $(Li_{0.62}K_{0.38})_2CO_3$ eutectics. *J. Power Sources* 137, 62–70.

23. Kim, Y.-S., Yi, C.-W., Choi, H.S. and Kim, K. 2011. Modification of Ni-based cathode material for molten carbonate fuel cells using Co_3O_4. *J. Power Sources* 196, 1886–1893.

24. Escudero, M.J., Mendoza, L., Cassir, M., Gonzalez, T. and Daza, L. 2006. Porous nickel MCFC cathode coated by potentiostatically deposited cobalt oxide II. Structural and morphological behavior in molten carbonate. *J. Power Sources* 160, 775–781.

25. Inaba, M., Todzuka, Y., Yoshida, H., Grincourt, Y. and Tasaka, A. 1995. Raman spectra of $LiCo_{1-y}Ni_yO_2$. *Chem. Lett.* 24, 889–890.

26. Takizawa, K. and Hagiwara, A. 2001. *Electrochem* (Japanese only, English title is unavailable.) 69, 692–698.

27. Durairajan, A., Colon-Mercado, H., Haran, B., White, R. and Popov, B. 2002. Electrochemical characterization of cobalt-encapsulated nickel as cathodes for MCFC. *J. Power Sources* 104, 157–168.

28. Yazici, M.S. and Selman, J.R. 1999. Oxidation-lithiation of nickel, iron and cobalt in contact with molten carbonate. *Solid State Ionics* 124, 149–160.

29. Sato, T., Hsien-Chang, C., Endo, T. and Shimada, M. 1986. Preparation and electrical properties of $Li_xM_{1-x}O$ (M = Ni, Co). *J. Mater. Sci. Lett.* 5, 552–554.

30. Moore, R.J. and White, J. 1974. Equilibrium relationships in the system NiO-CoO-O_2. *J. Mater. Sci.* 9, 1393–1400.

31. Pérez, F.J., Duday, D., Hierro, M.P., Gómez, C. et al. 2000. Analysis by electrochemical impedance spectroscopy of new MCFC cathode materials. *J. Power Sources* 86, 309–315.

32. Kuk, S.T., Song, Y.S. and Kim, K. 1999. Properties of a new type of cathode for molten carbonate fuel cells. *J. Power Sources* 83, 50–56.

33. Veldhuis, J.B.J., Eckes, F.C. and Plomp, L. 1992. The dissolution properties of $LiCoO_2$ in molten 62:38 mol% Li:K carbonate. *J. Electrochem. Soc.* 139, L6–L8.

34. Veldhuis, J.B.J., van der Molen, S.B., Makkus, R.C. and Broers, G.H.J. 1990. *Ber. Bunsenges Phys. Chem.* 94, 947.

35. Bloom, I., Lanagan, M.T., Krumpelt, M. and Smith, J.L. 1999. The development of $LiFeO_2$–$LiCoO_2$–NiO cathodes for molten carbonate fuel cells. *J. Electrochem. Soc.* 146, 1336–1340.

36. Lundblad, A., Schwartz, S. and Bergman, B. 2000. Effect of sintering procedures in development of LiCoO2- cathodes for the molten carbonate fuel cell. *J. Power Sources* 90, 224–230.

37. Han, J., Kim, S.-G., Yoon, S.P., Nam, S.W. et al. 2002. Performance of $LiCoO_2$-coated NiO cathode under pressurized conditions. *J. Power Sources* 106, 153–159.

38. Li, F., Chen, H.-Y., Wang, C.-M. and Hu, K. 2002. A novel modified NiO cathode for molten carbonate fuel cells. *J. Electroanal. Chem.* 531, 53–60.

39. Simonetti, E. and Lo Presti, R. 2006. Characterization of Ni porous electrode covered by a thin film of $LiMg_{0.05}Co_{0.95}O_2$. *J. Power Sources* 160, 816–820.

40. Kim, S.-G., Yoon, S.P., Han, J., Nam, S.W., Lim, T.-H., Hong, S.-A. and Lim, H.C. 2002. A stabilized NiO cathode prepared by sol-impregnation of $LiCoO_2$ precursors for molten carbonate fuel cells. *J. Power Sources* 112, 109–115.

41. Ringuedé, A., Wijayasinghe, A., Albin, V., Lagergren, C., Cassir, M. and Bergman, B. 2006. Solubility and electrochemical studies of $LiFeO_2$–$LiCoO_2$–NiO materials for the MCFC cathode application. *J. Power Sources* 160, 789–795.

42. Kudo, T., Kihara, K., Hisamitsu, Y., Yu, Q., Mohamedi, M. and Uchida, I. 2002. Electrochemical stability and solubility of Ni–Fe–Co ternary alloy oxides in Li/Na carbonate eutectic as an alternative material for MCFC cathodes. *J. Mater. Chem.* 12, 2496–2500.

43. Park, E., Hong, M., Lee, H., Kim, M. and Kim, K. 2005. A new candidate cathode material as (Co/Mg)-coated Ni powder for molten carbonate fuel cell. *J. Power Sources* 143, 84–92.

44. Cho, J.P. 2000. $LiNi_{0.74}Co_{0.26-x}Mg_xO_2$ cathode material for a Li-ion cell. *Chem. Mater.* 12, 3089–3094.

45. Levin, E.M., McMurdie, H.F. and Robbins, C.R. 1964. Phase diagrams for ceramists: Volume 11, *The American Ceramic Society*, Columbus, Ohio, pp. 52.

46. Daza, L., Rangel, C.M., Baranda, J., Casais, M.T., Martınez, M.J. and Alonso, J.A. 2000. Modified nickel oxides as cathode materials for MCFC. *J. Power Sources* 86, 329–333.

47. Kim, M.H., Hong, M.Z., Kim, Y.-S., Park, E., Lee, H., Ha, H.-W. and Kim, K. 2006. Cobalt and cerium coated Ni powder as a new candidate cathode material for MCFC. *Electrochim. Acta* 51, 6145–6151.

48. Albin, V., Mendoza, L., Goux, A., Ringuedé, A., Billard, A., Briois, P. and Cassir, M. 2006. Morphological, structural and electrochemical analysis of sputter-deposited ceria and titania coatings for MCFC application. *J. Power Sources* 160, 821–826.

49. Ota, K.-I., Matsuda, Y., Matsuzawa, K., Mitsushima, S. and Kamiya, N. 2006. Effect of rare earth oxides for improvement of MCFC. *J. Power Sources* 160, 811–815.

50. Baranda, J. 1997. Ph.D. thesis, Universidad Autónoma de Madrid, Spain.

51. Liu, G., Li, M., Zhou, Y. and Zhang, Y. 2003. Corrosion behavior and strength degradation of Ti_3SiC_2 exposed to a eutectic K_2CO_3 and Li_2CO_3 mixture. *J. Eur. Ceram. Soc.* 23, 1957–1962.

52. Zeng, C.L. and Wu, W.T. 2002. Corrosion of Ni–Ti alloys in the molten $(Li,K)_2CO_3$ eutectic mixture. *Corros. Sci.* 44, 1–12.

53. Perez, F.J., Duday, D., Hierro, M.P. and Gomez, C. 2002. Hot corrosion study of coated separator plates of molten carbonate fuel cells by slurry aluminides. *Surf. Coat. Technol.* 161, 293–301.

54. Chauvaut, V. and Cassir, M. 1999. Behaviour of titanium species in molten $Li_2CO_3+Na_2CO_3$ and $Li_2CO_3+K_2CO_3$ in the anodic conditions used in molten carbonate fuel cells: II. Electrochemical intercalation of Li^+ in Li_2TiO_3 at 600 and 650°C. *J. Electroanal. Chem.* 474, 9–15.

55. Liu, B., Zhang, Y. and Zhang, L. 2009. Oxygen reduction mechanism at Ba0.5Sr0.5Co0.8Fe0.2O3–δ cathode for solid oxide fuel cell. *Int. J. Hydrogen Energy* 34, 1008–1014.

56. Lee, K.T. and Manthiram, A. 2006. Comparison of $Ln_{0.6}Sr_{0.4}CoO_{3-\delta}$ (Ln = La, Pr, Nd, Sm, and Gd) as cathode materials for intermediate temperature solid oxide fuel cells. *J. Electrochem. Soc.* 153, A794–A798.

57. Huang, S., Peng, C. and Zong, Z. 2008. A high-performance $Gd_{0.8}Sr_{0.2}CoO_3$–$Ce_{0.9}Gd_{0.1}O_{1.95}$ composite cathode for intermediate temperature solid oxide fuel cell. *J. Power Sources* 176, 102–106.

58. Chen, J., Liang, F., Liu, L., Jiang, S., Chi, B., Pu, J. and Li, J. 2008. Nano-structured (La,Sr) (Co,Fe) O3+YSZ composite cathodes for intermediate temperature solid oxide fuel cells. *J. Power Sources* 183, 586–589.

59. Song, S.A., Jang, S.-C., Han, J., Yoon, S.P., Nam, S.W., Oh, I.-H. and Lim, T.-H. 2011. Enhancement of cell performance using a gadolinium strontium cobaltite coated cathode in molten carbonate fuel cells. *J. Power Sources* 196, 9900–9905.

60. Wiemhofer, H.-D., Bremes, H.-G. and Nigge, U. 2004. Mixed conduction and electrode properties of doped cobaltites and chromites. *Solid State Ionics* 175, 93–98.

61. Yuh, C.Y. and Selman, J.R. 1991. The polarization of molten carbonate fuel cell electrodes: I. Analysis of steady-state polarization data. *J. Electrochem. Soc.* 138, 3642–3648.

62. Ganesan, P., Colon, H., Haran, B. and Popov, B.N. 2003. Performance of La0.8Sr0.2CoO3 coated NiO as cathodes for molten carbonate fuel cells. *J. Power Sources* 115, 12–18.

63. Yuh, C.Y. and Selman, J.R. 1988. Characterization of fuel cell electrode processes by AC impedance. *AIChE J.* 34 (12), 1949–1958.

64. Fukui, T., Ohara, S., Okawa, H., Naito, M. and Nogi, K. 2003. Synthesis of metal and ceramic composite particles for fuel cell electrodes. *J. Eur. Ceram. Soc.* 23, 2835–2840.

65. Okawa, H., Lee, J.-H., Hotta, T., Ohara, S., Takahashi, S., Shibahashi, T. and Yamamasu, Y. 2004. Performance of NiO/MgFe$_2$O$_4$ composite cathode for a molten carbonate fuel cell. *J. Power Sources* 131, 251–255.

66. Clarke, S.H., Dicks, A.L., Pointon, K., Smith, T.A. and Swann, A. 1997. Catalytic aspects of the steam reforming of hydrocarbons in internal reforming fuel cells. *Catal. Today* 38, 411–423.

67. Lv, P., Chang, J., Xiong, Z., Huang, H., Wu, C. and Chen, Y. 2003. Biomass air-steam gasification in a fluidized bed to produce hydrogen-rich gas. *Energy Fuels* 17, 677–682.

68. Sime, R., Kuehni, J., D'Souza, L., Elizondo, E. and Biollaz, S. 2003. The redox process for producing hydrogen from woody biomass. *Int. J. Hydrogen Energy* 28, 491–498.

69. Garcia, L., Benedicto, A., Romeo, E., Salvador, M.L., Arauzo, J. and Bilbao, R. 2002. Hydrogen production by steam gasification of biomass using Ni-Al coprecipitated catalysts promoted with magnesium. *Energy Fuels* 16, 1222–1230.

70. Asadullah, M., Miyazawa, T., Ito, S.-I., Kunimori, K. and Tomishige, K. 2003. Catalyst performance of Rh/CeO$_2$/SiO$_2$ in the pyrogasification of biomass. *Energy Fuels* 17, 842–849.

71. Li, Z., Devianto, H., Yoon, S.P., Han, J., Lim, T.-H. and Lee, H.-I. 2010. Electrolyte effect on the catalytic performance of Ni-based catalysts for direct internal reforming molten carbonate fuel cell. *Int. J. Hydrogen Energy* 35, 13041–13047.

72. Park, D.S., Li, Z., Devianto, H. and Lee, H.-I. 2010. Characteristics of alkali-resistant Ni/MgAl$_2$O$_4$ catalyst for direct internal reforming molten carbonate fuel cell. 2010. *Int. J. Hydrogen Energy* 35, 5673–5680.

73. Park, K., Kim, K.Y., Lu, L., Lim, T.-H., Hong, S.-A. and Lee, H.-I. 2007. Structural characteristics of (NiMgAl)O$_x$ prepared from a layered double hydroxide precursor and its application in direct internal reforming molten carbonate fuel cells. *Fuel Cells* 7, 211–217.

74. Frusteri, F. and Freni, S. 2007. Bio-ethanol, a suitable fuel to produce hydrogen for a molten carbonate fuel cell. *J. Power Sources* 173, 200–209.

75. Frusteri, F., Freni, S., Spadaro, L., Chiodo, V., Bonura, G., Donato, S. and Cavallaro, S. 2004. H$_2$ production for MC fuel cell by steam reforming of ethanol over MgO supported Pd, Rh, Ni and Co catalysts. *Catal. Comm.* 5, 611–615.

76. Berger, R.J., Doesburg, E.B.M., van Ommen, J.G. and Ross, J.R.H. 1996. Nickel catalysts for internal reforming in molten carbonate fuel cells. *Appl. Catal., A* 143, 343–365.

77. Zhou, J., Zhang, X., Zhang, J., Liu, H., Zhou, L. and Yeung, K. 2009. Preparation of alkali-resistant, Sil-1 encapsulated nickel catalysts for direct internal reforming-molten carbonate fuel cell. *Catal. Comm.* 10, 1804–1807.

78. Kim, H.-W., Kang, K.-M., Kwak, H.-Y. and Kim, J.H. 2011. Preparation of supported Ni catalysts on various metal oxides with core/shell structures and their tests for the steam reforming of methane. *Chem. Eng. J.* 168, 775–783.

79. Fatsikostas, A.N., Kondarides, D.I. and Verykios, X.E. 2002. Production of hydrogen for fuel cells by reformation of biomass-derived ethanol. *Catal. Today* 75, 145–155.

80. Llorca, J., Homs, N., Sales, J. and de la Piscina, P.R. 2002. Efficient production of hydrogen over supported cobalt catalysts from ethanol steam reforming. *J. Catal.* 209, 306–317.

81. Freni, S., Cavallaro, S., Mondello, N., Spadaro, L. and Frusteri, F. 2003. Production of hydrogen for MC fuel cell by steam reforming of ethanol over MgO supported Ni and Co catalysts. *Catal. Commun.* 4, 259–268.

82. Matsumura, M. and Hirai, C. 1998. Deterioration mechanism of direct internal reforming catalyst. *J. Chem. Eng. Jpn.* 31, 734–740.

83. Rostrup-Nielsen, J.R. and Christiansen, L.J. 1995. Internal steam reforming in fuel cells and alkali poisoning. *Appl. Catal. A: Gen.* 126, 381–390.

84. Dicks, A.L. 1998. Advances in catalysts for internal reforming in high temperature fuel cells. *J. Power Sources* 71, 111–122.

85. Sugiura, K., Yamauchi, M., Tanimoto, K. and Yoshitani, Y. 2005. Evaluation of volatile behaviour and the volatilization volume of molten salt in DIR-MCFC by using the image measurement technique. *J. Power Sources* 145, 199–205.

86. Sugiura, K. and Ohtake, K. 1995. Deterioration of a catalyst's activity by liquid phase MC poisoning in DIR-MCFC. *J. Chem. Eng. Jpn.* 21 (6), 1170–1178.

87. Miyake, Y., Nakanishi, N., Nakajima, T., Itoh, Y., Saitoh, T., Saiai, A. and Yanaru, H. 1995. A study on degradation phenomena of reforming catalyst in DIR-MCFC. *J. Chem. Eng. Jpn.* 21 (6), 1104–1109.

88. Passaacqua, E., Frini, S. and Barone, F. 1998. Alkali resistance of tape-cast SiC porous ceramic membranes. *Mater. Lett.* 34, 257–262.

89. Heidebrecht, P. and Sundmacher, K. 2005. Optimization of reforming catalyst distribution in a cross-flow molten carbonate fuel cell with direct internal reforming. *Ind. Eng. Chem. Res.* 44, 3522–3528.

90. Berger, R.J., Doesburg, E.B.M., van Ommen, J.G. and Ross, J.R.H. 1996. Investigation of alkali carbonate transport toward the catalyst in internal reforming MCFCs. *J. Electrochem. Soc.* 143, 3186–3191.

91. Matsumura, M. and Hirai, C. 1998. Transport mechanism of electrolyte vapor to reforming catalyst. *Ind. Eng. Chem. Res.* 37, 1793–1798.

92. Sugiura, K., Yodo, T., Yamauchi, M. and Tanimoto, K. 2006. Visualization of electrolyte volatile phenomenon in DIR-MCFC. *J. Power Sources* 157, 739–744.

93. Mas, V., Kipreos, R., Amadeo, N. and Laborde, M. 2006. Thermodynamic analysis of ethanol/water system with the stoichiometric method. *Int. J. Hydrogen Energy* 31, 21–28.

94. Sasaki, K., Watanabe, K. and Teraoka, Y. 2004. Direct-alcohol SOFCs: Current–voltage characteristics and fuel gas compositions. *J. Electrochem. Soc.* 151, A965–A970.

95. Devianto, H., Li, Z.L., Yoon, S.P., Han, J., Nam, S.W., Lim, T.-H. and Lee, H.-I. 2010. The effect of Al addition on the prevention of Ni sintering in bio-ethanol steam reforming for molten carbonate fuel cells. *Int. J. Hydrogen Energy* 35, 2591–2596.

96. Cavallaro, S., Chiodo, V., Freni, S., Mondello, N. and Frusteri, F. 2003. Performance of Rh/Al$_2$O$_3$ catalyst in the steam reforming of ethanol: H$_2$ production for MCFC. *Appl. Catal. A: Gen.* 249, 119–128.

97. Roh, H.-S., Jung, Y., Koo, K.Y., Jung, U.H., Seo, Y.-S. and Yoon, W.L. 2010. Steam reforming of methane over highly active and KOH-resistant Ni/Al$_2$O$_3$ catalysts for direct internal reforming (DIR) in a molten carbonate fuel cell (MCFC). *Appl. Catal. A: Gen.* 383, 156–160.

98. Frusteri, F., Arena, F., Calogero, G., Torre, T. and Parmaliana, A. 2001. Potassium-enhanced stability of Ni/MgO catalysts in the dry-reforming of methane. *Catal. Commun.* 2, 49–56.

99. Frusteri, F., Freni, S., Chiodo, V., Spadaro, L., Di Blasi, O., Bonura, G. and Cavallaro, S. 2004. Steam reforming of bio-ethanol on alkali-doped Ni/MgO catalysts: Hydrogen production for MC fuel cell. *Appl. Catal. A: Gen.* 270, 1–7.

100. Klouz, V., Fierro, V., Denton, P., Katz, H., Lisse, J.P., Bouvout-Mauduit, S. and Mirodatos, C. 2002. Ethanol reforming for hydrogen production in a hybrid electric vehicle: Process optimization. *J. Power Sources* 105, 26–34.

101. Freni, S., Cavallaro, S., Mondello, N., Spadaro, L. and Frusteri, F. 2002. Steam reforming of ethanol on Ni/MgO catalysts: H_2 production for MCFC. *J. Power Sources* 108, 53–57.

102. Moon, H.-D., Lim, T.-H. and Lee, H.-I. 1999. Chemical poisoning of Ni/MgO catalyst by alkali carbonate vapor in the steam reforming reaction of DIR-MCFC. *Bull. Korean Chem. Soc.* 20, 1413–1417.

103. Moon, H.-D., Kim, J.-H., Ha, H.Y., Lim, T.-H., Hong, S.-A. and Lee, H.-I. 1999. Poisoning of the Ni/MgO catalyst by alkali carbonates in a DIR-MCFC. *J. Korean Ind. Eng. Chem.* 10, 754–759.

104. Choi, J.-S., Kwon, H.-H., Lim, T.-H., Hong, S.-A. and Lee, H.-I. 2004. Development of nickel catalyst supported on MgO–TiO_2 composite oxide for DIR-MCFC. *Catal. Today* 93–95, 553–560.

105. Ruckenstein, E. and Hu, Y.H. 1999. Methane partial oxidation over NiO/MgO solid solution catalysts. *Appl. Catal. A: Gen.* 183, 85–92.

106. Frusteri, F., Freni, S., Chiodo, V., Spadaro, L., Blasi, O.D., Bonura, G. et al. 2004. Steam reforming of bio-ethanol on alkali-doped Ni/MgO catalysts: Hydrogen production for MC fuel cell. *Appl. Catal. A* 270, 1–7.

107. Choi, J.-S., Yun, J.-S., Kwon, H.-H., Lim, T.-H., Hong, S.-A. and Lee, H.-I. 2005. Effect of lithium carbonate on nickel catalysts for direct internal reforming MCFC. *J. Power Sources* 145, 652–658.

108. Parmaliana, A., Frusteri, F., Tsiakaras, P. and Giordano, N. 1988. Out of the cell performance of reforming catalysts for direct molten carbonate fuel cells (DMCFC). *Int. J. Hydrogen Energy* 13, 729–734.

109. Wee, J.-H. 2006. Performance of a unit cell equipped with a modified catalytic reformer in direct internal reforming-molten carbonate fuel cell. *J. Power Sources* 156, 288–293.

110. Arena, F., Frusteri, F. and Parmaliana, A. 1999. Alkali promotion of Ni/MgO catalysts. *Appl. Catal. A: Gen.* 187, 127–140.

111. Antolini, E. 2001. Formation of ternary lithium oxide nickel oxide magnesium oxide solid solution from the Li/Ni/MgO system. *Mater. Lett.* 51, 385–388.

112. Sugiura, K., Soga, M., Yamauchi, M. and Tanimoto, K. 2008. Volatilization behavior of Li/Na carbonate as an electrolyte in MCFC. *ECS Trans.* 12, 355–360.

113. Li, Z., Devianto, H., Kwon, H.-H., Yoon, S.P., Lim, T.-H. and Lee, H.-I. 2010. The catalytic performance of Ni/$MgSiO_3$ catalyst for methane steam reforming in operation of direct internal reforming MCFC. *J. Ind. Eng. Chem.* 16, 485–489.

114. Noda, S., Nishioka, M. and Sadakta, M. 1999. Gas-phase hydroxyl radical emission in the thermal decomposition of lithium hydroxide. *J. Phys. Chem. B* 103, 1954–1959.

115. Ham, H.C., Maganyuk, A.P., Han, J., Yoon, S.P., Nam, S.W., Lim, T.H. et al. 2007. Preparation of Ni–Al alloys at reduced temperature for fuel cell applications. *J. Alloy Compd.* 446, 733–737.

116. Xu, Y., Kameoka, S., Kishida, K., Demura, M., Tsai, A. and Hirano, T. 2005. Catalytic properties of alkali-leached Ni_3Al for hydrogen production from methanol. *Intermetallics* 13 (2), 151–155.

117. Takehira, K., Shishido, T., Wand, P. and Kosaka, T. 2004. Autothermal reforming of CH_4 over supported Ni catalysts prepared from Mg–Al hydrotalcite-like anionic clay. *J. Catal.* 221, 43–54.
118. Gadalla, A.M. and Bower, B. 1998. The role of catalyst support on the activity of nickel for reforming methane with CO_2. *Chem. Eng. Sci.* 43, 3049–3062.

7

Nanostructured Nitrogen–Carbon–Transition Metal Electrocatalysts for PEM Fuel Cell Oxygen Reduction Reaction

Gang Wu, Zhongwei Chen, and Jiujun Zhang

CONTENTS

7.1 Introduction 195
7.2 Roles of Transition Metals 198
7.3 Roles of Nitrogen 201
7.4 Roles of Supporting Templates 204
    7.4.1 Carbon Black Nanoparticles.. 204
    7.4.2 Carbon Nanotubes... 205
    7.4.3 Graphene and Graphene Oxide 207
    7.4.4 Noncarbon Supports 207
7.5 *In Situ* Formed Graphitized Carbon Nanostructures......................... 209
    7.5.1 Nitrogen Precursors 210
    7.5.2 Transition Metals 214
    7.5.3 Heating Temperature 215
7.6 Conclusion 216
References.. 217

7.1 Introduction

The oxygen reduction reaction (ORR) represents one of the most important electrochemical reactions, which is crucial to a variety of electrochemical energy storage and conversion technologies, particularly low-temperature polymer electrolyte membrane fuel cells (PEMFCs). The ORR in acidic and alkaline media can be correspondingly described by the following electrochemical reactions and standard reaction potentials ($E°$):

$$O_2 + 4H^+ + 4e^- \rightarrow 2H_2O \quad E° = 1.229 \text{ V (acidic media)} \tag{7.1}$$

$$O_2 + 2H_2O + 4e^- \rightarrow 4OH^- \quad E° = 0.401 \text{ V (alkaline media)} \tag{7.2}$$

Unfortunately, the ORR electrode kinetics is quite sluggish with a high over-potential, which has been the main obstacle to its effective usage in PEMFCs. In order to speed up the ORR kinetics, electrocatalysts are necessary. In practice, platinum (Pt)-based catalysts are the state of the art in terms of activity and durability for the ORR. However, prohibitive cost and scarcity have limited widespread implementation of Pt metal-based catalysts [1]. To fully realize these technologies, such as PEMFCs, highly active, durable, and inexpensive catalysts based solely on earth-abundant elements are desperately needed for the ORR. In the most recent decade, some promising nonprecious metal catalysts (NPMCs) for ORR have been studied intensively, including organometallic components, nonprecious metal chalcogenides, and nitrogen-doped carbon nanotubes (CNTs) and graphene catalysts [2–4]. Although some of the best performance NPMCs have exhibited ORR catalytic activity close or even superior to Pt-based catalysts in alkaline electrolyte [5–7], the grand challenge of developing an NPMC for Nafion membrane and ionomer-based acidic PEMFCs has remained for decades due to both insufficient activity and durability. Recent progresses in the development of high-performance NPMCs for PEMFCs have suggested that a type of catalyst synthesized from Fe, Co, N, and C (abbreviated as M–N–C with M=Fe or Co) has the potential to efficiently catalyze ORR, showing approaching activity and durability to Pt catalysts [8,9].

Such promising M–N–C catalysts are obtained from simultaneously heat-treating precursors of nitrogen, carbon, and transition metals at 700°C–1000°C. Pioneer explorations were started with the pyrolysis of transition metal–containing macrocycles by Jasinski [10] and Gupta et al. [11]. The resulting catalysts had improved ORR activity compared to their non-pyrolyzed counterparts; however, they still demonstrated insufficient performance in an acidic environment. Later studies replaced the expensive macrocycle precursors with a wide variety of common and cost-effective nitrogen precursors (ammonia, acetonitrile, amines, etc.), transition metal inorganic salts (sulfates, nitrates, acetates, hydroxides, and chlorides), and carbon supports (Vulcan XC-72, Ketjenblack, BlackPearls, etc.) [9,12–17]. Although a heat treatment of almost any mixture of nitrogen, carbon, and metal species can result in a material that is ORR active, the resulting ORR activity is greatly dependent on the selection of the employed precursors (nitrogen, transition metal, and supports) as well as synthesis conditions (i.e., heat-treatment duration, temperature, preparation technique) [3,4,18–20].

Regarding the type of metals used for NPMC synthesis, the experimental data indicate that Fe and Co are more efficient than other transition metals in producing active catalysts for the ORR [15]. However, it is worth noting that these two transition metals seem to contribute differently to the formation of active sites. The active sites generated in the presence of Co species appear to have electrochemical properties similar to that exhibited by metal-free nitrogen-doped carbon (N-C) catalysts, forming abundant CN_x structures (pyridinic, quaternary nitrogens) as potential active sites [12].

Unlike Co, there is increasing evidence to support an assumption that Fe species directly participate in the active-site by coordinating with nitrogen and carbon, forming a Fe-Nx/C moiety with higher intrinsic activity for ORR [9,18,20]. In addition, electrochemical kinetic analysis indicated that the values of the Tafel slopes are different between polyaniline (PANI) derived Co-N-C (67 mV dec^{-1}) and Fe–N–C (87 mV dec^{-1}) catalysts [8]. As the Fe–N–C catalyst is more active and exhibits a more than 100 mV positive shift in ORR onset potential relative to the Co–N–C catalyst, this significant difference

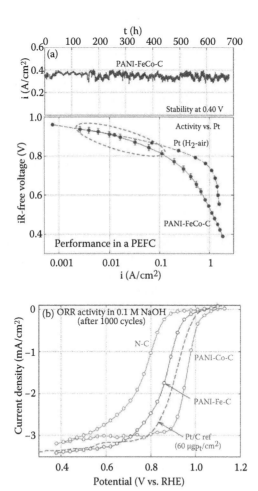

FIGURE 7.1

Activity and stability of NPMC ORR catalysts developed in Los Alamos in (a) Nafion-based PEMFCs. (From Wu, G. et al. *Science*, 332, 443–447, 2011. Reprinted with permission of AAAS). (b) Alkaline aqueous solution using PANI as a nitrogen–carbon precursor. (Reproduced from Wu, G. et al. *ECS Transactions*, 41, 1709–1717. Copyright 2011. The Electrochemical Society. With permission.)

in Tafel slope implies that the nature of their active sites is likely to be different. According to theoretical calculation of Tafel slopes for ORR catalysts [21,22], reaction rates on the Co–N–C catalyst are primarily determined by the migration of adsorbed oxygen intermediates. A more complicated ORR mechanism on the Fe–N–C catalysts is likely, simultaneously involving both intermediate migration and charge transfer as the rate-determining steps [6,8].

Regarding nitrogen precursors used for NPMC synthesis, larger molecular polymer and interlinked complexes may represent a favorable nitrogen–carbon template compound when compared to small molecular nitrogen precursors, such as NH_3 and amine [13]. In particular, the PANI-derived catalysts exhibited much higher activity and stability relative to ethylene diamine (EDA)–derived ones [13]. Because of the similarity between the structures of aromatic PANI and graphite, the heat treatment of PANI could facilitate the incorporation of nitrogen-containing active sites into the graphitized carbon matrix. For example, Wu et al. [6] showed that although several catalysts gave promising oxygen reduction activity, only PANI-derived formulations appeared to combine high ORR activity and four-electron selectivity with much improved performance durability as shown in Figure 7.1.

The exact nature of the active site(s) created during the heat treatment for this type of promising NPMC remains unknown due to a lack of direct evidence about whether the transition metal participates in the active site or simply catalyzes its formation. However, in either case, the nitrogen–carbon species (e.g., in-plane, edge, subsurface) embedded within the carbon nanostructures are likely critical to the active-site performance. In this chapter, nanoparticles and nanostructures associated with ORR catalysis on NPMCs are reviewed and discussed with focus on nitrogen doping, supports, and *in situ* formed graphitized carbon materials.

7.2 Roles of Transition Metals

Although metal-free nitrogen-doped carbon materials show some extent of activity toward ORR, the addition of transition metals is indispensable in increasing both catalyst activity and durability in such types of heat-treated NPMCs [15,23]. As discussed above, the type of metals can have a strong effect on NPMC ORR activity. For example, the influence of different transition metal ions on the ORR activity of polyacrylonitrile-based materials prepared by a heat-treatment process was studied using oxygen (air) cathodes in alkaline and acidic solutions [24]. These results indicated that the identity of the metallic center in the precursor plays a critical role in activity improvements after heat treatment. All experimental evidence clearly shows that compared to other metal centers, iron and cobalt centers exhibit

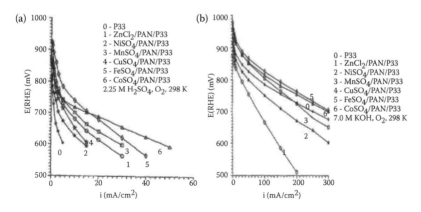

FIGURE 7.2
Polarization curves of oxygen electrodes made from different materials in (a) 2.25 M H₂SO₄ and (b) 7.0 M KOH at 25°C. (Reprinted from *Journal of Power Sources*, 38, Ohms, D. et al. Influence of metal ions on the electrocatalytic oxygen reduction of carbon materials prepared from pyrolyzed polyacrylonitrile, 327–334, Copyright 1992, with permission from Elsevier.)

the highest activity toward the ORR in both acid and alkaline electrolyte (Figure 7.2). In particular, under the same experimental conditions in acidic media, the best results were obtained using Co, followed by Fe and Mn. However, Fe-containing catalysts showed more positive onset potentials than Co catalysts, suggesting higher intrinsic ORR activity associated with Fe. Importantly, the Fe-containing catalysts exhibited more predominantly four-electron selectivity for the ORR. On the other hand, Fe and Co electrocatalysts showed similar performance in alkaline electrolyte.

Regarding the structures of the catalyst active site in Fe- and Co-containing NPMCs, it seems that the catalyst active site structure in the Fe-based catalyst is different from that in the Co-based catalyst. In the Co-derived catalyst, abundant CN_x structures (pyridinic, quaternary nitrogens) formed via catalysis of Co are likely the potential active sites [12]. On the contrary, an Fe–N_x moiety may be a type of potential active site with higher intrinsic activity for ORR [9,18,20]. For example, the average environments of these two metals in PANI-derived Co and Fe catalysts are completely different as analyzed by *ex situ* x-ray absorption fine structure (XAFS) for the coordination environment of noncrystalline transition metal species as shown in Figure 7.3a. The Fe spectrum displays only a single, low-amplitude peak at short R (R = distance from Fe atom) and no long-range order, whereas the Co exhibits an extended order and a nearest neighbor at a significantly longer distance. After fitting, O/N–Fe is dominant in the Fe catalysts (signals from O and N in the local environment of the metal are equivalent in EXAFS). Consistent with the XRD pattern of the Co catalyst, the Co EXAFS is well fit by a series of neighbor shells that correspond well with those of Co_9S_8 that

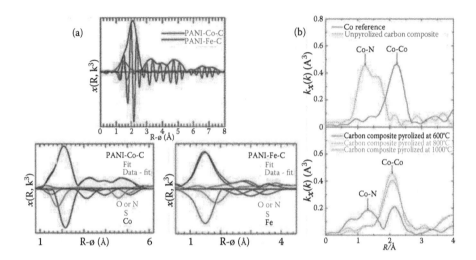

FIGURE 7.3

EXAFS spectra for (a) PANI-derived Fe and Co catalysts. (Wu, G. et al. *Journal of Materials Chemistry*, 21, 11392–11405, 2011. Reproduced by permission of The Royal Society of Chemistry.) (b) EDA-derived Co catalyst. (Reprinted from *Journal of Power Sources*, 183, Nallathambi, V. et al. Development of high performance carbon composite catalyst for oxygen reduction reaction in PEM proton exchange membrane fuel cells, 34–42, Copyright 2008, with permission from Elsevier.)

perhaps is a byproduct of catalyst synthesis as a sulfur-containing oxidant was used to polymerize aniline. However, there are small fractions of Co species with O shell at 1.5 Å and the possible S shell at 1.8 Å that does not belong to Co_9S_8. These forms are likely to be important because Co_9S_8 cannot account for the observed degree of ORR activity. In the case of EDA-derived Co catalysts, EXAFS spectra for the unpyrolyzed sample indicated that there is one major peak centered at R values of about 1.2 Å corresponding to the Co–N which confirms the presence of the Co species coordinated with nitrogen groups on the carbon surface (Figure 7.3b). The Co–Co peak becomes strong with increasing the heat-treatment temperature, and only a Co–Co peak is observed when the catalyst was heated at 800°C and 1000°C. This means that the Co–N_x chelate complexes decompose at high pyrolysis temperatures above 800°C, resulting in the formation of the metallic Co species. Hence, both heat-treated PANI and EDA-derived Co catalysts contain much less Co–N species compared to the Fe catalysts.

Regarding the induction of nitrogen doping into carbon structures, independently of the choice of a nitrogen precursor used for catalyst preparation (i.e., EDA, HAD, or PANI), the Fe–N–C has relatively higher peak intensity for quaternary nitrogen compared to the resulting Co–N–C. This indicates that nitrogen doping is favored at the interior, rather than at the

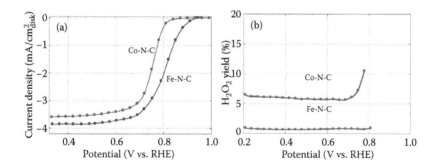

FIGURE 7.4
(a) Steady-state ORR polarization plots and (b) four-electron selectivity (H_2O_2 yield) measured with PANI-derived Co and Fe catalysts. (Wu, G. et al. *Journal of Materials Chemistry*, 21, 11392–11405, 2011. Reproduced by permission of The Royal Society of Chemistry.)

edge planes of the graphene layers in the presence of Fe [18,25]. Generally, during the pyrolysis of carbon-containing nitrogen species, the ratio of pyridinic and quaternary nitrogen would reach equilibrium [26]. The addition of Fe seems to be favorable for shifting the equilibrium toward quaternary nitrogen [25].

Probably due to the differences of the physicochemical properties and ORR mechanisms associated with Fe- and Co-based catalysts, both rotating ring-disk electrode (RRDE) tests have clearly indicated that the ORR activity and four-electron selectivity of the Co catalysts is generally lower than those of Fe catalysts as shown in Figure 7.4 using PANI-derived catalysts as an example.

Although the intrinsic activity derived from Co-associated active sites is lower than that of Fe, relative to Fe-derived catalysts, Co species usually lead to highly graphitized carbon nanostructures in catalysts [8]. Generally, this structural configuration is a key to electrocatalysis, providing both enhanced electronic conductivity and corrosion resistance [27]. Thus, through simultaneously taking advantage of the catalytic properties of Co and Fe, binary CoFe-based NPMCs have exhibited enhancements in both activity and durability as demonstrated in EDA–CoFe [12,28] and HDA–CoFe [25] as well as PANI–FeCo catalysts [8].

7.3 Roles of Nitrogen

There has been increasing evidence showing that the electrochemical and physical properties of carbon-based catalysts are extremely sensitive to heteroatom (N, B, S, and P) doping into the carbon structure, leading to a desirable way to modify the ORR catalytic properties of the carbon material [5,29–32]. Among them, nitrogen doping has been generally believed to be the most important to ORR. According to the literature [14,26], in nitrogen-containing carbon materials pyrolyzed above 700°C, the nitrogen atoms can incorporate into the graphene layers to replace carbon atoms at different sites (Figure 7.5).

In doing so, they show various binding energies in the XPS spectra, including (1) pyridinic N (398.6 ± 0.3 eV), (2) pyrrolic N (400.5 ± 0.3 eV), and (3) quaternary N (401.3 ± 0.3 eV). The pyridinic form comprises the nitrogen atoms doped at the edges of the graphitic carbon layers, and the quaternary form comprises nitrogen atoms doped inside the basal graphitic carbon plane. The pyrrolic form is assigned to nitrogen atoms in a pentagon structure but is considered indistinguishable from pyridone (pyridinic-N next to an OH group) [26]. However, both of these pyrrolic nitrogen and pyridone species have been shown to decompose at temperatures above 600°C, forming pyridinic-N and quaternary-N [26]. Importantly, due to difficulties in identifying and distinguishing various nitrogen species by XPS, the presence of other nitrogen functionalities may be falsely identified as quaternary-N. Quaternary nitrogen could also include protonated pyridine species in addition to the "graphitic" nitrogen in which nitrogen is within a graphite plane and bonded to three carbon atoms. In addition, the high-binding energy assignments (402–405 eV) in the N 1s region may be related to the interaction of graphitic nitrogen with other nitrogen atoms or to the differences in the binding energy of graphitic nitrogen depending on its position in the graphite plane [14,26]. Primarily, in the M–N–C catalysts, although positive species identification has proven difficult by XPS, nitrogen can be viewed as an *n*-type carbon dopant that assists in the formation of disordered carbon nanostructures and donates electrons to the carbon, theoretically facilitating the ORR.

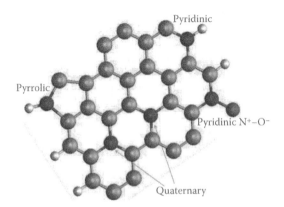

FIGURE 7.5
Scheme of nitrogen-doped carbon structures in different forms.

Nitrogen doping will lead to subtle changes of graphitic structures. The N-doped graphitic carbon nanostructure shows more disorder (curvature and dislocations) in the graphene stacking due to the propensity of substituted nitrogen for the formation of pentagonal defects in the graphene sheets [33]. Using Raman analysis, Maldonado et al. [34] found that the disorder in N-doped CNTs (N-CNTs) linearly increased with nitrogen content, providing direct evidence that the surface disorder is related to nitrogen doping. Additionally, the N-doped carbon structure appears more compact than the nondoped analogues. The shorter compartment distances caused by nitrogen doping were also detected in the CN_x films [35] and N-doped multiwalled CNTs (MWNTs) grown by chemical vapor deposition (CVD) [36] as explained by the fact that N inclusion encourages nanocrystallite formation and suppresses surface diffusion of the carbon during graphitization. Therefore, the disordered nanostructure and defects in the edge and interior sites in the N-doped graphitic layer may serve as active sites or facilitate adjacent carbon atoms as the sites for adsorption and reduction of oxygen.

Therefore, it is reasonable to expect that the doped nitrogen atoms in the carbon structure behave as reactive sites and show increased ORR reactivity due to favorable morphology modification and high electron density generated from the incorporation of nitrogen. Therefore, it is generally believed that nitrogen doping sites on graphitic carbon layers can enhance the ORR. These doped N–C structures in M–N–C catalysts were found to be controllable through tuning catalyst synthesis conditions in terms of maximizing catalytic activity [18].

The content and relative ratios of different types of nitrogen species are highly dependent on the heat-treatment temperature as shown in Figure 7.6. The N 1s XPS spectra of the catalysts indicate that raising the heat-treatment temperature leads to a decrease in total nitrogen content. The ratio of quaternary to pyridinic nitrogen rises with an increasing heating temperature from 600°C to 1000°C [18,37]. It is of note, however, that the decreased nitrogen content with the heat-treatment temperature does not lead to a commensurate drop in ORR activity. This may suggest that the ORR activity is not dependent on the total amount of incorporated nitrogen, but more on how the nitrogen is incorporated into doped nitrogen–carbon hybrid nanostructures. A similar trend was also found for Co-based catalysts and the TiO_2 supported Fe catalysts [6,37]. Although the pyridinic nitrogen has been believed to facilitate oxygen reduction [14], these data may indicate that the presence of quaternary species and optimal ratios of quaternary to pyridinic nitrogen are also important factors in providing a geometric and electronic structure for ORR active sites. However, because the total nitrogen content is decreased with heating temperatures, the absolute content of quaternary nitrogen does not actually increase. Thus, there is no exclusive correlation between any type of nitrogen content and the active-site density yet.

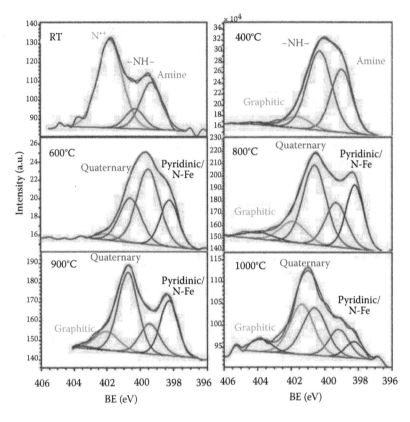

FIGURE 7.6

N 1s spectra of PANI-derived Fe–N–C catalyst as a function of heat-treatment temperature. (Wu, G. et al. *Journal of Materials Chemistry, 21*, 11392–11405, 2011. Reproduced by permission of The Royal Society of Chemistry.)

7.4 Roles of Supporting Templates

7.4.1 Carbon Black Nanoparticles

Support nanoparticles are key components in NPMCs to provide high specific surface areas and good electronic conductivity for low polarization and improved mass transport in the catalyst layer. Similar to carbon-supported Pt catalysts, the electrochemical properties of NPMCs can also be significantly affected by the supporting material [38,39]. The most effective support materials for NPMCs are carbon black–based ones, including Vulcan XC-72, Ketjenblack, and Black Pearls, which have a good balance of electron conductivity, surface area, corrosion resistance, and cost for optimum active-site distribution and catalyst utilization. With respect to this, a variety of carbon supports with different native disordered carbon contents was compared for Fe–N–C catalyst synthesis in which NH_3 was used for a nitrogen source. After heat treatments, the activity of the obtained catalyst was found to correlate well with the content of the native disordered carbon in the supports. It was assumed that the high activity should be linked to a high density of Fe–N–C catalytic sites hosted in the micropores formed by NH_3 etching of the disordered carbon phase [16]. In particular, Black Pearls 2000 (Cabot) (micropore surface area 934 $m^2\,g^{-1}$) was used as the optimal microporous carbon support to synthesize an Fe-based catalyst with much improved ORR activity for PEMFCs [9]. A substantial increase in ORR active site density was obtained when a mixture of carbon nanoparticle support, 1, 10-phenanthroline, and ferrous acetate was ball-milled and then pyrolyzed twice, first in argon and then in ammonia. The active site structures in this class of catalyst are believed to contain iron cations coordinated by pyridinic nitrogen functionalities in the interstices of graphitic crystallites within the micropores of carbon nanoparticles.

Although a correlation between catalyst microporosity and activity has been found, it also remains unclear that different families of active NPMCs should necessarily have the same active site structures and formation mechanisms. In a different study, it was found that the microporosity of NPMC does not dictate the activity for PANI-derived catalysts fabricated using Ketjenblack EC-300J nanoparticles (~20 nm) as supports (Figure 7.7) [8]. Surface area values (apparent BET and microporous) of the final catalyst after heat treatment and acid leaching were much lower than the original carbon support, attributed to pore blocking by PANI decomposition products [40]. In particular, the second heat treatment following acid leaching, which was believed to be a crucial step in the NPMC synthesis, was found to lead to a decline in both the apparent BET and microporous surface area while conversely increasing the density of ORR active sites with much improved activity. It was interpreted that the carbon phases derived from the polymer itself might be capable of hosting a significant number of active sites without the need of microporous supports. The carbon support chosen will furthermore have a much larger impact on the NPMC durability, which is a subject of ongoing study [39].

7.4.2 Carbon Nanotubes

It was known that CNTs could provide some overwhelming advantages as compared with conventional carbon black nanoparticles for fuel cell electrocatalyst supports, such as beneficial mass and electron transport, a specific interaction between active site and CNTs, and a few impurities as well as high corrosion resistance, which altogether leads to a higher overall fuel cell performance [27]. In NPMC development, a recent work was shown that

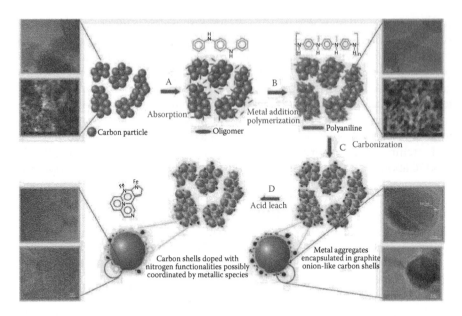

FIGURE 7.7
(See color insert.) Schematic diagram of the synthesis of PANI–M–C catalysts supported by carbon black nanoparticles. (From Wu, G. et al. *Science*, 332, 443–447, 2011. Reprinted with permission of AAAS.)

Fe–N–C catalysts derived from MWCNTs, and PANI exhibited a higher stability for ORR than those based on CB during a 500 h fuel cell life test [38].

As shown in Figure 7.8, in the Fe–N–MWNT catalyst, the presence of MWNTs can still be observed after the heat treatment in the catalyst synthesis. Furthermore, the dominant graphene sheet–like structures can be observed in a MWNT–supported Fe catalyst, which is different from a

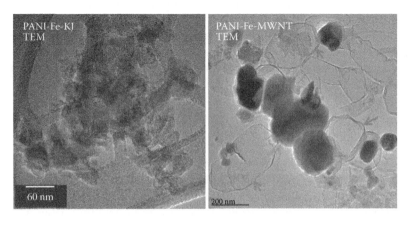

FIGURE 7.8
Morphology comparison between (left) KJ- and (right) MWNT-supported Fe–N–C catalysts.

KJ–supported Fe–N–KJ catalyst [18]. Owing to the unique properties of graphene sheets, such as high surface area, good conductivity, and a graphitized basal-plane structure [41], the presence of graphene structures presumably contributes to the increased catalytic performance of the Fe–N–MWNTs catalyst relative to the CB-supported Fe–N–KJ. However, the mechanistic understanding of how MNWTs contribute to the formation of graphene-like structures is still unknown. In the case of Co–N–C catalysts, while graphene sheet–like structures are found in both CB (KJ)- and MWNT-supported catalysts, the morphology of Co–N–MWNTs appears to be less agglomerated than that of Co–N–KJ, exhibiting an enhanced activity due to improved mass transport. Being consistent with the observed catalyst morphology, MWNT supports provide a higher surface area with abundant mesopores in the Co–N–MWNTs catalyst ($278 \text{ m}^2 \text{ g}^{-1}$) in comparison to the KJ-supported catalysts ($101 \text{ m}^2 \text{ g}^{-1}$).

7.4.3 Graphene and Graphene Oxide

Graphene is a novel form of carbon, which has been explored recently in nanostructured electrodes for electrochemical capacitors and biosensing because of its very high surface area and flexible surface functionalization chemistry [42]. The use of graphene and its oxide (graphene oxide, GO) in fuel cells began with its application as a catalyst support. The 2-D planar structure of the carbon sheet allows both the edge planes and basal planes to interact with the catalyst nanoparticles. The fast electron transport mechanism offered by graphene, especially N-doped graphene, can particularly facilitate the ORR. The disorders and defects introduced in the graphene stack due to the incorporation of nitrogen are known to act as active sites for the ORR [41].

Recently, chemically reduced graphene oxide (rGO), a derivative of graphene, has been employed in the development of NPMC through a heat treatment with Fe salt and graphitic carbon nitride (g-C_3N_4), as shown in Figure 7.9 [43]. The reported Fe–N–rGO catalyst exhibits an ORR mass activity approaching that of state-of-the-art NPMC reported to date [3,4], which highlights the opportunities of utilizing unique surface chemistry of rGO to create active Fe–N_x sites for highly active ORR catalysts [43].

7.4.4 Noncarbon Supports

Carbon corrosion is thermodynamically favorable above 0.20 V versus SHE in acidic media, which is located within the cathode potential range of PEMFC operation. As a result, many carbon black–supported ORR electrocatalysts, especially precious-metal ones, suffer from performance loss caused by carbon corrosion [27]. Oxide supports widely used in heterogeneous catalysis are inherently more stable than carbon in oxidizing environments. Thus, replacing carbon supports with these materials possessing higher intrinsic

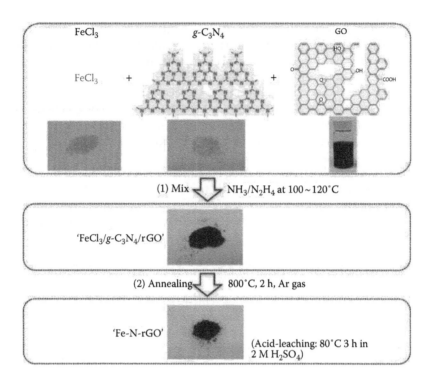

FIGURE 7.9
Schematic illustration of Fe–N–rGO synthesis and optical images of powder samples at various stages of synthesis. (Reprinted with permission from Byon, H.R. et al. *Chemistry of Materials, 23,* 3421–3428. Copyright 2011 American Chemical Society.)

stability is likely to offer longer fuel cell operating lifetimes. Specifically, a TiO_2-supported Fe-based NMPC derived from PANI was explored in Los Alamos [37]. The catalyst was synthesized by heat-treating a hybrid precursor material containing PANI and an Fe precursor that was polymerized *in situ* onto TiO_2 particles instead of conventional carbon black. After heat treatment, the samples were preleached in sulfuric acid to remove unstable and inactive species from the catalysts. Compared to traditional carbon black–supported NPMCs, the high ORR activity of the TiO_2-supported PANI-derived catalyst (PANI–Fe–TiO_2) has been verified by rotating disk electrode (RDE) and fuel cell tests (Figure 7.10a). It was found that, besides nitrogen-doped carbon structures, nitrogen atoms could also be doped into TiO_2, possibly forming Ti–N–O/Ti–N–Ti structures (Figure 7.10b). The presence of Ti–N species might be one of the reasons for the observed promotional role of TiO_2 in ORR performance in PANI–Fe–TiO_2 catalysts. In the meantime, the structure of PANI–Fe–TiO_2 is dominated by graphene sheet–like structures but not in PANI–TiO_2 or PANI–Fe–C samples (Figure 7.10c and d). This significant difference in catalyst nanostructure among the PANI-derived catalysts implies that a combination of TiO_2 and Fe species may effectively catalyze

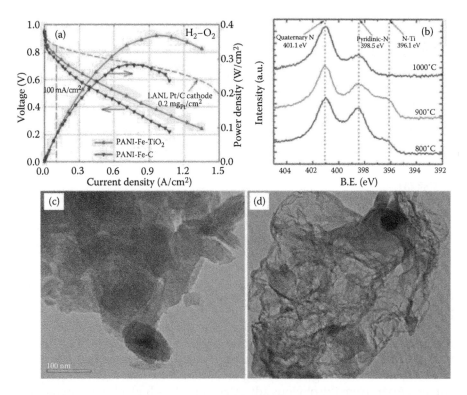

FIGURE 7.10
(a) Fuel cell polarization plots of C- and TiO_2-supported PANI–Fe catalysts. (b) N 1s XPS spectra of PANI–Fe–TiO_2 catalysts treated at various heating temperatures. HR-TEM images of (c) PANI–TiO_2 and (d) PANI–Fe–TiO_2 catalysts. (Wu, G. et al. *Chemical Communications, 46,* 7489–7491, 2010. Reproduced by permission of The Royal Society of Chemistry.)

PANI decomposition to form highly graphitic, N-doped graphene sheets at high temperatures [37]. Thanks to the unique properties of graphene sheets, such as high surface area, good conductivity, and a graphitized basal-plane structure, the presence of graphene sheet–like structures presumably contributes to the increased catalytic performance of the PANI–Fe–TiO_2 catalyst relative to the PANI–Fe–C.

7.5 *In Situ* Formed Graphitized Carbon Nanostructures

In the M–N–C catalyst synthesis, graphitic carbon, including nanotubes, onion-like, and graphene sheets (Figure 7.11) can be formed *in situ*, exhibiting a high diversity in crystallinity, morphology, porosity, and texture—factors

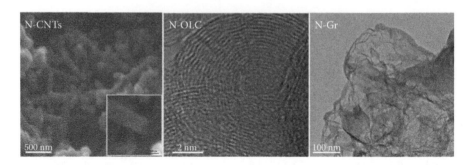

FIGURE 7.11

N-doped carbon nanostructures formed in non-PGM catalysts, including nanotube, onion-like carbon. (Reprinted with permission from Wu, G. et al. *Langmuir*, 24, 3566–3575. Copyright 2008 American Chemical Society) and graphene (Wu, G. et al. *Chemical Communications*, 46, 7489–7491, 2010. Reproduced by permission of The Royal Society of Chemistry.)

that play a crucial role in determining and optimizing the electrochemical performance of NPMC [44].

As transition metals are widely used as catalysts to generate carbon structures, such as nanotubes in other fields of research [45], the role of the metal in these M–N–C catalysts appears to be associated with forming the new carbon structures through catalyzing graphitization of nitrogen–carbon precursors. Apart from the obvious advantages, such as high electronic conductivity and corrosion resistance in electrocatalysis, these highly graphitized carbon nanostructures may serve as a matrix for hosting ORR–active nitrogen–metal moieties. A more specific hypothesis to explain the role of nanostructure in ORR catalysis is that nitrogen-containing carbon samples with more edge plane exposure (pyridinic nitrogen) could be more active for ORR [14]. The efforts in developing such a type of catalyst suggest that the presence of the graphitized carbon phase in an active catalyst may, at least, play a role in enhancing the stability of the NPMCs. Also, the formation of different carbon nanostructures could be greatly dependent on the type of nitrogen precursor, transition metal, heating temperatures, and supports used in catalyst synthesis [18,25,37,39], parameters that are directly correlated with catalyst performance. In this section, we briefly review the synthesis–structure–performance correlation for these graphitized carbon nanostructures observed in this type of NPMCs.

7.5.1 Nitrogen Precursors

Nitrogen precursors used during synthesis are a key to catalyst nanostructures. When PANI was used as a carbon–nitrogen precursor, graphene sheet–like structures are observed in the Co–N–C catalyst, surrounding the formed Co_9S_8 particles as shown in Figure 7.12.

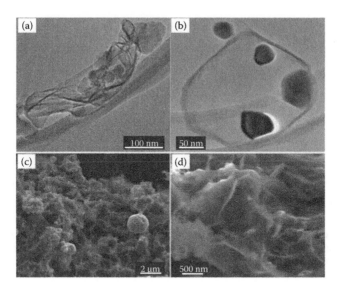

FIGURE 7.12
Microscopy images of Co–N–C catalyst derived from PANI: HR-TEM (a, b) and SEM (c, d). (Reproduced from Wu, G. et al. *ECS Transactions*, 41, 1709–1717. Copyright 2011. The Electrochemical Society. With permission.)

Just as important, the *in situ* formation of nitrogen-doped graphene sheets derived from PANI is unique, probably due to the structural similarities between the aromatic structure in PANI and graphene as shown in Figure 7.13 [46]. Traditionally, graphene with single and few layers has been grown epitaxially via CVD of hydrocarbons on metal substrates or by thermal decomposition of SiC. The developed new route to synthesize graphene layers from polymer offers controllability in morphology and nitrogen functionality for NPMC with enhanced oxygen reduction catalytic activity [46]. MWNTs proved to be an effective supporting template for the formation of graphene as compared to traditional carbon black supports. Unlike the agglomerated bubble-like graphene structures observed in the KJ-supported Co–N–KJ catalyst [6], a substantial fraction of the extended multilayered graphene sheets are observed in the MWNT-supported catalyst (Co–N–MWNTs) as shown as a TEM image (right) in Figure 7.13. This lower level of agglomeration is likely due to the unique surface properties and morphologies of the MWNTs [46].

However, no such graphene-sheet structures could be found in similar Co-based catalysts synthesized using other nitrogen–carbon precursors, such as ethylenediamine (EDA), hexamethylene diamine (HDA), and cyanamide. In particular, the morphology of the HDA–Co sample is dominated by other carbon nanostructures, including nanoshells, and onion-like carbon nanoshells (Figure 7.14) [19,25], resulting from the carbonization of the interlinked [-Co(HDA)$_4$-]$_n$ complexes. Such a well-defined graphitized carbon

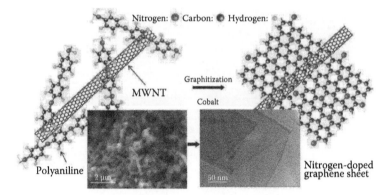

FIGURE 7.13
(See color insert.) Scheme of formation of nitrogen-doped graphene sheet derived from heteroatom polymers (polyaniline) and transition metals (Co). (Reprinted with permission from Wu, G. et al. *ACS Nan, 6,* 9764–9776. Copyright 2012 American Chemical Society.)

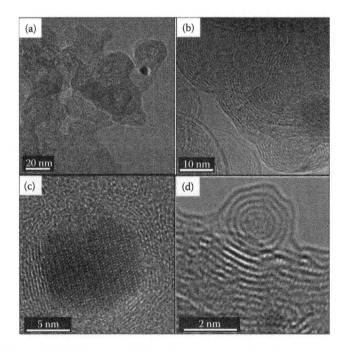

FIGURE 7.14
Nanoshells and onion-like carbon particles observed in NPMC derived from HAD. (Wu, G. et al. *Journal of Materials Chemistry, 20,* 3059–3068, 2010. Reproduced by permission of The Royal Society of Chemistry.)

shell surrounding metal-rich particles was also observed with iron(III) tetramethoxyphenyl porphyrin chloride (FeTMPP-Cl) when the heat-treatment temperature was raised to 1000°C [40]. The graphite shell formation was correlated to an increase in the catalyst open-circuit potential in an oxygen-saturated solution.

A possible growth mechanism for the formation of metal-encapsulating online-like carbon may be based on a vapor–liquid–solid model as shown in Figure 7.15. Briefly, during thermal treatment in an inert atmosphere at 900°C, HDA–CoFe complexes were decomposed into gaseous carbon, nitrogen, and metal species simultaneously. Metal atoms were first congregated into active metallic nanoparticles, and then gaseous carbon and nitrogen species were gradually captured by these metal nanoparticles and immediately decomposed into carbon and nitrogen atoms due to the catalytic functions of Co and Fe metals. Meanwhile, these atoms were precipitated and cooled down to form small nitrogen-doped carbon fragments around the encapsulated metal particles. The structural defects, especially the dangling bonds at the edges of these fragments, possibly act as a nucleus for further assembly and rearrangement, resulting in the formation of a layered structure on the surface of the metal particles. It is worth noting that small graphitic layers coalesced from the dangling bonds at the edges of carbon fragments are not definitely concentric to each other, thereby usually forming discontinuous graphitic layers, such as dislocations and irregular curvature. Moreover, the intrusion of nitrogen atoms suppresses surface diffusion of carbon atoms during graphitization, leading to more disorder and compact structures in these graphite sheets [19].

Alternatively, as a rule, amorphous carbon can transform into graphite at temperatures of 2500°C–3500°C. However, in the presence of such catalysts

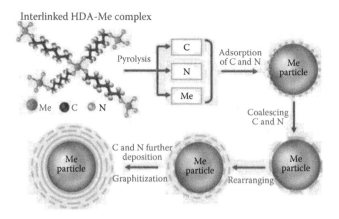

FIGURE 7.15
Growth model of magnetic N–Me–C onion-like graphitic carbon. (Wu, G. et al. *Journal of Materials Chemistry*, 20, 3059–3068, 2010. Reproduced by permission of The Royal Society of Chemistry.)

as Fe, Co, and Ni, the graphitization process can be substantially accelerated and takes place at a much lower temperature (600°C–900°C) [47]. These metal–carbon–nitrogen species can also interact with amorphous carbon particles, leading to the rearrangement of their surface [48]. Also, the energy minimization principle could explain why the carbon species are rearranged into closed concentric graphitic shells. While some metallic nanoparticles do not have enough time to agglomerate and were encapsulated into carbon shells, most of the catalytic particles leave the shells and accumulate in large spherical aggregates outside [25].

The different carbon structures and morphology observed in these catalysts can contribute to significant variations in the catalytic activity and durability for the ORR, implying that the corresponding carbon nanostructures (tubes, graphene-like, shells, etc.) may be critical to producing a highly active catalyst with improved performance.

7.5.2 Transition Metals

Besides the nitrogen precursors, the formed carbon nanostructures in M–N–C catalysts are greatly dependent on the transition metals (Co and Fe) used in synthesis. In spite of using the PANI as the same nitrogen–carbon precursors in both Co and Fe catalyst synthesis, graphene sheet–like structures are only abundant in the Co–N–C catalyst, while several layers of graphitized carbon coating the metal particles are observed in both types of catalysts (Figure 7.16). These layers of graphitized carbon and the graphene sheet structures may or may not directly relate to ORR active sites, but a connection between these structures and activity should be considered given that the presence of in-plane Fe–N_4 centers embedded in a graphene-type matrix, as determined by Mössbauer spectroscopy, has been previously correlated to the catalytic activity [20,49]. The significant morphological differences between the two catalysts demonstrate the strong effect of transition metal precursor selection on the carbon–nitrogen structures during the carbonization of polymers. It seems that Co is a more effective catalyst for the decomposition of PANI, and these decomposed carbon and nitrogen species would rearrange and coalesce together, forming highly graphitic carbon structures. In a different way, perhaps the complex could be broken into smaller species or moieties under the catalysis of Fe. Although Co species could facilitate the formation of such highly graphitized carbon structures that would be beneficial for electrocatalysis, the Co-based catalyst might be less active for ORR than the Fe-based one. This is probably due to the differences in the nature of the ORR active sites that are derived from different transition metals. A binary catalyst involving Co and Fe may simultaneously provide the benefits of graphitic carbon-rich morphologies and high concentrations of Fe-based active sites with excellent intrinsic ORR activity [8]. The research at Los Alamos to date has provided strong evidence of a correlation

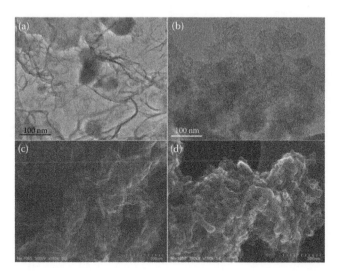

FIGURE 7.16
HR-TEM (a, b) and SEM (c, d) images for PANI-derived Co–N–C (a, c) and Fe–N–C (b, d) catalysts. (Wu, G. et al. *Journal of Materials Chemistry*, 21, 11392–11405, 2011. Reproduced by permission of The Royal Society of Chemistry.)

between the appearance of graphene sheets in the catalyst and enhanced performance durability [2,8,38,43]. The direct (structural properties) and indirect (promotional roles in catalytic properties) impact of the presence of graphene in these catalysts needs to be further elucidated.

7.5.3 Heating Temperature

The active sites in the M–N–C catalysts are generally believed to be formed during the heat-treatment step, and the resulting activity is strongly dependent on the temperature [10]. In particular, the ORR activity of the PANI-derived Fe–N–C catalysts has been found to increase with increasing heat-treatment temperatures from 400°C to 900°C and then decrease at 1000°C [18]. The corresponding carbon nanostructures and morphology in each catalyst synthesized at different temperatures may be one factor involved in determining the catalyst performance. As shown in Figure 7.17a through c for PANI-derived Fe–N–C catalysts, the characteristic appearance of PANI nanofibers gradually disappears as the heat-treatment temperature increases to 400°C, and spherical particle formation begins at 600°C. A higher degree of graphitization is observed at 900°C, including the graphitic shells that cover the metal particles identified using TEM imaging (not shown here). At higher temperatures, the morphology becomes nonuniform with the formation of larger agglomerated particles compared with the original carbon black, accompanied by significant losses in the catalyst

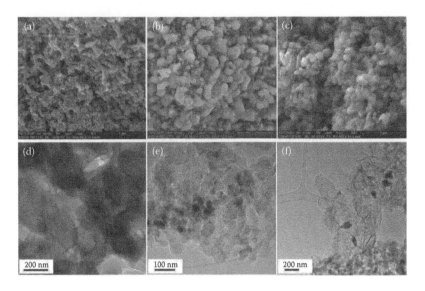

FIGURE 7.17
Morphology of PANI-derived Fe–N–C (a to c, SEM) and Co–N–C (d to f, TEM) catalysts as a function of heating temperature: (a, d) before heating, (b, e) 600°C, and (c, f) 900°C. (Wu, G. et al. *Journal of Materials Chemistry*, 21, 11392–11405, 2011. Reproduced by permission of The Royal Society of Chemistry.)

surface area. Likewise, in the case of PANI-derived Co–N–C catalysts, the morphology of these Co–N–C catalysts also greatly depends on the heat treatment in the synthesis as shown in Figure 7.17d through f. The graphene sheet–like structures grown over the solid particles are only observed when the heat-treatment temperature reaches 900°C, corresponding to maximum catalyst performance. The observed morphology with graphene sheet–like structures is in good agreement with the increase in the BET surface areas due to the higher surface area of graphene materials compared to other carbon structures [50]. The marked changes in the carbon structure that corresponds to large performance differences point to the importance of the metal-catalyzed transformation of the precursors into new carbon forms in such NPMCs.

7.6 Conclusion

The M–N–C catalysts synthesized from earth abundant elements have been widely considered the most promising NPMCs to replace Pt for the ORR, primarily due to recent substantial improvements in activity and stability [3,4]. The nanostructures related to nitrogen doping and graphitized carbon are controllable in the catalyst synthesis, which can be closely correlated to both catalyst activity and durability. The supports used, such as carbon black nanoparticles, nanotubes, graphene, and noncarbon oxides, are crucial to tuning catalyst morphology and enhancing catalyst performance. Understanding the detailed synthesis chemistry of how supports affect the catalytic activity and durability of active sites is still required. Although the vigorous debate is ongoing regarding whether metal atoms participate directly in active sites or merely catalyze their formation, the presence of nitrogen dopants in carbon structures is imperative for practical ORR activity. Also a graphitized carbon nanostructure and micropores in the carbon phase also have possible importance, such as serving as a host matrix for these active nitrogen moieties, regardless of if they are bound to metal centers or not. Although catalysts with a certain degree of activity for the ORR can be prepared without any detectable metal contents [5], the presence of metal is essential to generate the most active and durable catalysts known to date. The metal/carbon–nitrogen precursor interaction during the heat treatments needs precise control in order to formulate optimal chemical and morphological properties for NPMC activity and stability. Future synthesis efforts may focus on varying nitrogen precursors, multimetals, precursor ratios, surface/structure modifications (i.e., NH_3, CO_2, and N_2), and heat-treatment and posttreatment conditions.

Importantly, minimizing the remaining performance gap in both intrinsic activity and durability between NPMCs and Pt for the ORR in PEMFCs will require exclusive determination of the active oxygen reduction site and a better understanding of the reaction mechanism, thus providing rational design guidance. Probed with NO on a catalyst, advanced inorganic spectroscopic techniques like nuclear resonance vibrational spectroscopy (NRVS) can be used for the characterization of transition metal sites [51]. Using the spectroscopic information, density functional theory is able to identify the potential Fe–N–C centers having promise accessibility and favorable electronic structure including coordination with graphene/graphitic edges, multiple metal centers, and binding enhancement at graphene defect sites [52].

References

1. Matthey, J. 2013. Available at www.platinum.matthey.com/pgm-prices/price-charts.
2. Rabis, A., Rodriguez, P. and Schmidt, T.J. 2012. Electrocatalysis for polymer electrolyte fuel cells: Recent achievements and future challenges. *ACS Catalysis* 2, 864–890.
3. Chen, Z., Higgins, D., Yu, A., Zhang, L. and Zhang, J. 2011. A review on nonprecious metal electrocatalysts for PEM fuel cells. *Energy & Environmental Science* 4, 3167–3192.

4. Jaouen, F., Proietti, E., Lefevre, M., Chenitz, R., Dodelet, J.P., Wu, G., Chung, H.T., Johnston, C.M. and Zelenay, P. 2011. Recent advances in non-precious metal catalysis for oxygen-reduction reaction in polymer electrolyte fuel cells. *Energy & Environmental Science* 4, 114–130.

5. Gong, K.P., Du, F., Xia, Z.H., Durstock, M. and Dai, L.M. 2009. Nitrogen-doped carbon nanotube arrays with high electrocatalytic activity for oxygen reduction. *Science* 323, 760–764.

6. Wu, G., Chung, H.T., Nelson, M., Artyushkova, K., More, K.L., Johnston, C.M. and Zelenay, P. 2011. Graphene-enriched co_9s_8-n-c non-precious metal catalyst for oxygen reduction in alkaline media. *ECS Transactions* 41, 1709–1717.

7. Chen, Z., Higgins, D., Tao, H., Hsu, R.S. and Chen, Z. 2009. Highly active nitrogen-doped carbon nanotubes for oxygen reduction reaction in fuel cell applications. *The Journal of Physical Chemistry C* 113, 21008–21013.

8. Wu, G., More, K.L., Johnston, C.M. and Zelenay, P. 2011. High-performance electrocatalysts for oxygen reduction derived from polyaniline, iron, and cobalt. *Science* 332, 443–447.

9. Lefevre, M., Proietti, E., Jaouen, F. and Dodelet, J.P. 2009. Iron-based catalysts with improved oxygen reduction activity in polymer electrolyte fuel cells. *Science* 324, 71–74.

10. Jasinski, R. 1964. New fuel cell cathode catalyst. *Nature* 201, 1212.

11. Gupta, S., Tryk, D., Bae, I., Aldred, W. and Yeager, E. 1989. Heat-treated polyacrylonitrile-based catalysts for oxygen electroreduction. *Journal of Applied Electrochemistry* 19, 19–27.

12. Nallathambi, V., Lee, J.-W., Kumaraguru, S.P., Wu, G. and Popov, B.N. 2008. Development of high performance carbon composite catalyst for oxygen reduction reaction in PEM proton exchange membrane fuel cells. *Journal of Power Sources* 183, 34–42.

13. Wu, G., Chen, Z.W., Artyushkova, K., Garzon, F.H. and Zelenay, P. 2008. Polyaniline-derived non-precious catalyst for the polymer electrolyte fuel cell cathode. *ECS Transactions* 16, 159–170.

14. Matter, P.H., Zhang, L. and Ozkan, U.S. 2006. The role of nanostructure in nitrogen-containing carbon catalysts for the oxygen reduction reaction. *Journal of Catalysis* 239, 83–96.

15. Bezerra, C.W.B., Zhang, L., Lee, K., Liu, H., Marques, A.L.B., Marques, E.P., Wang, H. and Zhang, J. 2008. A review of Fe-N/C and Co-N catalysts for the oxygen reduction reaction. *Electrochimica Acta* 53, 4937–4951.

16. Charreteur, F., Jaouen, F., Ruggeri, S. and Dodelet, J.-P. 2008. Fe/N/C non-precious catalysts for pem fuel cells: Influence of the structural parameters of pristine commercial carbon blacks on their activity for oxygen reduction. *Electrochimica Acta* 53, 2925–2938.

17. Nallathambi, V., Wu, G., Subramanian, N., Kumaraguru, S., Lee, J.W. and Popov, B.N. 2007. Highly active carbon composite electrocatalysts for PEM fuel cells. *ECS Transactions* 11, 241–247.

18. Wu, G., Johnston, C.M., Mack, N.H., Artyushkova, K., Ferrandon, M., Nelson, M., Lezama-Pacheco, J.S., Conradson, S.D., More, K.L., Myers, D.J. and Zelenay, P. 2011. Synthesis-structure-performance correlation for polyaniline-me-c non-precious metal cathode catalysts for oxygen reduction in fuel cells. *Journal of Materials Chemistry* 21, 11392–11405.

19. Wu, G., Dai, C.S., Wang, D.L., Li, D.Y. and Li, N. 2010. Nitrogen-doped magnetic onion-like carbon as support for pt particles in a hybrid cathode catalyst for fuel cells. *Journal of Materials Chemistry* 20, 3059–3068.

20. Ferrandon, M., Kropf, A.J., Myers, D.J., Artyushkova, K., Kramm, U.I., Bogdanoff, P., Wu, G., Johnston, C.M. and Zelenay, P. 2012. Multi-technique characterization of a polyaniline-iron-carbon oxygen reduction catalyst. *The Journal of Physical Chemistry C* 30, 16001–16013.

21. Coutanceau, C., Croissant, M.J., Napporn, T. and Lamy, C. 2000. Electrocatalytic reduction of dioxygen at platinum particles dispersed in a polyaniline film. *Electrochimica Acta* 46, 579–588.

22. Wu, G., Cui, G.F., Li, D.Y., Shen, P.K. and Li, N. 2009. Carbon-supported co(1.67) te(2) nanoparticles as electrocatalysts for oxygen reduction reaction in alkaline electrolyte. *Journal of Materials Chemistry* 19, 6581–6589.

23. Bezerra, C.W.B., Zhang, L., Liu, H., Lee, K., Marques, A.L.B., Marques, E.P., Wang, H. and Zhang, J. 2007. A review of heat-treatment effects on activity and stability of PEM fuel cell catalysts for oxygen reduction reaction. *Journal of Power Sources* 173, 891–908.

24. Ohms, D., Herzog, S., Franke, R., Neumann, V., Wiesener, K., Gamburcev, S., Kaisheva, A. and Iliev, I. 1992. Influence of metal ions on the electrocatalytic oxygen reduction of carbon materials prepared from pyrolyzed polyacryloni-trile. *Journal of Power Sources* 38, 327–334.

25. Wu, G., Nelson, M., Ma, S.G., Meng, H., Cui, G.F. and Shen, P.K. 2011. Synthesis of nitrogen-doped onion-like carbon and its use in carbon-based cofe binary non-precious-metal catalysts for oxygen-reduction. *Carbon* 49, 3972–3982.

26. Pels, J.R., Kapteijn, F., Moulijn, J.A., Zhu, Q. and Thomas, K.M. 1995. Evolution of nitrogen functionalities in carbonaceous materials during pyrolysis. *Carbon* 33, 1641–1653.

27. Shao, Y., Liu, J., Wang, Y. and Lin, Y. 2009. Novel catalyst support materials for PEM fuel cells: Current status and future prospects. *Journal of Materials Chemistry* 19, 46–59.

28. Choi, J.-Y., Hsu, R.S. and Chen, Z. 2010. Nanoporous carbon-supported Fe/Co-N electrocatalyst for oxygen reduction reaction in PEM fuel cells. *ECS Transactions* 28, 101–112.

29. Yang, L., Jiang, S., Zhao, Y., Zhu, L., Chen, S., Wang, X., Wu, Q., Ma, J., Ma, Y. and Hu Z. 2011. Boron-doped carbon nanotubes as metal-free electrocatalysts for the oxygen reduction reaction. *Angewandte Chemie International Edition* 50, 7132–7135.

30. Maciel, I.O., Campos-Delgado, J., Cruz-Silva, E., Pimenta, M.A., Sumpter, B.G., Meunier, V., López-Urías, F., Muñoz-Sandoval, E., Terrones, H., Terrones, M. and Jorio, A. 2009. Synthesis, electronic structure, and raman scattering of phosphorus-doped single-wall carbon nanotubes. *Nano Letters* 9, 2267–2272.

31. Wohlgemuth, S.-A., Vilela, F., Titirici, M.-M. and Antonietti, M. 2012. A one-pot hydrothermal synthesis of tunable dual heteroatom-doped carbon micro-spheres. *Green Chemistry* 14, 741–749.

32. Herrmann, I., Kramm, U.I., Radnik, J., Fiechter, S. and Bogdanoff, P. 2009. Influence of sulfur on the pyrolysis of cotmpp as electrocatalyst for the oxygen reduction reaction. *Journal of the Electrochemical Society* 156, B1283–B1292.

33. Pollak, E., Salitra, G., Soffer, A. and Aurbach, D. 2006. On the reaction of oxygen with nitrogen-containing and nitrogen-free carbons. *Carbon* 44, 3302–3307.

34. Maldonado, S., Morin, S. and Stevenson, K.J. 2006. Structure, composition, and chemical reactivity of carbon nanotubes by selective nitrogen doping. *Carbon* 44, 1429–1437.

35. Fuge, G.M., Rennick, C.J., Pearce, S.R.J., May, P.W. and Ashfold, M.N.R. 2003. Structural characterisation of cnx thin films deposited by pulsed laser ablation. *Diamond and Related Materials* 12, 1049–1054.

36. Jang, J.W., Lee, C.E., Lyu, S.C., Lee, T.J. and Lee, C.J. 2004. Structural study of nitrogen-doping effects in bamboo-shaped multiwalled carbon nanotubes. *Applied Physics Letters* 84, 2877–2879.

37. Wu, G., Nelson, M.A., Mack, N.H., Ma, S.G., Sekhar, P., Garzon, F.H. and Zelenay, P. 2010. Titanium dioxide-supported non-precious metal oxygen reduction electrocatalyst. *Chemical Communications* 46, 7489–7491.

38. Wu, G., Artyushkova, K., Ferrandon, M., Kropf, A.J., Myers, D. and Zelenay, P. 2009. Performance durability of polyaniline-derived non-precious cathode catalysts. *ECS Transactions* 25, 1299–1311.

39. Wu, G., More, K.L., Xu, P., Wang, H.-L., Ferrandon, M., Kropf, A.J., Myers, D., Ma, S., Johnston, C.M. and Zelenay, P. 2013. A carbon-nanotube-supported graphene-rich non-precious metal oxygen reduction catalyst with enhanced performance durability. *Chemical Communications* 49, 3291–3293.

40. Gojkovic, S.L., Gupta, S. and Savinell, R.F. 1998. Heat-treated iron(iii) tetramethoxyphenyl porphyrin supported on high-area carbon as an electrocatalyst for oxygen reduction. *Journal of the Electrochemical Society* 145, 3493–3499.

41. Qu, L., Liu, Y., Baek, J.-B. and Dai, L. 2010. Nitrogen-doped graphene as efficient metal-free electrocatalyst for oxygen reduction in fuel cells. *ACS Nano* 4, 1321–1326.

42. Pumera, M. 2010. Graphene-based nanomaterials and their electrochemistry. *Chemical Society Reviews* 39, 4146–4157.

43. Byon, H.R., Suntivich, J. and Shao-Horn, Y. 2011. Graphene-based non-noble-metal catalysts for oxygen reduction reaction in acid. *Chemistry of Materials* 23, 3421–3428.

44. Wu, G., Li, D.Y., Dai, C.S., Wang, D.L. and Li, N. 2008. Well-dispersed high-loading pt nanoparticles supported by shell-core nanostructured carbon for methanol electrooxidation. *Langmuir* 24, 3566–3575.

45. Sinnott, S.B., Andrews, R., Qian, D., Rao, A.M., Mao, Z., Dickey, E.C. and Derbyshire, F. 1999. Model of carbon nanotube growth through chemical vapor deposition. *Chemical Physics Letters* 315, 25–30.

46. Wu, G., Mack, N.H., Gao, W., Ma, S., Zhong, R., Han, J., Baldwin, J.K. and Zelenay, P. 2012. Nitrogen-doped graphene-rich catalysts derived from heteroatom polymers for oxygen reduction in nonaqueous lithium–o_2 battery cathodes. *ACS Nano* 6, 9764–9776.

47. Krivoruchko, O.P., Maksimova, N.I., Zaikovskii, V.I. and Salanov, A.N. 2000. Study of multiwalled graphite nanotubes and filaments formation from carbonized products of polyvinyl alcohol via catalytic graphitization at 600–800°C in nitrogen atmosphere. *Carbon* 38, 1075–1082.

48. Lian, W., Song, H., Chen, X., Li, L., Huo, J., Zhao, M. and Wang, G. 2008. The transformation of acetylene black into onion-like hollow carbon nanoparticles at 1000°C using an iron catalyst. *Carbon* 46, 525–530.

49. Koslowski, U.I., Abs-Wurmbach, I., Fiechter, S. and Bogdanoff, P. 2008. Nature of the catalytic centers of porphyrin-based electrocatalysts for the ORR: A correlation of kinetic current density with the site density of fe–n4 centers. *The Journal of Physical Chemistry C* 112, 15356–15366.

50. Gao, W., Singh, N., Song, L., Liu, Z., Reddy, A.L.M., Ci, L., Vajtai, R., Zhang, Q., Wei, B. and Ajayan, P.M. 2011. Direct laser writing of micro-supercapacitors on hydrated graphite oxide films. *Nature Nanotechnology* 6, 496–500.

51. Holby, E.F., Wu, G., Zelenay, P. and Taylor, C.D. 2013. Metropolis monte carlo search for non-precious metal catalyst active site candidates. *ECS Transactions* 50, 1839–1845.

52. Holby, E.F. and Taylor, C.D. 2012. Control of graphene nanoribbon vacancies by fe and n dopants: Implications for catalysis. *Applied Physics Letters* 101, 064102.

8

Advances of the Nanostructured Carbon-Based Catalysts for Low-Temperature Fuel Cells

Fengping Hu and Pei Kang Shen

CONTENTS

8.1   Introduction...223
8.2   Nanostructured Carbon-Based Catalysts for Low-Temperature
     Fuel Cells..224
     8.2.1   Carbon Black–Supported Catalysts...224
     8.2.2   Fuel Cell Reactions with Ordered Mesoporous Carbon–
          Supported Catalysts 225
     8.2.3   Carbon Gel-Supported Catalysts...227
     8.2.4   CNT-Supported Catalysts...229
     8.2.5   CNF-Supported Catalysts...232
     8.2.6   CNH- and CNC-Supported Catalysts...234
     8.2.7   Graphene Oxide Counterpart (GO)-Supported Catalysts........235
     8.2.8   Boron-Doped Diamond-Supported Catalysts 236
     8.2.9   Hollow Carbon Sphere (HCS)-Supported Catalysts.................238
8.3   Conclusions..241
References..242

8.1 Introduction

Scientists are looking for a commercial technique to produce electricity by direct electrochemical conversion of hydrogen/methanol/ethanol and oxygen into water and carbon dioxide [1,2]. Fuel cells, especially low-temperature fuel cells, such as polymer electrolyte membrane fuel cells (PEMFCs), direct methanol fuel cells (DMFCs), and direct ethanol fuel cells (DEFCs), are one of these environmentally friendly techniques. Because of the high conversion efficiency, low pollution, light weight, and high power density, the low-temperature fuel cells have attracted lots of attention. However, the high catalyst cost of this kind of fuel cell is a major barrier to the commercialization due to the requirement of platinum-based catalysts for both the anode and cathode in the fuel cells at present [3,4]. In order to reduce the cost, lots of work have been tried to reduce the usage of the Pt by enhancing the Pt utilization efficiency, and the basic idea is to make the Pt catalyst with a small size and high surface area. Nowadays, the popular method to obtain a high active surface area is to deposit the Pt particles on a porous support. Due to the high specific surface area, flexible porous structure, high conductivity, and low cost, carbon materials have become a favorite catalyst support. Compared with any pure metals, the carbon-supported platinum has a higher catalyst activity for the oxygen reduction reaction (ORR) when it is used as electrocatalyst for a cathode [5]. And it has been found that the support materials have a strong influence on the electrocatalyst performance, durability, efficiency, and, finally, on the overall performance of the fuel cells. For the porous carbon support materials, the high specific surface area makes the uniform dispersion of metal particles possible, and the proper pore structure also improves the contact among the electronic conductors (carbon and metal), ionic conductors (polymer electrolyte), and the reactants [6]. Because of the good electrical and mechanical properties and their versatility in pore size and pore distribution tailoring, the novel nanostructured carbon materials have gained a lot of interest. As we know, the fuel cells' performances are mainly determined by the electrocatalysts that they used. The activity of a catalyst is determined by the active surface area of the catalyst, and the specific activity of the metal nanoparticles also decreases with the particle size due to the particle-size effect [7,8]. In order to solve this problem, catalysts are loaded on a porous support, and lots of experiments have proved that the supported metal catalysts have an improved stability and high activity. The ideal support is supposed to offer good electrical conductivity, large surface area, suitable mesoporous structure, good corrosion resistance, and good catalyst–support interaction [9]. The properties of the support material are very important to the performance of the catalyst. Sharma and Pollet [10] and Antolini [3] reviewed and compared the advantages and disadvantages of the various carbon materials as catalyst supports for low-temperature fuel cells. In this chapter, we discuss the performance, potential, and issues associated with the use of various nanostructured carbon materials, including carbon nanofibers (CNFs), carbon nanotubes (CNTs), carbon nanohorns (CNHs), and carbon nanocoils (CNCs), as support of catalysts in low-temperature fuel cells. Their electrochemical activity and the long-term stability in low-temperature fuel cells are also evaluated.

8.2 Nanostructured Carbon-Based Catalysts for Low-Temperature Fuel Cells

8.2.1 Carbon Black–Supported Catalysts

Due to their low cost and high availability, carbon blacks are widely used as the supporting materials for metal catalysts, and they are produced by oil-furnace processes and acetylene processes. Many works have been done to study their performance as electrocatalyst support [11–14]. It has been found that the metal catalyst deposits on the carbon by impregnation methods, the specific surface area of the carbon support, has little influence on Pt dispersion [11]. In the case of Pt/C catalysts prepared by colloidal methods, the effect of the specific surface area of different carbon on Pt particle size of Pt/C catalysts was evaluated by Uchida et al. [15]. Compared to oil-furnace black supports, acetylene black has a higher amount of pores with a diameter of 3–8 nm. Such pores are useful for fuel diffusion. Because the particles of ionomers are larger than the pore diameters and the Pt cannot contact the ionomer, the Pt in these pores does not contribute to the reaction for the fuel cells. The fuel diffusion becomes difficult when the pore is less than 3 nm and the activity of the catalyst is limited [16–18]. Rao et al. [19] investigated carbon material as supports for the anode catalysts of DMFCs. The specific surface area of the support varied in a wide range from 6 to 415 $m^2 g^{-1}$. PtRu catalysts were supported on these materials by a chemical route and showed higher catalyst utilization. The authors explained that this high catalyst utilization is due to the compatibility between the sizes of the pores in carbon supports. Mass activity and specific activity of PtRu anode catalysts change dramatically with the specific surface area of the support, increasing with the decrease in the latter. The results of this work proved the detrimental effect of pores with size <20 nm on the specific activity of PtRu/C electrocatalysts in methanol oxidation.

Zhou et al. [20] reported several Pt-based anode catalysts supported on XC-72R carbon, which, prepared by a novel method, resulted in uniform nanosized particles with sharp distribution, and Pt lattice parameter decreases with the addition of Ru or Pd. The graphite is an interesting support material for precious metal catalysts [21], and graphitized carbon is another support material of interest to catalyst manufacturers. This high surface area material is obtained by recrystallization of the spherical carbon particles at 2500°C–3000°C. The catalyst supported on Vulcan carbon shows a higher active surface area than that of the catalyst supported on HSAG.

In summary, the advantages of the activated carbon as catalyst support are (1) resistance to acid/alkaline media, (2) possibility to control porosity and surface chemistry, (3) chemical inertness, (4) ease of recovery of precious metal and large specific surface area, and (5) ease of obtaining. On the other hand, the disadvantages are (1) narrow microporosity (<2 nm), (2) inconsistent quality with traces of impurities, and (3) low mechanical and thermal stability.

8.2.2 Fuel Cell Reactions with Ordered Mesoporous Carbon–Supported Catalysts

The ordered mesoporous carbons (OMCs) are very interesting due to their potential use as catalytic supports in fuel cell electrodes, and they have controllable pore sizes, high surface areas, and large pore volumes [22]. It has been found that nanoporous carbons with 3D ordered pore structures can improve the mass transport of the reactants and products during fuel cell operation [22,23]. As we know, carbon supports have various parameters, such as high surface area for finely dispersing catalytic metal particles, high electrical conductivity for providing electrical pathways, mesoporosity for facile diffusion of reactants and byproducts, and water handling capability for removing water generated at the cathode. The electrocatalysts that employ such OMC supports exhibited promising catalytic activities toward methanol oxidation. From the viewpoint of pore structure, the OMCs are intriguing as carbon supports for fuel cell applications. The OMCs are composed of periodic arrays of carbon nanorods or nanopipes with uniform mesopores existing between them. They exhibit high specific surface areas up to 2000 $m^2 g^{-1}$; uniform pore diameters in the range of 2–10 nm; and high thermal, chemical, and mechanical stabilities. Therefore, their potential application as fuel cell catalyst supports has received a lot of attention [24–28]. Ryoo's group [24] first used OMCs as a carbon support and reported that Pt nanoparticles supported on CMK-5 mesoporous carbon exhibited superior electrocatalytic mass activity toward oxygen reduction in a half-cell configuration compared with the catalyst dispersed on a conventional carbon black support. The performance of the DMFC single cell using the Pt supported on CMK-3 carbon as a cathode catalyst was tested [27]. Joo et al. [28] reported that Pt particles supported on CMK-3-type OMCs exhibited better performance in DMFC single cell tests compared to the Pt black catalyst. Although aforementioned works demonstrated the possibility of OMCs as a new type of carbon support for DMFC, there is still space to improve the performance of DMFCs. In particular, because of the importance of the graphitic character (i.e., the electrical conductivity) of carbon supports for DMFC applications [29], the performance for fuel cell can be improved even more if the frameworks of OMCs can be graphitized. CMK-3 synthesized by Jun et al. [30] exhibited a large adsorption capacity, with a nitrogen BET–specific surface area of about 1500 $m^2 g^{-1}$ and total pore volume of about 1.3 $cm^3 g^{-1}$. CMK-3 has a primary pore size of about 4.5 nm, accompanied by micropores and some secondary mesopores. CMK-3 has been tested as support for fuel cell catalysts, and their metal dispersion and catalytic activity has been compared with that of catalysts supported on carbon blacks. Generally, all OMC-supported metals presented higher metal dispersion and higher catalytic activity, both for oxygen reduction and methanol oxidation than carbon black–supported metals. Compared to the commercial Pt catalyst from E-TEK, the catalytic performance of the mesoporous carbon-supported Pt catalyst in room temperature for methanol oxidation was higher. Calvillo et al. [31] prepared functionalized OMCs with a specific area of 570 $m^2 g^{-1}$. Transmission electron microscopy (TEM) images (Figure 8.1a) confirmed that the structure of CMK-3 is highly ordered. Pt supported electrocatalysts are well dispersed on the carbon support as can be seen for Pt/CMK3-Nc0.5h, leading to nanoparticles

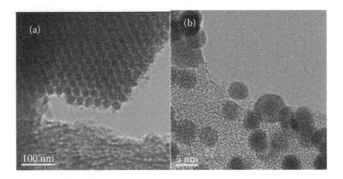

FIGURE 8.1
TEM image of CMK-3 carbons, (a) TEM image of platinum supported on CMK-3-Nc0.5h, (b). (Reproduced from Calvillo, L. et al. *Journal of Power Sources*, 169, 59–64, 2007. With permission.)

in the 3–5 nm range (Figure 8.1b). An OMC-supported Pt electrocatalyst was prepared using the impregnation method followed by the reduction of Pt precursor with sodium borohydride. According to the authors' description, platinum was well dispersed over the functionalized mesoporous support, and its catalytic performance toward methanol oxidation, compared with carbon Vulcan XC-72, was improved. However, OMCs contain a small amount of oxygen surface groups, which is disadvantageous for many applications. The functionalization of OMCs has not been studied to a large extent because their ordered structure could collapse during the process.

8.2.3 Carbon Gel–Supported Catalysts

The high purity of the carbon supports is essential to the electrochemical applications [32], and many works were done to prepare porous carbon materials with special attention to controlling the textural properties. Carbon gels, that is, carbon xerogels, aerogels, and cryogels, prepared by drying and pyrolysis of organic gels, have been extensively studied [33–35]. The surface area of the carbon gel, pore volume, and pore-size distribution can be tuned in the synthesis and processing process, and a wide spectrum of materials with unique properties can be obtained [36]. They are suitable for application as new carbonaceous supporting materials. Their mesoporosity could be controlled by varying the amount of catalyst used in the sol-gel polycondensation [37]. These materials have a great versatility both at the nanoscopic level in terms of their pore texture and at the macroscopic level in terms of their forms (e.g., microsphere, powder, or thin film). According to different synthesis methods, there are three types of carbon gels: carbon aerogels, carbon xerogels, and carbon cryogels. Moreno-Castilla and Maldonado-Hodar [36] reviewed the preparation of metal-doped carbon aerogels, their surface properties, and their applications as catalysts in various reactions. Little work has been done to characterize the electrocatalytic properties of gel-supported catalysts for use in low-temperature fuel cells. Kim et al. [38] investigated the preparation of highly dispersed platinum nanoparticles on carbon cryogel microspheres. For DMFC application, a PtRu alloy was supported on carbon xerogels and activated carbons using the formaldehyde reduction method [39]. For a well-dispersed PtRu catalyst to form the three-phase boundary, the support with large pore size and high surface area (especially, a meso-macropore area) was preferred. A microporous framework, which resulted from the destruction of structural integrity, was insufficient for high dispersion of PtRu particles. The catalysts with higher metal dispersion and structural integrity exhibited higher catalytic activities in the methanol electrooxidation and the DMFC performance test. Compared with traditional activated carbons or carbon blacks, the well-adapted structure of the support has the following advantages: (1) in the gas phase reaction, diffusional limitations encountered with activated carbon supports are completely eliminated; (2) in fuel cells, Pt-containing catalysts enable the decreasing effect of diffusion in the catalytic layer on potential loss and lead to increasing drastically the global metal utilization ratio. Marie et al. [40] compared two carbon aerogels with different nanopore-size distributions but both with high surface area, high nanoporous volume, and low bulk density as platinum support. The 3D structure of the carbon aerogel structures is shown in Figure 8.2. The observed pore sizes on the pictures are 25–66 nm for CA #1 and 15–40 nm for CA #2. As for CA #1, its porous volume would be essentially constituted by pores in the 50–66 nm range.

The structural differences between the carbon aerogels did not yield any difference in platinum deposits in terms of Pt-surface area and ORR activity. According to the authors, the similarity of the platinum deposit kinetic activity on the two carbon aerogels may allow people to make new catalytic layers based on Pt-doped carbon aerogels with different structures but identical platinum deposit in terms of surface area and intrinsic activity. This should be beneficial in developing the structural improvement (pore-size

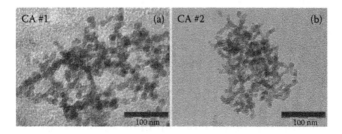

FIGURE 8.2
TEM images showing the voids between the primary carbon particles of carbon aerogels (a) CA #1 and (b) CA #2. (Reproduced from Marie, J. et al. *Journal of Applied Electrochemistry*, 37; 147–153, 2007. With permission.)

distribution optimization) of new PEMFC catalytic layers based on carbon aerogels. Conversely, the ORR mass activity of the high Pt-surface area samples, obtained by the cationic insertion technique, leading to the oxidation of carbon gel surface (oxCA), was several times lower than that of the samples obtained by the anionic technique. This result could be ascribed to (1) the size of platinum particles being too small on Pt/oxCA samples (negative particle-size effect) and (2) the platinum particles, due to their smallness, being located more deeply in the porous network of the carbon aerogel, which implies a more difficult access to oxygen and thus a decrease in the ORR performance.

8.2.4 CNT-Supported Catalysts

Compared with other forms of carbon, the tubular structure of CNTs is pretty unique and is also exploited as an alternative material for catalyst support in heterogeneous catalysis [41] and in fuel cells due to the high surface area, excellent electronic conductivity, and high chemical stability [42–46]. Conventional CNTs are made of seamless cylinders of hexagonal carbon networks and are synthesized as single-walled (SWCNTs) or multi-walled CNTs (MWCNTs). As we know, SWCNTs are a single graphene sheet rolled into a cylinder, and MWCNTs consist of several coaxially arranged graphene sheets rolled into a cylinder. Typical characteristics of CNTs for use as catalyst support are an outer diameter of 10–50 nm, inside diameter of 3–15 nm, and length of 10–50 mm. Compared to the activated carbons, CNTs have good mesoporosity (2–50 nm), which can improve mass transfer, high purity that can avoid the self-poisoning of the catalyst, and high mechanical and thermal stability. As reported by Serp et al. [47], pores in MWNTs can be mainly divided into inner hollow cavities of small diameter (narrowly distributed, mainly 3–6 nm) and aggregated pores (widely distributed, 20–40 nm) formed by interaction of isolated MWNTs. Typically, total surface area of as-grown SWNTs ranged between 400 and 900 $m^2\,g^{-1}$ whereas, for as-produced MWNTs, values ranging between 200 and 400 m^2 g^{-1} are often reported. Wildgoose et al. [48] reviewed the recent development in CNT-supported catalysts by exploring various techniques to load the CNTs with metals and other nanoparticles and the diverse applications of the resulting materials. More specifically, Lee et al. [49] reviewed the synthesis of CNT- and nanofiber-supported Pt electrocatalysts for PEM fuel cells, especially focusing on cathode nano-electrocatalyst preparation methods. Without surface modifications, however, most CNTs have sufficient binding sites to anchor the precursor metal ions or metal nanoparticles, which usually leads to poor dispersion and aggregation of metal nanoparticles, especially at high loading conditions. Indeed, highly dispersed and high loading metal nanoparticles have been obtained on carbon blacks; however, only less than 30 wt% Pt/MWCNTs catalysts can be achieved because high Pt loading unfunctionalized CNTs tend to aggregate [50]. Therefore, functionalization of CNTs should be done before the practical applications. Analogously to carbon blacks, in order to introduce more binding sites and surface anchoring groups, an acid oxidation process was adopted by treating CNTs in a refluxed, mixed-acid aqueous solution, commonly H_2SO_4/HNO_3 solution, at temperatures in the range of 90°C–140°C [51–53]. This treatment introduces surface-bound polar hydroxyl and carboxylic acid groups for subsequent anchoring and reductive conversion of precursor metal ions to metal nanoparticles. Hu et al. [54] reported the high-stability Pd/MWCNT catalyst, which used HF to treat the MWCNT support and got higher performance and stability for ethanol oxidation. The HRTEM images (Figure 8.3) of the Pd catalysts supported on the as-received and HF-treated MWCNTs (Pd/MWCNT and Pd/MWCNTHF) were compared in Figure 8.3. It shows that the Pd nanoparticles on the HF-treated MWCNTs are smaller than those on the as-received MWCNTs, and the nanoparticle number per unit length on the HF-treated MWCNTs (445 µm) is much larger than that on the as-received MWCNTs (125 µm). The size of nanoparticles dispersed on the HF-treated MWCNTs ranges from 2.2 to 4.2 nm with an average size of 3.15 ± 0.05 nm, which is more uniform than that on the as-received MWCNTs.

Most papers are indeed very optimistic regarding the potential interest of CNTs because of the presumably high activity of CNT-supported metals. In

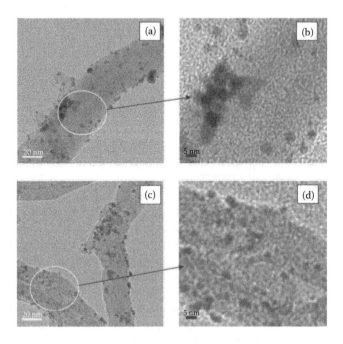

FIGURE 8.3
TEM images of (a) Pd/MWCNT and (c) Pd/MWCNTHF and HRTFM images of (b) Pd/MWCNT and (d) Pd/MWCNTHF. (Reproduced from Hu, F.P. et al. *Fuel Cells*, 8, 429–435, 2008. With permission.)

some cases, the authors overvalue their results, for example, by comparing CNT-supported catalysts with bad CB-supported catalysts. As reported in a lot of papers [51–53,55–60], when used as anode and/or cathode materials in low-temperature fuel cells, Pt and Pt-M (M = second metal) catalysts supported on CNTs presented higher catalytic activity than that of the same catalysts supported on carbon blacks. The higher activity of CNT-supported metal with respect to carbon black–supported metal was ascribed to different factors:

(1) The crystalline nature of CNTs [61] allows them to act as a good conductive substrate (the higher conductivity of CNTs is considered to contribute to the high performance of the CNT-supported metal electrodes).

(2) The hollow cavity and graphitic layer interspaces give more access to the gases than conventional supports. The Vulcan carbon support has randomly distributed pores of varying sizes, which may make fuel and product diffusion difficult, whereas the tubular three-dimensional morphology of the CNTs makes the fuel diffusion easier [52].

(3) The chemical difference between CNTs and carbon black induces flat disposition for Pt on the surface of CNTs [55].

(4) The architecture of the CNTs can give rise to specific sites (edge sites) where the Pt crystallites are anchored, and these sites may be more active than the conventional sites obtainable in carbon blacks [62].

(5) A low degree of alloying for MWCNT-supported PtRu with respect to PtRu supported on carbon black.

(6) The presence of different Pt crystallite phases on the MWCNTs and on the carbon. It is believed that the existence of the distinctive Pt crystallite phases, that is, Pt (110), on the PtRu particles supported on the MWCNTs could be the reason for enhancing the activity for the methanol oxidation reaction.

In order to understand which one is a better supporting carbon material for electrocatalysts in DMFCs, Wu and Xu [63] gave a detailed comparison between MWCNTs and SWCNTs. Pt particles were electrodeposited on MWCNT/Nafion and SWCNT/Nafion electrodes so as to investigate the effects of the carbon materials on the physical and electrochemical properties of the Pt catalyst. CO-stripping voltammograms showed that the onset and peak potentials on Pt-SWCNT/Nafion were significantly lower than those on the Pt-MWCNT/Nafion catalyst, revealing a higher tolerance to CO poisoning of Pt in Pt-SWCNT/Nafion. In the methanol electrooxidation reaction, the Pt-SWCNT/Nafion catalyst was characterized by a significantly higher current density, lower onset potentials, and lower charge transfer resistance. Therefore, SWCNT presents many advantages over MWCNT and

would emerge as an interesting supporting carbon material for fuel cell electrocatalysts. The enhanced electrocatalytic properties were discussed based
on the higher utilization and activation of Pt metal on the SWCNT/Nafion
electrode. The remarkable benefits from SWCNT were further explained by
its higher electrochemically accessible area and easier charge transfer at the
electrode–electrolyte interface due to the sound graphitic crystallinity of
the SWCNT, richness in oxygen-containing surface functional groups, and
highly mesoporous 3D structure. The methanol oxidation reaction activity
was in the order of PtRu/MWCNT > PtRu/C > PtRu/SWCNT. Long-term stability of supported catalysts is an important parameter for practical applications. The long-term performance of PtRu particles supported on MWCNTs
and on carbon black toward the methanol oxidation reaction was compared
by Prabhuram et al. [64]. They carried out chronoamperometry tests in a
0.5 M H_2SO_4 solution containing methanol for 3000 s. The close observation of the chronoamperometry curves revealed that potentiostatic current
decreases rapidly for MWCNT-supported PtRu. According to the authors,
this might be due to the higher deactivation of the Pt (110) crystallite phase
by the CO_{ads} species during the methanol oxidation reaction. Maiyalagan et
al. [65] investigated the durability of various electrodes by chronoamperometry measurements in H_2SO_4/CH_3OH at 0.6 V. The nitrogen-containing CNT
electrodes were the most stable for direct methanol oxidation. The increasing order of stability of various electrodes was Pt < Pt/Vulcan < Pt/N-CNTs.
According to the authors, the tubular morphology and the nitrogen functionality of the support have influence on the dispersion as well as the stability of
the electrode. Summarizing the results regarding the CNT stability are very
promising, but they are scarce and carried out in acidic solution. Further
tests, particularly in a single fuel cell, have to be performed to confirm the
good long-term performance of the CNTs as a support for fuel cell catalysts.

8.2.5 CNF-Supported Catalysts

Compared to the traditional carbon black and CNT catalyst supports, graphitized CNFs (G-CNFs) show superior thermal stability and corrosion resistance in the low-temperature fuel cell environment. This is because fibers
offer flexibility, which does not apply to the usual powdery or granular
materials. Fibrous catalytic packs offer the advantages of an immobile catalyst and a short diffusion distance. Another advantage of fibrous catalysts
is their low resistance to the flow of liquid and gases through a bundle of
fibers. Thus, they can be used as an alternative in fuel cells. However, CNFs
have an inert surface with only a very limited amount of surface defects
for the anchorage of Pt-catalyst nanoparticles, which may assist the particle
agglomeration. Modification of the fibrous surface is therefore needed. This
effect can be at least partly avoided by creating various surface functional
groups that platinum particles can use as anchoring sites [66]. Rodriguez et
al. [67] reported the use of CNFs as a catalyst support for Fe–Cu particles.

Since then, they have been extensively studied as fuel cell supports [68]. The basic difference between CNTs and CNFs is that CNFs either have a very thin or no hollow cavity. The diameters of CNFs are much larger than those of CNTs and may go up to 500 nm, while the length can be up to a few millimeters. Based on the orientation of the nanofibers with respect to the growth axis, there are three types of CNFs: ribbon-like CNFs, platelet CNFs, and herringbone (or stacked-cup) CNFs. Herringbone CNFs have intermediate characteristics between parallel and platelet types, thereby exhibiting higher catalytic activity than the parallel and better durability than the platelet forms [69]. Li et al. [70] prepared Pt nanoparticles (5–30 wt%) of 2–4 nm supported on stacked-cup CNFs (SC-CNFs) using the polyol process. The Pt/SC-CNFs displayed higher PEMFC performance than that of the commercial CB (E-TEK) catalyst. CNF-supported catalysts were prepared for use in fuel cells, and their metal dispersion and catalytic activity was compared with that of other carbon supports [71–73]. Gangeri et al. [72] deposited Pt by incipient wetness impregnation on CNFs. The TEM images of CNFs are similar to those of CNTs. The structure of CNFs was reported in Figure 8.4. A low magnification image confirmed the lack of hollow cavity in some parts and evidenced that no residual metallic particles, coming from the CNF production process, could be observed because they were encapsulated by the carbon. In the high magnification TEM image, it was evident that CNFs were herringbone (CNF-H), which means graphene layers are stacked obliquely with respect to the growth axis and regularly spaced by a distance of about 0.34 nm.

Tests in PEMFC indicated that the cells with Pt/CNFs as an anode material have a better performance than those with Pt/Vulcan. Compared to CNTs,

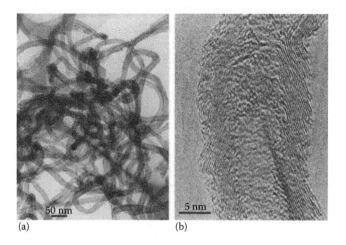

(a)   (b)

FIGURE 8.4

TEM images of CNFs at low and high magnification. (Reproduced from Gangeri, A. et al. *Catalysis Today*, 102, 50–57, 2005. With permission.)

Yuan and Ryu [74] showed that CNFs give better performance as a catalyst support material for a PEMFC. CNF-supported catalysts were prepared for use in fuel cells, and their metal dispersion and catalytic activity was compared with that of other carbon supports. Park et al. [75] prepared CNF-supported PtRu catalysts by the borohydride reduction method. Generally, it is difficult to obtain high-loaded and well-dispersed PtRu metal catalysts on CNFs by conventional methods. However, they obtained highly dispersed PtRu particles on CNFs and the herringbone structure of CNFs. Although CNFs have a small surface area for metal loading, the catalytic activities of CNF-supported PtRu nanoparticles were higher than those of Vulcan XC-72 carbon-supported catalysts. Compared to the Vulcan XC-72 carbon-supported catalyst, the electrochemical results indicated that the CNF-supported catalyst has a similar value in the mass-normalized currents and an increased value in the area-normalized currents. The authors thought that the enhancement in catalytic activity of the CNF-supported catalyst is the result of interactions between metal particles and CNFs. In particular, CNFs might modify the geometric characteristic of the supported catalysts.

8.2.6 CNH- and CNC-Supported Catalysts

CNHs and CNCs are a new class of carbon nanomaterials with properties that differ significantly from other forms of carbon. These materials have also been tested as support for fuel cell metal catalysts. The high catalytic activity of CNH-/CNC-supported catalysts demonstrates the suitability of their application in fuel cell technology. Single-walled CNH (SWCNH) aggregates can be produced by CO_2 laser vaporization of carbon, and a single aggregate can take either a "dahlia-like" or "bud-like" form. Yudakasa et al. [76] fabricated single-walled CNHs 30–50 nm long and 2–3 nm thick, forming aggregates that resemble dahlia flowers (diameter: 80 nm). Park et al. [77] employed CNCs with variable surface areas and crystallinity as the supports for 60 wt% Pt/Ru catalysts. The catalysts supported on all the CNCs exhibited better electrocatalytic performance compared to the catalyst supported on Vulcan XC-72 carbon. In particular, the PtRu alloy catalyst supported on the CNCs, which has both good crystallinity and a large surface area, showed a superior electrocatalytic performance. Sevilla et al. [78] synthesized highly graphitic CNCs by the catalytic graphitization of carbon spherules obtained by the hydrothermal treatment of different saccharides (sucrose, glucose, and starch). These CNCs were used as a support for PtRu nanoparticles, which were well dispersed over the carbon surface. They tested PtRu/CNCs as an electrocatalyst for methanol electrooxidation in an acid medium and found that the CNC-supported PtRu nanoparticles exhibit a high catalytic activity, which is even higher than that of PtRu supported on Vulcan XC-72R carbon. The high electrocatalytic activity of the PtRu/CNCs catalyst is due to the combination of a good electrical conductivity, derived from their graphitic structure, and a wide porosity that allows the diffusion

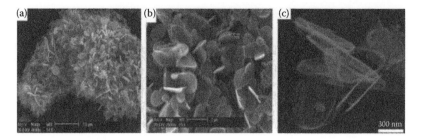

FIGURE 8.5
SEM micrographs of the as-prepared CNCs before (a) and after (b) the treatment by dilute HCl and (c) TEM image of the HCl-treated CNCs. (Reproduced from Yuan, D.S. et al. *Electrochemistry Communications*, 9, 2473–2478, 2007. With permission.)

resistances of reactants/products to be minimized. Yuan et al. [79] treated CNCs by HCl, which resulted in higher performance of Pt/CNCs. The scanning electron microscopy (SEM) and TEM images of CHC are shown in Figure 8.5. The as-synthesized CHC was covered by small powders, which were identified to be $MgCO_3$ and Ni. The unique monodispersed coin-like shape (Figure 8.5b) was clearly identified after the dissolution of $MgCO_3$ and Ni. The coin-like carbon is sized at 1–3 μm in diameter and has a thickness less than 150 nm. On the other hand, the TEM image also proved that the coin-like carbon products were hollow structured.

8.2.7 Graphene Oxide Counterpart (GO)-Supported Catalysts

Graphene is an atomical thin sheet of hexagonally arranged carbon atoms, and it has attracted a lot of interest since its discovery by Geim et al. in 2004 [80]. It offers high conductivity and one of the fastest available electron transfer capabilities. It is widely studied for various applications, including fuel cell catalyst support. The use of GO as catalyst support material in PEMFCs and DMFCs is one of the latest applications of GO and has shown promising results [81–84]. Oxygen groups are introduced into the graphene structure during the preparation of GO and then create defect sites on the surface as well as edge planes. These defect sites act as nucleation centers and anchoring sites for growth of metal nanoparticles. In comparison with CNTs, graphene nanosheets (GNSs) possess not only similar stable physical properties but also large surface area (theoretical specific area of $2620 \, m^2 \, g^{-1}$), which can be considered as a flat CNT. In addition, the production cost of GNS mass scale is cheaper than that of CNTs [85]. Also, GNS has been reported to have good dispersion stability and large surface area [86,87]. Sharma et al. recently reported the synthesis of Pt nanoparticles on GO using the microwave-assisted polyol method. The process allowed simultaneous partial reduction of GO and growth of Pt nanoparticles on reduced GO (RGO) support. Ha et al. reported the use of Pt nanoparticles embedded on reduced GO for ORR

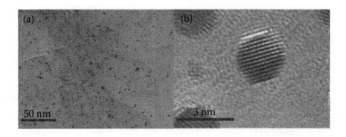

FIGURE 8.6
TEM image (a) and HRTEM image (b) of Pt/RGO composite. (Reproduced from Ha, H.W. et al. *Electrochemical and Solid State Letters*, 14, B70–B73, 2011. With permission.)

in DMFCs. The authors used a modified polyol process to deposit ~2.9 nm Pt nanoparticles on a reduced GO to produce 70 wt% Pt/RGO.

ORR studies and single cell polarization studies demonstrated an 11% higher maximum power density for Pt/RGO in comparison to commercial (75 wt% Pt/C [88]). TEM images of the Pt/RGO composite are shown in Figure 8.6. Small Pt nanoparticles uniformly embedded on the surface of reduced graphene oxide sheets were observed. The HRTEM of Pt nanoparticles shows their size to be about 3 nm, relatively close to that obtained by the Scherrer analysis discussed above. Other precious metal nanoparticles and bimetallic systems have also been grown on GO support and have exhibited superior catalytic activity [89]. A graphene nanosheet was synthesized with modified Hummer's method and characterized by Raman spectroscopy [90]. A graphene nanosheet–supported catalyst confirmed that it has uniformly dispersed metal particles of about 2 nm onto graphene in spite of a high metal content of 80 wt%. Their electrochemical data toward methanol oxidation exhibited that Pt-Ru/GNS had better catalytic activity than that of Pt-Ru/Vulcan carbon. These results indicate that a graphene nanosheet could be a good candidate as supporting material of high-loading metal catalyst in fuel cells.

8.2.8 Boron-Doped Diamond-Supported Catalysts

Polycrystalline boron-doped diamond (BDD) has ideal properties for the application as an electrocatalyst support for fuel cells. For example, the material possesses superior morphological stability and corrosion resistance. Compared to conventional sp^2 carbon support materials, it can withstand current densities in the scale of 1 A cm^{-2} for days, in both acidic and alkaline conditions, without any evidence of structural degradation [91,92]. The electrical conductivity of diamond remarkably increases after boron doping. Fischer and Swain [93] prepared BDD powders by coating insulated diamond powders with a thin boron-doped layer using microwave plasma-assisted chemical vapor deposition. Bennett et al. [94] reported the pulsed galvanostatic deposition of nanometer-sized Pt particles on electrically conducting microcrystalline and nanocrystalline diamond thin-film electrodes. The deposition of catalyst particles using 10 pulses at a current density of 1.25 mA cm^{-2} produced the smallest nominal particle size and the highest particle coverage on both diamond surfaces. Considering the electrochemically active Pt area normalized to the estimated metal loading, deposition under these conditions resulted in the most efficient utilization of the metal catalyst for H$^+$ adsorption. Although BDD appeared to be a possible alternative to carbon as a support material for fuel cell electrocatalysts, these catalysts showed low long-term stability. Siné et al. prepared bimetallic binary Pt–Sn [95] and ternary Pt–Ru–Sn[96] nanoparticles supported on a BDD substrate. These nanoparticles showed high activity toward methanol and/or ethanol oxidation. They believed that this substrate could promote the activation of the ethanol C–C bond scission or increase the turnover frequency of product formation. Spataru et al. [97] prepared a Pt/BDD powder electrocatalyst via electrochemical deposition of platinum and compared the electrochemical behavior with that of Pt/graphite powder. The BDD-deposited Pt particles were rather uniform in size (5–15 nm), although they form particle clusters. Electrodes were prepared by coating polycrystalline BDD films with these electrocatalysts with Nafion solution as a binder, and the activities for methanol oxidation were found to be comparable. The use of BDD/Nafion resulted in a much higher stability of the catalyst under severe anodic conditions. Figure 8.7 showed the BDD powder at low magnification. A small amount of Teflon suspension added cover on BBD powder.

Steady-state and long-time methanol oxidation polarization measurements demonstrated that the platinum on BDD powder was less sensitive

FIGURE 8.7
SEM image of the BDD powder mixed with a small amount of Teflon suspension. (Reproduced from Spataru, N. et al. *Journal of the Electrochemical Society*, 155, B264–B269, 2008. With permission.)

to deactivation, presumably due to CO poisoning, than that of platinum on graphite powders.

8.2.9 Hollow Carbon Sphere (HCS)-Supported Catalysts

Due to the high specific surface area and porous structure, the use of hollow carbon materials [98–101] is also used as electrocatalyst supports and is supposed to improve the dispersion of noble metal and the mass transfer in the electrochemical processes. The porous hollow carbon spheres (PHCSs) are different from the honeycomb-like carbon in shape [102,103]. PHCSs have good contact with each other to reduce resistance, which could further increase the performance of the electrocatalyst. Ethylene glycol was used to reduce the Pd salt in Pd nanoparticles as what was used by Xia's group [104,105] to prepare noble metals. Pd nanoparticles supported on PHCSs show an enhanced catalytic activity for alcohol oxidation. The porous graphitized carbon (PGC), as a new class of carbon material, has high surface area, low density, high pore volume and chemical inertia, and excellent electronic conductivity, which can be extensively useful for energy and gas storage [106,107] and so on. The use of the PGC as catalyst support can also lead to an improved catalytic activity [108].

Yan et al. [109] reported a method to synthesize PGC using an ion-exchange resin (D314) as a carbon source and $K_4[Fe(CN)_6]$ as a catalyst precursor to graphitize the resin at low temperatures during the heat treatment. The PGC was further tested as an electrocatalyst support. The Pt nanoparticles supported on the PGC showed an increased catalytic activity for the ORR and the methanol oxidation reaction, relative to the commercial Pt/C electrocatalyst. The typical SEM micrograph of the PGC formed at 1000°C with an iron catalyst (PGC-1000) is shown in Figure 8.8. The catalytically formed graphitized carbon was a foam-structured carbon (Figure 8.8a) with a large amount of macropores in different sizes. Those pores were formed during

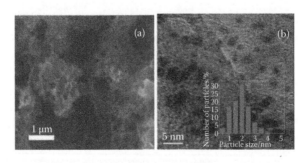

FIGURE 8.8
(a) SEM micrograph of the graphitized D314 resin catalyzed by iron at 1000°C and (b) TEM image of Pt/PGC-600 electrocatalyst. The inset corresponds to the Pt nanoparticle size distribution. (Reproduced from Yan, Z. et al. *Journal of Materials Chemistry*, 22, 2133–2139, 2012. With permission.)

the heat treatment due to the evaporation and decomposition of the resin. The corresponding TEM images and the particle size distribution (Figure 8.8b) reported that the average Pt particle size of the Pt/PGC-600 was 1.9 nm. It is clear that the porous structured graphitized resin is a suitable catalyst support for dispersing Pt particles uniformly with a smaller Pt particle size.

A higher exchange current density and lower reaction activation free energy were observed on Pt/HCS catalysts, indicating improved kinetics. It is recognized that the hollow structure with open microspores and nano-channels, which results in a larger three-phase interface, is responsible for the high catalytic activity. This novel HCS material can be a potential candidate as a catalyst support for DMFCs. The design of the three-dimensional electrode structure [110] and macroporous conducting support [102] resulted in a significant improvement in the electrode kinetics. Shen's group [111] reported a novel method to prepare HCSs by using polystyrene spheres (PSs) as templates along with the decomposition of glucose in a poly(ethylene glycol)-block-poly(propylene glycol)-block-poly(ethylene glycol) (P123) surfactant-containing solution under hydrothermal conditions. An intermittent microwave heating (IMH) technique was adopted to synthesize the HCSs rapidly. With the removal of the P123 molecules trapped during the IMH treatment, a large amount of nanopores and open nanochannels formed on the walls of the HCSs. Pt supported on HCSs, which were heated for 3 min in a microwave oven, gives the best performance for methanol oxidation. At the same loadings, Pt/HCS shows almost three times higher electrochemically active surface area and results in a higher catalytic activity for methanol oxidation compared with that of the commercial Pt/C catalyst.

Table 8.1 summarizes the main characteristics of nanostructured carbon supported catalysts for low-temperature fuel cells. Typical nanostructured

TABLE 8.1

Main Characteristics of Nanostructured Carbon Supported Catalysts

Nanostructured Carbon	Material Properties (Acted as Support)	Loaded Metal	Electrode	Supported Catalyst Properties	Ref.
Vulcan XC-72R	Graphitized carbon, low cost, high availability	Pt, Pd, Ag, Au, Pt–Ru, Pt–Pd, etc.	Anode, cathode	Good dispersion, low gas flow	[11–21]
OMCs	High surface area, good pore structure and size distribution, O-functional groups	Pt, Pd, Pt–Ru, Pt–Pd, Fe, etc.	Anode, cathode	High dispersion, high gas flow, low metal accessibility	[22–31]

(*continued*)

TABLE 8.1 (Continued)

Main Characteristics of Nanostructured Carbon Supported Catalysts

Nanostructured Carbon	Material Properties (Acted as Support)	Loaded Metal	Electrode	Supported Catalyst Properties	Ref.
Carbon gels	High surface area, good pore structure, hydrophobic	Pt, Pt–Ru, etc.	Anode	High dispersion, high gas flow, high metal accessibility	[32–40]
CNTs	Hydrophobic, high conductivity, good pore structure and size distribution, high stability	Pt, Pt–Ru, Pt–Co, Pt–Fe, Pt–Ni, Pt–WO$_3$, Pt–Ru–Ir, Pt–Ru–Ni, Ru–Se, etc.	Anode, cathode	Larger catalyst particles without surface treatment, good dispersion, high metal stability	[41–65]
CNF	High conductivity, high durability, N-doping for ORR	Pt, Pt–Ru, etc.	Anode, cathode	High dispersion, high gas flow, high metal stability	[66–75]
CNH, CNC	Large surface area, hollow pore structure	Pt, Pd, etc.	Anode	High dispersion, high gas flow	[76–79]
GO	Variable conductivity, hydrophilic, O-functional groups	Pt, Pt–Ru, etc.	Anode	O-defects, high loading with small particle size, O-species	[80–90]
BDD	Low surface area, high stability, low conductivity, doped, high resistance to poisoning	Pt, Pt–Ru, Pt–RuO$_2$, Pt–SnO$_2$, Pt–Ta$_2$O$_5$, etc.	Anode, cathode	Low metal stability and dispersion, high metal stability on BDD/Nafion	[91–97]
HCSs	Large surface area, good pore structure and size distribution, O-functional groups	Pt, Pd, Pt–Ru, Pt–Pd Au, etc.	Anode	High dispersion, high gas flow	[98–111]

carbon supported catalysts were compared by support properties, catalyst metal loading, and catalyst properties.

8.3 Conclusions

According to the International Union of Pure and Applied Chemistry (IUPAC), depending on pore width, pores are classified as micropores (<2 nm), mesopores (2–50 nm), and macropores (>50 nm). Generally, the high specific surface area of the carbon blacks is mainly contributed by micropores less than 1 nm, and therefore, it is more difficult to fully access. Indeed, when the average diameter of the pores is less than 2 nm, the supply of a fuel may not occur smoothly, and the activity of the catalyst may be limited. Compared with carbon blacks, mesoporous carbons presented a higher surface area and lower amount or absence of micropores. In a mesoporous carbon-supported catalyst, the metal catalyst particles are distributed and supported on the surface or in the pores of the mesoporous carbon. A large mesoporous surface area, particularly with a pore size of >20 nm, gives rise to a high dispersion of Pt particles, which results in a large effective surface area of Pt with a high catalytic activity. The mesoporous structure facilitated the smooth mass transportation, which provides high limited currents. Therefore, the use of carbon materials with high surface area, good electrical conductivity, and suitable porosity as catalyst supports for precious metals increases quickly due to the continuous development of fuel cells. The high availability and low cost make carbon blacks the most popular support for fuel cell catalysts. Among the new carbon materials, CNTs are the most investigated as catalyst support for low-temperature fuel cells. The high crystallinity of CNTs makes these materials have high conductivity, high surface area, and a high amount of mesopores, which results in a high metal dispersion and a good reactant flux in tubular structure. Moreover, CNTs have a positive effect on Pt structure, resulting in a higher catalytic activity and a higher stability than that of carbon blacks. Few works have been carried out on CNHs, CNCs, and carbon fibers as fuel cell catalyst support; tests in fuel cells of these materials showed promising results. New structures, synthesis, and surface modification techniques provide the possibilities to further improve the corrosion resistance and water handling capabilities while still maintaining good conductivity. An amalgamation of these novel electrocatalyst supports and improved catalyst loading techniques could bring about revolutionary changes in the quest for high-performance, long-lasting, low-temperature fuel cells.

References

1. Antolini, E. 2009. Carbon supports for low-temperature fuel cell catalysts. *Applied Catalysis B-Environmental* 88, 1–24.
2. Tang, S.H., Sun, G.Q., Qi, J. et al. 2010. New carbon materials as catalyst supports in direct alcohol fuel cells. *Chinese Journal of Catalysis* 31, 12–17.
3. Antolini, E. 2003. Formation of carbon-supported PtM alloys for low temperature fuel cells: A review. *Materials Chemistry and Physics* 78, 563–573.
4. Vengatesan, S., Kim, H.J., Kim, S.K. et al. 2008. High dispersion platinum catalyst using mesoporous carbon support for fuel cells. *Electrochimica Acta* 54, 856–861.
5. Costamagna, P. and Srinivasan, S. 2001. Quantum jumps in the PEMFC science and technology from the 1960s to the year 2000. Part I. Fundamental scientific aspects. *Journal of Power Sources* 102, 242–252.
6. Ren, X.M., Zelenay, P., Thomas, S. et al. 2000. Recent advances in direct methanol fuel cells at Los Alamos National Laboratory. *Journal of Power Sources* 86, 111–116.
7. Yahikozawa, K., Fujii, Y., Matsuda, Y. et al. 1991. Electrocatalytic properties of ultrafine platinum particles for oxidation of methanol and formic-acid in aqueous-solutions. *Electrochimica Acta* 36, 973–978.
8. Kabbabi, A., Gloaguen, F., Andolfatto, F. et al. 1994. Particle-size effect for oxygen reduction and methanol oxidation on Pt/C inside a proton-exchange membrane. *Journal of Electroanalytical Chemistry* 373, 251–254.
9. McBreen, J., Olender, H., Srinivasan, S. et al. 1981. Carbon supports for phosphoric-acid fuel-cell electrocatalysts alternative materials and methods of evaluation. *Journal of Applied Electrochemistry* 11, 787–796.
10. Sharma, S. and Pollet, B.G. 2012. Support materials for PEMFC and DMFC electrocatalysts: A review. *Journal of Power Sources* 208, 96–119.
11. Fraga, M.A., Jordao, E., Mendes, M.J. et al. 2002. Properties of carbon-supported platinum catalysts: Role of carbon surface sites. *Journal of Catalysis* 209, 355–364.
12. Gharibi, H., Mirzaie, R.A., Shams, E. et al. 2005. Preparation of platinum electrocatalysts using carbon supports for oxygen reduction at a gas-diffusion electrode. *Journal of Power Sources* 139, 61–66.
13. Antolini, E., Cardellini, F., Giacometti, E. et al. 2002. Study on the formation of Pt/C catalysts by non-oxidized active carbon support and a sulfur-based reducing agent. *Journal of Materials Science* 37, 133–139.
14. Takasu, Y., Kawaguchi, T., Sugimoto, W. et al. 2003. Effects of the surface area of carbon support on the characteristics of highly-dispersed Pt-Ru particles as catalysts for methanol oxidation. *Electrochimica Acta* 48, 3861–3868.
15. Uchida, M., Aoyama, Y., Tanabe, M. et al. 1995. Influences of both carbon supports and heat-treatment of supported catalyst on electrochemical oxidation of methanol. *Journal of the Electrochemical Society* 142, 2572–2576.
16. Gruver, G.A. 1978. Corrosion of carbon-black in phosphoric-acid. *Journal of the Electrochemical Society* 125, 1719–1720.
17. Wang, J.J., Yin, G.P., Shao, Y.Y. et al. 2007. Effect of carbon black support corrosion on the durability of Pt/C catalyst. *Journal of Power Sources* 171, 331–339.

18. Stonehart, P. 1984. Carbon substrates for phosphoric-acid fuel-cell cathodes. *Carbon* 22, 423–431.
19. Rao, V., Simonov, P.A., Savinova, E.R. et al. 2005. The influence of carbon support porosity on the activity of PtRu/Sibunit anode catalysts for methanol oxidation. *Journal of Power Sources* 145, 178–187.
20. Zhou, W.J., Zhou, Z.H., Song, S.Q. et al. 2003. Pt based anode catalysts for direct ethanol fuel cells. *Applied Catalysis B-Environmental* 46, 273–285.
21. Antolini, E. 2012. Graphene as a new carbon support for low-temperature fuel cell catalysts. *Applied Catalysis B-Environmental* 123, 52–68.
22. Chang, H., Joo, S.H. and Pak, C. 2007. Synthesis and characterization of mesoporous carbon for fuel cell applications. *Journal of Materials Chemistry* 17, 3078–3088.
23. Su, F.B., Zeng, J.H., Bao, X.Y. et al. 2005. Preparation and characterization of highly ordered graphitic mesoporous carbon as a Pt catalyst support for direct methanol fuel cells. *Chemistry of Materials* 17, 3960–3967.
24. Joo, S.H., Choi, S.J., Oh, I. et al. 2001. Ordered nanoporous arrays of carbon supporting high dispersions of platinum nanoparticles. *Nature* 412, 169–172.
25. Ding, J., Chan, K.Y., Ren, J.W. et al. 2005. Platinum and platinum-ruthenium nanoparticles supported on ordered mesoporous carbon and their electrocatalytic performance for fuel cell reactions. *Electrochimica Acta* 50, 3131–3141.
26. Zeng, J.Q., Francia, C., Dumitrescu, M.A. et al. 2012. Electrochemical performance of Pt-Based catalysts supported on different ordered mesoporous carbons (Pt/OMCs) for oxygen reduction reaction. *Industrial & Engineering Chemistry Research* 51, 7500–7509.
27. Nam, J.H., Jang, Y.Y., Kwon, Y.U. et al. 2004. Direct methanol fuel cell Pt-carbon catalysts by using SBA-15 nanoporous templates. *Electrochemistry Communications* 6, 737–741.
28. Joo, S.H., Pak, C., You, D.J. et al. 2006. Ordered mesoporous carbons (OMC) as supports of electrocatalysts for direct methanol fuel cells (DMFC): Effect of carbon precursors of OMC on DMFC performances. *Electrochimica Acta* 52, 1618–1626.
29. Park, K.W., Sung, Y.E., Han, S. et al. 2004. Origin of the enhanced catalytic activity of carbon nanocoil-supported PtRu alloy electrocatalysts. *Journal of Physical Chemistry B* 108, 939–944.
30. Jun, S., Joo, S.H., Ryoo, R. et al. 2000. Synthesis of new, nanoporous carbon with hexagonally ordered mesostructure. *Journal of the American Chemical Society* 122, 10712–10713.
31. Calvillo, L., Lazaro, M.J., Garcia-Bordeje, E. et al. 2007. Platinum supported on functionalized ordered mesoporous carbon as electrocatalyst for direct methanol fuel cells. *Journal of Power Sources* 169, 59–64.
32. Alegre, C., Calvillo, L., Moliner, R. et al. 2011. Pt and PtRu electrocatalysts supported on carbon xerogels for direct methanol fuel cells. *Journal of Power Sources* 196, 4226–4235.
33. Pekala, R.W. 1989. Organic aerogels from the polycondensation of resorcinol with formaldehyde. *Journal of Materials Science* 24, 3221–3227.
34. Al-Muhtaseb, S.A. and Ritter, J.A. 2003. Preparation and properties of resorcinol-formaldehyde organic and carbon gels. *Advanced Materials* 15, 101–114.
35. Job, N., Pirard, R., Marien, J. et al. 2004. Porous carbon xerogels with texture tailored by pH control during sol-gel process. *Carbon* 42, 619–628.

36. Moreno-Castilla, C. and Maldonado-Hodar, F.J. 2005. Carbon aerogels for catalysis applications: An overview. *Carbon* 43, 455–465.
37. Yamamoto, T., Endo, A., Ohmori, T. et al. 2005. The effects of different synthetic conditions on the porous properties of carbon cryogel microspheres. *Carbon* 43, 1231–1238.
38. Kim, S.I., Yamamoto, T., Endo, A. et al. 2006. Preparation of platinum nanoparticles supported on resorcinol-formaldehyde carbon cryogel microspheres. *Journal of Industrial and Engineering Chemistry* 12, 769–776.
39. Zhou, Z.H., Zhou, W.J., Wang, S.L. et al. 2004. Preparation of highly active 40 wt.% Pt/C cathode electrocatalysts for DMFC via different routes. *Catalysis Today* 93–95, 523–528.
40. Marie, J., Berthon-Fabry, S., Chatenet, M. et al. 2007. Platinum supported on resorcinol-formaldehyde based carbon aerogels for PEMFC electrodes: Influence of the carbon support on electrocatalytic properties. *Journal of Applied Electrochemistry* 37, 147–153.
41. Guo, D.J. and Li, H.L. 2006. Electrocatalytic oxidation of methanol on Pt modified single-walled carbon nanotubes. *Journal of Power Sources* 160, 44–49.
42. Che, G.L., Lakshmi, B.B., Fisher, E.R. et al. 1998. Carbon nanotubule membranes for electrochemical energy storage and production. *Nature* 393, 346–349.
43. Liu, Z.L., Lin, X.H., Lee, J.Y. et al. 2002. Preparation and characterization of platinum-based electrocatalysts on multiwalled carbon nanotubes for proton exchange membrane fuel cells. *Langmuir* 18, 4054–4060.
44. Knupp, S.L., Li, W.Z., Paschos, O. et al. 2008. The effect of experimental parameters on the synthesis of carbon nanotube/nanofiber supported platinum by polyol processing techniques. *Carbon* 46, 1276–1284.
45. Xing, Y.C. 2004. Synthesis and electrochemical characterization of uniformly-dispersed high loading Pt nanoparticles on sonochemically-treated carbon nanotubes. *Journal of Physical Chemistry B* 108, 19255–19259.
46. Wang, C., Waje, M., Wang, X. et al. 2004. Proton exchange membrane fuel cells with carbon nanotube based electrodes. *Nano Letters* 4, 345–348.
47. Serp, P., Corrias, M. and Kalck, P. 2003. Carbon nanotubes and nanofibers in catalysis. *Applied Catalysis A-General* 253, 337–358.
48. Wildgoose, G.G., Banks, C.E. and Compton, R.G. 2006. Metal nanoparticles and related materials supported on carbon nanotubes: Methods and applications. *Small* 2, 182–193.
49. Lee, K., Zhang, J.J., Wang, H.J. et al. 2006. Progress in the synthesis of carbon nanotube- and nanofiber-supported Pt electrocatalysts for PEM fuel cell catalysis. *Journal of Applied Electrochemistry* 36, 507–522.
50. Matsumoto, T., Komatsu, T., Arai, K. et al. 2004. Reduction of Pt usage in fuel cell electrocatalysts with carbon nanotube electrodes. *Chemical Communications* 840–841.
51. Bekyarova, E., Itkis, M.E., Cabrera, N. et al. 2005. Electronic properties of single-walled carbon nanotube networks. *Journal of the American Chemical Society* 127, 5990–5995.
52. Zhang, B., Chen, L.J., Ge, K.Y. et al. 2005. Preparation of multiwall carbon nanotubes-supported high loading platinum for vehicular PEMFC application. *Chinese Chemical Letters* 16, 1531–1534.

53. Wei, Z.D., Yan, C., Tan, Y. et al. 2008. Spontaneous reduction of Pt(IV) onto the sidewalls of functionalized multiwalled carbon nanotubes as catalysts for oxygen reduction reaction in PEMFCs. *Journal of Physical Chemistry C* 112, 2671–2677.

54. Hu, F.P., Shen, P.K., Li, Y.L. et al. 2008. Highly stable Pd-based catalytic nanoarchitectures for low temperature fuel cells. *Fuel Cells* 8, 429–435.

55. Yoo, E., Okada, T., Kizuka, T. et al. 2008. Effect of carbon substrate materials as a Pt-Ru catalyst support on the performance of direct methanol fuel cells. *Journal of Power Sources* 180, 221–226.

56. Ke, K. and Waki, K. 2007. Fabrication and characterization of multiwalled carbon nanotubes-supported Pt/SnOx nanocomposites as catalysts for electro-oxidation of methanol. *Journal of the Electrochemical Society* 154, A207–A212.

57. Wang, X., Waje, M. and Yan, Y.S. 2005. CNT-based electrodes with high efficiency for PEMFCs. *Electrochemical and Solid State Letters* 8, A42–A44.

58. Wang, J.J., Yin, G.P., Zhang, J. et al. 2007. High utilization platinum deposition on single-walled carbon nanotubes as catalysts for direct methanol fuel cell. *Electrochimica Acta* 52, 7042–7050.

59. Liao, S.J., Holmes, K.A., Tsaprailis, H. et al. 2006. High performance PtRuIr catalysts supported on carbon nanotubes for the anodic oxidation of methanol. *Journal of the American Chemical Society* 128, 3504–3505.

60. Paschos, O., Knupp, S.L., Choi, P. et al. 2007. Carbon nanotube-supported platinum electrode for ORR in phosphoric acid solution: Effect of PTFE content and annealing temperature. *Electrochemical and Solid State Letters* 10, B147–B149.

61. Zhang, W.H., Shi, J.L., Wang, L.Z. et al. 2000. Preparation and characterization of ZnO clusters inside mesoporous silica. *Chemistry of Materials* 12, 1408–1413.

62. Rajesh, B., Thampi, K.R., Bonard, J.M. et al. 2003. Carbon nanotubes generated from template carbonization of polyphenyl acetylene as the support for electrooxidation of methanol. *Journal of Physical Chemistry B* 107, 2701–2708.

63. Wu, G. and Xu, B.Q. 2007. Carbon nanotube supported Pt electrodes for methanol oxidation: A comparison between multi- and single-walled carbon nanotubes. *Journal of Power Sources* 174, 148–158.

64. Prabhuram, J., Zhao, T.S., Tang, Z.K. et al. 2006. Multiwalled carbon nanotube supported PtRu for the anode of direct methanol fuel cells. *Journal of Physical Chemistry B* 110, 5245–5252.

65. Maiyalagan, T., Viswanathan, B. and Varadaraju, U. 2005. Nitrogen containing carbon nanotubes as supports for Pt: Alternate anodes for fuel cell applications. *Electrochemistry Communications* 7, 905–912.

66. Guha, A., Lu, W.J., Zawodzinski, T.A. et al. 2007. Surface-modified carbons as platinum catalyst support for PEM fuel cells. *Carbon* 45, 1506–1517.

67. Rodriguez, N.M., Kim, M.S. and Baker, R.T.K. 1994. Carbon nanofibers: Unique catalyst support medium. *Journal of Physical Chemistry* 98, 13108–13111.

68. Lin, Z., Ji, L.W., Krause, W.E. et al. 2010. Synthesis and electrocatalysis of 1-aminopyrene-functionalized carbon nanofiber-supported platinum-ruthenium nanoparticles. *Journal of Power Sources* 195, 5520–5526.

69. Zheng, J.S., Zhang, X.S., Li, P. et al. 2007. Effect of carbon nanofiber microstructure on oxygen reduction activity of supported palladium electrocatalyst. *Electrochemistry Communications* 9, 895–900.

70. Li, W.Z., Waje, M., Chen, Z.W. et al. 2010. Platinum nanopaticles supported on stacked-cup carbon nanofibers as electrocatalysts for proton exchange membrane fuel cell. *Carbon* 48, 995–1003.
71. Boskovic, B.O., Stolojan, V., Khan, R.U.A. et al. 2002. Large-area synthesis of carbon nanofibres at room temperature. *Nature Materials* 1, 165–168.
72. Gangeri, A., Centi, G., La Malfa, A. et al. 2005. Electrocatalytic performances of nanostructured platinum-carbon materials. *Catalysis Today* 102, 50–57.
73. Wang, X.Z., Fu, R., Zheng, J.S. et al. 2011. Platinum nanoparticles supported on carbon nanofibers as anode electrocatalysts for proton exchange membrane fuel cells. *Acta Physico-Chimica Sinica* 27, 1875–1880.
74. Yuan, F.L. and Ryu, H.J. 2004. The synthesis, characterization, and performance of carbon nanotnbes and carbon nanofibres with controlled size and morphology as a catalyst support material for a polymer electrolyte membrane fuel cell. *Nanotechnology* 15, S596–S602.
75. Park, I.S., Park, K.W., Choi, J.H. et al. 2007. Electrocatalytic enhancement of methanol oxidation by graphite nanofibers with a high loading of PtRu alloy nanoparticles. *Carbon* 45, 28–33.
76. Yudasaka, M., Iijima, S. and Crespi, V.H. 2008. Single-wall carbon nanohorns and nanocones. *Carbon Nanotubes* 111, 605–629.
77. Hyeon, T., Han, S., Sung, Y.E. et al. 2003. High-performance direct methanol fuel cell electrodes using solid-phase-synthesized carbon nanocoils. *Angewandte Chemie-International Edition* 42, 4352–4356.
78. Sevilla, M., Lota, G. and Fuertes, A.B. 2007. Saccharide-based graphitic carbon nanocoils as supports for PtRu nanoparticles for methanol electrooxidation. *Journal of Power Sources* 171, 546–551.
79. Yuan, D.S., Xu, C.W., Liu, Y.L. et al. 2007. Synthesis of coin-like hollow carbon and performance as Pd catalyst support for methanol electrooxidation. *Electrochemistry Communications* 9, 2473–2478.
80. Novoselov, K.S., Geim, A.K., Morozov, S.V. et al. 2005. Two-dimensional gas of massless Dirac fermions in graphene. *Nature* 438, 197–200.
81. Li, Y.J., Gao, W., Ci, L.J. et al. 2010. Catalytic performance of Pt nanoparticles on reduced graphene oxide for methanol electro-oxidation. *Carbon* 48, 1124–1130.
82. Sharma, S., Ganguly, A., Papakonstantinou, P. et al. 2010. Rapid microwave synthesis of CO tolerant reduced graphene oxide-supported platinum electrocatalysts for oxidation of methanol. *Journal of Physical Chemistry C* 114, 19459–19466.
83. Wang, S.Y., Jiang, S.P. and Wang, X. 2011. Microwave-assisted one-pot synthesis of metal/metal oxide nanoparticles on graphene and their electrochemical applications. *Electrochimica Acta* 56, 3338–3344.
84. Qiu, J.D., Wang, G.C., Liang, R.P. et al. 2011. Controllable deposition of platinum nanoparticles on graphene as an electrocatalyst for direct methanol fuel cells. *Journal of Physical Chemistry C* 115, 15639–15645.
85. Lee, J.H., Shin, D.W., Makotchenko, V.G. et al. 2009. One-step exfoliation synthesis of easily soluble graphite and transparent conducting graphene sheets. *Advanced Materials* 21, 4383–4387.
86. Pumera, M. 2009. Electrochemistry of graphene: New horizons for sensing and energy storage. *Chemical Record* 9, 211–223.
87. Hummers, W.S. and Offeman, R.E. 1958. Prepartion of graphitic oxide. *Journal of the American Chemical Society* 80, 1339–1339.

88. Ha, H.W., Kim, I.Y., Hwang, S.J. et al. 2011. One-pot synthesis of platinum nanoparticles embedded on reduced graphene oxide for oxygen reduction in methanol fuel cells. *Electrochemical and Solid State Letters* 14, B70–B73.

89. Chen, X.M., Wu, G.H., Chen, J.M. et al. 2011. Synthesis of "clean" and well-dispersive Pd nanoparticles with excellent electrocatalytic property on graphene oxide. *Journal of the American Chemical Society* 133, 3693–3695.

90. Bong, S., Kim, Y.R., Kim, I. et al. 2010. Graphene supported electrocatalysts for methanol oxidation. *Electrochemistry Communications* 12, 129–131.

91. Xu, J.S., Granger, M.C., Chen, Q.Y. et al. 1997. Boron-doped diamond thin-film electrodes. *Analytical Chemistry* 69, A591–A597.

92. Chen, Q.Y., Granger, M.C., Lister, T.E. et al. 1997. Morphological and microstructural stability of boron-doped diamond thin film electrodes in an acidic chloride medium at high anodic current densities. *Journal of the Electrochemical Society* 144, 3806–3812.

93. Fischer, A.E. and Swain, G.M. 2005. Preparation and characterization of boron-doped diamond powder: A possible dimensionally stable electrocatalyst support material. *Journal of the Electrochemical Society* 152, B369–B375.

94. Bennett, J.A., Show, Y., Wang, S.H. et al. 2005. Pulsed galvanostatic deposition of Pt particles on microcrystalline and nanocrystalline diamond thin-film electrodes I. Characterization of as-deposited metal/diamond surfaces. *Journal of the Electrochemical Society* 152, E184–E192.

95. Sine, G., Foti, G. and Comninellis, C. 2006. Boron-doped diamond (BDD)-supported Pt/Sn nanoparticles synthesized in microemulsion systems as electrocatalysts of ethanol oxidation. *Journal of Electroanalytical Chemistry* 595, 115–124.

96. Sine, G., Smida, D., Limat, M. et al. 2007. Microemulsion synthesized Pt/Ru/Sn nanoparticles on BDD for alcohol electro-oxidation. *Journal of the Electrochemical Society* 154, B170–B174.

97. Spataru, N., Zhang, X.T., Spataru, T. et al. 2008. Platinum electrodeposition on conductive diamond powder and its application to methanol oxidation in acidic media. *Journal of the Electrochemical Society* 155, B264–B269.

98. Hu, F.P., Wang, Z., Li, Y. et al. 2008. Improved performance of Pd electrocatalyst supported on ultrahigh surface area hollow carbon spheres for direct alcohol fuel cells. *Journal of Power Sources* 177, 61–66.

99. Lim, K.H., Oh, H.S. and Kim, H. 2009. Use of a carbon nanocage as a catalyst support in polymer electrolyte membrane fuel cells. *Electrochemistry Communications* 11, 1131–1134.

100. Nam, K., Jung, D., Kim, S.K. et al. 2007. Operating characteristics of direct methanol fuel cell using a platinum-ruthenium catalyst supported on porous carbon prepared from mesophase pitch. *Journal of Power Sources* 173, 149–155.

101. Yan, Z.X., Hu, Z.F., Chen, C. et al. 2010. Hollow carbon hemispheres supported palladium electrocatalyst at improved performance for alcohol oxidation. *Journal of Power Sources* 195, 7146–7151.

102. Chen, S.X., Zhang, X. and Shen, P.K. 2006. Macroporous conducting matrix: Fabrication and application as electrocatalyst support. *Electrochemistry Communications* 8, 713–719.

103. Yan, Z., He, G., Zhang, G. et al. 2010. Pd nanoparticles supported on ultrahigh surface area honeycomb-like carbon for alcohol electrooxidation. *International Journal of Hydrogen Energy* 35, 3263–3269.

104. Lim, B., Yu, T. and Xia, Y. 2010. Shaping a bright future for platinum-based alloy electrocatalysts. *Angewandte Chemie-International Edition* 49, 9819–9820.
105. Yu, T., Zeng, J., Lim, B. et al. 2010. Aqueous-phase synthesis of Pt/CeO$_2$ hybrid nanostructures and their catalytic properties. *Advanced Materials* 22, 5188–5192.
106. Reshetenko, T.V., Avdeeva, L.B., Ismagilov, Z.R. et al. 2004. Catalytic filamentous carbon as supports for nickel catalysts. *Carbon* 42, 143–148.
107. Guha, A., Zawodzinski, T.A., Jr. and Schiraldi, D.A. 2007. Evaluation of electrochemical performance for surface-modified carbons as catalyst support in polymer electrolyte membrane (PEM) fuel cells. *Journal of Power Sources* 172, 530–541.
108. Yan, Z., Meng, H., Shi, L. et al. 2010. Synthesis of mesoporous hollow carbon hemispheres as highly efficient Pd electrocatalyst support for ethanol oxidation. *Electrochemistry Communications* 12, 689–692.
109. Yan, Z., Cai, M. and Shen, P.K. 2012. Low temperature formation of porous graphitized carbon for electrocatalysis. *Journal of Materials Chemistry* 22, 2133–2139.
110. Xie, F.Y., Tian, Z.Q., Meng, H. et al. 2005. Increasing the three-phase boundary by a novel three-dimensional electrode. *Journal of Power Sources* 141, 211–215.
111. Wu, J., Hu, F., Hu, X. et al. 2008. Improved kinetics of methanol oxidation on Pt/hollow carbon sphere catalysts. *Electrochimica Acta* 53, 8341–8345.

9

Atomic Layer Deposition of Metals and Metal Oxides for Fuel Cell Applications

Gaixia Zhang, Shuhui Sun, and Xueliang Sun

CONTENTS

9.1 Introduction 250
9.2 Fundamentals of ALD 252
    9.2.1 Principle of ALD 253
    9.2.2 Characteristics of ALD 254
    9.2.3 Key Deposition Factors of ALD 255
        9.2.3.1 Substrate 255
        9.2.3.2 Temperature 256
        9.2.3.3 Precursor 256
        9.2.3.4 Time 257
9.3 Materials Fabricated by ALD for Use as Fuel Cell Components 258
    9.3.1 Metals 258
        9.3.1.1 Platinum 258
        9.3.1.2 Palladium 259
    9.3.2 Mixed Metals 260
    9.3.3 Metal Oxides 262
    9.3.4 Coatings 263
9.4 ALD for Fuel Cells 265
    9.4.1 ALD for SOFCs 265
        9.4.1.1 ALD for Electrolytes 266
        9.4.1.2 ALD for Catalysts and Electrodes 268
        9.4.1.3 ALD for Other Components 271
    9.4.2 ALD for PEMFCs 272
        9.4.2.1 Reaction Mechanisms for PEMFCs 272
        9.4.2.2 Cathode Catalysts for ORR 273
        9.4.2.3 Anode Catalysts for DHFCs 275
        9.4.2.4 Anode Catalysts for DMFCs 275
9.5 Concluding Remarks and Outlooks 277
References 279

9.1 Introduction

Due to rising energy demands, depletion of fossil fuel reserves, and environmental pollution problems, a growing demand for exploring clean and sustainable energy sources as a replacement for combustion-based energy sources has sparked significant interest. As a promising candidate for environmentally benign electric power generation technology, fuel cells have drawn a great deal of attention because of their high efficiency, high energy density, lack of noise, and low or zero emissions.

A fuel cell is an electrochemical device that, with the help of catalysts, converts chemical energies of the fuel and oxidant directly into electrical energy. The primary components of a fuel cell are an ion-conducting electrolyte sandwiched between two electrodes, the cathode and anode. Together, these three are often referred to as the membrane-electrode assembly or simply a single-cell fuel cell. Generically, most fuel cells consume hydrogen (or hydrogen-rich fuels) and oxygen (or air) to produce electricity with by-products of only water and heat. Thus, fuel cells are very environmentally friendly.

Another selling point of fuel cells is their high efficiency [1]. Under favorable circumstances, this efficiency can be up to 60% or more (80% with heat recovery), which is two to three times that of combustion engines using fossil fuels (normally in the range of 20%–30%). This is because fuel cells convert chemical energy directly to electrical energy without the need of inefficient combustion as an intermediate step. Thus, fuel cells are not constrained by Carnot limit, giving them much higher conversion efficiency.

Further, compared to the limited capacity of batteries, fuel cells can continuously produce electricity as long as the fuel and oxidant inputs are supplied. Thus, the capability of continuously generating electricity makes fuel cells particularly suitable for automobile applications. In addition, fuel cells can provide power ranging from volts to megavolts; thus fuel cells have broad potential applications ranging from portable electronics to transportation and stationary uses [2–6].

While fuel cell technology is, by itself, very simple in principle, and there are huge advantages in many aspects of its implementation, it may still be considered in its infancy for applications in the wide consumer market because there are many reasons for the relatively slow development in "market-ready" fuel cells, such as high cost, low durability, and performance, which remain problems that are regarded as key challenges to the commercialization of fuel cell technology.

In recent years, the advent of nanomaterials and nanotechnology has shed light on almost every field of science and technology, particularly in the development of renewable energies, including fuel cells. Nanomaterials are structures with at least one dimension in the range of 1–100 nm. They can be classified into three categories according to their dimensions: zero-dimensional (0D) (nanoparticles, quantum dots, and fullerene), one-dimensional (1D) (nanotubes, nanowires, and nanobelts), and two-dimensional (2D) (nanofilms, nanowalls, and graphene) nanomaterials. One of the reasons for the considerable interest in nanomaterials is that such materials frequently display unusual physical (structural, electronic, magnetic, and optical) and chemical (catalytic) properties that vary from their bulk counterparts [7,8].

To date, many studies have reported nanostructured materials to be used as catalysts, electrodes, electrolytes, etc., which show extraordinary properties and are considered highly promising for fuel cell applications. Such materials can be synthesized by many methods, such as the template method [9–13], solution methods with or without surfactant [14–17], electrochemical depositions [18,19], sol-gel [20–25], physical or chemical vapor deposition (CVD) [26–32], and sputtering techniques [33–35]. Noble metals, especially platinum, are currently the most efficient electrocatalysts in fuel cells due to their outstanding catalytic and electrical properties and superior resistant characteristics to corrosion. However, they are rare and expensive. Therefore, the design of a highly efficient electrocatalyst is much desired. Many approaches have been developed to prepare Pt nanostructures in different shapes with improved performance, such as multioctahedrons [36], nanocubes [37], nanowires [16,38], and nanotubes [39,40], in order to reduce Pt loading or maximize its utility. Liang et al. [41] reported a free-standing Pt NW membrane catalyst that shows improved catalytic activity and remarkably high durability. In addition, bimetallic nanostructures have attracted increasing interest for fuel cells. Lim et al. [42] reported Pd–Pt nanodendrites that consist of dense arrays of Pt branches on a Pd core. Zhao et al. [43] reported Pt/CeO_2 deposited on hollow carbon spheres used as an anode electrocatalyst for DMFC. Other metals (e.g., Pd, Ru, Ir, Au), metal alloys, and composites (e.g., metals on carbon supports or metal oxide supports) have also shown improved catalytic activity and stability [44–48]. Yamaki et al. [49] demonstrated the preparation of nanostructured membranes for proton exchange membrane fuel cells (PEMFCs) by ion beam irradiation; such membranes exhibited less water uptake and lower methanol conductive regions. Many more studies demonstrated that a nanoscale membrane electrolyte presents opportunities for solid oxide fuel cells (SOFCs) [50–53]. Gao et al. [51] reported a composite electrolyte based on nanostructured $Ce_{0.8}Sm_{0.2}O_{1.9}$ (SDC) for low-temperature SOFCs, which showed a maximum power density of 900 mW cm^{-2} at 600°C. Despite much progress made in the past, translating nanostructures or nanofilms to real and useful catalysts, electrodes, or electrolytes is still challenging. There are some drawbacks of the above-mentioned methods and the as-synthesized materials, such as relative complexity and expensiveness, limited scalability and flexibility, impurity or large length scales of the products, undesirable reactions of the nanostructures due to the high surface area, and so on.

Atomic layer deposition (ALD) has recently been demonstrated as an effective method for synthesizing various materials from high-κ oxides (e.g., HfO_2, ZrO_2, Al_2O_3, Ta_2O_5), ternary oxides (e.g., $LaCoO_3$, $LaNiO_3$, $SrTiO_3$, $BaTiO_3$), metals (e.g., W, Mo, Pt, Pd, Ru, Rh, Ir, Os), fluorides, Li compounds, selenides/tellurides, and antimonides to organic and hybrid organic–inorganic polymers, via either a direct way to prepare nanomaterials when combined with templates or surface modification (e.g., coatings of metals, metal oxides, and other materials) to fine-tune the properties of the materials. For details, please refer to the reviews and other literature [54–59]. ALD provides the capability, via self-limiting chemical reactions, to achieve atomic-level control over film thickness and composition without the need for line-of-site access to the precursor sources. The as-synthesized materials are demonstrated with many meritorious properties. As a result, ALD could play a key role in achieving the breakthroughs in the syntheses of materials that can be used in many fields, including electronics, batteries, photovoltaics, fuel cells, capacitors, sensors, etc.

Recently, there have been increasing activities on the ALD synthesis of various materials for fuel cell components, showing the emerging applications of ALD for fuel cells [54–57,60–64]. However, very few reviews focus on ALD techniques in the fuel cell field. Very recently, Cassir et al. [65] have given an excellent review focused on ALD applications in SOFCs. Instead, the intent of this chapter is to give general concepts of fuel cells and ALD techniques as well as make a detailed survey on recent progress and directions in the field of ALD for fuel cells. It covers five parts in total: following the first part of introduction, the second part presents the fundamentals (e.g., principle, unique characteristics, and the key deposition factors) of ALD. We then illustrate the ALD technique for synthesizing specific materials (e.g., metals, mix-metals, and metal oxides) as well as coatings used for fuel cell components (e.g., catalysts, electrodes, electrolytes, current collectors) in the third part. And the fourth part highlights and details the applications of ALD for different types of fuel cells. In the end, we give concluding remarks and outlooks for future studies. We hope that the concepts and results presented in this chapter will prompt new researchers to join this field and help broaden the scope and impact of ALD technology for fuel cell applications.

9.2 Fundamentals of ALD

ALD, also called atomic layer epitaxy (ALE, before the year 2000), is a thin film deposition technique first demonstrated in Finland in the mid-1970s. Distinguished from other deposition techniques, such as CVD, physical vapor deposition, and other solution-based methods, ALD as a chemical vapor method operates with a unique mechanism relying on two gas–solid half-reactions to produce an ultimate material. In recent years, ALD has been attracting extensive and increasing interest in many fields for various applications.

9.2.1 Principle of ALD

ALD can be defined as a film deposition technique for atomic layer control and conformal deposition using sequential, self-limiting surface reactions. Normally, one growth cycle consists of four steps, taking Al_2O_3 ALD as a model system (Figure 9.1) [56], which are specified as follows:

(1) Exposure of the first precursor (trimethylaluminum, TMA)

(2) A purge of the reaction chamber to remove the nonreacted reactants (TMA) and the gaseous reaction by-products (CH_4)

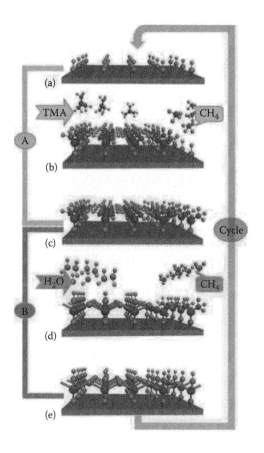

FIGURE 9.1

(See color insert.) A model ALD process for depositing Al_2O_3 using TMA and water as precursors: (a) the initial substrate covered by hydroxyl groups; (b) TMA molecules reacting with hydroxyl groups, leading to a new intermediate layer; (c) the substrate covered with a newly created intermediate layer; (d) H_2O molecules reacting with the intermediate layer, leading to new hydroxyl groups; and (e) the substrate again covered by hydroxyl groups. (Meng, X. et al.: *Advanced Materials*. 2012, 24, 3589–3615. Copyright Wiley-VCH Verlag GmbH & Co. KGaA. Reproduced with permission.)

(3) Exposure of the second precursor (water, H_2O), that is, another treatment to reactivate the surface for the reaction of the first reactant

(4) A further purge to remove the nonreacted reactants (H_2O) and the
gaseous by-product (CH_4)

The chemical reactions involved in the first and second self-limiting reactions and the overall reaction of a cycle deposition are given as follows:

First self-limiting reaction

$$AlOH^* + Al(CH_3)_3(g) \rightarrow Al-O-Al(CH_3)_2^* + CH_4(g) \qquad (9.1)$$

Second self-limiting reaction

$$Al\text{-}O\text{-}Al(CH_3)_2^* + 2H_2O(g) \rightarrow Al-O-Al(OH)_2^* + 2CH_4(g) \qquad (9.2)$$

Overall reaction

$$2Al(CH_3)_3 + 3H_2O \rightarrow Al_2O_3 + 6CH_4 \qquad (9.3)$$

where the asterisks (*) designate species attached on substrates by chemi-
sorptions and (g) the gaseous phase of the precursors or by-products. The
working temperature of ALD is usually lower than the decomposition of
the used precursors, typically lower than 400°C even down to room temperature. In addition, the growth cycles can be repeated as many times as
required to achieve a desired film thickness.

9.2.2 Characteristics of ALD

Ideally each exposure and purge step in ALD is complete. The precursor
molecules chemisorb or react with the surface groups saturatively, and after
the formation of the chemisorbed layer, no further adsorption takes place.
Under these saturative reaction conditions, the film growth is self-limiting;
that is, the amount of film material deposited in each reaction cycle is constant. Therefore, the self-limiting growth mechanism brings several advantages to ALD:

(1) Atomic level deposition
(2) Excellent uniformity and conformity
(3) Precise control of thickness and composition

(4) Large batches with larger or multiple substrates, thus easily scaled up

(5) Product with high purity and reproducibility

(6) Relatively low-temperature deposition

(7) Precursor variability from gases or volatile liquids to solids

In addition, the ALD processing window is often wide, which makes the process insensitive to small changes in temperature and precursor flows, and allows the processing of different materials to multilayer structures in a continuous process.

9.2.3 Key Deposition Factors of ALD

Just like all the other methods, the experimental parameters have very important effect on the as-synthesized materials. For an ALD process, as shown in Figure 9.1, there are several key deposition factors jointly determining the deposition features and, finally, the morphology and property of the synthesized materials. They are temperature, substrate, and precursor as well as pressure, time, and saturation/unsaturation of the reactions, etc. The effect of the experimental conditions on self-terminating gas–solid reactions must be well understood to be able to compare the results of different investigations.

9.2.3.1 Substrate

As illustrated in Figure 9.1, the first half-reaction of the ALD process relies on the interaction between the surface reactive sites and the first precursor. Due to the self-limiting, the new layer provides reactive sites for a following half-reaction. Similarly, the second self-limiting half-reaction creates another new layer with new functional groups for the next cycle of reaction. Therefore, both the initial coverage of reactive sites and the inherent properties of a substrate affect the growth of deposited materials. Differences in the surface characteristics of the substrates can cause differences in ALD experiments carried out for the same reactants and reaction temperature. The number and type of reactive surface sites may sometimes differ inherently on two substrates even if the general chemical compositions of the substrates are the same. An example is shown in Figure 9.2 [66], Pt ALD on Si substrates with HF-last and oxide-last surface treatments. On the oxide-last surface, Pt growth is already clearly detectable after 100 cycles of deposition, and an ~10 nm thick and relatively continuous Pt layer is found after 300 cycles. In contrast, on the HF-last surface, Pt growth is significantly retarded, and only discrete Pt particles are formed even after 300 cycles of deposition.

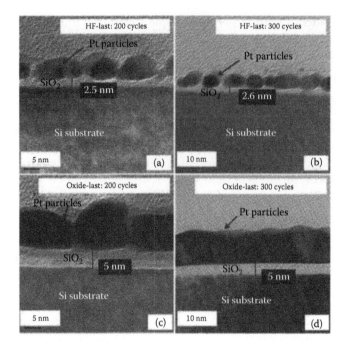

FIGURE 9.2

Cross-sectional TEM micrographs of Pt ALD on Si substrates (a, b) HF-last and (c, d) oxide-last. (The source of the material, Ge, L. et al. Influence of surface preparation on atomic layer deposition of Pt films, *Journal of Semiconductors*, 2012, is acknowledged. Reprinted with permission, copyright 2012 IOP Science.)

9.2.3.2 Temperature

The processing temperature is another crucial factor for sustaining ALD characteristics. To achieve ALD growth, the temperature must be held low enough so that the precursor does not decompose during surface adsorption, but the temperature must be high enough to thermally activate the reaction and/or avoid surface condensation. This leads to a range of temperatures, commonly referred to as the "ALD window" [54], as shown in Figure 9.3, in which the temperature is optimized to produce one monolayer of growth during each ALD cycle.

9.2.3.3 Precursor

The precursors for ALD could vary from gases and liquids to solids. The vapor pressure must be high enough for effective mass transportation; the solid and liquid precursors need to be heated. One must ensure that enough precursor is delivered to achieve full saturation; otherwise, the growth will be nonideal and nonuniform. On the other hand, we have to realize that for

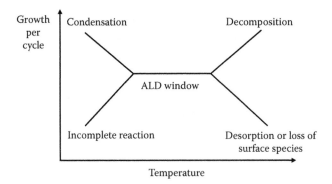

FIGURE 9.3
Schematic of possible behavior for the ALD growth per cycle versus temperature showing the "ALD window." (Reprinted with permission from George, S.M., *Chem. Rev.*, 110, 111–131. Copyright 2010 American Chemical Society.)

truly saturating, irreversible reactions, partial pressures are not expected to affect the amount of material adsorbed at saturation. Thus, a "pressure gap" has been hypothesized to exist between ALD growth at typical pressures of about 0.1–10 kPa. More studies are expected on this area.

9.2.3.4 Time

As related to the above, another factor, time, is important in order to make sure that the reactions of precursor and treatment gas go to completion, respectively, as well as the purge of oversupplied reactants and by-products to avoid an unexpected continuous process. In fact, the time for completing the self-terminating reactions may very well differ in orders of magnitude between different types of experiments. It has to be determined experimentally from deposition rate, substrate, etc. Obviously, the use of high-surface-area substrates increases the processing times compared to flat substrates.

Other factors, such as treatment gas and number of ALD cycles, may also affect the morphology of the product and thus the property, too. An excellent and comprehensive study on various deposition factors has been reported by Setthapun et al. [67]. Pt ALD was conducted on high-surface-area Al_2O_3, TiO_2, and $SrTiO_3$ supports. An *in situ* x-ray absorption fine structure technique was utilized to monitor the changes in the Pt species during each step of the synthesis. Through changing deposition parameters, different particle sizes, from 1–2 and 2–3 to 5 nm, were obtained, respectively. But under each set of synthesis conditions, the particles are fairly uniform in size and evenly distributed on the support. In addition, the ALD Pt catalysts exhibit excellent catalytic activity and stable WGS reaction rates that are comparable to those of conventional incipient wetness impregnation catalysts.

9.3 Materials Fabricated by ALD for Use as Fuel Cell Components

As mentioned above, ALD has been demonstrated to be an excellent method for synthesizing various materials for many applications in electronics, batteries, photovoltaics, fuel cells, capacitors, sensors, etc. This section is not a detailed review but rather a set of examples illustrating the promising synthetic ability of ALD in various nanomaterials, such as metals, mixed metals, and metal oxides as well as coatings (surface modifications). These materials could serve as fuel cell components, including catalysts, electrodes, electrolytes, current collectors, and others (e.g., interlayers via ALD coatings) and, therefore, hold very promising potential for fuel cell applications.

9.3.1 Metals

Metals (especially noble metals), mixed metals, and their composites are promising materials for catalysis in fuel cells [68–78]. They can serve as an electrode for current collection and a catalyst for the electrochemical reaction. Many studies have demonstrated the success of metal ALD on various supports with morphologies from thin films to nanoparticles [79–83].

9.3.1.1 Platinum

Pt is one of the commonly used electrode-catalyst materials for low-temperature fuel cells. It can serve as an electrode for current collection and a catalyst for the oxygen reduction reaction (ORR) and hydrogen oxidation reaction (HOR). Pt ALD is typically conducted by using alternating exposures of (trimethyl) methylcyclopentadienyl platinum (IV) (MeCpPtMe$_3$) and oxygen at 200°C–300°C with dose and purge times of several seconds required to coat planar surfaces. To prepare Pt nanoparticles on high surface area supports, a longer dose and purge times of several minutes are needed to allow the complete saturation and the removal of oversupplied precursor. The mechanism of Pt ALD is described by the following:

$$2(MeCp)PtMe_3 + 3O^* \rightarrow 2(MeCp)PtMe_2^* + CH_4 + CO_2 + H_2O \qquad (9.4)$$

$$2(MeCp)PtMe_2 + 24O_2 \rightarrow 2Pt^* + 3O^* + 16CO_2 + 13H_2O \qquad (9.5)$$

Aaltonen et al. [84] first reported the success of Pt thin films by ALD with excellent uniformity via using the above precursors. A growth rate of 0.45 Å per cycle was obtained with a 4 s total cycle time. The film thickness was found to linearly depend on the number of the reaction cycles. Noble-metal nanoparticles supported on metal oxide or carbon surfaces can serve as effective heterogeneous catalysts for accelerating the electrochemical process

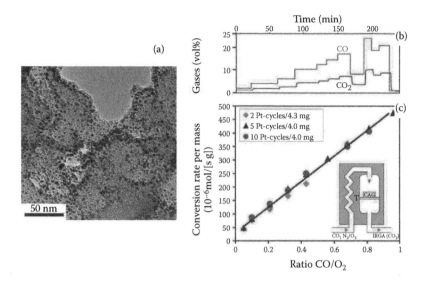

FIGURE 9.4

(a) TEM image of Pt-CA nanocomposites. (b, c) CO oxidation data for Pt-CA nanocomposites. (b) Time series of CO (inlet) and the resulting CO_2 signal at the reactor outlet for 4.0 mg of the 10 ALD cycle Pt-CA. (c) Conversion rate dependence of 2, 5, and 10 ALD cycle Pt-CA on the ratio of CO to O_2 (concentrations at the inlet). (Reprinted with permission from King, J.S. et al. *Nano Lett.*, 8, 2405–2409. Copyright 2008 American Chemical Society.)

[85–87]. Steven and Christensen et al. [85] studied the nucleation and growth of Pt nanoparticles on $SrTiO_3$ single crystals and nanocubes. Contrary to expectations, the Pt growth during the nucleation period on $SrTiO_3$ (001) was ~2× higher than in the steady-state regime. The higher initial growth was attributed to a greater density of reactive sites on $SrTiO_3$ compared to Pt. King et al. [88] demonstrated that Pt nanoparticles could be deposited on the inner surfaces of carbon aerogels (CAs) by ALD as shown in Figure 9.4. The Pt/CA materials exhibited nearly 100% conversion efficiency for the oxidation of CO in the 150°C–250°C temperature range with an ultralow Pt loading of only ~0.047 mg Pt/cm², and the total conversion rate seemed to be limited only by the thermal stability of the CA support in ambient oxygen.

9.3.1.2 Palladium

Palladium has very similar properties (same group of the periodic table, same fcc crystal structure, similar atomic size) as Pt. However, it is at least 50 times more abundant on earth and much cheaper than Pt. Therefore, Pd has been tested in fuel cells as a Pt co-catalyst (as anode and cathode material in acid media) and as a Pt-free catalyst (as anode material in alkaline media and in direct formic acid fuel cells [DFAFCs]). For detailed information, please refer to Antolini's review [89]. Here we only list a few researches on Pd ALD.

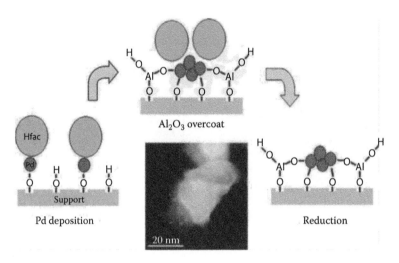

FIGURE 9.5
Subnanometer Pd particles on Al_2O_3 synthesized by ALD. (Reprinted with permission from Feng, H. et al. *ACS Catal.*, 1, 665–673. Copyright 2011 American Chemical Society.)

Feng et al. [90] demonstrated Pd ALD on different substrates (e.g., Al_2O_3, ZnO) using Pd(II) hexafluoroacetylacetonate (Pd(hfac)$_2$) and formalin as precursors. They revealed that the Pd growth begins with a relatively slow nucleation process and accelerates once an adequate amount of Pd has deposited on the surface. Furthermore, the Pd nucleation is faster on ALD ZnO surfaces and the catalyst possesses higher activity and hydrogen selectivity at relatively low temperatures compared to Pd ALD Al_2O_3 surfaces. By applying one cycle of ALD Al_2O_3 on top of the Pd–ZnO catalyst, the activity was enhanced, and the catalyst deactivation was mitigated. In addition, using Al_2O_3 as a substrate, they tuned the size of the ALD Pd particles by adjusting the preparation conditions. Conventional ALD conditions produced Pd particles with an average size of 1.4 nm. The removal of surface hydroxyls from the alumina support by a chemical treatment using TMA before performing Pd ALD led to nanoparticles larger than 2 nm. Ultra-small Pd particles were synthesized using low-temperature metal precursor exposures followed by applying protective ALD alumina overcoats as shown in Figure 9.5.

Other catalyst metals, such as Ru [91–93] and Ir [94–97], have also been studied with the ALD technique. Due to the space limitation, we will not give details here. Readers are recommended to refer to the related references.

9.3.2 Mixed Metals

Bimetallic catalysts offer the possibility of combining the unique advantages of each component, allowing the catalytic properties to be tuned by adjusting the nanoparticle composition and structure. However, the design and economical synthesis of catalysts at the atomic scale represents a scientific grand challenge. ALD offers a potential solution to this challenge. As with the pure Pt and Pd nanoparticles described in the literature, the size of bimetallic particles should be controlled by the total number of ALD cycles performed. Moreover, the composition will be dictated by the relative number of ALD cycles used for each component. Finally, the structure of the bimetallic nanoparticles might be controlled by the order in which the individual cycles are executed.

To demonstrate the viability of mixed-metal ALD, Christensen and Elam [98] and Comstock et al. [99] successfully synthesized thin-film mixtures of iridium and platinum by controlling the ratio between the iridium(III) acetylacetonate/oxygen cycles for Ir ALD and the (trimethyl)methylcyclopentadienyl platinum(IV)/oxygen cycles for Pt ALD. Christensen and Elam employed *in situ* quartz crystal microbalance and quadrupole mass spectrometry measurements to explore the growth mechanisms for the pure iridium and platinum metals. Accordingly, via adjusting the relative ALD cycle ratio of $Ir(acac)_3/O_2$ and $Pt(MeCp)Me_3/O_2$, the Ir/Pt ratio in the films could be controlled precisely. Comstock et al. further realized the synthesis of PtIr alloy films with controlled composition and morphology by template ALD as shown in Figure 9.6.

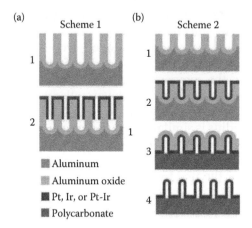

FIGURE 9.6
Templating schemes for the synthesis of nanostructured Pt/Ir films by ALD. (a) Scheme 1: (1) high-aspect ratio AAO template and (2) ALD deposition that does not fully infiltrate the AAO template. (b) Scheme 2: (1) low-aspect ratio AAO template; (2) ALD deposition that fully infiltrates the AAO template; (3) film is inverted, bonded to a polycarbonate substrate, and the aluminum substrate is etched away; and (4) AAO template is etched away to expose the templated ALD film. (From Comstock, D.J. et al.: *Advanced Functional Materials.* 2010. 20. 3099–3105. Copyright Wiley-VCH Verlag GmbH & Co. KGaA. Reproduced with permission.)

Some other mixed metals that have the potential to be used as fuel cell electrode catalysts are also developed, such as Pt–Ru [100–102], Pt–Ni [103], and others [104].

9.3.3 Metal Oxides

Metal oxide nanostructures, as well as their components with other materials, such as carbon, often show extraordinary properties and are considered highly promising for numerous future applications, such as electrochemical power sources, photocatalysis, flexible piezoelectric generators, and sensors. Recently, metal oxide ALD has attracted increasing interest for fuel cells because these materials can serve as electrolytes, catalysts, and even catalyst supports.

The effort put into solid oxide electrolytes by ALD so far includes zirconia-based oxides, ceria-based oxides, and lanthanum gallate–based oxides. They are used for SOFCs to increase the electrolyte ionic conductivity. Zirconia-based electrolytes, such as yttria-stabilized-ZrO_2 (YSZ), are the most popular materials employed as the electrolyte in SOFCs because of its attractive ionic conductivity, stability in both oxidizing and reducing environments, and compatibility with the electrode materials. Several groups have conducted YSZ ALD using different precursors, at different temperatures, on various supports, with different film thickness (from several tens of nanometers to a few micrometers) [105–108]. However, at lower temperatures, the ionic conductivity of YSZ is much lower than the other two types of electrolytes. The ceria-based electrolytes have an ionic conductivity superior to that of zirconia-based electrolytes. Gadolinia-doped ceria (GDC) electrolytes are good candidates due to their high ionic conductivity and are already used in SOFCs. However, the composition of GDC films deposited by ALD has proven to be difficult to optimize, and the growth rate is generally very low, that is, 0.4–0.5 Å/cycle [109]. Yttria could be a good substitute for gadolinia, as (1) the Y^{3+} ionic radius is similar to that of Gd^{3+}, (2) ALD processes for Y_2O_3 deposits are well established with more satisfactory growth rates, and (3) the lower cost is due to the fact that yttrium is more abundant than gadolinium. Ballée et al. [109], for the first time, processed the YDC electrolyte by ALD at 300°C with β-diketonate complexes $Y(thd)_3$ and $Ce(thd)_4$ used as yttrium and cerium precursors, respectively, whereas ozone was used as the oxygen source. It seems that no lanthanum gallate–based oxides have been produced by ALD so far, but $LaGaO_3$ and Ga_2O_3 have been reported by Nieminen et al. [110] and Dezelah et al. [111], respectively.

Some other metal oxides, such as NiO/CNT hybrid materials, have been studied for use as advanced electrocatalysts due to their unique hybrid structure, large surface area, and high electrochemical activity [112–116]. However, it is still difficult to permit precise control of the size of the deposited NiO nanoparticles at the subnanometer level while preserving the homogeneity. Very recently, Tong et al. [117] developed a simple and effective route for the

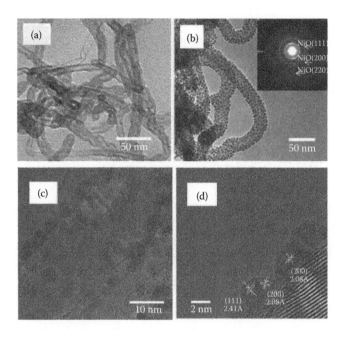

FIGURE 9.7
TEM images of (a) pristine CNTs and (b) NiO nanoparticles on CNTs grown after 400 ALD cycles (inset shows the SAED pattern). (c) HRTEM image of a NiO/CNT hybrid nanostructure after 400 ALD cycles and of (d) the area boxed region in (c) at higher magnification showing the {200} and {111} planes of cubic NiO. (Tong, X. et al.: *Small*. 2012. 8. 3390–3395. Copyright Wiley-VCH Verlag GmbH & Co. KGaA. Reproduced with permission.)

synthesis of NiO/CNT by ALD as shown in Figure 9.7. The sizes of the NiO nanoparticles range from 1.5 to 6.3 nm and can be precisely controlled by varying the number of ALD cycles. Compared to commercial NiO nanopowder and the results reported by Asgari et al. [115], the obtained NiO/CNT hybrid nanostructures show greatly enhanced electrochemical catalytic ability and stability for a methanol oxidation reaction.

9.3.4 Coatings

In addition to the direct way to prepare nanomaterials when combined with templates, ALD also allows surface modification and fabrication to fine-tune the properties of the material. Many papers have reported the synthesis of various materials via surface modification by ALD. Bae et al. gave an excellent review that addresses the recent progress in the field of nanostructured materials and devices that have been surface-modified and fabricated by ALD. Such materials have potential applications in high-performance transistors, sensors, and green energy conversion (e.g., batteries). Here, we give a few examples to demonstrate the coating ability of ALD and to show the enhanced catalytic activity and other properties of the surface-modified materials.

For example, Biener et al. [118] demonstrated that ALD could be used to stabilize and functionalize nanoporous metals as shown in Figure 9.8. Specifically, the effect of nanometer-thick ALD films of alumina and titania on thermal stability, mechanical properties, and the catalytic activity of nanoporous gold (np-Au) have been studied. The results demonstrated that even 1-nm-thick oxide films could stabilize the nanoscale morphology of np-Au up to 1000°C while simultaneously making the material stronger and

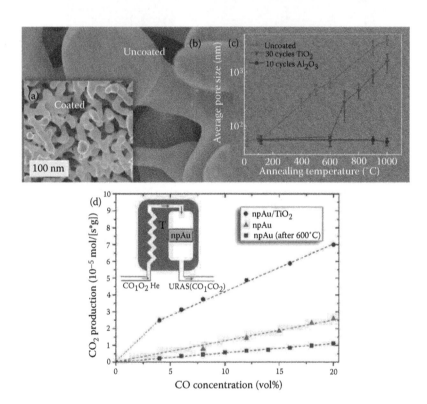

FIGURE 9.8

SEM micrographs obtained from (a) Al_2O_3-coated (10 cycles) and (b) uncoated np-Au after annealing at 600°C. (c) Thermal stability of Al_2O_3- and TiO_2-coated np-Au: development of average ligament size of Al_2O_3-coated (10 cycles), TiO_2-coated (30 cycles), and uncoated np-Au vs. annealing temperature. (d) Comparison of the catalytic activity of a TiO_2-modified (10 ALD cycles, annealed at 600°C) np-Au sample with an uncoated np-Au sample before and after annealing at 600°C in He. Shown is the vol% O_2 in a He carrier gas. Titania-coating causes a threefold increase in the catalytic activity with respect to the uncoated, not annealed np-Au sample, and annealing of the uncoated np-Au sample at 600°C decreases its reactivity by more than 50%. (Reprinted with permission from Biener, M.M. et al. *Nano Lett.*, 11, 3085–3090. Copyright 2011 American Chemical Society.)

stiffer. The catalytic activity of np-Au could be drastically increased by TiO_2 ALD coatings. These results showed much potential in high-temperature sensors, actuators, and catalysis applications and functionalized electrodes for energy storage and harvesting applications. Chen et al. [119] reported the ALD assisted $Pt-SnO_2$ hybrid catalysts on nitrogen-doped CNTs (NCNTs) with enhanced electrocatalytic activities for low-temperature fuel cells. The sample is prepared by an ALD coating of SnO_2 on NCNTs, followed by Pt nanoparticle deposition by the ethylene glycol reduction method. Readers will see more surface coatings by ALD somewhere in the next section for fuel cell applications. In any case, it is worth pointing out that ALD coating presents a powerful tool for fabricating nanomaterials with controlled composition and physical properties in an exceptionally accurate manner.

9.4 ALD for Fuel Cells

Traditionally, fuel cells can be classified, according to the types of electrolyte employed, into five categories: (1) alkaline fuel cells (AFCs), (2) phosphoric acid fuel cells (PAFCs), (3) molten carbonate fuel cells (MCFCs), (4) SOFCs, and (5) PEMFCs. In addition, direct methanol fuel cells (DMFCs) can be classified as PEMFCs because they use proton exchange membranes (PEMs) as the electrolyte, too. Generally, almost all fuel cells (with the exception of DMFCs) consume hydrogen and oxygen to produce electrical current as mentioned above; however, different types of fuel cells have their own strengths and weaknesses and, as a result, they have their application niches [2–6]. There are also other types of fuel cells that are employed less but may later find a specific application. Examples are the air-depolarized cells, sodium amalgam cells, biochemical fuel cells, inorganic redox cells, microbial fuel cells, regenerative cells, alkali metal–halogen cells, etc. We are not going to give details for each type of fuel cells (see Chapter 1 for general information on various types of fuel cells).

9.4.1 ALD for SOFCs

SOFCs have been intensively investigated because of their high energy conversion efficiency and flexible fuel choice as well as extremely low emissions. They are very promising energy devices to be used in large, high-power applications, such as full-scale industrial stations and large-scale electricity-generating stations. However, the high operating temperatures (800°C to 1000°C) of SOFCs not only pose significant challenges regarding structural stability and material integrity [120,121] but also reduce their practicality for mobile applications [122,123]. Lowering the operating temperature adversely affects the rate of the ORR and HOR at the electrode/electrolyte interfaces, thereby reducing the exchange current density and causing a drop in fuel cell performance.

To improve SOFC performance at low temperatures, recently, ALD has emerged as a promising new technology to be adopted. As a consequence, there are ongoing research activities by several groups to develop SOFC materials and processes by ALD. The choice of materials for each component in a SOFC device depends on the functional properties required. In the following, we will go through a brief description of materials synthesized by ALD, which can be used for different SOFC components as well as their performances for fuel cell applications.

9.4.1.1 ALD for Electrolytes

Currently, electrolyte by ALD may have been far more frequently studied than other components in fuel cells. In the previous section, we have given a brief introduction on the metal oxide electrolyte by ALD for fuel cells. Following that, we give a few typical examples to show their promising properties for fuel cells.

As mentioned above, the Prinz group [106,124] has developed YSZ by ALD for serving as an electrolyte in SOFCs designed to operate at low temperatures, via using Tetrakis-(dimethylamido)zirconium and tris(methylcyclopentadienyl) yttrium as precursors and distilled water as the oxidant. To evaluate ALD YSZ films as oxide ion conductors, freestanding 60 nm films were prepared with porous platinum electrodes on both sides of the electrolyte as shown in Figure 9.9. Maximum power densities of 28, 66, and 270 mW/cm^2 were observed at 265°C, 300°C, and 350°C, respectively. The high performance of SOFCs based on thin film ALD electrolytes is related to low electrolyte resistance and fast electrode kinetics. The exchange current density at the electrode–electrolyte interface was approximately four orders of magnitude higher compared to reference Pt-YSZ values.

The Prinz group [107] also presented a method for fabricating a thin-film SOFC featuring a three-dimensional corrugated catalyst/electrolyte membrane to increase the electrochemically active surface area. Membrane fabrication is carried out by employing a sequence of microelectromechanical systems (MEMS) processing steps shown in Figure 9.10. The corrugated YSZ electrolyte structure fabricated in this manner and their fuel cell performance are shown in Figure 9.11. The corrugated electrolyte membranes released from the silicon substrate showed an increase in power density relative to membranes with planar electrolytes. Maximum power densities of the corrugated fuel cells of 677 and 861 mW/cm^2 were obtained at 400°C and 450°C, respectively.

The yttria doped ceria (YDC) ALD was processed for the first time by Ballée et al. [109]. As shown in Figure 9.12, SEM observations showed that the films were relatively homogeneous and well covering. Their conductivities were significantly lower than those reported in the literature for bulk YDC.

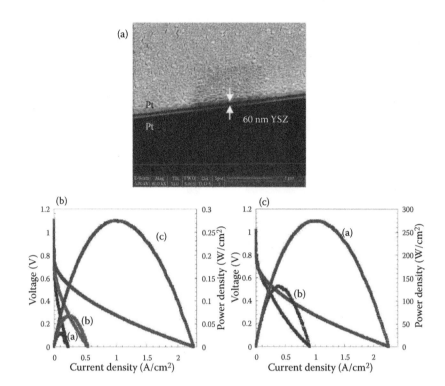

FIGURE 9.9

(a) SEM image of freestanding ALD YSZ films with porous Pt cathode–anode layers. Polarization curves and power density of YSZ fuel cells. (Left) Performance of ALD YSZ fuel cells at (a) 265°C, (b) 300°C, and (c) 350°C [125]. (Right) Performance comparison of (b) ALD YSZ fuel cells to (c) reference fuel cells with 50 nm RF-sputtered YSZ at 350°C. (Reprinted with permission from Shim, J.H. et al. *Chem. Mater.*, 19, 3850–3854. Copyright 2007 American Chemical Society.)

Nevertheless, the electrical properties of YDC thin films seem to be superior to those of the YSZ thin films, characterized under the same conditions, in the temperature range of 500°C–750°C. Thus, YDC by ALD represents a new and promising SOFC generation.

In addition, ALD coating provides another promising technique in fuel cell electrolytes. Still, we take the Prinz group's studies as examples. The surface modification by ALD adding a 1-nm-thin, high-yttria concentration (14–19 M%) YSZ film on YSZ electrolyte membranes led to an increase in the maximum power density of a low-temperature SOFC (LT-SOFC) by a factor of 1.50 at 400°C [126]. The enhanced performance can be attributed to an increased oxide ion incorporation rate on the surface of the modified electrolyte. Very recently, they also reported the significantly enhanced performance of SOFCs through ALD surface modification of YSZ electrolyte by doped ceria interlayers [127]. The doping concentrations of Y_2O_3 were

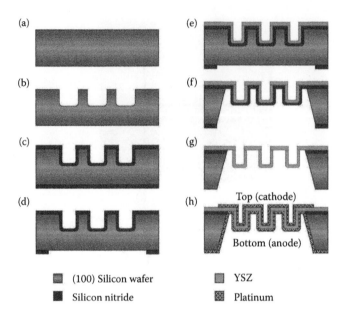

FIGURE 9.10

Process flow diagram for fabrication of the corrugated thin-film SOFC. The (100) silicon substrate is etched by DRIE to generate the template for pattern transfer (a and b). A 100-nm-thick silicon nitride layer is deposited with LPCVD on both sides of wafer (c). The backside of silicon nitride is patterned with openings (d), followed by ALD deposition of YSZ onto the template (e). The silicon template is etched in KOH (f), and the silicon nitride etch stop is removed by plasma etching (g). The free-standing corrugated electrolyte is deposited with porous platinum on top (cathode) and bottom (anode) sides acting as both electrode and catalyst (h). (Reprinted with permission from Su, P.C. et al. *Nano Lett.*, 8, 2289–2292. Copyright 2008 American Chemical Society.)

finely tuned by changing the Ce/Y pulsing ratios. YDC layers were inserted between a bulk YSZ electrolyte and a porous Pt cathode. The performance of SOFCs with different doping concentrations of YDC interlayers was measured using the linear sweep voltammetry method at operating temperatures between 300°C and 500°C. The fuel cell performance was enhanced by a factor of 2–3.6 within the Y_2O_3 doping range of 12–17 mol%. The enhanced performance was attributed to an increased oxide ion incorporation rate at the electrode–electrolyte interface.

9.4.1.2 ALD for Catalysts and Electrodes

So far, most of the contributions for ALD coatings in catalysts and catalyst–electrodes for SOFCs were made by Prinz's group. They conducted so many fuel cells tests via several strategies [124,128–132]. First, ultrathin Pt layers were deposited by ALD on various sputtered porous metals (e.g., Au, W, Ag, Pd,

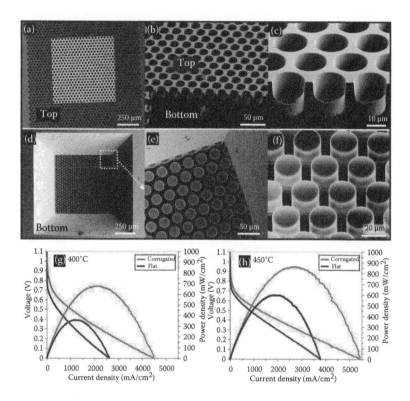

FIGURE 9.11

(See color insert.) The upper part shows the images of corrugated YSZ membrane, view from top (a through c) and bottom (d through f). (a) Optical microscopy image of free-standing corrugated YSZ membrane. The white square in the middle is the free-standing YSZ membrane released after KOH etching. The black dots represent cup-shaped trenches. (b and c) Cross section of cups in corrugated membrane. (d) Bottom view of membrane after silicon removal. (e) Bottom corner view of membrane; (f) perspective after tilting 40°. The lower part shows the fuel cell performance of corrugated (red line) and flat (black line) YSZ electrolyte at (g) 400°C and (h) 450°C. The membrane size is 0.0036 cm² (600 μm × 600 μm). The size of the cups embedded is 15 μm in diameter and 20 μm in depth. Maximum power densities of corrugated YSZ are slightly less than two times that of flat ones. (Reprinted with permission from Su, P.C. et al. *Nano Lett.*, 8, 2289–2292. Copyright 2008 American Chemical Society.)

Ru) as shown in Figure 9.13 [128]. Such materials served as cathodes, together with 200-μm-thick 8% YSZ electrolytes and 80-nm-thick sputtered platinum anodes to make the fuel cell tests. ALD Pt clearly enhanced the performance of the fuel cells and preserved the porous structure of the base metals after exposure to ambient air at elevated temperatures, while the bare porous metals showed poor performance presumably due to either a morphological change or thermal oxidation. All of the tested metals with ALD Pt coatings performed similar to fuel cells with pure Pt cathodes except the Au–ALD Pt.

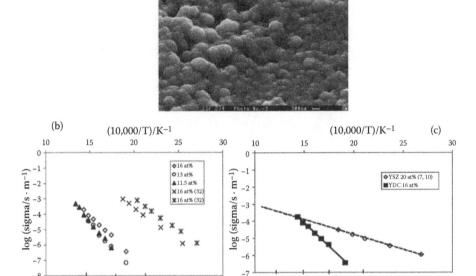

FIGURE 9.12

(a) SEM micrograph of yttria-doped ceria layers deposited by ALD at 300°C on LSF (425 nm thick, 16 at% Y). (b) Arrhenius plots of the YDC layer conductivity for samples (11.5, 13, and 16 at% Y, respectively) deposited by ALD as well as those for the YDC bulk [133]. (c) Arrhenius plots of YDC (16 at% Y) and YSZ (8 at% Y) [109,134] with a thickness of 425–450 nm, processed by ALD at 300°C on stainless steel. (Reprinted with permission from Ballée, E. et al. *Chem. Mater.*, 21, 4614–4619. Copyright 2009 American Chemical Society.)

Second, in order to improve the bonding between the electrolyte and the electrodes, they deposited a 1 nm platinum layer *in situ* on top of the electrolyte [132]. An 80% improvement in maximum power density was observed at 450°C. This confirms that *in situ* deposition of a platinum catalyst by ALD generates better interface quality, which reduces interfacial resistance in fuel cell operations. Third, they integrated the ALD technique into the electrode–catalyst–fabrication process with different ALD Pt cycles and further identified the optimal deposition conditions that delivered the highest triple-phase-boundary (TPB, where the electrode catalyst, gas, and electrolyte meet) density [130]. The catalyst structure that yielded the highest TPB density coincided with the structure that yielded the highest electrochemical performance. Jiang et al. also explored ALD for depositing various components of SOFCs. We are not going to give details about the catalyst/electrode layers here, but we will introduce their work for current collectors in the following section.

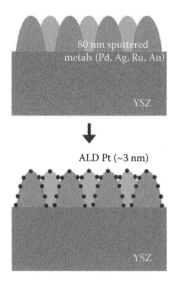

FIGURE 9.13
Schematics of ALD Pt surface coating on sputtered porous metal meshes as catalysts of ceramic fuel cells with YSZ electrolytes. (Reprinted with permission from Shim, J.H. et al. *J. Electrochem. Soc.*, 157, B793–B797, Copyright 2010, The Electrochemical Society.)

9.4.1.3 ALD for Other Components

Jiang et al. developed the area-selective ALD technique, a method for depositing materials in a well-controlled pattern on the surface [135–137] to fabricate a current collector consisting of a Pt grid pattern. The fabrication of a micropatterned Pt structure for the current collector grids of the fuel cells is shown in Figure 9.14 [87], in which a self-assembled monolayer, in particular, octadecyltrichlorosilane, is used as the ALD resist to modify the chemical properties of the substrate surface for patterning. Excitingly, an improvement of fuel cell performance (YSZ-based SOFCs) by a factor of 10 was observed when using the Pt current collector grid/patterned catalyst that integrated onto cathodic $La_{0.6}Sr_{0.4}Co_{0.2}Fe_{0.8}O_{3-\delta}$ (LSCF). Given the catalytic nature of Pt, it would be interesting to deposit another noncatalytic metal, such as Au, as the current collector to separate the influence of catalytic and electronic effects on LSCF cathode improvement. The above studies also suggest that the ALD-deposited Pt thin films and micropatterned structures can potentially be integrated onto any type of fuel cell as a catalyst layer to improve the electrode reaction and as a current collector to improve the current collection efficiency. To make full use of the advantages of ALD-deposited Pt films, such as high conformity and low resistivity for films as thin as 5 nm, better catalyst-support geometries should be designed for coating Pt through ALD.

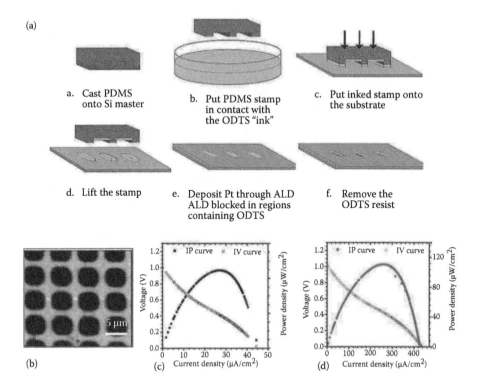

FIGURE 9.14

(See color insert.) (a) Schematic outline of the procedure to fabricate patterned Pt thin films using microcontact printing and selective ALD. (b) Auger electron spectroscopy elemental maps for platinum on the micropatterned grid structure after area-selective ALD of Pt on YSZ. (c and d) *I–V* and *I–P* characteristics of fuel cells without and with Pt current collector grids, respectively. (Reprinted with permission from Jiang, X. et al. *Chem. Mater.*, 20, 3897–3905. Copyright 2008 American Chemical Society.)

9.4.2 ALD for PEMFCs

Among the multitude of fuel cell technologies available, PEMFCs have become increasingly important because they operate at relatively low temperatures and have short start-up and transient-response times compared to other types of fuel cells that operate at higher temperatures (200°C to 800°C).

9.4.2.1 Reaction Mechanisms for PEMFCs

At the cathode of a PEMFC, oxygen is electrochemically reduced on catalyst surfaces through the ORR. On the anode side, various chemicals, including hydrogen, methanol, formic acid, and other alcohols, can be fed as fuels; accordingly, PEMFCs can be further classified as direct hydrogen fuel cells (DHFCs), DMFCs, DFAFCs, and direct alcohol fuel cells (DAFCs).

In PEMFCs, the slow reaction kinetics contributes greatly to the overall overpotential in a cell, and designing highly efficient and durable electrocatalysts is pivotal in advancing fuel cell technology. Platinum-based nanomaterials are the most widely used PEMFC electrocatalysts, and reducing the high cost of catalysts is one of the major challenges for widespread commercialization of PEMFCs [138]. In the following sections, we will give a few examples to demonstrate the recent development and progress of the catalysts in PEMFCs made by ALD technique.

9.4.2.2 Cathode Catalysts for ORR

One monolayer of Pt on tungsten monocarbide (WC) has been theoretically and experimentally demonstrated in the literature to be a promising system for fuel cell and electrolysis applications by allowing a reduction in Pt loading while maintaining Pt-like activity. WC is known as a promising material for fuel cell applications because it is inexpensive, and it has "platinum-like" catalytic properties [139–141] and exhibits good stability in acidic environments over a wide potential range [141–143]. WC has also shown great promise as a support material for platinum particles [144], leading to improved catalyst stability [145] and synergistic activity in a variety of electrochemical systems [146]. The common catalytic properties and good adhesion of Pt on WC are attributed to similarities in the electronic structures of these two materials [144]. Hsu et al. [140] reported, for the first time, the use of the ALD technique to investigate Pt on WC substrates with various cycles as shown in Figure 9.15. ALD Pt was found to deposit onto WC substrates following an island growth mechanism. When a few Pt ALD cycles were used, discrete Pt particles first formed and dispersed over the WC substrate. At least 100 ALD cycles were required for the WC substrate to be totally covered with Pt. CV measurements suggested that the low-loading ALD Pt on WC showed excellent ORR activity that was comparable to that for Pt foil. Qualitatively, ORR activity is detected for films with as few as 20 ALD cycles, but quantitative measurements of ORR activities are needed to compare the ALD Pt/WC samples with commercial Pt catalysts.

In addition, Hoover and Tolmachev [103] also demonstrated an ALD fabrication of Pt coatings on non-noble metals (e.g., Ni) with (sub)monolayer thickness, using MeCpPtMe$_3$ and H$_2$ as precursors. They found that the ALD results in micrometer-size Pt islands up to about 3.7 monolayers of Pt and continues more uniformly at higher loadings. The island-like growth at low Pt loadings is attributed to the presence of adsorbed O on atomically rough facets of polycrystalline Ni substrate. The coating becomes virtually pinhole free at Pt real-area loading over 8 ML. The Pt thin films show similar enhancement in ORR to PtNi alloys as shown in Figure 9.16. The thin films of Pt on Ni show enhancement factors in ORR similar to what has been reported for PtNi alloys. There is a great potential to employ the above methodology to design the third-generation PEMFC catalytic layers with ultralow

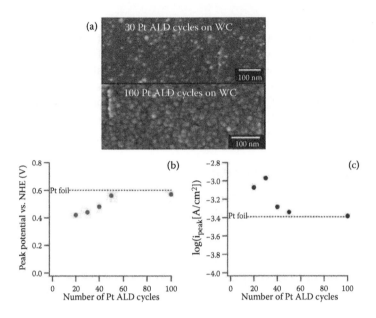

FIGURE 9.15
(a) SEM images of Pt-WC foil surfaces after 30 and 100 Pt ALD cycles. (b) ORR peak potentials vs. number of Pt ALD cycles and (c) ORR peak current vs. number of Pt ALD cycles. (Reprinted with permission from Hsu, I.J. et al. *J. Phys. Chem. C*, 115, 3709–3715. Copyright 2011 American Chemical Society.)

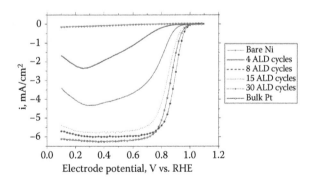

FIGURE 9.16
RDE oxygen-reduction polarization curves for Pt and Pt/Ni electrodes. 0.10 M $HClO_4$ saturated at 1.00 atm O_2, 1600 rpm, 20 mV/s anodic scan. The current is normalized per the geometric area of the disk electrode. (Reprinted with permission from Hoover, R.R., and Tolmachev, Y.V., *J. Electrochem. Soc.*, 156, A37–A43, Copyright 2009, The Electrochemical Society.)

Pt loading. It is also believed that a more uniform Pt coating at (sub)mono-layer loadings can be achieved using high-vacuum ALD.

9.4.2.3 Anode Catalysts for DHFCs

Less attention has been put into anode catalysts for DHFCs so far. Liu et al. [147] demonstrated the good dispersion of Pt nanoparticles on acid-treated carbon supports (CNTs and carbon cloth) by ALD. The particle density and loading of Pt could be controlled by the condition of substrate surface and the cycle number. Conformal Pt nanoparticle coating on acid-treated CNTs was achieved as shown in Figure 9.17, and it showed higher catalyst utilization efficiency in PEMFCs than commercial E-TEK electrodes.

9.4.2.4 Anode Catalysts for DMFCs

Although hydrogen is an ideal fuel for low-temperature fuel cells (PEMFCs), grand challenges remain in the production and storage of H_2. To bypass these disadvantages, the DMFC (or DAFC) has been developed with aqueous methanol (or alcohol) as the fuel. Methanol, as a liquid at room temperature, makes the fuel storage and transport much easier. Methanol also contains a significantly higher energy density (5 kWh/L for methanol vs. 3 kWh/L for hydrogen compressed at 5000 psi) [102], involving six electrons in the complete oxidation to CO_2 [148]. Potentially, this provides energy densities that are 10 times higher than that of Li-ion batteries, and it makes DMFCs an attractive power source for portable military and commercial applications [149–152]. However, unlike in hydrogen fuel cells, the anodic half-cell reactions have

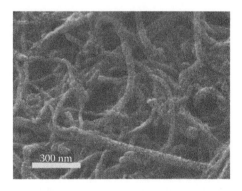

FIGURE 9.17
SEM cross-section view of the CNTs on Si wafer after ALD of Pt for 100 cycles. The CNTs were acid treated for 6 h. (Liu, C. et al.: *Small.* 2009. 5. 1535–1538. Copyright Wiley-VCH Verlag GmbH & Co. KGaA. Reproduced with permission.)

relatively large overpotential and slow kinetics in DMFCs [153]. The CO-like intermediates, generated during the MOR process, interact strongly with platinum atoms and can only be oxidized at potentials much higher than the working range for DMFCs [154,155]. The CO-like molecules can accumulate on and block the active sites and subsequently reduce the activity of platinum catalysts. Therefore, the removal of adsorbed poisoning intermediates from active sites is necessary to achieve reasonable activity for MOR.

Among many types of catalysts reported, mixed-metal catalysts are currently favored for methanol oxidation [86,156]. These catalysts contain platinum and other metals that can give the help of preventing CO poising of the anode. Jiang et al. [100] demonstrated an ALD approach for the synthesis and investigation of Pt–Ru catalyst toward the oxidation of stoichiometric (1:1) methanol solutions in advanced DMFCs. Two types of thin-film materials were investigated as catalysts for methanol oxidation: Pt–Ru films with various ruthenium contents that were codeposited by ALD and Pt skin catalysts made by ALD depositing porous platinum layers of different thicknesses on sputtered ruthenium films with total 10 samples summarized in Table 9.1. The catalytic activity and stability of such samples are shown in Figure 9.18. It is worth mentioning that the skin catalysts displayed superior catalytic activity over pure platinum.

Another trend to enhance the alcohol oxidation is to use alkaline media [157–160] and noble-metal catalysts. In acidic media, platinum is most commonly applied as an electrocatalyst by itself or as an alloy; however, in alkaline media, palladium has also shown high activity, especially for ethanol and isopropanol oxidation [161,162]. Rikkinen et al. [163] reported the ALD preparation of Pd nanoparticles on porous carbon support, and their activity and stability for alcohol oxidation were investigated in alkaline media

TABLE 9.1

Summary of the Ten Types of Samples Used in the Study

Sample	Description	Film Thickness (nm)	Composition (%)			Surface Area (cm²)
			Ru	Pt	O	
S1	ALD Pt:ALD Ru = 1:0	10.0	0.0	97.7	2.3	0.4071
S2	ALD Pt:ALD Ru = 8:4	10.2	12.3	86.9	1.8	0.8621
S3	ALD Pt:ALD Ru = 4:8	10.2	28.7	78.6	2.7	1.2821
S4	ALD Pt:ALD Ru = 4:16	10.8	50.8	46.3	2.9	1.3315
S5	ALD Pt:ALD Ru = 4:64	13.6	59.6	38.4	2.0	1.6427
S6	0 cycles of Pt ALD on sputtered Ru	0	100	0.0	0	2.4664
S7	10 cycles of Pt ALD on sputtered Ru	0.5	67.8	30.9	1.3	2.1303
S8	20 cycles of Pt ALD on sputtered Ru	1.0	48.3	50.0	1.7	2.6954
S9	30 cycles of Pt ALD on sputtered Ru	1.5	40.7	57.7	1.6	2.8081
S10	50 cycles of Pt ALD on sputtered Ru	2.5	0.0	97.4	2.6	0.9797

Source: Reprinted with permission from Jiang, X., Gür, T.M., Prinz, F.B. and Bent, S.F., *Chem. Mater.*, 22, 3024–3032. Copyright 2010 American Chemical Society.

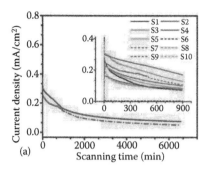
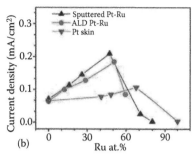

FIGURE 9.18

(a) Chronoamperometry (CA) analysis of samples S4 and S7 up to 6550 min in 16.6 M $CH_3OH/0.5$ M H_2SO_4 at 0.6 V vs. RHE. (b) Methanol oxidation activity plots, extracted from CA curves of co-sputtered Pt–Ru catalysts [164] (data denoted by upright solid triangles), ALD codeposited Pt–Ru catalysts (data denoted by solid circles), and the Pt skin catalysts (data denoted by inverted solid triangles), as a function of the ruthenium composition measured in 16.6 $CH_3OH/0.5$ M H_2SO_4, at 0.6 V vs. RHE at $t = 900$ min. The inset in panel (a) is the CA analysis of samples S1–S10 up to 900 min; the *x*-axis and *y*-axis titles are "scanning time (min)" and "current density (mA/cm^2)," respectively. (Reprinted with permission from Jiang, X. et al. *Chem. Mater*, 22, 3024–3032. Copyright 2010 American Chemical Society.)

as shown in Figure 9.19. The Pd particles have smaller average sizes on the support material resulting in ~50 mV lower onset potential and 2.5 times higher mass activity for alcohol oxidation compared with a commercial catalyst. The results indicate that the use of ALD allows the preparation of a noble metal nanoparticle catalyst, and this catalyst can provide similar mass activity with lower catalyst loading than state-of-the-art commercial fuel cell catalysts. This would significantly reduce the cost of the cell and provide a competitive advantage compared with other power sources.

9.5 Concluding Remarks and Outlooks

Due to energy shortage and environmental pollution problems, fuel cells have drawn a great deal of attention because of their high efficiency and low emissions. In recent years, ALD has emerged as a very powerful, unique tool for advancing energy technologies because it provides a unique combination of characteristics, including subnanometer precision, the capability to tailor surface properties, unparalleled conformity over high aspect ratio and nanoporous structures, and the ability to control chemical composition and size. In this book chapter, we first gave general concepts on fuel cells and ALD technique. Further, we made a comprehensive summary on ALD for fuel

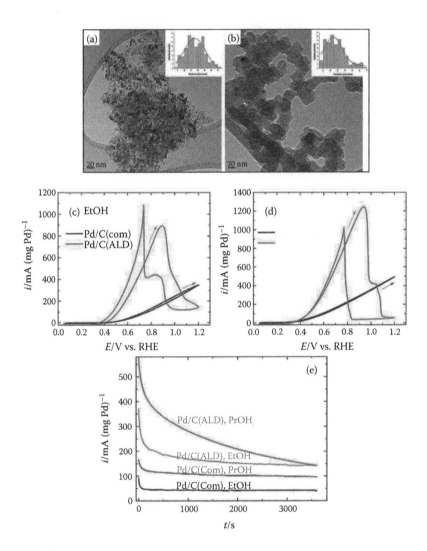

FIGURE 9.19
TEM images of Pd/C(Com) (a) and Pd/C(ALD) (b) with the histograms. Ethanol (c) and isopropanol (d) oxidation in solution containing 0.1 M NaOH and 1 M of the studied alcohol with 1800 rpm rotation and 10 mV s⁻¹. The third scan is presented, and the arrows indicate the forward scan of the voltammograms. (e) Chronoamperometric curves for 1 M alcohol in 0.1 M NaOH electrolyte at 0.7 V versus RHE potential. (Reprinted with permission from Rikkinen, E. et al. *J. Phys. Chem. C*, 115, 23067–23073. Copyright 2011 American Chemical Society.)

cells. It is clear from the literature that ALD appears as a very promising technique for processing various materials, which can serve as different components in fuel cells. Such ALD-fabricated components showed promising potential to further enhance the fuel cells' performance. We emphasize our summary of ALD for fuel cells as follows:

(1) Two general ways of ALD preparation: direct deposition and surface modification.

(2) New reactor designs to extend the capabilities of ALD and to improve the speed and simplicity of ALD coatings, such as direct-write (area-selected deposition), plasma, atmospheric pressure, and roll-to-roll processing [107,165–168].

(3) ALD can synthesize various materials for FCs: metals, mixed metals, and metal oxides.

(4) ALD can fabricate various nanostructures from 0D nanoparticles, 1D nanowires, and 2D ultrathin layers to 3D mesoporous structures or patterned structures.

(5) The above ALD-synthesized materials can serve different components, such as catalysts, electrodes, electrolyte, and current collectors for fuel cells.

(6) To date, most ALDs for fuel cell studies focus on SOFCs and PEMFCs, but such materials can also be easily extended to other fuel cell devices.

The future prospects for ALD in fuel cells are very promising. Despite some initial progresses that have been achieved, this field is still quite open. Generally, the ALD expansion can follow two directions: (1) introducing existing ALD processes into other fuel cell devices and (2) developing new processes for existing and new areas. Much more materials with various nanostructures are still waiting for discovery by ALD, and such materials may play a significant role in future fuel cell applications. Beyond the applications discussed here, ALD is also being investigated for other energy areas, such as supercapacitors, lithium-ion batteries, solar cells, catalysis, and sensing. The development of nanomaterials for energy applications is rapidly emerging as a major emphasis of ALD research, and future ALD applications in clean energy will undoubtedly continue to expand in the coming years.

References

1. Vishnyakov, V.M. 2006. Proton exchange membrane fuel cells. *Vacuum* 80, 1053–1065.
2. Hui, S.R. et al. 2007. A brief review of the ionic conductivity enhancement for selected oxide electrolytes. *Journal of Power Sources* 172, 493–502.
3. Du, Z., Li, H. and Gu, T. 2007. A state of the art review on microbial fuel cells: A promising technology for wastewater treatment and bioenergy. *Biotechnology Advances* 25, 464–482.
4. Steele, B.C.H. and Heinzel, A. 2001. Materials for fuel-cell technologies. *Nature* 414, 345–352.

5. Tawfik, H., Hung, Y. and Mahajan, D. 2007. Metal bipolar plates for PEM fuel cell: A review. *Journal of Power Sources* 163, 755–767.
6. Stambouli, A.B. and Traversa, E. 2002. Solid oxide fuel cells (SOFCs): A review of an environmentally clean and efficient source of energy. *Renewable and Sustainable Energy Reviews* 6, 433–455.
7. Lieber, C.M. 2003. Nanoscale science and technology: Building a big future from small things. *MRS Bulletin* 28, 486–491.
8. Daniel, M.C. and Astruc, D. 2004. Gold nanoparticles: Assembly, supramolecular chemistry, quantum-size-related properties, and applications toward biology, catalysis, and nanotechnology. *Chemical Reviews* 104, 293–346.
9. Tao, F. et al. 2006. An easy way to construct an ordered array of nickel nanotubes: The triblock-copolymer-assisted hard-template method. *Advanced Materials* 18, 2161–2164.
10. Bunz, U.H.F. 2006. Breath figures as a dynamic templating method for polymers and nanomaterials. *Advanced Materials* 18, 973–989.
11. Luo, Z. et al. 2008. Controllable nanonet assembly utilizing a pressure-difference method based on anodic aluminum oxide templates. *Angewandte Chemie International Edition* 47, 8905–8908.
12. Hurst, S.J., Payne, E.K., Qin, L. and Mirkin, C.A. 2006. Multisegmented one-dimensional nanorods prepared by hard-template synthetic methods. *Angewandte Chemie International Edition* 45, 2672–2692.
13. Zhang, G. et al. 2010. Controlled growth/patterning of Ni nanohoneycombs on various desired substrates. *Langmuir* 26, 4346–4350.
14. Cheng, C.L., Lin, J.S. and Chen, Y.F. 2008. Fabrication and growth mechanism of metal (Zn, Sn) nanotube arrays and metal (Cu, Ag) nanotube/nanowire junction arrays. *Materials Letters* 62, 1666–1669.
15. Liu, J., Maaroof, A.I., Wieczorek, L. and Cortie, M.B. 2005. Fabrication of hollow metal "nanocaps" and their red-shifted optical absorption spectra. *Advanced Materials* 17, 1276–1281.
16. Sun, S. et al. 2011. A highly durable platinum nanocatalyst for proton exchange membrane fuel cells: Multiarmed starlike nanowire single crystal. *Angewandte Chemie International Edition* 50, 422–426.
17. Zhang, G., Sun, S., Li, R. and Sun, X. 2010. New insight into the conventional replacement reaction for the large-scale synthesis of various metal nanostructures and their formation mechanism. *Chemistry—A European Journal* 16, 10630–10634.
18. Sun, S., Jaouen, F. and Dodelet, J.P. 2008. Controlled growth of Pt nanowires on carbon nanospheres and their enhanced performance as electrocatalysts in PEM fuel cells. *Advanced Materials* 20, 3900–3904.
19. Hou, Y., Kondoh, H., Ohta, T. and Gao, S. 2005. Size-controlled synthesis of nickel nanoparticles. *Applied Surface Science* 241, 218–222.
20. Jia, F.L., Zhang, L.Z., Shang, X.Y. and Yang, Y. 2008. Non-aqueous sol–gel approach towards the controllable synthesis of nickel nanospheres, nanowires, and nanoflowers. *Advanced Materials* 20, 1050–1054.
21. Yu, T., Joo, J., Park, Y.I. and Hyeon, T. 2005. Large-scale nonhydrolytic sol–gel synthesis of uniform-sized ceria nanocrystals with spherical, wire, and tadpole shapes. *Angewandte Chemie International Edition* 44, 7411–7414.
22. Boury, B., Corriu, R.J.P., Le Strat, V., Delord, P. and Nobili, M. 1999. Nanostructured silica-based organic–inorganic hybrid materials: Evidence for

self-organization of a xerogel prepared by sol–gel polymerization. *Angewandte Chemie International Edition* 38, 3172–3175.

23. Aida, T. and Tajima, K. 2001. Photoluminescent silicate microsticks containing aligned nanodomains of conjugated polymers by sol–gel-based in situ polymerization. *Angewandte Chemie International Edition* 40, 3803–3806.

24. Liu, B., Sun, T., He, J. and Dravid, V.P. 2010. Sol–gel-derived epitaxial nanocomposite thin films with large sharp magnetoelectric effect. *ACS Nano* 4, 6836–6842.

25. Moore, J.G., Lochner, E.J. and Stiegman, A.E. 2007. Unusual nanoparticle structures from the silica sol–gel-mediated self-assembly of a prussian-blue analogue and the formation of templated graphite regions. *Angewandte Chemie International Edition* 46, 8653–8655.

26. Kim, C. et al. 2008. Copper nanowires with a five-twinned structure grown by chemical vapor deposition. *Advanced Materials* 20, 1859–1863.

27. Hong, S.W., Banks, T. and Rogers, J.A. 2010. Improved density in aligned arrays of single-walled carbon nanotubes by sequential chemical vapor deposition on quartz. *Advanced Materials* 22, 1826–1830.

28. Song, L. et al. 2006. Large-scale synthesis of rings of bundled single-walled carbon nanotubes by floating chemical vapor deposition. *Advanced Materials* 18, 1817–1821.

29. Fontcubertai Morral, A., Arbiol, J., Prades, J.D., Cirera, A. and Morante, J.R. 2007. Synthesis of silicon nanowires with wurtzite crystalline structure by using standard chemical vapor deposition. *Advanced Materials* 19, 1347–1351.

30. Zhang, G., Sun, S., Bostetter, M., Poulin, S. and Sacher, E. 2010. Chemical and morphological characterizations of CoNi alloy nanoparticles formed by co-evaporation onto highly oriented pyrolytic graphite. *Journal of Colloid and Interface Science* 350, 16–21.

31. Zhang, G., Yang, D. and Sacher, E. 2007. X-ray photoelectron spectroscopic analysis of Pt nanoparticles on highly oriented pyrolytic graphite, using symmetric component line shapes. *Journal of Physical Chemistry C* 111, 565–570.

32. Zhang, G., Yang, D. and Sacher, E. 2007. Structure and morphology of Co nanoparticles deposited onto highly oriented pyrolytic graphite. *Journal of Physical Chemistry C* 111, 17200–17205.

33. Yoshida, H. et al. 2010. Nanoparticle-dispersed liquid crystals fabricated by sputter doping. *Advanced Materials* 22, 622–626.

34. Fu, G.D., Zhang, Y., Kang, E.T. and Neoh, K.G. 2004. Nanoporous ultra-low-k fluoropolymer composite films via plasma polymerization of allylpentafluorobenzene and magnetron sputtering of poly(tetrafluoroethylene). *Advanced Materials* 16, 839–842.

35. Wei, Q. et al. 2009. Self-assembly of ordered semiconductor nanoholes by ion beam sputtering. *Advanced Materials* 21, 2865–2869.

36. Lim, B. et al. 2008. Facile synthesis of highly faceted multioctahedral Pt nanocrystals through controlled overgrowth. *Nano Letters* 8, 4043–4047.

37. Wang, C., Daimon, H., Lee, Y., Kim, J. and Sun, S. 2007. Synthesis of monodisperse Pt nanocubes and their enhanced catalysis for oxygen reduction. *Journal of the American Chemical Society* 129, 6974–6975.

38. Sun, S. et al. 2009. Ultrathin single crystal Pt nanowires grown on N-doped carbon nanotubes. *Chemical Communications* 45, 7048–7050.

39. Chen, Z., Waje, M., Li, W. and Yan, Y. 2007. Supportless Pt and PtPd nanotubes as electrocatalysts for oxygen-reduction reactions. *Angewandte Chemie* 119, 4138–4141.

40. Alia, S.M. et al. 2010. Porous platinum nanotubes for oxygen reduction and methanol oxidation reactions. *Advanced Functional Materials* 20, 3742–3746.

41. Liang, H.W. et al. 2011. A free-standing Pt-nanowire membrane as a highly stable electrocatalyst for the oxygen reduction reaction. *Advanced Materials* 23, 1467–1471.

42. Lim, B. et al. 2009. Pd-Pt bimetallic nanodendrites with high activity for oxygen reduction. *Science* 324, 1302–1305.

43. Zhao, Y., Wang, F., Tian, J., Yang, X. and Zhan, L. 2010. Preparation of Pt/CeO$_2$/HCSs anode electrocatalysts for direct methanol fuel cells. *Electrochimica Acta* 55, 8998–9003.

44. Xiao, L., Zhuang, L., Liu, Y., Lu, J. and Abruña, H.D. 2009. Activating Pd by morphology tailoring for oxygen reduction. *Journal of the American Chemical Society* 131, 602–608.

45. Chu, Y.H. and Shul, Y.G. 2010. Combinatorial investigation of Pt–Ru–Sn alloys as an anode electrocatalysts for direct alcohol fuel cells. *International Journal of Hydrogen Energy* 35, 11261–11270.

46. Di Noto, V. and Negro, E. 2010. A new Pt–Rh carbon nitride electrocatalyst for the oxygen reduction reaction in polymer electrolyte membrane fuel cells: Synthesis, characterization and single-cell performance. *Journal of Power Sources* 195, 638–648.

47. Tayal, J., Rawat, B. and Basu, S. 2011. Bi-metallic and tri-metallic Pt–Sn/C, Pt–Ir/C, Pt–Ir–Sn/C catalysts for electro-oxidation of ethanol in direct ethanol fuel cell. *International Journal of Hydrogen Energy* 36, 14884–14897.

48. Lee, Y. et al. 2012. Virus-templated Au and Au–Pt core–shell nanowires and their electrocatalytic activities for fuel cell applications. *Energy & Environmental Science* 5, 8328.

49. Yamaki, T. et al. 2006. Nano-structure controlled polymer electrolyte membranes for fuel cell applications prepared by ion beam irradiation. *ECS Transactions* 3, 103–112.

50. Michel, M. et al. 2007. High-performance nanostructured membrane electrode assemblies for fuel cells made by layer-by-layer assembly of carbon nanocolloids. *Advanced Materials* 19, 3859–3864.

51. Gao, Z., Mao, Z., Wang, C., Huang, J. and Liu, Z. 2009. Composite electrolyte based on nanostructured Ce$_{0.8}$Sm$_{0.2}$O$_{1.9}$ (SDC) for low-temperature solid oxide fuel cells. *International Journal of Energy Research* 33, 1138–1144.

52. Sholklapper, T.Z., Kurokawa, H., Jacobson, C.P., Visco, S.J. and De Jonghe, L.C. 2007. Nanostructured solid oxide fuel cell electrodes. *Nano Letters* 7, 2136–2141.

53. Liu, Y., Zha, S. and Liu, M. 2004. Novel nanostructured electrodes for solid oxide fuel cells fabricated by combustion chemical vapor deposition (CVD). *Advanced Materials* 16, 256–260.

54. George, S.M. 2010. Atomic layer deposition: An overview. *Chemical Reviews* 110, 111–131.

55. Leskelä, M., Ritala, M. and Nilsen, O. 2011. Novel materials by atomic layer deposition and molecular layer deposition. *MRS Bulletin* 36, 877–884.

56. Meng, X., Yang, X.Q. and Sun, X. 2012. Emerging applications of atomic layer deposition for lithium-ion battery studies. *Advanced Materials* 24, 3589–3615.

57. Bae, C., Shin, H. and Nielsch, K. 2011. Surface modification and fabrication of 3D nanostructures by atomic layer deposition. *MRS Bulletin* 36, 887–897.

58. Elliott, S.D. and Nilsen, O. 2011. (Invited) Reaction mechanisms in ALD of ternary oxides. *ECS Transactions* 41, 175–183.

59. Kosola, A., Putkonen, M., Johansson, L.S. and Niinistö, L. 2003. Effect of annealing in processing of strontium titanate thin films by ALD. *Applied Surface Science* 211, 102–112.

60. Elam, J.W., Dasgupta, N.P. and Prinz, F.B. 2011. ALD for clean energy conversion, utilization, and storage. *MRS Bulletin* 36, 899–906.

61. Leskelä, M. and Ritala, M. 2003. Atomic layer deposition chemistry: Recent developments and future challenges. *Angewandte Chemie International Edition* 42, 5548–5554.

62. Peng, Q., Lewis, J.S., Hoertz, P.G., Glass, J.T. and Parsons, G.N. 2012. Atomic layer deposition for electrochemical energy generation and storage systems. *Journal of Vacuum Science & Technology A: Vacuum, Surfaces, and Films* 30, 010803.

63. Parsons, G.N., George, S.M. and Knez, M. 2011. Progress and future directions for atomic layer deposition and ALD-based chemistry. *MRS Bulletin* 36, 865–871.

64. Puurunen, R.L. 2005. Surface chemistry of atomic layer deposition: A case study for the trimethylaluminum/water process. *Journal of Applied Physics* 97, 121301 (52 pp.).

65. Cassir, M., Ringuedé, A. and Niinistö, L. 2010. Input of atomic layer deposition for solid oxide fuel cell applications. *Journal of Materials Chemistry* 20, 8987–8993.

66. Ge, L. et al. 2012. Influence of surface preparation on atomic layer deposition of Pt films. *Journal of Semiconductors* 33, 083003 (5 pp.).

67. Setthapun, W. et al. 2010. Genesis and evolution of surface species during Pt atomic layer deposition on oxide supports characterized by in situ XAFS analysis and water–gas shift reaction. *Journal of Physical Chemistry C* 114, 9758–9771.

68. Lu, J., Lei, Y. and Elam, W.J. 2012. Atomic layer deposition of noble metals: New developments in nanostructured catalysts. In *Noble Metals*, Su, Y.-H. (Ed.), ISBN 978-953-307-898-4, 426 pp., InTech Publisher.

69. Sun, S. et al. 2010. Direct growth of single-crystal Pt nanowires on Sn@CNT nanocable: 3D electrodes for highly active electrocatalysts. *Chemistry—A European Journal* 16, 829–835.

70. Wen, Z., Liu, J. and Li, J. 2008. Core/shell Pt/C nanoparticles embedded in mesoporous carbon as a methanol-tolerant cathode catalyst in direct methanol fuel cells. *Advanced Materials* 20, 743–747.

71. Xu, C.W., Wang, H., Shen, P.K. and Jiang, S.P. 2007. Highly ordered Pd nanowire arrays as effective electrocatalysts for ethanol oxidation in direct alcohol fuel cells. *Advanced Materials* 19, 4256–4259.

72. Okamoto, M., Fujigaya, T. and Nakashima, N. 2009. Design of an assembly of poly(benzimidazole), carbon nanotubes, and Pt nanoparticles for a fuel-cell electrocatalyst with an ideal interfacial nanostructure. *Small* 5, 735–740.

73. Xia, B.Y., Ng, W.T., Wu, H.B., Wang, X. and Lou, X.W. 2012. (David) Self-supported interconnected Pt nanoassemblies as highly stable electrocatalysts for low-temperature fuel cells. *Angewandte Chemie International Edition* 51, 7213–7216.

74. Gao, M.R. et al. 2011. A methanol-tolerant Pt/CoSe$_2$ nanobelt cathode catalyst for direct methanol fuel cells. *Angewandte Chemie International Edition* 50, 4905–4908.

75. Tada, M. et al. 2007. In situ time-resolved dynamic surface events on the Pt/C cathode in a fuel cell under operando conditions. *Angewandte Chemie International Edition* 46, 4310–4315.

76. Srivastava, R., Mani, P., Hahn, N. and Strasser, P. 2007. Efficient oxygen reduction fuel cell electrocatalysis on voltammetrically dealloyed Pt–Cu–Co nanoparticles. *Angewandte Chemie International Edition* 46, 8988–8991.

77. Deng, W. and Flytzani-Stephanopoulos, M. 2006. On the issue of the deactivation of Au–ceria and Pt–ceria water–gas shift catalysts in practical fuel-cell applications. *Angewandte Chemie International Edition* 45, 2285–2289.

78. Cao, L. et al. 2006. Novel nanocomposite Pt/RuO$_2$ x H$_2$O/carbon nanotube catalysts for direct methanol fuel cells. *Angewandte Chemie International Edition* 45, 5315–5319.

79. Thambidurai, C., Kim, Y.G., Jayaraju, N., Venkatasamy, V. and Stickney, J.L. 2009. Copper nanofilm formation by electrochemical ALD. *Journal of the Electrochemical Society* 156, D261–D268.

80. Bethge, O., Pozzovivo, G., Henkel, C., Abermann, S. and Bertagnolli, E. 2012. Fabrication of highly ordered nanopillar arrays and defined etching of ALD-grown all-around platinum films. *Journal of Micromechanics and Microengineering* 22, 085013.

81. Banga, D.O. et al. 2008. Formation of PbTe nanofilms by electrochemical atomic layer deposition (ALD). *Electrochimica Acta* 53, 6988–6994.

82. Tupala, J., Kemell, M., Härkönen, E., Ritala, M. and Leskelä, M. 2012. Preparation of regularly structured nanotubular TiO$_2$ thin films on ITO and their modification with thin ALD-grown layers. *Nanotechnology* 23, 125707.

83. Sivakov, V.A. et al. 2010. Silver coated platinum core–shell nanostructures on etched si nanowires: Atomic layer deposition (ALD) processing and application in SERS. *A European Journal of Chemical Physics and Physical Chemistry* 11, 1995–2000.

84. Aaltonen, T., Ritala, M., Sajavaara, T., Keinonen, J. and Leskelä, M. 2003. Atomic layer deposition of platinum thin films. *Chemistry of Materials* 15, 1924–1928.

85. Steven, T. and Christensen, J.W.E. 2009. Controlled growth of platinum nanoparticles on strontium titanate nanocubes by atomic layer deposition. *Small* 5, 750–757.

86. Liu, H. et al. 2006. A review of anode catalysis in the direct methanol fuel cell. *Journal of Power Sources* 155, 95–110.

87. Jiang, X., Huang, H., Prinz, F.B. and Bent, S.F. 2008. Application of atomic layer deposition of platinum to solid oxide fuel cells. *Chemistry of Materials* 20, 3897–3905.

88. King, J.S. et al. 2008. Ultralow loading Pt nanocatalysts prepared by atomic layer deposition on carbon aerogels. *Nano Letters* 8, 2405–2409.

89. Antolini, E. 2009. Palladium in fuel cell catalysis. *Energy & Environmental Science* 2, 915.

90. Feng, H., Libera, J.A., Stair, P.C., Miller, J.T. and Elam, J.W. 2011. Subnanometer palladium particles synthesized by atomic layer deposition. *ACS Catalysis* 1, 665–673.

91. Schaekers, M., Swerts, J., Altimime, L. and Tőkei, Z. 2011. ALD Ru and its application in DRAM MIM-capacitors and interconnect. *ECS Transactions* 34, 509–514.

92. Thambidurai, C., Kim, Y.G. and Stickney, J.L. 2008. Electrodeposition of Ru by atomic layer deposition (ALD). *Electrochimica Acta* 53, 6157–6164.

93. Choi, S.H. et al. 2011. Thermal atomic layer deposition (ALD) of Ru films for Cu direct plating. *Journal of the Electrochemical Society* 158, D351–D356.

94. Szeghalmi, A. et al. 2011. Atomic layer deposition of iridium thin films and their application in gold electrodeposition. *Proceedings of SPIE* 8186, 81680K.

95. Hämäläinen, J. et al. 2009. Atomic layer deposition of iridium thin films by consecutive oxidation and reduction steps. *Chemistry of Materials* 21, 4868–4872.

96. Knapas, K. and Ritala, M. 2011. In situ reaction mechanism studies on atomic layer deposition of Ir and IrO2 from Ir(acac)3. *Chemistry of Materials* 23, 2766–2771.

97. Vuori, H. et al. 2011. The effect of iridium precursor on oxide-supported iridium catalysts prepared by atomic layer deposition. *Applied Surface Science* 257, 4204–4210.

98. Christensen, S.T. and Elam, J.W. 2010. Atomic layer deposition of Ir–Pt alloy films. *Chemistry of Materials* 22, 2517–2525.

99. Comstock, D.J., Christensen, S.T., Elam, J.W., Pellin, M.J. and Hersam, M.C. 2010. Tuning the composition and nanostructure of Pt/Ir films via anodized aluminum oxide templated atomic layer deposition. *Advanced Functional Materials* 20, 3099–3105.

100. Jiang, X., Gür, T.M., Prinz, F.B. and Bent, S.F. 2010. Atomic layer deposition (ALD) co-deposited Pt–Ru binary and Pt skin catalysts for concentrated methanol oxidation. *Chemistry of Materials* 22, 3024–3032.

101. Christensen, S.T. et al. 2010. Supported Ru–Pt bimetallic nanoparticle catalysts prepared by atomic layer deposition. *Nano Letters* 10, 3047–3051.

102. Jiang, X., Prinz, F. and Bent, S. 2008. Pt-Ru alloys deposited by sputtering as catalysts for methanol oxidation. *ECS Transactions* 16, 605–612.

103. Hoover, R.R. and Tolmachev, Y.V. 2009. Electrochemical properties of Pt coatings on Ni prepared by atomic layer deposition. *Journal of the Electrochemical Society* 156, A37–A43.

104. Hsu, I.J., McCandless, B.E., Weiland, C. and Willis, B.G. 2009. Characterization of ALD copper thin films on palladium seed layers. *Journal of Vacuum Science Technology A: Vacuum, Surfaces, and Films* 27, 660–667.

105. Bernay, C., Ringuedé, A., Colomban, P., Lincot, D. and Cassir, M. 2003. Yttria-doped zirconia thin films deposited by atomic layer deposition ALD: A structural, morphological and electrical characterization. *Journal of Physics and Chemistry of Solids* 64, 1761–1770.

106. Shim, J.H., Chao, C.C., Huang, H. and Prinz, F.B. 2007. Atomic layer deposition of yttria-stabilized zirconia for solid oxide fuel cells. *Chemistry of Materials* 19, 3850–3854.

107. Su, P.C., Chao, C.C., Shim, J.H., Fasching, R. and Prinz, F.B. 2008. Solid oxide fuel cell with corrugated thin film electrolyte. *Nano Letters* 8, 2289–2292.

108. Brahim, C., Ringuedé, A., Cassir, M., Putkonen, M. and Niinistö, L. 2007. Electrical properties of thin yttria-stabilized zirconia overlayers produced by atomic layer deposition for solid oxide fuel cell applications. *Applied Surface Science* 253, 3962–3968.

109. Ballée, E., Ringuedé, A., Cassir, M., Putkonen, M. and Niinistö, L. 2009. Synthesis of a thin-layered ionic conductor, CeO2–Y2O3, by atomic layer deposition in view of solid oxide fuel cell applications. *Chemistry of Materials* 21, 4614–4619.

110. Nieminen, M., Lehto, S. and Niinistö, L. 2001. Atomic layer epitaxy growth of LaGaO3 thin films. *Journal of Materials Chemistry* 11, 3148–3153.

111. Dezelah, N.J., Arstila, K.N.L. and Winter, C.H. 2006. Atomic layer deposition of Ga2O3 films from a dialkylamido-based precursor. *Chemistry of Materials* 18, 471–475.

112. Zhang, W.D., Chen, J., Jiang, L.C., Yu, Y.X. and Zhang, J.Q. 2010. A highly sensitive nonenzymatic glucose sensor based on NiO-modified multi-walled carbon nanotubes. *Microchimica Acta* 168, 259–265.

113. Deo, R.P., Lawrence, N.S. and Wang, J. 2004. Electrochemical detection of amino acids at carbon nanotube and nickel/carbon nanotube modified electrodes. *The Analyst* 129, 1076.

114. Xu, C., Sun, J. and Gao, L. 2011. Large scale synthesis of nickel oxide/multi-walled carbon nanotube composites by direct thermal decomposition and their lithium storage properties. *Journal of Power Sources* 196, 5138–5142.

115. Asgari, M., Maragheh, M.G., Davarkhah, R. and Lohrasbi, E. 2011. Methanol electrooxidation on the nickel oxide nanoparticles/multi-walled carbon nanotubes modified glassy carbon electrode prepared using pulsed electrodeposition. *Journal of the Electrochemical Society* 158, K225–K229.

116. Lee, J.Y., Liang, K., An, K.H. and Lee, Y.H. 2005. Nickel oxide/carbon nanotubes nanocomposite for electrochemical capacitance. *Synthetic Metals* 150, 153–157.

117. Tong, X. et al. 2012. Enhanced catalytic activity for methanol electro-oxidation of uniformly dispersed nickel oxide nanoparticles: Carbon nanotube hybrid materials. *Small* 8, 3390–3395.

118. Biener, M.M. et al. 2011. ALD functionalized nanoporous gold: Thermal stability, mechanical properties, and catalytic activity. *Nano Letters* 11, 3085–3090.

119. Chen, Y. et al. 2011. Atomic layer deposition assisted Pt-SnO$_2$ hybrid catalysts on nitrogen-doped CNTs with enhanced electrocatalytic activities for low temperature fuel cells. *International Journal of Hydrogen Energy* 36, 11085–11092.

120. McEvoy, A. 2000. Thin SOFC electrolytes and their interfaces: A near-term research strategy. *Solid State Ionics* 132, 159–165.

121. Minh, N.Q. 2004. Solid oxide fuel cell technology: Features and applications. *Solid State Ionics* 174, 271–277.

122. Steele, B.C.H. 2001. Material science and engineering: The enabling technology for the commercialization of fuel cell systems. *Journal of Materials Science* 36, 1053–1068.

123. Litzelman, S.J., Hertz, J.L., Jung, W. and Tuller, H.L. 2008. Opportunities and challenges in materials development for thin film solid oxide fuel cells. *Fuel Cells* 8, 294–302.

124. Prinz, F. 2008. Opportunities of ALD for thin film solid oxide fuel cells. *ECS Transactions* 16, 15–18.

125. Huang, H. et al. 2007. High-performance ultrathin solid oxide fuel cells for low-temperature operation. *Journal of the Electrochemical Society* 154, B20–B24.

126. Fan, Z., Chao, C.C., Hossein-Babaei, F. and Prinz, F.B. 2011. Improving solid oxide fuel cells with yttria-doped ceria interlayers by atomic layer deposition. *Journal of Materials Chemistry* 21, 10903–10906.

127. Chao, C.C., Kim, Y.B. and Prinz, F.B. 2009. Surface modification of yttria-stabilized zirconia electrolyte by atomic layer deposition. *Nano Letters* 9, 3626–3628.

128. Shim, J.H., Jiang, X., Bent, S.F. and Prinz, F.B. 2010. Catalysts with Pt surface coating by atomic layer deposition for solid oxide fuel cells. *Journal of the Electrochemical Society* 157, B793–B797.

129. Shim, J.H., Jiang, X., Bent, S.F. and Prinz, F.B. 2009. Metal alloy catalysts with Pt surface coating by atomic layer deposition for intermediate temperature ceramic fuel cells. *ECS Transactions* 25, 323–332.

130. Chao, C.C., Motoyama, M. and Prinz, F.B. 2012. Nanostructured platinum catalysts by atomic-layer deposition for solid-oxide fuel cells. *Advanced Energy Materials* 2, 651–654.

131. Shim, J.H. and Prinz, F. 2008. Platinum coated fuel cell electrodes by atomic layer deposition. *ECS Transactions* 11, 27–33.

132. Chao, C.C. and Prinz, F.B. 2009. Solid oxide fuel cells with atomic layer deposited platinum catalyst. *ECS Transactions* 25, 855–858.

133. Wang, D.Y., Park, D.S., Griffith, J. and Nowick, A.S. 1981. Oxygen-ion conductivity and defect interactions in yttria-doped ceria. *Solid State Ionics* 2, 95–105.

134. Brahim, C. 2006. Ph.D. thesis, ENSCP, University of Paris 6, Paris.

135. Jiang, X. and Bent, S.F. 2007. Area-selective atomic layer deposition of platinum on YSZ substrates using microcontact printed SAMs. *Journal of the Electrochemical Society* 154, D648–D656.

136. Jiang, X. and Bent, S.F. 2007. Atomic layer deposition of platinum for solid oxide fuel cells. *ECS Transactions* 3, 249–259.

137. Jiang, X., Chen, R. and Bent, S.F. 2007. Spatial control over atomic layer deposition using microcontact-printed resists. *Surface and Coatings Technology* 201, 8799–8807.

138. Bing, Y., Liu, H., Zhang, L., Ghosh, D. and Zhang, J. 2010. Nanostructured Pt-alloy electrocatalysts for PEM fuel cell oxygen reduction reaction. *Chemical Society Reviews* 39, 2184–2202.

139. Esposito, D.V. et al. 2010. Low-cost hydrogen-evolution catalysts based on monolayer platinum on tungsten monocarbide substrates. *Angewandte Chemie International Edition* 49, 9859–9862.

140. Hsu, I.J., Hansgen, D.A., McCandless, B.E., Willis, B.G. and Chen, J.G. 2011. Atomic layer deposition of Pt on tungsten monocarbide (WC) for the oxygen reduction reaction. *Journal of Physical Chemistry C* 115, 3709–3715.

141. Weigert, E.C., Stottlemyer, A.L., Zellner, M.B. and Chen, J.G. 2007. Tungsten monocarbide as potential replacement of platinum for methanol electrooxidation. *Journal of Physical Chemistry C* 111, 14617–14620.

142. Weigert, E.C., Esposito, D.V. and Chen, J.G. 2009. Cyclic voltammetry and X-ray photoelectron spectroscopy studies of electrochemical stability of clean and Pt-modified tungsten and molybdenum carbide (WC and Mo_2C) electrocatalysts. *Journal of Power Sources* 193, 501–506.

143. Bozzini, B., Pietro De Gaudenzi, G., Fanigliulo, A. and Mele, C. 2004. Electrochemical oxidation of WC in acidic sulphate solution. *Corrosion Science* 46, 453–469.

144. Shao, Y., Liu, J., Wang, Y. and Lin, Y. 2009. Novel catalyst support materials for PEM fuel cells: Current status and future prospects. *Journal of Materials Chemistry* 19, 46–59.

145. Chhina, H., Campbell, S. and Kesler, O. 2007. Thermal and electrochemical stability of tungsten carbide catalyst supports. *Journal of Power Sources* 164, 431–440.

146. Jeon, M.K. et al. 2008. Investigation of Pt/WC/C catalyst for methanol electrooxidation and oxygen electro-reduction. *Journal of Power Sources* 185, 927–931.

147. Liu, C., Wang, C.C., Kei, C.C., Hsueh, Y.C. and Perng, T.P. 2009. Atomic layer deposition of platinum nanoparticles on carbon nanotubes for application in proton-exchange membrane fuel cells. *Small* 5, 1535–1538.

148. Salazar-Banda, G.R., Suffredini, H.B., Avaca, L.A. and Machado, S.A.S. 2009. Methanol and ethanol electro-oxidation on Pt–SnO$_2$ and Pt–Ta$_2$O$_5$ sol–gel-modified boron-doped diamond surfaces. *Materials Chemistry and Physics* 117, 434–442.

149. Icardi, U.A., Specchia, S., Fontana, G.J.R., Saracco, G. and Specchia, V. 2008. Compact direct methanol fuel cells for portable application. *Journal of Power Sources* 176, 460–467.

150. Thompsett, D. and Hogarth, M. 2003. *Fuel Cell Technology Handbook*. Boca Raton: CRC Press, p. 61.

151. O'Hayre, R., Cha, S., Colella, W. and Prinz, F.B. 2006. *Fuel Cell Fundamentals*. New York: John Wiley and Sons.

152. Williams, M.C. 2004. *Fuel Cell Handbook*, 7th ed. Morgantown, WV: U. S. Department of Energy, pp. 1–3.

153. Olah, G.A., Goeppert, A. and Prakash, G.K.S. 2006. *Beyond Oil and Gas: The Methanol Economy*. Weinheim: Wiley-VCH.

154. Park, S., Xie, Y. and Weaver, M.J. 2002. Electrocatalytic pathways on carbon-supported platinum nanoparticles: Comparison of particle-size-dependent rates of methanol, formic acid, and formaldehyde electrooxidation. *Langmuir* 18, 5792–5798.

155. Beden, B., Kadirgan, F., Lamy, C. and Leger, J.M. 1981. Electrocatalytic oxidation of methanol on platinum-based binary electrodes. *Journal of Electroanalytical Chemistry and Interfacial Electrochemistry* 127, 75–85.

156. Jiang, X., Gür, T.M., Prinz, F.B. and Bent, S.F. 2010. Sputtered Pt–Ru alloys as catalysts for highly concentrated methanol oxidation. *Journal of the Electrochemical Society* 157, B314–B319.

157. Jiang, L., Hsu, A., Chu, D. and Chen, R. 2010. Ethanol electro-oxidation on Pt/C and PtSn/C catalysts in alkaline and acid solutions. *International Journal of Hydrogen Energy* 35, 365–372.

158. Dimos, M.M. and Blanchard, G.J. 2010. Evaluating the role of Pt and Pd catalyst morphology on electrocatalytic methanol and ethanol oxidation. *Journal of Physical Chemistry C* 114, 6019–6026.

159. Lai, S.C.S. and Koper, M.T.M. 2009. Ethanol electro-oxidation on platinum in alkaline media. *Physical Chemistry Chemical Physics* 11, 10446–10456.

160. Fujiwara, N. et al. 2008. Direct ethanol fuel cells using an anion exchange membrane. *Journal of Power Sources* 185, 621–626.

161. Santasalo-Aarnio, A. et al. 2011. Comparison of methanol, ethanol and isopropanol oxidation on Pt and Pd electrodes in alkaline media studied by HPLC. *Electrochemistry Communications* 13, 466–469.

162. Xu, C., Cheng, L., Shen, P. and Liu, Y. 2007. Methanol and ethanol electrooxidation on Pt and Pd supported on carbon microspheres in alkaline media. *Electrochemistry Communications* 9, 997–1001.

163. Rikkinen, E. et al. 2011. Atomic layer deposition preparation of Pd nanoparticles on a porous carbon support for alcohol oxidation. *Journal of Physical Chemistry C* 115, 23067–23073.

164. Lipkowski, J. and Ross, P.N. 1998. *Electrocatalysis*. New York: Wiley-VCH.

165. Kessels, W.M.M. and Putkonen, M. 2011. Advanced process technologies: Plasma, direct-write, atmospheric pressure, and roll-to-roll ALD. *MRS Bulletin* 36, 907–913.

166. Ten Eyck, G.A., Pimanpang, S., Bakhru, H., Lu, T.M. and Wang, G.C. 2006. Atomic layer deposition of Pd on an oxidized metal substrate. *Chemical Vapor Deposition* 12, 290–294.
167. Chao, C.C., Hsu, C.M., Cui, Y. and Prinz, F.B. 2011. Improved solid oxide fuel cell performance with nanostructured electrolytes. *ACS Nano* 5, 5692–5696.
168. Ten Eyck, G.A. et al. 2005. Plasma-assisted atomic layer deposition of palladium. *Chemical Vapor Deposition* 11, 60–66.

10

Noncarbon Material–Supported Electrocatalysts for Proton Exchange Membrane Fuel Cells

Vladimir Neburchilov and Jiujun Zhang

CONTENTS

10.1 Introduction 291
10.2 Metal Oxide-Based Supported Catalysts 294
    10.2.1 Magneli Phase Titania-Supported Catalysts 294
    10.2.2 Metal-Doped Titania-Supported Catalysts 295
    10.2.3 Nanostructured Titanium Dioxide 299
    10.2.4 Sulfated Zirconia (S-ZrO$_2$)-Supported Catalysts 300
    10.2.5 SnO$_2$-Based Material-Supported Catalysts 300
    10.2.6 WO$_x$-Supported Catalysts 301
    10.2.7 SiO$_2$-Supported Catalysts 301
10.3 Boride-Supported Catalysts 302
10.4 Carbide-Supported Catalysts 303
    10.4.1 Tungsten Carbide (WC)-Supported Catalysts 303
    10.4.2 Titanium Carbide (TiC)-Supported Catalysts 305
    10.4.3 Silicon Carbide (SiC)-Supported Catalysts 306
10.5 Nitride-Supported Catalysts 307
10.6 Effect of a Metal–Support Interaction on Catalyst Performance 308
10.7 Conclusions 308
References 310

10.1 Introduction

The proton exchange membrane (PEM) fuel cell is one of the most advanced energy devices, which has high power or energy densities, high efficiency of energy conversion, and low or zero emissions when it is used in automobile, portable, and stationary applications. However, two major challenges of PEM fuel cell technology, such as high cost and low durability, have been identified as the barriers hindering its commercialization. It has been recognized that the electrocatalysts, which are used for catalyzing anode fuel oxidation reactions, such as the hydrogen oxidation reaction (HOR) and cathode oxygen reduction reaction (ORR), are the major contributors to these two challenges. At the current technology state, the most practical catalysts are carbon-supported platinum (Pt)–based catalysts. However, these Pt-containing catalysts are expensive and, at the same time, also not stable enough at PEM fuel cell operating conditions. Regarding the cost, the cost portion of Pt catalyst/ink is about 34% of the total cost of a fuel cell stack as shown in Figure 10.1 [1]. Regarding the catalyst durability, carbon support corrosion and Pt dissolution are the major degradation modes.

To overcome the low durability issue of the catalysts, in particular, the issue associated with the corrosion or oxidation of carbon support, new support materials that have high resisting properties to oxidation and, at the same time, their supported Pt catalysts, which have the same as or even higher activity than those carbon-supported ones, have to be developed.

Generally, a catalyst support can provide high catalyst dispersion, specific surface area, narrow distribution of particle size, and interaction with catalyst components, forming a highly active catalyst. In this respect, the most popular catalyst support in PEM fuel cells today is carbon black, which has a sufficient level of the required properties, such as high conductivity, surface area, porosity, and reasonable interaction with catalyst particles. Unfortunately, its electrochemical stability is relatively limited due to the oxidation under the fuel cell operation conditions, such as acidic media and the presence of O_2, leading to a low durability of the fuel cells.

Regarding the carbon corrosion mechanism, for example, in an automobile application in which the fuel cells work in frequent start-up/shutdown cycles, the electrode potential can achieve more than 1.0 V versus SHE in fuel cell start-up [2], leading to significant oxidation of the carbon support. This is because the standard potential of electrochemical carbon oxidation to CO_2 [3], which is about 0.2 V versus SHE, is less than this electrode potential. The

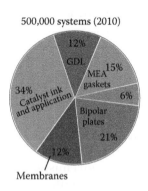

500,000 systems (2010)

FIGURE 10.1
Breakdown of the cost of PEM fuel cells. (From B. James, and J. Kalinisky, DTI, Inc., *DOE Hydrogen Program Review*, Washington, DC, June 9, 2010.)

electrochemical oxidation reaction and its standard electrode potential (E) are as follows:

$$C + 2H_2O \leftrightarrow CO_2 + 4H^+ + 4e^-, \quad E = 0.2V \text{ vs. SHE [4]} \quad (10.1)$$

Although the kinetics of reaction 10.1 is slow, the presence of Pt and fuel impurities [5], acidic media, O_2, high temperature (~80°C) [6], and low humidity [6] can accelerate this thermodynamically favorable reaction, leading to the carbon corrosion. As shown in Figure 10.2, due to the carbon corrosion, Pt dissolution and then agglomeration can happen, resulting in the losses of Pt active electrochemical surface area (ECSA) and the interaction between Pt catalysts and carbon support [7–10], causing a reduction in catalyst activity.

Recently, high-temperature PEM fuel cells (95°C–250°C) have attracted considerable attention because of their several advantages over those operated at 70°C–80°C, such as fast ORR kinetics, a fast heat rejection rate, and easier water management as well as simple flowfield design [11]. However, at the high temperatures $T > 95°C$, degradation of carbon support will be accelerated, causing fast degradation of fuel cell performance through the mechanisms shown in Figure 10.2 for a carbon-supported catalyst and its associated catalyst layer (Figure 10.3) [11,12].

Therefore, insufficient corrosion resistance of the carbon supports in a wide range of temperatures and humidities in PEM fuel cells has to be improved, or carbon supports should be replaced with more durable noncarbon materials. These noncarbon materials must meet the following main requirements:

(1) High electrochemical stability
(2) High chemical stability in acidic media
(3) High electrochemical stability in acidic medium
(4) High thermal stability
(5) High electronic conductivity

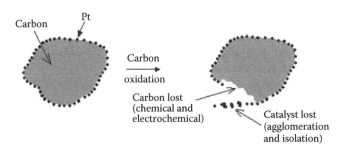

FIGURE 10.2
Schematic process of carbon corrosion. (From J. Zhang et al. *J. Fuel Cell Sci. Technol.* 8, 051006-1–051006-5, 2011.)

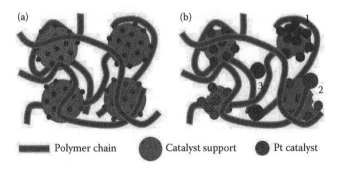

FIGURE 10.3
(See color insert.) Effect of carbon corrosion on (1) agglomeration, (2) coalescence, and (3) loss of Pt particles in catalyst layer during operation of PEM fuel cells: (a) normal (corrosion-resistant) electrode and (b) corroded electrode. (Reprinted with permission from S.-Y. Huang et al. *J. Am. Chem. Soc.*, 131, 13898. Copyright 2009, American Chemical Society.)

(6) High specific surface area for maximal catalyst dispersion

(7) High porosity for efficient mass transport

(8) Strong metal–support interaction (SMSI) to avoid catalyst migration, agglomeration, and coalescence

To meet these requirements, metal oxides are considered to be possible candidates for catalyst support materials and have been extensively explored in the most recent years [13–15].

In this chapter, progress in the development of metal oxide support materials and their supported Pt or/and Pt alloy catalysts will be reviewed, and also some challenges and perspectives will be discussed.

10.2 Metal Oxide-Based Supported Catalysts

10.2.1 Magneli Phase Titania-Supported Catalysts

In the most recent years, suboxide titania and metal-doped titania have been widely explored as catalyst supports due to their unique properties, such as relative high conductivity and high corrosion resistibility at PEM fuel cell operation conditions. For example, the mixture of titanium suboxides, Ti_nO_{2n-1} where $4 \leq n \leq 10$ (the Magneli phase), was one kind of the first ceramic materials that were explored as the support for PEM fuel cell catalysts [16–18].

Normally, the Magneli phase consists of TiO_6 octahedra and *n*-layers of titanium atoms separated by oxygen-deficient layers. Two typical materials are Ti_4O_7 and Ti_5O_9, both of which have high electric conductivities, such as 10^3 and $10^2\,S\,cm^{-1}$, respectively [19]. However, they alone do not have catalytic activity for HOR and ORR [20–22] and show high electrochemical stabilities at 0–2 V versus RHE in 1 M H_2SO_4 [19]. The commonly explored material is Ti_4O_7, which has a specific surface area of 0.95 m^2 g^{-1}. This material is synthesized by the reduction of TiO_2 at 1050°C in H_2 for 6 h. Ti_4O_7 has a higher onset potential of the corrosion current than that of carbon black Vulcan XC-72 at PEM fuel cell operation conditions, which demonstrates its higher stability. The 5% Pt/Ti_4O_7 catalyst was synthesized by impregnating Ti_4O_7 by Pt salt then being reduced at 250°C in N_2/H_2. The ORR mass activity of this catalyst was slightly lower than that of 20% Pt/C, which was attributed to the larger catalyst particle size and smaller ECSA [17]. Another 20% Pt/Ti_4O_7 nanofiber catalyst was also synthesized and gave a low ORR mass activity due to a support degradation during an *ex situ* durability test at the potential cycling (1000 cycles) at $E = 0.05–1.2$ V versus RHE in 0.1 M $HClO_4$. The XPS analysis of this catalyst before and after the durability test showed a formation of electronically insulating surface oxides. Ti_4O_7 nanofibers (Ti_4O_7) were also synthesized by the reduction of TiO_2 nanofibers (electrospinning method) [16].

The first commercial porous suboxide titania Ebonex (Altraverde, Ltd.) Ti_4O_7 + Ti_5O_9 has a lower specific surface area of 2–3 m^2/g than carbon black [23,24]. Ebonex demonstrated a good electrochemical stability in both alkaline and acidic solutions even with the presence of reduced titanium [20]. This support has a good interaction with deposited Pt as a catalyst, which can positively affect the ORR activity [7,25–27]. For example, PtCo/Ebonex catalysts [24] could prevent Pt from agglomeration [24]. However, the Pt_4Ru_4Ir/Ebonex catalyst showed an ORR current drop after 7 h of conditioning at 1.6 V versus RHE in 0.5 M H_2SO_4 due to the partial Ebonex oxidation [27].

In general, Ti_4O_7-supported catalysts have a lower ORR mass activity when compared to carbon-supported catalysts, which seems not to be in favor for PEM fuel cell applications.

10.2.2 Metal-Doped Titania-Supported Catalysts

To improve the ORR mass activity, some metal-doped TiO_2 materials were explored as alternative catalyst supports. For example, after Nb doping, the resistive TiO_2 could become conductive due to the appearance of defects in the TiO_2. The charge compensation at the Nb^{5+} substitution for Ti^{4+} requires one titanium cation vacancy per four Nb atoms in the oxidative conditions and low Nb doping level (Equation 10.2) or a generation of Ti^{3+} ions at the reduction of stoichiometric TiO_2 in the reductive conditions and high Nb doping level (Equation 10.3) [28]:

$$Nb_2O_5 \xrightarrow{Ti_2O_4} 2Nb_{Ti} + 2Ti'_{Ti} + 4O_o^X + \frac{1}{2}O_{2(g)} \tag{10.2}$$

$$2Nb_2O_5 \xrightarrow{Ti_5O_{10}} 4Nb_{Ti}^{\cdot} + V'''_{Ti} + 10O_o^X \tag{10.3}$$

Due to the similarity of the ionic radius of Ti^{4+} (0.61 Å) and Nb^{5+} (0.64 Å), Nb can incorporate into a TiO_2 lattice up to 30% and generates donor states with the same energy, which provide donor states during the reduction of stoichiometric TiO_2 [29].

Regarding the Nb-doped TiO_2 support and catalyst synthesis, Chhina et al. [29,30] developed a $Nb_{0.1}Ti_{0.9}O_2$ support using the sol-gel method with calcinations at 500°C and subsequent reduction at 700°C in H_2 for 2 h, and then deposited the Pt catalyst onto this support to form a 10% $Pt/Nb_{0.1}Ti_{0.9}O_2$ catalyst. This catalyst showed a better durability (no changes in Pt surface area after 60 h) compared to that of the 40% Pt/C catalyst at a high potential of 1.4 V as shown in Figure 10.4.

The conductivity of the Nb-doped TiO_2 (abbreviated as $NbTiO_2$) support with the rutile structure, that is, 610 μS cm^{-1}, is higher than that of its anatase phase (0.12 μS cm^{-1}) [7]. Due to this, $NbTiO_2$ with a rutile phase is more favored than that of a TiO_2 anatase phase in catalyst supporting. The transition from rutile to anatase titania for 10% $NbTiO_2$ happened in the range of $T = 600°C–700°C$. At the temperature of 500°C or 600°C, the dominating phase was found to be an anatase phase (13–16 nm) and to be a rutile phase (39–44 nm) at the range of 700°C or 900°C. Regarding the specific surface area, normally, $NbTiO_2$ has a higher specific area than that of suboxide titania. However, synthesis conditions have a great effect on the specific surface area. For example, at the conditions of calcination followed by a reduction at 500°C, the material with anatase phase could be formed with a specific surface area of 150 $m^2 g^{-1}$. However, if under the conditions of direct reduction at 600°C, the obtained material with a mixed anatase–rutile phase could have a specific surface area of 65.8 $m^2 g^{-1}$; if at the conditions of direct reduction

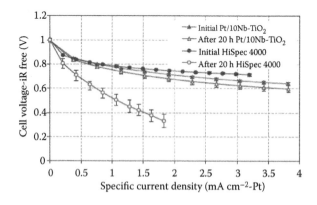

FIGURE 10.4
Single-cell polarization curves for 10 wt% $Pt/Nb_{0.1}Ti_{0.9}O_2$ and 40% Pt/C (HiSpec 4000) at the initial state and after a potential hold at 1.4 V for 20 h under N_2. Ambient pressure with humidified H_2 and O_2 at 50°C and cell temperature of 85°C. (From H. Chhina et al. *J. Electrochem. Soc.,* 156, 10, B1232–B1237, 2009.)

at 700°C, the formed material with a rutile phase only had a specific area of 13.5 m^{2}g^{-1}.

The morphology of Pt catalyst particles supported on NbTiO$_2$ was found to be strongly affected by the support phase. For example, on anatase-phased support, the Pt particles were spheres; on rutile-phased support, it became flattened, indicating a SMSI [28]. Park and Seoleao [31] synthesized a 40% Pt/ NbTiO$_2$ catalyst in which the support was synthesized by a thermal hydrolysis method at 120°C with a subsequent annealing at 400°C for 2 h in H$_2$. This support had a conductivity of 0.1 Ω^{-1}cm^{-1} and a particle size of 10 nm. The Pt catalyst (particle size ~3 nm) was deposited on this support using an impregnation–reduction (borohydride) method. The developed catalyst showed higher ORR activity probably due to the strong interaction between support and catalyst. XANES analysis showed that the intensity of the white line for the 40% Pt/NbTiO$_2$ catalyst (Figure 10.5) is lower than that of the 40% Pt/C, corresponding to less Pt oxidation in the Pt/NbTiO$_2$ catalyst, which may explain its higher ORR activity and durability.

Huang et al. [32] developed a 33.8 wt% Pt/NbTiO$_2$ catalyst using a more conductive support (1.1 S cm^{-1}). This support was synthesized via a template-assisted method in ethanol with hydrolyzing titanium isopropoxide and niobium ethoxide in the presence of 1-octadecylamine at 400°C–900°C in H$_2$. The Pt catalyst was deposited by a modified borohydride reduction method using, additionally, sodium dodecylsulfate. This catalyst was used for a durability test using a 20% Pt/C as the baseline. The ECSA of this 20% Pt/C after an *ex situ* durability test was decreased significantly due to two effects: Pt particle migration and sintering through the Ostwald ripening effect [33,34]. However, the Pt/Nb$_x$Ti$_{(1-x)}$O$_2$ catalyst showed a higher stability in the *in situ* accelerated durability test (continuous potential cycling between 0.6 and 1.4 V [RHE] at 50 mV s^{-1}) and also a tenfold higher ORR activity. The 20% Pt/C was completely degraded in a PEM fuel cell after 1000 cycles during the

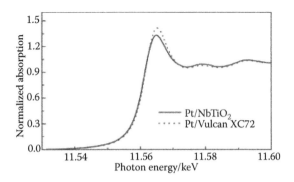

FIGURE 10.5
XANES spectra of PtL edge for 40% Pt/C (Vulcan XC72) and 40% Pt/NbTiO$_2$. (From K.-W. Park and K.-S. Seoleao, *Electrochemistry Communications*, 9, 2256–2260, 2007.)

test due to the carbon corrosion and the loss of its support–catalyst interaction, while 33.8% Pt/NbTiO$_2$ demonstrated only a small voltage loss, 0.11 V at 0.6 A cm^{-2}, after 3000 test cycles under the same conditions (Figure 10.6) [32].

A new approach was used for the synthesis of NbTiO$_2$ support and its supported 20% Pt/NbTiO$_2$ catalysts by Do et al. [35]. The method consisted of titanium and niobium hydroxide deposition on a surface of nanotemplate (polystyrene latex) particles by a sol-gel technique with the subsequent template removal at 500°C in air and then reduction of NbTiO$_2$ at 1050°C in H$_2$ to achieve the conductive rutile phase. The obtained NbTiO$_2$ support had a specific surface area of 116 m^2 g^{-1} and a porosity of 0.22 cm^{-1} g^{-1}. An *ex situ* test of the developed catalyst, 16% Pt/NbTiO$_2$, showed both higher mass and ORR-specific activity at 0.9 V (RHE) than that of the 20% Pt/C catalyst as listed in Table 10.1.

The similar wet-chemical method was also used for the synthesis of mesoporous NbTiO$_2$ hollow spheres [36]; the obtained ORR mass activity was about 80 mA mg$_{Pt}^{-1}$. However, the durability test of this hollow-sphered

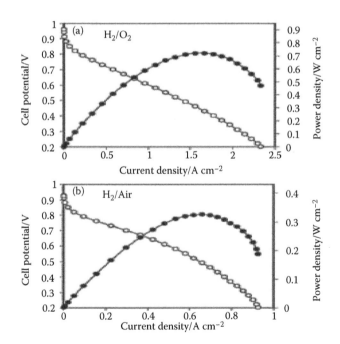

FIGURE 10.6
Polarization and power density curves of the PEM fuel cells with the Pt/Nb$_x$Ti$_{(1-x)}$O$_2$ cathode electrocatalysts: (a) H$_2$/O$_2$ and (b) H$_2$/air at 75°C with fully humidified reactants and flow rates were 150/150/500 mL min^{-1} for H$_2$/O$_2$/air without backpressure. Pt loading is 0.5 mg$_{Pt}$ cm^{-2} and 0.4 mg$_{Pt}$ cm^{-2} for anode and cathode, respectively. (From S.-Y. Huang et al. *Applied Catalysis B: Environmental*, 96, 224–231, 2010.)

TABLE 10.1

Comparison of Pt/NbTiO$_2$ and Pt/C Catalyst Performance in 0.1 M HClO$_4$, 5 mV s^{-1}, 1600 rpm, GC RDE

Catalyst	Loading, μg_{Pt} cm^{-2}	Specific Surface Area, m^2 g$_{Pt}^{-1}$	Mass Activity, mA mg$_{Pt}^{-1}$	Specific Activity, μA cm$_{Pt}^{-2}$
16% Pt/NbTiO$_2$	13	90	160	180
16% Pt/NbTiO$_2$	16	94	160	170
20% Pt/C (Vulcan XC72)	14	86	110	130

Source: T.B. Do et al. *Electrochimica Acta*, 55, 27, 8013–8017, 2010.

Pt/NbTiO$_2$ during 30,000 cycles in 0.1 M HClO$_4$ at a potential cycling regime (0.6–1.1 V) showed a slower degradation rate than that of 40% Pt/C (E-TEC). The improved durability of the hollow-sphered Pt/NbTiO$_2$ catalyst was attributed to its possible advantages over carbon-supported Pt catalysts, such as (1) higher corrosion resistance of NbTiO$_2$ supports compared to carbon and (2) stronger Pt–metal oxide support interaction, which prevented Pt agglomeration [37].

The ORR Pt$_4$Ru$_4$Ir catalysts on two types of TiO$_2$-based supports, namely, titania suboxide Ti$_4$O$_7$ produced by the reduction of TiO$_2$ and Nb$_{0.1}$Ti$_{0.9}$O$_2$ produced by Nb doping of TiO$_2$, were developed by Macak et al. [38]. The Nb$_{0.1}$Ti$_{0.9}$O$_2$ support was prepared by heat treatment over the course of 5 days in a sealed quartz tube under vacuum. The heating procedure was at 650°C for 1 day, 950°C for 2 days, and 1000°C for 2 days. The Pt$_4$Ru$_4$Ir catalyst was deposited by borohydride reduction of the corresponding salts. The durability tests showed that Pt$_4$Ru$_4$Ir/Ti$_4$O$_7$ had less durability than Pt$_4$Ru$_4$Ir/ Nb$_{0.1}$Ti$_{0.9}$O$_2$ due to the decrease in conductivity by at least five orders of magnitude induced by the oxidation of Ti$_4$O$_7$.

10.2.3 Nanostructured Titanium Dioxide

Nanostructured titanium dioxide (NSTO) in the form of titanium dioxide nanotubes (TONT), nanorods, nanosheets, and nanofibers have recently been used for fuel cell, mainly direct methanol fuel cell (DMFC), catalyst supports [38–41]. One of the most popular NSTOs as a catalyst support is TONT. Synthesized TONTs on Ti substrates by an electrochemical anodization method had a length of 1 μm and a diameter of 120 nm. These self-ordered supports were used for magnetron sputtering ORR PtNi (particle size of 5–10 nm) and PtCo (3–4 nm) catalysts. The PtNi/TONT catalyst showed a high ORR activity only after heat treatment in hydrogen at 400°C, indicated by a 120 mV increase in half-wave potential under the simulated PEM fuel cell operational conditions (O$_2$-saturated 0.5 M HClO$_4$ with a scan rate of 5 mV s^{-1}). This effect was achieved due to the formation of a PtNi alloy. This alloying could modify the electronic structure of Pt, favoring the adsorption of oxygen and also a SMSI [40,41]. An even higher performance was achieved for $Pt_{0.7}Co_{0.3}$ on the same TONT support, indicated by a positive shift of 200 mV of onset potential. This was attributed to Pt electronic modification by an alloying effect, which may shorten the Pt–Pt interatomic distance for oxygen adsorption, a necessary step for ORR.

10.2.4 Sulfated Zirconia (S-ZrO$_2$)-Supported Catalysts

Sulfated zirconia is the crystalline solid acid with high proton conductivity and hydrophilicity. Its usage as a catalyst support could reduce the content of the electrolyte ionomer in the catalyst layer, leading to an improvement in mass transport. PEM fuel cell cathode with 53% Pt/S-ZrO$_2$ in the absence of the ionomer in the catalyst layer showed less degradation; only 16% cell voltage drop was observed, which was less than that of PtC (33%). Commercial (Walco Pure Chemicals) S-ZrO$_2$ has a specific surface area of 80 m^2 g^{-1} and a particle size of 50–100 nm [42].

10.2.5 SnO$_2$-Based Material-Supported Catalysts

Tin oxide (SnO$_2$) is an n-type semiconductor and amphoteric oxide. As a support material, SnO$_2$ is attractive because it has an electronic effect and interaction with a Pt catalyst. However, both durability and resistivity are not sufficient and need to be improved mainly by doping, using metals, such as Sb [43] and Ru [44,45]. For example, antimony (Sb^{5+})-doped SnO$_2$ (2 mol% Sb–SnO$_2$) with particle sizes of 8–12 nm demonstrated less resistivity of 10^{-2}–10^{-3} versus 10–10^6 Ω cm [43] than undoped SnO$_2$.

Catalyst-supporting SnO$_2$ nanowires (SnO$_2$–NW) grown on carbon paper using the vapor deposition method (VPD) were explored as the support for a Pt catalyst [46]. This VPD method included the evaporation of tin at 800°C–900°C for 2 h in Ar and the deposition of Sn vapor in the form of SnO$_2$–NW on carbon paper in the same chamber. Pt was then electrochemically deposited on the SnO$_2$–NW/carbon paper. The ESCA of Pt/SnO$_2$–NW/carbon paper was 62% higher than that of 30% Pt/C (E-TEC), which indicated a higher Pt utilization. A significant 50 mV positive shift of the O$_2$ onset potential was observed in comparison with the 30% Pt/C (E-TEK) catalyst in 0.5 M H$_2$SO$_4$. For this Pt/SnO$_2$ catalyst, it was also found that SnO$_2$ could suppress the adsorption/desorption of oxygen and formation/reduction of platinum oxides at $E > 0.6$ V (RHE) and stimulate a SMSI, hindering catalyst degradation [47,48] even though both the electrochemical stability and conductivity of SnO$_2$ are still relatively low, requiring further improvement [47]. It was found that the combination of SnO$_2$ with indium oxide (indium–tin oxide, ITO) could give both improved stability and conductivity. The resulting ITO could have a high conductivity as $>10^3$ S cm^{-1}. Its stability to oxidation was also better than that of Vulcan XC72R. For example, the 40% Pt/ITO catalyst (Pt particle size of 13 nm) had less ESCA loss during accelerated durability tests (potential cycling) than that of the 40% Pt/C catalyst [49]. If an ITO-graphene (hybrid support, MeO+carbon)–supported Pt catalyst is used, the improvement of both durability and ORR activity could be further improved [50]. In 22 h of an *ex situ* durability test, 66.7% loss of ORR mass activity (from 78 to 26 mA mg$_{Pt}^{-1}$) for a Pt/graphene catalyst was observed. However, for the Pt/ITO–graphene catalyst, the lost was only 22.2% (from 108 to 84 mA mg$_{Pt}^{-1}$). This improvement was attributed to the formation of a triple-junction structure (Pt-ITO-graphene), which could stabilize Pt and limit its agglomeration [50].

10.2.6 WO$_x$-Supported Catalysts

Similar to SnO$_2$, tungsten oxide (WO$_x$) is an n-type semiconductor. Its nonstoichiometric oxide has high conductivity due to oxygen-vacancy defects [51]. The 40% Pt/WO$_3$ catalyst (WO$_3$ commercial powder) was synthesized by an impregnation–reduction method using the reduction of Pt (H$_2$PtCl$_6$) by hydrogen in 0.5 M H$_2$SO$_4$ at 75°C [52]. The accelerated durability test of 40% Pt/WO$_3$ (potential steps oxidation cycles: 0.6 V for 60 s and at 1.8 V vs. RHE for 20 s at 30°C) showed that it was stable than 40% Pt/C. However, ORR catalytic activity of 40% Pt/WO$_3$ was less than that of the 40% Pt/C due to its smaller specific surface area and conductivity [52].

WO$_3$ nanorods (WO$_3$–NR), another possible catalyst support material, showed a higher electrochemical stability than bulk WO$_3$ during the durability tests in 1 M H$_2$SO$_4$ at $E = -0.2$–1.1 V [53]. WO$_3$–NR was synthesized by the pyrolysis of tetrabutylammonium decatungstate at 450°C (length: 130–480 nm and diameter: 18–56 nm). The surfactant, tetrabutylammonium bromide, was used before pyrolysis for the avoidance of an irregular arrangement. Regarding its supported Pt catalyst, the reduction–impregnation method was used for synthesis of the 20% Pt/WO$_3$–NR catalyst (Pt particle size: 4–6 nm), but it was only tested for the methanol oxidation [53].

10.2.7 SiO$_2$-Supported Catalysts

SiO$_2$ is one of the most durable ceramic supports. The Pt/SiO$_2$ catalyst was prepared by both the reduction–impregnation method [54,55] and the microwave-assisted polyol method [56]. Uniform distribution of the Pt catalyst on the surface was obtained at the Pt:SiO$_2$ ratios of 1:1 and 2:1, and the ORR mass activity of formed catalysts was comparable to that of Pt/C. The normalized power density of 2% Pt/SiO$_2$ at the 2:1 ratio showed the highest power density. The Pt/SiO$_2$ catalyst had the similar ohmic loss to that of commercial Pt/C, demonstrating the absence of the support effect [56].

10.3 Boride-Supported Catalysts

Titanium boride (TiB$_2$) has high thermal stability and corrosion resistance in both acidic media and also good electric conductivity (10^5 S cm^{-1}) and thermal conductivity (65 W m^{-1} K^{-1}) [57–59]. Its supported Pt catalyst (Pt/TiB$_2$) showed some promise in terms of both ORR mass activity and electrochemical stability [60,61]. TiB$_2$ was synthesized by a self-propagating high-temperature synthesis (SHS) method with perfluorosulfonic acid to stabilize the TiB$_2$ surface (Figure 10.7) [61]. This could increase the hydrophilicity, surface area, and uniform distribution of Pt nanoparticles (Figure 10.7).

The 19% Pt/TiB$_2$ catalyst showed a higher ORR mass activity of 300 mA mg$_{Pt}^{-1}$ as well as a lower ECSA loss rate (3×10^{-3} m^2 g^{-1} cycle^{-1}) than that of the commercial 20% Pt/C catalyst (JM) (1.5×10^{-3} m^2 g^{-1} cycle^{-1}). This low ECSA loss rate of Pt/TiB$_2$ could be translated to a five times higher lifetime for the Pt/TiB$_2$ catalyst. The high durability of Pt/TiB$_2$ was explained by a strong Pt-support interaction and its dependence on Nafion electron-rich –SO$_3$ groups on the electron-deficient support surface. This strong interaction anchors Pt particles on the support, enabling the avoidance of Pt agglomeration and its accelerated degradation. Two models of Pt migration on TiB$_2$ were suggested as shown in Figure 10.8. Pt migration begins when the electric field forces (f_1) affecting Pt particles during testing are stronger than the friction resistance (force f_2) between the Pt and TiB$_2$ surfaces with –SO$_3$ groups. At $f_1 < f_2$ (Figure 10.8a), migration and aggregation of Pt particles starts, but agglomeration begins only when the Nafion film on TiB$_2$ is destroyed during a long durability test, and Pt migration does not occur (Figure 10.8b).

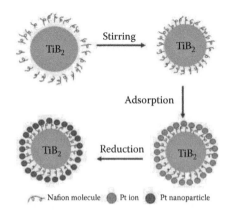

FIGURE 10.7
Synthesis route for Pt/TiB$_2$ catalysts. (From Y. Shibin et al. *J. Power Sources*, 196, 7931–7936, 2011.)

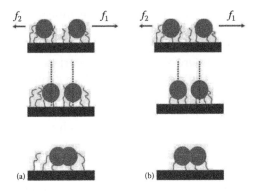

FIGURE 10.8

Migration models of Pt nanoparticles moving on the catalyst support: (a) migration of Pt colloid particles driven by an electric field force and (b) migration of Pt nanoparticles coupled with Nafion polymer oxidization. (From Y. Shibin et al. *J. Power Sources*, 196, 7931–7936, 2011.)

10.4 Carbide-Supported Catalysts

Carbides of transition metals, having similar electronic structure to Pt (close Fermi level, electronic density states), are very promising candidates for catalyst supports due to their satisfaction of several key requirements for supports [15,62–67]. Among carbides of transition metals, two main carbide supports, tungsten carbide [62,64,68] and titanium carbide [69–72], were used as catalyst supports for PEM fuel cell catalysts.

10.4.1 Tungsten Carbide (WC)-Supported Catalysts

Tungsten carbide (WC) has several unique properties, such as high chemical, thermal, and electrochemical stabilities as well as catalytic activity for several electrochemical reactions, including ORR [73]. A synergetic effect using WC support for ORR Pt catalysts could be observed [74,75]. WC could promote catalytic ORR activity of Pt/WC due to the uniform catalyst distribution of WC [75]. The main forms of tungsten carbide, WC and W_2C, have different durability under PEM fuel cell operating conditions. Normally, the thermal and electrochemical stabilities of W_2C are less than those of WC. WC is stable at $E < 0.6$ V in 0.5 H_2SO_4, in contrast to W_2C, due to the formation of substoichiometric W_xO_y. Regarding a WC-supported Pt catalyst, Zhu et al. [64] synthesized and tested a 20% Pt/WC catalyst where WC had a high specific surface area. WC was synthesized using a polymer precursor route modified based on Ganesan's procedure [76], which was performed by refluxing the precursor at 1000°C for 2 h to form WC. For Pt deposition on WC, both intermittent microwave heating and an assisted polyol method were used with impregnation and reduction H_2PtCl_6 at pH 8 in a 600 W microwave for 60 s (six cycles with 10 s on/10 s off). The technical characteristics of the synthesized WC (syn) and commercial WC (Alfa) are given in Table 10.2 [64].

The synthesized WC (35 nm) has significantly higher specific surface area (89 m^2 g^{-1}) than the commercial WC specific surface area (1.6 m^2 g^{-1}) and higher conductivity (3.0 vs. 0.87 S cm^{-1}) as well as comparable chemical stability at 200°C in 0.5 M H$_2$SO$_4$. A durability test of 20% Pt/WC (Syn) (Pt particle size: 5 nm) in 0.1 M HClO$_4$ at 30°C showed its degradation due to its oxidation to WO$_3$ at $E > 0.7$ V versus RHE, which was confirmed by x-ray photon spectroscopy measurements. The intensity of the observed Pt oxidation peak was decreased with increasing cycle number due to the isolation of the Pt catalyst by the deposited nonconductive WO$_3$. It was also found that Pt could also inhibit the WC oxidation, signaling by the disappearance of oxidation current at $E = 0.7$ V versus RHE for the Pt/WC catalyst with respect to WC. This observation agrees well with those reported in the literature [77,78]. The ORR mass activity of 20% Pt/WC (Syn) was relatively low, 3.3 mA cm^{-2} at 1600 rpm, due to the low specific surface area of 9 m^2 g$_{Pt}^{-1}$ [64]. According to the Pt particle size of 5 nm, the theoretical external surface area of the Pt catalyst should be around 56 m^2 g$_{Pt}^{-1}$, only 16% of the external electrochemically active surface.

In terms of stability, a degradation of the Pt/WC catalyst was observed in the accelerated durability test in 0.5 M H$_2$SO$_4$ (cycling –1.8 V for 20 s and 0.6 V for 60 s at 30°C) due to WC oxidation and WC core encapsulating by WO$_x$ [70]. However, compared to the 40% Pt/C catalyst, the 40% Pt/WC catalyst showed a better stability. In addition, the electrochemical stability of WC could be improved with the addition of Ta to WC [78]. The formed WC-Ta was also used for supporting the Pt catalyst.

Pt/WC catalysts were also developed for HOR [79–81]. The WC (20 nm) used for a 10% Pt/WC HOR catalyst was prepared by carburization of W$_2$N, WS$_2$, and WP. The maximal HOR mass activity was obtained for the catalyst 10% Pt/WC (WC precursor, WS$_2$) of 120.7 A/g$_{cat}$ at 450 mV versus RHE was less than that of the commercial E-TEK Pt/C catalyst (163.8 A/g$_{cat}$ at 450 mV) due to its low specific surface area of 42.4 m^2 g^{-1} [79]. WC with the higher specific surface area of 80 m^2 g^{-1} was also obtained via a modification of the

TABLE 10.2

Properties of WC

	Specific Surface Area, m^2 g^{-1}	Solubility (wt%)		Conductivity, S cm^{-1}		Thermal Stability
		95°C	200°C	95°C	200°C	
WC (Alfa)	1.6	0.15	0.63	0.75	0.87	Stable up to 200°C
WC (Syn)	89	0.16	0.27	2.4	3.0	Stable up to 200°C

Source: W. Zhu et al. *Electrochimica Acta,* 61, 198–206, 2012.

carburization method. The resulting Pt/WC catalyst showed a higher HOR activity than that of commercial Pt/C, which was attributed to a synergetic effect between support and catalyst. In the suggested mechanism of HOR on a Pt/WC catalyst, Pt is responsible for the acceleration of a rate-determining stage of the dissociative adsorption of H_2 on WC, and WC takes over for the other steps [80].

10.4.2 Titanium Carbide (TiC)-Supported Catalysts

Like WC, titanium carbide (TiC) has high thermal stability, chemical stability, and electronic conductivity. TiC also has a superior corrosion resistance in PEM fuel cells, phosphoric acid fuel cells, and water and hydrogen chloride cells [70,71,82–84]. Unfortunately, its irreversible electrochemical oxidation in acidic electrolyte at $E = 0$–1.2 V versus RHE compromises its usage as a catalyst support [69]. The modification of TiC by forming a more durable support with core-shell design TiC@TiO$_2$ to form Pt/TiC@TiO$_2$ catalyst could significantly improve the durability. The TiC@TiO$_2$ core-shell support was synthesized via thermal oxidation of TiC in air at 330°C [83]. In addition, 20% Pt/TiC and 20% Pt$_3$Pd/TiC@TiO$_2$ were synthesized by a microwave-assisted polyol process. However, the conductivity of TiC@TiO$_2$ at 95°C was less than that of TiC due to the formation of an insulating layer of TiO$_2$ (Table 10.3) [69].

Similar to WC, both TiC and TiC@TiO$_2$ are not electrochemically stable. A large oxidation peak was observed with the onset potential at 0.95 V (RHE), and the intensity of this peak was increased with increasing cycle number in the accelerated durability test (1000 CVs at 0.05–1.2 V vs. RHE, 20 mV s^{-1}). This could be explained by the following reaction [85–94]:

$$TiC + 5H_2O \rightarrow TiO_2 + CO_3^{2-} + 10H^+ + 8e^- \qquad (10.4)$$

Regarding the stability, both 20% Pt/TiC and 20% Pt$_3$Pd/TiC catalysts showed a loss of ECSA 78% and 94%, respectively, after 500 cycles in an accelerated durability test (0.1 M HClO$_4$, 0.05–1.2 V vs. RHE, 20 mV s^{-1}) as shown in Figure 10.9. Both of these two ECSA losses were higher than that of Pt/C [95]. In addition, a 20% Pt/TiC@TiO$_2$ catalyst had less loss of ECSA (20%) and

TABLE 10.3

Electrical Conductivities of TiC, TiC@TiO$_2$, and Carbon at 670 kPa

Conductivity (S m^{-1})	TiC	TiC–TiO$_2$ (330°C)	Carbon
95°C	0.28	0.09	46
200°C	0.9	0.12	225

Source: A. Ignaszak et al. *Electrochimica Acta*, 69, 397–405, 2012.

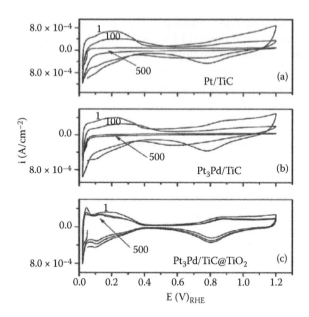

FIGURE 10.9
Cyclic voltammograms of (a) Pt/TiC, (b) Pt$_3$Pd/TiC, and (c) Pt$_3$Pd/TiC@TiO$_2$ catalysts, recorded during durability tests (1, 100, and 500 cycles). Scan rate: 20 mV s^{-1}. (From A. Ignaszak et al. *Electrochimica Acta*, 69, 397–405, 2012.)

TABLE 10.4

Electrochemical Performance of 20% Pt/TiC, 20% Pt$_3$Pd/TiC, and 20% Pt$_3$Pd/TiC@TiO$_2$

Catalyst	ECA (m^2 g$_{Pt}^{-1}$)	i_{ma} (mA mg$_{(Pt+Pd)}^{-1}$) at 0.9 V	i_{spec} (μA cm^{-2}) at 0.9 V
Pt–TiC	40.0	9.7	24.0
Pt$_3$Pd–TiC	41.6	34.7	83.7
Pt$_3$Pd–TiC–TiO$_2$	37.6	33.3	88.3

Source: A. Ignaszak et al. *Electrochimica Acta*, 69, 397–405, 2012.

also ORR mass activity (45%) after 500 cycles of accelerated test than 20% Pt$_3$Pd/TiC (94%), showing TiC@TiO$_2$ was a promising support material. The ORR mass activity of 20% Pt$_3$Pd/TiC and 20% Pt$_3$Pd/TiC@TiO$_2$ (38–42 mA mg$_{Pt}^{-1}$) catalysts was three times higher than that of Pt/TiC (Table 10.4) [69].

10.4.3 Silicon Carbide (SiC)-Supported Catalysts

Commercially available (SiC) silicon carbide has both high chemical and thermal durability and both relatively low conductivity and specific surface area [96,97]. However, a thermal plasma method seemed to allow the SiC

TABLE 10.5

Physical and Electrochemical Properties of 20% Pt/SiC Catalysts

Catalysts	Pt Loading, wt%	Particle Size, nm	Specific Surface Area, $m^2 g^{-1}$	ECSA, $m^2 g_{Pt}^{-1}$	ORR Mass Activity at 0.85 V (NHE), mA mg_{Pt}^{-1}
Pt/SiC	18.4	3.5	76	78	125
Pt/C (E-TEC)	19.7	3.7	81	85	141

Source: V. Rao et al. *Indian J. Chem.*, 47A, 1619, 2008.

vapor deposition on cold walls of the reactor chamber and achieve a specific surface area of 35 $m^2 g^{-1}$ and a high conductivity of 1×10^6 S cm^{-1} [98,99].

Regarding a SiC-supported Pt catalyst, 20% Pt/SiC was synthesized by the hydrogen method. As shown in Table 10.5, the mass activity of 20% Pt/SiC was 125 mA mg_{Pt}^{-1}, which is comparable with the ORR mass activity of the commercial 20% Pt/C (E-TEK) catalyst of 146 mA mg_{Pt}^{-1} at 0.85 V versus NHE (Table 10.5). It was found that an equal distribution of low-index planes of Pt on the support surface could be obtained, indicating the electronic interaction between Pt and SiC. This interaction might inhibit the formation of OH$^-$ species and improve the ORR activity [98]. Degradation of the initial ORR mass activity after 22 h of durability test at 0.85 V (NHE) in 0.5 M H_2SO_4 for 20% Pt/SiC and 20% Pt/C (E-TEK) were 26% and 45%, respectively, indicating the 20% Pt/SiC was more stable than 20% Pt/C.

10.5 Nitride-Supported Catalysts

Transition metal nitrides have both high conductivity and stability and are considered candidates for catalyst supports. One of them, titanium nitride, has been used for the synthesis of Pt/TiN catalysts [100]. Titanium nitride is a triple-bond transition metal compound with corrosion resistance attributed to the formation of surface oxide/oxynitrides [101–103], high electrical conductivity (4000 S cm^{-1}) [104], and nonzero electron density at the Fermi level [101]. These properties of TiN made it a possible catalyst support material for PEM fuel cell applications [17,105,106]. The 20% Pt/TiN catalyst showed a higher ORR mass activity than that of commercial 20% Pt/C but low durability in $HClO_4$ [107]. Regarding the stability, Pt/TiN catalyst showed a strong dependence, similar to TiN, on temperature and acid concentration as well as the type of acids. For example, the durability of the Pt/TiN catalyst in 0.1 M $HClO_4$ acid was dependent on temperature [107]. The formation of an oxide film on TiN in Pt/TiN catalyst could lead to a significant dropping of ECSA in 0.1 M $HClO_4$ at 60°C.

10.6 Effect of a Metal–Support Interaction on Catalyst Performance

Compared to carbon-supported catalysts, both the higher ORR/HOR activities and durability of noncarbon-supported Pt catalysts, such as $Pt/NbTiO_2$ [31], Pt/TiO_2-NT [40], Pt/TiB_2 [61], and Pt/SiC [98], could be attributed to the stronger metal–support interaction [13–15]. Such a metal–support interaction could favor the anchoring of Pt nanoparticles on a support surface, promoting uniform catalyst dispersion and preventing them from migration, coalescence, aggregation, and loss (see Figures 10.3 and 10.8). The electronic interaction between support and Pt could favor the electron transfer from Pt to adsorbed oxygen atoms, resulting in high ORR activity [108]. Thus, the advantages of noncarbon-supported catalysts in terms of both their higher durability and activity are mainly realized through strong metal–support interaction.

10.7 Conclusions

In this chapter, various noncarbon support materials, such as metal oxides, borides, nitride, and carbides, and their supported Pt-based catalysts for ORR in PEM fuel cells were presented and reviewed in terms of both their ORR activity and stability as summarized in Table 10.6. Some of these noncarbon material–supported Pt catalysts have both higher ORR mass activity and stability, higher electrical conductivity, stronger metal–support interaction, better catalyst dispersion, and lower specific surface area than carbon-supported catalysts.

The most durable and active catalysts, such as Pt/TiB_2 (ORR mass activity: 300 mA mg_{Pt}^{-1} and ECSA loss: 19%) and $Pt/NbTiO_2$ (ORR mass activity: 180 mA mg_{Pt}^{-1}) catalysts, showed the optimal balance among chemical, thermal, and electrochemical stability and ORR mass activity. Other ceramic-supported catalysts, such as Pt/TiO_{2-x} [14], Pt/WC [63], Pt/TiC [69], Pt/SiC [98], as well as Pt/TaN [108] had higher initial ORR mass activities but faster degradation due to the electrooxidation of their supports when compared to that of carbon-supported Pt catalyst.

It should be noted that one of the main current trends in PEM fuel cell catalyst development is the use of hybrid (carbon + metal oxide)–supported catalysts. Hybrid supports combine the main advantages of noncarbon supports (high stability and strong metal–support interaction) and carbon supports (high specific surface area and high conductivity) to give both high ORR activity and stability without compromising the main required support properties [50,109–111].

TABLE 10.6

Characteristics of Noncarbon-Supported Catalysts for PEM Fuel Cells

Catalyst	Electrical Conductivity (Support), S cm⁻¹	Pt Particle Size, nm	Specific Surface Area m² g⁻¹/ECSA (m² g_{Pt}^{-1})	ORR Mass/Specific Activity (mA mg_{Pt}^{-1})/ (μA cm_{Pt}^{-1})	Loss of ECSA/Mass Activity after Durability Test, %	Reference
20% Pt/C (JM)	–	–	–/61	110/130	–	[61]
19% Pt/TiB₂	–	3.1	–/38	300/–	19/–	[61]
Pt/NbTiO₂	*Resistivity 14 Ω cm (C–0.22 Ω cm)	10	90/–	160/180	–	[31,32,35]
5%–20% Pt/TiO₂₋ₓ	–	–	1–2/–	Low	Low	[16]
20% Pt/WO₃	–	–	–	ORR MA Pt/WO₃ < ORR MA Pt/C	Durable at 1.8 V (100 cycles)	[52]
1% Pt/SiO₂	PEMFC with Pt/SiO₂ has higher power density vs. Pt/C					[56]
20% Pt/TiC	0.28 (95°C)	4	41/34	35/84	94/–	[69]
20% Pt/TiC + TiO₂	–	4	76/78	33/88	45/–	[69]
20% Pt/SiC	–	3.5	–	–	–	[39]
20% Pt/WC	2.4 (95°C)	5	9/–	3.3/–	–	[107]
20% Pt/TiN	–	–	–	–	Low durability	[98]
20% Pt/SnO₂	–	3–4	76/78	125 (at 0, 85 V) –	–	[42]
53% Pt/S-ZrO₂	–	–	–	–	16% dropping of cell voltage vs. 33% for Pt/C	

References

1. James, B. and Kalinisky, J. 2010. DTI, Inc., *DOE Hydrogen Program Review*. Washington, DC, June 9.
2. Mathias, M., Makharia, R., Gasteiger, H., Conley, J., Fuller, T., Gittleman, C., Kocha, S., Miller, D., Mittesteadt, C., Xie, T., Yan, S. and. Yu, P. 2005. *Interface (USA)* 14, 24.
3. Pourbaix, M. 1979. *Atlas of Electrochemical Equilibria in Aqueous Solution*. National Association of Corrosion Engineers, Houston, TX.
4. Kinoshita, K. 1988. *Carbon: Electrochemical and Physicochemical Properties*. Wiley, New York, p. 319.
5. Zhu, W., Zheng, J.P., Liang, R., Wang, B., Zhang, C., Au, G. and Plichta, E.J. 2009. *J. Electrochem. Soc.* 156 (9), B1099.
6. Borup, R.L., Davey, J.R., Garzon, F.H., Wood, D.L. and Inbody, M.A. 2006. *J. Power Sources* 163, 76.
7. Ioroi, T., Senoh, H., Yamazaki, S., Siroma, Z., Fujiwara, N. and Yasuda, K. 2008. *J. Electrochem. Soc.* 155, B321.
8. Gruver, G.A. 1978. *J. Electrochem. Soc.* 125, 1719.
9. Stonehart, P. 1984. *Carbon* 22, 423.
10. Kangasniemi, K.H., Condit, D.A. and Jarvi, T.D. 2004. *J. Electrochem. Soc.* 151, E125.
11. Zhang, J., Song, C. and Zhang, J. 2011. *J. Fuel Cell Sci. Technol.* 8, 051006-1–051006-5.
12. Huang, S.-Y., Ganesan, P., Park, S. and Popov, J. 2009. *Am. Chem. Soc.* 131, 13898.
13. Yu, X. and Ye, S. 2007. *J. Power Sources* 172, 133–144.
14. Tauster, S.J., Fung, S.C. and Garten, R.L. 1978. *J. Am. Chem. Soc.* 100 (1), 170–175.
15. Wang, Y., Wilkinson, D.P. and Zhang, J. 2011. *Chem. Rev.* 12, 7625–7651.
16. Keerthi, S., Hui, R., Campbell, S., Ye, S. and Zhang, J. 2011. *Electrochim. Acta* 59, 538–547.
17. Ioroi, T., Siroma, Z., Fujiwara, N., Yamazaki, S. and Yasuda, K. 2005. *Electrochem. Comm.* 7, 183.
18. Marezio, M. and Dernier, P.D. 1971. *J. Solid State Chem.* 3, 340.
19. Bartholomew, R. and Frankl, D. 1967. *Phys. Rev.* 187, 828.
20. Farndon, E.E. and Pletcher, D. 1997. *Electrochim. Acta* 42, 1281.
21. Przyluski, J. and Kolbrecka, K.J. 1993. *J. Appl. Electrochem.* 23, 1063.
22. Kolbrecka, K. and Przylski, J. 1994. *J. Electrochim. Acta* 39, 1591.
23. Smith, R., Walsh, F.C. and Clarke, R.L. 1998. *J. Appl. Electrochem.* 28, 1021.
24. Slavcheva, E., Nikolova, V., Petkova, T., Lefterova, E., Dragieva, I., Vitanov, T. and Budevski, E. 2005. *Electrochim. Acta* 50, 5444.
25. Adamczyk, L., Kulesza, P.J., Miecznikowski, K., Palys, B., Chojak, M. and Krawczyk, D. 2005. *J. Electrochem. Soc.* 152, E98.
26. Vracar, L.M., Krstajic, N.V., Radmilovic, V.R. and Jaksic, M.M. 2006. *J. Electroanal. Chem.* 587, 99.
27. Chen, Z., Bare, G. and Mallouk, S.R. 2003. *J. Electrochem. Soc.* 149, A1092.
28. Ruiz, A., Dezanneau, G., Arbiol, J., Cornet, A. and Morante, J. 2004. *Chem. Mater.* 16, 862.
29. Chhina, H., Campbell, S. and Kesler, O. 2009. *J. N. Electrochem. Systems* 12, 177–185.

30. Chhina, H., Campbell, S. and Kesler, O. 2009. *J. Electrochem. Soc.* 156 (10), B1232–B1237.
31. Park, K.-W. and Seoleao, K.-S. 2007. *Electrochem. Comm.* 9, 2256–2260.
32. Huang, S.-Y., Ganesan, P. and Popov, B.N. 2010. *Appl. Catal. B: Environ.* 96, 224–231.
33. Chen, Z.W., Waje, M., Li, W.Z. and Yan, Y.S. 2007. *Angew. Chem. Int. Ed.* 46, 4060–4063.
34. Popov, B.N., Gojkovic, S.L., Zecevic, S.K. and Savinell, R.F. 1998. *J. Electrochem. Soc.* 145, 3713–3720.
35. Do, T.B., Cai, M., Ruthkosky, M.S. and Moylan, T.E. 2010. *Electrochim. Acta* 55 (27), 8013–8017.
36. Sun, S., Sun, X., Cai, M. and Ruthkosky, M. 2012. Hindawi Publishing Corporation. *J. Nanotechnol.* 2012, 1–8.
37. Sasaki, K., Zhang, L. and Adzik, R.R. 2008. *Phys. Chem. Chem. Phys.* 10, 159–167.
38. Macak, M., Barczuk, P.J., Tsuchiya, H., Nowakowska, M.Z., Ghicov, A., Chojak, M., Bauer, S., Virtanen, S., Kulesza, P.J. and Schmuki, P. 2005. *Electrochem. Comm.* 7, 1417.
39. Wang, S.H.M., Guo, D. and Li, H. 2005. *J. Solid State Chem.* 178, 1996.
40. Kang, S.H., Jeon, T.-Y., Kim, H.-S., Sung, Y.-E. and Smyrl, W.H. 2008. *J. Electrochem. Soc.* 155, B1058.
41. Kang, H., Sung, Y.E. and Smyrl, W.H. 2008. *J. Electrochem. Soc.* 155, B1128.
42. Suzuki, Y., Ishihara, A., Mitsushima, S., Kamiya, N. and Ota, K. 2007. *Electrochem. Solid State Lett.* 10, B105.
43. Santos, A.L., Profeti, D. and Olivi, P. 2005. *Electrochim. Acta* 50, 2615.
44. Pang, H.L., Zhang, X.H., Zhong, X.X., Liu, B., Wei, X.G., Kuang, Y.F. and Chen, J.H. 2008. *J. Coll. Int. Sci.* 319, 193.
45. Safonova, O.V., Delabouglise, G., Chenevier, B., Gaskov, A.M. and Labeau, M. 2002. *Mater. Sci. Eng. C* 21, 105.
46. Saha, S., Li, R.Y., Cai, M. and Sun, X.L. 2007. *Electrochem. Solid State Lett.* 10, B130–B133.
47. Nakada, M., Ishihara, A., Mitsushima, S., Kamiya, N. and Ota, K. 2007. *Electrochem. Solid State Lett.* 10, F1–F4.
48. Ota, K., Ishihara, A., Mitsushima, S., Lee, K., Suzuki, Y., Horibe, N., Nakagawa, T. and Kamiya, N. 2005. *J. New Mat. Electrochem. Syst.* 8, 25–35.
49. Chhina, H., Campbell, S. and Kesler, O. 2006. *J. Power Sources* 161, 893–900.
50. Kou, R., Shao, Y., Mei, D., Nie, Z., Wang, D., Wang, C., Viswanathan, V.V., Park Ilhan, S., Aksay, A., Lin, Y., Wang, Y. and Liu, J. 2011. *J. Am. Chem. Soc.* 133, 2541–2547.
51. Nakajima, H. and Honma, I. 2002. *Solid State Ionics* 48, 607–610.
52. Chhina, H., Campbell, S. and Kesler, O. 2007. *J. Electrochem. Soc.* 154, B533.
53. Rajeswari, J., Viswanathan, B. and Varadarajan, T.K. 2007. *Mater. Chem. Phys.* 106, 168.
54. Zhu, X., Zhang, H., Liang, Y., Zhang, Y. and Yia, B. 2006. *Electrochem. Solid State Lett.* 9, A49.
55. Wang, L., Xing, D.M., Liu, Y.H., Cai, Y.H., Shao, Z.G., Zhai, Y.F., Zhonga, H.X., Yi, B.L. and Zhang, H.M. 2006. *J. Power Sources* 161, 61.
56. Seger, B., Kongkanand, A., Vinodgopal, K. and Kamat, P.V. 2008. *J. Electroanal. Chem.* 621, 198.
57. Zhang, J.Y., Fu, Z.Y. and Wang, W.M. 2005. *J. Mater. Sci. Technol.* 21, 841.

58. Fu, Z.Y., Wang, H., Wang, W.M., Zheng, Q.J. and Yuan, R.Z. 2002. *Key Eng. Mater.* 217, 41.
59. Basu, B., Raju, G.B. and Suri, A.K. 2006. *Int. Mater. Rev.* 51, 352.
60. Yin, S., Mu, S., Lv, H., Cheng, N., Pan, M. and Fu, Z. 2010. *Appl. Catal. B* 93, 233.
61. Shibin, Y., Shichun, M., Mu, P. and Zhengyi, F. 2011. *J. Power Sources* 196, 7931–7936.
62. Chhina, H., Campbell, S. and Kesler, O. 2007. *J. Power Sources* 164, 431–440.
63. Zellner, M.B. and Chen, J.G.G. 2005. *Catal. Today* 99, 299–307.
64. Zhu, W., Ignaszak, A., Song, C., Bakera, R., Hui, R., Zhang, J., Nan, F., Botton, G., Ye, S. and Campbell, S. 2012. *Electrochim. Acta* 61, 198–206.
65. Meng, H. and Shen, P.K. 2005. *J. Phys. Chem., B* 109, 22705.
66. Antolini, E. and Gonzalez, E.R. 2009. *Solid State Ionics* 180, 746–763.
67. Shao, Y., Liu, J., Wang, Y. and Lin, Y. 2009. *J. Mater. Chem.* 19, 46–59.
68. Levy, R.B. and Boudart, M. 1973. *Science* 181, 547–549.
69. Ignaszak, A., Song, C., Zhua, W., Zhang, J., Bauer, A., Baker, R., Neburchilov, V., Ye, S. and Campbell, S. 2012. *Electrochim. Acta* 69, 397–405.
70. Ma, L., Sui, S. and Zhai, Y. 2008. *J. Power Sources* 177, 470.
71. Ou, Y., Cui, X., Zhang, X. and Jiang, Z. 2010. *J. Power Sources* 195, 1365.
72. Jalan, V. and Frost, D.G. 1989. US Patent 4,795,684.
73. Hwu, H.H. and Chen, J.G. 2003. *J. Phys. Chem. B* 107, 2029–2039.
74. Nie, M., Shen, P.K., Wu, M., Wei, Z.D. and Meng, H. 2006. *J. Power Sources* 162, 173–176.
75. Meng, H. and Shen, P.K. 2006. *Electrochem. Comm.* 8, 588–594.
76. Ganesan, R. and Lee, J.S. 2005. *Angew. Chem. Int. Ed.* 44, 6557.
77. Weigert, E., Esposito, D. and Chen, J.G. 2009. *J. Power Sources* 193, 501.
78. Lee, K., Ishihara, A., Mitsushima, S., Kamiya, N. and Ota, K. 2004. *Electrochim. Acta* 49, 3479.
79. Hara, Y., Minami, N. and Itagaki, H. 2007. *Appl. Catal. A* 323, 86.
80. Hara, Y., Minami, N., Matsumoto, H. and Itagaki, H. 2007. *Appl. Catal. A* 332, 289.
81. Ham, D.J., Kim, Y.K., Han, S.H. and Lee, J.S. 2008. *Catal. Today* 132, 17.
82. LaConti, A.B., Griffith, A.E., Cropley, C.C. and Kosek, J.A. 2000. US Patent 6,083,641.
83. Stott, S.J., Mortimer, R.J., Dann, S.E., Oyama, M. and Marken, F. 2006. *Phys. Chem. Chem. Phys.* 8, 5437.
84. Lee, H.Y., Choi, J.W., Hwang, G.H. and Kang, S.G. 2006. *Met. Mater. Int.* 12 (2), 147.
85. Ou, Y., Cui, X., Zhang, X. and Jiang, Z. 2010. *J. Power Sources* 195, 1365.
86. Cocino, B.S. and Alman, D.E. 2002. DOE/ARC-02-011.
87. Cowling, R.D. and Hintermann, H.E. 1970. *J. Electrochem. Soc.* 117 (11), 1447.
88. Beverskog, B., Carlsson, J.O., Delblank Bauer, A., Deshpandey, C.V., Doerr, H.J., Bunshah, R.F. and O'Brien, B.P. 1990. *Surf. Coat. Technol.* 41, 221.
89. Delblanc Bauer, A. and Carlsson, J.O. 1991. *J. Phys., IV*, 1, C2–C641.
90. Brynza, A.P., Kosolapova, T.Y., Khmelovskaya, S.A., Fedorus, V.B. and Simonova, E.K. 1971. *Poroshkovaya Metallurgiya* 8, 67.
91. Hintermann, H.E., Riddiford, A.C., Cowling, R.D. and Malyszko, J. 1972–1973. *Electrodepos. Surf. Treat.* 1, 59.
92. Verkhoturov, A.D., Kuzenkova, M.A., Lebukhova, N.V. and Podchernyaeva, I.A. 1988. *Poroshkovaya Metallurgiya* 2 (302), 81.

93. Morancho, R., Petit, J.A., Dabosi, F. and Constant, G. 1982. *J. Electrochem. Soc.* 129, 854.

94. Alloca, C.M., Williams, E.S. and Kaloyeros, A.E. 1987. *J. Electrochem. Soc.* 134, 3170.

95. Zhang, Y., Huang, Q., Zou, Z., Yang, J., Vogel, W. and Yang, H. 2010. *J. Phys. Chem. C* 114, 6860.

96. Li, H., Sun, G., Li, N., Su, D. and Xin, Q. 2007. *J. Phys. Chem. C* 111, 5605.

97. Ignaszak, A., Ye, S. and Gyenge, E. 2009. *J. Phys. Chem. C* 113, 298.

98. Rao, V., Singh, S.K. and Viswanathan, B. 2008. *Indian J. Chem.* 47A, 1619.

99. Jalan, V., Taylor, E.T., Frost, D. and Morriseau, D.B. 1983. *National Fuel Cell Semin. Abstr.* 127.

100. Avasarala, B., Murray, T., Li, W. and Haldar, P. 2009. *J. Mater. Chem.* 19, 1803.

101. Milosev, I., Strehblow, H., Navinsek, B. and Metikos-Hukovic, M. 1995. *Surf Interface Anal.* 23 529–539.

102. Saha, N.C. and Tompkins, H.G. 1992. *J. Appl. Phys.* 72, 3072.

103. Ernsberger, C., Nickerson, J., Smith, J.T., Miller, T.A.E. and Banks, D. 1986. *J. Vac. Sci. Technol A* 4 (6), 2784–2788.

104. Oyama, S.T. 1996. *The Chemistry of Transition Metal Carbides and Nitrides*, 1st ed. Springer.

105. Ham, D.J. and Lee, J.S. 2009. *Energies* 2 (4), 873–899.

106. Lo, C.P., Kumar, A. and Ramani, V. 2010. Abstract #973, 218th ECS Meeting, The Electrochemical Society.

107. Avasarala, B. and Haldar, B.P. 2010. *Electrochim. Acta* 55 (28), 9024–9034.

108. Bogotski, V.S. and Snudkin, A.A. 1984. *Electrochim. Acta* 29, 757.

109. Bauer, A., Song, C., Ignaszak, A., Hui, R., Zhang, J., Checallier, L., Jones, D. and Rozière, J. 2010. *Electrochim. Acta.* 55, 8365–8370.

110. Ignaszak, A., Song, C., Zhu, W., Zhang, J., Bauer, A., Baker, R., Neburchilov, V., Ye, S. and Campbell, S. 2012. *Electrochim. Acta* 75 (30), 220–228.

111. Bing, Y., Neburchilov, V., Song, C., Baker, R., Guest, A., Ghosh, D., Ye, S., Campbell, S. and Zhang, J. 2012. *Electrochim. Acta* 77, 225–231.

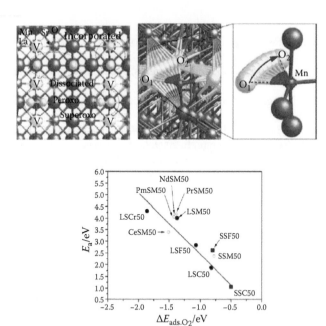

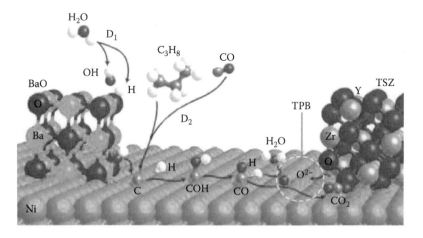

FIGURE 2.5
Comparison of adsorption energies versus diffusion barriers of oxygen ions through the MIEC bulk phases. (Reprinted from *J Power Sources*, 195, Choi, Y. et al. Rational design of novel cathode materials in solid oxide fuel cells using first-principles simulations, 1441–1445, Copyright 2010, with permission from Elsevier).

FIGURE 2.10
Proposed mechanism for water-mediated carbon removal on the anode with BaO/Ni interfaces. (Reprinted by permission from Macmillan Publishers Ltd. *Nature Communications*, Yang, L. et al. Promotion of water-mediated carbon removal by nanostructured barium oxide/nickel interfaces in solid oxide fuel cells, 2, 357–367, Copyright 2011.)

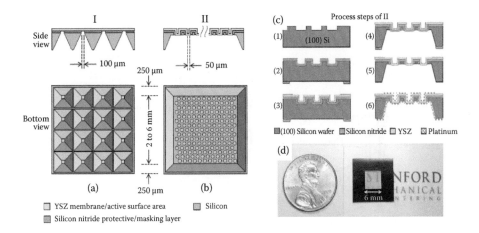

FIGURE 4.15
(a and b) Illustration of schematic comparison of μSOFC free-standing membrane active areas (yellow) of two different fabrication processes. (a) Membrane array structure fabricated by common KOH through-wafer etching, (b) membrane array structure with high surface utilization by the modified method. The upper pictures in (a and b) are cross-section views of a membrane array on a (100) silicon substrate, and the bottom pictures present the bottom side view of the array. (c) Modified process steps of fabrication of planar μSOFC array on silicon substrate; (d) A 6 mm by 6 mm YSZ membrane electrolyte array on a silicon chip (Su, P.-C., and Prinz, F.B., *Electrochem. Comm.*, 16, 77–79, 2012).

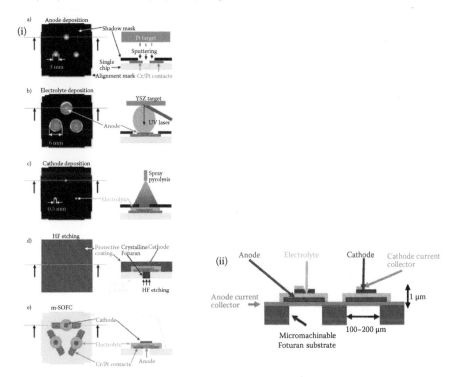

FIGURE 4.17
(i) Schematic showing the fabrication steps of a planar μSOFC on Foturan substrate. Top views on the left and cross sections on the right. (ii) Schematic view of a μSOFC cell on Foturan substrate showing all the components (From Muecke, U.P., Beckel, D., Bernard, A., Bieberle-Hütter, A., Graf, S., Infortuna, A., Müller, P., Rupp, J.L.M., Schneider, J., and Gauckler, L.J., *Adv. Funct. Mater.*, 18, 3158–3168, 2008; Rupp, J.L.M., Muecke, U.P., Nalam, P.C., and Gauckler, L.J., *J. Power Sources*, 195, 2669–2676, 2010).

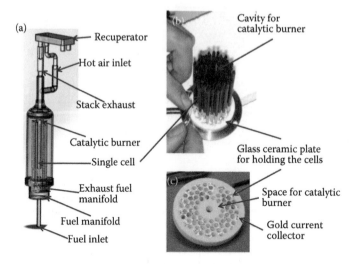

FIGURE 4.39
(a) CAD drawing of thermally sustainable 60-cell anode-supported microtubular SOFC system, (b) photograph of AITF–built 60-cell stack, and (c) machined glass-ceramic holding plate showing gold current collector, space for catalytic burner.

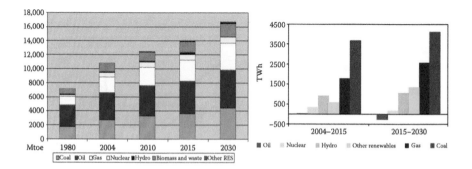

FIGURE 5.1
Primary world energy consumption is set to march on with the majority from fossil sources (left) contributing to CO_2 emission. Coal is expected to dominate future electricity production (right). (Reproduced from Malzbender, J. et al. *Solid State Ionics*, 176, 2201–2203, 2005. With permission.)

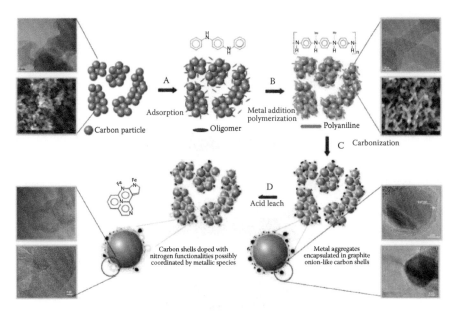

FIGURE 7.7
Schematic diagram of the synthesis of PANI–M–C catalysts supported by carbon black nanoparticles. (From Wu, G. et al. *Science*, 332, 443–447, 2011. Reprinted with permission of AAAS.)

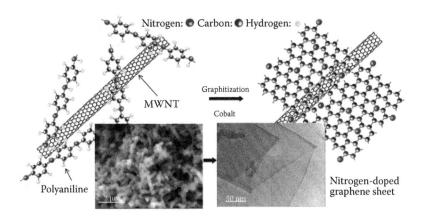

FIGURE 7.13
Scheme of formation of nitrogen-doped graphene sheet derived from heteroatom polymers (polyaniline) and transition metals (Co). (Reprinted with permission from Wu, G. et al. *ACS Nan, 6*, 9764–9776. Copyright 2012 American Chemical Society.)

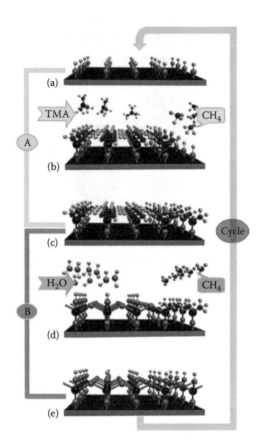

FIGURE 9.1
A model ALD process for depositing Al_2O_3 using TMA and water as precursors: (a) the initial substrate covered by hydroxyl groups; (b) TMA molecules reacting with hydroxyl groups, leading to a new intermediate layer; (c) the substrate covered with a newly created intermediate layer; (d) H_2O molecules reacting with the intermediate layer, leading to new hydroxyl groups; and (e) the substrate again covered by hydroxyl groups. (Meng, X. et al.: *Advanced Materials*. 2012. 24. 3589–3615. Copyright Wiley-VCH Verlag GmbH & Co. KGaA. Reproduced with permission.)

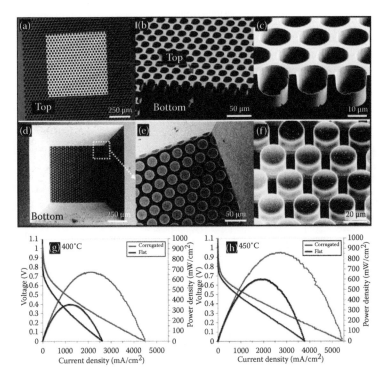

FIGURE 9.11
The upper part shows the images of corrugated YSZ membrane, view from top (a through c) and bottom (d through f). (a) Optical microscopy image of free-standing corrugated YSZ membrane. The white square in the middle is the free-standing YSZ membrane released after KOH etching. The black dots represent cup-shaped trenches. (b and c) Cross section of cups in corrugated membrane. (d) Bottom view of membrane after silicon removal. (e) Bottom corner view of membrane; (f) perspective after tilting 40°. The lower part shows the fuel cell performance of corrugated (red line) and flat (black line) YSZ electrolyte at (g) 400°C and (h) 450°C. The membrane size is 0.0036 cm² (600 μm × 600 μm). The size of the cups embedded is 15 μm in diameter and 20 μm in depth. Maximum power densities of corrugated YSZ are slightly less than two times that of flat ones. (Reprinted with permission from Su, P.C., Chao, C.C., Shim, J.H., Fasching, R., and Prinz, F.B., *Nano Lett.*, 8, 2289–2292. Copyright 2008 American Chemical Society.)

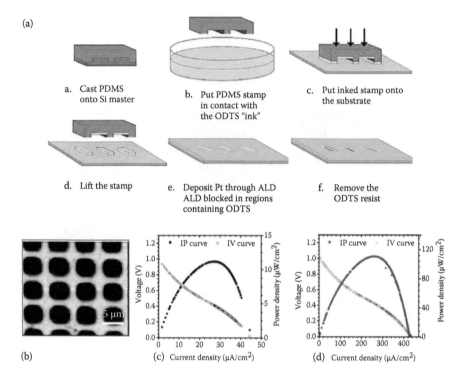

FIGURE 9.14
(a) Schematic outline of the procedure to fabricate patterned Pt thin films using microcontact printing and selective ALD. (b) Auger electron spectroscopy elemental maps for platinum on the micropatterned grid structure after area-selective ALD of Pt on YSZ. (c and d) *I–V* and *I–P* characteristics of fuel cells without and with Pt current collector grids, respectively. (Reprinted with permission from Jiang, X., Huang, H., Prinz, F.B., and Bent, S.F., *Chem. Mater.*, 20, 3897–3905. Copyright 2008 American Chemical Society.)

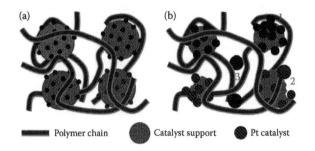

FIGURE 10.3
Effect of carbon corrosion on (1) agglomeration, (2) coalescence, and (3) loss of Pt particles in catalyst layer during operation of PEM fuel cells: (a) normal (corrosion-resistant) electrode and (b) corroded electrode. (Reprinted with permission from S.-Y. Huang et al. *J. Am. Chem. Soc.*, 131, 13898. Copyright 2009 American Chemical Society.)

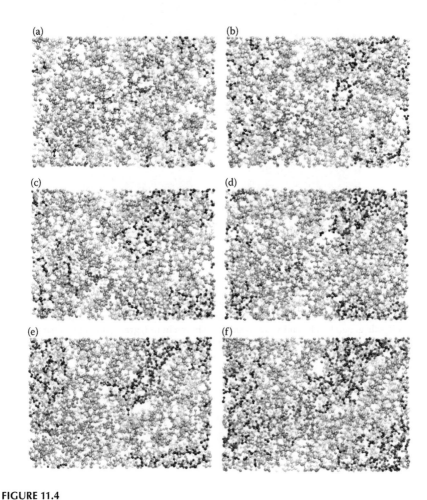

FIGURE 11.4
Orthographic projection of hydrated Nafion for the following λ values: (a) 3, (b) 5, (c) 7, (d) 9, (e) 11, and (f) 13.5. Water molecules, hydronium ions, sulfonate groups, and the rest of the membrane are represented in blue, red, yellow, and gray, respectively. (Reprinted with permission from Devanathan, R., Venkatnathan, A., and Dupuis, M. *Journal of Physical Chemistry B*, 111, 8069–8079. Copyright 2007 American Chemical Society.)

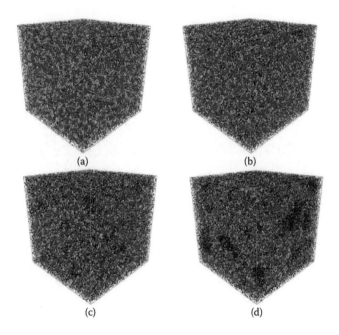

FIGURE 11.5
3D images of hydrated SSC PFSA membrane morphologies at λ values of (a) 5, (b) 7, (c) 11, and (d) 16. The simulation box is a cube with dimensions of 32.4 nm. The fluorocarbon beads, terminal ionic side group beads, and water beads are shown in red, green, and blue, respectively. (Reprinted with permission from Wu, D.S., Paddison, S.J., and Elliott, J.A. *Macromolecules*, 42, 3358–3367. Copyright 2009 American Chemical Society.)

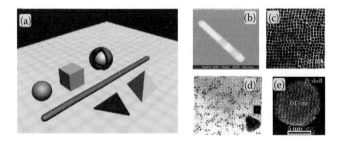

FIGURE 13.3
Some example shapes and structures of nanoparticles, which had been synthesized. (a) Some ideal shapes. (b) Field emission scanning electron microscopy (FESEM) image of five-segmented nanorods, Pt-RuNi(2)-Pt-RuNi(2)-Pt. (From Fang, L. et al. *J Electrochem Soc*, 153: A2133–A2138, 2006.) (c) Transmission electron microscopy (TEM) image of Pt nanocubes. (From Yang, H.Z. et al. *Angew Chem-Int Edit*, 49: 6848–6851, 2010.) (d) TEM image of Pt tetrahedral nanoparticles. (From Ahmadi, T.S. et al. *Science*, 272: 1924–1925, 1996.) (e) High resolution TEM (HRTEM) image of Pd-Pt core-shell nanoparticle. (From Long, N.V. et al. *Int J Hydrog Energy*, 36: 8478–8491, 2011.)

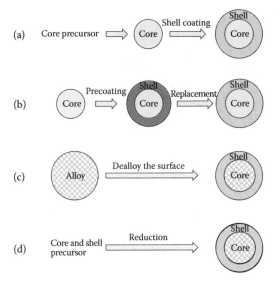

FIGURE 13.16
Scheme of four ways to synthesize core-shell structure. (a) A two-stage procedure method, (b) shell replacement method, (c) dealloy method, and (d) one-step synthesis method.

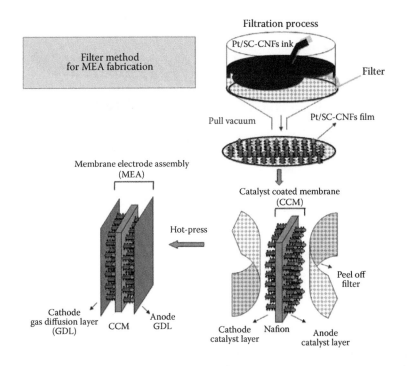

FIGURE 14.10
Schematic illustration of filtration method for fabricating SC-CNFs based MEA. (Reprinted from Li, W.M. et al. *Carbon*, 48, 995–1003. Copyright 2010. With permission.)

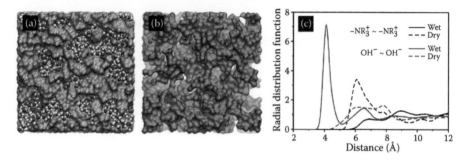

FIGURE 15.14

Molecular dynamics (MD) simulations for the structure of hydrated aQAPS ($\lambda = 30$). (a) A view of the simulated hydrophilic/hydrophobic domain separation: The hydrophilic domains that contain water molecules (represented by red-white balls) and OH$^-$ (represented by green-white balls) are intercalated in the hydrophobic network formed by the polysulfone backbone and the alkyl side-chains (represented by the blue continuum). (b) Another view of (a) with the hydrophobic network being hidden and the hydrophilic domains being represented in continuum mode so as to see the interconnections between aqueous clusters. (c) The radial distribution function (RDF) of the $-NR_3^+$ and the OH$^-$ in wet and dry aQAPS membranes, based on which the aggregation of OH$^-$ upon hydration is evident. (Reprinted with permission from Pan, J., Chen, C., Zhuang, L., and Lu, J., *Acc. Chem. Res.*, 45, 473–481. Copyright 2012 American Chemical Society.)

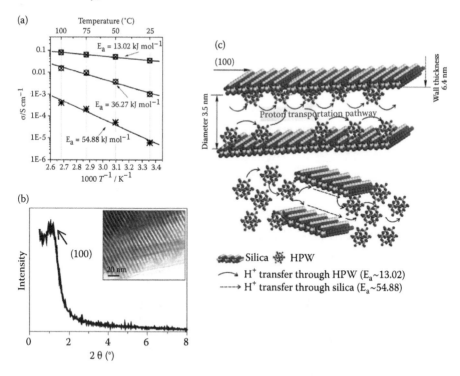

FIGURE 16.7

(a) Impedance curves of the cell measured at a current density of 100 and 600 mA cm^{-2}. (b) Small angle XRD pattern of the self-assembled 25 wt% HPW/meso-silica electrolyte membrane, (c) the proposed proton transportation pathways of the self-assembled HPW/meso-silica electrolyte membrane and the mixed HPW/meso-silica composite electrolyte membrane, respectively. The insert in (b) is the HRTEM micrograph of the self-assembled 25 wt% HPW/meso-silica electrolyte membrane. (From Tang, H.L. et al. *Chem. Commun.*, 46, (24), 4351–4353, 2010.)

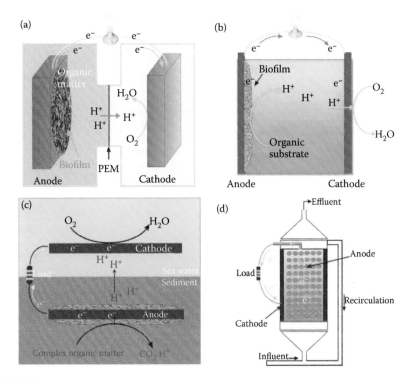

FIGURE 17.1
Configurations of MFCs. (a) Dual-chamber MFC; (b) single-chamber MFC; (c) sediment MFC; and (d) tubular MFC.

11

Understanding the Nanostructures in Nafion Membranes

Hongwei Zhang and Pei Kang Shen

CONTENTS

11.1 Introduction 315
11.2 Multiscale Computational Simulations 317
11.3 Morphology Models of Nafion 321
11.4 Structural Evolution and Proton Transport of Nafion with
Increasing Water Content 326
11.5 Conclusions 329
References 329

11.1 Introduction

Although Nafion membranes were primarily developed for the chloro-alkali processes by DuPont in the 1960s, they did demonstrate unexpected applicability to polymer electrolyte membrane fuel cells (PEMFCs) because of their high proton conductivity and inherent chemical, thermal, and oxidative stability. Moreover, they were widely used as polymer electrolyte membranes (PEMs), a key component in PEMFCs, and considered as the *de facto* benchmark of PEMs. The PEM functions as a protonic conductor, an electronic insulator, and a separator for the two-reactant gases. Because Nafion membranes play such important roles in PEMFCs, it is necessary to gain an insight into the nanostructures in Nafion membranes, which can be useful in understanding the relationships among their chemical structure, morphology, and macroscopic physicochemical properties (Figure 11.1).

It is well known that Nafion membranes are composed of a polytetrafluoroethylene (PTFE) backbone and perfluorinated vinyl ether side chains terminating with sulfonic acid groups (Figure 11.2). When Nafion membranes are fully hydrated, they are separated into hydrophobic regions and water-filled hydrophilic domains. The former endows them with superior

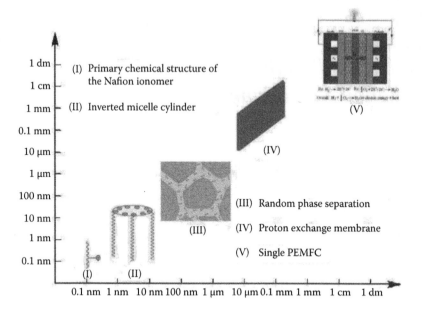

FIGURE 11.1
Schematic depiction of the dimensional evolution of PEMFC.

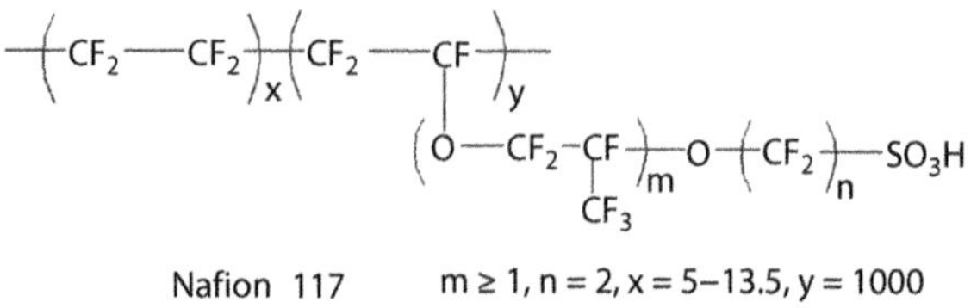

Nafion 117 $m \geq 1, n = 2, x = 5\text{–}13.5, y = 1000$

FIGURE 11.2
Chemical structures of Nafion membrane.

properties of structural integrity, and the latter imparts them with desirable proton conductivity.

In this chapter, both computational simulations and morphology models of Nafion membranes will be reviewed, but the relative literatures are not confined to Nafion membranes. The literature about short-side-chain (SSC) perfluorosulfonic acid (PFSA) membranes (such as the Dow membrane and Hyflon Ion membrane) are also included because they have similar structures and no distinct differences in water and proton transport and in the hydrophobic/hydrophilic separation as a function of water volume fraction compared to Nafion membrane of the same equivalent weight (EW) [1].

11.2 Multiscale Computational Simulations

An abundance of morphological information on Nafion has been obtained from small-angle x-ray scattering (SAXS) [2–5], small-angle neutron scattering (SANS) [6,7], differential scanning calorimetry (DSC) [8], infrared and Raman spectra [9,10], nuclear magnetic resonance (NMR) [11,12], dielectric relaxation spectroscopy (DRS) [13], dynamic mechanical analysis (DMA) [14], electron spin resonance spectroscopy (ESRS) [15], scanning electron microscopy (SEM) [16], atomic force microscopy (AFM) [17], and positron annihilation spectroscopy (PAS) [18]. But these experimental studies of Nafion do not draw a clear conclusion on the molecular-level morphology of Nafion because the experimental studies are difficult to directly probe the nanoscale morphological changes and transport phenomena. Consequently, multiscale computational simulations are employed to undertake these tasks.

The main object of multiscale computational simulations is to provide a profound understanding of structures on the molecular scale, larger-scale morphological structures, and chemical functionalities of hydrated Nafion. Among a variety of different simulation techniques, the atomistic and molecular simulations are valuable tools to give a visual molecular picture about the nanoscale morphological changes with water molecules and to analyze proton transfer in Nafion. Various atomistic computer simulation techniques are adopted to carry out the object, including classical molecular dynamics (MD) simulations [19–22], *ab initio* MD simulations [23], molecular orbital calculations [24], statistical mechanical models [25,26], and empirical valence bond (EVB) simulations [27,28]. All of these works reveal the multiscale nature of transport processes in Nafion membranes.

Devanathan et al. [29,30] used classical MD simulation with the DREIDING force field to characterize the changes in the nanostructure of Nafion membrane brought about by systematically changing the hydration level. A simulation cell of hydrated Nafion with a projection of 4.2×3.0 nm was used. They found that hydronium (H_3O^+) ions were trapped by multiple sulfonate groups at low λ (λ stands for the hydration level, which is the ratio of water molecules per sulfonic acid group in Nafion), while only a few H_3O^+ ions were coordinated by multiple sulfonate groups at high λ ($\lambda = 20$). There existed an abrupt structural change above $\lambda = 5$, which was probably related to the changes in hydration number and confined configurations of H_3O^+ ions. The relative proportions of bound, weakly bound, and free water molecules in Nafion at different λ are shown in Figure 11.3.

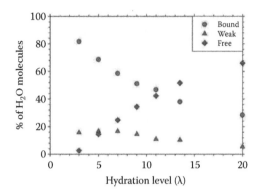

FIGURE 11.3

Relative proportions of bound, weakly bound, and free water molecules in Nafion as a function of λ. (Reprinted with permission from Devanathan, R., Venkatnathan, A., and Dupuis, M. *Journal of Physical Chemistry B*, 111, 8069–8079. Copyright 2007 American Chemical Society.)

They also revealed that the distances between sulfonate groups become larger with increasing membrane hydration. Furthermore, the mean residence time of H_2O molecules near sulfonate groups decreased with increasing hydration level due to the changes in membrane nanostructure. Figure 11.4 presented a visual representation of their simulated Nafion nanostructure [29]. Depth was hinted at by indicating the atoms closer to the top in a darker shade of blue, red, yellow, and gray, respectively, for water molecules, hydronium ions, sulfonate groups, and the rest of the membrane. From the bottom right of Figure 11.4a, it can be clearly seen that H_3O^+ ions are surrounded by multiple sulfonate groups at $\lambda = 3$. When λ increases to 7, H_2O molecules are more likely to be found away from the sulfonate groups and form a continuous channel. And a network for transporting protons has been well developed at $\lambda = 13.5$.

Their group further investigated the temperature effect on H_3O^+ ion mobility by classical MD simulation and proton hopping in Nafion by the quantum hopping (Q-HOP) MD simulation [31,32]. They concluded that temperature had a significant effect on the absolute value of the diffusion coefficients for both water and H_3O^+ ions. The mean residence time of the proton on a H_2O molecule decreases by two orders of magnitude when the λ value was increased from 5 to 15.

Their work provided insights into nanoscale phase separation in hydrated Nafion and an explanation for the changes observed experimentally with increasing hydration level. In addition, a bridge between Nafion membrane nanostructure and the dynamics of H_2O molecules and H_3O^+ ions was established.

On a larger scale, mesoscale morphology evolution of Nafion often involves dissipative particle dynamics (DPD) simulation [33–36] and self-consistent

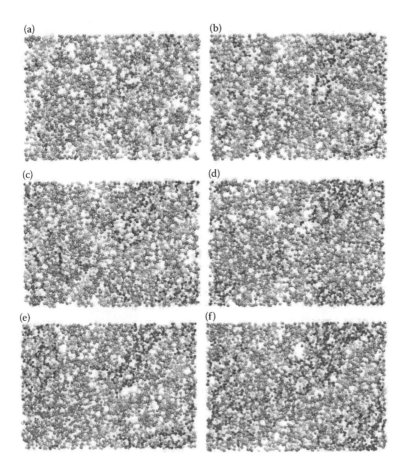

FIGURE 11.4

(See color insert.) Orthographic projection of hydrated Nafion for the following λ values: (a) 3, (b) 5, (c) 7, (d) 9, (e) 11, and (f) 13.5. Water molecules, hydronium ions, sulfonate groups, and the rest of the membrane are represented in blue, red, yellow, and gray, respectively. (Reprinted with permission from Devanathan, R., Venkatnathan, A., and Dupuis, M. *Journal of Physical Chemistry B*, 111, 8069–8079. Copyright 2007 American Chemical Society.)

mean-field (SCMF) simulation [37,38]. These simulations aim at investigating the mesoscopic structure of Nafion at the scales, which are greater and longer than atomistic and MD simulations.

Wu et al. [39] also addressed the nanostructure changing with the hydration level and the effect of molecular weight (MW) on the hydrated morphology of the SSC PFSA membrane. They used more explicit polymer chains in their DPD simulations. The three-dimensional (3D) images of hydrated SSC PFSA morphologies and two-dimensional (2D) cross sections for the corresponding 3D images are shown in Figures 11.5 and 11.6, respectively.

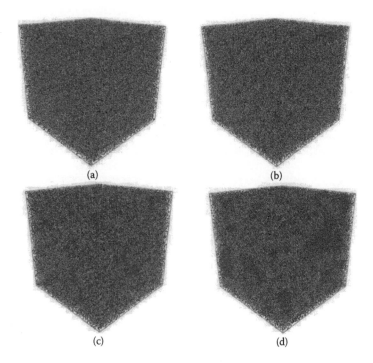

FIGURE 11.5

(See color insert.) 3D images of hydrated SSC PFSA membrane morphologies at λ values of (a) 5, (b) 7, (c) 11, and (d) 16. The simulation box is a cube with dimensions of 32.4 nm. The fluorocarbon beads, terminal ionic side group beads, and water beads are shown in red, green, and blue, respectively. (Reprinted with permission from Wu, D.S., Paddison, S.J., and Elliott, J.A. *Macromolecules*, 42, 3358–3367. Copyright 2009 American Chemical Society.)

From Figure 11.5, it can be seen that the ion clusters consisted of water, and the terminal ionic side group becomes more and more distinct with an increase in the hydration level. And the partial ion clusters are gradually coalesced in the process; the long-range connectivity of the water domains occurs at the highest water content. The phenomena are broadly consistent with what is observed in MD simulations of Nafion studies.

A visual inspection of 2D slices in Figure 11.6 confirms that the water clusters are isolated at the lower water contents, and a channel morphology is formed when the water is increased. Furthermore, the morphologies of water clusters at all levels of hydration appear to be self-similar in shape despite of their different sizes.

In addition, they found that there is a strong influence of MW on both the shape and size of water-rich ionic aggregates formed as a function of water content. The increase in MW induces aggregation of the fluorocarbon backbone in order to minimize chain-bending forces while maintaining a phase-separated structure and results in larger, more elongated water domains, especially at high EW [39].

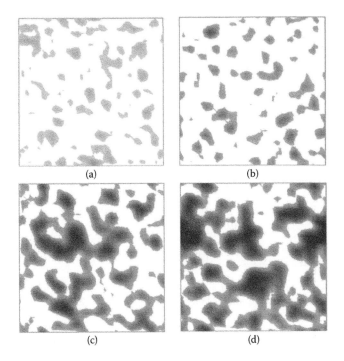

FIGURE 11.6
Contour plots of the density of the water as a 2D cross section of the hydrated 3D morphology images at λ values of (a) 5, (b) 7, (c) 11, and (d) 16. The darkness of the gray level is linearly proportional to the water density over the range from 0.00 to 3.25. (Reprinted with permission from Wu, D.S., Paddison, S.J., and Elliott, J.A. *Macromolecules*, 42, 3358–3367. Copyright 2009 American Chemical Society.)

11.3 Morphology Models of Nafion

In order to better link structural information with the macroscopic properties of Nafion membrane, a rational and predictive morphology model for Nafion is needed. But all the existing morphology models for Nafion only are phenomenological, and they are proposed on the basis of the information gathered from SAXS and wide-angle diffraction studies [40].

The SAXS profile of hydrated Nafion (Figure 11.7) [41] gives at least the following four basic pieces of information [42–44]: The first is an upturn of the intensity $I(q)$ at the very low angles ($q < 0.01$ Å^{-1}), which is strongly related to the ionic aggregates. This is generally associated with the long-range heterogeneities in the spatial distribution of the ionic clusters. The second is the presence of a peak, centered at $q \approx 0.05$ Å^{-1}, called the "matrix peak." This peak generally appears as a shoulder, and the intensity of this maximum strongly depends on the crystallinity of the Nafion films or the presence of

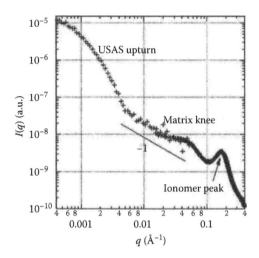

FIGURE 11.7

Typical x-ray scattering spectrum of a water swollen Nafion-1100. (Reprinted with permission from Diat, O., Rubatat, L., and Gebel, G. *Macromolecules*, 37, 7772–7783. Copyright 2004 American Chemical Society.)

crystallites in the fluorocarbon hydrophobic phase. The third is the well-known "ionomer peak" centered at $q \approx 0.2$ Å^{-1}. The ionomer peak originates from either the shape of the ionic clusters (acidic groups) or their spatial distribution. The angular position of the ionomer peak has been proposed to correspond to the mean center-to-center spacing between the ionic clusters. However, the intensity and position of the signal and the structure responsible for this maximum are still not clearly understood. Finally, the tail region of the SAXS profile shows an asymptotic scattering behavior in the high q-region and follows Porod's law $I(q) \alpha\ q^{-4}$, which is an indication that the polymer–solvent interface is sharp and also confirms the phase separation.

The widely popular morphological model of Nafion membrane is the cluster-network model (Figure 11.8) [45], which is extended from the inverted spherical micelle model proposed by Gierke et al. [46]. In a dry state, the diameter of these inverted micelles is ~2 nm, and they are disconnected from each other. With water-filling, the hydrated spherical ionic clusters with a diameter of ~4–5 nm, inside which the SO_3^- groups and water molecules are confined, are connected by water-filled cylindrical channels of around 1 nm in diameter, thus forming a connected cluster-network structure. The well-connected cluster network is embedded in the surrounding, sponge-like fluorocarbon phases. Approximately every fully hydrated spherical ionic cluster contains 70 SO_3^- groups fixed at the inner surface and 1000 water molecules. However, the aqueous necks in this model are factitiously hypothesized for the coherence of the cluster network while they cannot be detected directly. Furthermore, the assumption of periodic distribution of spherical clusters in this model is also too simplistic, and the computational simulations also indicate that water

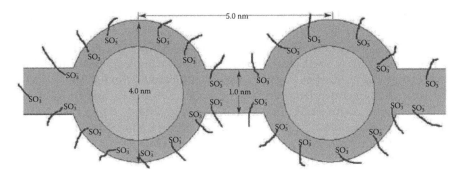

FIGURE 11.8

Cluster-network model of Nafion. (Reprinted from *Journal of Membrane Science*, 13, Hsu, W.Y., and Gierke, T.D., Ion transport and clustering in Nafion perfluorinated membranes, 307–326, Copyright 1983, with permission from Elsevier.)

domains appear as an elongated, nonspherical aggregated structure, instead of a simple shape, such as spherical clusters.

Under the premise of the recognition that ionic groups aggregate to form a network of clusters, a variety of morphology models for Nafion are proposed. Because the cluster-network model is unsuitable to describe the structural evolution of membranes over a wide range of humidification conditions, Gebel proposed a conceptual qualitative model for structural evolution of Nafion membranes from membrane to solution during the dissolution process [47]. In dehydrated Nafion, the dry cluster diameter of ~1.5 nm is significantly smaller than the intercluster distance of ~2.7 nm, which explains the extremely low ionic conductivity observed at the anhydrous state. With the absorption of water, a modification of the cluster structure occurs, which results in the formation of spherical water pools with the ionic groups at the polymer water interface in order to minimize the interfacial energy. The water pool diameter of ~2.0 nm is still lower than the interaggregate distance of ~3.0 nm, which is evidenced by the low ionic conductivity at low water content. When the water volume fraction, φ, is larger than 0.2, the large increase in the ionic conductivity happens, indicating a percolation of the ionic aggregates. The origin of percolation can be probably interpreted as a combination of the effect of the interfacial energy and of the limitation of the swelling due to the polymer chain elastic energy. As the water content increases to between $\varphi = 0.3$ and 0.5, the structure of spherical ionic domains connected with cylinders of water dispersed in the polymer matrix formed. The ionic domain diameter increases from 4 to 5 nm, and the increase in ionic conductivity with the water content increasing reveals that both the connectivity and the diameter increase. At φ values larger than 0.5, an inversion of the structure occurs and the membranes correspond to a connected network of rod-like polymer aggregates. For $\varphi = 0.3$–0.9, the connected rod-like network swells, and the radius of the rod is about 2.5 nm. The structure of highly swollen membranes is then very close to that of the Nafion solution.

A local-order model was developed by Dreyfus et al. [48] in which ionic domains have a tetrahedral-like packing arrangement but with short-range order. As an interparticle model, it can be used to define the spatial distribution of spherically shaped ionic clusters in Nafion. The core-shell model is proposed by Fujimura et al. [49], which is an intraparticle model. It suggested that ionic clusters are surrounded by a fluorocarbon phase, and these core-shell particles are embedded in a matrix of fluorocarbon chains containing fluorocarbon polymer and nonclustered ions. But the model cannot well fit the experimental data. A lamellar model proposed by Litt [50] describes the ionic domains as hydrophilic micelles separated by thin, lamellar PTFE-like crystallites and the Nafion structure as a lamellar organization of planar clusters. The lamellar model gives a convenient and simple explanation for the swelling behavior of Nafion. But it ignores the low angle maximum attributed to the crystalline, interlamellar long-range spacing. Gebel and Lambard [51] proposed a model that was based on elongated (cylindrical or ribbonlike) polymer aggregates with a diameter of about 4 nm and a length larger than 100 nm. These aggregates are surrounded by ionic groups and packed in bundles with an ordered orientation. The ionomer peak in SAXS represents the average distance between the aggregates in a typical bundle. The bundles are randomly arranged at the mesoscale. The sandwich-like model is proposed by Haubold et al. [52]. It assumes the structure of Nafion to be a core region sandwiched between shells made of hydrophilic groups and polymer side chains. The lateral dimensions are between 1.5 and 4.5 nm with a total thickness of 6 nm. In a dry state, the core is empty, and after being hydrated, the core is filled with water. The model provides information on the basic structural unit while ignoring the mesoscale structure. The three-phase model proposed by Yeager and Steck [53] depicted Nafion as a water/ionomer mixture without regular structure, which consists of a fluorocarbon polymer region, an interfacial region, and an ionic cluster region.

All the morphology models for Nafion mentioned above cannot match the experimental SAXS data, while the model based on parallel cylindrical water channels (Figure 11.9) shows a well fit to the experimental SAXS data [54]. The parallel inverted-micelle cylinder model is proposed by Schmidt-Rohr et al. which features long parallel water channels in cylindrical inverted micelles and parallel cylindrical water channels with diameters between 1.8 and 3.5 nm. The cylindrical inverted micelles are lined with hydrophilic groups and are stabilized by the stiff polymer backbone (Figure 11.9a), and these cylindrical inverted micelles are locally parallel to their neighbors. At 20 vol% water, the water channels have diameters of between 1.8 and 3.5 nm, with an average of 2.4 nm, which is larger than that in a cluster-network model. The model also considers the crystallinity, which is elongated and parallel to the water channels, with cross sections of ~5 nm². These Nafion crystallites (~10 vol%) form physical crosslinks that are crucial for the mechanical properties of the Nafion membrane. Based on this model, many important features of Nafion, such as fast diffusion of water and protons through Nafion and its

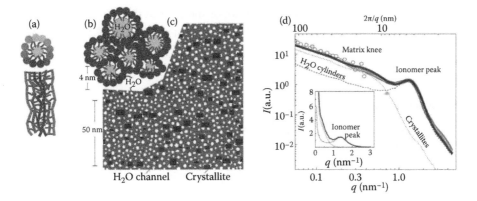

FIGURE 11.9

Parallel water-channel (inverted-micelle cylinder) model of Nafion. (a) Two views of an inverted-micelle cylinder with the polymer backbones on the outside and the ionic side groups lining the water channel. Shading is used to distinguish chains in front and in the back. (b) Schematic diagram of the approximately hexagonal packing of several inverted-micelle cylinders. (c) Cross sections through the cylindrical water channels (white) and the Nafion crystallites (black) in the noncrystalline Nafion matrix (gray). (d) Small-angle scattering data (circles) in a log(I) versus log(q) plot for Nafion at 20 vol% of H_2O, and the model curve is in black (solid line). The inset shows the ionomer peak in a linear plot of $I(q)$. Simulated scattering curves from the water channels and the crystallites by themselves (in a structureless matrix) are shown dashed and dotted, respectively. (Reprinted by permission from Macmillan Publishers, Ltd. *Nature Materials,* Schmidt-Rohr, K., and Chen, Q., 7, 75–83. Copyright 2008.)

persistence at low temperatures, can be satisfactorily explained. The characteristic "ionomer peak" in SAXS is attributed to inverted-micelle cylinders, which are parallel but otherwise randomly packed.

However, when this model is applied at length scales beyond 20 nm, it must be modified because channels bend and merge. Remaining the central recognition of cylindrical water channels and crystallinity elongated and parallel to the water channels, the revised parallel cylindrical water channels at mesoscale are proposed as Figure 11.10. In the left figure, some

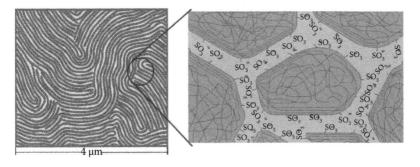

FIGURE 11.10

Revised parallel cylindrical water channel model at mesoscopic scale.

cylindrical water channels emerge, bend, or merge, but they are still parallel to their new neighbors. In the right figure, lines parallel to the water channels stand for crystallinity.

11.4 Structural Evolution and Proton Transport of Nafion with Increasing Water Content

For convenience of understanding, the approximate relations among λ, water content, and relative humidity (RH) are listed in Table 11.1 [29,55].

Generally, the water in fully hydrated Nafion is classified as three types: (1) tightly bound water, which is strongly associated with ionic and polar sites in the polymer chain through the solvation effect and also defined as nonfreezing bound water; (2) loosely bound water, which is weakly bound to the polymer chain or interacts weakly with nonfreezable water, also called freezable bound water; and (3) free water, which is not intimately bound to the polymer chain through ion–water interactions and exhibits the same behavior as bulk water. The three kinds of water can be determined from the melting behavior of the water in Nafion via DSC.

It is universally accepted that proton transport in Nafion may be accomplished by two mechanisms: (1) vehicle mechanism and (2) Grotthus mechanism (structural diffusion). The former involves the diffusion of protons on a vehicle, such as H_3O^+ ions, with the diffusion constant of water, while the latter involves the proton relay–relay transport through a sequence of consecutive Eigen $\left(H_9O_4^+\right)$-to-Zundel $\left(H_5O_2^+\right)$ and Zundel-to-Eigen transitions in the hydrogen bonding network, which is tightly associated with the forming and breaking of hydrogen bonds in the neighborhood of the proton exchange

TABLE 11.1

Approximate Relations among λ, Water Content, and RH

λ	Water Content (wt.%)	RH (%)
1	1.54	3
3	4.50	18
5	7.28	68
7	9.90	83
9	12.38	92
11	14.72	96
13.5	17.48	99
20	23.89	In water

Sources: Devanathan, R. et al. *Journal of Physical Chemistry B*, 111, 8069–8079, 2007; Costamagna, P. et al. *Journal of Power Sources*, 178, 537–546, 2008.

position. This determines that the Grotthus mechanism cannot take place at low water content. The initial driving force for the two mechanisms is the concentration gradient of proton, and the reorganization and stabilization of the side chains induced by the hydrogen bonding of the SO_3^- groups with a H_2O molecule and a H_3O^+ ion, respectively [23,56]. Vehicle mechanism occurs on slower time (nanoseconds) scales, whereas the Grotthus mechanism happens on faster time scales (up to a few picoseconds), which makes Grotthus mechanism more effective than vehicle mechanism.

In order to better describe the structural evolution of the Nafion membrane from the dry state to the fully humidified state, the process mesoscopic scale and molecular scales must be taken into account at the same time. According to the parallel cylindrical water channel model, the structural evolution of Nafion membrane at two scales can be depicted as Figures 11.11 and 11.12.

The structural evolution of Nafion with water content at the mesoscopic scale can be described as follows: In the anhydrous Nafion (Figure 11.11a), the inverted-micelle cylinders are still existing, but their diameters are reduced. And because these inverted-micelle cylinders are empty, the proton conductivity is nearly zero due to the absence of proton carriers. When λ is lower than 2.0, the water is associated strongly with the acidic groups, and these solvated clusters are isolated reciprocally (Figure 11.11b), which results in fairly poor proton conductivity. With λ increasing, these clusters grow and form interconnections with each other. When λ is around 2.0, the percolation threshold is achieved, and the proton conductivity is still comparatively low (Figure 11.11c). The proton conductivity increases but remains

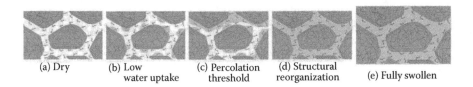

(a) Dry   (b) Low water uptake   (c) Percolation threshold   (d) Structural reorganization   (e) Fully swollen

FIGURE 11.11
Evolution of Nafion structure with water content at mesoscopic scale.

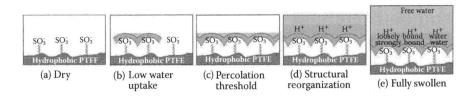

(a) Dry   (b) Low water uptake   (c) Percolation threshold   (d) Structural reorganization   (e) Fully swollen

FIGURE 11.12
Structural evolution of Nafion membrane from the anhydrous state to the swollen state at the molecular level.

low with λ values from 2.0 to 5.0. When λ approaches 5.0, the nanochannels in Nafion cannot accommodate more water, so a structural reorganization occurs, and the continuous phase, which is favorable for proton transport, begins to develop. The channels also become broader (Figure 11.11d). The conductivity displays an abrupt increase. Then the continuous phase continues to develop gradually with λ increasing. Finally, a complete water channel network is formed, and the resultant considerably high proton conductivity is obtained in fully hydrated Nafion (Figure 11.11e).

The structural evolution in a single inverted-micelle cylinder of Nafion can be given as follows: In the case of dry Nafion, the SO_3^- groups locate inside the nanochannels, and no proton conductivity can be detected (Figure 11.12a). When a small quantity of water is absorbed in the nanochannels, they firstly fulfill the tightly bound water, which is necessary but little attributed to proton conductivity. These water molecules mainly serve as a plasticizer to plasticize polymer and decrease resistance to the next water molecules (Figure 11.12b). When λ is lower than 3.0, most water molecules remain inside the first hydration shell of the nearest sulfonate group [29], ion cages [57] formed by hydronium ions hydrogen-bonded to multiple sulfonate groups, leading to only 2.7% free water in water, which not only prohibits the structural transport of protons due to the lack of water molecules solvating the proton to allow the formation of the extended networks [58] but also impedes vehicular transport of protons through steric hindrance. As a result, low proton conductivity can be anticipated because it follows the vehicle mechanism through H_3O^+ hopping from acid side to acid side. With the water content increasing, more water becomes available to solvate the H_3O^+ ions, which leads to a significant decrease in the formation of $SO_3^- - H_3O^+ - SO_3^-$ bridges. When λ approaches 5.0, the percentage of free water is near 16% in water. At this stage, a reorganizational transition of the interfacial region happens, and the diffusion of H_3O^+ ions displays an abrupt increase. Both of them are related to changes in the hydration number and bridging configurations of H_3O^+ ions between SO_3^- groups [30]. Moreover, the structural transition opens the possibility of structural diffusion (Figure 11.12c). As λ increases from 5.0 to 9.0, the average distance of the H_3O^+ ions from the nearest SO_3^- group further increases and remains beyond the first hydration shell radius at last. The average hydration number of H_3O^+ is approaching 3 [29]. The proton transport mechanism changes from the prevailing of vehicle mechanism to the coexisting of the two mechanisms step by step (Figure 11.12d). When λ reaches 11.0 or higher, the percentage of free water is higher than 40%, and the most likely SO_3^- coordination number of H_3O^+ is zero; at the same time, the hydration number of H_3O^+ starts to saturate at about $\lambda = 11.0$ [29]. The structural diffusion dominates proton transport, and the proton conductivity exhibits a significant increase because it is a faster means of proton transport. Finally, the proton conductivity achieves a value similar to that of bulk water when Nafion is fully hydrated (Figure 11.12e). And at this stage, the vehicular diffusion accounts for only about one fifth of the total proton transport [57].

In Figure 11.12e, the water in Nafion is still simplified as three water regions: the tightly bound water region, loosely bound water region, and free water region. In terms of the activation of proton hops [59] and the permittivity (dielectric constant) of the water [60] in a Nafion membrane pore, the thickness of the tightly bound water region is around 4.3 Å, which is nearly the first solvation shell radius of a SO_3^- group, and the thickness of the loosely bound water region is approximately 2.5 Å. Proton transport by vehicle mechanism primarily takes place in the free water region, while the one by the Grotthus mechanism mainly occurs in the loosely bound water region. Secundum, the structure and transport mechanism of three water regions, it is reasonable to conclude that the methanol crossover mostly passes through the same channels as the vehicle mechanism occurs. Based on this conclusion, a PEM with high methanol resistance may be developed.

11.5 Conclusions

The main issue addressed in this chapter is the nanostructures in the Nafion membrane. It can be found that the understanding of Nafion membranes has been substantially advanced because of systematic experiments in conjunction with computational simulations. And a series of qualitative morphology models of Nafion has also been proposed, but a widely accepted model that can quantitatively predict new phenomena is still the subject of debates due to the complexity and multihierarchy levels of time and length scales in Nafion (from microscopic mechanisms of proton transfer to conductance in the single channel, from the conductivity network of Nafion membrane to macroscopical engineering properties). To take the most successful parallel water–channel (inverted-micelle cylinder) model as example, it is questionable whether its prediction about proton conductance is isotropic.

Hence, many more efforts still have to be made before an ideal model of Nafion is established. Once such a model is developed, a much deeper understanding to the relationships between microstructure and properties of Nafion will be yielded. In turn, this can also facilitate the searching of new materials with rapid proton conductivity, optimum fuel permeability, and excellent durability.

References

1. Kreuer, K.D., Schuster, M., Obliers, B., Diat, O., Traub, U., Fuchs, A., Klock, U., Paddison, S.J. and Maier, J. 2008. Short-side-chain proton conducting perfluorosulfonic acid ionomers: Why they perform better in PEM fuel cells. *Journal of Power Sources* 178, 499–509.

2. Kreuer, K.D. 2001. On the development of proton conducting polymer membranes for hydrogen and methanol fuel cells. *Journal of Membrane Science* 185, 29–39.

3. Rubatat, L., Rollet, A.L., Diat, O. and Gebel, G. 2002. Characterization of ionomer membrane structure (NAFION) by small-angle X-ray scattering. *Journal de Physique* IV 12, 197–205.

4. Diat, O., Rubatat, L., Rollet, A.L. and Gebel, G. 2002. Evidence of elongated polymeric aggregates in Nafion. *Macromolecules* 35, 4050–4055.

5. Manthiram, A. and Yang, B. 2006. Comparison of the small angle X-ray scattering study of sulfonated poly (etheretherketone) and Nafion membranes for direct methanol fuel cells. *Journal of Power Sources* 153, 29–35.

6. Paciaroni, A., Casciola, M., Cornicchi, E., Marconi, M., Onori, G., Donnadio, A., Sganappa, M. and De Francesco, A. 2007. Low-frequency dynamics of water absorbed in Nafion membranes as a function of temperature. *Philosophical Magazine* 87, 477–483.

7. Lyonnard, S., Perrin, J.C. and Volino, F. 2007. Quasielastic neutron scattering study of water dynamics in hydrated Nafion membranes. *Journal of Physical Chemistry C* 111, 3393–3404.

8. Thompson, E.L., Capehart, T.W., Fuller, T.J. and Jorne, J. 2006. Investigation of low-temperature proton transport in Nafion using direct current conductivity and differential scanning calorimetry. *Journal of the Electrochemical Society* 153, A2351–A2362.

9. Korzeniewski, C., Basnayake, R., Peterson, G.R. and Casadonte, D.J. 2006. Hydration and interfacial water in Nafion membrane probed by transmission infrared spectroscopy. *Journal of Physical Chemistry B* 110, 23938–23943.

10. Moilanen, D.E., Piletic, I.R. and Fayer, M.D. 2007. Water dynamics in Nafion fuel cell membranes: The effects of confinement and structural changes on the hydrogen bond network. *Journal of Physical Chemistry C* 111, 8884–8891.

11. Goward, G.R., Ye, G. and Hayden, C.A. 2007. Proton dynamics of Nafion and Nafion/SiO$_2$ composites by solid state NMR and pulse field gradient NMR. *Macromolecules* 40, 1529–1537.

12. Goward, G.R., Ye, G. and Janzen, N. 2006. Solid-state NMR study of two classic proton conducting polymers: Nafion and sulfonated poly(ether ether ketone)s. *Macromolecules* 39, 3283–3290.

13. Lu, Z.J., Polizos, G., Macdonald, D.D. and Manias, E. 2008. State of water in perfluorosulfonic ionomer (Nafion 117) proton exchange membranes. *Journal of the Electrochemical Society* 155, B163–B171.

14. Bauer, F., Denneler, S. and Willert-Porada, M. 2005. Influence of temperature and humidity on the mechanical properties of Nafion (R) 117 polymer electrolyte membrane. *Journal of Polymer Science Part B: Polymer Physics* 43, 786–795.

15. Schlick, S. and Bosnjakovic, A. 2004. Naflon perfluorinated membranes treated in Fenton media: Radical species detected by ESR spectroscopy. *Journal of Physical Chemistry B* 108, 4332–4337.

16. McGrath, J.E., Kim, Y.S., Wang, F., Hickner, M. and Zawodzinski, T.A. 2003. Fabrication and characterization of heteropolyacid (H$_3$PW$_{12}$O$_{40}$)/directly polymerized sulfonated poly(arylene ether sulfone) copolymer composite membranes for higher temperature fuel cell applications. *Journal of Membrane Science* 212, 263–282.

17. McGrath, J.E., Kim, Y.S., Wang, F., Hickner, M., McCartney, S., Hong, Y.T., Harrison, W. and Zawodzinski, T.A. 2003. Effect of acidification treatment and morphological stability of sulfonated poly(arylene ether sulfone) copolymer proton-exchange membranes for fuel-cell use above 100°C. *Journal of Polymer Science Part B: Polymer Physics* 41, 2816–2828.

18. Lu, G.Q., Ladewig, B.P., Knott, R.B., Hill, A.J., Riches, J.D., White, J.W., Martin, D.J. and da Costa, J.C.D. 2007. Physical and electrochemical characterization of nanocomposite membranes of Nafion and functionalized silicon oxide. *Chemistry of Materials* 19, 2372–2381.

19. Neimark, A.V. and Vishnyakov, A. 2000. Molecular simulation study of Nafion membrane solvation in water and methanol. *Journal of Physical Chemistry B* 104, 4471–4478.

20. Neimark, A.V. and Vishnyakov, A. 2001. Molecular dynamics simulation of microstructure and molecular mobilities in swollen Nafion membranes. *Journal of Physical Chemistry B* 105, 9586–9594.

21. Metiu, H., Blake, N.P., Petersen, M.K. and Voth, G.A. 2005. Structure of hydrated Na-Nafion polymer membranes. *Journal of Physical Chemistry B* 109, 24244–24253.

22. Spohr, E., Commer, P. and Kornyshev, A.A. 2002. Enhancing proton mobility in polymer electrolyte membranes: Lessons from molecular dynamics simulations. *Journal of Physical Chemistry B* 106, 10560–10569.

23. Paddison, S.J., Eikerling, M., Pratt, L.R. and Zawodzinski, T.A. 2003. Defect structure for proton transport in a triflic acid monohydrate solid. *Chemical Physics Letters* 368, 108–114.

24. Paddison, S.J., Eikerling, M. and Zawodzinski, T.A. 2002. Molecular orbital calculations of proton dissociation and hydration of various acidic moieties for fuel cell polymers. *Journal of New Materials for Electrochemical Systems* 5, 15–23.

25. Paddison, S.J., Paul, R. and Kreuer, K.D. 2002. Theoretically computed proton diffusion coefficients in hydrated PEEKK membranes. *Physical Chemistry Chemical Physics* 4, 1151–1157.

26. Paul, R. and Paddison, S.J. 2005. Effects of dielectric saturation and ionic screening on the proton self-diffusion coefficients in perfluorosulfonic acid membranes. *The Journal of Chemical Physics* 123, 224704–224717.

27. Seeliger, D., Hartnig, C. and Spohr, E. 2005. Aqueous pore structure and proton dynamics in solvated Nafion membranes. *Electrochimica Acta* 50, 4234–4240.

28. Voth, G.A., Petersen, M.K., Wang, F., Blake, N.P. and Metiu, H. 2005. Excess proton solvation and delocalization in a hydrophilic pocket of the proton conducting polymer membrane narion. *Journal of Physical Chemistry B* 109, 3727–3730.

29. Devanathan, R., Venkatnathan, A. and Dupuis, M. 2007. Atomistic simulation of Nafion membrane: I. Effect of hydration on membrane nanostructure. *Journal of Physical Chemistry B* 111, 8069–8079.

30. Devanathan, R., Venkatnathan, A. and Dupuis, M. 2007. Atomistic simulation of Nafion membrane. 2. Dynamics of water molecules and hydronium ions. *Journal of Physical Chemistry B* 111, 13006–13013.

31. Venkatnathan, A., Devanathan, R. and Dupuis, M. 2007. Atomistic simulations of hydrated Nafion and temperature effects on hydronium ion mobility. *Journal of Physical Chemistry B* 111, 7234–7244.

32. Devanathan, R., Venkatnathan, A., Rousseau, R., Dupuis, M., Frigato, T., Gu, W. and Helms, V. 2010. Atomistic simulation of water percolation and proton hopping in nation fuel cell membrane. *Journal of Physical Chemistry B* 114, 13681–13690.

33. Yamamoto, S. and Hyodo, S.A. 2003. A computer simulation study of the mesoscopic structure of the polyelectrolyte membrane Nafion. *Polymer Journal* 35, 519–527.

34. Groot, R.D. and Rabone, K.L. 2001. Mesoscopic simulation of cell membrane damage, morphology change and rupture by nonionic surfactants. *Biophysical Journal* 81, 725–736.

35. Grafmuller, A., Shillcock, J. and Lipowsky, R. 2007. Pathway of membrane fusion with two tension-dependent energy barriers. *Physical Review Letters* 98, 218101–218104.

36. Wu, D.S., Paddison, S.J. and Elliott, J.A. 2008. A comparative study of the hydrated morphologies of perfluorosulfonic acid fuel cell membranes with mesoscopic simulations. *Energy & Environmental Science* 1, 284–293.

37. Khalatur, P.G., Talitskikh, S.K. and Khokhlov, A.R. 2002. Structural organization of water-containing Nafion: The integral equation theory. *Macromolecular Theory and Simulations* 11, 566–586.

38. Qi, Y., Wescott, J.T., Subramanian, L. and Capehart, T.W. 2006. Mesoscale simulation of morphology in hydrated perfluorosulfonic acid membranes. *Journal of Chemical Physics* 124, 134702–134715.

39. Wu, D.S., Paddison, S.J. and Elliott, J.A. 2009. Effect of molecular weight on hydrated morphologies of the short-side-chain perfluorosulfonic acid membrane. *Macromolecules* 42, 3358–3367.

40. Mauritz, K.A. and Moore, R.B. 2004. State of understanding of Nafion. *Chemical Reviews* 104, 4535–4585.

41. Diat, O., Rubatat, L. and Gebel, G. 2004. Fibrillar structure of Nafion: Matching Fourier and real space studies of corresponding films and solutions. *Macromolecules* 37, 7772–7783.

42. Mistry, M.K., Choudhury, N.R., Dutta, N.K., Knott, R., Shi, Z.Q. and Holdcroft, S. 2008. Novel organic-inorganic hybrids with increased water retention for elevated temperature proton exchange membrane application. *Chemistry of Materials* 20, 6857–6870.

43. Sekhon, S.S., Park, J.S., Cho, E., Yoon, Y.G., Kim, C.S. and Lee, W.Y. 2009. Morphology studies of high temperature proton conducting membranes containing hydrophilic/hydrophobic ionic liquids. *Macromolecules* 42, 2054–2062.

44. Diat, O., van der Heijden, P.C. and Rubatat, L. 2004. Orientation of drawn Nafion at molecular and mesoscopic scales. *Macromolecules* 37, 5327–5336.

45. Hsu, W.Y. and Gierke, T.D. 1983. Ion transport and clustering in Nafion perfluorinated membranes. *Journal of Membrane Science* 13, 307–326.

46. Gierke, T.D., Munn, G.E. and Wilson, F.C. 1981. The morphology in Nafion perfluorinated membrane products, as determined by wide- and small-angle X-ray studies. Journal of Polymer Science: *Polymer Physics Edition* 19, 1687–1704.

47. Gebel, G. 2000. Structural evolution of water swollen perfluorosulfonated ionomers from dry membrane to solution. *Polymer* 41, 5829–5838.

48. Dreyfus, B., Gebel, G., Aldebert, P., Pineri, M., Escoubes, M. and Thomas, M. 1990. Distribution of the "micelles" in hydrated perfluorinated ionomer membranes from SANS experiments. *Journal of Physics (Paris)* 51, 1341–1354.

49. Fujimura, M., Hashimoto, R. and Kawai, H. 1982. Small-angle X-ray scattering study of perfluorinated ionomer membranes. 2. Models for ionic scattering maximum. *Macromolecules* 15, 136–144.
50. Litt, M.H. 1997. A reevaluation of Nafion morphology. *Polymeric Preprints* 38, 80–81.
51. Gebel, G. and Lambard, J. 1997. Small-angle scattering study of water-swollen perfluorinated ionomer membranes. *Macromolecules* 30, 7914–7920.
52. Haubold, H.G., Vad, T., Jungbluth, H. and Hiller, P. 2001. Nano structure of Nafion: a SAXS study. *Electrochimica Acta* 46, 1559–1563.
53. Yeager, H.L. and Steck, A. 1981. Cation and water diffusion in Nafion ion exchange membrane-influence of polymer structure. *Journal of the Electrochemical Society* 128, 1880–1884.
54. Schmidt-Rohr, K. and Chen, Q. 2008. Parallel cylindrical water nanochannels in Nafion fuel-cell membranes. *Nature Materials* 7, 75–83.
55. Costamagna, P., Grosso, S. and Di Felice, R. 2008. Percolative model of proton conductivity of Nafion (R) membranes. *Journal of Power Sources* 178, 537–546.
56. Tsuda, M., Arboleda, N.B. and Kasai, H. 2006. Initial driving force for proton transfer in Nafion. *Chemical Physics* 324, 393–397.
57. Hristov, I.H., Paddison, S.J. and Paul, R. 2008. Molecular modeling of proton transport in the short-side-chain perfluorosulfonic acid ionomer. *Journal of Physical Chemistry B* 112, 2937–2949.
58. Cui, S.T., Liu, J.W., Selvan, M.E., Keffer, D.J., Edwards, B.J. and Steele, W.V. 2007. A molecular dynamics study of a nafion polyelectrolyte membrane and the aqueous phase structure for proton transport. *Journal of Physical Chemistry B* 111, 2208–2218.
59. Eikerling, M. and Kornyshev, A.A. 2001. Proton transfer in a single pore of a polymer electrolyte membrane. *Journal of Electroanalytical Chemistry* 502, 1–14.
60. Paul, R. and Paddison, S.J. 2004. The phenomena of dielectric saturation in the water domains of polymer electrolyte membranes. *Solid State Ionics* 168, 245–248.

12

Advances in Proton Exchange Membranes for Direct Alcohol Fuel Cells

Haekyoung Kim and Doo-Hwan Jung

CONTENTS

12.1 Introduction 335
12.2 Roles and Requirements of PEM 337
12.3 Technical Approaches 342
    12.3.1 Perfluorinated and Partially Fluorinated Membranes 342
    12.3.2 Hydrocarbon Membranes...348
    12.3.3 Composite and Other Modifications of Ionomers....................358
12.4 Conclusions...368
References...369

12.1 Introduction

Portable devices for telecommunications with technical advances and user-friendly functions call upon an energy source greater than the limit of conventional technology. Because mobile devices perform more functions of conversation and exchanging information via the Internet and have digital mobile broadcasting (DMB) systems, cameras, etc., the requirements of more power and energy are getting increased. Mobile devices require nearly 3800 mAh per day, which needs three 5-h charges with a conventional power source for a 1-day use. Direct methanol fuel cells (DMFCs) have been becoming attractive as an alternative to Li-ion batteries in portable electronics for higher power and energy. Also, there is a demand for lighter and longer-operation power sources for mobile informative devices. Of all the system characteristics of DMFCs, energy efficiency, energy density, power density, and cost have to be competitive with rechargeable batteries. The requirements for the portable DMFC suggested by DOE are the following [1]:

- Power density, 100 W·l^{-1}
- Energy density, 1000 Wh·l^{-1}
- Lifetime, 5000 h
- Cost, $3 W^{-1}

These requirements are somewhat different from other types of fuel cells for residential or vehicle application, which can utilize more freely auxiliary components for managing fuel, water, heat, and electricity, such as a compressor, pump, humidifier, heat exchanger, and so on. The components have to be miniaturized effectively or eliminated for portable power sources. Many technical advances have been achieved, and mobile devices with DMFCs can be expected to be on the market. A fuel cell system is a complicated electrochemical device, which should come up to the commercial level with the technological integration of catalyst, membrane, stack, system energy balance, fluid dynamics, and hybrid circuit design [2]. Among the various technologies, the technology gaps for DMFCs between the current status and the status for commercialization are impossible to bridge without significant improvements in the economic and technical efficiency of polymer electrolyte membranes. In recent years, significant progress has been made in the development of polymer electrolyte membranes for DMFCs in terms of cost reduction and improvement of performance together with other associated technological advancements. Cheap and durable membranes, such as hydrocarbon membranes, have been produced by PolyFuel, Inc. (5000 h lifetime in a passive DMFC). High performance nonplatinum or low-platinum anode catalysts with a loading of <0.2 mg·cm^{-2} have been developed, and high-performance nonplatinum cathode catalysts with low precious metal loading (0.2–0.5 mg·cm^{-2}), such as palladium alloys, have been developed. The noncarbon cathode supports that are more resistant to oxidation, such as porous titanium, are used for DMFCs.

With regard to the DMFC membrane, the technical goal has been well defined, that is, maintaining the ionic conductivity at the current technical level and reducing the methanol crossover as much as possible, preferably down to zero. The requirements for DMFC membranes are high ionic conductivity, electrical insulation, gas and liquid (especially methanol) tightness, and chemical and mechanical stability. The performance curve of a DMFC is shown in Figure 12.1, which shows that ohmic polarization is mainly due to the ionic resistance of membranes. To increase the DMFC performance, Figure 12.1 shows that a proton exchange membrane (PEM) exhibits the high ionic conductivity for reducing ohmic resistance. The open circuit potential in a DMFC, as shown in Figure 12.1, is lower than 0.8 V due to the voltage drop by a mixed potential from methanol crossover through the membrane. The reduction of methanol crossover is also achieved to increase DMFC performances. The low cost of material and processing should be considered in terms of commercialization. There are several reasons why the perfluorinated sulfonic acid ionomer (PFSI) membrane is still the best even though it permeates significant amounts of methanol during operation. First, its chemical and mechanical stability with high ionic conductivity is far better than the other types of membranes that are still struggling to reach 0.1 S cm^{-1} at ambient temperatures with rigidity. Second, the processes for membrane electrode assembly (MEA) have been designed for utilizing PFSI-based membranes, whose critical functions as a MEA are simple and

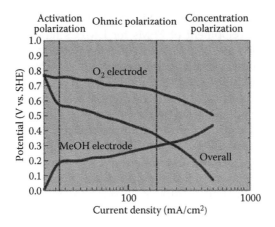

FIGURE 12.1
Performance curve of DMFC.

the most important. However, alternative PEMs are under development to exhibit the requirements, and researchers are focused on three types: perfluorinated and partially fluorinated membranes, hydrocarbon membranes, and composite membranes. Based on the alternative PEM, ionomer and inorganic materials for PEM are also developed. Especially for a passive type MEA in small powered mobile applications, a PEM needs to be designed for managing the mass (methanol and water) balance through PEM.

12.2 Roles and Requirements of PEM

Proton exchange membranes (PEMs) in DMFCs should transport protons as an electrolyte and prevent mixing of the fuel and oxidant as a separator. The proton transport capacity affects the resistance and performance of fuel cells. The ability as a separator influences the long-term stability and fuel efficiency. The insufficiency of function in separation, which is called methanol crossover, leads to the generation of a mixed potential and, thus, deteriorated cathode catalysts, which induce the decrease in the performance and efficiency of DMFCs. The proton transports in PEM have been explained by the vehicular and Grotthus mechanisms [3–5]. The proton transports take place with water molecules, and the water, which the membrane contains, accelerates the proton transport. Ionic conductivity is a function of mobility (stiffness) and capability of absorbing ions. Highly concentrated ionic sites and lower activation energy for proton transfer in PEMs give high proton conductivity.

The design of ionic site concentration and stiffness of the polymer matrix are issues for PEM development. The strategies for reducing methanol crossover must be considered in designing PEMs for DMFCs. Based on the conducting mechanisms, the development of PEMs for DMFCs has the contradiction in ionic conductivity and methanol crossover. For high ionic conductivity, the PEM should be swollen enough with water. However, for low methanol crossover, it should not be swollen with water. Methanol crossover occurs by the absorption of fuel (methanol) by membranes, diffusion of methanol through the membranes, and desorption of methanol from the membranes. To diminish methanol crossover, three stages of methanol crossover are considered. The diffusion through the membrane relates to electro-osmotic water and methanol drag with proton transport [6–9]. The electro-osmotic drag should be reduced by new electrolyte design [10]. A rigid polymer matrix and small free volume of PEM can reduce the electro-osmotic methanol diffusion through the membrane. Besides ionic conductivity and methanol crossover, the criteria for PEMs are composed of low electronic conductivity, good mechanical properties in dry and hydrated status, capability for improvement with MEA fabrications, chemical and hydrolysis stability, and low cost. However, PEMs, which satisfy both high ionic conductivity and low methanol crossover for DMFCs, are hard to develop easily.

With the current technology in PEM, the priority of criteria can be chosen, followed by the operation methods of DMFC systems. The ionic conductivity and the methanol crossover of PEM are dependent on the operation temperature and fuel supply methods [11,12]. The ionic conductivity and the methanol crossover increase with the temperature. The concentration and the rate of supplied fuel/oxidant affect the methanol crossover [13]. As mentioned, DMFC systems can be applied to automobile, stationary, and mobile energy sources [14]. Depending on the applications, the differences in the operational temperature and the fuel supply method lead to the strategic development of materials and designs of components. Table 12.1 shows the DMFC systems, which research groups have reported. DMFC systems can be categorized by the method of supplying the fuel and oxidant into active, semipassive, and passive systems. The active system is considered to be applied to several 10 W DMFC systems and which has active components, such as a pump for methanol fuel, a compressor for the oxidant, a methanol sensor, etc. A highly concentrated methanol solution can be located in active DMFC systems, and an adjusting dilute methanol solution by methanol sensor is supplied into the DMFC stack. The highly concentrated methanol solution is located in a system for achieving high energy density, and then the diluted methanol fuel (3–10 wt%) with water produced at the cathode is supplied into the stack. In the active system, it is easy to control the supply of methanol fuel and air, which exhibits the higher power density. However, power for active components is needed, and the noise from the active components is an issue. A bipolar-type stack design is preferred in an active DMFC system, which is considered for Note PCs with several 10 W powers. However, it is hard to apply to phone, which requires smaller power and lower noise. For smart phones with smaller power and near zero noise, a passive type of DMFC system can be proposed. While the passive DMFC does not include the active

TABLE 12.1

Examples of DMFC for Portable Devices

Motorola	MTI micro-fuel cell	Toshiba	Hitachi

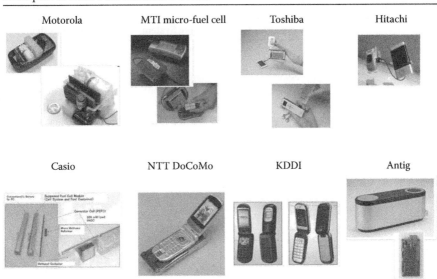

Casio	NTT DoCoMo	KDDI	Antig

components, the methanol fuel is supplied by capillary force, and the oxidant is supplied by natural convection. The produced water in the passive DMFC is dried out in ambient air. The passive DMFC is simple and small, and the power is lower than that of the active DMFC. In the passive DMFC, the materials and components should play roles as the active components.

A semipassive system takes advantage of the active system and the passive system, and its design is more flexible than other systems. The systems are summarized in Table 12.2. The passive DMFC system needs the high-performance MEA, catalysts, and membranes. Table 12.3 shows the management of the methanol solution and water for each system. DMFC systems consist of a stack, fuel tank, BOP (balance of power), and electric circuit. The efficiency of the DMFC system, the energy density, is affected by the volume of the fuel tank, BOP, and electric circuit. With the view of system efficiency, it is induced that there are contradictions of water and heat in DMFC systems. The contradictions are from the fact that the water is a product in the cathode side and a reactant in the anode side at the same time. The water in a DMFC is mass that should be supplied and removed. Furthermore, the water controls the concentration of the methanol solution. Due to the methanol crossover and lower catalytic activity, a lower concentrated methanol solution than 2 M is preferred, and the performances with a 0.5 or 1 M methanol solution show the highest value. The highly concentrated methanol should be located for achieving the higher energy density, and the lower concentrated methanol fuel should be supplied to the MEA. PEMs should contain much water for higher ionic conductivity, while, for lower methanol crossover, PEMs

TABLE 12.2

Advantages and Disadvantages for Direct Methanol Fuel Cell with Fuel Supply Method

	Active System	Semipassive System	Passive System
Anode (methanol, water)	Pump	Pump	Capillary
Cathode (air)	Compressor	Blower (fan), breathing (natural convection)	Breathing (natural convection)
Stack	Bipolar	Bipolar, monopolar	Monopolar
Strong	*High power	*Low noise	*No noise
	*Easy to control system	*Low BOP electric consume	*No BOP electric
	*Water recycling	*Design flexibility	*Flat design (design flexibility)
Weak	*Noise	*Low power	*Lowest power
	*BOP volume	*Stack volume	*Stack volume
	*BOP electric consume	*Difficulty of water recycling	*Environment
			*System control
			*Difficulty of water recycling
			*Slow start-up

TABLE 12.3

Requirements in Materials for DMFC with Fuel Supply Method

		Active System	Semipassive System	Passive System
	Anode	Pump	Pump	Capillary
	Cathode	Compressor	Blower (fan), breathing (natural convection)	Breathing natural convection
	Stack	Bipolar	Bipolar, monopolar	Monopolar
Membrane	Ionic conductivity/conductance		Maximize	
	Methanol crossover		Minimize	
	Water swelling	Minimize	Minimize	Maximize
Electrode	Cathode diffusion	Remove water hydrophobic	Remove water hydrophobic	Remove water/air supply hydrophilic and hydrophobic
Catalyst	Activity/loading		Maximize/minimize	

should contain less water. Heat is helpful for catalytic activity while it promotes the methanol crossover. The MEA in a DMFC needs to be humidified to lower the resistance while the humidity of the MEA hinders the air supply of the cathode. These contradictions can be defined as the water and heat, which are the technical issues to be solved for DMFC commercialization. For solving the issues, a breakthrough in materials and components is needed.

The higher flow rate of the fuel and oxidant promotes the methanol crossover due to the higher concentration gradient. In particular, the passive condition of DMFC application, in which the fuel and the oxidant are supplied without active components, such as a pump or a compressor, tends to make PEM developments complicated. The supplied method and amount of water will decide the fuel cell efficiency and the system volume. The development of the PEM will be designed considering the supply methods of the fuel, oxidant, and water. There have been many approaches for solving problems that are induced by methanol cross-over. Methanol-tolerant catalysts and design of electrodes are examples of these approaches [15–18]. The combined approaches to the membrane, electrode, and diffusion layer achieve the best fuel cell performance. For higher fuel cell effi-ciency, DMFC systems should have a small volume and high performance. The approaches for water management are another way to resolve the issue among methanol concentration, power density, and fuel efficiency in the DMFC system. The water produced at the cathode can be used to dilute the anode fuel concen-tration to achieve the high energy density [19] and the long-term stability [20–22]. The reuse method of water from the cathode is one of the important factors for higher DMFC efficiency. These relationships are described in Figure 12.2.

PEMs for DMFCs are mainly characterized by ionic conductivity and methanol crossover. Ionic conductivity is tested with the four-point probe method. Through-plane and in-plane ionic conductivity is characterized with temperature and humidity [23]. Methanol crossover can be measured with the diffusion cell method, pervaporation of methanol solution, dictating

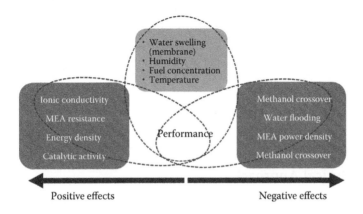

FIGURE 12.2
Contradictions in DMFC.

the CO_2 amount from the cathode out stream [24]. An oxidant-impermeable property is also one of the important factors in PEMs for DMFCs. The properties of PEMs in DMFCs are also characterized with an interfacial resistance, an electro-osmotic drag, methanol crossover through the MEA, and so on. In this chapter, developments of PEMs for DMFCs are discussed.

Although varying for different applications, common requirements for a PEM in DMFC applications include the following:

- Low methanol crossover ($<10^{-6}$ mol·min^{-1} cm^{-1}) or low methanol diffusion coefficient in the membrane ($<5.6 \times 10^{-6}$ cm^2·s^{-1} at $T = 25°C$) [25]
- High ionic conductivity (>0.008 S·cm^{-1}) [25]
- High chemical and mechanical durability, especially at $T > 80°C$ (for increased CO tolerance)
- Low cost ($<\$10$ k·W^{-1} based on a PEMFC) [26]

An analysis of the DMFC market reveals three main types of membranes, which are or could be used in commercial DMFCs. Approaches for PEMs in DMFCs are classified as fluorinated and partially fluorinated polymers, hydrocarbon polymers, and modification of polymer materials, such as composite membranes. The fluorinated commercial membranes and partially fluorinated membranes, including Nafion [27], Dow Chemical (USA) [28], Asahi Glass Engineering (Japan) (Flemion R, IEC 1.0 mequiv·g^{-1} dry resin, 50 μm dried film thickness) [29], Asahi Kasei (Japan) (Aciplex-S, based on a weak functional acid [–COOH] instead of the SO_3H groups in Nafion) [30], W.L. Gore & Associates (USA) (Gore-Tex, Gore-Select) [31,32], and 3P Energy (Germany) [33], were not specifically developed for DMFC applications. Nevertheless, Nafion is now one of the main membranes for commercial DMFCs despite many inherent disadvantages for this application. The hydrocarbon membrane groups contain membranes that are in various stages of development. Some examples are still in the laboratory stage, while a few others have been, to a degree, commercialized. Improvements to the structure and functionality of membranes are usually made by adding inorganic–organic and/or acidic–basic constituents to produce composite membranes. The water uptake of the membranes typically determines the proton conductivity and methanol permeation based on the ionomer microstructure, cluster, and channel size [34].

12.3 Technical Approaches

12.3.1 Perfluorinated and Partially Fluorinated Membranes

By virtue of the exceptional chemical inertness of the fluoroalkyl group, fluorine-containing ionomers can survive during the harsh fuel cell operating conditions and therefore are mostly preferred in DMFCs as well as PEMFCs. On the other hand, the environmental and safety issues in the manufacturing process remain serious problems. The fluorine-containing polymer membranes can be classified into two major categories based on the chemical structure of their backbone. Conceptually, the perfluorinated membranes consist of a tetrafluoroethylene (PTFE)-like backbone, which is copolymerized with perfluorinated vinyl ethers with an acid terminal group. Other types of fluorine-containing ionomers, which include hydrocarbon moiety in their structure, fall into the category of partially fluorinated membranes.

As often represented by Nafion, which was commercialized by Dupont in the late 1960s, the perfluorinated membrane has been the most commonly used ionomer material due to its exceptional stability. It has been reported that the membrane is durable up to 60,000 h [35]. Its first application was not for a fuel cell system but as a permselective membrane in chlor-alkali cells. Its usefulness as a fuel cell membrane was verified in 1966 for NASA's space mission project [36]. In the perfluorinated ionomers, a PTFE-based polymeric backbone offers the chemical stability from the radical or acid–base species, which causes the hydrolytic degradation of the polymer chain. The ionic conductivity is provided by a pendant acidic moiety in a carboxylate or sulfonate form. There are some reports on perfluorinated carboxylic acid (PFCA) materials, most of which are derived from Nafion [37–40]. However, a PFCA is not suitable for a fuel cell application due to its lower proton conductivity. The perfluorosulfonic acid (PFSA) is the most favored choice among not only perfluorinated membranes but also all other ionomers in fuel cell applications. The sulfonic acid form of Nafion is a representative PFSA and thus has been intensively studied since the 1960s. The reported chemical structure of the Nafion membrane is given in Figure 12.3. Such a nonionic form of Nafion is further treated with a base followed by an acid to give a highly ion-conducting sulfonic acid ($-SO_3H$) form (>0.1 S cm^{-1} at an ambient temperature under humid conditions). By changing the comonomer ratio with ionic moiety, the ion exchange capacity (IEC = H$^+$ mequiv·g^{-1} of polymer) of the membrane can be varied over a wide range. However, too much ionic content will cause the excessive swelling of the membrane, which deteriorates its mechanical strength. Generally, Nafion with an equivalent weight of 1100 is popularly used in the most applications, although the equivalent weights

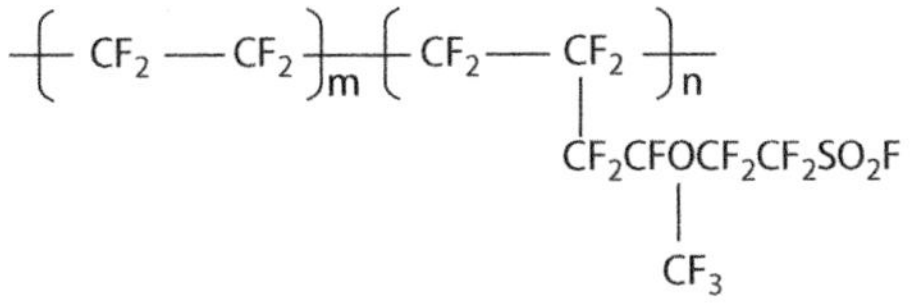

FIGURE 12.3
Chemical structure of Nafion.

(EWs, grams of dry polymer per mole of ion exchange sites) of 900 and 1200 are commercially available. The neat PFSA ionomer has the undesirable great tendency to swell when it is wet, which results in poor mechanical properties. The swelling problem is particularly serious in a fuel cell operating at elevated temperatures. In 1995, W.L. Gore & Associates developed the Gore-Select membrane, which has a micro-reinforced composite structure of PTFE fabrics and PFSA ionomers. A similar reinforced PFSA was also reported by Asahi Glass, and these membranes are as thin as 5 μm with good mechanical strength.

Nafion has also stimulated the mass production of other PFSAs, such as Flemion (Asahi Glass) [29], Aciplex-S (Asahi Kasei Corp., formerly Asahi Chemicals) [30], Dow membranes (Dow Chemical), and Gore & Associates [31,32]. They are similar in structure to Nafion because the PTFE backbone and PFSA side chains are the main constituents for all of them [41]. The differences come from the details of the side chains as shown by comonomer structures as shown in Table 12.4. Those membranes have been recommended by their manufacturers for PEMFC, but they may also be modified for use in DMFC applications. The Dow membrane is prepared by the copolymerization of tetrafluoroethylene with a vinylene monomer. With its short side chain, the Dow membrane functions very differently from Nafion. The specific conductance of the 800 and 850 EW membranes is 0.2 and 0.12 S·cm^{-1}, respectively [42]. In terms of ionic current, tests showed approximate values of methanol crossover to be 2.7×10^{-10} A·cm^{-2} for Nafion membranes [43]. Performance testing of a Dow membrane in a six-cell Ballard Power Systems PEMFC MK4 stack in 1988 showed better performance than with Nafion (at E [potential] = 0.5 V, j [current density] = 5000 and 1400 A·ft^{-2}, respectively) [42]. The German firm 3P Energy also developed a perfluorinated sulfonic acid (PFSA) membrane. The 3P membranes have 20 times lower methanol crossover than the commercially available Nafion membrane. A DMFC fabricated with the 3P membrane could operate with a higher methanol concentration, resulting in a higher power density [30].

Because Nafion has received the greatest interest as a proton-conducting membrane for fuel cells, researchers are trying to understand the water transport phenomena within the membrane from the microstructural point of view. There has been astonishing progress in the theoretical modeling of transport mechanisms related to Nafion's nanoscale morphology. In the

TABLE 12.4

Typical Functional Monomers of Commercial PFSA

Membrane	Comonomer Structure
Nafion, Flemion	$CF_2 = CFOCF_2CF(CF_3)OCF_2CF_2SO_2F$
Aciplex	$CF_2 = CFOCF_2CF(CF_3)OCF_2CF_2CF_2SO_2F$
Dow	$CF_2 = CFOCF_2CF_2SO_2F$

early 1980s, Gierke and Hsu [44] proposed a cluster network model that was thought to be the basis of other theories. They described that the polymeric membrane has an inverted micellar structure in which the sulfonyl ionic sites are spatially separated from the fluorocarbon backbone to form spherical clusters that are interconnected. According to this model, a dry Nafion membrane consists of clusters having a diameter of 1.8 nm containing 26 sulfonyl groups distributed on the inner pore surface. Upon swelling with water, the pore size becomes 4 nm, and each pore contains 70 sulfonyl groups surrounding 1000 water molecules in the pore. Gierke and Hsu also adopted a percolation theory to explain a nonlinear relationship between the ionic conductivity and the swelling ratio of Nafion membrane. Eikerling et al. [43] developed a random network model that assumes that the hydrated regions are randomly distributed in the polymer matrix to facilitate quicker transport of protons upon the movement of pendant sulfonyl groups. Basically, their work was a modified version of the cluster network model. Some reports verified random network models by experimental techniques, such as a small angle x-ray scattering (SAXS) or an atomic force microscopy (AFM) using Nafion membranes [45,46].

Meanwhile, transport theories from the microstructural point of view are being developed to understand the general transport mechanism of water molecules and protons in the fuel cells. There have been efforts to explain methanol transport in DMFC by semi-empirical methods [24,47–54]. Verbrugge [55] developed a simple diffusion model for methanol through an ionomer membrane, assuming the dilute solution. He validated his model by experimentally showing that the methanol diffusion through the membrane occurs nearly as readily as through water. Cruickshank and Scott [56] presented a simple model to describe the methanol crossover in a vapor feed DMFC and its effect on the cathodic overpotential. The measured permeation rates of water and methanol through a Nafion 117 membrane under varied pressure differentials across the PEM were used to determine the essential parameters in the model. This model has also been extended to include a one-dimensional model of the potential distribution and concentration distribution of methanol in the anode catalyst layer for a vapor feed DMFC [56]. Ren and Gottesfeld [57] used a DMFC to experimentally determine the electro-osmotic drag of water in the membrane. Although Nafion-based PFSA membranes still dominate most PEM in the fuel cell systems, there are critical drawbacks, such as a high cost (US\$700 m^{-2}), a limited temperature range, and a high methanol crossover. Particularly, the methanol crossover issue hampers its practical application as a DMFC membrane for long-term operation. The methanol permeability of the Nafion membrane is about 2×10^{-6} cm^2·s^{-1} at room temperature, which gets more serious when the operation temperature increases. The methanol crossover through the membrane irreversibly contaminates the Pt catalyst at the cathode, consequently aggravating cell performance. The methanol crossover limits the fuel concentration supplied to an anode, which is as low as 1 M in DMFC experiments. Ren et al. [58] have reported that the methanol crossover rate can be significantly diminished by designing the anode to adjust the concentration of methanol between the flow field and the membrane surface. They performed a 3000 h lifetime test on a DMFC single cell at 75°C with 0.3 M methanol fuel, in which a 40% loss of initial current was shown. There are efforts to modify Nafion-based membranes to improve their properties. Various composite materials to improve disadvantageous properties of the PFSA ionomer will be introduced later in another section.

Partially fluorinated membrane research has been actively conducted for the past 30 years to search for alternatives to perfluorinated membranes for PEMFC and DMFC applications. Partially fluorinated ionomers are of great interest among other various candidates in such efforts. Like perfluorinated membranes, partially fluorinated ionomers also have a PTFE-like polymer backbone as a main part to resist chemical attacks. Due to its availability and easy sulfonation, styrene is a reasonable choice for ionomeric material. The ionic groups are attached to styrenic moieties instead of perfluorinated side chains. In the late 1990s, Ballard Power Systems introduced a partially fluorinated low-cost membrane for fuel cell applications [35,59,60]. The Ballard Advanced Materials (BAM) membrane is a family of sulfonated styrenic copolymers of α, β, β-trifluorostyrene and substituted α, β, β-trifluorostyrene comonomers by emulsion polymerization. The chemical structure of the BAM membrane is shown in Figure 12.4. Due to the excellent chemical stability of the fluorinated backbone structure of the BAM membrane, its performance is reported to be comparable to perfluorinated membranes, and more than 100,000 h of PEMFC operation is possible [60]. Similarly, by copolymerizing trifluorostyrene and substituted vinyl monomers, Kim [61] has improved the mechanical properties of partially fluorinated membranes and confirmed their performance in PEMFC and DMFC operation. Another approach toward partially fluorinated membranes is the use of the radiation-grafting technique to cross-link styrenic polymers on the fluorinated base polymer film followed by sulfonation of the phenyl group. This approach has the advantage of (1) easy modification of both surface property and bulk property and (2) flexibility of introducing a variety of monomers with suitable microstructural properties that would be generally difficult to obtain by classical synthesis routes. Gubler et al. [62] developed radiation-grafted

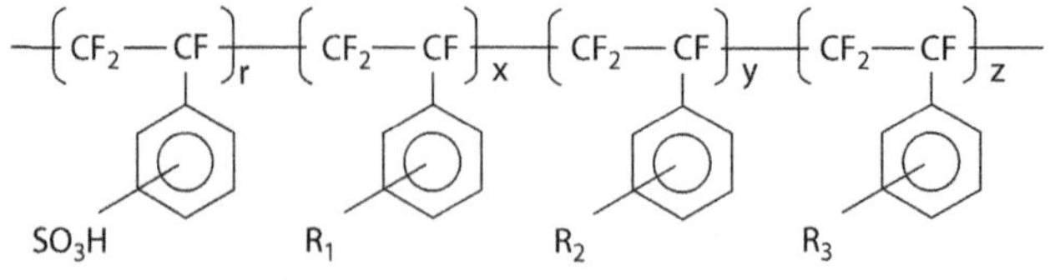

FIGURE 12.4
Chemical structure of BAM membrane. R_1, R_2, and R_3 are alkyl, halogen, or $CF = CF_2$, CN, NO_2, OH.

poly(tetrafluoro-ethylene-cohexa-fluoropropylene)-g-polystyrene (FEP-g-PS) membranes and conducted PEMFC performance using those membranes. The preparation of the membrane starts with an irradiation of base polymer film of FEP (DuPont) with an electron beam, followed by grafting PS in the presence of a divinylbenzene (DVB) cross-linker. Subsequently, the grafted films were sulfonated with chlorosulfonic acid. Using sulfonated FEP-sPS (sulfonated polystyrene) membranes, they performed a PEMFC single-cell durability test over 7900 h at a cell temperature of 80°C–85°C, which resulted in 500 mA·cm^{-2} of current density. Under this condition, the cell showed a negligible degradation during the first 4000 h. Other research groups developed a FEP-g-sAA (sulfonated acrylic acid) and tested it in PEMFCs [63]. Because acrylic acid itself has low ionic conductivity, further sulfonation at the α-carbon of the carboxylic group was performed to give comparable ion conductivity to the membrane. They reported successful PEMFC MEA performance with the membrane having a maximum 32% sulfonation degree. A radiation-grafted membrane has been also applied for a DMFC system. While maintaining conductivity, a substantial decrease in methanol crossover was achieved by using a low-cost ethylene-tetrafluoroethylene (ETFE)–based polymer membrane [64]. However, a poor membrane–electrode interface and delamination were the problems to be solved. Shen et al. [65] reported the performance of a DMFC with a radiation grafted polymer electrolyte membrane. The membranes used polyethylene-tetrafluoroethylene (ETFE), polyvinylidene fluoride (PVDF), and low-density polyethylene (LDPE) as base polymer films. The base polymer films were grafted with polystyrene sulfonic acid (PSSA) as proton-conducting groups. They reported that varying the intrinsic properties of the copolymer by increasing the degree of grafting (DOG) could increase the conductivity, methanol diffusion coefficient, and surface expansion rate of the membrane. As membrane thickness increases, the resistance of the membrane also increases, which reduces the performance of DMFC. However, at the same time, it limits methanol crossover, which counteracts the DMFC performances. Also reported were the effects of the substrate and DOG for composite membrane in DMFC performance. Saarinen et al. [66] investigated membranes that are based on 35 μm thick commercial poly(ethylene-*alt*-tetrafluoroethylene) (ETFE) films. These membranes showed exceptionally low water uptake and excellent dimensional stability. The methanol permeation through the membranes was examined at different temperatures (30°C, 50°C, and 70°C), resulting in 90% lower values for the ETFE-SA than that for Nafion 115, which was more than three times thicker than ETFE-SA. Arico et al. [67] investigated durability of ETFE-based membranes during DMFC operation. They performed DMFC operation at temperatures of 90°C–130°C. Their MEA showed an initial performance of 92 mW cm^{-2} at 110°C with 1 M methanol and 2 atm air as anode and cathode fuel, respectively. After 20 days, the performance still remained the same as the initial value, and the authors confirmed a good adhesion of electrodes to the membrane by observation of the MEA cross-section image using optical microscopy. Grandfield University developed membranes based on copolymers of PVDF (50 μm thick) and LDPE (125 μm thick) with styrene. The membranes were produced through radiation grafting with a cobalt source (Co-60). The radiation was carried out in air and resulted in the formation of peroxy radicals on the polymer backbone. The membranes were then polymerized with styrene in an inert gas, sulfonated, and then hydrolyzed in hot water. The grafting process incorporates the styrene or its derivatives into substrates with subsequent conversion into sulfonate or other functional groups. These membranes have a factor of 10 higher resistance than Nafion but have lower cost and lower methanol diffusion coefficients at 2 M methanol and $T = 20°C$ ($<0.05 \times 10^6$ cm$^2 \cdot$s^{-1} compared to 0.11×10^6 for Nafion) [66,68]. Lifetime and mechanical durability data, however, were not found. Although perfluorinated or partially fluorinated ionomers are mostly favored as fuel cell membrane materials due to their unique resistance to chemical attack from their surroundings, the fundamental solution to overcome the inherent shortcomings of Nafion and other fluorinated membranes would be the development of chemically different types of ionomers. Alternative membranes for PFSA can reduce methanol crossover as low as two orders of magnitude. Such membranes are discussed in the following sections.

12.3.2 Hydrocarbon Membranes

As discussed before, PFSA-based membranes are widely used due to their stability against electrochemical reactions and high acidity to confer high proton conductivity. Alternatives to PFSA, however, are required for the DMFC application because the fluorinated polymers have a few obstacles for commercialization: high cost of the membranes, safety of fluorinated monomers, methanol crossover through the membranes, and low mechanical properties, including high swelling ratio in organic solvents. The high cost and safety of fluorinated monomers might be the main stimulus for producing nonfluorinated membranes because the methanol crossover and low mechanical properties would not be major concerns if anodic or cathodic catalyst efficiencies would increase. Hydrocarbon membranes are defined as polymeric materials consisting of nonfluorinated main chains and side chains containing functional groups. The functional groups are mainly a sulfonic acid relating to proton conductivity. Such hydrocarbon polymers have merits on cost and structural diversity to fluorinated polymer membranes, but the hydrocarbon polymers should meet with other requirements for real applications because sulfonated polystyrene–DVB copolymer membranes adopted for the fuel cells for the Gemini space program in the early 1960s had short lifetimes due to an oxidative degradation. The membranes are usually fabricated via casting polymer solutions on substrates or coating polymer melts on substrates using coating machines. In the casting polymer solution methods, membranes are prepared by dissolving copolymers or homopolymers in effective solvents (e.g., dimethylacetamide), filtering the polymer solutions to remove undissolved polymers or impurities, and casting the solutions onto clean glass substrates. The polymer membranes are subjected to acid treatment to convert to the required acid form if the fabricated membranes are in the salt form. During film formation, the dissolved polymer chains adhere to the substrates experiencing shear and tensile stresses [69]. The residual stresses of the fabricated membranes cause cracks or deformation on membranes in long-term performance. Therefore, the stresses in the fabricated films need to be removed through annealing or thermal treatment. The factors involved in film castings are a solubility parameter of solvents, a solvent concentration, a rate of evaporation, and annealing times, whereas those of coating methods are rheological behaviors, such as viscosity of polymer solutions, gap size of extruding dies, and pressure of extruders.

Polymer membranes have low proton conductivity until they contain enough proton conductive functional groups, for example, sulfonic acid, in polymer backbones or side chains [23]. The sulfonic acid in polymer chains shows a high dissociation constant, resulting in high proton conductivity in aqueous environments compared with other acids; therefore, most DMFC membranes use the sulfonic moiety as the proton transfer medium. Sulfonation methods of polymer membranes are generally classified as post-sulfonation in the presence of polymers and *in situ* sulfonation through copolymerization of sulfonated monomers and nonsulfonated monomers. The degree of sulfonation via post-sulfonation relates with the electrophilic characteristics of pristine polymers and sulfonating agents. For example, aromatic polymers, the most widely researched alternatives to fluorinated polymer membranes, are sulfonated using concentrated sulfuric acid, fuming sulfuric acid, chlorosulfonic acid, sulfur trioxide (or complexes thereof), a trimethylsilylchlorosulfonate sulfonating agent, or a sulfur trioxide–triethyl phosphate complex. Because the sulfonation reactions are based on an electrophilic substitution reaction, the reaction depends on the substituents present on the aromatic rings. The electrophilic substitution reactions are carried out favorably with electron-donating substituents but not with electron-withdrawing substituents. The sulfonation degree is also affected by the activated position on the aromatic rings. For example, in bisphenol A–based polymers, more than one sulfonic acid group cannot be attached in a repeat unit due to the stability of the aromatic ring, the charge of existing sulfonic acid, and steric hindrances. Other methods for post-sulfonation are methylation (Li), sulfination by SO_2 gas, and oxidation. For lithium-based sulfonation, the choice of oxidant to convert the lithium sulfinate to sulfonic acid is a critical step. The disadvantages of post-sulfonation reactions are lack of precise control over the degree and location of functionalization, the possibility of side reactions, and degradation of the polymer backbone. Sulfonated organic molecules have been a subject of research for their flame-retardant properties because the sulfonated organic molecules show enhanced thermal characteristics. The acidities and thermal stabilities of sulfonic acid groups covalently bound to electron-deficient aromatic rings might be higher than those connected to electron-donating aromatic rings. Therefore, monomer sulfonations are generally carried out with aromatic monomers containing electron-deficient groups (dihalide, chloride, fluoride) using fuming sulfonic acids and bases (NaOH, KOH) for neutralization. *In situ* sulfonations are preceded by copolymerization of sulfonated monomers and neat monomers under similar conditions of condensation polymerizations and nucleophilic substitution reactions. The resultant copolymers afford random copolymers, which have hydrophilic or hydrophobic portions. The hydrophilic portions are originated from the sulfonated monomers and the hydrophobic portions are from the nonsulfonated monomers. The degree of the sulfonation could be controlled by sulfonated monomer amounts.

Types of hydrocarbon membranes are classified as novel hydrocarbons, cross-linked hydrocarbons (covalent cross-linking, ionic cross-linking through acid-base interaction), polymer blends, and grafted polymers. More rigorously, the methods for synthesizing membranes are categorized as nucleophilic substitution reactions (condensation polymerizations) and radical polymerization using vinyl monomers and radical initiators (grafted polymerizations). The nucleophilic substitution or condensation polymerization rates depend on intermediate nucleophilic substitution rates that involve the acidity/basicity of monomers, stoichiometric balance of monomers, and removal of impurities. The radical polymerization rates are affected by an initiator concentration, a monomer concentration, polarity of monomers (structures), and a radical transfer constant of solvents. The grafted polymerizations depend on the radical formation rates of polymers, grafting initiator types, and catalyst types.

The sulfonated phenol-formaldehyde polymer is the first hydrocarbon-based polymer membrane in the literature [70]. The phenolic polymer membrane was prepared by condensation polymerization of sulfonated phenol with formaldehyde, but the sulfonated phenolic polymer had low chemical and mechanical stability for fuel cell applications. The first type of hydrocarbon membrane for fuel cell applications was the sulfonated polystyrene–DVB copolymer membranes equipped for the power source in NASA's Gemini space flights, but the sulfonated polystyrene had low chemical stability for long-term applications because the proton on the tertiary carbons and benzylic bonds are easily dissociated in an oxygen environment, forming hydroperoxide radicals. Because a styrene monomer is easily copolymerized with other vinyl monomers via radical polymerization methods, various styrenic polymers were researched intensively. Two commercial polystyrene-based or -related membranes are available: BAM (Ballard) and Dais Analytic's sulfonated styrene–ethylene–butylene–styrene (SEBS) membrane. Dais membranes are produced using well-known commercial block copolymers containing SEBS blocks. Due to the low chemical stability of the styrene block, Dais membranes are designed for portable fuel cell power sources of 1 kW or less and <60°C operation temperatures. Polyphosphazene-based polymers have high chemical and thermal stabilities and structural diversity through attaching various side chains and cross-linking onto the polymer backbones, so they could be potential materials for fuel cell applications. A typical synthesis of polyphosphazene is shown in Figure 12.5. The strong point of polyphosphazenes is related to the synthetic and technological concerns because the structures of the side groups can be diversified to unlimited molecular structures. Polyphosphazene films show low mechanical properties under hydrated conditions, so the mechanical property should be enhanced through modifications or polymer blends [71]. Arylene-based polymers have been intensively developed because they have high mechanical and chemical stabilities next to fluorinated polymers [72]. Typical polymerizations of arylene-based polymers are carried out via nucleophilic substitution reactions in the presence of sulfonated monomers and nonsulfonated monomers (Figure 12.6). The degree of sulfonation of the polymers can be controlled by varying the amounts of sulfonated monomers. The arylene-based polymers are categorized as the following: poly(phenylene ethers), poly(2,6-dimethyl-1,4-phenylenether), poly(2,6-diphenyl-1,4-phenylenether), poly(ether sulfones), poly(etherketones), poly(phenylenesulfides), and poly(phenyl-quinoxalines). Among the arylene-based polymers, poly(arylene ether) materials, such as poly(arylene ether ether ketone), poly(arylene ether sulfone), and their derivatives, attract huge attention because they have good chemical stability under harsh conditions. The synthetic procedure for the most studied arylene-based copolymer is carried out via step polymerization. Los Alamos National Laboratory and Virginia Polytechnic Institute reported the biphenol-based poly(arylene ether sulfone) (PES) membrane BPSH-40 (40 refers to the molar fraction of sulfonic acid by percent) and a poly(arylene ether benzonitrile) membrane 6FCN-35 (Figure 12.7) [73]. The membranes were prepared by direct aromatic nucleophilic substitution and

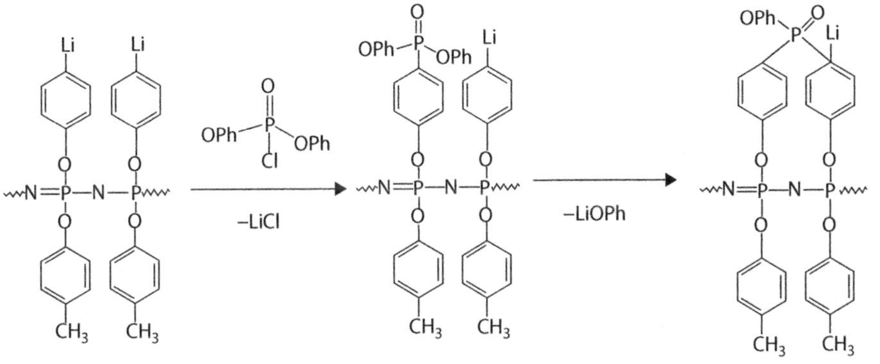

FIGURE 12.5
Intermolecular coupling of polyphosphazene with diphenyl chlorophosphate.

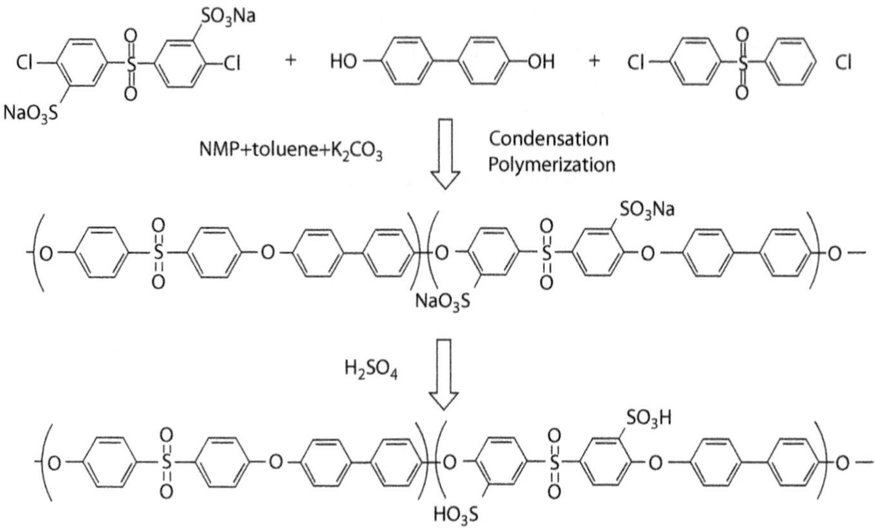

FIGURE 12.6
Synthesis of directly copolymerized wholly aromatic sulfonated poly(arylene ether sulfone), BPSH-xx, where xx is the ratio of sulfonated/unsulfonated activated halide.

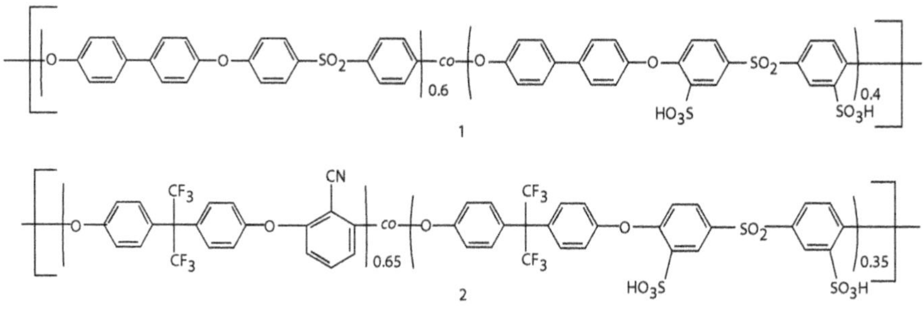

FIGURE 12.7
Structure of 1: BPSH-40 and 2:6FCN-35.

polycondensation of 4,4-hexafluoroisopropylidene diphenol (6F), 2,6-dichlorobenzonitrile, and 3,3-disulfonate-4,4-dichlorodiphenylsulfone under basic conditions in N-methyl-2-pyrrolidinonone at 200°C. As is typically the case, water uptake correlated proportionally with IEC with both decreasing in the order of BPSH-40, 6FCN-35, and Nafion. However, the methanol permeability, which is typically proportional to water uptake and IEC, of Nafion was two times higher than those of the other two membranes. Kim et al. [73] explain the situation by the presence of less loosely bonded and/or free water in BPSH membranes than Nafion membranes due to structural differences even at the higher total water uptake. Another interesting feature of the 6FCN-35 membranes was fluorine enrichment of the air-contacting surface (38%), which had more than twice the fluorine concentration of the bulk material (17%). It was hypothesized that the surface fluorine enrichment was the source of high adhesion to Nafion bonded electrodes [73], which can increase the fuel cell performances due to lowered interfacial resistance. The methanol crossover of 6FCN-35 and BPSH-40 are nearly equal (81 × 10⁻⁸ and 87 × 10⁻⁸ cm²·cm⁻¹) with 0.5 M methanol at T = 30°C. At a slightly higher (10%) membrane resistance, the methanol crossover is a factor of two times lower than for Nafion (167 × 10⁻⁸ cm²·cm⁻¹) (Figure 12.8) [73]. The performance of the membrane in terms of methanol crossover and ionic conductivity properties results in a 50% improvement in selectivity, regardless of membrane thickness, and a DMFC with this membrane has shown a performance of j = 200 mA·cm⁻² (compared to 150 mA·cm⁻² for Nafion) at 0.5 V, T = 80°C, and ambient air pressure [74]. Another prospective DMFC membrane (originally developed for PEMFCs) composed of modified PES is sulfonated poly(arylene thioether sulfone) (sPTES) [75]. This membrane was synthesized at the Air Force Research Laboratory (AFRL) by dissolving the sodium for 6F-sPTES-50 copolymer (50% sulfonation) in DMAc followed by direct casting on a glass dish. Sulfonation increases the acidity and hydrophilicity, improving the conductivity but also increasing the water uptake. The latter problem was resolved by incorporation of [–CF₃] groups into the backbone of the polymer resulting in a new sPTES modification with comparable water uptake in the backbone to Nafion. The German company FuMA-Tech GmbH developed

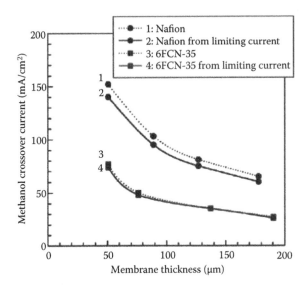

FIGURE 12.8
Effect of thickness on methanol crossover for Nafion and 6FCN-35 membrane: dotted line, electro-osmotic drag corrected value; solid line, uncorrected value.

cost-effective sulfonated poly(arylether ketone) membranes (FKE Series, 50 μm thick) that have higher mechanical stability while also demonstrating increased efficiency and significantly higher power density due to the considerably lower methanol permeability than Nafion membranes. FKE membranes operate in the 100°C–160°C temperature range. Inorganic or polymer materials will be blended with these membranes to increase stability [76].

In the case of polyamide-based membranes, the first step in the synthesis involves preparation of short sequences of 4,4-diamino-2,2-biphenyl disulfonic acid condensed with 1,4,5,8-tetracarboxylic dianhydride. An adjusted ratio of these two monomers allows one to create different block lengths of the sulfonated sequence [77]. In the second step of polymerization, the degree of sulfonation can be precisely controlled by regulating the molar ratio of BDA and the unsulfonated diamine, which is 4,4'-oxydianiline (ODA) in SPI. The degree of sulfonation of polyamides should be carefully controlled because highly sulfonated polyamides became excessively swollen or dissolved in aqueous solutions. The stabilities of sulfonated polyamides also involve the structures. Five-membered ring polyamides, having high performances in nonsulfonated forms, showed expected quick degradation in sulfonated forms because phthalic imides tend to hydrolyze, leading to chain scissions, but naphthalenic polyamides showed much higher stability in fuel cell applications. Due to sulfonic acid group in 2-acrylamido-2-methylpropanesulfonic acid (AMPS) molecules, homo-poly (AMPS) or (AMPS)-based copolymers would be applicable to proton-conductive membranes for fuel cell applications [78]. The amphiphilic monomer can be easily polymerized with radical initiators, and the price of the monomer is reasonable compared with other monomers. A research result showed that proton conductivity of semisolid poly-AMPS showed higher than that of partially hydrated. The conductivity of poly-AMPS increases with water content only up to about six molecules per equivalent and then levels off, but Nafion is hydrated to 15 water molecules per sulfonic acid group. The conductivity illustrates that poly-AMPS may have high proton conductivity under low water conditions. Because poly-AMPSs are highly swollen in aqueous media and become gels that are too fragile for DMFC membranes, the hydrophilicity of poly-AMPSs need to be controlled by introducing hydrophobic portions into the polymer backbones via a copolymerization of the hydrophobic monomer. Chitosan is an inexpensive and abundant natural material with high molecular weight, but it is water soluble or highly swelling in aqueous solutions. Therefore, chitosan needs to be modified in solubility by blending or cross-linking using ionic characters. Recently, chitosan-based membranes were prepared by the spontaneous formation of a polyion complex due to the occurrence of ionic cross-linking [79]. Covalently cross-linked ionomer membranes or blend membranes are expected to have dimensional and chemical stability to reduce methanol crossover. Cross-linking agents, such as DVB, sulfonyl *n*-imidazolide, and 4,4'-diaminodiphenylsulfone, can imbibe in the polymer main chain during polymerizations. In 2005, PolyFuel, Inc., developed novel hydrocarbon membranes (the exact composition is unavailable) for passive DMFCs with power densities of 60 and 80 mW·cm^{-2} for thicknesses of 62 and 45 μm, respectively (at 0.28 V, $T = 40°C$, and 8 mg·cm^{-2} catalyst loading). The lifetime for a nearly constant runtime is 5000 h, which is longer than the lifetime required for a commercially viable portable fuel cell (2000–3000 h). In comparison, the runtime of a 4 h Li ion laptop battery decreases to 2.5 h after 3000 h of operation [80]. PolyFuel had worked with six major corporations that are developing DMFC systems, including NEC and SANYO Electric.

The basic concept for using block copolymer in the DMFC application is that the ordered hydrophilic/hydrophobic phase separations offer a route for the selective transport of protons with the reduced methanol crossover in the hydrophilic domains because block copolymers can be selectively sulfonated using post-sulfonation methods, and the block copolymers can be verified over a wide range of structures during anionic polymerization. For example, methanol transport behaviors of a triblock copolymer ionomer, sulfonated poly (styrene-isobutylene-styrene) (S-SIBS), were compared with Nafion to determine that the sulfonated block copolymer could serve as a viable alternative membrane for application to DMFC [81]. The S-SIBS membranes show approximately 5–10 times more methanol selectivity than Nafion 117, although the S-SIBS membranes exhibited low conductivity compared with Nafion 117.

Cross-linked polymer membranes show reduced methanol crossover; however, it is questionable whether the cross-linking bridges are stable in the strongly acidic environment of the fuel cell. Another problem is that the cross-linked polymers become very brittle upon drying out, which is a severe problem when these membranes are used as fuel cell membranes. The brittleness is possibly caused by the inflexibility of the covalent networks. Flexible ionomer networks can be built up via mixing polymeric acids and polymeric bases, obtaining networks that contain ionic cross-links formed by a proton transfer from the polymeric acid onto the polymeric base. It was observed that upon drying, acid–base blend membranes decrease brittleness, compared with un-cross-linked or covalently cross-linked ionomer membranes, which is possibly caused by the flexibility of the ionic network. The developed acid–base blend membranes show outstanding thermal stabilities determined by DSC and TGA. Graft copolymers, in which ion-containing polymer grafts are attached to a hydrophobic backbone, could be suitable structures for studying structure–property relationships in ion-conducting membranes if the length of the graft and number density of graft chains can be controlled. In principle, the grafting length would determine the size of the ionic domains, whereas the number density of graft chains would determine the number of ionic domains per unit volume. Thermally, a cross-linkable sulfonated poly(arylene ether sulfone) has been developed and showed that the cross-linked membranes had a lower swelling ratio relative to their un-cross-linked counterparts [82–84].

Collectively, the size and number density of ionic aggregates/clusters are expected to control the degree of connectivity between ionic domains. Recently, several researchers have shown that it is possible to synthesize graft copolymers possessing ionic grafts bound to hydrophobic backbones using macromonomers formed by stable free radical polymerization (SFRP) techniques [85]. The detailed synthesis and characterization of this class of copolymer that comprises a styrenic main chain and sodium styrene sulfonate graft chains was reported by Holdcroft et al. PS-g–PSSNa was prepared by (1) pseudo-living, tempo-mediated free radical polymerization of sodium styrenesulfonate (SSNa) and (2) termination with DVB. The macromonomer, *mac* PSSNa, serves as both the comonomer and emulsifier in the emulsion copolymerization with styrene. During polymerization, the DVB terminus is located in the core of the micellar particles and is incorporated into the growing polystyrene (PS) as graft chains. From the results of intensive research, a number of polymer membranes were reported with high proton conductivity and low methanol permeation and water uptake, but many researchers assert that additional properties should be considered. The additional properties would be (1) the relationship of molecular weights and mechanical properties (e.g., tensile modulus in dried and wet states); (2) the dependence of acid treatment methods after sulfonation on proton conductivity; and (3) introducing an optimized way for the MEA, such as proper binders for new developed polymer membranes, pressure range, temperature, and catalyst layer configurations [23,86].

The PBI-based membranes were cast from a solution of high molecular weight PBI in dimethylacetamide (DMAc) followed by immersion into an 11 M solution of phosphoric acid. PBI is a basic polymer (pKa = 5.5), and doping with acid forms a single-phase polymer electrolyte that has a good oxidative and thermal stability and mechanical flexibility at elevated temperatures $T <$ 200°C [87]. At $T > 100$°C, these membranes have higher ionic conductivity, lower electro-osmotic drag (about 0 compared to 0.6 for Nafion), and lower methanol crossover (it of 80 µm thick is 1/10 of 210 µm thick Nafion) than Nafion. The major disadvantage is leaching of the low molecular weight acid (H_3PO_4, shown in Figure 12.9) in hot methanol solutions. These problems were solved by the addition of high molecular weight phosphotungstic acid as a replacement of the low molecular weight acid. At $T > 130$°C, the conductivity of the doped PBI membrane is similar to Nafion (30 mS·cm^{-1} at 130°C and 80 mS·cm^{-1} at 200°C). The methanol crossover at different methanol concentrations is as shown in Table 12.5 [87]. The maximum methanol crossover from Table 12.5 (10.7 mA·cm^{-2} at 100% methanol) is in good agreement with Wainright et al. [88]. The DMFC used by Wainright et al. achieved a power density of 210 mW·cm^{-2} at $T = 200$°C, atmospheric pressure, a water/methanol ratio of 2:1, and $j = 500$ mA·cm^{-2}. Recently, a PBI-based membrane was commercialized by PEMEAS (USA) for DMFC applications. The Celtec V membrane has operating temperatures that range from $T = 60$°C to 160°C without humidification, can be produced much more cheaply than Nafion,

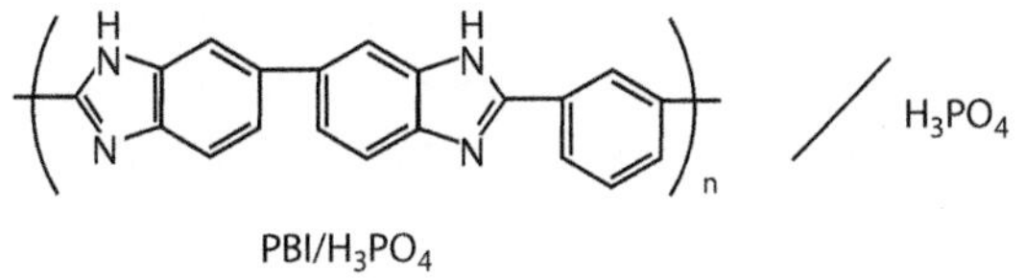

PBI/H₃PO₄

FIGURE 12.9
Phosphoric acid–doped PBI structure.

TABLE 12.5

Effects of Methanol Concentration on Methanol Crossover in DMFC

Water/Methanol	Methanol Crossover (mA·cm^{-2})
3	4.5
2	6.4
1	10.4
0 (pure methanol)	10.7

Note: $E = 0.9$ V, $T = 180°C$, PBI doped in 5 M H₃PO₄, air flow, anode feed (24.7 ml·min^{-1}), anode (Pt/C, 0.5 mg·cm^{-2}), and cathode (Pt/C, 4mg·cm^{-2}).

has low methanol crossover, and possesses good mechanical properties [89]. However, quantitative data on the lifetime and durability of PBI-based membranes were not found in the literature.

The National University of Singapore developed a novel asymmetric membrane (Figure 12.10) composed of a three-component acrylic polymer blend (TCPB). The membrane was prepared by the polymerization of an acrylic polymer blend consisting of 4-vinylphenol-co-methyl methacrylate (P(4-VP-MMA)), poly(butyl methacrylate) (PBMA), and Paraloid B-82 acrylic copolymer resin [90]. To maintain membrane homogeneity, the hydrophilic monomers, 2-acrylamido-2-methyl propanesulfonic acid (AMPS), 2-hydroxyethyl methacrylate (HEMA), and poly(ethylene glycol) dimethylacrylate (PEGDMA), are polymerized only after they have been embedded in the TCPB matrix. The resulting membrane has a novel asymmetric structure in which the hydrophilic network is sandwiched between two layers of matrixes containing high percentages of TCPB. The two exterior layers in this asymmetric membrane primarily provide methanol-blocking with some proton-conducting functionalities, while the middle layer supplies protons and

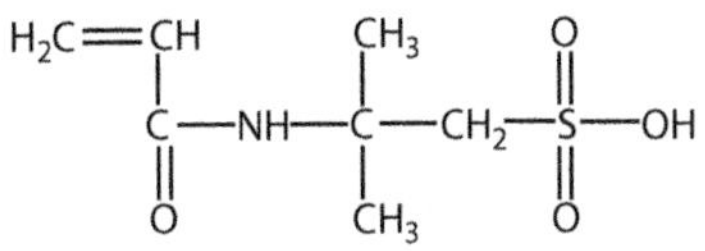

FIGURE 12.10
Structure of asymmetric acrylic membranes.

conserves water [90]. The proton conductivity of AMPS-based membranes (1.2 mS·cm^{-1}) is lower than Nafion (80 mS·cm^{-1}) due to lower sulfonic group content (0.7 mmol·g^{-1} compared to 0.91 mmol·g^{-1} for Nafion). The methanol permeability, however, of the membranes, is lower (10^{-8} cm^2·s^{-1}). The clusters strain the carbon bonds and link them with polymer main chains. The high water uptake of this membrane is determined by the PEGDMA component in the hydrophilic network (it is responsible for 75% of the water uptake). These membranes have high thermal stability (up to 270°C) and good mechanical properties due to the association of the AMPS sulfonic acid groups via ionic interaction, which forms heavy clusters. Lifetime data, however, were not found in the literature.

12.3.3 Composite and Other Modifications of Ionomers

In order to diminish the methanol crossover through the membrane, main directions can be followed: a development of new polymer electrolytes, a modification of the structure of the conventional membranes, or a modification of new polymer electrolytes. Much research has been carried out to improve the performance of the ionic conductivity and methanol crossover of PEM. Hydrocarbon-based new electrolytes have been described, and sulfonated polyarlysulfone [23] shows better results as an electrolyte for DMFC. However, many kinds of membranes have contradictions in ionic conductivity and methanol crossover up to now. Ionic conductivity decreases with a reduction of methanol crossover. For application of the DMFC system, this contradiction has to be solved. To overcome this drawback, modification of a conventional membrane or a new membrane has been approached. Inorganic materials have been adapted for reducing fuel crossover and increasing water sustainability. Morphology, size, and properties of inorganic materials can affect PEM performance in a DMFC system. Here, modified composite membranes will be classified as (1) multilayer composite membranes with methanol impermeable substrate, (2) composite membranes with various kinds of organic materials and blends, and (3) composite membranes with inorganic materials.

Approach for modification has been achieved by various methods, such as physical or chemical modifications and arrangements. Plasma etching [91] or palladium film [92] were tried to reduce methanol crossover by physical modification of the perfluorinated polymer membrane itself. Yoon et al. [93] carried out the modification of Nafion membrane with Pd film by sputtering. Pd film, which is thinner than 300 nm, acts as a barrier to methanol crossover, and at the same time, it also reduces proton conductivity. The nano silica layer has also been deposited by the plasma-enhanced chemical vapor deposition (PECVD) technique by Kim et al. [94]. They deposited the nanoscale layer of silica (10, 32, and 68 nm) on Nafion membranes. The TiO$_2$ layer has also been coated for the methanol barrier [95]. The ionic conductivity of Nafion/silica composite membranes is reduced by 7%–22% to the unmodified Nafion membrane, but its methanol permeability was reduced by 40%–70%. Layered double hydroxides (LDHs) were incorporated into polyelectrolyte membranes in order to investigate the electrochemical reaction processes affected by transport rates of the methanol and proton in DMFC applications by Lee et al. [96]. Depending on different ion exchange capacities and the composition of LDH, the polyelectrolyte membranes gave different diffusion coefficients of methanol and proton conductivities, which were correlated with the maximum power density of DMFC. They observed an optimal property condition in which the proton conductivity and diffusion coefficient were balanced in a desirable manner, which exhibited a 24% increase in the maximum power density compared with the pristine Nafion membrane at 5 M of methanol feed concentration. The developed nanocomposite technique identified the effects on the diffusion coefficient and proton conductivity of polymer membranes in the electrochemical reactions of DMFC. Ren et al. [97] modified the hydrocarbon polymer, SPEEK (sulfonated poly ether ether ketone), with a Nafion ionomer for decreasing methanol crossover with SPEEK and low interfacial resistance with the electrode with the Nafion ionomer.

Multilayer composite membranes using a methanol impermeable substrate have been achieved to reduce methanol crossover for DMFC. A porous substrate, impermeable to methanol, has been impregnated with an ionomeric material, usually Nafion ionomer. In this approach, by reducing resistance with a thinner membrane, the performance of DMFC can be improved with lowering fuel crossover. With a thin porous substrate, a composite membrane can reduce the internal resistance of the membrane and reduce the material cost. A methanol impermeable porous membrane, such as Teflon, PVDF, PTFE, and Polyethylene-terephthalate, has been used for a multilayer composite membrane. Multilayer composite membranes have also been used in a PEMFC as the self-humidified membranes for water management. A water management of thin membranes can be used in DMFC systems for particular applications. An impregnation of ionomer to substrate is affected by the kinds of ionomer and substrate. A composite membrane with a thin porous substrate was first fabricated by Nafion ionomer with a porous PTFE [19] or micro PTFE fibril [98] for PEMFC applications. These approaches can also be applied in DMFCs. Lin found that a Nafion/PTFE composite membrane caused the reduction of not only the methanol crossover by diffusion but also the electro-osmosis of methanol through the membrane. To improve membrane performance, an impregnated ionomer and a substrate should be well connected for a good interface formation. In DMFCs, a methanol permeable ionomer can be delaminated from a methanol impermeable substrate in a highly concentrated fuel. The delaminating problem can be solved by using chemical bonding between the ionomer and substrate. Grafted ionomers to a fluoropolymer backbone can also fall into a partially fluorinated polymer. This was reviewed in the previous section. Yamaguchi et al. [99,100] reported the pore-filling membranes, which reduce the methanol crossover in a wide range of methanol concentrations due to the suppression effect of the substrate matrix. They used porous cross-linked high-density polyethylene (CLPE) on which proton conductivity was given by acrylamide tert-butyl sulfonate sodium. The MEA with pore-filling membrane successfully generated electricity and showed excellent fuel cell performance with a high concentration of methanol fuel. To reduce the methanol crossover using a new type of membrane, high-concentration methanol of 32 wt% (10 M) fuels can be used for the operation. Multilayer composite membranes have low resistance by thinner thickness and low methanol crossover by impermeable substrate. The correlation between impregnated ionomer and substrate are the factors that control performance of PEM in DMFC. Process is also an important design factor to obtain good fuel cell performance.

Other approaches to develop PEM in DMFC are composite membranes, which are composed of organic–organic or organic–inorganic systems. Each system has been achieved by various processes or materials. Organic–organic systems consist of an ionic conducting phase and nonconducting phase or methanol-permeable and -impermeable phase. An organic–organic system can control the morphology, properties, and size of each phase. The process of membrane fabrication is one of the important factors to control performance of PEM for DMFC. The ratio and structure of phases control ionic conductivity and methanol crossover. These properties are driving the development of a block copolymer for DMFC. A block copolymer can be applied for good control of morphology in polymer membranes. More studies should be carried out to develop morphology of PEM for controlling methanol-water diffusion and electro-osmosis effects. In recent years, several polymer blends have been developed as effective PEM alternatives for DMFCs [101–104]. Ren et al. [105] investigated polymer blends of sulfonated poly(ether ether ketone) (SPEEK) with PVDF. The liquid uptake of the 9/1 SPEEK/PVDF blend membrane is about 20%, and the value of proton conductivity of this membrane is $1.75 \times 10^{-3}\,S\,cm^{-1}$, which is about 9% of Nafion 115. The methanol permeability of the 9/1 SPEEK/PVDF blend membrane is about 1/20 that of Nafion 115 under the same testing conditions. A composite membrane with SPEEK/polyaniline (PANI) was also investigated by Li et al. [106]. Qiao et al. [107] investigated a new type of chemically cross-linked polymer blend membranes consisting of poly(vinyl alcohol) (PVA), 2-acrylamido-2-methyl-1-propanesulfonic acid (PAMPS), and poly(vinyl pyrrolidone) (PVP) as proton-conducting polymer electrolytes. Through chemical cross-linking reactions between the hydroxyl groups (–OH) of the polyhydroxy polymer PVA and the aldehyde groups (–CHO) of glutaraldehyde cross-linkers, the swelling property of PVA was effectively controlled with the increased cross-linking density. The resultant PVA–PAMPS–PVP blend membranes are thus capable of possessing all the required properties of a PEM, namely, reasonable swelling. The proton conductivity of the membranes was investigated as a function of cross-linking time, blending composition, water content, and ion exchange capacity. The membranes attained $0.088\,S\,cm^{-1}$ of the proton conductivity and 1.63 mequiv·g^{-1} of IEC at 25°C for a polymer composition PVA–PAMPS–PVP of 1:1:0.5 in mass, and a methanol permeability of 6.1 × 10^{-7} cm^{2}·s^{-1}, showed a comparable proton conductivity to Nafion 117, but only one third of Nafion 117's methanol permeability under the same measuring conditions. Polymer blending with PVDF has been carried out by many researchers because it is impermeable to methanol [108–110]. Nafion membranes were modified by the *in situ* electrodeposition of polypyrrole inside the membrane pores on the anode side only in order to prevent the crossover of methanol in the DMFC [111]. The modified membranes were studied in terms of morphology, electrochemical characteristics, and methanol permeability. FTIR and SEM confirmed the presence of the polypyrrole on the anode side of the Nafion membrane. SEM showed the polymer to be present both on the membrane surface and inside the membrane pores. It was found to be deposited as small grains with two distinct sizes; the smallest particles had a diameter of around 100 nm whereas the larger particles had diameters of around 700 nm. Methanol permeability was determined electrochemically and was shown to be effectively reduced. Controlling the phase through blend or block copolymer will be a good approach in PEM development for satisfying requirement in various DMFC systems. Nafion-polyfurfuryl alcohol (PFA) nanocomposite membranes can be synthesized by *in situ* polymerization of furfuryl alcohol within commercial Nafion membranes. Furfuryl alcohol is miscible with mixtures of water and alcohols (it penetrates into the hydrophilic channels of Nafion) and becomes hydrophobic following polymerization via acid catalysis. The chemically stable PFA is responsible for the low methanol crossover through homogenous Nafion–PFA nanocomposite membranes at PFA concentrations varying from 3.9% to 8%. The Nafion 8 wt% PFA membrane has a methanol crossover of 1.72 × 10^{-6} mol·min^{-1}·cm^{-1} and a proton conductivity of 70.4 mS·cm^{-1} at room temperature. The corresponding properties of Nafion 115 are 4.66 × 10^{-6} mol·min^{-1}·cm^{-1} and 95.3 mS·cm^{-1} [112]. The 26% lower conductivity of Nafion 8% PFA is offset by almost three times lower methanol crossover. The Nafion 4.7% PFA (290 mV) has much higher cell performance than the plain Nafion 115 (58 mV) at $j = 40$ mA·cm^{-2}, $T = 60$°C, 1 M methanol, and ambient air. The OCV for Nafion 4.7% PFA is 705 mV compared to 624 mV for Nafion.

The composite membranes (Figure 12.11) have been prepared by blending main-chain polymers, sulfonated PEEK Victex or sulfonated PSU Udel (sPSU), with basic polymers, poly(4-vynylpyridine) (P4VP) or PBI. The membranes show comparable performance to Nafion 105 (500 mV) at $j = 0.5$ A·cm^{-2} and $T = 110$°C. However, these membranes have higher methanol crossover than Nafion 117 (150 mA·cm^{-2}), as shown in Table 12.6. The testing parameters were 5.3 mg·cm^{-2} Pt black cathode, 5.2 mg·cm^{-2} Pt/RuOx E-TEK anode, 4 ml·min^{-1} of 1 M methanol at 2.5 bar, and 1.5 ml·min^{-1} of air at 4 bar [113].

Using inorganic materials is another effective way for reducing methanol crossover. PEM in fuel cells is very permeable or prone to swelling in methanol fuel. A high swelling of a membrane in fuel solution reinforces

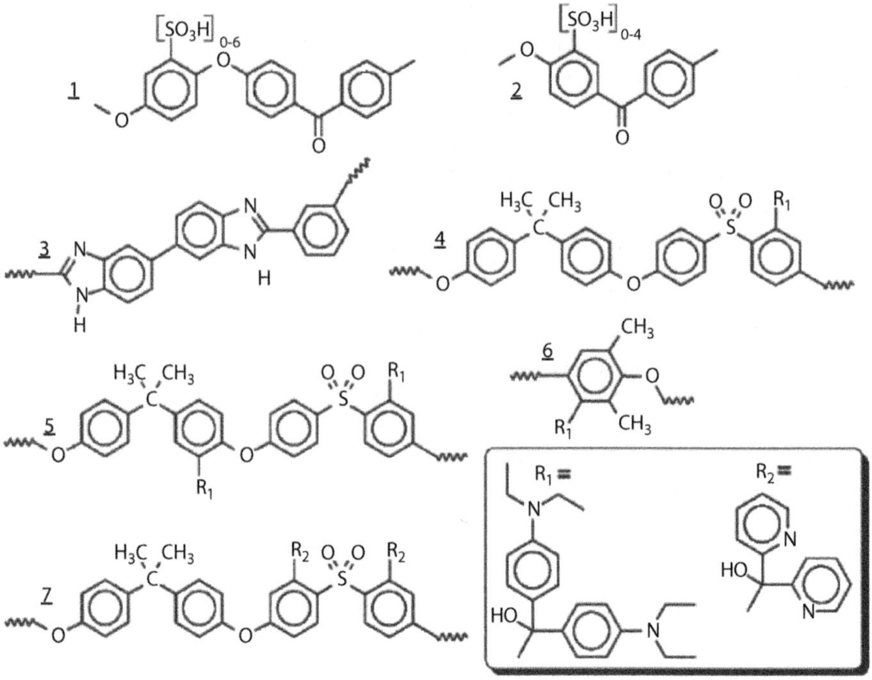

FIGURE 12.11
Arylene-chain polymers used for the preparation of an acid–base blend membrane.

TABLE 12.6

Comparison of Nafion and sPEEK + sPSU + PBI Membranes

Parameters DMFC (Current Density, mA·cm⁻²)	Methanol Permeation Equivalent (mA·cm⁻²)		
	Nafion 117	Nafion 105	sPEEK + sPSU + PBI (E504)
0	163	343	195
200	103	257	150
500		135	

or accelerates fuel crossover. The transport or crossover of fuel through the membrane is progressed by adsorption of fuel to the membrane, diffusion through the membrane, and desorption from the membrane. The efforts to reduce methanol crossover can be achieved to suppress the phenomena of these three stages. Organic-inorganic systems have been tried to reduce methanol crossover through suppression of fuel crossover in three stages. Inorganic material hinders the swelling of PEM, which reduces the diffusion rate of methanol. Inorganic materials play a role as a barrier or an obstacle for diffusion of fuel through the membrane in the diffusion and electro-osmotic drag of methanol in DMFC. Electro-osmotic drag of methanol in DMFC cannot be easily reduced by just inorganic materials. To reduce electro-osmotic drag effectively, the ionic cluster or electrostatic forces inside the ionic channel and the interaction of ionic and mobile phase should be controlled. Nanosized metal oxides, hetero-polyacid, layered inorganic materials, and modified inorganic materials have been used and well dispersed for the composite membranes in a polymer matrix.

An organic-inorganic system was first reported by Watanabe [114] in 1994. Watanabe reported that a hygroscopic metal oxide enhanced the water-absorbing ability of PEMFCs. Thereafter, many articles were published and patents granted. Antonucci et al. [115] tested oxide-containing membranes in a DMFC at 145°C, and obtained 350 mA cm^{-2} at 0.5 V. Mauritz [116] developed a sol-gel process to introduce SiO_2 into the fine hydrophilic channels (50 nm of diameter). Detailed investigations on the microstructure and fundamental properties of the composite membranes by a sol-gel process have been carried out [117–120]. The modification of a fabrication method is proposed by using a Nafion solution instead of a Nafion membrane, mixed with tetraethyl orthosilicate (TEOS) or TMDES [121,122]. The modified inorganic materials have been also introduced into Nafion membranes by a sol-gel process [123]. Nunes et al. [124] reported the results of composite membranes with various inorganic materials. They developed inorganic modification by *in situ* hydrolysis of different alkoxides of Si, Ti, Zr, Sr, and organically modified silanes with basic groups (I-silane) [125]. Modification of Nafion through the addition of silica is a common approach utilized for the improvement of membrane performance in DMFC applications. Nafion-silica membranes have been prepared according to several methods by casting mixtures, such as silica powder [126], dithenylsilicate (DPS) [127], sol-gel reaction with tetra-ethylorthosilicate (TEOS) followed by solution casting of the Nafion solution [128], phosphotungstic acid (PWA)-dopes composite silica/Nafion/PWA [129], and silica oxide [115,130]. SPEEK-based membranes were developed to contain heteropolyacids and an oxide phase that was either produced by hydrolysis of amino-modified silanes or by dispersion of surface-modified fumed silica. The degree of sulfonation ranged from 65% to 66%. The heteropolyacid was based on lacunary divacant $[-SiW_{10}O_{36}]^{-8}$ and contained epoxy groups. The reaction provided a covalent bond between the heteropolyacid and the insoluble oxide phase, resulting in its fixation within the membrane [131]. The organic–inorganic materials are mechanically more stable than membranes without inorganic compounds in alcohol solutions. Heteropolyacid has good proton conductivity, but degradation was shown to be an issue for DMFC applications due to its dissolution in water. The stability has usually been increased through silica modification, which increases the strength of the covalent bonds or columbic interactions but reduces the acid strength. Heteropolyacid bleeding can also be reduced through the addition of silanes and the dispersion of surface-modified, fumed silica [131]. The silane-modified sPEEK membrane has lower methanol crossover but also lower conductivity than that modified by silica. The inorganic phase decreases the water and methanol crossover in addition to fixing of the heteropolyacid to the membrane. The stability of these membranes is higher than unmodified sPEEK due to their higher stability in alcohol solution [132]. A remarkable reduction of methanol and water permeability was achieved by an inorganic modification of SPEK and SPEEK membranes [133,134]. With zirconium phosphate and $Zr(OPr)_4$/ACAC as inorganic compounds, the water and methanol crossover was reduced drastically without diminishing conductivity to the same extent. Nafion-zirconium membranes can be prepared by starting with an extruded film, such as Nafion 115. The film is then impregnated with zirconium phosphate (ZrP) via an exchange reaction involving zirconium ions followed by precipitation of zirconium phosphate by immersion of the membrane in H_3PO_4 solution. The result is an insoluble ZrP entrapped in the pores of the Nafion membrane. A Nafion zirconium membrane is stable at $T = 150°C$ with a dry oxidant. The membrane resistance was measured as $0.08\ \Omega$ cm^2 and maximum power densities of 380 and $260\ mW\cdot cm^{-2}$ for a DMFC with this membrane were achieved with oxygen and air feeds, respectively. The ZrP additive enhanced water retention characteristics, raised the maximum working temperature, and increased the dry weight and thickness of the membrane by 23% and 30%, respectively [135]. The membrane conductivity decreased from $0.12\ \Omega\cdot cm^2$ at 90°C to $0.08\ \Omega\cdot cm^2$ at 140°C–150°C. The OCV of the DMFC cell was measured between 0.86 and 0.87 V during the operation between 120°C and 150°C with oxygen or air [136]. Crystallinity and surface morphology play an important role in determining the conductivity of ZrP–modified membranes. The distribution of ZrP particles inside the membrane is uniform, and the particle size is 1161 nm, which is larger than the pore size of the bare Nafion membrane under complete hydration. The surface area of the Nafion membrane increases by two orders of magnitude when modified by ZrP. Nafion-zirconium membranes also have comparable proton conductivity to that of Nafion (10^{-2} $S\cdot cm^{-1}$) at room temperature and 100% relative humidity [135]. Nafion-silica membranes show good performance at $T > 100°C$ due to low levels of dehydration. Nafion-silica membranes were prepared by mixing Nafion ionomer (5%) with 3% SiO_2 followed by a regular membrane casting procedure. In the final stage, the membranes were heat-treated at 160°C for 10 min to achieve both a high crystallinity and high mechanical stability [115]. A DMFC utilizing these membranes was tested under galvanostatic conditions at $500\ mA\cdot cm^{-2}$. The voltage initially decreased from 0.42 to 0.36 V but then remained stable for 8 h. The performance decrease is due to adsorption of poisoning species, which appears to be a reversible process at 145°C (removed by short circuit discharging in the presence of water). The effect of the operating temperature on the performance of DMFCs with this membrane (conditions as mentioned above) is given in Figure 12.12, demonstrating higher performance with increasing temperature [129]. Nafion membranes with 10–20 wt% DPS have a nanolayered microstructure that results in low methanol crossover. The proton conductivity increases with increasing DPS

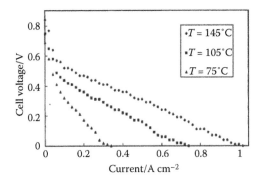

FIGURE 12.12
Effect of operating temperature on DMFC performance with Nafion–SiO$_2$ membrane.

content as the phenyl group can impact hydrophobic characteristics and reduces water adsorption. The increase in the hydrophobic properties of the membrane reduces flooding at the DMFC cathode [127]. DMFC performance with this membrane is dominated by methanol crossover at high methanol concentrations and by resistance at low concentrations (1 M). The methanol crossover current density (mA·cm^{-2}) and conductivity ($\times 10^{-2}$ S·cm^{-1}) for the different membranes were 329 and 2.89 for Nafion, 183 and 2.39 for Nafion/10 wt% DPS, and 160 and 2.16 for Nafion/20 wt% DPS [127]. Nafion/5 wt% silica membranes utilizing TEOS, fabricated according to the sol-gel method, have a methanol permeability of 4.17×10^{-7} cm^2·s^{-1} in a two glass compartment cell at 20°C in 1 M methanol compared to 9.7×10^{-7} cm^2·s^{-1} for Nafion. The membranes with 3 and 5 wt% silica have a MCO of 215–245 mA·cm^{-2} in a DMFC at 60°C and fuelled with 3 ml·min^{-1} of 1 M methanol versus 273 mA·cm^{-2} for Nafion. The conductivity of a Nafion/10% SiO$_2$ membrane was found to be slightly lower than for pure Nafion 112 [128]. A DMFC with the Nafion–SiO$_2$ membrane (80 μm thick) could be operated at $T = 145$°C and reach a maximum power density of 240 and 150 mW·cm^{-2} with an oxygen and air feed supply, respectively. Methanol crossover was 4×10^{-6} mol·min^{-1}·cm^{-1} at $j = 0.5$ A·cm^{-2} with 2 M methanol. A Pt–Ru/C catalyst with a loading of 2 mg·cm^{-2} was used at the anode, and a 20% Pt/C catalyst with a 2 mg·cm^{-2} loading was used at the cathode. The operating pressure at the anode was 3.5, and 5.5 atm O$_2$ was supplied to the cathode. Figure 12.12 shows the performance of this membrane at various operating temperatures [115]. The modification of Nafion membranes through the addition of molybdophosphoric acid (MoPh-a) has been shown to increase the proton conductivity two to two and a half times but with slightly increased methanol crossover [126]. As shown in Table 12.7, Nafion membranes modified by 3.3% MoPh-a and 4.3% SiO$_2$ have slightly higher MCO than Nafion 117 but much higher conductivity. A polyaniline coating was used to create Polyaniline-Nafion-Silica nanocomposite membranes (PaniNC), a modified membrane with decreased methanol crossover.

TABLE 12.7

Effects of the Composition of Nafion Composite Membranes on Conductivity and Methanol Crossover

Membrane	Conductivity (S·cm⁻¹)		Equivalent Permeation (mol·s⁻¹·m⁻¹) in 1.5 M Methanol at $T = 65°C$
	$T = 60°C$	$T = 90°C$	
Nafion 117 + 4.3% SiO_2	0.23	0.29	7
Nafion 117 + 3.3% $H_3PO_4 \cdot 12MoO_3 \cdot H_2O$ (MoPh-a)	0.27	0.39	7.5
Nafion 117	0.09	0.15	5.9

Polyaniline, an electronically conductive polymer, modifies the membrane structure and correspondingly reduces the MCO while the silica nanocomposite improves the conductivity. One approach to prepare a PaniNC membrane is through the sol-gel method. This method embeds the silica in the hydrophilic clusters of Nafion first. Polyaniline was then deposited on the silica-Nafion membranes by redox polymerization [137]. Dimitrova et al. [126] also used inorganic materials for DMFC. They studied composite membranes that have different kinds of inorganic materials. They found that ionic conductivity of composite membranes had no correlation with change of crystallinity. An increment of crystallinity in composite membrane improves the mechanical properties. They concluded that addition of inorganic materials to the polymer matrix increased the water-absorbing ability and lowered methanol crossover by increasing crystallinity.

The modified PVDF membranes (24% PVDF–16% SiO_2 with 60% –3 M H_2SO_4 [vol%]) are nanoporous proton-conducting membranes (NP-PCM). The membranes, which utilize PVDF as the polymer binder to implant SiO_2 powder and acid into the polymer matrix to provide proton conductivity, can have very high surface area and thus two to four times higher ionic conductivity than Nafion. Meanwhile, the methanol crossover of PVDF membranes are two to four times lower than Nafion due to smaller pore sizes (1.5–3 nm compared to 3 nm for Nafion) [137]. Impregnation of the NP-PCM pores with Na_2SiO_3 solution and hydrolyzing the silicate in sulfuric acid (forming silica gel) decreases the MCO further to almost an order of magnitude lower than Nafion. Table 12.8 shows a comparison between two modified PVDF membranes and Nafion [137]. These membranes operate in a wider temperature range than Nafion (from 0°C to over 90°C). They are also less sensitive to iron impurities (Fe > 500 ppm) than Nafion, which allows for the use of Pt–Fe catalysts or stainless steel fuel cell hardware. MEAs utilizing these membranes have achieved power densities of 85 mW·cm⁻² (at $T = 80°C$, 1 M methanol, and a Pt–Ru anode with a loading of 4–6 mg·cm⁻²). At about $4 m⁻², the cost of modified PVDF is significantly lower than Nafion [137]. Modified PVDF membranes have good mechanical properties and can be bent 180° without breaking (at membrane thicknesses of 30–1000 μm). They

TABLE 12.8

Effect of Temperature and Composition on (a) Methanol Crossover ($mA \cdot cm^{-2}$) and (b) Conductivity ($mS \cdot cm^{-1}$)

Temperature (°C)	PVDF + 16% SiO$_2$ (300 µm)	PVDF + Silica Gel (225 µm)	Nafion 117
60	53	22	125–150
80	74	37	
90		42	250–300
25	200	70	

can be manufactured with 50%–90% (by volume) acid solution, have a high thermal stability (from subzero to over 100°C) and have stable dimensions, which do not change with absorption of water.

Nanocomposite membranes have been tested as a barrier of methanol fuel [137–139]. Jung et al. [140] studied nanocomposite membranes with the layered structure of inorganic materials. In their work, they made a nanocomposite membrane by the melt intercalation method with the various contents of montmorillonite (MMT) and the modified montmorillonite (m-MMT) as an internal mixer. Nanocomposite membranes were characterized by XRD, and intercalation or exfoliation of layered structures was checked. The ionic conductivity of the composite membranes was measured to be $8.9–6.7 \times 10^{-2}$ $S \cdot cm^{-1}$ at 110°C. They confirmed that the nanocomposite membranes have lower methanol crossover than Nafion membranes, which perform the higher MEA performance at high temperature. The inorganic materials in the composite membrane decrease not only methanol crossover but also ionic conductivity. The inorganic materials, which have ionic sites, have been investigated for reducing methanol crossover without decreasing the ionic conductivity. Hetero-polyacids and solid acids have been used as composite membranes. Also, inorganic material modifications have been carried out for not decreasing ionic conductivity. Ponce et al. [141] have used an organic matrix of sulfonated polyetherketone (SPEK), different hetero polyacids, and an inorganic network of ZrO_2 or $RSiO_{3/2}$. The inorganic oxide network decreased methanol and water permeability across the membrane as well as decreasing the bleeding out of the heteropolyacid. Hybrid membranes with SPEK and ZrO_2–TPA had the highest conductivity values ($0.110–0.086$ $S \cdot cm^{-1}$). Rhee et al. [142] reported nanocomposite membranes using modified montmorillonite. A modification method is given in Figure 12.13. In their work, nanocomposite membranes decreased methanol crossover without reducing ionic conductivity. They developed the method of modification of inorganic materials to have ionic capability. The novel sodium titanate nanotube had been reported as an additive for DMFC membranes, which showed the improved DMFC performances [143].

Modified inorganic materials need to be more developed, and ionic capability in polymer and inorganic materials will be maximized. And the ionic

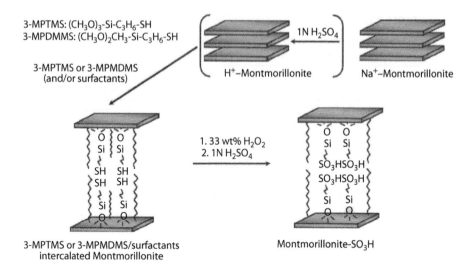

FIGURE 12.13
Representation of the processes of preparing Nafion/sulfonated Montmorillonite composite.

capability in inorganic materials should be freer for the transport of proton in composite membranes. Development in PEM seems to be very difficult for getting high ionic conductivity and zero methanol crossover. The development of enhancing the Grotthus mechanism at high temperatures should be considered for the selection of an ionomer and an inorganic material in designing a composite membrane. The vehicular mechanism of ionic sites in inorganic materials should be reinforced for low-temperature application. Composite membranes have been investigated for optimizing ionic conductivity and methanol crossover. Both properties, the proton conductivity and the methanol crossover, can be represented as the ratio, and the ratio can be expressed as selectivity for the membrane. Compositions of polymer and inorganic materials are the design factors in composite membranes. More subtle research needs to be advanced to examine the correlation between composite membranes and DMFC performance.

12.4 Conclusions

Analysis of the various DMFC membranes showed that PEMs are categorized as partially fluorinated membranes, hydrocarbon membranes, and composite membranes. Composite fluorinated and nonfluorinated (hydrocarbon) DMFC membranes have been reported with low cost, methanol and ruthenium crossover (for Pt–Ru anodes), wider temperature range

(80°C–180°C) and higher ionic conductivity in comparison to Nafion membranes. The traditional Nafion membranes for DMFC applications do not satisfy all of the DMFC requirements. In hydrogen PEMFC applications, unlike DMFCs, thinner membrane materials are preferred as they offer reduced ionic resistance and increased MEA performance. In DMFCs, however, thin membranes (such as Nafion 112) result in a high methanol crossover. These disadvantages exceed the advantage of low ionic resistance, and thus thicker membranes like Nafion 117 are typically used. However, Nafion 117 in a DMFC at $j > 0.3$ A cm^{-2} has a very low cell voltage. Hydrocarbon membranes are the main candidates for the replacement of the expensive Nafion membranes commonly used. The improvement of Nafion-based membranes through the addition of inorganic compounds (SiO$_2$, silanes, Zr, MoPha, etc.) and acidic–basic composites (polyaryl) decrease the methanol crossover but do not reduce cost. The hydrocarbon membranes are cheaper and more technically effective for DMFC than Nafion membranes. They have lower methanol crossover and higher conductivity and stability.

The technical approaches of PEM for MFC are to increase ionic conductivity and reduce methanol crossover. PEM needs to be developed in considering the applications. DMFC for mobile applications required the MEA performances with the design of stack and system. With high-performance PEM, the MEA needs to be designed with an electrode and diffusion layer for applications.

References

1. Lightner, V. 2005. *Small Fuel Cells 2005 Conferences.* Washington DC, USA, April 27–29.
2. Park, Y.C., Peck, D.H., Dong, S.K., Kim, S.K., Lim, S., Jung, D.H., Jang, J.H. and Lee, D.Y. 2001. *Int. J. Hydrogen Energy* 36, 1853.
3. Kreuer, K.D., Paddison, S.J., Spohr, E. and Schuster, M. 2004. *Chem. Rev.* 104, 4637.
4. Tang, H. and Pintauro, P.N. 2001. *J. Appl. Polym. Sci.* 79, 49.
5. Wycisk, R. and Pintauro, P.N. 1996. *J. Membr. Sci.* 119, 155.
6. Ren, X., Springer, T.E., Zawodzinski, T.A., Jr. and Gottesfeld, S. 2000. *J. Electrochem. Soc.* 147, 466.
7. Janssen, G.J.M. and Overvelde, M.L.J. 2001. *J. Power Sources* 101, 117.
8. Smitha, B., Suhanva, D., Sridhar, S. and Ramakrishna, M. 2002. *J. Membr. Sci.* 241, 1.
9. Edmondson, C.A. and Fontanella, J.J. 2002. *Solid State Ionics* 152–153, 355.
10. Cha, S.Y., Tran, N., Duong, A.T., Hou, G., Lefebvre, M. and Attia, A. 2003. PBFC 1st International Conference of Polymer for Battery and Fuel Cell.
11. Mauritz, K.A. and Moore, R.B. 2004. *Chem. Rev.* 104, 4535.

12. Park, Y., Peck, D., Kim, S., Lim, S., Lee, D.Y., Ji, H. and Jung, D.H. 2010. *Electrochim. Acta* 55, 4512.
13. Kim, H., Cho, J., Yoon, J. and Chang, H. 2001. ECS Conference.
14. Chang, H., Kim, J.R., Cho, J.H., Kim, H.K. and Choi, K.H. 2002. *Solid State Ionics* 148, 601.
15. Guo, J., Sun, G., Wang, O., Wang, G., Zhou, Z., Tang, S., Jiang, L., Zhou, B. and Xin, O. 2006. *Carbon* 44, 152.
16. Makino, K., Furukawa, K., Okajima, K. and Sudoh, M. 2005. *Electrochim. Acta* 51, 961.
17. Baranton, S., Coutanceau, C., Leger, J.M., Roux, C. and Capron, P. 2005. *Electrochim. Acta* 51, 517.
18. Dillon, R., Srinivasan, S., Arico, A.S. and Antonucci, V. 2004. *J. Power Sources* 127, 112.
19. Gore, W.L., Bahar, B., Hobson, A.R., Kodle, J.A. and Zuckerbrod, A. 1996. W.L. Gore & Associates, Inc., US patent 5547551.
20. Xie, C., Bostaph, J. and Pavio, J. 2004. *J. Power Sources* 136, 55.
21. Kim, K.H., Choi, S.J. and Chang, H. 2003. 203rd ECS Paris Meeting.
22. Lu, G.Q., Liu, F.Q. and Wang, C.Y. 2005. *Electrochem. Solid-State Lett.* 8, 1099.
23. Hickner, M.A., Ghassemi, H., Kim, Y.S., Einsla, B.R. and McGrath, J.E. 2004. *Chem. Rev.* 104, 4587.
24. Drake, J.A., Wilson, W. and Killeen, K. 2004. *J. Electrochem. Soc.* 151, A413.
25. Aricòa, A.S., Srinivasan, S. and Antonuccia, V. 2001. *Fuel Cells* 1, 133–161.
26. Kalhammer, F.R., Prokopius, P.R., Roan, V.P. and Voecks, G.E. 1998. Status and Prospects of Fuel Cells as Automobile Engines, A Report of the Fuel Cell Technical Advisory Panel, July.
27. DuPont website: http://www.dupont.com/fuelcells.
28. Küver, A. and Potje-Kamloth, K. 1998. *Electrochim. Acta* 43, 2527.
29. Yoshitake, M., Kunisa, Y., Endoh, E. and Yanagisawa, E. US Patent 6,933,071.
30. Miyake, N., Wakizoe, M., Honda, E. and Ohta, T. 2004. Proceedings of the Fourth International Symposium on Proton Conducting Membrane Fuel Cells, October 3–8, Abs# W-1880.
31. Penner, R.M. and Martin, C.R. 1985. *J. Electrochem. Soc.* 132, 514.
32. Bahar, B., Hobson, A.R., Kolde, J.A. and Zuckerbrod, D. US. Patent 5,547,551.
33. 3P Energy website: http://www.3p-energy.de.
34. Miyake, N., Wainright, J.S. and Savinell, R.F. 2001. *J. Electrochem. Soc.* 148, A905.
35. Steck, A.E. 1995. Proceedings of 1st International Symposium on New Material Fuel Cell Systems, 74.
36. Grot, W.G. 1994. *Chem. Ind.* 82, 161.
37. Yeager, H.L., Kipling, B. and Dotson, R.L. 1980. *J. Electrochem. Soc.* 127, 303.
38. Robertson, M.A.F. and Yeager, H.L. 1996. *Macromolecules* 29, 5166.
39. Doyle, M., Lewittes, M.E., Roelofs, M.G., Perusich, S.A. and Lowrey, S.A. 2001. *J. Membr. Sci.* 184, 257.
40. Perusich, S.A., Avakian, P. and Keating, M.Y. 1993. *Macromolecules* 26, 4756.
41. Rajendran, R.G. 2005. *MRS Bull.* 30, 587.
42. Savadogo, O. 1998. *J. New Mater. Electrochem. Syst.* 1, 47.
43. Eikerling, M., Kornyshev, A.A. and Stimming, U. 1997. *J. Phys. Chem. B* 101, 10807.
44. Gierke, T.D. and Hsu, W.Y. 1982. Perfluorinated ionomer membranes. In A. Eisenberg, and H.L. Yeager, Eds., ACS Symposium Series No. 180, *American Chemical Society*, Washington, DC, p. 283.

45. Haubold, H.G., Vad, T., Jungbluth, H. and Hiller, P. 2001. *Electrochim. Acta* 46, 1559.
46. James, P.J., McMaster, T.J., Newton, J.M. and Miles, M.J. 2000. *Polymer* 41, 4223.
47. Every, H.A., Hickner, M.A., McGrath, J.E. and Zawodzinski, T.A., Jr. 2005. *J. Membr. Sci.* 250, 183.
48. Choi, P., Jalani, N.H. and Datta, R. 2005. *J. Electrochem. Soc.* 152, A1548.
49. Argyropoulos, P., Scott, K., Shukla, A.K. and Jackson, C. 2003. *J. Power Sources* 123, 190.
50. Barragan, V.M. and Heinzel, A. 2002. *J. Power Sources* 104, 66.
51. Divisek, J., Fuhrmann, J., Gartner, K. and Jung, R. 2003. *J. Electrochem. Soc.* 150, A811.
52. Murgia, G., Pisani, L., Shukla, A.K. and Scott, K. 2003. *J. Electrochem. Soc.* 150, A1231.
53. Wang, Z.H. and Wang, C.Y. 2001. Proceedings of International Symposium on Direct Methanol Fuel Cells, 286.
54. Wang, Z.H. and Wang, C.Y. 2003. *J. Electrochem. Soc.* 150, A508.
55. Verbrugge, M.W. 1989. *J. Electrochem. Soc.* 136, 417.
56. Cruickshank, J. and Scott, K. 1998. *J. Power Sources* 70, 40.
57. Ren, X. and Gottesfeld, S. 2001. *J. Electrochem. Soc.* 144, L267.
58. Ren, X., Wilson, M.S. and Gottesfeld, S. 1996. *J. Electrochem. Soc.* 143, L12.
59. Wei, J., Stone, C. and Steck, A.E. 1995. Power Systems, Inc., US Patent 5422411.
60. Steck, A.E. and Stone, C. 1997. Proceedings of 2nd International Symposium On New Material Fuel-Cell and Modern Battery Systems II, 792.
61. Kim, H. US Patent, 6774150B2.
62. Gubler, L., Kuhn, H., Schmidt, T.J., Scherer, G.G., Brack, H.P. and Simbeck, K. 2004. *Fuel Cells* 4, 196.
63. Patri, M., Hande, V.R., Phadnis, S., Somaiah, B., Roychoudhury, S. and Deb, P.C. 2004. *Polym Adv. Technol.* 15, 270.
64. Scott, K., Taama, W.M. and Argyropoulos, P. 2000. *J. Membr. Sci.* 171, 119.
65. Shen, M., Roy, S., Kuhlmann, J.W., Scott, K., Lovell, K. and Horsfall, J.A. 2005. *J. Membr. Sci.* 251, 121.
66. Saarinen, V., Kallio, T., Paronen, M., Tikkanen, P., Rauhala, E. and Kontturi, K. 2005. *Electrochim. Acta* 50, 3453.
67. Arico, A.S., Baglio, V., Creti, P., Di Blasi, A., Antonucci, V., Brunea, J., Chapotot, A., Bozzi, A. and Schoemans, J. 2003. *J. Power Sources* 123, 107.
68. Reeve, R.W. Update on status of direct methanol fuel cells, ETSUF/03/00232/REP, DTI/Pub URN 02/592.
69. Tirumkudulu, M.S. and Russel, W.B. 2005. *Langmuir* 21, 4938.
70. Zaidi, S.M.J. and Matsuura, T. 2008. *Polymer Membranes for Fuel Cells.* Springer.
71. Allcock, H.R., Hofmann, M.A., Ambler, C.M. and Morford, R.V. 2002. *Macromolecules* 35, 3484.
72. Wang, F., Hickner, M., Kim, Y.S., Zawodzinski, T.A. and McGrath, J.E. 2002. *J. Membr. Sci.* 197, 231.
73. Kim, Y.S., Summer, M.J., Harrison, W.L., Siffle, J.S., McGrath, J.E. and Pivovar, B.S. 2004. *J. Electrochem. Soc.* 151, A2150.
74. Piela, P., Eickes, C., Brosha, E., Garzon, F. and Zelenay, P. 2004. *J. Electrochem. Soc.* 151, A2053.
75. Lufrano, F., Squadrito, G., Patti, A. and Passalacqua, E. 2000. *J. Appl. Polym. Sci.* 77, 1250.

76. FuMA-Tech GmbH website: http://www.fumatech.com.
77. Genies, C., Mercier, R., Sillion, B., Cornet, N., Gebel, G. and Pineri, M. 2001. *Polymer* 42, 359.
78. Walker, C.W., Jr. 2002. *J. Power Sources* 110, 144.
79. Smitha, B., Sridhar, S. and Khan, A.A. 2004. *Macromolecules* 37, 2233.
80. Polyfuel website: http://www.polyfuel.com.
81. Elabd, Y.A., Napadensky, E., Sloan, J.A., Crawford, D.M. and Walker, C.W. 2003. *J. Membr. Sci.* 217, 227.
82. Feng, S., Shang, Y., Xie, X., Wang, Y. and Xu, J. 2009. *J. Membr. Sci.* 335, 13.
83. Gil, S.C., Kim, J.C., Ahn, D., Jang, J., Kim, H., Jung, J.C., Lim, S., Jung, D. and Lee, W. 2012. J. *Membrane Science* 417–418, 2.
84. Jung, D.H., Myoung, Y.B., Cho, S.Y., Shin, D.R. and Peck, D.H. 2001. *Int. J. Hydrogen Energy* 26, 1263.
85. Ding, J., Chuy, C. and Holdcroft, S. 2002. *Macromolecules* 35, 1348.
86. Kim, Y.S., Wang, F., Hickner, M., McCartney, S., Hong, Y.T., Harrison, W., Zawodzinski, T.A. and McGrath, J.E. 2003. *J. Polym. Sci. B* 41, 2816.
87. Wang, J.T., Wainright, J.S., Savinell, R.F. and Litt, M. 1996. *J. Appl. Electrochem.* 26, 751.
88. Wainright, J.S., Wang, J. and Litt, M.H. 1997. In O. Savadogo and P.R. Roberge, Eds., Proceedings of the Second International Symposium on New Materials for Fuel Cell and Modern Battery Systems, Montréal, Canada, July 6–10.
89. PEMEAS website: http://www.pemeas.com.
90. Pei, H., Hong, L. and Lee, J.Y. 2006. *J. Membr. Sci.* 270, 169.
91. Choi, W.C., Kim, J.D. and Woo, S.I. 2001. *J. Power Sources* 96, 411.
92. Smotkin, E., Mallouk, T., Wardchael, M. and Ley, K. US Patent 5846669.
93. Yoon, S.R., Hwang, G.H., Cho, W.I., Oh, I.H., Hong, S.A. and Ha, H.Y. 2002. *J. Power Sources* 106, 215.
94. Kim, D., Scibioh, M.A., Kwak, S., Oh, I.H. and Ha, H.Y. 2004. *Electrochem. Commun.* 6, 1069.
95. Cho, S.A., Cho, E.A., Oh, I., Kim, H., Ha, H.Y., Hong, S.A. and Ju, J.B. 2006. *J. Power Sources* 155, 286.
96. Lee, K., Nam, J.H., Lee, J.H., Lee, Y., Cho, S.M., Jung, C.H., Choi, H.G., Chang, Y.Y., Kwon, J.U. and Nam, J.D. 2005. *Electrochem. Commun.* 7, 113.
97. Ren, S., Li, C., Zhao, X., Wu, Z., Wang, S., Sun, G., Xin, Q. and Yang, X. 2005. *J. Membr. Sci.* 247, 59.
98. Higuchi, Y., Terada, N., Shimoda, H. and Hommura, S. 2001. Ashai Glass Co., EP1139472.
99. Yamaguchi, T., Miyata, F. and Nakao, S. 2003. *Adv. Mater.* 15, 1198.
100. Yamaguchi, T., Kuroki, H. and Miyata, F. 2005. *Electrochem. Commun.* 7, 730.
101. Kerres, J.A., Ullrich, A., Meier, F. and Haering, T. 2002. *J. Membr. Sci.* 206, 443.
102. Kerres, J.A. and Ullrich, A. 2001. *Sep. Purif. Technol.* 22–23, 1.
103. Lin, J., Ouyang, M., Fenton, J.M., Kunz, H.R., Koberstein, J.T. and Cutlip, M.B. 1998. *J. Appl. Polym. Sci.* 70, 121.
104. Hasiotis, C. 2001. *Electrochim. Acta* 46, 2401.
105. Ren, S., Sun, G., Li, C., Wu, Z., Jin, W. and Chen, W. 2006. *Mater. Lett.* 60, 44.
106. Li, X., Chen, D., Xu, D., Zhao, C., Wang, Z.H., Lu, H. and Na, H. 2006. *J. Membr. Sci.* 275, 134.
107. Qiao, J., Hamaya, T. and Okada, T. 2005. *Polymer* 46, 10809.

108. Cho, K.-Y., Eom, J.-Y., Ung, H.-Y., Choi, N.-S., Lee, Y.M., Park, J.-K., Choi, J.-H., Park, K.-W. and Sung, Y.-E. 2004. *Electrochim. Acta* 50, 583.
109. Kim, H.J., Kim, H.J., Shul, Y.G. and Han, H.S. 2004. *J. Power Sources* 135, 66.
110. Prakash, G.K.S., Smart, M.C., Wang, Q.-J., Atti, A., Pleynet, V., Yang, B., McGrath, K., Olah, G.A., Narayanan, S.R. and Chun, W. 2004. *J. Fluorine Chem.* 125, 1217.
111. Smit, M.A., Ocampo, A.L., Espinosamedina, M.A. and Sebastian, P.J. 2003. *J. Power Sources* 124, 59.
112. Liu, J., Wang, H., Cheng, S. and Chan, K.Y. 2004. *Chem. Commun.* 728.
113. Jörissen, L., Gogel, V., Kerres, J. and Garche, J. 2002. *J. Power Sources* 105, 267.
114. Watanabe, M. 1994. Paper presented at The Electrochemical Society Meeting PV 94–2, Pennington, NJ.
115. Antonucci, P.L., Arico, A.S., Creti, P., Ramunni, E. and Antonucci, V. 1999. *Solid State Ionics* 125, 431.
116. Mauritz, K.A. 1998. *Mater. Sci. Eng. C* 6, 121.
117. Deng, Q., Moore, R.B. and Mauritz, K.A. 1995. *Chem. Mater.* 7, 2259.
118. Deng, Q., Hu, Y., Moore, R.B., McCormick, C.L. and Mauritz, K.A. 1997. *Chem. Mater.* 9, 36.
119. Deng, Q., Wilkie, C.A., Moore, R.B. and Mauritz, K.A. 1998. *Polymer* 39, 5961.
120. Jung, D.H., Cho, S.Y., Peck, D.H., Shin, D.R. and Kim, J.S. 2002. *J. Power Sources* 106, 173.
121. Zoppi, R.A., Yoshida, I.V.P. and Nunes, S.P. 1997. *Polymer* 39, 1309.
122. Zoppi, R.A. and Nunes, S.P. 1997. *Electroanal. Chem.* 445, 39.
123. Kim, H., Cho, J., Yoon, J. and Chang, H. 2004. Korea Patent KR-P0413801.
124. Nunes, S.P., Ruffmann, B., Rikowski, E., Vetter, S. and Richau, K. 2002. *J. Membr. Sci.* 203, 215.
125. Kang, S., Peck, D.H., Park, Y.C., Jung, D.H., Jang, J.H. and Lee, H.R. 2008. *J. Phys. Chem. Solids* 69, 1280.
126. Dimitrova, P., Friedrich, K.A., Stimming, U. and Vogt, B. 2002. *Solid State Ionics* 150, 115.
127. Liang, Z.X., Zhao, T.S. and Prabhuram, J. 2006. *J. Membr. Sci.* 283, 219.
128. Jiang, R.C., Kunz, H.R. and Fenton, J.M. 2006. *J. Membr. Sci.* 272, 116.
129. Staiti, P., Arico, A.S., Baglio, V., Lufrano, F., Passalacqua, E. and Antonucci, V. 2001. *Solid State Ionics* 145, 101.
130. Adjemian, K.T., Lee, S.J., Srinivasan, S., Benziger, J. and Bocarsly, A.B. 2002. *J. Electrochem. Soc.* 149, A256.
131. Ponce, M.L., de A. Prado, L.A.S., Silva, V. and Nunes, S.P. 2004. *Desalination* 162, 383.
132. Silva, V.S., Weisshaar, S., Reissner, R., Ruffmann, B., Vetter, S., Mendes, A., Madeira, L.M. and Nunes, S. 2005. *J. Power Sources* 145, 485.
133. Jaafar, J., Ismaila, A.F., Matsuuraa, T. and Nagai, K. 2011. *J. Membr. Sci.* 382, 202.
134. So, S.Y., Hong, Y.T., Kim, S.C. and Lee, S. 2010. *J. Membr. Sci.* 346, 131.
135. Vaivars, G., Maxakato, N.W., Mokrani, T., Petrik, L., Klavins, J., Gericke, G. and Linkov, V. 2004. *Mater. Sci.* 10, 162.
136. Yang, C., Srinivasan, S., Aricò, A.S., Creti, P. and Baglio, V. 2001. *Electrochem. Solid-State Lett.* 4, A31.
137. Peled, E., Duvdevani, T., Aharon, A. and Melman, A. 2000. *Electrochem. Solid-State Lett.* 3, 525.
138. Maiti, J., Kakati, N., Lee, S.H., Jee, S.H. and Yoon, Y.S. 2011. *Solid State Ionics* 201, 21.

139. Mohtar, S.S., Ismail, A.F. and Matsuura, T. 2011. *J. Membr. Sci.* 371, 10.
140. Jung, D.H., Cho, S.Y., Peck, D.H., Shin, D.R. and Kim, J.S. 2003. *J. Power Sources* 118, 205.
141. Ponce, M.L., Prado, L., Ruffmann, B., Richau, K., Mohr, R. and Nunes, S.P. 2003. *J. Membr. Sci.* 217, 5.
142. Rhee, C., Kim, H., Lee, J.S. and Chang, H. 2005. *Chem. Mater.* 17, 1691.
143. Wei, Y., Matar, S., Shen, L., Zhang, X., Guo, Z., Zhu, H. and Liu, H. 2012. *Int. J. Hydrogen Energy* 37, 1857.

13

Nanostructure Advances in Catalysts for Direct Alcohol Fuel Cells

XiaoChun Zhou and Hui Yang

CONTENTS

13.1 Introduction 376
13.2 Nanocube Catalysts 380
    13.2.1 Structure Characteristics 380
        13.2.1.1 TEM, XRD Characteristics of Pt and Pt Alloy
                 Nanocubes 380
        13.2.1.2 SEM, XRD, SAED, and HRTEM Characteristics of
                 Pd Nanocubes 382
        13.2.1.3 Element Analysis 383
    13.2.2 Synthesis Methods 383
        13.2.2.1 Pt and Pt-M(Co, Fe, Ni, Mn, etc.) Nanocube Synthesis ... 383
        13.2.2.2 Pd Nanocube Synthesis 385
    13.2.3 Electrochemical Performance 385
        13.2.3.1 Hydrogen Adsorption/Desorption on Pt and Pt
                 Alloy Nanocubes 386
        13.2.3.2 Alcohol Oxidation on Pt Nanocubes 386
        13.2.3.3 Alcohol Oxidation on Pt Alloy Nanocubes 387
        13.2.3.4 Oxygen Reduction on Pt and Pt Alloy Nanocubes 388
        13.2.3.5 Alcohol Oxidation on Pd Nanocubes 390
13.3 Core-Shell and Decorated Structure Catalysts 391
    13.3.1 Structure Characteristics 391
        13.3.1.1 TEM and HRTEM Characteristics of Core-Shell
                 Structure 391
        13.3.1.2 XRD Characteristics of Core-Shell Structure 392
        13.3.1.3 XPS Characteristics of Core-Shell Structure 393
        13.3.1.4 EDS Characteristics of Core-Shell Structure 394
    13.3.2 Synthesis Methods 394
    13.3.3 Electrochemical Performance 396
        13.3.3.1 Carbon Monoxide (CO) and CO-Like Oxidation
                 on Ru@Pt 396
        13.3.3.2 Methanol Oxidation on Core-Shell Catalysts 398
        13.3.3.3 Ethanol Oxidation on Core-Shell Catalysts 400

        13.3.3.4 Ethylene Glycol Oxidation on Ru@Pt............................401
        13.3.3.5 Oxygen Reduction Reaction.....................................401
13.4 Hollow Structure Catalysts...405
    13.4.1 Structure Characteristics...405
    13.4.2 Synthesis Methods...406
    13.4.3 Electrochemical Performance of Hollow Structure
           Nanoparticles...407
        13.4.3.1 Methanol Oxidation...408
        13.4.3.2 Ethanol Oxidation..409
        13.4.3.3 Oxygen Reduction...409
13.5 One-Dimensional Structure Catalysts..................................410
    13.5.1 Pure Metal Nanowires...412
    13.5.2 Multi-Segment Nanowires..419
    13.5.3 Alloy Nanowires...421
13.6 Hierarchical Structure Catalysts...424
13.7 Other Shape Catalysts...426
13.8 Summary and Perspectives...428
References...430

13.1 Introduction

Direct alcohol fuel cells (DAFCs) have been studied since 1962 [1]. They are thought to be the replacement candidates for H_2/O_2 fuel cells in some application fields, such as portable devices, signal devices, emergency power generators, etc., due to their unique advantages of high safety and high energy density. Moreover, DAFCs are receiving more and more attention because of more and more personal portable devices and more serious environment and energy problems [2]. Figure 13.1 shows that the number of papers related to direct methanol fuel cells (DMFCs) obviously flies up from 1990, and the number of published papers even rises over 2000 during the period from 2009 to 2010. The total number has already been more than 2000 during the period from 2011 to October 2012.

As we know, methanol is the most important fuel of all alcohols due to its high reactivity, wide resources, high energy density, low price, etc. However, methanol still has some shortcomings, such as toxicity, low boiling point, permeating through proton exchange membrane (PEM), etc. So some other alcohol fuels, such as ethanol, isopropanol, ethylene glycol (EG), and glycerol, are also studied as candidates. Especially, ethanol has a higher theoretical mass energy than methanol (8 kWh kg^{-1} vs. 6.1 kWh kg^{-1}) [3], and is considered as a "green" energy for its massive production from renewable biofuels [4], low toxicity and low price.

Since the beginning of DAFCs, Pt electrocatalysts were recognized as very effective catalysts for the alcohol electrooxidation reaction (AEOR) and oxygen

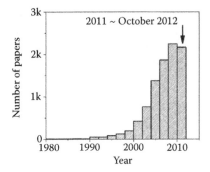

FIGURE 13.1
Literature number distribution versus years for DMFCs. The number of papers was obtained by searching *Web of Science* with keywords "fuel cell" and "methanol" every two years from 1960 to 2012. The last column is the number of papers from January 2011 to October 2012.

reduction reaction (ORR). However, Pt electrocatalysts are always poisoned by some poisonous intermediates from the decomposition of alcohols. The poisonous intermediates can be removed at high potential, but the potential is meaningless for real fuel cells. In order to lower the potential, promoters (e.g., Ru, Mo, W) were added into Pt electrocatalysts because the promoters could provide the promotion effect of the bifunctional effect [5] or giving electronic effect [6] or both of them. A large number of di- and tri- Pt-based catalysts, such as PtRu, PtSn, PtW, etc. [7–10], were synthesized. Among the large numbers of Pt-based catalysts, the PtRu catalyst is the most famous and widely used one because of its high activity and stability [11]. During the same period, synthesis methods, such as the impregnation method [12,13], colloidal method [14,15], spray pyrolysis method [16], etc., were developed successfully. The preparation methods and compositions of Pt-based catalysts for methanol [7–10] and ethanol [17] electrooxidation have been well reviewed.

Besides the anode with fuel, DAFCs also include a cathode with oxidant, which usually is pure oxygen or oxygen in air. So ORR is another important reaction for DAFCs. Currently, the most practical catalysts in DAFCs are highly dispersed Pt-based nanoparticles supported on a carbon black surface. But pure Pt catalysts have some drawbacks, such as high cost, sensitivity to contaminants, no tolerance to alcohol oxidation, fewer completed four-electron reduction reactions, Pt dissolution, etc. In order to solve these problems, transition metal–doped Pt-based alloy catalysts, including PtCo [18–21], PtFe [22], PtCr [23], PtVFe [24], PtPd [25], etc., were synthesized. Traditional surface science revealed that Pt (100) planes are more active than Pt (111) [26,27] for ORR in acidic media. So the surface structure has a strong effect on the activity of Pt-based catalysts.

Recently, the effect of nanostructure on fuel cell catalysts is attracting more and more attentions due to the outstanding advantages and huge achievements in the nanoscience field. Some nanostructures, such as core-shell structures, nanorods, nanowires, nanotubes, and nanocubes, are extensively studied because of their special characteristics. Figure 13.2 shows the number of published papers of these nanostructures applied in DAFCs from the year 2000 to 2011. The plots clearly exhibit that there are very few papers in the year 2000, but more and more nanostructure applications have been done in DAFCs after 2000.

Nanoparticles with different shapes and structures usually show different activity and selectivity. Sometimes, special shapes and structures can enhance the activity and selectivity of nanoparticles to an outstanding stage. The enhancement is basically attributed to four reasons: First, nanoparticles

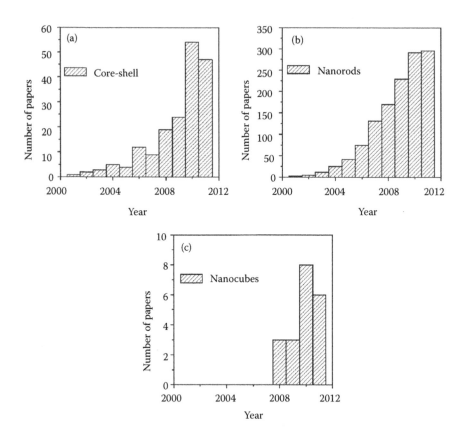

FIGURE 13.2
Literature number distributions versus year for (a) core-shell structures; (b) nanorods, nanowires, and nanotubes; (c) nanocubes. The results were obtained by searching *Web of Science* with keywords core shell and methanol or ethanol or propanol or EG or glycerol and fuel cell for (a); nanowires or nanotubes or nanorods and methanol or ethanol or propanol or EG or glycerol and fuel cell for (b); and nanocube or nanocubes and methanol or ethanol or propanol or EG or glycerol and fuel cell for (c).

with different shapes can expose specific facets or surface sites, some of which have higher activity for AEOR [28] or ORR [29]. Second, the special nanostructure (e.g., core-shell structure) can reduce the noble metal loading and thus increase the mass activity. Third, special nanostructures (e.g., core-shell structures) can show electronic or quantum effect to promote the catalysis process [30,31]. Fourth, special nanostructures (e.g., nanowire structures) can facilitate mass and electron transfer [32,33].

Before 1996, many studies on colloidal particles were focused on the control of particle sizes, their growth kinetics and the relationship between their size and catalytic activity. In 1996, Ahmadi et al. [34] first succeeded in synthesizing nanoparticles with multiple shapes, such as tetrahedral- and cubo-octahedral-shaped Pt nanoparticles, whose distribution was dependent on the concentration ratio of the capping polymer material (sodium polyarylate) to the platinum cation. However, the mechanism of shape- or morphology-dependent synthesis of colloidal nanoparticles was not known at that time.

After 1996, researchers realized the special characteristics of well-defined nanostructures in optics, catalysis, biology, and so on. More and more shapes of nanoparticles, some of which are shown in Figure 13.3 [34–37], were synthesized and studied afterwards. The shape-controlled synthesis of diverse palladium and silver nanostructures was well reviewed by Xiong and Xia [38,39], and Pt-based nanostructures were reviewed by Chen and Holt-Hindle [40]. Nanoparticles with some well-defined structures were also studied in AEOR and ORR.

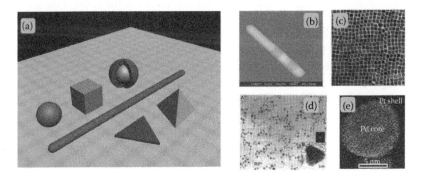

FIGURE 13.3
(See color insert.) Some example shapes and structures of nanoparticles, which had been synthesized. (a) Some ideal shapes. (b) Field emission scanning electron microscopy (FESEM) image of five-segmented nanorods, Pt-RuNi(2)-Pt-RuNi(2)-Pt. (From Fang, L. et al. *Journal of the Electrochemical Society* 153: A2133–A2138, 2006.) (c) Transmission electron microscopy (TEM) image of Pt nanocubes. (From Yang, H.Z. et al. *Angewandte Chemie International Edition* 49: 6848–6851, 2010.) (d) TEM image of Pt tetrahedral nanoparticles. (From Ahmadi, T.S. et al. *Science*, 272: 1924–1925, 1996.) (e) High resolution TEM (HRTEM) image of Pd-Pt core-shell nanoparticle. (From Long, N.V. et al. *International Journal of Hydrogen Energy* 36: 8478–8491, 2011.)

13.2 Nanocube Catalysts

It has been extensively shown that platinum surface facets can strongly affect the methanol electrooxidation on Pt catalyst. For example, Housmans et al. [28] found that the reactivity of the methanol oxidation reaction (MOR) on Pt(111), Pt(110), and Pt(100) increases in the order Pt(111) < Pt(110) < Pt(100). Pt, Pt alloys, and Pd nanocubes, which are preferential to expose (100) facet on their six faces, are such kinds of specially shaped nanoparticles. They possibly show some distinct activity and selectivity toward AEOR and ORR, which are primary reactions in DAFCs, as compared to spherical nanoparticles.

Currently, researchers have already synthesized some nanocubes, such as Pt [36], PtFe [41], Pt_3Fe [42], Pt_3Co [36], PtCo [43], Pd [44], etc. Some of them have been tested in AEOR and ORR, and showed very promising performance. However, the mass-specific activity of nanocubes sometimes is slightly lower than small-size spherical nanoparticles due to the bigger size of nanocubes. A problem is how to preserve the improved activity by cubic shape but reduce the mass consumption of precious metals. Maybe one of the solutions is to replace the inner part of nanocubes by some cheaper materials to form a core-shell structure, which will be introduced in next section.

13.2.1 Structure Characteristics

13.2.1.1 TEM, XRD Characteristics of Pt and Pt Alloy Nanocubes

In order to take the TEM images of nanocubes, the particles need to be dispersed in a solvent (e.g., water, ethanol, etc.) under sonication. The particle suspension will be deposited on a carbon-coated copper grid. Then, the copper grids are installed into a TEM machine to take some TEM images.

TEM technology is widely used to measure the shape and size of nanoparticles [29,45–48]. The typical platinum nanocubes are shown in Figure 13.4a and b [36]. Figure 13.4a shows high yield and well mono-dispersed Pt nanocubes, which assemble to a very ordered pattern. A HRTEM image in Figure 13.4b reveals highly crystalline cubes with clearly resolved lattice fringes. The measured d spacings 1.96 Å for Pt are consistent with those lattice planes in other literature [44] and demonstrate that Pt nanocubes are perfect (100)–orientated structures. The typical Pt alloy (Pt_3Co) nanocubes, which are as good as Pt nanocubes, are shown in Figure 13.4c and d [36]. The measured d spacings 1.92 Å for Pt_3Co in Figure 13.4d not only demonstrates that Pt_3Co nanocubes are perfect (100)–orientated structures but also indirectly verifies the composition of Pt_3Co.

X-ray diffraction (XRD) is another commonly used technique [29,45,46,49–52], which can check the composition and crystal phase of the obtained samples. For example, Zhang and Fang [49] prepared their XRD sample by depositing the nanocubes on a surface-polished Si wafer. The resultant XRD pattern of each sample shows that as-synthesized nanocubes possess a highly

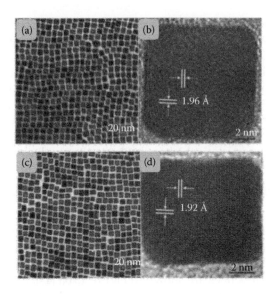

FIGURE 13.4
(a) TEM image of Pt nanocubes, (b) HRTEM image of a single Pt nanocube, (c) TEM image of Pt₃Co nanocubes, and (d) HRTEM image of a single Pt₃Co nanocube.

crystalline face-centered cubic Pt phase shown in Figure 13.5A (e through h). There is a greatly enhanced (200) peak as compare to the XRD patterns of the same Pt₃M cubic samples randomly deposited on a regular XRD holder/ substrate as in Figure 13.5A (a through d). This enhancement of the (200) peak in the XRD observation indicates that these Pt₃M nanocubes align perfectly flat on the surface of the substrates with (100) texture, and further supports that the Pt₃M nanocubes have a (100)-dominated cubic morphology with

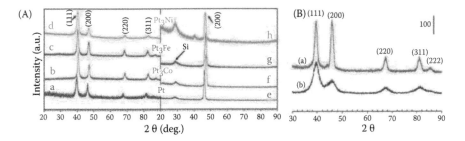

FIGURE 13.5
XRD patterns of (A) Pt₃M nanocubes: (a,e) Pt, (b,f) Pt₃Co, (c,g) Pt₃Fe, and (d,h) Pt₃Ni. (a–d) Samples were randomly deposited on a zero-background Si sample holder. (e–h) Samples were assembled on a surface-polished 25 mm Si <100> wafer. (From Zhang, J., and Fang, J., *Journal of the American Chemical Society* 131, 18543–18547, 2009.) (B) XRD patterns for the as-synthesized Pt nanocubes (a, top curve) and Pt nanoparticles (b, bottom curve). (From Loukrakpam, R. et al. *Chemical Communications*, 46, 7184–7186, 2010.)

very narrow shape distributions. However, no diffraction signal of pure M and/or pure W was detected from all of the patterns in Figure 13.5A, which means the formation of alloys and no single-phase metals.

A XRD comparison was done between as-synthesized Pt nanocubes and the spherical nanoparticles as shown in Figure 13.5B [50]. The lattice parameter was found to be 0.3940 and 0.3947 nm for the nanocubes and spherical nanoparticles, respectively. Interestingly, the flat alignment of the nanocubes on the substrate induced the relative peak intensity (200) increase one time compare to Pt nanoparticles, which is similar to the result above.

13.2.1.2 SEM, XRD, SAED, and HRTEM Characteristics of Pd Nanocubes

In order to take SEM images, the precipitate is dispersed into a solvent (e.g., water, ethanol) under sonication. The suspension was dropped on the cleaned substrate (e.g., ITO glass, glassy carbon, etc.) and dried in air for SEM characterization [53].

The typical Pd nanocubes are shown in Figure 13.6a [44]. Nearly monodisperse single-crystalline Pd nanocubes were successfully prepared in aqueous solution at room temperature. The sample consists of cubic particles and has eight sharp corners with a very uniform size of about 60 nm. The XRD pattern

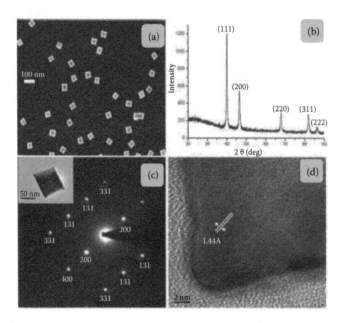

FIGURE 13.6
(a) SEM image of Pd nanocubes; (b) XRD pattern of the as-prepared Pd nanocube; (c) TEM image (inset) and the corresponding SAED pattern taken with the incident beam perpendicular to one facet of the cube; (d) typical HRTEM image of a selected corner of an individual Pd nanocube. (From Fan, F.-R. et al. *Crystal Growth & Design*, 9, 2335–2340, 2009.)

of the as-prepared Pd nanocubes is shown in Figure 13.6b. All of the peaks can be indexed to fcc bulk Pd metal. The absence of other diffraction peaks other than Pd metal indicates that this material was synthesized successfully as a pure phase. SAED pattern of the corresponding nanoparticle (inset in Figure 13.6c), which was obtained by directing the electron beam to one of the facets of the cube, infers that the particle is a single crystal enclosed by six (100) faces. Figure 13.6d shows the HRTEM image of a selected corner of an individual Pd nanocube and the lattice fringes of the image were measured to be 0.144 nm in agreement with the (220) lattice spacing of the fcc crystal, which also indicates that the individual cube is a single crystal [44].

13.2.1.3 Element Analysis

For the composition measurement of alloy nanocubes, inductively coupled plasma mass spectrometry (ICP-MS), x-ray spectroscopy (EDS), x-ray photoelectron spectroscopy (XPS), and XRD were commonly used. Yang et al. [36] used ICP-MS analysis to measure the molar ratio (ca. 3.02:1.00) between Pt and Co in this Pt_3Co sample, which is in good agreement with the average result from energy-dispersive EDS evaluation (about 76.8:23.2). Further support for 3:1 Pt/Co molar ratio can be found in XRD patterns.

13.2.2 Synthesis Methods

13.2.2.1 Pt and Pt-M(Co, Fe, Ni, Mn, etc.) Nanocube Synthesis

Essentially, Pt nanocubes are synthesized by reducing the Pt ion to Pt metal with the protection of capping ligands. The early synthesis of Pt nanocrystals with different shapes and crystal facets was reported by El-Sayed's group [34] in 1996. They synthesized Pt nanocubes by reducing K_2PtCl_4 with H_2 using sodium polyacrylate as capping ligands. In detail, they prepared a water solution of 250 ml 1×10^{-4} M K_2PtCl_4 to which they added 0.2 ml of 0.1 M sodium polyacrylate. The Pt ions were reduced by H_2 for 12 h, and the solution color turned lightly golden. The TEM image showed that the sample contained 80% cubic particles [34].

After 1996, some other methods, such as the colloidal method [34,54], thermal decomposition [52], polyol method [55], liquid–liquid phase transfer method [43], etc., were developed for Pt nanocube synthesis. For all methods, capping ligands are necessary for shape control and stabilization because the facet (100) on platinum nanocubes has large surface energy at bare status. Originally, El-Sayed's group used sodium polyacrylate as capping polymer material [34] and was followed by other researchers [54]. Lately, some other stabilizers, such as oleylamine + oleic acid [36], polyvinylpyrrolidone (PVP) [48,52], 1-octadecene + oleic acid + oleylamine [56], 1,2-hexadecanediol + adamantanecarboxylic acid + hexadecylamine + diphenyl ether [45], cetyltrimethylammonium bromide (CTAB) [57], etc., were found to be effective in making Pt nanocubes. In addition, the exposure of facets can also be controlled by surfactants and additives (Fe^{3+}, Ag^+, etc. [58,59]) in the chemical process [48]. The frequently used reductants are H_2 [34], CO [41,46,49,60], $NaBH_4$ [61], etc.

Pt-M (Co, Fe, Ni, Mn, etc.) alloy nanocubes are now being vigorously studied because of their special chemical and magnetic properties. Chen et al. [60] synthesized monodisperse 6.9 nm PtFe nanocubes at 205°C by controlling decomposition of $Fe(CO)_5$ and reduction of $Pt(acac)_2$ under the protection of oleic acid and oleylamine. In detail, the $Pt_{50}Fe_{50}$ nanocubes were synthesized by mixing oleic acid and $Fe(CO)_5$ with benzyl ether/octadecene solution of $Pt(acac)_2$ and heating the mixture to 120°C for about 5 min before oleylamine was added. Then the mixture was heated at 205°C for 2 h. The two key conditions are the temperature 205°C and adding oleic acid first.

Later, Zhang and Fang [49] developed a facile, reliable, general, and robust synthetic method for preparation of high-quality, (100)-terminated Pt_3M nanocubes (M = Pt or 3d-transition metals Co, Fe, and Ni). They found that the addition of $W(CO)_6$ is crucial to control the nucleation process when the metallic precursors are reduced. The $W(CO)_6$ might induce the formation of a large number of critical nuclei in a short interval of time, followed by the simultaneous and steady growth of those nuclei. Figure 13.7 illustrates the mechanism of this method. In addition, monodisperse crystalline Pt_9Co [62], PtFeCo [56], and PtMn [29] nanocubes were prepared by other researchers.

It is broadly accepted that the evolution of nanocubes in a solution system consists of a nucleation stage and a subsequent Ostwald ripening growth on the existing seeds (or nuclei). The rates of nucleation and subsequent nanocube growth are the keys to shape control [39,63]. It is reasonable that freshly formed tiny nanocrystallites have diverse crystallographic planes on their surfaces, which are possible to grow. Ideally, a large number of critical nuclei should be formed in a short interval of time, followed by the simultaneous and steady growth of those nuclei [64]. A relatively slow rate of crystal growth usually results in growth much more selectively in crystal directions and benefits the shape control of nanocubes [57]. A solution-based evolution environment with the surfactant-binding ability [39] may tune the surface

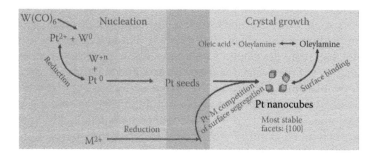

FIGURE 13.7
Illustration of both nucleation and crystal growth processes for Pt_3M nanocubes (M-Co, Fe, and Ni). (From Zhang, J., and Fang, J., *Journal of the American Chemical Society* 131, 18543–18547, 2009.)

energy of certain facets to be lower than others; then other planes will be eliminated in later growth. Therefore, oleic acid + oleylamine [36,49,60], PVP [48,52], CTAB [57], etc., are necessary to form Pt and PtM (M = Co, Fe, Ni, etc.) nanocubes. These surfactants are the key point to enabling the lowest total surface energy on (100) facets in order to develop cubic nanocrystals.

13.2.2.2 Pd Nanocube Synthesis

The principle of Pd nanocube formation is similar to that of Pt nanocubes. Pd ions are reduced to a metallic particle by some reductants (e.g., EG) under the protection of capping ligands (e.g., EG/PVP). The capping ligands can protect the facet (100) to induce the formation of Pd nanocubes.

Pd nanocube synthesis was first succeeded by Xia's group in 2005 [65]. They synthesized monodisperse Pd nanocubes in EG/PVP organic system through a seed etching process by reducing $PdCl_4^{2-}$ with EG. They found that the cube dimension is dependent on the concentration of the etchant $FeCl_3$, and produced 8, 25, and 50 nm Pd nanocubes.

In 2007, Chang et al. [53] reported a facile method to synthesize monodisperse 85 nm Pd nanocubes in aqueous solution at room temperature. In detail, a solution of 0.5 ml 0.01 M K_2PdCl_6 was added in 18 ml of 0.1 M CTAB aqueous solution with stirring at 25°C. Then, 0.1 ml of 0.1 M ascorbic acid was added, followed by the addition of 0.12 ml of 0.1 M NaOH. The solution was just shaken to mix it and then left for 24 h without stirring. The color of the solution gradually changed from light brown to brown black. After 24 h, the solution was centrifuged at 3500 rpm for 15 min to obtain black precipitate. The CTAB acts as a capping reagent, and ascorbic acid works as a reductant.

Fan et al. [44] also prepared nearly monodisperse single-crystalline palladium (Pd) nanocubes and nanodendrites using a similar method. They found that the morphology of the obtained Pd nanocrystals had eight sharp corners with a very uniform size of about 60 nm and could be tuned by the addition of foreign halide ions (Cl⁻ and Br⁻).

13.2.3 Electrochemical Performance

Because the size, shape, and structure of metal nanostructures have a significant effect on the catalytic reactivity of a nanocatalyst, the well-defined nanostructures are essential in preparing high-performance nanocatalysts for DAFCs. The Pt-based or Pd-based cubic nanoparticles are the promising high active catalysts for AEOR and ORR due to their exposure of (100) facet, which has high surface energy and consequently higher activity.

The Pt nanocubes were usually deposited on carbon black (e.g., Vulcan XC-72R) to form a carbon-supported catalyst (cubic Pt/C) to prevent the aggregation of nanocubes [52]. The supported catalysts are dispersed into a solvent (e.g., water, ethanol) with some binders (e.g., Nafion) under sonication. Then the suspension will be deposited onto an electrochemical electrode (e.g., glassy carbon). At last, the catalyst is usually tested by electrochemical methods, such as cyclic voltammetry (CV) and linear scan.

13.2.3.1 Hydrogen Adsorption/Desorption on Pt and Pt Alloy Nanocubes

It has been established that (100) facets of a Pt crystal surface lead to a distinct hydrogen desorption peak between −0.05 and −0.02 V (vs. Ag/AgCl [saturated KCl]) while the surface defects or corners/edges or (110) facets will show the peak between −0.15 and −0.10 V (vs. Ag/AgCl [saturated KCl]) [27]. Figure 13.8 shows the hydrogen desorption region curves of cubic Pt/C and spherical Pt/C [52]. The cubic Pt/C exhibits a relatively intense peak at −0.031 V while spherical Pt/C exhibits a relatively intense peak at −0.125 V. The relative difference between these two peaks further proves that cubic Pt/C with dominantly exposed (100) facets has distinct electrochemical activity as compare to spherical Pt/C with polycrystalline structure.

13.2.3.2 Alcohol Oxidation on Pt Nanocubes

The size of Pt nanocubes can be controlled in ~4.5 nm [52], ~8 nm [50,66,67], ~9 nm [36,57], ~10 nm [54,68], ~20 nm [69], and >20 nm [70], which is usually larger than spherical Pt nanoparticles of typically ~3 nm in diameter. The shape-controlled Pt nanoparticles thus exhibit smaller electrochemical active surface areas (EASAs) in comparison to the spherical Pt nanoparticles.

The activity of Pt nanocubes toward the alcohol electrooxidation is found to be higher than that of Pt spherical nanoparticles because Pt nanocubes have dominantly exposed (100) facets, which are different from the spherical nanoparticles with multitudinous facets. Han et al. [48] found that the onset potential for methanol electrooxidation on the Pt nanocube catalyst shifted to 0.17 V compared to 0.24 V of the polycrystalline Pt nanocatalyst. Moreover, the

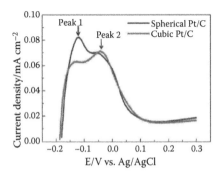

FIGURE 13.8
Hydrogen desorption region curves of cubic Pt/C and spherical Pt/C in Ar-saturated 0.1 M HClO$_4$ at a scan rate of 50 mV s^{-1} at 25°C. (From Lee, Y.-W. et al. *Chemical Communications* 47, 6296–6298, 2011.)

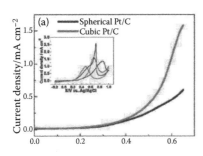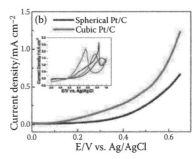

FIGURE 13.9
CVs of cubic Pt/C and spherical Pt/C in Ar-saturated (a) 0.1 M $HClO_4$ + 2.0 M CH_3OH; (b) 2.0 M C_2H_5OH at a scan rate of 50 mV s^{-1} at 25°C. (From Lee, Y.-W. et al. *Chemical Communications* 47, 6296–6298, 2011.)

ratio of forward scan peak and backward scan peak (I_f/I_b) of the Pt nanocube catalyst is higher than that of the polycrystalline Pt nanocatalyst. Higher I_f/I_b ratio implies excellent oxidation of methanol during the reverse anodic scan and less accumulation of residues on the catalyst according to Goodenough's research [71]. Lee et al. [52] tested the specific activity of cubic Pt/C for methanol, ethanol, or formic acid and compared it to spherical Pt/C. The cubic Pt/C exhibits about 1.8–2.4 times higher electrochemical activity for all three molecules than spherical Pt/C at 0.6 V. For the ethanol electrooxidation in Figure 13.9b, the onset potential of cubic Pt/C (+0.05 V) is much lower than that of spherical Pt/C (+0.16 V). Lu et al. [54] found that Pt nanocrystals were nearly three times higher than that on a commercial available Pt black catalyst.

13.2.3.3 Alcohol Oxidation on Pt Alloy Nanocubes

The Pt catalyst is very active for alcohol electrooxidation, but it is easily inhibited by the poisonous intermediates (e.g., CO). The dual- or tri-Pt-based alloys are widely used in alcohol electrooxidation and DAFCs because the second or third elements can facilitate the removal of the poisonous intermediates at lower potential due to the bifunctional effect [5] or electronic effect [6] or both of them. In addition, the edge of stepped (100) faces in the Pt nanocube catalyst was preferable to the breakage of CH_3OH and CH_3CH_2OH [48], which is critical in alcohol oxidation.

Kang and Murray [29] synthesized PtMn nanocubes from platinum acetylacetonate and manganese carbonyl in the presence of oleic acid and oleylamine. For methanol oxidation, the cubic PtMn nanoparticles exhibit better activity than commercial E-TEK Pt while spherical PtMn nanoparticles are higher in activity to E-TEK Pt for methanol oxidation. These results suggest that the (100) surface of PtMn is more active for methanol oxidation than the (111) surface of PtMn.

Although PtM nanocubes show higher activity of (100) facet [29,72], their mass specific current density is sometimes lower than that of the sphere-like nanoparticles [45]. Moreover, their mass specific current density decreases with the increasing size. Teng and Yang [45] found that the mass current densities decreased from 32 ± 2.4 to 30 ± 2.5 and to 20 ± 0.9 mA mg^{-1} metal for PtCo nanocubes when the edge length of cubic NPs increased from 9.4 to 10.5 and to 13.2 nm, respectively.

The capping ligands on nanoparticles should be removed before the catalytic activity measurement because PtM nanocubes are usually synthesized under the protection of capping ligands. Yang et al. [36] used argon plasma treatment and potential cycling between 0.05 and 1.0 V to remove organic residuum and surfactant. The cyclic voltammogram shows that Pt$_3$Co nanocubes have two pairs of peaks at about 0.20 V and 0.30 V, which are attributed to hydrogen adsorption/desorption on Pt(100) surface sites, suggesting the particle surface is clean. The oxidation current peaks of methanol at 0.87 V on Pt$_3$Co nanocubes, which is about 30 mV more negative than that on Pt nanocubes. The overall current density on the positive potential sweep is higher on Pt$_3$Co nanocubes [51].

Currently, Pt, PtMn [29], PtNi [49], PtCo [43,51,62], PtFe [41,42,60,73], PtFeCo [56], PtCu [72,74], and PtPd [75] have been synthesized and tested in alcohol oxidation. Unfortunately, the most interesting PtRu nanocubes are still not found to be successfully synthesized. The reason may be due to the proper capping ligand is still not successfully discovered yet.

13.2.3.4 Oxygen Reduction on Pt and Pt Alloy Nanocubes

Previous studies revealed that Pt (100) planes are more active than Pt (111) [26,27] for ORR in acidic media. Notably, all six faces of Pt and Pt alloy nanocubes expose (100) facet, which makes it possible to make Pt-based nanocubes much more active than polyhedral and truncated cubic Pt nanoparticles [76]. The activities of catalysts for ORR are usually measured with a rotating disk electrode (RDE) in O$_2$-saturated H$_2$SO$_4$ or HClO$_4$ solution. A linear scan is carried out from high to low potential. For example, Choi et al. [62] compared the activities of Pt/C (JM), spherical Pt$_9$Co/C, cubic Pt/C and cubic Pt$_9$Co/C in Figure 13.10. The onset potential of the Pt$_9$Co nanocubes has a more positive value than those of the other catalysts, and the current density at half-wave potential (about 0.68 V vs. NHE) of the Pt$_9$Co nanocubes is about 4, 1.6, and 1.9 times higher than that of the commercial Pt/C catalyst, the spherical Pt$_9$Co, and the cubic Pt nanoparticles, respectively [62]. In the kinetically controlled region, the current density on cubic Pt$_9$Co is higher than those three catalysts [62]. A similar improvement was observed by other researchers in Pt nanocubes [50,57,77], PtMn nanocubes [29].

Wang et al. [76] compared the performance of nanocubes, polyhedral and truncated cubic Pt nanoparticles. They used Levich–Koutecky plots shown

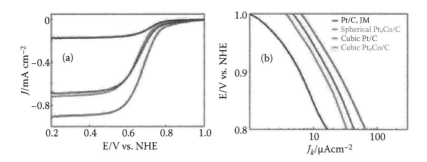

FIGURE 13.10
(a) The polarization curves in O_2 saturated 0.5 M H_2SO_4 at a scan rate of 10 mV s^{-1} and a rotation rate of 1600 rpm. (b) J_k for the corresponding nanocatalysts. (From Choi, S.-I. et al. *Chemical Communications* 46, 4950–4952, 2010.)

in Figure 13.11 to study the relationship between disk electrode current density (J) and the rotation rate in the following equation [78]:

$$\frac{1}{J} = \frac{1}{J_k} + \frac{1}{J_{diff}} = \frac{1}{J_k} + \frac{1}{B\omega^{1/2}} \tag{13.1}$$

in which

$$B = \frac{0.62nFC_0D_0^{2/3}}{\eta^{1/6}}$$

where J_k is the kinetic current density and J_{diff} the diffusion limiting current density, n is the overall number of electrons transferred, F is the Faraday constant, C_0 is the O_2 concentration in the electrolyte, D_0 is the diffusion coefficient of O_2 in the H_2SO_4 electrolyte, and η is the viscosity of the electrolyte. In the J^{-1} versus $w^{-1/2}$ plot, the slope is $1/B$. The so-called B factor can be applied to obtain the number of electrons involved in the ORR. Then the calculated number n for the 7 nm platinum nanocubes is 3.6 while it is 0.7 for the polyhedral and truncated cubic Pt nanoparticles. So the ORR on platinum nanocubes is nearly a four-electron process, which is an ideal process for ORR.

However, the mass activity of the Pt nanocube catalyst is usually comparable with that for the spherical Pt catalyst although the specific area activity of the nanocube catalyst is higher. The low mass activity of the Pt nanocube is because as-prepared samples were usually relative bigger Pt nanocubes (>5 nm) as compared to spherical particles. Consequently, a large portion of Pt atoms are wasted in the center of Pt nanocubes. Therefore, the center of Pt nanocubes is better to be empty or replaced by other cheaper materials in order to lower Pt consumption, reserve the activity at the mean time, and

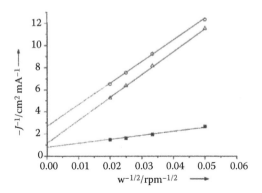

FIGURE 13.11
Koutecky–Levich plots at 0.6 V for the platinum catalysts (△ polyhedron, ○ truncated cube, ■ cube). The potential was applied with scanning rate of 10 mV s⁻¹. (Bard, A.J., and Faulkner, L.R.: *Electrochemical Methods: Fundamentals and Applications.* 2000. Copyright Wiley-VCH Verlag GmbH & Co. KGaA. Reproduced with permission.)

consequently improve the mass activity. These structures are hollow structure and core-shell structure, which will be introduced in next sections.

13.2.3.5 Alcohol Oxidation on Pd Nanocubes

A Pd catalyst almost has no activity for AEOR in acidic conditions, but it can show comparable or higher activity as a Pt catalyst in basic condition. Ding et al. [79] found that the current density of Pd nanocubes for methanol oxidation is evidently higher than that of the Pd spherical nanoparticles shown in Figure 13.12. The onset potential of Pd nanocubes shifts 24 mV more negative than that of 8 nm Pd spherical nanoparticles. The reason is

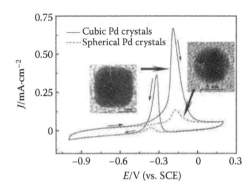

FIGURE 13.12
Cyclic voltammograms of methanol oxidation on Pd nanocubes (solid curve) and spherical Pd nanoparticles (dashed curve) in 0.5 M NaOH +1.0 M methanol solution (scan rate: 20 mV s⁻¹). The electrode potentials are versus saturated calomel electrode (SCE). (From Ding, H. et al. *Chin Phys B*, 19, 2010.)

that the cubic-shaped Pd nanocubes are abundant in (100) facets, which are highly catalytic active.

13.3 Core-Shell and Decorated Structure Catalysts

13.3.1 Structure Characteristics

Currently, the most effective catalysts in DAFC are Pt-based catalysts, which are too expensive to be accepted by commercial application. The Pt loading (i.e., general ~1 mg_{Pt} cm^{-2} for both anode and cathode) should be reduced to an acceptable level. Core-shell structure is one of the most promising ways to reduce Pt loading while reserving similar or higher performance. Core-shell structure can generally keep the outer shape and morphology of nanoparticles while replacing the high cost core with cheaper materials. As a heterogeneous catalysis, only the Pt on catalyst surface is effective for AEOR and ORR. Then core-shell structure Pt-based catalysts principally show similar activities as common Pt-based nanoparticle catalysts. Recently, more and more attention has been focused on the core-shell structure with Pt shell and low cost element core for the sake of lower Pt loading and possible novel performance originated from electric or synergistic effect.

Decorated structure is a special core-shell structure with uncompleted shell [80,81]. The decorated structure is much easier to synthesize than the completed core-shell structure. Furthermore, it is possible that the decorated particles could be very small [82] and have synergistic effect with the core. Actually, the core here acts as more like a support. Here, we only discuss some decorated structure with metal core. More information about support materials for PEMFC and DMFC can refer to the reviews [83,84].

13.3.1.1 TEM and HRTEM Characteristics of Core-Shell Structure

Both TEM and HRTEM can easily identify the big core-shell structure shown in Figure 13.13a and b, in which the particle size is ~20 nm [37]. The clear boundary between core and shell can be easily recognized in the TEM and HRTEM images. The Pd shells were well grown with similar low-index planes of (111), (100), and (110) or Pd (hkl), which are the same as those of the Pt cores. However, there is a very small lattice mismatch between Pt core and Pd shell.

For the small core-shell structure, TEM and HRTEM still can identify the structure directly in some cases [20,85]. But the core-shell structure information also can be revealed by the particle size in other cases [86,87]. Principally, a thicker shell will cause larger average size in TEM images. Experimentally, Figure 13.13c proves that the particle size increases clearly with more and more shell material deposition [87]. Besides diamond and spherical core-shell

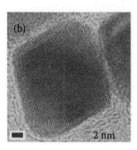
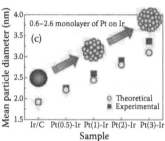

FIGURE 13.13

(a) TEM and (b) HRTEM images of Pt@Pd core-shell nanoparticles. (From Long, N.V. et al. *International Journal of Hydrogen Energy* 36, 8478–8491, 2011.) (c) Relation between shell coverage and mean particle diameter. (From Lee, K.S. et al. *Langmuir*, 27, 3128–3137, 2011.)

structure nanoparticles, triangular [88], cubic [88], and rod-shaped [89,90] core-shell structures were also synthesized.

13.3.1.2 XRD Characteristics of Core-Shell Structure

XRD is a bulk analysis tool, which can reveal the crystal structure, lattice constants, and crystal orientation of the core-shell structure. For example, the results in Figure 13.14 show a crystalline face centered cubic (fcc) phase for prepared nanoparticles. For Ru@Pt/C and Ru@PtPd/C, there was a diffraction peak at 39.0°, which is apparently associated with Ru(101). This implies that Pt or PtPd was not alloyed with Ru. The XRD patterns of Ru@Pt/C and Ru@PtPd/C are quite different from that of alloyed PtRu/C in which the main diffraction peak is fcc (111) at ~39° [91]. Therefore, the XRD diffraction peaks of core are usually easy to identify while the shell will vary a lot depending on the thickness of the shell [20,80,81,86,87].

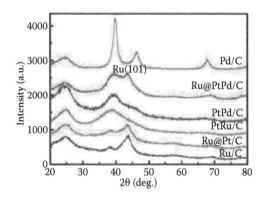

FIGURE 13.14

XRD patterns of Ru@PtPd/C, Ru@Pt/C, PtPd/C, PtRu/C, Pd/C, and Ru/C (A). (From Gao, H.L. et al. *Journal of Power Sources* 196, 6138–6143, 2011.)

By using one XRD peak, the average size of the Pt nanocrystallites can be calculated by the width of the reflection according to the Debye–Scherrer equation:

$$D = 0.9\lambda/(\beta\cos\theta) \tag{13.2}$$

where λ is the full width at half maximum (FWHM) of the peak, β is the angle of diffraction, and θ is the wavelength of the x-ray radiation. Then, the average sizes of Pd/C, PtPd/C, Ru/C, and Ru@PtPd/C catalysts are calculated to be ca. 5.0, 2.6, 2.5, and 2.9 nm, respectively [91].

13.3.1.3 XPS Characteristics of Core-Shell Structure

From XPS, we can get the binding energies (BEs) of both core and shell for core-shell structure. Then the interaction between core and shell [91] and the chemical species [80] can be revealed. Figure 13.15 shows the typical XPS spectrums of core-shell structure. The Pt BEs of Ru@PtPd/C catalyst are higher than those of PtPd/C, indicating an interaction between the Ru core

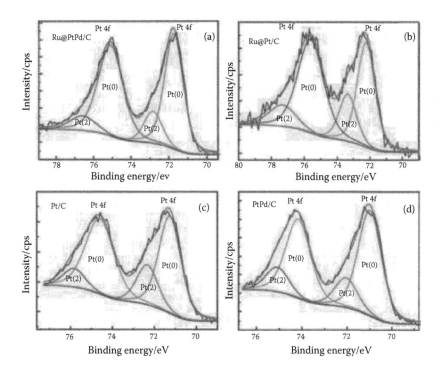

FIGURE 13.15
XPS survey spectrums of (a) the Pt_{4f} core level XPS spectra of Ru@PtPd/C, (b) Ru@Pt/C, (c) Pt/C, and (d) PtPd/C. (From Gao, H.L. et al. *Journal of Power Sources* 196, 6138–6143, 2011.)

and the PtPd shell [91]. The metallic and oxide states and their relative ratios were also addressed by XPS [80,87]. The information from XPS can help us understand the relationship between catalytic activity of core-shell structure nanoparticles and surface species.

13.3.1.4 EDS Characteristics of Core-Shell Structure

Although TEM and HRTEM can clearly identify the core-shell structure, it is difficult to get the element composition of core or shell. EDS is a very powerful tool to get the element composition of the prepared core-shell structure catalysts [91]. Moreover, the composition of a single nanoparticle also can be obtained [87]. For example, STEM-EDS line spectra can give the element distribution in a single core-shell structure nanoparticle [92]. STEM-EDS line spectra of Ru from Ru@Pt NPs showed a single Gaussian distribution across the particle, which was due to the monometallic core. STEM-EDS line spectra of Pt showed a bimodal distribution implying a Pt shell. The Pt shell and Ru core structure was further proved.

13.3.2 Synthesis Methods

It is highly possible that the concept of core-shell structure was originated from the surface decoration by "irreversible" adsorption or template replacement [93,94]. Wieckowski and coworkers [93,94] prepared Ru adlayers on a Pt(hkl) surface by immersing a Pt surface in a Ru ion-containing solution to do spontaneous deposition. Brankovic et al. [95] prepared Pt decorated Ru(0001) single crystal surface by immersing Ru(0001) in the $[PtCl_6]^{2-}$ solution for a certain time. On the other hand, Cu adlayer can be used as a template to decorate Pt, Pd, and Ag on the Au(111) surface [96].

Currently, there are mainly four methods for core-shell synthesis shown in Figure 13.16. The widely used method for core-shell structure synthesis is a two-stage procedure shown in Figure 13.16a. First, the core is synthesized; then, the core is coated by a shell. For example, Alayoglu et al. [20,92] used a sequential polyol process to synthesize Ru@Pt catalyst. First, Ru(acac)$_3$ (acac = acetylacetonate) was reduced in refluxing EG in the presence of PVP stabilizer. Second, the resulting Ru cores were coated with Pt by reducing PtCl$_2$ in the Ru/EG colloid at temperature 200°C. Some other core-shell nanoparticles, such as Ru-oxide@Pt [80], Ru@Pt [20,92,97–99], Ru@PtPd/C [91], Au@Pt [86], IrNi@PtRu [99], were synthesized by the same strategy. However, the reduction agencies are quite different for these shell coatings. The most used reductants include NaBH$_4$ [99], hydrogen [80,81,97], ascorbic acid [86,87,100], etc. Very few researchers use other ways, such as radiolytic process [98], to form the metal shell.

For monolayer or submonolayer core-shell structure synthesis, the effective way is the galvanic displacement method shown in Figure 13.16b. For example, Karan et al. [101] prepared a Pt/Pd/IrRe core-shell catalyst by

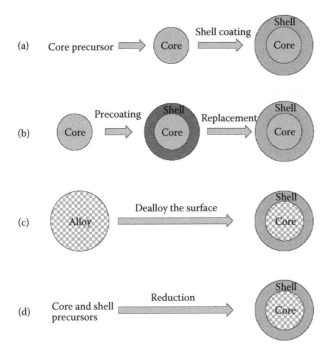

FIGURE 13.16
(See color insert.) Scheme of four ways to synthesize core-shell structure. (a) A two-stage procedure method, (b) shell replacement method, (c) dealloy method, and (d) one-step synthesis method.

galvanic displacement of an underpotentially deposited Cu monolayer by Pd or Pt. The Pd or Pt ions were reduced by Cu monolayer, and Pd or Pt atoms were deposited on the IrRe core. This method has been widely used by many other researchers [101–104].

Additionally, core-shell structure can be achieved by chemical leaching or by electrochemical dealloying of the non-noble element in the catalysts [105–108] shown in Figure 13.16c. For example, Mani et al. [106] firstly prepared PtCu alloy by an impregnation/freeze drying route followed by annealing, and then PtCu alloy was dealloyed by treatment in 1 M H_2SO_4 at 80°C for 36 h. The dealloying also can be done electrochemically by CV scanning 200 times at 100 mV s^{-1} between 0.5 and 1.0 V/RHE in deoxygenated condition at 80°C.

The one-step synthesis method was also successfully developed. For example, Chen et al. [85] found that one-step reaction of $Fe(CO)_5$ and $Pt(acac)_2$ could synthesize PtFe particles with benzyl ether as solvent and oleic acid and oleylamine as stabilizers. Core-shell FePt/Fe_3O_4 nanoparticles were separated by controlled addition of an excess of $Fe(CO)_5$ into the reaction mixture and air oxidation. Size, composition, and shape of the particles can be controlled by varying the synthetic parameters, such as the molar ratio of stabilizers to metal precursor, heating temperature, heating duration, etc. Huang et al. [109] synthesized a carbon-supported PdAu@Au core-shell nanostructure by a simultaneous reduction method without using any stabilizer.

13.3.3 Electrochemical Performance

The core-shell structure can reduce the cost of noble metal elements by replacing the nanoparticle's core with some other cheaper materials. The activity and stability may also be improved by the cooperation between core and shell. Usually, a core-shell structure shows higher mass activity [37,81,87,99] or apparent activity [86,97] but sometimes shows lower intrinsic catalytic activity [80].

13.3.3.1 Carbon Monoxide (CO) and CO-Like Oxidation on Ru@Pt

CO and CO-like species are the main poisons for the electrooxidation of alcohol on precious metal electrodes. The poisonous species are formed by dehydrogenation of alcohol and adsorb on the electrode surface strongly. They can be removed at high potential, but high potential is meaningless for fuel cells. Figure 13.17A is the comparison of adsorbed CO and CO-like species of methanolic residues for Pt-decorated unsupported Ru catalyst. The peak potentials for the removal of adsorbed CO and methanolic residues are almost same, indicating that the adsorbed methanolic species are essentially CO-like species. Figure 13.17B shows that the electrooxidation of CO-like species is temperature dependent. Both the onset potential and peak potential negatively shift with temperature increase. The decorated catalyst shows more positive stripping potentials with respect to the 60% PtRu/C and close to the bare, unsupported Ru. This indicates a lower intrinsic catalytic activity of the decorated catalyst with respect to the PtRu alloy. However, the stripping charge is significantly larger in the decorated catalyst than in the PtRu/C at high temperatures.

CO-stripping voltammetry is a very common method to investigate the activity and measure the active area of Pt-based catalysts. Ru@Pt/C was compared with Pt/C and Ru/C in Figure 13.17C by Ochal et al. [92]. The CO-stripping voltammetries of Pt/C and Ru/C agreed well with the literature data. The CO stripping voltammetry of Ru@Pt in Figure 13.17C shows two peaks. The first one at ~0.6 V, which is 0.2 V and 0.05 V lower than pure Pt and pure Ru, may come from either PtRu alloy particles or from Ru@Pt NPs. The second one at about 0.8 V is similar as that of Pt.

CO-stripping voltammetry of Pt decorated Ru catalysts with different Pt coverage are also shown in Figure 13.17D [97]. The activity of all catalysts toward methanol oxidation increases with the increase in the platinum uptake as shown both in the voltammetric and chronoamperometric experiment. The best result was obtained under 40% Ru@Pt/C catalyst

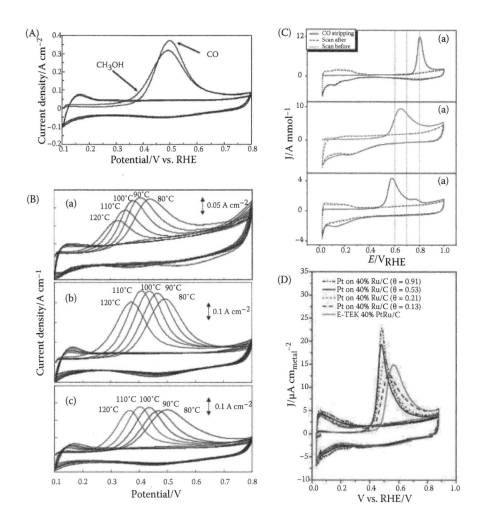

FIGURE 13.17
(A) Comparison of adsorbed CO and methanolic residues stripping profiles for Pt-decorated unsupported Ru catalyst, at 80°C under a DMFC configuration. Anode: 1 M methanol or 2% CO in argon, 1 atm. rel. adsorbed for 30 min at 0.1 V; cathode: H_2 feed 1 atm. rel. (From Arico, A.S. et al. *Journal of Electroanalytical Chemistry* 576: 161–169, 2005.) (B) *In situ* methanol residues stripping voltammetry for (a) commercial PtRu/C, (b) Pt-decorated and (c) bare, unsupported Ru catalysts at various temperatures under DMFC configuration. Anode: 1 M methanol, 1 atm rel. adsorbed for 30 min; cathode: H_2 feed 1 atm rel. (From Arico, A.S. et al. *Electrochemistry Communications* 6: 164–169, 2004.) (C) CO stripping voltammetry of Pt/XC-72R reference NPs (a), Ru/XC-72R reference NPs (b) and Ru@Pt/XC-72R NPs (c). Conditions: 10 mV s^{-1}, 0.5 M HClO$_4$; adsorption at 50 mV. The current has been normalized to the total number of moles of catalyst. (From Ochal, P. et al. *Journal of Electroanalytical Chemistry* 655: 140–146, 2011.) (D) CO stripping voltammograms on the Ru/Pt catalyst different Pt coverage, scan rate 5 mV s^{-1}, 0.5 M H$_2$SO$_4$ supporting electrolyte. The catalyst composition is given in the legend. (From Kuk, S.T., and Wieckowski, *Journal of Power Sources* 141: 1–7, 2005.)

with packing density of 0.53, which is higher than that of the commercial carbon-supported 40% Pt-Ru/C catalyst of the 50:50 Pt:Ru ratio.

Sun's group [110] investigated two different core-shell, Ru@Pt and Au@Pt/C, metal nanoparticles (NPs) by electrochemical *in situ* surface-enhanced infrared reflection absorption spectroscopic (SEIRAS). They found that the most active sites for the COR and MOR were the Pt islands on Ru core sites, where the onset potential was as low as ~0.1 V for the COR. However, the Au@Pt/C NPs had a much higher onset potential (0.45 V) for the COR in SEIRAS data.

13.3.3.2 Methanol Oxidation on Core-Shell Catalysts

The area-specific activity of core-shell catalysts has an improvement as compared to the pure or alloyed catalysts. For example, Zhang et al. [100] found that onset potential of the MOR at the Pd@Pt catalyst was 0.16 V, which was much lower than 0.27 V of Pt catalyst in Figure 13.18A. On the other hand, the current density of the positive sweep for Pd@Pt at 0.65 V is about sixfold higher than that for the Pt catalyst. This improvement may be due to the generation of oxygen-containing species, such as PdO and PdO_x, which assist the removal of CO-like species. Similar results were also observed by other researchers in core-shell catalysts, such as Pd@Pt [37], Pt@Pd [37,88], Ir@Pt [87], Au@Pt [86], etc.

One of the main advantages for core-shell structure is the improvement of mass specific activity because the core is replaced by cheaper materials, and less precious metal is consumed. Kaplan et al. [99] found that IrNi@PtRu catalyst show ~50% higher mass activity for methanol than the alloy-based catalyst, and the Ru@Pt catalyst shows ~66% higher mass activity for

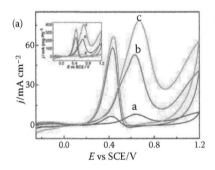
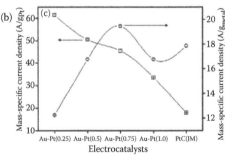

FIGURE 13.18

(A) CVs of the oxidation of 0.5 M methanol at GC electrodes modified with (a) Pt, (b) Pd@Pt, and (c) graphene-supported Pd@Pt in 0.5 M H_2SO_4 solution containing 0.5 M methanol. Scan rate: 50 mV s⁻¹. (From Zhang, H. et al. *Electrochim Acta*, 56: 7064–7070, 2011.) (B) Electrocatalytic activity of Pt-modified Au nanoparticles on carbon support for methanol electrooxidation: mass-specific current densities at 0.6 V vs. NHE. Current densities vs. potential plots were measured in 0.5 M H_2SO_4 +1M CH_3OH at room temperature. (From Park, I.S. et al. *Electrochim Acta*, 52, 5599–5605, 2007.)

EG oxidation. Park et al. [86] found that Pt-modified Au nanoparticles could show two times higher current densities for $Pt\left(Ag_{Pt}^{-1}\right)$ and 10% higher current densities for total metal $\left(Ag_{Pt+Au}^{-1}\right)$ shown in Figure 13.18B.

We must emphasize that the mass specific activity may be enhanced even more greatly if the core-shell structure is applied into well-defined shaped nanoparticles [37,88]. As we know from the previous section, cubic nanoparticles show very high area-specific activity but don't show outstanding mass specific activity due to the problem of bigger size. If the core of a cubic nanoparticle is replaced by cheaper materials, the cubic nanoparticle catalyst will have higher mass-specific activity and be promising in real applications. The stability of the core-shell structure is increased greatly as compared to a pure metal nanocatalyst. A Pt@Pd catalyst retained 30% current after 2 h polarization while Pt catalyst only had 3.67% left [88].

For the decorated catalyst, the coverage is an important factor for the catalytic activity. Park et al. [86] investigated the effect of atomic Pt/Au ratio on the activity of methanol oxidation. They found that the current densities $\left(Ag_{Pt}^{-1}\right)$ for Pt decorated Au decrease with the ratio increase while current densities (A/g_{Pt}) have a maximum value at the ratio 0.75. Lee et al. [87] and Kuk et al. [97] also found a similar behavior in Ir@Pt [87] and Ru@Pt [97] catalysts.

However, the decorated catalyst showed lower intrinsic catalytic activity with respect to the PtRu alloy. But the decorating particles promote methanol dehydrogenation and adsorption of methanolic residues on the Ru support allowing the reaction to occur at significant rates even in the presence of small Pt amounts [80]. A high DMFC power density of 150 mW cm^{-2} was achieved with the decorated catalyst under air feed operation at 130°C (see Figure 13.19 [81]). The anode had ultra-low Pt loading (0.1 mg cm^{-2}), which was 20 times reduction while the power density only lost 35% [81].

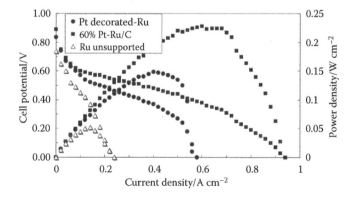

FIGURE 13.19
DMFC single cell polarizations at 130°C for commercial PtRu/C, Pt-decorated, and bare, unsupported Ru catalysts. Anode: 1 M methanol, 2 atm.; cathode: air feed 2.5 atm. (From Arico, A.S. et al. *Electrochemistry Communications* 6, 164–169, 2004.)

TABLE 13.1

Specific Activity of Core-Shell Catalysts with the Test Conditions

Catalysts	Activity	Temperature	Potential (vs. Ag/AgCl)	Reference
Ru@Pt	1.0 µA cm^{-2}	25°C	0.2 V	[111]
Pt@Ru	4.1 µA cm^{-2}			
PtRu (1:1)	2.8 µA cm^{-2}			
Pt/C (E-Tek)	51.0 mA (mg$_{Pt}$)$^{-1}$	22°C	0.4 V	[100]
Pd@Pt	300 mA (mg$_{Pt}$)$^{-1}$			
Pt/C(JM)	18 A g^{-1}	RT $^{(1)}$	0.4 V	[86]
Au@Pt	62 A g^{-1}			
Pt@Pd	1.29 × 10^{-3} A cm^{-2}	RT	0.5 V	[88]
Pt	4.33 × 10^{-4} A cm^{-2}			
PtPd	1.4 × 10^{-3} A cm^{-2}	RT	0.5 V	[37]
Pt@Pd (15 min)	1.5 × 10^{-3} A cm^{-2}			
Pd@Pt (15 min)	1.4 × 10^{-3} A cm^{-2}			
Pt@Pd (6 h)	0.94 × 10^{-3} A cm^{-2}			
Pd@Pt (6 h)	1.0 × 10^{-3} A cm^{-2}			
Ir@Pt	200 A g$_{Pt}^{-1}$	RT	0.5 V	[87]
Pt	100 A g$_{Pt}^{-1}$			
PtRu(JM)	3 mA (mg$_{metal}$)$^{-1}$	25°C	0.2 V	[97]
Ru@Pt	0.8 mA (mg$_{metal}$)$^{-1}$			
PtRu/C(E-TEK)	2.5 mA (mg$_{metal}$)$^{-1}$			
Ru@Pt	6.0 mA (mg$_{metal}$)$^{-1}$			

Note: (1) room temperature.

However, it is very difficult to compare the specific activity for core-shell catalysts appearing in the different literature because the specific activity can be affected by many parameters, such as temperature, reactant concentration, potential, etc. Unfortunately, the structure and experiment parameters are quite different in the literatures. So the comparison is somewhat reliable for the experiments from the same researcher. Nevertheless, it is useful to list all of these comparisons for later researchers. Here, we present some specific activity of core-shell catalysts in Table 13.1 not to critically compare the activity between them but to provide a general reference. Some important parameters are presented as well.

13.3.3.3 Ethanol Oxidation on Core-Shell Catalysts

The core-shell structure nanoparticles also show improved catalytic performance towards ethanol oxidation. Gao et al. [91] found that the catalyst Ru@PtPd/C is 1.3, 3, 1.4, and 2.0 times more active than PtPd/C, PtRu/C, Pd/C, and Pt/C in basic condition, respectively, indicating high utilization of Pt

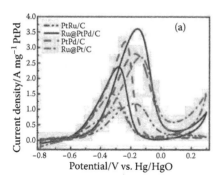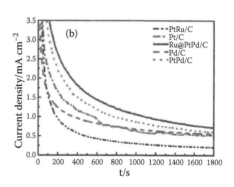

FIGURE 13.20

(a) CVs of Ru@PtPd/C, Ru@Pt/C, PtRu/C and PtPd/C catalysts in 1.0 M KOH solution + 1.0 M ethanol (room temperature, scanning rate of 30 mV s^{-1}). (b) CAs of the catalysts at −0.4 V in 1.0 M ethanol + 1.0 M KOH at room temperature. (From Gao, H.L. et al. *Journal of Power Sources* 196, 6138–6143, 2011.)

and Pd shown in Figure 13.20a. In addition, they also found that the stability of Ru@PtPd/C is higher than that of Pt/C and PtPd/C as evidenced by the chronoamperometric evaluations shown in Figure 13.20b.

13.3.3.4 Ethylene Glycol Oxidation on Ru@Pt

EG is a promising candidate to replace methanol because it has a much higher boiling point than methanol (198°C vs. 64.7°C) and greater volumetric capacity (4.8 Ah ml^{-1} vs. 4.0 Ah ml^{-1}) [112]. Furthermore, fuel crossover to the cathode can be much lower because EG is a much larger molecule. The application of core-shell structure in EG oxidation has been carried out by Kaplan et al. [99,112]. They found that the mass-specific activity of core-shell structure PtRu/IrNi/XC-72 and Pt/Ru/XC-72 was higher than the commercial catalyst while their surface activity is lower. This is yet another demonstration of the importance of high platinum utilization in the electrocatalysis process.

13.3.3.5 Oxygen Reduction Reaction

ORR is very important for all fuel cells using oxygen as the oxidant. Core-shell structure is a promising nano-technique to find better ORR catalysts due to the advantages as follows. Therefore, some core-shell structure nanoparticles with different cores and shells, such as Pt$_{ML}$@Pd$_{ML}$@Ir$_2$Re$_1$ [101], Pt$_{ML}$@Pd$_{2layers}$@Ir$_2$Re$_1$ [101], and Pt$_{ML}$@Pd$_{2layers}$@Ir$_7$Re$_3$ [101], Pt$_{ML}$/IrNi/C [103], Pd$_3$Co@Pt/C [113], Au@Pt/C [114,115], Co@Pt/C, etc., have been prepared for ORR to find high performance catalysts.

First, a thin precious-metal shell can dramatically reduce the loading of metal elements and lower the cost. The ideal shell is a monolayer or sub-monolayer shell because the precious-metal usage could be extremely high.

This kind of monolayer structure is usually synthesized by the galvanic displacement method. For example, Karan et al. [101] prepared a core-shell structure that consists of an iridium-rhenium nanoparticle core and a platinum monolayer shell or platinum and palladium bilayer shell. The monolayer concept facilitates the use of much less platinum than in traditional approaches. The activities of the $Pt_{ML}@Pd_{ML}@Ir_2Re_1$, $Pt_{ML}@Pd_{2layers}@Ir_2Re_1$, and $Pt_{ML}@Pd_{2layers}@Ir_7Re_3$ catalysts were, in fact, better than that of conventional Pt electrocatalysts. In addition, a Pt monolayer on an IrNi core was synthesized by Kuttiyiel et al. [103] and showed higher mass specific activity than that of the commercial Pt/C electrocatalyst. Wang et al. [113] found that the electrocatalytic activity of the $Pd_3Co@Pd/C$ nanoparticles for the ORR was enhanced by spontaneously depositing a nominal monolayer of Pt. The addition of Pt not only increased the onset potential for the ORR but also modified the electronic structure of Pd, which they believe is responsible, at least in part, for the higher activity.

Second, the core of the core-shell structure can lead to improved activity due to the geometric, electronic, and segregation effects on the shell, where the ORR happens. The activities of the different catalysts for the ORR could be correlated with the oxygen adsorption energy and the d-band center of the catalyst surface, which can be calculated by density functional theory (DFT) [113]. Karan et al. [101] found that Ir and Re mixed-metal cores facilitate the decreasing of OH poisoning on Pt according to their DFT calculations. Their calculations revealed that the molar ratio of Ir to Re affects the binding strength of adsorbed OH and, thereby, the O_2 reduction activity of the catalysts. The maximum specific activity was found for an intermediate OH BE with the corresponding catalyst on the top of the volcano plot, where the

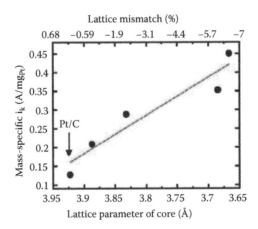

FIGURE 13.21
Variations in the Pt mass specific activity at 0.9 V for $Pt@Pd_xCu_y/C$ with the lattice parameter of the Pd_xCu_y/C core and the percentage mismatch between the lattice parameter of Pt and the core. PdCu/C was not included because it showed two distinct phases.

core composition is Ir_2Re_1. These catalysts also exhibited a remarkable stability during the accelerated test involving 30,000 cycles of CVs.

Cochell and Manthiram [102] also provided that the activity enhancements over Pt/C are due to the delay in the onset of Pt-OH formation caused by interactions between the Pd_xCu_y-rich core and the Pt-rich shell. The relationship between the activity and lattice parameter shown in Figure 13.21 illustrates this connection of the core-shell structure to ORR activity.

Kuttiyiel et al. [103] showed that the activity of Pt monolayer electrocatalysts for the ORR is strongly substrate metal-dependent, and the IrNi core-shell structured substrate is a very suitable core for Pt monolayer electrocatalysts. Their DFT calculations using a sphere-like model shown in Figure 13.22

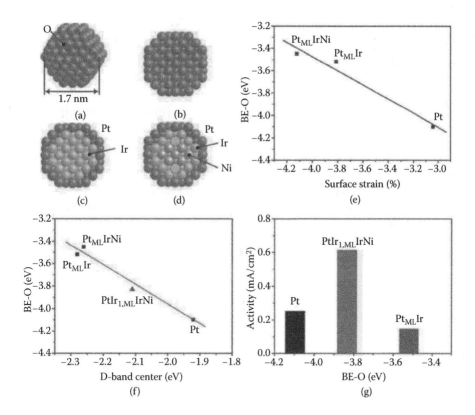

FIGURE 13.22

(a) A schematic of a sphere-like nanoparticle considered in DFT calculations representing adsorption of atomic oxygen at the fcc site. Cross-sectional views of nanoparticle models of (b) Pt, (c) $Pt_{ML}Ir$, and (d) $Pt_{ML}IrNi$ with 1.7 nm. The dotted circles represent Ir in the sub-core. (e) The DFT calculated BE of oxygen (BE–O) as a function of strain on $Pt_{ML}IrNi$, $Pt_{ML}Ir$, and Pt using the nanoparticle models. (f) The DFT calculated BE of oxygen (BE–O) as a function of the average d-band center of metals interacting with O on $PtIr_{1,ML}IrNi$, $Pt_{ML}IrNi$, $Pt_{ML}Ir$, and Pt. (g) The Pt specific activity against BE–O on $PtIr_{1,ML}IrNi$, $Pt_{ML}IrNi$, $Pt_{ML}Ir$, and Pt. "ML" is the monolayer. (From Kuttiyiel, K.A. et al. *Energy & Environmental Science* 5, 5297–5304, 2012.)

clearly demonstrate that mixing Ni with Ir (Pt_{ML}/IrNi/C) induces geometric, electronic, and segregation effects and thus weakens the BE of oxygen, resulting in higher activity than pure Pt/C and Pt_{ML}/Ir/C electrocatalysts.

Third, the core-shell structure also can lead to new catalysts, which have high ORR activity and high alcohol tolerance. Alcohol, permeating from the anode, will cause mixture potential and decrease the performance of the cathode greatly. So alcohol-tolerant cathode catalysts are desired. Wang et al. [113] found that $Pd_3Co@Pd/C$ had an activity comparable to Pt/C and exhibited a much higher tolerance to methanol in the ORR shown in Figure 13.23. The electronic properties of Pd were modulated by alloying with different amounts of Co, which affects the ORR activity. The similar catalyst, the Pd_4Co core–shell structure, was found to be much more active than alloyed NPs and be comparable to Pt catalyst by Jang et al. [116].

The high performance of the core-shell structure was also proved by other researchers. Wu et al. [117] found that Co@Pt/C is more active than pure Pt/C. Liu et al. [118] found that both the ORR activities and long-term durability of Pt@Au nanorods are better than that of commercial Pt/C nanoparticles.

The coverage of shell was studied by Alia et al. [104]. They controlled the Pt coatings to 9 ($PtPd_9$), 14 ($PtPd_{14}$), and 18 ($PtPd_{18}$) wt% and was estimated to have a thickness of 1.1, 1.7, and 2.2 Pt atoms, respectively. $PtPd_9$, $PtPd_{14}$, and $PtPd_{18}$ produce dollar activities of 10.4, 9.4, and 8.7 A $\$^{-1}$, respectively; $PtPd_9$ exceeds the DOE (USA) dollar activity target (9.7 A $\$^{-1}$) by 7%. Pt/Pd nanotubes further exceed the DOE area activity target by 40%–43%.

However, not all core-shell structures show a positive effect on ORR reaction. For example, Lin et al. [114] found that the activity of Au@Pt/C is lower than the Pt/C (JM) catalyst, and the activity of Au@Pt (2:4)/C decreased several times after 500 cycles CV test. Moreover, the half-wave potential $E_{1/2}$ of

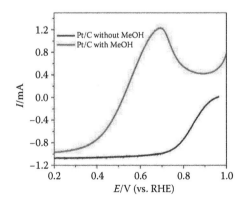

FIGURE 13.23
Electrochemical test results of Pd_xCo/C and Pt/C catalysts in O_2-saturated 0.1 M H_2SO_4 with and without the presence of 1 M MeOH. (From Wang, D. et al. *Chemistry of Materials* 24, 2274–2281, 2012.)

Au@Pt (2:4)/C is negatively shifted by about 40 mV vs. RHE in comparison with the Pt/C (JM) catalyst. But they revealed that the Au@Pt/C catalyst with a ratio of 2:4 for Au to Pt (atomic ratio) exhibited the best catalytic activity toward ORR, and Au@Pt(2:4)/C led to an approximately four-electron pathway in acid solution.

13.4 Hollow Structure Catalysts

13.4.1 Structure Characteristics

Hollow structure is a special structure, which has an ordinary shape outside but is concave inside. The hollow structure has increased surface area, low density, and saving of material due to the empty inside of particle. It is possible to find some new interesting properties of catalysis because the inside of the hollow structure has a special physical and chemical environment. The typical hollow structure is show in Figure 13.24 [68,119–121]. We

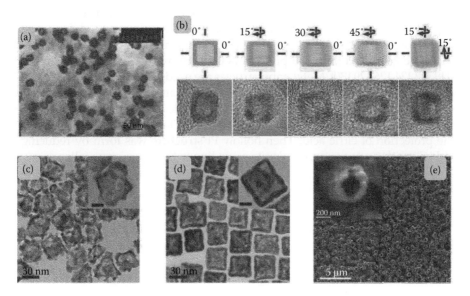

FIGURE 13.24
(a) Representative TEM images of hollow Pt/C electrocatalysts. (From Zhao, J. et al. *J Power Sources*, 162: 168–172, 2006.) (b) Individual Pt hollow nanocubes under various tilting angles with respect to the direction of the imaging beam. (From Peng, Z. et al. *Nano Letters* 10: 1492–1496, 2010.) (c) PdPt octahedral nanocages (ONCs). (From Hong, J.W. et al. *ACS Nano*, 6: 2410–2419, 2012.) (d) PdPt cubic nanocages (CNCs). High-magnification TEM images are shown in the inset. Scale bars in the insets indicate 10 nm. (From Hong, J.W. et al. *ACS Nano*, 6: 2410–2419, 2012.) (e) Ni@Pt core-shell nanotube arrays after etching ZnO nanorods. (From Ding, L.-X. et al. *Chemistry (Weinheim an der Bergstrasse, Germany)*, 18: 8386–8391, 2012.)

can clearly see the hollow structure with a dark edge and grey center, which is the projection characteristic of the hollow structure. Besides the hollow sphere in Figure 13.24a, there are some other hollow structures, such as nanobox, nanocage, and nanotube, shown in Figure 13.24b through 13.24e. Currently, most of hollow structure nanoparticles are an assembly of small nanoparticles [119,121–127], but there are few examples of nanocrystal hollow nanoparticles [68,120].

13.4.2 Synthesis Methods

The hollow structures were usually synthesized by templating methods [1–4,9,12,13] shown in Figure 13.25. The templates are usually metal, polymer, SiO_2 spheres, etc., which define the shape and size of the nanoshells. For example [128], Pd hollow spheres were synthesized by using silica spheres as a template. In detail, the surfaces of the silica spheres were functionalized with mercaptopropylsilyl (MPS) groups to adsorb the palladium precursor, palladium acetylacetonate ($Pd(acac)_2$). Once the silica spheres were removed by HF at 250°C, Pd metal hollow structure NPs formed. The similar method can be used to synthesize hollow spheres [123,125], nanotubes [121], etc.

The synthesis method above includes two procedures, which are complicated and hard to control. Currently, the best known method for metal deposition in such cases is the galvanic replacement method, in which metal shells are generated from salt precursors through oxidation of the metal cores [119,122,124,126,129,130]. Metal nanoparticles are used as both template and support for depositing targeted metal elements. For example [119], hollow Pt nanospheres were synthesized by using Co nanoparticles as template. In detail, Co nanoparticles were synthesized by reducing $CoCl_2 \cdot 6H_2O$ with $NaBH_4$ under the protection of citric acid. Then hollow Pt structure was form by reducing H_2PtCl_6 with Co nanoparticles *in situ* according to the equation as follows:

$$2Co + PtCl_6^{2-} = Pt + 2Co^{2+} + 6Cl^- \quad (13.3)$$

The similar method can be used to synthesize hollow nanospheres [119,122,124,126,127,129], hollow nanocages [120], and hollow nanoboxes [68].

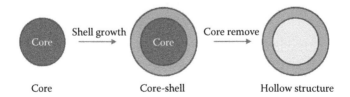

FIGURE 13.25
Schematic of procedure of hollow structure nanoparticle synthesis.

13.4.3 Electrochemical Performance of Hollow Structure Nanoparticles

Hollow structure nanoparticles have been found to obviously enhance performance for MOR, EOR, and ORR. The enhanced performance can be attributed to several possible reasons:

First, most hollow structure particles are raspberry-like hollow spheres, which are composed of many smaller NPs. These small NPs themselves usually have urchin-like nanostructure. So the hollow structure particles usually have relatively large surface area [125]. The hollow cavity and nano-channels between particles can increase the surface-to-volume ratio greatly [122,125]. So one of the reasons for improvement in electrocatalytic activity is due to the fact that the hollow Pt structure particles possess a higher active surface area [119].

Second, a relatively high density of defects, particularly vacancies, is formed at the surface of the shell, which might result in a product with a rough surface rather than a smooth one. For example, the surface rough-ness could be ascribed to the poor epitaxial growth of Pt on Ag as caused by the relatively large difference (~4.5%) in lattice constant between Ag and Pt. These defects (vacancies) with atomic scale and nanochannels were created by removing silver from a hollow Ag-Pt alloy; such defects may have high activity as catalysts and provide more bonding sites for chemical reaction. The nanochannels may give rise to the activation of the inner surface and provide a route for the transport of reactant and product [127].

Third, the nanochannels between particles in a porous shell facilitate the mass transfer during the reaction [131]. Especially, the nanotubes can obvi-ously enhance the diffusion of reaction species and facilitate the reaction kinetics on the electrode surfaces [121].

Fourth, the curvature radii of the inner surface of the hollow structure are smaller than that of the outer surface, so the inner surface has a much higher activity than the outer surface. This is another reason that the catalytic activ-ity of PtRu nanosphere for methanol oxidation is much higher [126].

Fifth, the active sites in the inner surface of the hollow shell are possibly formed [127] when nanochannels are formed after the treatment of chemical etching or assembly of nanoparticles. The mesoporous structure shell pro-vides very large fraction of edges and corner atoms, endowing the hollow structure nanoparticles with good electrocatalytic performance [124].

Sixth, the unique core-shell nanostructure of bimetallic nanoparticles may allow greatly enhanced electrocatalytic activity because of the electronic, strain, and synergistic effects of the various metals [121]. The unique hol-low nanotubes with double-layer walls and highly dispersed Pt nanocrystals lead to a high electrocatalyst utilization ratio [121].

Due to the reasons above, larger surface area, more and new active sites, and easier mass transfer, fast electron transport is possessed by hollow structure nanoparticles. Consequently, higher electrochemical activity, higher stabil-ity and lower charge transfer resistance [129] are achieved. The applications of hollow structure nanoparticles in methanol oxidation, ethanol oxidation, and oxygen reduction will be introduced as follows.

13.4.3.1 Methanol Oxidation

High-performance hollow structure nanoparticles of methanol oxidation have already been developed by many researchers [119,126,129,132]. Hu et al. [124] found that the peak current density of the hollow mesoporous PtNi nanospheres (HMPNNs) was higher than that of the solid PtNi nanospheres and the commercial Pt catalyst, which were shown in Figure 13.26A and B [124]. The onset potential in Figure 13.26C of MOR catalyzed by the HMPNNs is ~150 mV (vs. a saturated calomel electrode, SCE), which is ~260 and 265 mV

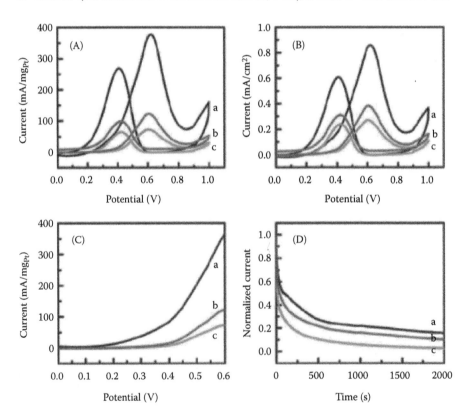

FIGURE 13.26
(A, B) Cyclic voltammetric, (C) linear-sweep voltammetric, and (D) chronoamperometric responses for the oxidation of 0.5 M methanol in H_2SO_4 solution (0.5 M). The currents in (A) and (C) were normalized with the mass of Pt loading on electrode surface; the current in (B) was normalized with ECSA. The current in (D) was normalized with the highest current obtained at the (a) HMPNNs, (b) solid PtNi nanoparticles, and (c) commercial Pt catalyst, respectively. The cyclic and linear-sweep voltammograms were recorded at 50 mV s^{-1}. (From Hu, Y. et al. *Electrochemistry Communications* 18, 96–99, 2012.)

negative as compare with those of PtNi nanospheres and the commercial Pt catalyst, respectively, indicating that the MOR was easier on the HMPNN electrocatalysts. The stability of HMPNN electrocatalysts was evaluated by chronoamperometric measurement, in which HMPNNs showed higher stability and activity. Similar enhanced electrochemical activity toward methanol oxidation was also found in hollow PtRu [126], Pt [119], PtCu [132], and PtCo [129] nanosphere electrocatalysts.

More complex hollow nanostructures have been studied also. Hollow Au@Pd nanoparticles were found to be more active than hollow Au nanospheres or Pd nanoparticles for methanol electrooxidation [122]. Pt cubic nanoboxes were found to be more active than hollow spheres and commercial Pt catalysts [68]. Ding et al. [121] found that the onset potential of the anodic peak of the porous Ni@Pt core-shell nanotube arrays is located at 0.406 V, which is similar to 0.394 V for the commercial Pt/C catalyst. However, the peak current density of the porous Ni@Pt core-shell nanotube arrays is almost twice of the commercial Pt/C catalyst. The stability of porous Ni@Pt core-shell nanotube arrays is higher. As a conclusion, the hollow nanostructures, including hollow nanosphere [122], hollow nanocage [120], hollow nanobox [68], hollow nanotube [121,133], etc. always show higher activity and stability for methanol electrooxidation compared to normal structures.

13.4.3.2 Ethanol Oxidation

Most of the catalytic research on hollow structures is concerned with MOR. Only very few papers reported the performance of hollow structure on EOR. But all of them showed that hollow structure can greatly improve the catalytic activity for EOR. Dai et al. [123] found that the Pd hollow nanosphere nanocatalyst showed a higher electrochemically active area, higher electrochemical activity, and long-term performance of ethanol oxidation than Pd film electrode in an alkaline condition. The onset potential for a Pd hollow sphere modified electrode is −0.547 V, which is 43 mV more negative than −0.504 V observed on the Pd film electrode implying enhanced kinetics of the ethanol oxidation reaction [123]. In addition, hollow Au@Pd nanoparticles were found to be more active than hollow Au nanospheres or Pd nanoparticles for ethanol electrooxidation [122].

13.4.3.3 Oxygen Reduction

Hollow structure nanoparticles, such as a hollow PdPt alloy [120], porous triangular AgPd nanoplates [134], hollow Pt_3Co/C [135], etc., exhibited excellent electrocatalytic activity and durability for the ORR [120,125,127,134,135]. Guo et al. [125] found that a raspberry-like hierarchical Au/Pt NP assembling a hollow sphere (RHAHS) modified electrode exhibited more positive potential (the half-wave potential at about 0.6 V), higher specific activity, and higher mass activity for ORR than that of commercial platinum black (CPB)

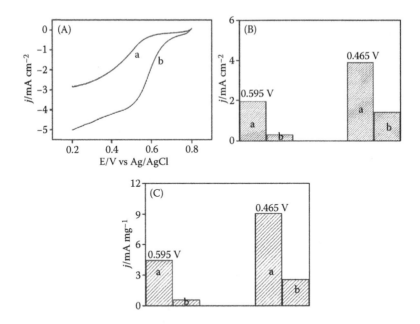

FIGURE 13.27
(A) Polarization curves for the ORR on (a) CPB and (b) RHAHS in air-saturated 0.5 M H₂SO₄ solutions at room temperature. Rotation rate: 500 rpm. Sweep rate: 50 mV s⁻¹. The Pt loading is 11.2 μg in the above two cases. (B) Comparison of specific activity for (b) CPB and (a) RHAHS at 0.595 and 0.465 V, respectively. (C) Comparison of mass activity for (b) CPB and (a) RHAH Sat 0.595 and 0.465 V, respectively. (From Guo, S. et al. *The Journal of Physical Chemistry C* 113, 5485–5492, 2009.)

shown in Figure 13.27. Excitingly, a four-electron reduction of O₂ to H₂O was demonstrated by rotating ring-disk electrode (RRDE) voltammetry for a RHAHS-modified electrode in a 0.5 M air-saturated H₂SO₄ solution. Chen et al. [127] also found that the hollow Pt shell with nanochannels showed more positive onset potential and higher activity than that of the commercial catalyst.

13.5 One-Dimensional Structure Catalysts

One-dimensional (1-D) nanostructure catalysts usually includes nanowires and nanorods shown in Figure 13.28a and b, respectively [136,137]. One of them is arranged randomly, another is aligned very well. Nanorods and nanowires are defined by their length and morphology, but there is no distinct edge between them. In order to simplify the statement, both nanorods and nanowires will be called "nanowires."

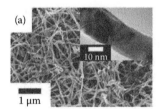

FIGURE 13.28

(a) SEM images of Pt nanowires. (From Kim, Y.S. et al. *Electrochemistry Communications* 11, 1026–1029, 2009.) (b) Typical SEM images of surface of Pd nanowire array with the diameters of 50 nm, cross-section of Pd nanowire array. (From Cheng, F. et al. *Electrochemistry Communications* 10, 798–801, 2008.)

Presently, the commonly used zero-dimensional (0-D) nanoparticle catalyst degrades obviously in its catalytic performance, which can be attributed to two reasons. One is the carbon support degrading due to chemical and electrochemical corrosion. Another one is the aggregation and dissolution of nanoparticles. As we all know, the most widely used catalyst in DAFCs consists of Pt-based nanoparticles (2–5 nm) supported on active carbon. The Pt-based nanoparticles with 0-D nanoparticle morphologies generally have a large number of low coordination atoms and defects on their surfaces, which will cause Ostwald ripening. In addition, carbon supports were known to undergo electrochemical corrosion that resulted in Pt nanoparticle migration, coalescence, and even detaching from the catalyst system in operational fuel cells. All the reasons above will make the supported 0-D Pt-based nanoparticle catalyst degrade in its catalytic performance.

One-dimensional Pt nanostructure catalysts are promising to overcome this shortcoming owing to their inherent structural characteristics [138]. Nanowires have elongated single crystalline segments with smooth crystal planes that are connected by grain boundaries forming a nanowire structural motif in Figure 13.29 [139]. Therefore, they possess less low coordination

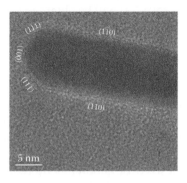

FIGURE 13.29

HRTEM images of gold nanorod with well-defined tip. The planes are indicated. (From Sreeprasad, T.S. et al. *Langmuir*, 23, 9463–9471, 2007.)

atoms and defects, which make them much more stable as compared with 0-D nanoparticles.

Moreover, nanowires could preferentially expose highly active crystal facets shown in Figure 13.29, which bring high electrocatalytic activity for alcohol oxidation and oxygen reduction. The structural anisotropy due to the large size in the longitudinal direction makes nanowires and nanorods have high conductivity [138,140]. Especially for the nanowires with less than 2 nm diameter, the electronic property of noble metal nanowires (Au, Pt, Pd, etc.) could be largely altered due to a surface contraction effect and a surface-stress-induced phase transformation, which could significantly enhance the catalytic performance in the ORR and MOR [31]. Furthermore, the 1-D nanostructure could facilitate effective mass transfer by networking the anisotropic morphology.

13.5.1 Pure Metal Nanowires

Originally, pure metal nanowires were synthesized due to the easier synthesis condition. The earliest literature on metal nanowires (here, gold nanowires) in *Web of Science* was published by Yu et al. [141]. However, gold is not an ideal catalyst for DAFCs as compared to Pt and Pd. So researchers took much effort to synthesize Ag [142,143], Pt [30,31,144–152], and Pd [152–157] nanowires shown in Figure 13.28.

The typical morphology of nanorods and nanowires in Figure 13.28a and b is quite different. The morphology difference is mainly dependent on the synthesis method. If the nanowires are synthesized by the hard template method, usually a well-aligned nanowire array will be created. This is due to the hard templates, such as anodic aluminum oxide (AAO), having well-oriented pores, inside which nanowires form. If the soft templates, such as PVP, are used, usually randomly arranged nanowires will be created. This is due to the fact that the soft templates are randomly dispersed in solution. However, if the soft templates are not in solution but oriented well on the substrate, very organized nanowire networks (NWNs) will be obtained [33]. The XRD patterns of as-prepared one-dimensional Ag, Au, and Pt nanostructures match well with the standard JCPDS diffraction patterns, which means that pure and crystal phases of the desired materials have been synthesized. Furthermore, there are no detectable impurities noted in any of the patterns [158].

Nanowires are usually synthesized by both soft template and hard template methods. For the soft template method, the templates include some organic molecules or polymers, such as oleyamine (OAm) [159–161], 1-octadecene (ODE) [159], triton X-114 [152], insulin powder [162], poly(vinyl pyrrolidone) (PVP) [136,143,144,146,163–167], etc., which are listed in Table 13.2. For example, Lee et al. [163] synthesized Pt nanowires by reducing H_2PtCl_6 with EG at 110°C in the presence of PVP, which serves as a stabilizer. They added $FeCl_3$ or $FeCl_2$ to lower the reduction rate of Pt(II) species to achieve uniform nanowires. The soft template method is very simple and robust because it can

TABLE 13.2

Summary of Nanowire Synthesis by Soft Template Procedure

Nanowires	Template	Length	Width	Ref.
Pt	Poly(vinyl pyrrolidone) (PVP)	~50 nm	~7 nm	[163]
Pt	PVP	100–200 nm	Several nm	[144]
Pt and PtRu	PVP	Tens μm	30 ± 10 nm	[136]
PtBi network	PVP	245 ± 81 nm	3.6 ± 0.5 nm	[164]
Ag	PVP	>10 μm	130, 80 nm	[143]
RhPt	NaBr, PVP	10–30 nm	5.0 ± 0.3 nm	[165]
PtRh and PtRu	Electrospinning, PVP	>10 μm	50 ± 10 nm	[166]
Pt	Electrospinning, PVP	>10 μm	100–150 nm	[146]
PtFe	Electrospinning, PVP tetrabutyl ammonium chloride	>10 μm	20–30 nm	[167]
Pd	CTAB	Tens nm	4 nm	[155]
Pt	CTAB	Cross net	2.5 ± 0.7 nm	[168]
PtRu	CTAB	Cross net	3.0 ± 0.5 nm	[169]
FePtPd	Oleyamine (OAm) or a mixture of OAm and 1-octadecene (ODE)	<100 nm	2.5 nm	[159]
PdFe	OAm	10–50 nm	3 nm	[160]
PtFe	OAm	~5 to ~200 nm	4.2 nm	[161]
Pt	Dimethylformamide, toluene, triethylamine	100 ± 25 nm	<2 nm	[30]
Pt	Triethylamine	6 ± 2 nm	2.4 ± 0.4 nm	[31]
Pt–Au core/shell	Formic acid (possible)	10–20 nm	~4 nm	[118]
Pt	Formic acid (possible)	100 nm	3.3 nm	[151]
Pd networks	Trisodium citrate (possible)	Tens nm	4 nm	[157]
Au, Pt and Pd	Triton X-114	Hundreds nm	2–3 nm	[152]
Pt	Insulin powder	2.1 μm	~1.8 nm	[162]

work in one pot and has large volume. Because the soft template is composed of some small molecules, very thin and high aspect ratio nanowires [144,163–165] can be synthesized by this method. The length of the nanowires can vary in large range from several nanometers to tens of micrometers. This method is a very common one, which has been used to synthesize Pd [155,157], Ag [143], Au [152], and Pt-based alloy [136,164–166] nanowires, etc.

The hard template method is another widely used one. The most popular hard template is a porous AAO template [35,147,170–172] shown in Table 13.3. Some other hard templates, such as SBA-15 [173,174], MCM-41 [149,174], polycarbonate template [158] etc., are also used. In this method, the metal atoms are deposited in the pores of the hard template to form a nanowire shape structure. Because the pore size of AAO [35,147,170–172] and polycarbonate [158] templates is large, the prepared nanowires are of a large diameter from tens of nm to more than 200 nm with the length from hundreds of nm to tens of μm. The nanowires synthesized by using SBA-15 [173,174],

TABLE 13.3

Summary of Nanowire Synthesis by Hard Template Procedure

Nanowires	Template	Length	Width	Ref.
Porous Pt arrays	Porous AAO templates	3 μm	36 nm	[170]
Pt	Porous AAO templates & PVP	~15 μm	200 nm	[147]
Mesoporous Pt array	Porous AAO templates	1.5 μm	40–50 nm	[171]
PtCo	Porous AAO templates	>300 nm	28 nm	[172]
Multisegment Pt–RuNi	Porous AAO templates	1.448 μm	209 ± 9 nm	[35]
Multisegment Pt-Ru, Pt-Ni, and Pt-RuNi	Porous AAO templates	1.453 μm	210 ± 10 nm	[176]
Multisegment Pt-Ni	Porous AAO templates	530 ± 60 nm	170 ± 20 nm	[177]
PtCoNi, PtCoAu, PtRuCoNi	Porous AAO templates	>1 μm	50, 48, 45 nm	[178]
PtCo	Porous AAO templates	3–10 μm	280 ± 50 nm	[179]
PtRu and Pt array	Porous AAO templates	~1 μm	30 nm	[150]
Pd	Porous AAO templates	850 nm	50 nm	[137]
Pd arrays	Porous AAO templates	850 nm	50 nm	[153]
Pd arrays	Porous AAO templates	800 nm	80 nm	[154]
Pd arrays	Porous AAO templates	800 nm	80 nm	[156]
NiCu	Porous AAO templates	5 μm	75 nm	[180]
PtRu network	SBA-15 molecular sieves	70–200 nm	7–8 nm	[173]
Pt	MCM-41 molecular sieves	Up to 100 nm	3 nm	[149]
PtRu	SBA-15 molecular sieves	10–20 nm	6–7 nm	[174]
	MCM-41 molecular sieves	60–90 nm	~2.0 nm	
Multilayered Pt-Ru	Ti/Si(111) or Si(111)	>500 nm	55 nm	[181]
Ag and AgMn	Polycarbonate track etched membranes	–	50 nm	[175]
Ag, Au, and Pt	Polycarbonate template	2 ± 1 μm	220 ± 30 nm	[158]
Ag	Polycarbonate membranes	10 μm	100 nm	[142]
Pt networks	Irradiated polymer foil	~35 μm	35 nm	[33]
Pt	Rack-etched polycarbonate (PCTE) membrane	0.5–6 μm	47 ± 9.8 nm	[148]
PdPt and PdAu	Te NWs sacrificial template	Tens μm	10.8 nm	[182]
Au/Pt and Au/Pt₃Ni	On Au nanowires	>1 μm	~30 nm	[140]
Pt	On Sn@CNT nanocable	–	~4 nm	[183]
Pt arrays	Glancing angle deposition no template	50–400 nm	5–100 nm	[145]

MCM-41 [149,174] templates are as thin as 2 nm with lengths from 10–200 nm. This method is a very common one too, which has been used to synthesize Pd [137,153,154,156], Ag [142,158,175], Au [158], Pt-based alloy [35,150,172,176] nanowires, etc.

The pure metal nanowires have been applied to MOR, EOR, ORR, etc. The researchers expect high performance of nanowires due to their 1-D nanostructure. The origin of the remarkable electrocatalytic performance of nanowires may be attributed to the factors as follows. First, 1-D nanostructured electrocatalysts possess improved electron transport characteristics due to the shape anisotropy as compared to nanoparticles. The contact possibility between nanowires increased extensively to form a nanowire network. The numerous contact points and parallel pathways for electron transfer result in larger electronic conductivity. Second, high porosity is achieved by a nanowire network, which facilitates the effective exposure of catalytically active sites and mass transportation. Third, nanowires could preferentially expose highly active crystal facets shown in Figure 13.29, which bring high electrocatalytic activity for alcohol oxidation and oxygen reduction. Fourth, the electronic property of noble metal nanowires (Au, Pt, Pd, etc.) could be largely altered for some nanowires with less than a 2 nm diameter due to a surface contraction effect and a surface-stress-induced phase transformation, which could significantly enhance the catalytic performance in the ORR and MOR [31].

However, the specific surface area of a nanowire is significantly lower than that of very small Pt nanoparticles. Consequently, the mass activity of pure metal nanowires is hard to be outstandingly higher than that of small Pt nanoparticles. But the mass activity may be improved by synthesizing alloy-based networks or porous nanowires or by using network structures of another material as a support.

The durability of the nanowire shown in Figure 13.30 was investigated by Rauber et al. [33]. A significant loss of ECSA is observed for PtB and Pt/C catalysts whereas the CV curves show only a slight decrease for Pt NWNs. The reasons for different durability are multiple degradation pathways, such as particle aggregation; loss of electrical contact; Ostwald ripening;

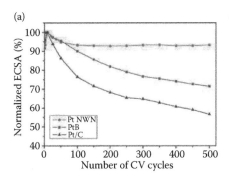 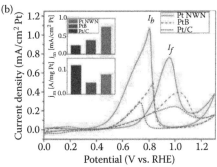

FIGURE 13.30
(a) Loss of ECSA of Pt NWNs, PtB, and Pt/C catalysts with number of CV cycles in 0.5 M H_2SO_4 solution (0–1.3 V, scan rate 50 mV). (b) CV curves for MOR catalyzed by Pt NWNs (r = 15 nm), PtB, and Pt/C in 0.5 M H_2SO_4, 0.5 M CH_3OH solution (0–1.3 V, scan rate 50 mV s^{-1}). Inset: Peak current densities j_a (top) and peak mass activities j_m (bottom) of the forward anodic peak I_f for Pt NWNs, PtB, and Pt/C. (From Rauber, M. et al. *Nano Letters* 11, 2304–2310, 2011.)

and dissolution of platinum; etc. The current densities of the forward scanning peaks (I_f) of Pt NWNs, PtB, and Pt/C are 0.76, 0.39, and 0.24 mA cm^{-2} in Figure 13.30b, respectively, implying the high activity of Pt NWNs. But the mass activity of Pt nanowire is 82 mA mg^{-1}, which is lower than 121 mA mg^{-1} of Pt/C.

But Choi et al. [148] found that the Pt nanowires also can show higher catalytic mass activities and stable methanol electrooxidation than the Pt nanoparticle catalysts over the electrode systems with high Pt content. The Pt nanowires synthesized by different researchers also show high activity and stability to MOR [146,149,151,152,162,163,168,171].

EOR on Pt nanowires—Pt nanowires have much higher activity than nanoparticles at peak position for the EOR, and the onset potential negatively shifted ~0.22 V [31], a high current density with at least a twofold enhancement in the potential region of 0.2–0.9 V as compared with associated 0-D counterparts. In order to improve the performance, porous Pt nanowire arrays were made by Zhang et al. [170]. They found that the anodic peak current on the Pt porous platinum nanowire array electrode is about 2.2 times than that on the Pt film and 1.5 times than that on the Pt NWA electrode. But they stressed that the increase of current is not related to the increase of active surface area because the current densities have already been normalized. The precise mechanism has not been determined yet.

ORR on Pt nanowires—The ORR activity of the nanowire shows higher area specific activity, more positive reduction peak potential, higher reaction rate constant, comparable activation energy in Figure 13.31 [145], and greater stability against electrochemically active surface area loss compared to conventional Pt/C electrodes [138,145]. The mass specific ORR activity of the Pt nanowires was found to be lower than that of the Pt/C catalyst, mainly due

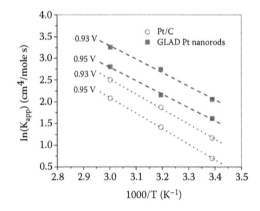

FIGURE 13.31
Arrhenius plots for the apparent rate constant k_{app} on Pt/C and GLAD Pt nanowires at 0.93 and 0.95 V vs. RHE. Ea for 20% Pt/C is 30.0 ± 0.5 kJ mol^{-1}; Ea for Pt nanowires is 25.8 ± 0.5 kJ mol^{-1}. (From Khudhayer, W.J. et al. *Journal of the Electrochemical Society* 158, B1029–B1041, 2011.)

to the large diameter of the nanowires [145]. Koenigsmann et al. [30] found that acid-treated ultra-thin nanowires displayed an electrochemical surface area activity of 1.45 mA cm^{-2}, which was nearly four times greater than that of analogous, unsupported platinum nanotubes and seven times greater than that of commercially supported platinum nanoparticles.

However, Pt nanowires also can show higher mass activity. Sun et al. [183] studied the single-crystal Pt nanowires on Sn@CNT nanocable and found that the Pt NW/Sn@CNT composite shows 1.2 times higher mass activity and 2.4-fold better area-specific activity for ORR than those of the commercial catalysts made of Pt nanoparticles on carbon black. Shimizu et al. [184] found that the nanowire network has high initial mass activity (110.5A g^{-1} Pt^{-1} at 0.85 V RHE) as compared to commercial Pt/C catalyst. These novel structures have the potential to possess high Pt utilization, high activity, and high durability for fuel cell applications.

Propanol and isopropanol oxidation on Pd nanowires—A highly ordered Pd nanowire array was prepared by template fabrication and studied in propanol electrooxidation [153]. The electrocatalytic performance of Pd nanowire electrode is shown in Figure 13.32. The onset potential for propanol oxidation on the Pd nanowire electrode is −0.67 V, which is 190 mV more negative than the −0.48 V observed on the Pd film electrode. The reduction in the onset anodic potential shows the significant enhancement in the kinetics of the propanol oxidation reaction. The peak current density is 10 and 102 mA cm^{-2} for the reaction on the Pd film and Pd NWA electrodes, respectively. The electrocatalytic activity of the Pd nanowires for propanol electrooxidation is obviously higher than that of conventional Pd film electrodes. The Pd nanowires show great potential as excellent electrocatalysts for propanol electrooxidation in alkaline media in DAFCs.

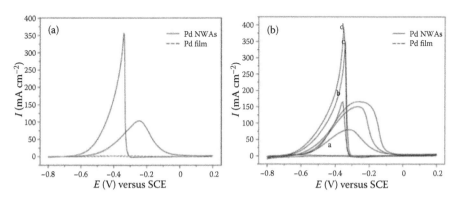

FIGURE 13.32
(a) CVs measured on Pd nanowire electrode with the diameters of about 50 nm and Pd film electrode in 1.0 M KOH + 1.0 M propanol solution at a scan rate of 50 mV s^{-1}. (b) CVs of propanol with different concentrations in 1 M KOH solution on Pd nanowire electrode. a: 0 M propanol; b: 0.2 M propanol; c: 0.5 M propanol; d: 1.0 M propanol. (From Sun, Z.H. et al. *J Autom Methods Manag Chem*, 5, 2009.)

Isopropanol is the smallest secondary alcohol with less toxicity and higher electrochemical reactivity. The direct isopropanol fuel cells show higher performance and lower crossover current than the DMFCs [185]. A Pd nanowire catalyst exhibits good electrocatalytic activity for isopropanol oxidation in alkaline media at room temperature [137]. The peak current density for isopropanol is 65 mA cm^{-2}, demonstrating the well-defined Pd nanowire structure possesses the high active sites for the electrooxidation reaction of isopropanol. Meanwhile, the peak current density for methanol is 111 mA cm^{-2}, which is higher than that for isopropanol. But the onset potential of isopropanol oxidation is ~240 mV more negative than that of methanol oxidation. The negative potential indicates that isopropanol oxidation has much higher catalytic activity than methanol electrooxidation on Pd nanowire electrodes in alkaline media and will directly improve the fuel cell efficiency.

EOR on Pd nanowire—The electrocatalytic oxidation of ethanol was selected as a test reaction in an alkaline medium, in which Pd is known to be among the best electrode materials [154–156]. Ksar et al. [155] found that Pd nanowires exhibit both a very important electrocatalytic activity for ethanol oxidation and a very high stability shown in Figure 13.33. Xu et al. [154] found that the electrocatalytic activity and stability of the Pd nanowires for ethanol electrooxidation are not only significantly higher than that of conventional Pd film electrodes, but also are higher than that of the well-established commercial E-TEK PtRu/C electrocatalysts. The Pd nanowires possess excellent electrocatalytic properties for ethanol electrooxidation in alkaline media and may be a great potential in direct ethanol fuel cells and ethanol sensors.

ORR on Ag nanowires—A catalyst composed of cheaper materials is essential for the commercial application of DAFCs. The price of silver is

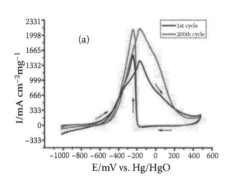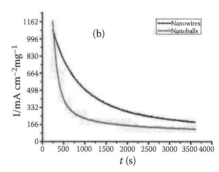

FIGURE 13.33

(a) Superposition of the first (black solid line curve) and the 200th (gray solid line curve) cyclic voltammetric runs associated with the electrocatalytic oxidation of 1 M EtOH in 1 M KOH with continuous cycling of the electrode potential. The working electrode was a glassy carbon disk modified with the Pd nanowires synthesized as described in the text. The reference electrode was a Hg/HgO (1 M KOH) electrode. The scan rate was 50 mV s^{-1}. (b) CA curves for ethanol electrooxidation at –0.230 V on a glassy carbon electrode modified with Pd nanowires (black curve) or Pd nanoballs (gray curve). (From Ksar, F. et al. *Chemistry of Materials* 21, 1612–1617, 2009.)

usually less than 30 USD oz^{-1} while the price of Pt is more than 1400 USD oz^{-1}. Fortunately, silver nanowires were proved to be active catalysts for the ORR through electrochemical testing [142]. The performance increases with the increasing loading of Ag. Furthermore, silver nanowires could act as alcohol-tolerant catalysts for the oxygen electrochemical reduction [143]. Cyclic voltammetry and Tafel results showed that silver nanowires with a high aspect ratio exhibited better activity and alcohol-tolerant stability for oxygen electroreduction in alkaline media than nanowires with a low aspect ratio. The effects of alcohol on the loss of ORR activity for silver nanowires follow the order: isopropanol > ethanol > methanol [143].

13.5.2 Multi-Segment Nanowires

Metal–metal junctions are possible to show the special properties of catalysis due to the interaction between two different metals. In order to improve the performance, multi-segment nanowires, which consist of many metal–metal junctions, were synthesized and tested. Figure 13.34 shows four sets of typical multi-segment nanowires with two metal species, Pt and Ru. The average diameter of the rods is 55 nm. The Pt and Ru segments can be clearly identified by the brightness differences in the SEM image. The brightness differences are due to the different atomic number of Pt and Ru, which have different abilities to scatter the electron. The Pt segment is brighter because of the higher atomic number of Pt. The element in each segment was confirmed by EDS.

FIGURE 13.34
FE-SEM images of segmented nanowires. (a) Pt nanowire, (b) Pt/Ru nanowire with three layers, (c) Pt/Ru nanowire with seven layers, and (d) Pt/Ru nanowire with 13 layers. (From Yoo, S.J. et al. *Physical Chemistry Chemical Physics* 12, 15240–15246, 2010.)

The XRD data in Figure 13.35 shows that the diffraction peaks of Pt-Ru nanowires with 13 layers could be superimposed on pure Ru and Pt without detectable shifts in the Bragg angles [181]. That is to say there is no alloying between Pt and Ru in the segmented nanowires. So multi-segment nanowires are a good metal–metal junction system to study the catalysis property on this kind of special structure [181]. In addition, more multi-segment nanowires, such as multi-segment Pt/Ru [176,181], PtNi nanowires [176,177], Pt–RuNi nanowires [35,176], etc., have been synthesized and studied with relation to DAFCs.

The hard template method is the main way to synthesize multi-segment nanowires. Anodic aluminum oxide (AAO) [35,176,177,186–189] membranes are commonly used as hard templates, and Si(111) [181], and Ti/Si(111) [181] membranes were used as well. The template has a large number of tunnels perpendicular to the membrane surface. Metals will be deposited into the tunnels in the designed order to form multi-segment nanowires. The most commonly used deposition method is the electrodeposition method due to its low cost, easy procedure, high productivity, etc. [35,176,177,186–189]. Before the electrodeposition, a layer of metal film, such as Cu [176,177,188,189] or Ag [186] film, was first deposited by vacuum evaporation onto one side of the membrane to form the working electrode. During the electrodeposition, different metals can be deposited into the AAO membrane from a solution of two or more metal ions and changing applied potential [190,191] or from different electrolyte solutions consecutively [35,176,177,186–189] to form multi-segment nanowires. After the electrodeposition, an aqueous solution of NaOH was then used to dissolve the alumina membrane. At last, the multi-segment nanowires are obtained.

Till now, researchers have synthesized many kinds of multi-segment nanowires with different segments and metal species, such as Pt-RuNi-Pt-RuNi-Pt [176], Pt-Ru-Pt-Ru-Pt [176], Pt-Ni-Pt-Ni-Pt [176], Ni-Pt [177], Ni-Pt-Ni [177], Ni-Pt-Ni-Pt [177], Ni-Pt-Ni-Pt-Ni [177], and etc. Even the multi-segment

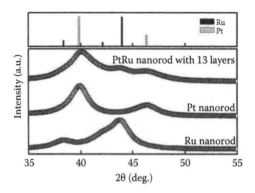

FIGURE 13.35
XRD profiles of Ru nanowire, Pt nanowire, and Pt/Ru nanowires with 13 layers. The upper figure with black and gray lines represents diffraction patterns of the fcc. Pt phase and hcp. Ru phase, respectively. (From Yoo, S.J. et al. *Physical Chemistry Chemical Physics* 12, 15240–15246, 2010.)

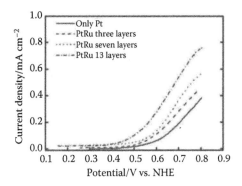

FIGURE 13.36
Current density vs. potential for Pt nanowires, Pt-Ru nanowires with three layers, Pt-Ru nanowires with seven layers, and Pt-Ru nanowires with 13 layers cycled between potentials of 0.05 and 1.2 V vs. NHE at 50 mV s^{-1}.

Pt-Ru nanowires with 13 layers [181] was also synthesized. Here, the sign "-" means the interface of two segments.

Currently, the multi-segment nanowires are mainly used in methanol oxidation. The performance of multi-segment nanowires could be affected by the number of segments [176,177], length ratio between segments [35], metal species, etc. [35,176,177,181]. Usually, the number of segments has a positive effect [177,181] shown in Figure 13.36 while the length ratio between segments [35] and metal species, etc. [35,176,177,181] have an optimized result. The stability was improved if the multi-segment nanowires had more segments [177,181] or the mid part was alloy [176].

13.5.3 Alloy Nanowires

Different from multi-segment nanowires, alloy nanowires are a homogenous mixture of two or more metals. The typical alloy nanowires, which are $Pd_{80}Pt_{20}$ nanowires, are shown in Figure 13.37 [166]. The morphology of alloy nanowires is very similar to that of pure metal nanowires shown in Figure 13.28a. The average diameter of these nanowires is ~50 nm while the length is too long to be determined by the SEM image. However, some nanowires show a 1-D structure under SEM image with high magnification while they are actually composed of small nanoparticles in their HRTEM image [166], which implies these nanowires are of a porous structure, which will be discussed in the following section.

Alloy nanowires are usually synthesized by both soft template and hard template methods. For the soft template method, the precursors are reduced under the protection of some organic ligands, such as oleylamine (OAm) or a mixture of OAm and 1-octadecene (ODE) [159], PVP (MW = 30,000) [165], PVP (MW = 58,000) [164], CTAB, etc. For example, in order to synthesize Pd-Fe nanowires, the precursor $Pd(acac)_2$ and $Fe(CO)_5$ were reduced under the protection of OAm at 160°C [160]. Besides Pd-Fe nanowires, other nanowires, such as ultrathin Fe-Pt-Pd nanowires

FIGURE 13.37
(a) TEM images of $Pd_{80}Pt_{20}$. (From Zhu, C.Z. et al. *Advanced Materials* 24, 2326–2331, 2012.) (b) FESEM images of as-spun Pt-Rh. (From Kim, Y.S. et al. *Electrochemistry Communications* 10, 1016–1019, 2008.) (c) FESEM images of Pt-Ru. (From Kim, Y.S. et al. *Electrochemistry Communications* 10, 1016–1019, 2008.)

[159], Pt-Rh nanowires [165], Pt-Ru nanowire [169], Pt-Bi nanowires [164], etc., have been synthesized. Therefore, crystal, very thin, very long, and very large amounts of nanowires could be obtained through the soft template method.

In the hard template method, the precursors are reduced and formed in the pore of hard templates, which can constrain the shape of the nanoparticles. For example, in order to synthesize Pd-Ru nanowires, Pt and Ru were electrodeposited into the porous AAO membrane in a solution of H_2PtCl_6 and $RuCl_3$. The commonly used templates include porous AAO membrane [150,179], Whatman's nuclepore polycarbonate track-etched membranes [175], SBA-15 [173,192], MCM-41 [192], etc.

Besides the soft and hard template methods, there are some other methods to synthesize alloy nanowires. One is the sacrificial template method, in which Pd-M (M = Pt, Au) bimetallic alloy nanowires were synthesized by using Te nanowires as a sacrificial template and reducing agent [182]. The precursors, such as H_2PtCl_6 and H_2PdCl_4, were reduced by Te nanowires and deposited on Te nanowires simultaneously. A Pd-Pt bimetallic alloy nanowire formed when the whole Te nanowire was consumed. Another one is the electrospin method, in which the mixture of precursors and PVP are electrospun to form hybrid nanowires. Then the formed hybrid nanowires are burned in air to remove the polymer and get the alloy nanowires. Through this method, Pt-Rh and Pt-Ru bimetallic nanowires have been synthesized [166].

Alloy nanowires show excellent activity and stability for methanol oxidation. Kim et al. [166] found that the mass-specific activity of Pt-Ru nanowires was superior to the other nanowires and even better than the commercial catalysts of the highly dispersed Pt-Ru nanoparticles on carbon (Figure 13.38). They attributed the enhancements to the one-dimensional features, which can outperform on the electrooxidations over the fuel cell electrodes [166]. The ratio between metal species has an optimized value, which is 1:1. The stability of the PtRu nanowire and PtRu/C is similar while the stability of the Pt-Rh nanowire is lower. In addition, the onset potential of Pt-Co nanowires is 0.33 ± 0.03 V, which is lower than 0.42 ± 0.06 V of PtCo thin

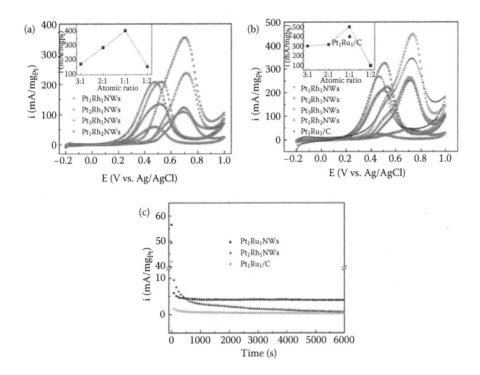

FIGURE 13.38

The cyclic voltammograms of electrospun bimetallic nanowires of (a) PtRh and (b) PtRu with compositional variation in 2 M CH_3OH containing 0.5 M H_2SO_4 solution at a scan rate of 50 mV s^{-1} and (c) chronoamperometric curves of PtRh nanowires and PtRu nanowires compared with PtRu/C (E-TEK) in 2 M CH_3OH containing 0.5 M H_2SO_4 solution for 6000 s.

film [179]. The improved performance was also found for nanowires of PtRu [150,169,173,192], $Pt_{90}Co_{10}$ [179], $Fe_{28}Pt_{38}Pd_{34}$ [159], PtFe [159], PtPd [182], etc.

The application of alloy nanowires into EOR is not studied as much as that into MOR. Yuan et al. [165] found that PtRh nanowires with higher ratios of Pt had higher activity, and the highest activity was higher than Pt black shown in Figure 13.39. Du et al. [164] found that the as-made $Pt_{95}Bi_5$ appeared to have superior electrocatalytic activity toward ethanol oxidation in comparison with commercial Pt/C. Zhu et al. [182] found that the as-prepared PtPd nanowires exhibited significantly enhanced activity and stability toward ethanol and methanol electrooxidation in an alkaline medium, demonstrating the potential of applying PtPd nanowires as effective electrocatalysts for DAFCs.

The alloy nanowires have been applied to ORR as well as MOR and EOR. The accelerated electrochemical cycling tests show that PtFe nanowires have better electrochemical durability than conventional Pt/C catalysts. Before the durability test, PtFe nanowires and conventional Pt/C have similar activity at low overpotential, while PtFe nanowires show clearly higher activity afterwards [161]. This similar performance improvement for ORR by other studies [160,175].

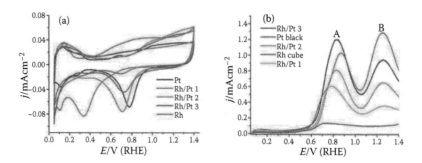

FIGURE 13.39
Cyclic voltammograms (CVs) of Rh cubes, commercial Pt black, and RhPt bimetal alloys in (a) a 0.1 M HClO$_4$ solution and (b) a 0.1 M ethanol and 0.1 M HClO$_4$ solution. (From Yuan, Q. et al. *Chemistry of Materials* 22, 2395–2402, 2010.)

13.6 Hierarchical Structure Catalysts

A hierarchical organization is an organizational structure in which every entity in the organization, except one, is subordinate to a single other entity. Hierarchical structure appearing in the literature includes nanoflowers [32], branched nanocubes [56], spongelike particles (SSPs) [193,194], etc. shown in Figure 13.40. The figure shows a clear trunk and branch structure in the nanoparticle. This kind of structure has advantages in mass and electron transfer and incorporation between different scales. So it is a potential structure in electrochemical catalysis. But it is very hard to define a clear structure model for a hierarchical structure. Here, we only present three examples and their application in DAFCs.

Hu et al. [32] synthesized Pt nanoflowers by electrodepositing a Pt overlayer on a BaTiO$_3$/Ti electrode in the deaerated H$_2$PtCl$_6$ (1 mM) + HCl (10 mM) +

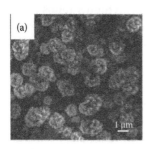
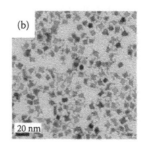

FIGURE 13.40
(a) FESEM images of Pt/BaTiO3/Ti catalyst. (From Hu, C. et al. *Journal of Power Sources* 195, 1594–1598, 2010.) (b) TEM images of PtFeCo branched nanocubes. (From Kim, S. et al. *Topics in Catalysis* 53, 686–693, 2010.) (c) TEM image of the Pd spherical spongelike particles. (From Shen, Q. et al. *The Journal of Physical Chemistry* C 113, 1267–1273, 2009.)

$HClO_4$ (0.1 mM) solution. The morphology of Pt nanoflowers is shown in Figure 13.40a. The CV current density of $Pt/BaTiO_3/Ti$ is much higher than that of Pt/Ti for electrochemical oxidation of both methanol and ethanol. The anodic peak of $Pt/BaTiO_3/Ti$ electrode is 0.69 V and 0.72 V while that of Pt/Ti electrode shift to 0.72 V and 0.76 V for methanol and ethanol, respectively.

Kim et al. [56] synthesized branched cubic PtFeCo nanoparticles by reducing platinum acetylacetonate, $Fe(CO)_5$, and cobalt acetylacetonate in 1-octadecene (Aldrich), oleic acid (Aldrich), and an oleylamine mixture solution under an N_2 environment without the use of a continuous N_2 flow. The morphology of Pt nanoflowers is shown in Figure 13.40b. CVs in Figure 13.41A for the electrocatalytic oxidation of methanol showed that PtFeCo branched nanocubes had the highest activity, although their H adsorption/desorption abilities were higher for PtFeCo nanocubes or nanoparticles. CA

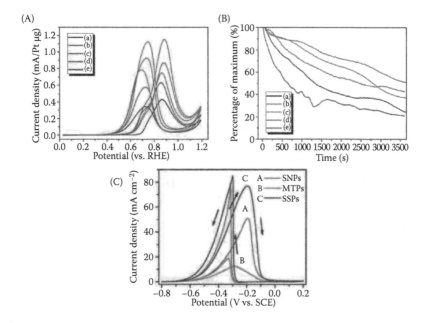

FIGURE 13.41
(A) MeOH oxidation cyclovoltammograms of carbon-supported nanoparticles in 1 M H_2SO_4 and 2 M MeOH solution with a scan rate of 20 m V s^{-1}. Current density was calculated based on unit Pt mass: (a) PtFe nanocubes, (b) PtFeCo nanocubes, (c) PtFeCo branched nanocubes, (d) PtFeCo nanoparticles with low cobalt content, (e) PtFeCo nanoparticles with high cobalt content. (From Kim, S. et al. *Topics in Catalysis* 53, 686–693, 2010.) (B) Chronoamperometry of carbon-supported nanoparticles in 1 M H_2SO_4 and 2 M MeOH solution at 0.6 V: (a) PtFe nanocubes, (b) PtFeCo nanocubes, (c) PtFeCo branched nanocubes, (d) PtFeCo nanoparticles with low cobalt content, (e) PtFeCo nanoparticles with high cobalt content. (From Kim, S. et al. *Topics in Catalysis* 53, 686–693, 2010.) (C) CVs of the SNPs (curve A), MTPs (curve B), and sSSPs (curve C) Pd-modified electrode in 1.0 M KOH solution containing 1.0 M ethanol. (From Shen, Q. et al. *The Journal of Physical Chemistry C* 113, 1267–1273, 2009.)

measurements in Figure 13.41B showed that PtFeCo branched nanocubes had the lowest loss of activity over time.

Shen et al. [193] synthesized Pd spherical spongelike particles by electro-chemically reducing H_2PdCl_4 in KNO_3, CTAB and a polydiallyldimethylam-monium chloride (PDDA) mixture solution. The morphology of Pt spherical spongelike particles is shown in Figure 13.41C. The catalytic activity of spherical spongelike particles (SSPs) in Figure 13.41C is six times higher than that of the multitwinned particles (MTPs) and ~1.5 times higher than that of the spherical nanoparticles (SNPs), which demonstrates that Pd SSPs have higher electrocatalytic activity for the oxidation of ethanol than SNPs and MTPs. CA curves also showed that Pd SSP modified electrodes exhibit a higher initial current and much slower current decay than those of the SNPs and MTPs modified electrodes.

13.7 Other Shape Catalysts

There are some other interesting nanostructured nanoparticles that have been studied in AEOR and ORR. The structures include polyhedral [76], truncated cubic [76], hexagonal [195], octahedral [195], or tetrahedral [195] shapes, and tetrahexahedral shapes [196], etc. Not all of these nanostructured nanoparticles showed outstanding higher performance in AEOR and ORR. But we can gain more knowledge about the relationship between nanostructure and catalytic activity. We will present a few novel nanostructures and their applications in DAFCs.

Polyhedral and truncated cubic Pt nanoparticles in Figure 13.42 were synthesized and studied by Wang et al. [76]. They controlled the size and the

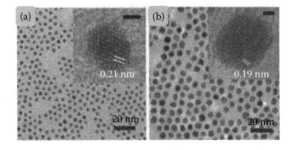

FIGURE 13.42

Representative TEM images of (a) the 3 nm polyhedral (b) the 5 nm truncated cubic and the insets are the representative HRTEM images of corresponding single particles, showing (a) Pt(111) and (b) Pt(100) lattice fringes. All scale bars in the insets correspond to 1 nm. (From Wang, C. et al. *Angewandte Chemie International Edition* 47, 3588–3591, 2008.)

shape of the Pt nanoparticles by tuning the reaction temperature at which Fe(CO)$_5$ was injected into the Pt(acac)$_2$ solution. They compared the performance of nanocube, polyhedral, and truncated cubic Pt nanoparticles, then found that the number of electrons involved in the ORR for the polyhedral and truncated cubic Pt nanoparticles is only 0.7, which is much lower than 3.6 for the 7 nm platinum nanocubes. They also found that the activity of the polyhedral and truncated cubic Pt nanoparticles is much lower than that of platinum nanocubes.

Solla-Gullón et al. [195] systematically studied the nanoparticles with variant facets. They synthesized polyoriented Pt nanoparticles by the water-in-oil method and synthesized preferentially (100), (111), and (100)–(111) oriented Pt nanoparticles by using a so-called colloidal method. The polyoriented, (100), (111), and (100)–(111) oriented Pt nanoparticles are spherical, cubic, hexagonal, and octahedral or tetrahedral shapes shown in Figure 13.43, respectively. Interestingly, (111) Pt nanoparticles showed the highest electrocatalytic activity toward MOR in the low potential range. The electrocatalytic activity follows the series, (111) > (100)–(111) > polyoriented > (100) Pt nanoparticles after 600 s at 0.6 V. But this result is somehow completely in conflict with the results in nanocube study in section 13.2.

Sun's group [196] prepared tetrahexahedral (THH) shape platinum nanocrystals shown in Figure 13.44 by an electrochemical treatment of Pt nanospheres. The single-crystal THH nanoparticle is enclosed by 24 high-index

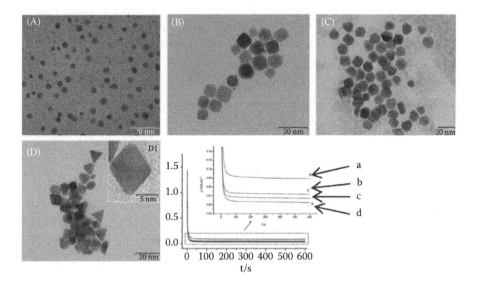

FIGURE 13.43
TEM pictures of (A) polyoriented, (B) (100), (C) (100)–(111), and (D) (111) Pt nanoparticles. (E) Chronoamperometric measurements at 0.6 V for (a) polyoriented, (b) (100), (c) (100)–(111), and (d) (111) Pt nanoparticles. Test solution: 0.5 M H$_2$SO$_4$ + 0.5 M CH$_3$OH. (From Solla-Gullon, J. et al. *Physical Chemistry Chemical Physics* 10, 3689–3698, 2008.)

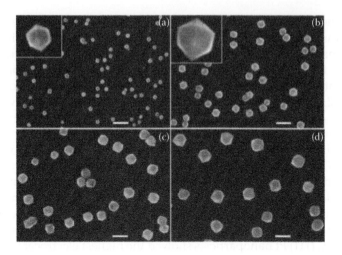

FIGURE 13.44
Size control of THH Pt nanoparticles and their thermal stability. SEM images of THH Pt nanoparticles grown at (a) 10, (b) 30, (c) 40, and (d) 50 min. The insets in (a) and (b) are the high-magnification SEM images that confirm the shape of THH. Scale bars, 200 nm. (From Tian, N. et al. *Science*, 316, 732–735, 2007.)

facets such as (730), (210), and/or (520) surfaces that have a large density of atomic steps and dangling bonds. These high-energy surfaces show much higher specific area catalytic activity for MOR and EOR. Outstandingly, the shape and high-energy surfaces can survive at high temperature up to 800°C.

13.8 Summary and Perspectives

Here, we introduced the structural characteristics, synthesis, and application in DAFCs of several nanostructures, such as nanocube, core-shell structure, hollow structure, 1D structure, hierarchical structure, etc. Most of them show higher performance as compared to conventional spherical nanoparticles in reactions of MOR, EOR, ORR, isopropanol reactions, etc., which are critical reactions for DAFCs. The improved performance is mainly due to the factors as follows. First, the novel nanostructure could bring the opportunity to modify the conditions of catalyst surface, which include facets, composition, electron-bonding energy, coverage, etc. The modifications of these surface conditions will affect the activity of catalysts in large range. The nanoparticles, such as nanocubes; 1D structures; tetrahexahedrals; hexagonal, octahedral, tetrahedral shapes, etc., synthesized by different ligands and conditions usually show different shapes and expose different facets.

The composition of the nanoparticle surface could be controlled by methods of alloying, dealloying, decorating, etc. With the variations of composition and size, the electron bonding energy will be tuned. Second, a special nanostructure could bring advantages in saving materials, improving mass transportation, improving electron transfer, increasing surface-to-volume ratio, etc. Core-shell structure and hollow structure can save precious material by removing the useless inner part. One-dimensional structures and hierarchical structures can facilitate mass transportation and electron transfer.

Although the application of nanostructures in DAFCs has been a huge achievement in recent years, it is just a promising start and needs be propelled further. At least, there are four works need to be done:

First, a number of variant nanostructures have been synthesized and tested in DAFC-related researches, but a systematical research even in one structure has not been done yet. The same nanostructure studied by different researchers usually shows different catalytic activities and sometime even shows confliction. Therefore, it is necessary to do some strict researches on one structure before we justify any nanostructure.

Second, how to assemble and operate the nanostructured particles to a DAFC electrode is a serious problem. The nanostructured particles are usually bare particles. They need to be supported stably on some supports, which are mainly carbon materials. Then, the capping ligands, which are added during the synthesis, need to be removed before any electrochemical reactions. The operation conditions need to be optimized and controlled thoroughly because most nanostructured particles have worse stability than spherical NPs.

Third, the advantage of one nanostructure is monotonous. Can we combine the advantages of several nanostructures together to achieve outstanding performance? Principally, we may realize this target by synthesizing a nanostructure with different structural characteristics. The hollow nanocube is a good example, which has the advantages of both nanocube and hollow structure.

Fourth, nanostructured particles with cheaper materials urgently need to be studied. Most of the nanoparticles in this chapter are precious metals, which are too expensive to be accepted. In addition, the activity of these catalysts does not satisfy the applications. However, the bio-catalysts, enzymes, are composed by cheaper elements, such as C, N, O, H, etc., but they can show extremely high activity. So it may be possible to synthesize highly active catalysts for DAFCs by cheaper materials, such as W_xC_y [197], NiMoC [198], NiWC [198], Ta-Ni-C [199], etc. Because nanostructures have so many advantages, it is possible to find some active catalysts in nanostructured particles with cheaper materials.

References

1. Boies, D.B. and Dravnieks, A. 1962. New high-performance methanol fuel cell electrodes. *Journal of the Electrochemical Society* 109, C198.
2. Cao, J., Yuan, T., Huang, Q., Zou, Z., Xia, B., Yang, H. and Feng, S. 2008. Development of micro direct methanol fuel cell systems and their possible applications. *ECS Transactions* 16, 1495–1506.
3. Zhou, W.J., Song, S.Q., Li, W.Z., Zhou, Z.H., Sun, G.Q., Xin, Q., Douvartzides, S. and Tsiakaras, P. 2005. Direct ethanol fuel cells based on PtSn anodes: The effect of Sn content on the fuel cell performance. *Journal of Power Sources* 140, 50–58.
4. Lai, S.C.S. and Koper, M.T.M. 2009. Ethanol electro-oxidation on platinum in alkaline media. *Physical Chemistry Chemical Physics* 11, 10446–10456.
5. Watanabe, M. and Motoo, S. 1975. Electrocatalysis by ad-atoms: Part II. Enhancement of the oxidation of methanol on platinum by ruthenium ad-atoms. *Journal of Electroanalytical Chemistry and Interfacial Electrochemistry* 60, 267–273.
6. Tong, Y.Y., Kim, H.S., Babu, P.K., Waszczuk, P., Wieckowski, A. and Oldfield, E. 2002. An NMR investigation of CO tolerance in a Pt/Ru fuel cell catalyst. *Journal of the American Chemical Society* 124, 468–473.
7. Zhao, X., Yin, M., Ma, L., Liang, L., Liu, C., Liao, J., Lu, T. and Xing, W. 2011. Recent advances in catalysts for direct methanol fuel cells. *Energy & Environmental Science* 4, 2736–2753.
8. Antolini, E. and Gonzalez, E.R. 2010. The electro-oxidation of carbon monoxide, hydrogen/carbon monoxide and methanol in acid medium on Pt-Sn catalysts for low-temperature fuel cells: A comparative review of the effect of Pt-Sn structural characteristics. *Electrochimica Acta* 56, 1–14.
9. Liu, H., Song, C., Zhang, L., Zhang, J., Wang, H. and Wilkinson, D.P. 2006. A review of anode catalysis in the direct methanol fuel cell. *Journal of Power Sources* 155, 95–110.
10. Burstein, G.T., Barnett, C.J., Kucernak, A.R. and Williams, K.R. 1997. Aspects of the anodic oxidation of methanol. *Catalysis Today* 38, 425–437.
11. Liu, J., Cao, J., Huang, Q., Li, X., Zou, Z. and Yang, H. 2008. Methanol oxidation on carbon-supported Pt–Ru–Ni ternary nanoparticle electrocatalysts. *Journal of Power Sources* 175, 159–165.
12. Alerasool, S. and Gonzalez, R.D. 1990. Preparation and characterization of supported Pt-Ru bimetallic clusters: Strong precursor support interactions. *Journal of Catalysis* 124, 204–216.
13. Arico, A.S., Poltarzewski, Z., Kim, H., Morana, A., Giordano, N. and Antonucci, V. 1995. Investigation of a carbon-supported quaternary Pt-Ru-Sn-W catalyst for direct methanol fuel-cells. *Journal of Power Sources* 55, 159–166.
14. Watanabe, M., Uchida, M. and Motoo, S. 1987. Preparation of highly dispersed Pt + Ru alloy clusters and the activity for the electrooxidation of methanol. *Journal of Electroanalytical Chemistry* 229, 395–406.
15. Paulus, U.A., Endruschat, U., Feldmeyer, G.J., Schmidt, T.J., Bonnemann, H. and Behm, R.J. 2000. New PtRu alloy colloids as precursors for fuel cell catalysts. *Journal of Catalysis* 195, 383–393.
16. Xue, X.Z., Lu, T.H., Liu, X.P. and Xing, W. 2005. Simple and controllable synthesis of highly dispersed Pt-Ru/C catalysts by a two-step spray pyrolysis process. *Chemical Communications*, 1601–1603.

17. Antolini, E. 2007. Catalysts for direct ethanol fuel cells. *Journal of Power Sources* 170, 1–12.
18. Guha, A., Lu, W.J., Zawodzinski, T.A. and Schiraldi, D.A. 2007. Surface-modified carbons as platinum catalyst support for PEM fuel cells. *Carbon* 45, 1506–1517.
19. He, T., Kreidler, E., Xiong, L., Luo, J. and Zhong, C.J. 2006. Alloy electrocatalysts: Combinatorial discovery and nanosynthesis. *Journal of the Electrochemical Society* 153, A1637–A1643.
20. Alayoglu, S., Nilekar, A.U., Mavrikakis, M. and Eichhorn, B. 2008. Ru-Pt core-shell nanoparticles for preferential oxidation of carbon monoxide in hydrogen. *Nature Materials* 7, 333–338.
21. Wang, W., Zheng, D., Du, C., Zou, Z., Zhang, X., Xia, B., Yang, H. and Akins, D.L. 2007. Carbon-supported Pd-Co bimetallic nanoparticles as electrocatalysts for the oxygen reduction reaction. *Journal of Power Sources* 167, 243–249.
22. Zhao, H., Li, L., Yang, J., Zhang, Y. and Li, H. 2008. Synthesis and characterization of bimetallic Pt-Fe/polypyrrole-carbon catalyst as DMFC anode catalyst. *Electrochemistry Communications* 10, 876–879.
23. Choi, J.-S., Chung, W.S., Ha, H.Y., Lim, T.-H., Oh, I.-H., Hong, S.-A. and Lee, H.-I. 2006. Nano-structured Pt–Cr anode catalyst over carbon support, for direct methanol fuel cell. *Journal of Power Sources* 156, 466–471.
24. Luo, J., Kariuki, N., Han, L., Wang, L., Zhong, C.-J. and He, T. 2006. Preparation and characterization of carbon-supported PtVFe electrocatalysts. *Electrochimica Acta* 51, 4821–4827.
25. Wang, W., Huang, Q., Liu, J., Zou, Z., Li, Z. and Yang, H. 2008. One-step synthesis of carbon-supported Pd–Pt alloy electrocatalysts for methanol tolerant oxygen reduction. *Electrochemistry Communications* 10, 1396–1399.
26. Kinoshita, K. 1990. Particle-size effects for oxygen reduction on highly dispersed platinum in acid electrolytes. *Journal of the Electrochemical Society* 137, 845–848.
27. Markovic, N.M., Gasteiger, H.A. and Ross, P.N. 1995. Oxygen reduction on platinum low-index single-crystal surfaces in sulfuric acid solution: Rotating ring-Pt(hkl) disk studies. *The Journal of Physical Chemistry* 99, 3411–3415.
28. Housmans, T.H.M., Wonders, A.H. and Koper, M.T.M. 2006. Structure sensitivity of methanol electrooxidation pathways on platinum: An on-line electrochemical mass spectrometry study. *Journal of Physical Chemistry B* 110, 10021–10031.
29. Kang, Y. and Murray, C.B. 2010. Synthesis and electrocatalytic properties of cubic Mn-Pt nanocrystals (nanocubes). *Journal of the American Chemical Society* 132, 7568.
30. Koenigsmann, C., Zhou, W.-P., Adzic, R.R., Sutter, E. and Wong, S.S. 2010. Size-dependent enhancement of electrocatalytic performance in relatively defect-free, processed ultrathin platinum nanowires. *Nano Letters* 10, 2806–2811.
31. Zhou, W.-P., Li, M., Koenigsmann, C., Ma, C., Wong, S.S. and Adzic, R.R. 2011. Morphology-dependent activity of Pt nanocatalysts for ethanol oxidation in acidic media: Nanowires versus nanoparticles. *Electrochimica Acta* 56, 9824–9830.
32. Hu, C., He, X. and Xia, C. 2010. Pt hierarchical structure catalysts on BaTiO3/Ti electrode for methanol and ethanol electrooxidations. *Journal of Power Sources* 195, 1594–1598.
33. Rauber, M., Alber, I., Muller, S., Neumann, R., Picht, O., Roth, C., Schokel, A., Toimil-Molares, M.E. and Ensinger, W. 2011. Highly-ordered supportless three-dimensional nanowire networks with tunable complexity and interwire connectivity for device integration. *Nano Letters* 11, 2304–2310.

34. Ahmadi, T.S., Wang, Z.L., Green, T.C., Henglein, A. and El-Sayed, M.A. 1996. Shape-controlled synthesis of colloidal platinum nanoparticles. *Science* 272, 1924–1925.

35. Fang, L., Jim Yang, L. and Weijiang, Z. 2006. Multi-segment Pt: RuNi nanorods for methanol electro-oxidation at room temperature. *Journal of the Electrochemical Society* 153, A2133–A2138.

36. Yang, H.Z., Zhang, J., Sun, K., Zou, S.Z. and Fang, J.Y. 2010. Enhancing by weakening: Electrooxidation of methanol on Pt(3)Co and Pt nanocubes. *Angewandte Chemie International Edition* 49, 6848–6851.

37. Long, N.V., Hien, T.D., Asaka, T., Ohtaki, M. and Nogami, M. 2011. Synthesis and characterization of Pt-Pd alloy and core-shell bimetallic nanoparticles for direct methanol fuel cells (DMFCs): Enhanced electrocatalytic properties of well-shaped core-shell morphologies and nanostructures. *International Journal of Hydrogen Energy* 36, 8478–8491.

38. Wiley, B., Sun, Y., Mayers, B. and Xia, Y. 2005. Shape-controlled synthesis of metal nanostructures: The case of silver. *Chemistry—A European Journal* 11, 454–463.

39. Xiong, Y. and Xia, Y. 2007. Shape-controlled synthesis of metal nanostructures: The case of palladium. *Advanced Materials* 19, 3385–3391.

40. Chen, A. and Holt-Hindle, P. 2010. Platinum-based nanostructured materials: Synthesis, properties, and applications. *Chemical Reviews* 110, 3767–3804.

41. Shukla, N., Liu, C. and Roy, A.G. 2006. Oriented self-assembly of cubic FePt nanoparticles. *Materials Letters* 60, 995–998.

42. Margeat, O., Tran, M., Spasova, M. and Farle, M. 2007. Magnetism and structure of chemically disordered FePt3 nanocubes. *Physical Review B* 75, 134410.

43. Demortière, A. and Petit, C. 2007. First synthesis by liquid-liquid phase transfer of magnetic CoxPt100-x nanoalloys. *Langmuir* 23, 8575–8584.

44. Fan, F.-R., Attia, A., Sur, U.K., Chen, J.-B., Xie, Z.-X., Li, J.-F., Ren, B. and Tian, Z.-Q. 2009. An effective strategy for room-temperature synthesis of single-crystalline palladium nanocubes and nanodendrites in aqueous solution. *Crystal Growth & Design* 9, 2335–2340.

45. Teng, X. and Yang, H. 2010. Synthesis and electrocatalytic property of cubic and spherical nanoparticles of cobalt platinum alloys. *Frontiers of Chemical Engineering in China* 4, 45–51.

46. Kang, Y., Ye, X. and Murray, C.B. 2010. Size- and shape-selective synthesis of metal nanocrystals and nanowires using CO as a reducing agent. *Angewandte Chemie International Edition* 49, 6156–6159.

47. Ren, J. and Tilley, R.D. 2007. Preparation, self-assembly, and mechanistic study of highly monodispersed nanocubes. *Journal of the American Chemical Society* 129, 3287–3291.

48. Han, S.-B., Song, Y.-J., Lee, J.-M., Kim, J.-Y. and Park, K.-W. 2008. Platinum nanocube catalysts for methanol and ethanol electrooxidation. *Electrochemistry Communications* 10, 1044–1047.

49. Zhang, J. and Fang, J. 2009. A general strategy for preparation of Pt 3d-transition metal (Co, Fe, Ni) nanocubes. *Journal of the American Chemical Society* 131, 18543–18547.

50. Loukrakpam, R., Chang, P., Luo, J., Fang, B., Mott, D., Bae, I.T., Naslund, H.R., Engelhard, M.H. and Zhong, C.J. 2010. Chromium-assisted synthesis of platinum nanocube electrocatalysts. *Chemical Communications* 46, 7184–7186.

51. Lu, Y., Jiang, Y., Wu, H. and Chen, W. 2013. Nano-PtPd cubes on graphene exhibit enhanced activity and durability in methanol electrooxidation after CO stripping–cleaning. *The Journal of Physical Chemistry C* 117, 2926–2938.
52. Lee, Y.-W., Han, S.-B., Kim, D.-Y. and Park, K.-W. 2011. Monodispersed platinum nanocubes for enhanced electrocatalytic properties in alcohol electrooxidation. *Chemical Communications* 47, 6296–6298.
53. Chang, G., Oyama, M. and Hirao, K. 2007. Facile synthesis of monodisperse palladium nanocubes and the characteristics of self-assembly. *Acta Materialia* 55, 3453–3456.
54. Lu, L.L., Yin, G.P., Wang, Z.B. and Gao, Y.Z. 2009. Electro-oxidation of dimethyl ether on platinum nanocubes with preferential {100} surfaces. *Electrochemistry Communications* 11, 1596–1598.
55. Ding, H., Shi, X.-Z., Shen, C.-M., Hui, C., Xu, Z.-C., Li, C., Tian, Y., Wang, D.-K. and Gao, H.-J. 2010. Synthesis of monodisperse palladium nanocubes and their catalytic activity for methanol electrooxidation. *Chinese Physics B* 19, 106104.
56. Kim, S., Kim, C. and Lee, H. 2010. Shape- and composition-controlled Pt–Fe–Co nanoparticles for electrocatalytic methanol oxidation. *Topics in Catalysis* 53, 686–693.
57. Yang, W., Wang, X., Yang, F., Yang, C. and Yang, X. 2008. Carbon nanotubes decorated with Pt nanocubes by a noncovalent functionalization method and their role in oxygen reduction. *Advanced Materials* 20, 2579.
58. Herricks, T., Chen, J. and Xia, Y.N. 2004. Polyol synthesis of platinum nanoparticles: Control of morphology with sodium nitrate. *Nano Letters* 4, 2367–2371.
59. Herrero, E., Franaszczuk, K. and Wieckowski, A. 1994. Electrochemistry of methanol at low-index crystal planes of platinum: An integrated voltammetric and chronoamperometric study. *Journal of Physical Chemistry* 98, 5074–5083.
60. Chen, M., Kim, J., Liu, J.P., Fan, H.Y. and Sun, S.H. 2006. Synthesis of FePt nanocubes and their oriented self-assembly. *Journal of the American Chemical Society* 128, 7132–7133.
61. Demortiere, A., Launois, P., Goubet, N., Albouy, P.A. and Petit, C. 2008. Shape-controlled platinum nanocubes and their assembly into two-dimensional and three-dimensional superlattices. *Journal of Physical Chemistry B* 112, 14583–14592.
62. Choi, S.-I., Choi, R., Han, S.W. and Park, J.T. 2010. Synthesis and characterization of Pt(9)Co nanocubes with high activity for oxygen reduction. *Chemical Communications* 46, 4950–4952.
63. Murray, C.B., Kagan, C.R. and Bawendi, M.G. 2000. Synthesis and characterization of monodisperse nanocrystals and close-packed nanocrystal assemblies. *Annual Review of Materials Science* 30, 545–610.
64. Sun, S.H. and Murray, C.B. 1999. Synthesis of monodisperse cobalt nanocrystals and their assembly into magnetic superlattices (invited). *Journal of Applied Physics* 85, 4325–4330.
65. Xiong, Y.J., Chen, J.Y., Wiley, B., Xia, Y.N., Yin, Y.D. and Li, Z.Y. 2005. Size-dependence of surface plasmon resonance and oxidation for pd nanocubes synthesized via a seed etching process. *Nano Letters* 5, 1237–1242.
66. Kim, C. and Lee, H. 2009. Change in the catalytic reactivity of Pt nanocubes in the presence of different surface-capping agents. *Catalysis Communications* 10, 1305–1309.

67. Fu, G., Wu, K., Jiang, X., Tao, L., Chen, Y., Lin, J., Zhou, Y., Wei, S., Tang, Y., Lu, T. and Xia, X. 2013. Polyallylamine-directed green synthesis of platinum nano-cubes. Shape and electronic effect codependent enhanced electrocatalytic activity. *Physical Chemistry Chemical Physics* 15, 3793–3802.
68. Peng, Z., You, H., Wu, J. and Yang, H. 2010. Electrochemical synthesis and catalytic property of sub-10 nm platinum cubic nanoboxes. *Nano Letters* 10, 1492–1496.
69. Mahmoud, M.A., Snyder, B. and El-Sayed, M.A. 2010. Polystyrene micro-spheres: Inactive supporting material for recycling and recovering colloidal nanocatalysts in solution. *Journal of Physical Chemistry Letters* 1, 28–31.
70. Mahmoud, M.A., Snyder, B. and El-Sayed, M.A. 2010. Surface plasmon fields and coupling in the hollow gold nanoparticles and surface-enhanced raman spectros-copy. Theory and experiment. *Journal of Physical Chemistry C* 114, 7436–7443.
71. Mancharan, R. and Goodenough, J.B. 1992. Methanol oxidation in acid on ordered NiTi. *Journal of Materials Chemistry* 2, 875–887.
72. Xu, D., Liu, Z., Yang, H., Liu, Q., Zhang, J., Fang, J., Zou, S. and Sun, K. 2009. Solution-based evolution and enhanced methanol oxidation activity of mono-disperse platinum-copper nanocubes. *Angewandte Chemie International Edition* 48, 4217–4221.
73. Zhang, J., Yang, H., Yang, K., Fang, J., Zou, S., Luo, Z., Wang, H., Bae, I.-T. and Jung, D.Y. 2010. Monodisperse Pt(3)Fe nanocubes: Synthesis, characterization, self-assembly, and electrocatalytic activity. *Advanced Functional Materials* 20, 3727–3733.
74. Xu, D., Bliznakov, S., Liu, Z., Fang, J. and Dimitrov, N. 2010. Composition-dependent electrocatalytic activity of Pt-Cu nanocube catalysts for formic acid oxidation. *Angewandte Chemie International Edition* 49, 1282–1285.
75. Yuan, Q., Zhou, Z., Zhuang, J. and Wang, X. 2010. Pd-Pt random alloy nano-cubes with tunable compositions and their enhanced electrocatalytic activities. *Chemical Communications* 46, 1491–1493.
76. Wang, C., Daimon, H., Onodera, T., Koda, T. and Sun, S.H. 2008. A general approach to the size- and shape-controlled synthesis of platinum nanoparticles and their catalytic reduction of oxygen. *Angewandte Chemie International Edition* 47, 3588–3591.
77. Wang, C., Daimon, H., Lee, Y., Kim, J. and Sun, S. 2007. Synthesis of monodis-perse Pt nanocubes and their enhanced catalysis for oxygen reduction. *Journal of the American Chemical Society* 129, 6974–6975.
78. Bard, A.J. and Faulkner, L.R. 2000. *Electrochemical Methods: Fundamentals and Applications.* New York: Wiley.
79. Ding, H., Shi, X.-Z., Shen, C.-M., Hui, C., Xu, Z.-C., Li, C., Tian, Y., Wang, D.-K. and Gao, H.-J. 2010. Synthesis of monodisperse palladium nanocubes and their catalytic activity for methanol electrooxidation. *Chin Phys B* 19.
80. Arico, A.S., Baglio, V., Di Blasi, A., Modica, E., Monforte, G. and Antonucci, V. 2005. Electrochemical analysis of high temperature methanol electro-oxidation at Pt-decorated Ru catalysts. *Journal of Electroanalytical Chemistry* 576, 161–169.
81. Arico, A.S., Baglio, V., Modica, E., Di Blasi, A. and Antonucci, V. 2004. Performance of DMFC anodes with ultra-low Pt loading. *Electrochemistry Communications* 6, 164–169.
82. Zhang, Y., Huang, Q., Zou, Z., Yang, J., Vogel, W. and Yang, H. 2010. Enhanced durability of Au cluster decorated Pt nanoparticles for the oxygen reduction reaction. *The Journal of Physical Chemistry C* 114, 6860–6868.

83. Sharma, S. and Pollet, B.G. 2012. Support materials for PEMFC and DMFC electrocatalysts: A review. *Journal of Power Sources* 208, 96–119.
84. Yuan, W., Tang, Y., Yang, X.J. and Wan, Z.P. 2012. Porous metal materials for polymer electrolyte membrane fuel cells: A review. *Applied Energy* 94, 309–329.
85. Chen, M., Liu, J.P. and Sun, S. 2004. One-step synthesis of FePt nanoparticles with tunable size. *Journal of the American Chemical Society* 126, 8394–8395.
86. Park, I.S., Lee, K.S., Jung, D.S., Park, H.Y. and Sung, Y.E. 2007. Electrocatalytic activity of carbon-supported Pt-Au nanoparticles for methanol electro-oxidation. *Electrochimica Acta* 52, 5599–5605.
87. Lee, K.S., Yoo, S.J., Ahn, D., Jeon, T.Y., Choi, K.H., Park, I.S. and Sung, Y.E. 2011. Surface structures and electrochemical activities of Pt overlayers on Ir nanoparticles. *Langmuir* 27, 3128–3317.
88. Long, N.V., Ohtaki, M., Hien, T.D., Randy, J. and Nogami, M. 2011. A comparative study of Pt and Pt-Pd core-shell nanocatalysts. *Electrochimica Acta* 56, 9133–9143.
89. Cheng, F., Dai, X., Wang, H., Jiang, S.P., Zhang, M. and Xu, C. 2010. Synergistic effect of Pd-Au bimetallic surfaces in Au-covered Pd nanowires studied for ethanol oxidation. *Electrochimica Acta* 55, 2295–2298.
90. Guo, S., Dong, S. and Wang, E. 2010. Ultralong Pt-on-Pd bimetallic nanowires with nanoporous surface: Nanodendritic structure for enhanced electrocatalytic activity. *Chemical Communications* 46, 1869–1871.
91. Gao, H.L., Liao, S.J., Liang, Z.X., Liang, H.G. and Luoa, F. 2011. Anodic oxidation of ethanol on core-shell structured Ru@PtPd/C catalyst in alkaline media. *Journal of Power Sources* 196, 6138–6143.
92. Ochal, P., de la Fuente, J.L.G., Tsypkin, M., Seland, F., Sunde, S., Muthuswamy, N., Ronning, M., Chen, D., Garcia, S., Alayoglu, S. and Eichhorn, B. 2011. CO stripping as an electrochemical tool for characterization of Ru@Pt core-shell catalysts. *Journal of Electroanalytical Chemistry* 655, 140–146.
93. Chrzanowski, W. and Wieckowski, A. 1997. Ultrathin films of ruthenium on low index platinum single crystal surfaces: An electrochemical study. *Langmuir* 13, 5974–5978.
94. Chrzanowski, W., Kim, H. and Wieckowski, A. 1998. Enhancement in methanol oxidation by spontaneously deposited ruthenium on low-index platinum electrodes. *Catalysis Letters* 50, 69–75.
95. Brankovic, S.R., McBreen, J. and Adzic, R.R. 2001. Spontaneous deposition of Pt on the Ru(0001) surface. *Journal of Electroanalytical Chemistry* 503, 99–104.
96. Brankovic, S.R., Wang, J.X. and Adzic, R.R. 2001. Metal monolayer deposition by replacement of metal adlayers on electrode surfaces. *Surface Science* 474, L173–L179.
97. Kuk, S.T. and Wieckowski, A. 2005. Methanol electrooxidation on platinum spontaneously deposited on unsupported and carbon-supported ruthenium nanoparticles. *Journal of Power Sources* 141, 1–7.
98. Yamamoto, T.A., Kageyama, S., Seino, S., Nitani, H., Nakagawa, T., Horioka, R., Honda, Y., Ueno, K. and Daimon, H. 2011. Methanol oxidation catalysis and substructure of PtRu/C bimetallic nanoparticles synthesized by a radiolytic process. *Applied Catalysis A-General* 396, 68–75.
99. Kaplan, D., Burstein, L., Rosenberg, Y. and Peled, E. 2011. Comparison of methanol and ethylene glycol oxidation by alloy and core-shell platinum based catalysts. *Journal of Power Sources* 196, 8286–8292.

100. Zhang, H., Xu, X.Q., Gu, P., Li, C.Y., Wu, P. and Cai, C.X. 2011. Microwave-assisted synthesis of graphene-supported Pd(1) Pt(3) nanostructures and their electrocatalytic activity for methanol oxidation. *Electrochimica Acta* 56, 7064–7070.

101. Karan, H.I., Sasaki, K., Kuttiyiel, K., Farberow, C.A., Mavrikakis, M. and Adzic, R.R. 2012. Catalytic activity of platinum mono layer on iridium and rhenium alloy nanoparticles for the oxygen reduction reaction. *ACS Catalysis* 2, 817–824.

102. Cochell, T. and Manthiram, A. 2011. Pt@PdxCuy/C core-shell electrocatalysts for oxygen reduction reaction in fuel cells. *Langmuir* 28, 1579–1587.

103. Kuttiyiel, K.A., Sasaki, K., Choi, Y., Su, D., Liu, P. and Adzic, R.R. 2012. Bimetallic IrNi core platinum monolayer shell electrocatalysts for the oxygen reduction reaction. *Energy & Environmental Science* 5, 5297–5304.

104. Alia, S.M., Jensen, K.O., Pivovar, B.S. and Yan, Y. 2012. Platinum-coated palladium nanotubes as oxygen reduction reaction electrocatalysts. *ACS Catalysis* 2, 858–863.

105. Strasser, P., Koh, S., Anniyev, T., Greeley, J., More, K., Yu, C., Liu, Z., Kaya, S., Nordlund, D., Ogasawara, H., Toney, M.F. and Nilsson, A. 2010. Lattice-strain control of the activity in dealloyed core-shell fuel cell catalysts. *Nature Chemistry* 2, 454–460.

106. Mani, P., Srivastava, R. and Strasser, P. 2008. Dealloyed Pt-Cu core-shell nanoparticle electrocatalysts for use in PEM fuel cell cathodes. *Journal of Physical Chemistry C* 112, 2770–2778.

107. Shao, M., Shoemaker, K., Peles, A., Kaneko, K. and Protsailo, L. 2010. Pt monolayer on porous PdAu alloys as oxygen reduction electrocatalysts? *Journal of the American Chemical Society* 132, 9253–9255.

108. Yang, R., Leisch, J., Strasser, P. and Toney, M.F. 2010. Structure of dealloyed PtCu3 thin films and catalytic activity for oxygen reduction. *Chemistry of Materials* 22, 4712–4720.

109. Huang, Y., Zhou, X., Yin, M., Liu, C. and Xing, W. 2010. Novel PdAu@Au/C coreshell catalyst: Superior activity and selectivity in formic acid decomposition for hydrogen generation. *Chemistry of Materials* 22, 5122–5128.

110. Chen, D.J., Hofstead-Duffy, A.M., Park, I.S., Atienza, D.O., Susut, C., Sun, S.G. and Tong, Y.Y.J. 2011. Identification of the most active sites and surface water species: A comparative study of CO and methanol oxidation reactions on core-shell M@Pt (M = Ru, Au) nanoparticles by in situ IR spectroscopy. *Journal of Physical Chemistry C* 115, 8735–8743.

111. Lewera, A., Zhou, W.P., Vericat, C., Chung, J.H., Haasch, R., Wieckowski, A. and Bagus, P.S. 2006. XPS and reactivity study of bimetallic nanoparticles containing Ru and Pt supported on a gold disk. *Electrochimica Acta* 51, 3950–3956.

112. Kaplan, D., Alon, M., Burstein, L., Rosenberg, Y. and Peled, E. 2011. Study of core-shell platinum-based catalyst for methanol and ethylene glycol oxidation. *Journal of Power Sources* 196, 1078–1083.

113. Wang, D., Xin, H.L., Wang, H., Yu, Y., Rus, E., Muller, D.A., DiSalvo, F.J. and Abruna, H.D. 2012. Facile synthesis of carbon-supported Pd-Co core-shell nanoparticles as oxygen reduction electrocatalysts and their enhanced activity and stability with monolayer Pt decoration. *Chemistry of Materials* 24, 2274–2281.

114. Lin, R., Zhang, H., Zhao, T., Cao, C., Yang, D. and Ma, J. 2012. Investigation of Au@Pt/C electro-catalysts for oxygen reduction reaction. *Electrochimica Acta* 62, 263–268.

115. Xu, C., Wang, L., Mu, X. and Ding, Y. 2010. Nanoporous PtRu alloys for electro-catalysis. *Langmuir* 26, 7437–7443.

116. Jang, J.-H., Pak, C. and Kwon, Y.-U. 2012. Ultrasound-assisted polyol synthesis and electrocatalytic characterization of PdxCo alloy and core-shell nanoparticles. *Journal of Power Sources* 201, 179–183.

117. Wu, H., Wexler, D., Wang, G. and Liu, H. 2012. Co-core-Pt-shell nanoparticles as cathode catalyst for PEM fuel cells. *Journal of Solid State Electrochemistry* 16, 1105–1110.

118. Liu, C.W., Wei, Y.C., Liu, C.C. and Wang, K.W. 2012. Pt-Au core/shell nanorods: Preparation and applications as electrocatalysts for fuel cells. *Journal of Materials Chemistry* 22, 4641–4644.

119. Zhao, J., Chen, W., Zheng, Y. and Li, X. 2006. Novel carbon supported hollow Pt nanospheres for methanol electrooxidation. *Journal of Power Sources* 162, 168–172.

120. Hong, J.W., Kang, S.W., Choi, B.-S., Kim, D., Lee, S.B. and Han, S.W. 2012. Controlled synthesis of Pd-Pt alloy hollow nanostructures with enhanced catalytic activities for oxygen reduction. *ACS Nano* 6, 2410–2419.

121. Ding, L.-X., Li, G.-R., Wang, Z.-L., Liu, Z.-Q., Liu, H. and Tong, Y.-X. 2012. Porous Ni@Pt core-shell nanotube array electrocatalyst with high activity and stability for methanol oxidation. *Chemistry (Weinheim an der Bergstrasse, Germany)* 18, 8386–8391.

122. Liu, Z., Zhao, B., Guo, C., Sun, Y., Xu, F., Yang, H. and Li, Z. 2009. Novel hybrid electrocatalyst with enhanced performance in alkaline media: Hollow Au/Pd core/shell nanostructures with a raspberry surface. *The Journal of Physical Chemistry C* 113, 16766–16771.

123. Dai, L., Jiang, L.-P., Abdel-Halim, E.S. and Zhu, J.-J. 2011. The fabrication of palladium hollow sphere array and application as highly active electrocatalysts for the direct oxidation of ethanol. *Electrochemistry Communications* 13, 1525–1528.

124. Hu, Y., Shao, Q., Wu, P., Zhang, H. and Cai, C. 2012. Synthesis of hollow mesoporous PtNi nanosphere for highly active electrocatalysis toward the methanol oxidation reaction. *Electrochemistry Communications* 18, 96–99.

125. Guo, S., Dong, S. and Wang, E. 2009. Raspberry-like hierarchical Au/Pt nanoparticle assembling hollow spheres with nanochannels: An advanced nanoelectrocatalyst for the oxygen reduction reaction. *The Journal of Physical Chemistry C* 113, 5485–5492.

126. Guo, D.-J., Zhao, L., Qiu, X.-P., Chen, L.-Q. and Zhu, W.-T. 2008. Novel hollow PtRu nanospheres supported on multi-walled carbon nanotube for methanol electrooxidation. *Journal of Power Sources* 177, 334–338.

127. Chen, H.M., Liu, R.-S., Lo, M.-Y., Chang, S.-C., Tsai, L.-D., Peng, Y.-M. and Lee, J.-F. 2008. Hollow platinum spheres with nano-channels: Synthesis and enhanced catalysis for oxygen reduction. *The Journal of Physical Chemistry C* 112, 7522–7526.

128. Kim, S.-W., Kim, M., Lee, W.Y. and Hyeon, T. 2002. Fabrication of hollow palladium spheres and their successful application to the recyclable heterogeneous catalyst for Suzuki coupling reactions. *Journal of the American Chemical Society* 124, 7642–7643.

129. Guo, D.-J. and Cui, S.-K. 2009. Hollow PtCo nanospheres supported on multi-walled carbon nanotubes for methanol electrooxidation. *Journal of Colloid and Interface Science* 340, 53–57.

130. Vasquez, Y., Sra, A.K. and Schaak, R.E. 2005. One-pot synthesis of hollow super-paramagnetic CoPt nanospheres. *Journal of the American Chemical Society* 127, 12504–12505.
131. Balgis, R., Anilkumar, G.M., Sago, S., Ogi, T. and Okuyama, K. 2012. Nanostructured design of electrocatalyst support materials for high-performance PEM fuel cell application. *Journal of Power Sources* 203, 26–33.
132. Yu, X., Wang, D., Peng, Q. and Li, Y. 2011. High performance electrocatalyst: Pt-Cu hollow nanocrystals. *Chemical Communications* 47, 8094–8096.
133. Bi, Y.P. and Lu, G.X. 2009. Control growth of uniform platinum nanotubes and their catalytic properties for methanol electrooxidation. *Electrochemistry Communications* 11, 45–49.
134. Lee, C.-L., Chiou, H.-P., Syu, C.-M., Liu, C.-R., Yang, C.-C. and Syu, C.-C. 2011. Displacement triangular Ag/Pd nanoplate as methanol-tolerant electrocatalyst in oxygen reduction reaction. *International Journal of Hydrogen Energy* 36, 12706–12714.
135. Dubau, L., Durst, J., Maillard, F., Guotaz, L., Chatenet, M., Andro, J. and Rossinot, E. 2011. Further insights into the durability of Pt3Co/C electrocatalysts: Formation of hollow Pt nanoparticles induced by the Kirkendall effect. *Electrochimica Acta* 56, 10658–10667.
136. Kim, Y.S., Kim, H.J. and Kim, W.B. 2009. Composited hybrid electrocatalysts of Pt-based nanoparticles and nanowires for low temperature polymer electrolyte fuel cells. *Electrochemistry Communications* 11, 1026–1029.
137. Cheng, F., Wang, H., Sun, Z., Ning, M., Cai, Z. and Zhang, M. 2008. Electrodeposited fabrication of highly ordered Pd nanowire arrays for alcohol electrooxidation. *Electrochemistry Communications* 10, 798–801.
138. Liang, H.-W., Cao, X., Zhou, F., Cui, C.-H., Zhang, W.-J. and Yu, S.-H. 2011. A free-standing Pt-nanowire membrane as a highly stable electrocatalyst for the oxygen reduction reaction. *Advanced Materials* 23, 1467–1471.
139. Sreeprasad, T.S., Samal, A.K. and Pradeep, T. 2007. Body- or tip-controlled reactivity of gold nanorods and their conversion to particles through other anisotropic structures. *Langmuir* 23, 9463–9471.
140. Tan, Y.M., Fan, J.M., Chen, G.X., Zheng, N.F. and Xie, Q.J. 2011. Au/Pt and Au/Pt3Ni nanowires as self-supported electrocatalysts with high activity and durability for oxygen reduction. *Chemical Communications* 47, 11624–11626.
141. Yu, Y.Y., Chang, S.S., Lee, C.L. and Wang, C.R.C. 1997. Gold nanorods: Electrochemical synthesis and optical properties. *Journal of Physical Chemistry B* 101, 6661–6664.
142. Kostowskyj, M.A., Gilliam, R.J., Kirk, D.W. and Thorpe, S.J. 2008. Silver nanowire catalysts for alkaline fuel cells. *International Journal of Hydrogen Energy* 33, 5773–5778.
143. Ni, K., Chen, L. and Lu, G. 2008. Synthesis of silver nanowires with different aspect ratios as alcohol-tolerant catalysts for oxygen electroreduction. *Electrochemistry Communications* 10, 1027–1030.
144. Du, S., Millington, B. and Pollet, B.G. 2011. The effect of Nafion ionomer loading coated on gas diffusion electrodes with in-situ grown Pt nanowires and their durability in proton exchange membrane fuel cells. *International Journal of Hydrogen Energy* 36, 4386–4393.

145. Khudhayer, W.J., Kariuki, N.N., Wang, X., Myers, D.J., Shaikh, A.U. and Karabacak, T. 2011. Oxygen reduction reaction electrocatalytic activity of glancing angle deposited platinum nanorod arrays. *Journal of the Electrochemical Society* 158, B1029–B1041.

146. Kim, J.M., Joh, H.-I., Jo, S.M., Ahn, D.J., Ha, H.Y., Hong, S.-A. and Kim, S.-K. 2010. Preparation and characterization of Pt nanowire by electrospinning method for methanol oxidation. *Electrochimica Acta* 55, 4827–4835.

147. Song, Y.-J., Han, S.-B. and Park, K.-W. 2010. Pt nanowire electrodes electrodeposited in PVP for methanol electrooxidation. *Materials Letters* 64, 1981–1984.

148. Choi, S.M., Kim, J.H., Jung, J.Y., Yoon, E.Y. and Kim, W.B. 2008. Pt nanowires prepared via a polymer template method: Its promise toward high Pt-loaded electrocatalysts for methanol oxidation. *Electrochimica Acta* 53, 5804–5811.

149. Park, I.-S., Choi, J.-H. and Sung, Y.-E. 2008. Synthesis of 3 nm Pt nanowire using MCM-41 and electrocatalytic activity in methanol electro-oxidation. *Electrochemical and Solid-State Letters* 11, B71–B75.

150. Zhao, G.-Y., Xu, C.-L., Guo, D.-J., Li, H. and Li, H.-L. 2006. Template preparation of PtRu and Pt nanowire array electrodes on a Ti/Si substrate for methanol electro-oxidation. *Journal of Power Sources* 162, 492–496.

151. Si, F.Z., Ma, L., Liu, C.P., Zhang, X.B. and Xing, W. 2012. The role of anisotropic structure and its aspect ratio: High-loading carbon nanospheres supported Pt nanowires with high performance toward methanol electrooxidation. *RSC Advances* 2, 401–403.

152. Liu, R., Liu, J.-F. and Jiang, G.-B. 2010. Use of Triton X-114 as a weak capping agent for one-pot aqueous phase synthesis of ultrathin noble metal nanowires and a primary study of their electrocatalytic activity. *Chemical Communications* 46, 7010–7012.

153. Sun, Z.H., Cheng, F.L. and Dai, X.C. 2009. Highly ordered Pd nanowire array by template fabrication for propanol electrooxidation. *Journal of Automated Methods and Management in Chemistry*, vol. 2009, Article ID 496281, 5 pp. doi:10.1155/2009/496281.

154. Xu, C.W., Wang, H., Shen, P.K. and Jiang, S.P. 2007. Highly ordered Pd nanowire arrays as effective electrocatalysts for ethanol oxidation in direct alcohol fuel cells. *Advanced Materials* 19, 4256–4259.

155. Ksar, F., Surendran, G., Ramos, L., Keita, B., Nadjo, L., Prouzet, E., Beaunier, P., Hagege, A., Audonnet, F. and Remita, H. 2009. Palladium nanowires synthesized in hexagonal mesophases: Application in ethanol electrooxidation. *Chemistry of Materials* 21, 1612–1617.

156. Wang, H., Xu, C., Cheng, F. and Jiang, S. 2007. Pd nanowire arrays as electrocatalysts for ethanol electrooxidation. *Electrochemistry Communications* 9, 1212–1216.

157. Wang, J., Chen, Y., Liu, H., Li, R. and Sun, X. 2010. Synthesis of Pd nanowire networks by a simple template-free and surfactant-free method and their application in formic acid electrooxidation. *Electrochemistry Communications* 12, 219–222.

158. Zhou, H., Zhou, W.-P., Adzic, R.R. and Wong, S.S. 2009. Enhanced electrocatalytic performance of one-dimensional metal nanowires and arrays generated via an ambient, surfactantless synthesis. *The Journal of Physical Chemistry C* 113, 5460–5466.

159. Guo, S.J., Zhang, S., Sun, X.L. and Sun, S.H. 2011. Synthesis of ultrathin FePtPd nanowires and their use as catalysts for methanol oxidation reaction. *Journal of the American Chemical Society* 133, 15354–15357.

160. Li, W. and Haldar, P. 2009. Supportless PdFe nanorods as highly active electrocatalyst for proton exchange membrane fuel cell. *Electrochemistry Communications* 11, 1195–1198.

161. Zhang, Z.Y., Li, M.J., Wu, Z.L. and Li, W.Z. 2011. Ultra-thin PtFe-nanowires as durable electrocatalysts for fuel cells. *Nanotechnology* 22, 5.

162. Zhang, L., Li, N., Gao, F., Hou, L. and Xu, Z. 2012. Insulin amyloid fibrils: An excellent platform for controlled synthesis of ultrathin superlong platinum nanowires with high electrocatalytic activity. *Journal of the American Chemical Society* 134, 11326–11329.

163. Lee, E.P., Peng, Z., Chen, W., Chen, S., Yang, H. and Xia, Y. 2008. Electrocatalytic properties of Pt nanowires supported on Pt and W gauzes. *ACS Nano* 2, 2167–2173.

164. Du, W.X., Su, D., Wang, Q., Frenkel, A.I. and Teng, X.W. 2011. Promotional effects of bismuth on the formation of platinum-bismuth nanowires network and the electrocatalytic activity toward ethanol oxidation. *Crystal Growth & Design* 11, 594–599.

165. Yuan, Q., Zhou, Z., Zhuang, J. and Wang, X. 2010. Seed displacement, epitaxial synthesis of Rh/Pt bimetallic ultrathin nanowires for highly selective oxidizing ethanol to CO_2. *Chemistry of Materials* 22, 2395–2402.

166. Kim, Y.S., Nam, S.H., Shim, H.-S., Ahn, H.-J., Anand, M. and Kim, W.B. 2008. Electrospun bimetallic nanowires of PtRh and PtRu with compositional variation for methanol electrooxidation. *Electrochemistry Communications* 10, 1016–1019.

167. Shui, J.I., Chen, C. and Li, J.C.M. 2011. Evolution of nanoporous Pt-Fe alloy nanowires by dealloying and their catalytic property for oxygen reduction reaction. *Advanced Functional Materials* 21, 3357–3362.

168. Wang, S., Jiang, S.P., Wang, X. and Guo, J. 2011. Enhanced electrochemical activity of Pt nanowire network electrocatalysts for methanol oxidation reaction of fuel cells. *Electrochimica Acta* 56, 1563–1569.

169. Li, B., Higgins, D.C., Zhu, S., Li, H., Wang, H., Ma, J. and Chen, Z. 2012. Highly active PtRu nanowire network catalysts for the methanol oxidation reaction. *Catalysis Communications* 18, 51–54.

170. Zhang, X., Lu, W., Da, J., Wang, H., Zhao, D. and Webley, P.A. 2009. Porous platinum nanowire arrays for direct ethanol fuel cell applications. *Chemical Communications* 195–197.

171. Zhong, Y., Xu, C.-L., Kong, L.-B. and Li, H.-L. 2008. Synthesis and high catalytic properties of mesoporous Pt nanowire array by novel conjunct template method. *Applied Surface Science* 255, 3388–3393.

172. Liu, L., Pippel, E., Scholz, R. and Goesele, U. 2009. Nanoporous Pt-Co alloy nanowires: Fabrication, characterization, and electrocatalytic properties. *Nano Letters* 9, 4352–4358.

173. Choi, W.C. and Woo, S.I. 2003. Bimetallic Pt-Ru nanowire network for anode material in a direct-methanol fuel cell. *Journal of Power Sources* 124, 420–425.

174. Sun, S., Yang, D., Zhang, G., Sacher, E. and Dodelet, J.P. 2007. Synthesis and characterization of platinum nanowire-carbon nanotube heterostructures. *Chemistry of Materials* 19, 6376–6378.

175. Kostowskyj, M.A., Kirk, D.W. and Thorpe, S.J. 2010. Ag and Ag-Mn nanowire catalysts for alkaline fuel cells. *International Journal of Hydrogen Energy* 35, 5666–5672.

176. Liu, F., Lee, J.Y. and Zhou, W.J. 2006. Segmented Pt/Ru, Pt/Ni, and Pt/RuNi nanorods as model bifunctional catalysts for methanol oxidation. *Small* 2, 121–128.

177. Liu, F., Lee, J.Y. and Zhou, W.J. 2004. Template preparation of multisegment PtNi nanorods as methanol electro-oxidation catalysts with adjustable bimetallic pair sites. *Journal of Physical Chemistry B* 108, 17959–17963.

178. Liu, L.F., Huang, Z.P., Wang, D.A., Scholz, R. and Pippel, E. 2011. The fabrication of nanoporous Pt-based multimetallic alloy nanowires and their improved electrochemical durability. *Nanotechnology* 22, 6.

179. Bertin, E., Garbarino, S., Ponrouch, A. and Guay, D. 2012. Synthesis and characterization of PtCo nanowires for the electro-oxidation of methanol. *Journal of Power Sources* 206, 20–28.

180. Tian, X.-K., Zhao, X.-Y., Zhang, L.-D., Yang, C., Pi, Z.-B. and Zhang, S.-X. 2008. Performance of ethanol electro-oxidation on Ni-Cu alloy nanowires through composition modulation. *Nanotechnology* 19, 215711.

181. Yoo, S.J., Jeon, T.-Y., Kim, K.S., Lim, T.-H. and Sung, Y.-E. 2010. Multilayered Pt/Ru nanorods with controllable bimetallic sites as methanol oxidation catalysts. *Physical Chemistry Chemical Physics* 12, 15240–15246.

182. Zhu, C.Z., Guo, S.J. and Dong, S.J. 2012. PdM (M = Pt, Au) bimetallic alloy nanowires with enhanced electrocatalytic activity for electro-oxidation of small molecules. *Advanced Materials* 24, 2326–2331.

183. Sun, S., Zhang, G., Geng, D., Chen, Y., Banis, M.N., Li, R., Cai, M. and Sun, X. 2010. Direct growth of single-crystal Pt nanowires on Sn@CNT nanocable: 3D electrodes for highly active electrocatalysts. *Chemistry—A European Journal* 16, 829–835.

184. Shimizu, W., Okada, K., Fujita, Y., Zhao, S. and Murakami, Y. 2012. Platinum nanowire network with silica nanoparticle spacers for use as an oxygen reduction catalyst. *Journal of Power Sources* 205, 24–31.

185. Wang, J., Wasmus, S. and Savinell, R.F. 1995. Evaluation of ethanol, 1–propanol, and 2–propanol in a direct oxidation polymer–electrolyte fuel cell. *Journal of the Electrochemical Society* 142, 4218–4224.

186. Reiss, B.D., Freeman, R.G., Walton, I.D., Norton, S.M., Smith, P.C., Stonas, W.G., Keating, C.D. and Natan, M.J. 2000. Electrochemical synthesis and optical readout of striped metal rods with submicron features. *Journal of Electroanalytical Chemistry* 522, 95–103.

187. Martin, B.R., Dermody, D.J., Reiss, B.D., Fang, M.M., Lyon, L.A., Natan, M.J. and Mallouk, T.E. 1999. Orthogonal self-assembly on colloidal gold-platinum nanorods. *Advanced Materials* 11, 1021–1025.

188. Bauer, L.A., Reich, D.H. and Meyer, G.J. 2003. Selective functionalization of two-component magnetic nanowires .*Langmuir* 19, 7043–7048.

189. Birenbaum, N.S., Lai, B.T., Chen, C.S., Reich, D.H. and Meyer, G.J. 2003. Selective noncovalent adsorption of protein to bifunctional metallic nanowire surfaces. *Langmuir* 19, 9580–9582.

190. Guo Wan, L.-J., Zhu, C.-F., Yang, D.-L., Chen, D.-M. and Bai, C.-L. 2003. Ordered NiRu nanowire array with enhanced coercivity. *Chemistry of Materials* 15, 664–667.

191. Zhang, H.M., Guo, Y.G., Wan, L.J. and Bai, C.L. 2003. Novel electrocatalytic activity in layered Ni-Cu nanowire arrays. *Chemical Communications*, 3022–3023.
192. Sun, S., Xu, H., Tang, S., Guo, J., Li, H., Cao, L., Zhou, B., Xin, Q. and Sun, G. 2006. Synthesis of PtRu nanowires and their catalytic activity in the anode of direct methanol fuel cells. *Chinese Journal of Catalysis* 27, 932–936.
193. Shen, Q., Min, Q., Shi, J., Jiang, L., Zhang, J.-R., Hou, W. and Zhu, J.-J. 2009. Morphology-controlled synthesis of palladium nanostructures by sonoelectrochemical method and their application in direct alcohol oxidation. *The Journal of Physical Chemistry C* 113, 1267–1273.
194. Xu, C., Li, Q., Liu, Y., Wang, J. and Geng, H. 2011. Hierarchical nanoporous PtFe alloy with multimodal size distributions and its catalytic performance toward methanol electrooxidation. *Langmuir* 28, 1886–1892.
195. Solla-Gullon, J., Vidal-Iglesias, F.J., Lopez-Cudero, A., Garnier, E., Feliu, J.M. and Aldaza, A. 2008. Shape-dependent electrocatalysis: Methanol and formic acid electrooxidation on preferentially oriented Pt nanoparticles. *Physical Chemistry Chemical Physics* 10, 3689–3698.
196. Tian, N., Zhou, Z.-Y., Sun, S.-G., Ding, Y. and Wang, Z.L. 2007. Synthesis of tetrahexahedral platinum nanocrystals with high-index facets and high electrooxidation activity. *Science* 316, 732–735.
197. Levy, R.B. and Boudart, M. 1973. Platinum-like behavior of tungsten carbide in surface catalysis. *Science* 181, 547–549.
198. Barnett, C.J., Burstein, G.T., Kucernak, A.R.J. and Williams, K.R. 1997. Electrocatalytic activity of some carburised nickel, tungsten and molybdenum compounds. *Electrochimica Acta* 42, 2381–2388.
199. Burstein, G.T., McIntyre, D.R. and Vossen, A. 2002. Relative activity of a base catalyst toward electro-oxidation of hydrogen and methanol. *Electrochemical and Solid State Letters* 5, A80–A83.

14

Carbon Nanotubes and Nanofibers and Their Supported Catalysts for Direct Methanol Fuel Cells

Madhu Sudan Saha and Jiujun Zhang

CONTENTS

14.1 Introduction 444
14.2 Commonly Used Catalyst Carbon Supports 445
14.3 CNT- and CNF-Based Catalyst Supports 446
    14.3.1 Significance of CNTs and CNFs 446
    14.3.2 Carbon Nanotubes (CNTs) 447
    14.3.3 Carbon Nanofibers (CNFs) 449
14.4 Synthetic Methods of CNT– and CNF–Supported Catalysts 450
    14.4.1 Synthesis of CNT–Supported Catalysts 450
        14.4.1.1 Functionalization of Carbon Nanotubes 450
        14.4.1.2 Impregnation Method 452
        14.4.1.3 Microwave-Heated Polyol Method 453
        14.4.1.4 Colloidal Method 455
        14.4.1.5 Electrochemical Method 455
        14.4.1.6 Sputtering Method 456
    14.4.2 Synthesis of Carbon Nanofiber–Supported Catalysts 457
        14.4.2.1 Functionalization of Carbon Nanofibers 457
        14.4.2.2 Impregnation Method 459
        14.4.2.3 Microwave-Heated Polyol Method 459
        14.4.2.4 Colloidal Method 460
        14.4.2.5 Electrochemical Method 461
        14.4.2.6 Sputtering Method 461
14.5 Fabrication Process of DMFC Electrodes and Their
    Corresponding DMFC Performance 462
14.6 Challenges to Nanocatalysts in DMFCs 466
    14.6.1 Control of Particle Size 466
    14.6.2 Distribution of Catalyst Nanoparticles 466
    14.6.3 Fabrication of DMFC Electrode 466
14.7 Conclusions and Prospects 467
References 467

14.1 Introduction

In the past two decades, proton exchange membrane fuel cells (PEMFCs), including direct methanol fuel cells (DMFCs), have been demonstrated to be feasible for a large variety of technical areas, such as portable devices, automotive, and stationary power systems. However, their commercial applications have been limited by their high cost and insufficient durability. In particular, DMFCs using liquid methanol rather than hydrogen gas as fuel have been demonstrated to be suitable technologies for low-power portable electronic device applications due to their high theoretical energy density and low working temperatures. The use of methanol as fuel has several advantages in comparison to hydrogen: it is a cheap liquid fuel that is easily handled, transported, and stored [1–3].

The operating principle of DMFCs involves the anode electrooxidation of methanol catalyzed by carbon-supported Pt–Ru alloy catalyst and the cathode electroreduction of oxygen catalyzed by carbon supported platinum catalysts [4]. Unfortunately, both catalyzed methanol electrooxidation and oxygen electroreduction are kinetically slow, resulting in relatively lower performance. As one of the main factors that limit and diminish the practical performance of the DMFCs, the sluggish methanol oxidation and the poisoning of the anode catalyst have attracted a lot of efforts in catalyst development. Platinum (Pt), the most practical catalyst for fuel cells, is a good catalyst for methanol oxidation. However, it can easily suffer from a quick poisoning by the CO, a major reaction intermediate [5–12]. The remedy has been to use Pt-based binary or ternary eletrocatalysts [13–18]. Some improved catalytic activities have been reported when using Pt-based alloys, such as Pt–Ru, Pt–Mo, Pt–Sn, Pt–Os, Pt–Ru–Os [15,19–21]. This enhancement effect of alloying has been explained by models, such as the "bi-functional mechanism" [22–25] and/or by the "electronic effect" [26–28], which indicates a promotional effect of the alloyed metal on Pt. Particularly, Pt–Ru has been the most investigated binary system and has shown the best catalytic activity [29–33].

The catalytic activity of Pt-based catalysts may be influenced by many factors, among which the catalyst-supporting material, such as a carbon particle, plays an important role in promoting catalyst activity. Carbon black particles are widely used as catalyst supports because of their relative stability in both acidic and basic media, good electric conductivity, and high specific surface area. However, the mesopores on carbon particles can result in part of the Pt nanoparticles getting buried deeply inside the pores and hence becoming inaccessible for the electrochemical reaction at the triple phase boundary. Further, the carbon particle can undergo corrosion (more rampant under peroxide intermediate formation conditions at fuel cell cathodes), resulting in the aggregation and dissolution as well as isolation of Pt nanoparticles [34–38]. To overcome these challenges in carbon support, many efforts have been made to search for new catalyst supports [39].

In this regard, nanomaterials, such as carbon nanotubes (CNTs) and carbon nanofibers (CNFs) have been explored as support materials for catalysts, in particular for fuel cell catalysts in the last few years. For example, a large amount of studies have shown that Pt (or Pt alloys) supported on CNTs and CNFs could exhibit better performance for the electro-oxidation of methanol [40–44], oxygen reduction reaction [45–55] and higher durability than that on Vulcan XC–72 carbon particles [49,56,57].

This chapter begins by briefly describing the processing techniques used to synthesize the carbon nanotubes and carbon nanofibers. Next, the methods by which metal catalysts can be deposited onto the CNTs and CNFs to form supported catalysts will be discussed along with the surface functionalization. Then, this is followed by a review of fabrication processes of DMFC membrane electrode assemblies (MEAs), which are containing CNT- or CNF-supported catalysts. Finally, we will discuss the challenges of using of nanocatalysts in DMFCs.

14.2 Commonly Used Catalyst Carbon Supports

Traditionally, carbon black is the most popular material used as support for fuel cell catalysts due to its advantages as follows: (1) high electrical and thermal conductivities; (2) available with a wide range of surface areas and porosities; (3) high stability in acid environments; (4) relatively stable in reducing and reasonably oxidizing environments; (5) available with low impurity levels; and (6) less expensive (less than or equal to a few dollars per kilogram) and available in at least multikilogram quantities.

A wide range of carbon blacks is available from a number of suppliers (e.g., Cabot, Columbian Chemicals, Azko Nobel, Denka, Timcal, Degussa, and Mitsubishi). These types of carbon blacks are characterized by particularly high surface area–to–volume ratios and have much lower impurity levels. However, despite the many black products available, mainly three types of carbon materials, such as Vulcan XC-72, BP2000, and Ketjen EC300J, were reported for fuel cell usage. The physical properties of these commercial carbon supports are shown in Table 14.1.

Within a DMFC electrode catalyst layer, in order to facilitate the electrocatalytic reaction, such as methanol oxidation, an ideal catalyst carbon support should simultaneously possess high specific surface area, good electric conductivity, suitable pore size, and good corrosion resistance and can also support necessary surface functional groups to guarantee the reaction occurring continuously without significant activity degradation. Normally, good electronic conductivity (graphitization degree) to reduce the electrode resistance and also to protect the carbon support from being oxidized is important for the electrocatalytical reaction. Suitable surface functional groups on

TABLE 14.1

Physical Properties of Various Commercial Carbon Supports

Carbon Materials	Supplier	Specific Surface Area (m²/g)	Pore Volume (cm³/g)	DBPᵃ/Unit
Vulcan XC-72R	Cabot	250	0.592	190
Black Pearls 2000	Cabot	1500	3.349	330
Ketjen Black EC-300J	Ketjen Black	800	0.480–0.501	360
Ketjen Black EC-600 JD	Ketjen Black	1400	–	495
Shawinigan	Chevron	80	–	–
Denka Black	Denka	65	–	165

ᵃ DBP: Dibutyl phthalate number (measure of carbon void volume).

the carbon support are also needed to act as anchors for the noble metal particles during catalyst preparation, which would improve the lifetime of the catalyst and the dispersion of the active components. In this regard, the commonly used activated carbon XC–72R seems still not to satisfy meeting these demands for an ideal electrocatalyst support even though it is the best commercial support currently. Therefore, the development of new carbon materials and modification of existing ones are necessary. Actually, the requirements for carbon support materials to have high conductivity and high specific surface area as well as a large amount of mesopores or macropores are contradictive to each other, there exists a trade-off. In order to meet these requirements as much as possible, since 1998, there have been a variety of new carbon materials, such as CNTs and CNFs, that have been investigated as electrocatalyst supports in DMFCs, and some impressive results have been achieved.

14.3 CNT- and CNF-Based Catalyst Supports

14.3.1 Significance of CNTs and CNFs

With the continuous progress of nanotechnology in materials science, different types of carbon nanomaterials have attracted considerable attention, especially carbon nanotubes (CNTs) and carbon nanofibers (CNFs) and their applications, particularly as fuel cell catalyst supports. The potential benefits of using CNTs and CNFs as fuel cell catalyst supports that have been suggested include higher utilization of active metal due to the lack of smaller porosity and higher corrosion resistance due to the (theoretically) inert surfaces. The use of CNTs for fuel cell catalyst applications has been recently experimented and largely reviewed [58]. It is believed that when catalyst is supported on nanotubes, the reactant accessibility could be improved, leading to higher catalyst utilization [59]. Better performance of a carbon nanotube–based catalyst electrode with respect to a carbon black–based one has also been correlated to the higher electronic conductivity of the former (104 S cm^{-1}) when compared to the latter (4.0 S cm^{-1}) in the case of VulcanXC-72 [60–62]. Regarding the stability of CNT as a catalyst support, Li et al. carried out some experiments to evaluate their electrochemical durability [63]. The comparison with carbon black Vulcan XC–72 demonstrated that CNTs are more resistant to electrochemical oxidation.

Other carbon nanostructures, in particular graphite nanofibers, have been largely studied as support for catalysts [64–67]. For example, Bessel et al. [68] reported about the deposition of Pt on various types of nanofibers, and their electrochemical activity toward methanol oxidation reaction was studied. The authors ascribed the enhanced performance of these catalysts compared to carbon black–supported catalysts to a specific crystallographic orientation of the metal particles deposited on highly tailored graphite nanofibers. Literature results suggest that the use of carbon nanostructures, which are different from carbon black, as catalyst supports is quite promising to prepare electrodes for both fuel cell anodes and cathodes.

14.3.2 Carbon Nanotubes (CNTs)

Among several new nanostructured carbon materials used for catalyst supports, such as CNTs [69,70], CNFs [71], ordered mesoporous carbon [72,73], carbon aero, and xerogels, CNTs and CNFs are the most well-known nanostructured carbon, which have shown promising results as catalyst supports for fuel cell applications due to their unique electrical and structural properties.

CNTs are a novel class of one-dimensional nanomaterials discovered by Iijima in 1991 [74]. The extraordinary mechanical properties and unique electrical properties of CNTs have stimulated extensive research activities worldwide, resulting in many special issues of journals [75–77] and several books [78–81]. CNTs are nanoscale cylinders of rolled up graphene sheets with an extensive range of variations, such as single-walled nanotubes (SWNTs), double-walled nanotubes (DWNTs), and multi-walled nanotubes (MWNTs) as shown in Figure 14.1 [82]. It can be seen that structures comprising one cylindrical tube are called SWCNTs. SWCNTs have a relatively smaller diameter, as low as 0.4 nm, and could be metallic or semiconducting in nature, depending on their structure. MWCNTs can be considered as concentric SWCNTs that have increasing diameter and are coaxially disposed (Figure 14.1B). The number of walls present can vary from two (double wall nanotubes) to several tens, so that the external diameter can reach to 100 nm. The concentric walls are regularly spaced by 0.34 nm similar to the intergraphene distance in turbostratic graphite materials.

The methods for synthesizing CNTs include arc discharge, laser ablation, and chemical vapor deposition (CVD) [59,83]. Although both the arc discharge and laser ablation methods are efficient in producing high-quality

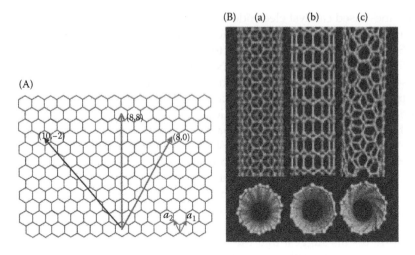

FIGURE 14.1

(A) Schematic honeycomb structure of a graphene sheet. Single-walled carbon nanotubes can be formed by folding the sheet along lattice vectors. The two basis vectors a_1 and a_2 are shown. (B) Schematic illustrations of the structures of carbon nanotubes: (a) armchair, (b) zigzag, and (c) chiral SWNTs. (Reprinted from Dai, H.J., *Acc. Chem. Res.*, 35, 1035–1044. Copyright 2002. With permission.)

nanotube material in large quantities; they do not offer control over the spatial arrangement of the produced nanostructures. Furthermore, some complex purification procedures are also required to remove amorphous carbon particles and entangled catalysts in order to obtain useful material. Among these three methods, only CVD allows controlled synthesis of CNTs. The advantage of the CVD method is allowing a scaled-up production to the industrial level and producing CNTs in a predictive fashion with controlled length, positions, and orientations on the substrates. In the synthesis, controlling the CNT structure and its growth can be achieved through changing experimental parameters, such as process temperature, gas composition, pressure, and flow rates as well as catalytic materials. There are various different approaches used for CVD methods, including general thermal CVD [84,85], floating catalyst CVD (FC–CVD) [86,87], aerosol-assisted CVD (AA–CVD) [88–90], plasma-enhanced CVD (PE–CVD) [91,92] as well as DC plasma-enhanced CVD (DC–PEVCD) [93].

Among these CVD methods, the FC–CVD and AA–CVD are two important ones in growing high-density and high-quality CNTs [86]. Compared to the traditional CVD method, the advantage of using the FC–CVD and AA–CVD methods is allowing the catalyst and carbon source to be introduced into the reaction chamber simultaneously to produce well-aligned CNTs on the substrates [87].

The AA–CVD involves pyrolysis of a mixed liquid aerosol, which consists of both a liquid hydrocarbon and a catalyst precursor [90,94]. In general,

FIGURE 14.2
SEM images of CNTs showing water effects (a) without water, (b) a little wet argon (10 sccm), (c) optimized amount of wet argon (150 sccm), and (d) excessive wet argon (700 sccm). All scale bars in inserted images are 500 nm. In each image, the bigger left image is low magnification SEM image, the upper-right corner one is high magnification SEM image and the lower-right corner one is the back-scattered electron image. (Reprinted from Liu, H. et al. *Appl. Surf. Sci.*, 256, 4692–4696. Copyright 2010. With permission.)

ferrocene, cobaltocene, cobalt nitrate, nickelocene, and iron pentacarbonyl are used as both catalysts and carbon sources. Additionally, acetylene, benzene, toluene, xylene, mesitylene, and tetrahydrofuran are also used as solvents for additional carbon sources. For example, Liu et al. [86] synthesized some aligned multiwalled carbon nanotubes with high purity on SiO_2 substrate by the AA-CVD method. In their method, an aerosol consisting of a liquid hydrocarbon source was decomposed over a catalyst for CNT growth. The aerosol droplets were produced by ultrasonication and transported by argon gas. Figure 14.2 shows cross-sectional SEM images of the as-synthesized CNTs. The lengths of the aligned CNTs are approximately 150, 300, 400, and 150 µm with wet argon flow rates of 0, 10, 150, and 700 sccm, respectively.

14.3.3 Carbon Nanofibers (CNFs)

The unique properties of carbon nanofibers (CNFs) have been explored in a number of applications, including selective absorption [95], energy storage [66,96], and polymer reinforcement [97] as well as catalyst supports [64,65,98–100]. CNFs have lengths on the order of micrometers while their diameter varies between some tens of nanometers up to several hundreds of nanometers. The mechanical strength and electric properties of CNFs are similar to that of CNTs while their size and graphite ordering can be well controlled [101]. The primary distinguishing characteristic of CNFs from CNTs is the stacking of graphene sheets of varying shapes, producing more edge sites on the outer wall of CNFs than CNTs [102]. In comparison with conventional supports, the use of CNFs as catalyst support could increase the catalyst activity in DMFCs [68,103].

CNFs can be divided into the platelet CNFs, the fish-bone CNFs and the tubular CNFs, according to the different arrangement of graphene layers. Regarding their applications in catalyst supporting, the microstructure controllable CNFs have been used as support for noble metal catalysts [64,104–106]. The CNF nanostructures consist of graphite sheets oriented in a direction dictated by the growth process. The fishbone CNFs can reveal their graphene layers stacking obliquely with respect to the fiber axis; the platelet CNFs consist of graphene layers oriented perpendicularly to the growth axis, and the ribbon CNFs display their graphene layers parallel to the growth axis.

Regarding CNF fabrication, the CVD method was normally used for the synthesizing CNFs [107–110], in which a carbon source was decomposed catalytically over a metal catalyst, such as Ni, Co, Fe, or their alloy. The catalytic decomposition of methane on Ni catalysts is one of the most extensively studied systems [111–113]. For example, Sebastian et al. [100] synthesized CNFs by thermal CVD in a quartz tube electric furnace with methane as a carbon source and $Ni:Cu:Al_2O_3$ as a catalyst. In their experiment, a fixed bed reactor was used, in which CNFs were grown for 10 h. In general, different reaction conditions, varying temperatures (550°C–750°C) and gas space velocity (CH_4, air products) were used to obtain CNFs with different properties.

14.4 Synthetic Methods of CNT– and CNF–Supported Catalysts

14.4.1 Synthesis of CNT–Supported Catalysts

14.4.1.1 Functionalization of Carbon Nanotubes

The inert surfaces of CNTs pose some difficulties for metal cluster deposition because no sites exist for deposition or stabilization of catalyst particles. To introduce more binding sites and surface anchoring groups, surface functionalization of CNTs is generally necessary. In this regard, much research has been conducted in the past decade in studying surface modification of CNTs [114]. Generally, CNTs could be functionalized by (a) covalent attachment of functional groups through forming covalent bonds to the π-conjugated skeleton of the CNTs or (b) noncovalent absorption such as π–π stacking, hydrophobic interaction, electrostatic attraction, or wrapping of various functional molecules or polymers [115,116]. In recent years, a large number of papers have reported the chemical modification of CNTs via covalent grafting of a wide range of functional groups and molecules onto the surface of CNTs. The most common covalent functionalization involves the addition of carbonyl and carboxyl groups onto the CNT surface via an aggressive oxidation treatment with a HNO_3 or HNO_3/H_2SO_4 mixture [117–122]. The carbonyl and carboxyl groups on the surface of CNTs can serve as the nucleation sites for the deposition and dispersion of noble metal nanoparticles. These noble metal nanoparticles include Pt, Pd, Pt-Ru, Au, Ag nanoparticles, and so on. This functionalization process can also be carried out by pretreatment of CNTs in HCl, HF, O_3, $KMnO_4$, or H_2O_2 [119–122]. The electrochemical method for surface modification of CNTs can also be used [123,124].

The surface functional groups have strong attraction forces toward metal ions that cling to the CNT surface. Some functional groups, such as carboxylic acid groups, even have ion-exchange capabilities. As a result, these surface functional groups serve as metal-anchoring sites to facilitate metal nuclei formation and electrocatalyst deposition. For example, Hernandez-Fernandez et al. [44] investigated the effect of the surface functionalization on the deposition of Pt–Pd nanoparticles for the electrooxidation of methanol. The functionalization treatment of the support entailed refluxing the CNTs in a sulfonitric mixture for several hours at various temperatures. Specifically, CNT–MT (MT, mild treatment) was treated in 5.5:3.0 M H_2SO_4:HNO_3 for 2 h at 60°C whereas for CNT–ST (ST, strong treatment), the concentration of the sulfonitric mixture was 18.3:15.5 M, the temperature and time were 110°C and 12 h, respectively. It was observed that Pt–Ru nanoparticles that are supported on CNTs with high amounts of functional groups had the best behavior in the methanol oxidation reaction compared with Pt–Ru supported on low amounts of functional groups.

Unfortunately, the chemical modification of the CNT surface by acid treatment could reduce considerably the mechanical and electronic performance of the tubes due to the introduction of large numbers of defects. To reduce this effect, doping CNTs with other elements (e.g., nitrogen) could be a particularly interesting way to modify their electrical and mechanical properties [125–127]. Recently, nitrogen-doped CNTs (N–CNTs) were used as support materials in fuel cell catalysts, and found that the dispersion of Pt nanoparticles on the N–CNT surface was improved [128–134]. N–CNTs contain nitrogenated sites (substitutional and pyridinic nitrogen) that are chemically active. Therefore, it should be possible to avoid functionalization processes that use strong acid treatments, and it is also relatively easier to deposit metal catalysts onto N–CNTs than onto acid treated CNTs [128–132]. For example, Maiyalagan et al. [131] used N–CNTs as supports for Pt electrocatalysts and applied them for the methanol oxidation reaction. Nitrogen-containing carbon nanotubes were synthesized by impregnating polyvinylpyrrolidone inside the alumina membrane template and subsequent carbonization of the polymer [131]. Platinum nanoparticles were deposited on the N–CNTs by the impregnation method. They observed that the Pt particles were homogeneously dispersed on the nanotubes with particle sizes of 3 nm.

14.4.1.2 Impregnation Method

The impregnation method is simple and straightforward in depositing a metal catalyst on the carbon support for the preparation of fuel cell catalysts [48,50,135,136]. This method involves the impregnation of support material with a salt solution containing the depositing metal and then followed by a reduction step. Common liquid phase reducing agents are $Na_2S_2O_3$, $NaBH_4$, $Na_4S_2O_5$, N_2H_4, and formic acid while H_2 is the predominant gas phase reducing agent. For depositing catalyst particles in CNT, the impregnation method is the most popular procedure. It has been found that the chemical modifications of a CNT surface and the synthetic conditions, including the process temperature, the choice of the species of metal precursor, and the reducing agent are essential to the impregnation efficiency [114]. Representative CNT–supported catalysts that were synthesized by the impregnation method [59,62,137–144] as reported in the literature.

Jha and coworkers [59] synthesized Pt–Ru nanoparticles supported on CNTs by the chemical reduction method and used as an anode catalyst for the oxidation of methanol. In their experiments, purified CNTs were ultrasonicated in acetone for 1 h and then magnetic stirred for 24 h after addition of 0.075 M H_2PtCl_6 and 0.15 M $RuCl_2$ solution. They achieved the uniformly dispersed Pt and Pt–Ru nanoparticles on the CNT surface. DMFC performance with Pt–Ru/CNT as the anode electrocatalyst with a fixed loading of 2.5 mg cm^{-2} at various temperatures showed that the power density was increased up to 40 mW cm^{-2} with increasing temperature.

Recently, core-shell nanoparticles with very thin noble metal shells have attracted great attention in catalyst development [141,145–147]. For example, Zhao et al. [141] prepared well-dispersed Ni@Pd core-shell nanoparticles on multiwalled carbon nanotubes by the co-reduction of Ni^{2+} and $PdCl_4^{2-}$ in sodium dodecyl sulfate, which was used as the capping and structure-directing agent. Briefly, the aqueous solution containing H_2PdCl_4, $NiSO_4$, trisodium citrate dehydrate, and sodium dodecyl sulfate was dispersed in ethanol solution, stirring for 30 min to form a mixed solution. Then a freshly prepared aqueous solution of $NaBH_4$ was added drop-wise into this mixed solution under vigorous stirring for 30 min to form another mixed solution into which MWCNTs were added. After stirring for an additional 10 h, the black solid was separated by centrifuging; it was washed with deionized water for several cycles and then dried overnight in an oven at 70°C. Based on the experimental process, they proposed a possible preparation mechanism of Ni@Pd/MWCNTs as shown in Figure 14.3. Chen et al. [148] used a sequential reduction approach to prepare Ni@Pt nanoparticles supported on MWCNTs. TEM images showed that the prepared core-shell particles had a mean diameter of ca. 5.0–5.4 nm with a narrow size distribution. These particles exhibited some structural characteristics of nearly fcc Ni nanocrystals as indicated by their XRD patterns. Both the TEM and XRD results indicated that the deposition of the Pt shell could result in a lattice expansion of the Ni core.

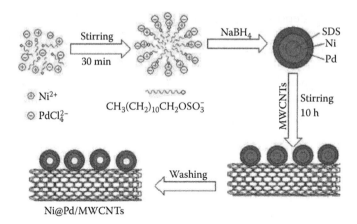

FIGURE 14.3
Schematic diagram of the synthesis of Ni@Pd electrocatalysts on MWCNTs. (Reprinted from Zhao, Y. et al. *Int. J. Hydrogen Energy,* 35, 3249–3257. Copyright 2010. With permission.)

14.4.1.3 Microwave-Heated Polyol Method

Microwave irradiation has been widely used for the catalyst synthesis due to its unique advantage in that it can transfer heat to the substance uniformly through the microwave-transparent reaction container (i.e., glass or plastic), leading to a more homogeneous nucleation and shorter crystallization time compared to conventional heating during which unavoidable temperature gradients occur, in particular when large volumes of solutions are used [149]. The microwave-assisted polyol method has showed some successes in preparing high purity Pt or Pt-Ru nanoparticles supported on CNTs [150–156]. Actually, the polyol process is normally used in preparing colloidal metal particles [157]. In this process, the polyol (most commonly ethylene glycol) solution containing the metal salt is refluxed at 120°C–170°C to decompose the polyol to yield an *in situ* generated reducing species for the reduction of metal ions to their elemental states [150,151,153]. For example, Hsieh et al. [155] used a microwave-assisted polyol process for synthesizing bimetallic Pt-Co electrocatalysts on CNTs. Their synthesis procedure can be outlined as follows: CNT/CP (carbon paper) composites were chemically oxidized by 1 M nitric acid at 85°C for 1 h. Subsequently, the oxidized CNT/CP composites were impregnated with $PtCl_4$ and $Co(NO_3)_2 \cdot 6H_2O$ precursor solution. The weight ratio of Pt to Co was set at 36:64 in the ionic solution. The pH value was around 11.7. Then the beaker was kept open and placed in the center of a household microwave oven under power of 720 W. The deposition period for each step of microwave heating was set at 6 min. The metal-coated CNT/CP samples were then separated from the Pt ionic solution and subsequently dried in a vacuum oven at 105°C overnight. To inspect the influence of deposition sequence on the activity of Pt-Co catalysts, three types of microwave heating are used to prepare different bimetallic catalysts: (1) simultaneous deposition of Pt and Co species; (2) deposition of Pt followed by Co species; and (3) deposition of Co followed by Pt species, which are designated as Pt–Co–m, Pt–Co, and Co–Pt, respectively. HR-TEM images of the resulting samples reflect that the small particles, attached to the outer surface of CNTs, have a uniform dispersion, as shown in Figure 14.4. However, the influence of the deposition sequence on the average size and uniformity seems to be minor [155].

It has been recognized that the pH value of the synthesis solution is the crucial factor in influencing catalyst particle size when using the polyol reduction method [158]. The pH effect on both the Pt particles' size and distribution were investigated by Li et al. [153]. At a lower pH range (pH 3.6–5.8), the Pt nanoparticles were agglomerated and not as well dispersed on the CNT surfaces. However, in the pH range of 7.4–9.2, a less agglomeration and a more uniform dispersion of Pt nanoparticles could be formed.

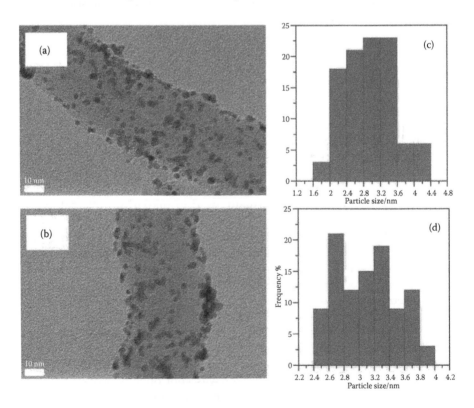

FIGURE 14.4
HR-TEM micrographs of (a) Pt–Co–m and (b) Pt–Co catalysts. Particle size distributions of (c) Pt–Co–m and (d) Pt–Co catalysts. (Reprinted from Hsieh, C.-T. et al. *J. Power Sources*, 199, 94–102. Copyright 2012. With permission.)

The use of surfactants as stabilizers is another important parameter in achieving uniform and stable nanoparticles. Preferably, water-soluble surfactants, that is, anionic or zwitterionic surfactants, have been used to disperse CNTs and stabilize the Pt nanoparticles [159]. With suitable surfactants, the particle size of Pt could be controlled by the surfactant content in the reaction medium [160]. Sakthivel et al. [156] investigated the effect of surfactants to Pt precursor weight ratio on stabilizing nanoparticles supported on carbon nanotubes. A uniform, small and spherical Pt nanoparticle–decorated CNT surface was obtained using 3-(N,N-dimethyldodecylammonio) propanesulfonate as a stabilizer.

14.4.1.4 Colloidal Method

The colloidal method is another effective technique to prepare CNT–supported catalysts with a narrow particle size distribution [40,161–165]. In this method, the catalyst precursors are first reduced into the colloids in the solution, followed by the adsorption of colloids onto the CNT surfaces. Normally, the size of metal nanoparticles can be largely controlled or stabilized by the protecting agents, such as ligands, surfactants, or polymers [166]. For example, Cao et al. [165] used a sodium dodecylbenzene sulfonate (SDBS) as the stabilizer and CO as the reducing agent to prepare Pt nanoparticle–supported CNTs.

The major drawback of the colloidal method is the requirement of a protecting agent, which may hinder the catalytic performance of the nanoparticles. Therefore, it would be preferable to use other alternative route to prepare colloidal nanoparticles without using protecting agents. For example, Yoshitake et al. [167] reported that the use of $NaHSO_3$ and H_2O_2 seemed to be suitable for the preparation of Pt oxide colloids without the presence of a protecting agent. The fuel cell performance catalyzed by the catalysts prepared using this method showed a 20% higher performance than that of conventional Pt/C catalysts. It was recently reported that the colloidal Pt nanoparticles could be protected by glycol, which served as both a solvent and the protecting agent [40,163]. For example, Kongkanand et al. [163] used an ethylene glycol as a reducing agent to prepare well-dispersed Pt catalysts supported by polymer-wrapped CNTs.

14.4.1.5 Electrochemical Method

The electrodeposition method used for metal catalyst deposition onto the CNT supports has been developed by many research groups to improve the Pt utilization and reduce Pt loading precursors [124,168–174]. Several options, including pulse current, direct current, and cyclic voltammetry, have been utilized to deposit catalysts. In the synthesis, the size of noble metal nanoparticles and their distribution on the sidewalls of CNTs can be controlled by the concentration of noble metal salts and various electrochemical

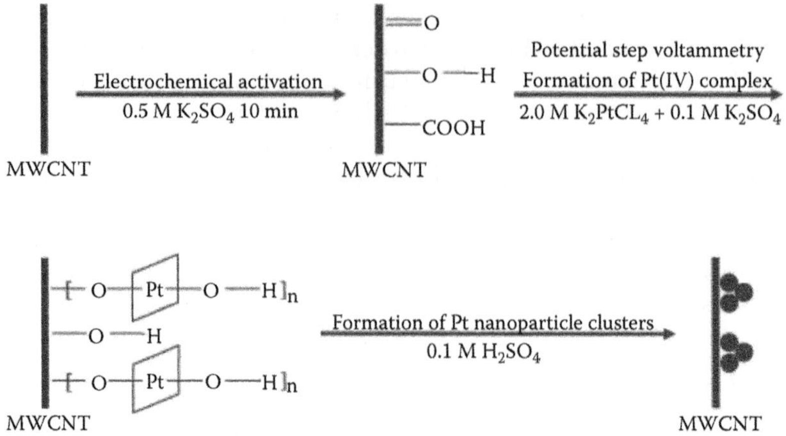

FIGURE 14.5
Illustration of the three-step process for electrochemical synthesis of 3D Pt/MWCNTs. (Reprinted from Zhao, Y. et al. *Adv. Funct. Mater.*, 17, 1537–1541. Copyright 2007. With permission.)

deposition parameters, including nucleation potential and deposition time [175–177]. For example, Zhao et al. [178] reported electrodeposition of Pt nanoparticles onto the multiwalled carbon nanotubes using a three-step protocol (see Figure 14.5) which was a slightly different process than that employed by Guo et al. [179]. Their synthesis procedure can be outlined as follows: (1) electrochemical treatment of CNTs to introduce various oxide functional groups such as quinonyl, carboxyl, or hydroxyl groups without further damage of the CNTs; (2) formation of the octahedral complexes of Pt(IV) with functional groups on the surface of CNTs, and (3) electrochemical transformation of the Pt(IV) complex to Pt nanoparticles on the surface of CNTs. TEM images revealed that the Pt nanoparticles were well dispersed in the CNTs with a 3 nm average size of the nanoparticles clusters [178].

Direct potential control electrodeposition has been used for the formation of metal nanoparticles that are supported on CNTs [174]. For example, Day et al. [180] studied the nucleation and growth mechanisms of metal electrodeposition, such as Ag and Pt onto nanostructures. In their study, CNT networks were used as a template, and the deposition of Pt was performed by stepping the potential from 0.0 V (at which point there was no electrode reaction) to −0.4 V for a period of 30 s and then back to an open circuit. This procedure allowed the density and size of metal nanoparticles to be controlled by careful choice of the applied potential and deposition time.

14.4.1.6 Sputtering Method

The sputtering deposition method has been recognized to have a great potential for reducing the Pt loading and enhance the catalyst utilization in fuel cells

[181,182]. For example, Hirano et al. [183] reported some promising performance results using a thin layer of Pt that was deposited on wet-proofed noncatalyzed gas diffusion electrode (equivalent to 0.01 mg$_{Pt}$ cm^{-2}) using sputter deposition.

Recently, the sputter-deposition technique has been employed to prepare a CNT–supported Pt catalyst by several research groups [128,130,183–185]. Using this method, uniform Pt nanoparticles could be deposited on the external wall surface of CNTs through controlling the current and sample exposure time. In the efforts of Chen et al. [183], the CNTs were first fabricated on the carbon cloth on which the Pt particles were then sputtered. The employed sputtering current and time were 10 mA and 30 s, respectively. For comparison, the electrode deposition method was also used to deposit Pt particles on the same CNTs. TEM results showed that the sputtering method could generate highly uniformed Pt nanoparticles on CNTs compared to that produced by the electroless deposition method. Soin et al. [184] reported one of representative examples to deposit Pt nanoparticles on CNTs by a direct current (DC) sputtering system. In their study, Pt nanoparticles with a particle size of 3–5 nm were deposited uniformly onto the surface of the vertically aligned carbon nanotubes. Sun et al. [128] tried to deposit Pt nanoparticles on nitrogen-containing CNTs (N–CNTs) for μDMFC application. For Pt deposition, a DC sputtering technique was employed. According to their approach, some well-separated Pt nanoparticles could be formed with an average diameter of 2 nm on the bamboo-like structure of N–CNTs. Cyclic voltammograms showed that a Pt/N–CNTs catalyst had electrochemical activity toward methanol oxidation. Their results suggested that the sputter-deposition technique could be a better way to deposit small and uniform Pt nanoparticles.

14.4.2 Synthesis of Carbon Nanofiber–Supported Catalysts

14.4.2.1 Functionalization of Carbon Nanofibers

Carbon supports with a poor surface chemistry, that is, with a low content of surface functional groups, need to be chemically treated to increase their hydrophilicity and, additionally, to improve the interaction between the support surface and the active phase [186]. Therefore, the functional groups are needed for an optimum metal deposition. Normally, CNFs contain only a small number of surface oxygen groups due to their inert nature. However, their surface chemistry can be modified by oxidation treatments in a gas or liquid phase in order to create functional groups [186–189]. These functional groups can significantly affect the performance of electrocatalysts, and they are responsible for both the acid-base and the redox properties of the CNFs. However, the effect of these groups on the dispersion and anchoring of the platinum is not well established yet. Some authors have determined that the presence of surface oxygen groups is necessary to achieve a good Pt dispersion [186] whereas others argue that they do not influence or have a negative effect on the dispersion of the active phase [190].

For example, Toebes et al. [188] found that a more severe treatment could lead to a considerable increase in the surface area and pore volume of CNFs, which was ascribed to an opening of the inner tubes of the nanofibers once the growth catalyst was removed. Calvillo et al. [187,191] investigated the effect of the functionalization of CNFs using concentrated HNO_3 or a HNO_3–H_2SO_4 mixture to optimize their ability to disperse active metal particles. They observed that a very severe treatment could destroy partially the structure of CNFs; thus careful selection of the oxidizing process should be mandatory. Oh et al. [189] examined the effect of chemical oxidation of CNFs on the electrochemical carbon

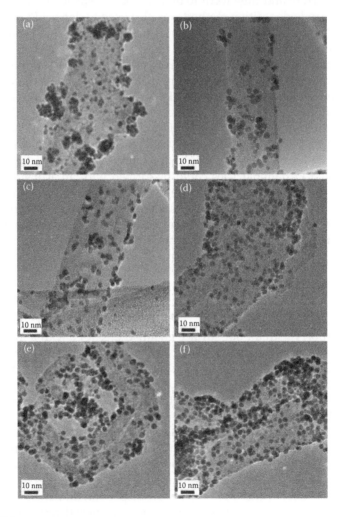

FIGURE 14.6
HR–TEM images of Pt particles on the CNFs with different acid treatment times: (a) 0 h, (b) 1 h, (c) 3 h, (d) 4 h, (e) 8 h, and (f) 16 h. (Reprinted from Oh, H.-S. et al. *Int. J. Hydrogen Energy*, 35, 701–708. Copyright 2010. With permission.)

corrosion in PEMFCs. They observed that with increasing time of chemical oxidation treatment using an acidic solution, more oxygen functional groups were formed on the surface of CNF, resulting in an increasingly hydrophilic carbon surface. This effect could contribute to the improvements in Pt loading and the distribution of Pt particles on carbon supports as well as shown in Figure 14.6. Guha et al. [192] investigated the electrochemical performance of Pt catalysts deposited on various functionalized carbon supports using a commercial catalyst as the baseline. Each carbon support sample was functionalized by treatment with nitric acid. They concluded that CNFs had potential for reducing fuel cell resistance without sacrificing fuel cell performance.

14.4.2.2 Impregnation Method

The impregnation method has been used for highly dispersing metal catalysts on the CNFs support in a PEMFC catalyst preparation. For example, Park et al. [193] prepared a highly dispersed 60 wt% Pt–Ru (1:1) on CNFs using borohydride as a reducing agent. In their experiment, the CNFs were dispersed in ultrapure water, and then an adequate amount of $H_2PtCl_6 \cdot xH_2O$ and $RuCl_3 \cdot xH_2O$ were added to the solution. After mixing for 1 h at room temperature, the metal salts were reduced by $NaBH_4$ while the solution was stirred vigorously. Sebastián et al. [100] compared two synthesis methods using different reduction agents for Pt catalyst deposition on the CNFs: One was the reduction of the Pt ion by sodium borohydride reduction, and the other was the reduction of the Pt ion by formic acid. They pointed out that Pt deposition using sodium borohydride reduction seemed to be better than formic acid reduction with respect to electrochemical activity, CO tolerance, and performance [100].

14.4.2.3 Microwave-Heated Polyol Method

The microwave-heated polyol process has been employed for the preparation of Pt or Pt–Ru catalysts on CNFs for fuel cells [149,194–197]. Tsuji et al. [194] prepared some PtRu alloy nanoparticles supported on CNFs using a microwave-polyol method. Three types of CNFs with very different surface structures, such as platelet, herringbone, and tubular, were used as supports. In their experiment, a three-necked flask was placed in the microwave oven and connected to the condenser. The solution containing $H_2PtCl_6.6H_2O$, $RuCl_3.2H_2O$, KOH/ethylene glycol (EG), and CNFs in EG was irradiated by microwave in a continuous wave mode for only 2.5 min. The solution was rapidly heated to the boiling point of EG (198°C) after about 1.5 min under MW irradiation and held on this temperature for another 1 min. Figure 14.7 shows the TEM images of Pt–Ru nanoparticles supported on platelet, herringbone, and tubular CNFs, respectively. It is clear that some well-dispersed Pt–Ru nanoparticles are loaded on each CNF. The average diameters of Pt–Ru nanoparticles were determined to be 3.4 ± 0.3, 3.5 ± 0.3, and 3.7 ± 0.5 nm for platelet, herringbone, and tubular CNFs, respectively. Khosravi et al. [197]

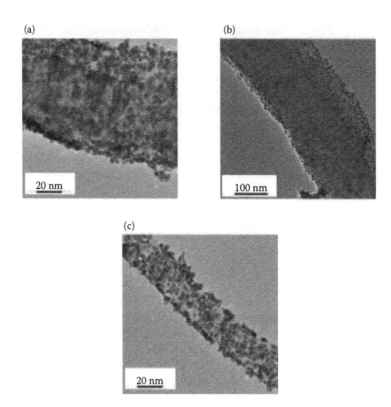

FIGURE 14.7
TEM images of PtRu nanoparticles (24 [1 wt%, Ru/Pt atomic ratios] 0.91–0.97) loaded on (a) platelet, (b) herringbone, and (c) tubular CNFs. These PtRu catalysts were prepared under MW heating for 2.5 min. (Reprinted from Tsuji, M. et al. *Langmuir*, 23, 387–390. Copyright 2006. With permission.)

developed a novel synthetic method for preparing uniformly dispersed Pt nanoparticles on CNFs, which were grown directly on the carbon paper. The CNFs were synthesized on Ni-deposited carbon paper utilizing the common laboratory ethanol flame as both heat and carbon source. The Pt electrocatalyst was deposited on both carbon paper and CNF/carbon paper by the microwave-assisted polyol method from a solution containing H_2PtCl_6 in ethylene glycol adjusted to pH 7.4 with 1 M KOH solution in the same solvent.

14.4.2.4 Colloidal Method

Colloid methods [167,198–200] have been used to deposit Pt or Pt alloy nanoparticles onto the CNFs, and the electrocatalytic activity of such prepared catalysts toward the methanol oxidation in DMFCs have also been studied. For example, Sebastián et al. [198] reported the colloidal method for the synthesis of Pt nanoparticle–supported fishbone carbon nanofiber catalysts. In their study, different temperatures were used for the synthesis of herringbone CNFs with varying textural and crystalline properties. Briefly, the platinum precursor (H_2PtCl_6) dissolved in aqueous solution was drop-wise added into the mixture of surfactant and n-heptane. The quantity of surfactant was 16.5% in volume whereas the ratio of water to surfactant was 33. Subsequently, 2-propanol was added until an optically transparent mixture was observed, indicating the formation of the microemulsion. After 4 h of stirring, the reducing agent ($NaBH_4$) was slowly added into the solution under continuous stirring. The suspension was stirred overnight and then slowly added to a suspension of CNFs in ethanol under sonication and left under continuous stirring for 16 h. The catalyst was then thoroughly washed with ethanol and water and finally dried overnight at 70°C. The Pt crystallite size values calculated from XRD patterns were between 3.0 and 4.2 nm. Highly graphitic CNFs displayed a better catalytic activity despite its lower surface area and pore volume compared to carbon black.

14.4.2.5 Electrochemical Method

The electrochemical method has also been employed for the deposition of a catalyst on CNFs. In the electrodeposition, CNFs are first cycled in a sulfuric acid solution between −0.7 and +1.2 V to oxidize the surface. During that process, various surface functional groups, such as quinoid (= O), hydroxy (–OH), and carboxyl (–COOH), can be generated, which can supply defect sites for the Pt nanoparticle deposition [201]. Tang et al. [202] used the cyclic voltammogram method to deposit Pt particles onto graphitic carbon nanofibers (GCNFs). According to their approach, Pt nanoparticles were electrodeposited on a GCNF/graphite working electrode from an acidic solution of H_2PtCl_6 by cyclic voltammetry under condition of deposition potential +0.1 to −0.25 V and sweep rate 15 mV s^{-1}. They reported that the Pt nanoparticles could be uniformly dispersed on the whole surface of the GCNFs with a diameter of about 40–50 nm, which was smaller than that of Pt/graphite catalyst. Li et al. [203] also used the cyclic voltammogram method to deposit Pt particles onto CNFs. The diameters of deposited Pt nanoclusters are 50–200 nm. Lin et al. [204] reported the preparation and characterization of Pt/CNFs by the electrodeposition of smaller Pt nanoparticles (≤55 nm) onto electrospun CNFs under different potentials.

14.4.2.6 Sputtering Method

Directly sputtering catalyst particles onto the carbon support layers [205–207] or the surface of membranes [181,208] has been proven to be another effective method to improve the performance and utilization of the catalyst. Sputtering Pt nanoparticles on the CNTs/CNFs promises to improve the Pt utilization by securing the electronic route from Pt to the supporting electrode in fuel cells [135,209–211]. Zhang et al. [212] developed Pt/VACNF

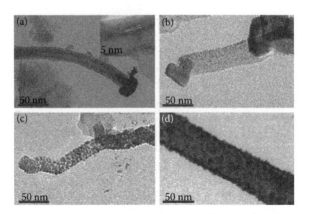

FIGURE 14.8
Transmission electron microscopy of (a) VACNFs directly grown on carbon paper, (b) Pt/
VACNFs-10, (c) Pt/VACNFs-25 and (d) Pt/VACNFs-50. The inset of (a) is a high-resolution
transmission electron microscopy image of the VACNFs. (Reprinted from Zhang, C. et al.
Electrochim. Acta, 56, 6033–6040. Copyright 2011. With permission.)

(vertically aligned carbon nanofiber) electrodes by direct RF plasma magnetron sputtering of Pt nanoparticles onto VACNFs. The Pt nanoparticles
were highly dispersed over VACNFs, and the particle size became larger
by increasing the Pt loading. TEM images of VACNFs and Pt/VACNFs are
shown in Figure 14.8. The Pt nanoparticles can be seen on the surface of the
CNFs, which have a high and homogeneous dispersion. Meanwhile, it can
be seen from Figure 14.8d that the Pt nanoparticles can agglomerate to clusters. The mean particle sizes are about 2.6, 4.2, and 8.1 nm for Pt/VACNFs
10, Pt/VACNFs 25 and Pt/VACNFs 50, respectively. The Pt/VACNFs with a
low metal loading were exhibiting excellent electrochemical activity of the
methanol oxidation according to CV measurement. The results showed that
the Pt/VACNFs had a significant improvement of Pt utilization than commercial Pt/C catalyst, indicating a potential application in DMFCs.

14.5 Fabrication Process of DMFC Electrodes and
Their Corresponding DMFC Performance

The MEA is the core of PEMFCs, including DMFCs, in which the chemical
energy of the fuel (i.e., hydrogen or methanol) is converted into electricity
through the electrochemical oxidation of fuel at the anode and the electrochemical reduction of oxygen at the cathode. A typical DMFC MEA consists of three basic components, for example, electrodes, polymer electrolyte
membrane, and gas diffusion layers (GDL), which are sandwiched together
to form a MEA as shown in Figure 14.9.

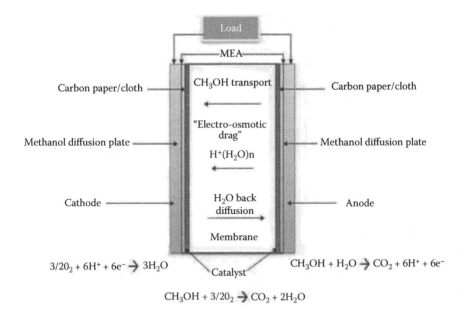

FIGURE 14. 9
Schematic illustration of the membrane electrode assembly (MEA) in a DMFC.

The anode in the MEA is a negatively charged electrode with the catalyst (e.g., carbon-supported Pt or Pt–Ru) dispersed. When methanol is fed into the electrode, it is oxidized on the surface of the catalyst to form protons and electrons, which is known as the methanol oxidation reaction (MOR). The protons flow through the electrolyte (PEM) to reach the cathode; at the same time, the electrons flow through the electrical circuit and reach the cathode, during which the electricity is produced. Among various catalyst supports, CNTs and CNFs have been widely investigated as the promising support materials in place of carbon black for catalysts in DMFCs due to their unique one-dimensional structure, high electronic and thermal conductivities, and good electrochemical stability [213,214].

The fabrication of a MEA is an intricate procedure, in which all details of the structure and preparation are important because the function of the electrodes is far more than just catalyzing a reaction. Generally, there are two model procedures to fabricate the MEA, that is, (1) the application of the catalyst layer onto the GDL followed by membrane addition and (2) the application of the catalyst onto the membrane followed by GDL addition [215].

The CNT– and CNF–based MEAs are usually prepared using a conventional process of pasting mixed CNT–supported Pt or Pt–Ru catalyst and Nafion ink onto a GDL [40,45,135,216,217]. The method consists of preparation of the catalyst ink by ultrasonicating an appropriate amount of the mixture of CNT–supported Pt or Pt–Ru catalyst, Nafion solution, and iso-propyl alcohol [217]. The ink is then coated on the GDL by the brush or spray method. The GDL of the electrodes consists of two layers, one is a subcarbon layer containing polytetrafluorethylene, and the other is a carbon cloth or paper substrate. After the catalyzed electrodes (anode and cathode) are pre-pared, the anode will placed onto one side of a PEM (Nafion) and the cath-ode will be placed onto the other side of this PEM to form a sandwich-like assembly, followed by hot pressing to form a MEA. For example, Li et al. [56] prepared some electrodes from a mixture of ink containing CNT–supported Pt catalyst, Nafion solution, and distilled H_2O. This ink was made by ultra-sonically blending the mixture for 1 h. The ink was then spread uniformly across the surface of a carbon paper substrate, which was previously hydro-phobically treated. Their MEA was then prepared by sandwiching a Nafion membrane between the anode and cathode. They observed a higher activity in the oxygen reduction reaction and a better performance in a direct metha-nol fuel cell when compared to that of a Pt catalyst supported on carbon particles [56].

Recently, electrophoretic deposition (EPD) has been investigated as a novel method for fabricating electrodes for fuel cells [218–220]. This method is generally conducted in nonaqueous suspensions under high-voltage con-ditions. At present, studies in this field are limited. However, EPD may be capable of producing an even and compact catalyst layer as well as main-taining the desired catalyst composition and morphology in the electrode. In addition, EPD can be conducted under low applied voltage conditions, yielding safer and more environmentally friendly operations. A Pt–Ru/C catalyst catalyzed electrode, the most active anode for methanol oxidation, was bonded directly onto a PEM using IR methodology and investigated in DMFCs [219]. Another method for the synthesis of Pt–Ru based electrodes includes vacuum filtration, followed by physical leveling [221]. Filtration has been achieved with a ceramic funnel connected to a device that produces a vacuum. During the procedure, a piece of Teflon-treated or waterproofed carbon cloth was used as a filtration medium and catalyst/Nafion solution was filtered through the funnel several times until the filtrate became clear.

Li et al. [222] prepared Pt nanoparticles (5–30 wt.%) of 2–4 nm supported on stacked-cup CNFs (SC–CNFs) using the polyol process. MEAs based on these Pt/SC–CNFs were prepared using a unique filtration process. The schematic illustration of the fabrication process is shown in Figure 14.10. Pt/SC–CNFs suspension in ethanol with a known catalyst quantity is drawn through a 0.2 μm pore polycarbonated filter paper. After filtration, a 5 wt.% Nafion solution is used to spay on the surface of the filtrated Pt/SCCNFs solid, the Nafion weight ratio is altered from 30 wt.% to 50 wt.% for the cathode and is constant at 50 wt.% for the anode. The filtrated Pt/SC–CNF catalysts are transferred onto a Nafion 112 membrane by hot-pressing two catalyst-coated sides of the filters onto the Nafion membrane to pro-duce a catalyst-coated membrane (CCM). The Pt/SC–CNF–based MEA with

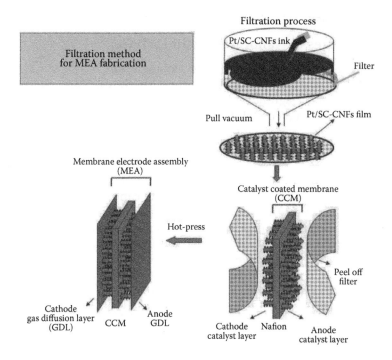

FIGURE 14.10
(See color insert.) Schematic illustration of filtration method for fabricating SC-CNFs based MEA. (Reprinted from Li, W.M. et al. *Carbon*, 48, 995–1003. Copyright 2010. With permission.)

50 wt.% Nafion displayed a higher PEMFC performance than the commercial CB (E–TEK)–based MEA with 30 wt.% Nafion content. The improved performance was attributed to the high aspect ratio (length to diameter ratio) of CNFs which might allow the formation of continuously conducting networks in the Nafion matrix [222].

Steigerwalt et al. [196] and Bessel et al. [68] demonstrated that CNF–supported catalysts showed some improved methanol oxidation activity as compared to CB. Bessel et al. [68] found that catalysts consisting of 5 wt.% Pt supported on "platelet" and "ribbon" type graphite nanofibers exhibited activities comparable to that displayed by about 25 wt.% Pt on Vulcan carbon. Furthermore, they observed that the graphite nanofiber–supported metal particles were significantly less susceptible to CO poisoning than the traditional catalyst systems. This improvement in performance was believed to depend on the fact that the metal particles could adopt specific crystallographic orientations when dispersed on the highly tailored graphitic nanofiber structures. Tsuji et al. [194] also showed that the DMFC activities of Pt–Ru/CNFs were 1.7–3.0 times higher than that of a standard Pt–Ru catalyst loaded on Vulcan XC–72R support.

14.6 Challenges to Nanocatalysts in DMFCs

There are several challenges to the development of CNT– and CNF–supported catalysts in term of fabrication and synthesis. Thus, this section highlighted the most significant limitation controlled by particle size, distribution, fabrication, and crystallinity.

14.6.1 Control of Particle Size

The size of Pt–Ru particles is a major challenge in the synthesis of CNT– and CNF–supported catalysts. Because the support can strongly affect catalyst utilization and activity, the synthesis of Pt nanoparticles supported by CNT and CNF is of fundamental importance. To control the particle size, uniform platinum nanotubes have been synthesized by directly mixing Ag nanowires and H_2PtCl_6 in a saturated solution of NaI at room temperature. The crystal structure of the resulting Pt nanotube was investigated in detail by field emission scanning electron microscopy, transmission electron microscopy, and x-ray diffraction [223]. Furthermore, catalyst particle size can also be controlled through CVD methods. There are various methods for CVD based on the different functions and chemicals used.

14.6.2 Distribution of Catalyst Nanoparticles

Due to the slow oxidation kinetics of methanol oxidation, catalyst loadings as high as 50 wt.% are required for DMFCs. Despite the high catalyst loading in DMFC cells, CO poisoning is prevalent. Thus, a considerable amount of research has been conducted in recent years to develop a method that distributes Pt/Ru evenly throughout the structure. Fujigaya et al. [224] described the fabrication of a CNT hybrid wrapped by pyridine-containing polybenzimidazole (PyPBI). Based on this method, highly homogeneous and remarkably efficient Pt loading onto the surface of CNTs through a coordination reaction between Pt and PyPBI has been achieved. They found that the wrapped PyPBI could serve as glue for immobilizing Pt nanoparticles onto the surface of CNTs without any strong oxidation process for the CNTs.

14.6.3 Fabrication of DMFC Electrode

In comparison with carbon black, CNT–supported electrocatalysts commonly exist in unusual shapes and have bulky specific volumes. As a result, CNT–supported electrocatalysts are difficult to fabricate into fuel cell electrodes by conventional means, such as painting, brushing, spraying, screenprinting, etc. Additionally, high catalyst loadings required for DMFCs are also difficult to achieve in CNT–supported electrocatalysts [225]. Common obstacles include thick catalyst layers, loose structures in coated layers, poor cell performance, and high ionomer content, leading to a high electrode resistance. These problems are exacerbated when catalysts with low amounts of metal and high amounts of CNT are used in the electrode catalyst layer.

14.7 Conclusions and Prospects

Electrocatalyst supports play a vital role in ascertaining the performance, durability, and cost of DMFC systems. Myriad nanostructured materials, including carbon nanotubes and nanofibers, have been exhaustively researched over the past few decades to improve existing, and also develop novel, PEMFC/DMFC catalyst supports. This article has reviewed the recent developments and investigations reported on various catalyst supports. Many developments and improvements can be seen in the structure, poisoning tolerance, and stability of various nanostructured carbonaceous supports over the recent years.

In conclusion, the uses of nanomaterials in fuel cells may significantly improve the electrocatalytic performance for high energy density and high power density while reducing the manufacturing cost. The prominent electrocatalytic behavior of the nanomaterials is contributed mainly from their unique physical–chemical properties such as sizes, shapes, pore structures/distribution, surface defects, and chemical properties. The great challenges to synthesize and further use various nanostructured catalysts, such as CNT– and CNF–supported catalysts are not only from chemistry but also from nanoengineering approaches. In general, the core-shell nanostructures could provide an economic and effective way to prepare precious metal catalysts for remarkably reducing the usage of the noble metals while the unique nanostructures, such as nanotubes and nanofibers are believed to provide high specific active surface area, superior conductivity, and better mass transport as well as high intrinsic catalytic activity. An amalgamation of these novel electrocatalyst supports and improved catalyst loading techniques could bring about revolutionary changes in the quest for high-performance, long-lasting DMFCs. However, more detailed investigations (MEA studies, continuous cycling, and accelerated degradation tests) are still required to understand the behavior of these materials under "real" fuel cell conditions.

References

1. Chu, D. and Jiang, R. 2002. Novel electrocatalysts for direct methanol fuel cells. *Solid State Ionics* 148, 591–599.
2. Aricò, A.S., Cretì, P., Antonucci, P.L., Cho, J., Kim, H. and Antonucci, V. 1998. Optimization of operating parameters of a direct methanol fuel cell and

physico-chemical investigation of catalyst-electrolyte interface. *Electrochim. Acta* 43, 3719–3729.

3. Reddington, E., Sapienza, A., Gurau, B., Viswanathan, R., Sarangapani, S. et al. 1998. Combinatorial electrochemistry: A highly parallel, optical screening method for discovery of better electrocatalysts. *Science* 280, 1735–1737.

4. Parsons, R. and VanderNoot, T. 1988. The oxidation of small organic molecules: A survey of recent fuel cell related research. *J. Electroanal. Chem.* 257, 9–45.

5. Schmidt, T.J., Paulus, U.A., Gasteiger, H.A. and Behm, R.J. 2001. The oxygen reduction reaction on a Pt/carbon fuel cell catalyst in the presence of chloride anions. *J. Electroanal. Chem.* 508, 41–47.

6. Matsumoto, T., Komatsu, T., Arai, K., Yamazaki, T., Kijima, M. et al. 2004. Reduction of Pt usage in fuel cell electrocatalysts with carbon nanotube electrodes. *Chem. Commun.* 840–841.

7. Yu, J.-S., Kang, S., Yoon, S.B. and Chai, G. 2002. Fabrication of ordered uniform porous carbon networks and their application to a catalyst supporter. *J. Am. Chem. Soc.* 124, 9382–9383.

8. Liang, H.-P., Zhang, H.-M., Hu, J.-S., Guo, Y.-G., Wan, L.-J. and Bai, C.-L. 2004. Pt hollow nanospheres: Facile synthesis and enhanced electrocatalysts. *Angew. Chem.* 43, 1540–1543.

9. Anderson, M.L., Stroud, R.M. and Rolison, D.R. 2002. Enhancing the activity of fuel-cell reactions by designing three-dimensional nanostructured architectures: Catalyst-modified carbon-silica composite aerogels. *Nano Lett.* 2, 235–240.

10. Chen, W.X., Lee, J.Y. and Liu, Z. 2002. Microwave-assisted synthesis of carbon supported Pt nanoparticles for fuel cell applications. *Chem. Commun.* 2588–2589.

11. Vinodgopal, K., Haria, M., Meisel, D. and Kamat, P. 2004. Fullerene-based carbon nanostructures for methanol oxidation. *Nano Lett.* 4, 415–418.

12. Lamy, C., Belgsir, E.M. and Lager, J.-M. 2001. Electrocatalytic oxidation of aliphatic alcohols: Application to the direct alcohol fuel cell (DAFC). *J. Appl. Electrochem.* 31, 799–809.

13. Gasteiger, H.A., Markovic, N., Ross, J., Philip, N. and Cairns, E.J. 1994. Temperature-dependent methanol electro-oxidation on well-characterized Pt-Ru alloys. *J. Electrochem. Soc.* 141, 1795–1803.

14. Lizcano-Valbuena, W.H., de Azevedo, D.C. and Gonzalez, E.R. 2004. Supported metal nanoparticles as electrocatalysts for low-temperature fuel cells. *Electrochim. Acta* 49, 1289–1295.

15. Watanabe, M., Uchida, M. and Motoo, S. 1987. Preparation of highly dispersed Pt-Ru alloy clusters and the activity for the electrooxidation of methanol. *J. Electroanal. Chem.* 229, 395–406.

16. Liu, L., Pu, C., Viswanathan, R., Fan, Q., Liu, R. and Smotkin, E.S. 1998. Carbon supported and unsupported Pt-Ru anodes for liquid feed direct methanol fuel cells. *Electrochim. Acta* 43, 3657–3663.

17. Liu, R., Iddir, H., Fan, Q., Hou, G., Bo, A. et al. 2000. Potential-dependent infrared absorption spectroscopy of adsorbed CO and X-ray photoelectron spectroscopy of arc-melted single-phase Pt, PtRu, PtOs, PtRuOs, and Ru electrodes. *J. Phys. Chem. B* 104, 3518–3531.

18. Chrzanowski, W. and Wieckowski, A. 1997. Ultrathin films of ruthenium on low index platinum single crystal surfaces: An electrochemical study. *Langmuir* 13, 5974–5978.

19. Ley, K.L., Liu, R., Pu, C., Fan, Q., Leyarovska, N. et al. 1997. Methanol oxidation on single-phase Pt-Ru-Os ternary alloys. *J. Electrochem. Soc.* 144, 1543–1548.
20. Gasteiger, H.A., Markovic, N., Ross, P.N. and Cairns, E.J. 1994. Carbon monoxide electrooxidation on well-characterized platinum-ruthenium alloys. *J. Phys. Chem.* 98, 617–625.
21. Gurau, B., Viswanathan, R., Liu, R., Lafrenz, T.J., Ley, K.L. et al. 1998. Structural and electrochemical characterization of binary, ternary, and quaternary platinum alloy catalysts for methanol electro-oxidation1. *J. Phys. Chem. B* 102, 9997–10003.
22. Markovic, N.M., Gasteiger, H.A., Ross, P.N., Jiang, X., Villegas, I. and Weaver, M.J. 1995. Electro-oxidation mechanisms of methanol and formic acid on Pt-Ru alloy surfaces. *Electrochim. Acta* 40, 91–98.
23. Gojkovic, S.L., Vidakovic, T.R. and Durovic, D.R. 2003. Kinetic study of methanol oxidation on carbon-supported PtRu electrocatalyst. *Electrochim. Acta* 48, 3607–3614.
24. Watanabe, M. and Motoo, S. 1975. Electrocatalysis by ad-atoms: Part II. Enhancement of the oxidation of methanol on platinum by ruthenium ad-atoms. *J. Electroanal. Chem.* 60, 267–273.
25. Friedrich, K.A., Geyzers, K.-P., Linke, U., Stimming, U. and Stumper, J. 1996. CO adsorption and oxidation on a Pt(111) electrode modified by ruthenium deposition: An IR spectroscopic study. *J. Electroanal. Chem.* 402, 123–128.
26. Christensen, P.A., Hamnett, A. and Troughton, G.L. 1993. The role of morphology in the methanol electro-oxidation reaction. *J. Electroanal. Chem.* 362, 207–218.
27. Tong Kim, H.S., Babu, P.K., Waszczuk, P., Wieckowski, A. and Oldfield, E. 2001. An NMR investigation of CO tolerance in a Pt/Ru fuel cell catalyst. *J. Am. Chem. Soc.* 124, 468–473.
28. Waszczuk, P., Wieckowski, A., Zelenay, P., Gottesfeld, S., Coutanceau, C. et al. 2001. Adsorption of CO poison on fuel cell nanoparticle electrodes from methanol solutions: A radioactive labeling study. *J. Electroanal. Chem.* 511, 55–64.
29. Hamnett, A. 1997. Mechanism and electrocatalysis in the direct methanol fuel cell. *Catal. Today* 38, 445–457.
30. Chu, D. and Gilman, S. 1996. Methanol electro-oxidation on unsupported Pt-Ru alloys at different temperatures. *J. Electrochem. Soc.* 143, 1685–1690.
31. He, C., Kunz, H.R. and Fenton, J.M. 1997. Evaluation of platinum-based catalysts for methanol electro-oxidation in phosphoric acid electrolyte. *J. Electrochem. Soc.* 144, 970–979.
32. Choi, J.-H., Park, K.-W., Kwon, B.-K. and Sung, Y.-E. 2003. Methanol oxidation on Pt/Ru, Pt/Ni, and Pt/Ru/Ni anode electrocatalysts at different temperatures for DMFCs. *J. Electrochem. Soc.* 150, A973–A978.
33. Wasmus, S. and Küver, A. 1999. Methanol oxidation and direct methanol fuel cells: A selective review. *J. Electroanal. Chem.* 461, 14–31.
34. Liu, J.G., Zhou, Z.H., Zhao, X.X., Xin, Q., Sun, G.Q. and Yi, B.L. 2004. Studies on performance degradation of a direct methanol fuel cell (DMFC) in life test. *Phys. Chem. Chem. Phys.* 6, 134.
35. Shao, Y., Yin, G., Zhang, J. and Gao, Y. 2006. Comparative investigation of the resistance to electrochemical oxidation of carbon black and carbon nanotubes in aqueous sulfuric acid solution. *Electrochim. Acta* 51, 5853–5857.
36. Shao, Y., Yin, G., Wang, Z. and Gao, Y. 2007. Proton exchange membrane fuel cell from low temperature to high temperature: Material challenges. *J. Power Sources* 167, 235–242.

37. Auer, E., Freund, A., Pietsch, J. and Tacke, T. 1998. Carbon as supports for industrial precious metal catalysts. *App. Catal. A: General* 173, 259–271.

38. Sun, X., Li, R., Villers, D., Dodelet, J.P. and Desilets, S. 2003. Composite electrodes made of Pt nanoparticles deposited on carbon nanotubes grown on fuel cell backings. *Chem. Phys. Lett.* 379, 99–104.

39. Borup, R., Meyers, J., Pivovar, B., Kim, Y.S., Mukundan, R. et al. 2007. Scientific aspects of polymer electrolyte fuel cell durability and degradation. *Chem. Rev.* 107, 3904–3951.

40. Ma, X.L., Li, Y. and Zhu, Y.L. 2003. Growth mode of the SnO2 nanobelts synthesized by rapid oxidation. *Chem. Phys. Lett.* 376, 794–798.

41. Che, G., Lakshmi, B.B., Martin, C.R. and Fisher, E.R. 1999. Metal-nanocluster-filled carbon nanotubes: Catalytic properties and possible applications in electrochemical energy storage and production. *Langmuir* 15, 750–758.

42. He, Z.J., Chen, D., Liu Zhou, H. and Kuang, Y. 2004. Electrodeposition of Pt–Ru nanoparticles on carbon nanotubes and their electrocatalytic properties for methanol electrooxidation. *Diamond Relat. Mater.* 13, 1764–1770.

43. Wu, G., Chen, Y.S. and Xu, B.Q. 2005. Remarkable support effect of SWNTs in Pt catalyst for methanol electrooxidation. *Electrochem. Commun.* 7, 1237–1243.

44. Hernández-Fernández, P., Nuño, R., Fatás, E., Fierro, J.L.G. and Ocón, P. 2011. MWCNT-supported PtRu catalysts for the electrooxidation of methanol: Effect of the functionalized support. *Int. J. Hydrogen Energy* 36, 8267–8278.

45. Liu, Z., Lin, X., Lee, J.Y., Zhang, W., Han, M. and Gan, L.M. 2002. Preparation and characterization of platinum-based electrocatalysts on multi-walled carbon nanotubes for proton exchange membrane fuel cells. *Langmuir* 18, 4054–4060.

46. Carmo, M., Paganin, V.A., Rosolen, J.M. and Gonzalez, E.R. 2005. Alternative supports for the preparation of catalysts for low-temperature fuel cells: The use of carbon nanotubes. *J. Power Sources* 142, 169–176.

47. Wang, X., Waje, M. and Yan, Y. 2005. CNT-based electrodes with high efficiency for PEMFCs. *Electrochem. Solid-State Lett.* 8, A42–A44.

48. Rajalakshmi, N., Ryu, H., Shaijumon, M.M. and Ramaprabhu, S. 2005. Performance of polymer electrolyte membrane fuel cells with carbon nanotubes as oxygen reduction catalyst support material. *J. Power Sources* 140, 250–257.

49. Steigerwalt, E.S., Deluga, G.A. and Lukehart, C.M. 2002. Pt-Ru/carbon fiber nanocomposites: Synthesis, characterization, and performance as anode catalysts of direct methanol fuel cells. A search for exceptional performance. *J. Phys. Chem. B* 106, 760–766.

50. Matsumoto, T., Komatsu, T., Nakanoa, H., Arai, K., Nagashima, Y. et al. 2004. Efficient usage of highly dispersed Pt on carbon nanotubes for electrode catalysts of polymer electrolyte fuel cells. *Catal. Today* 90, 277–281.

51. Villers, D., Sun, S.H., Serventi, A.M. and Dodelet, J.P. 2006. Characterization of Pt nanoparticles deposited onto carbon nanotubes grown on carbon paper and evaluation of this electrode for the reduction of oxygen. *J. Phys. Chem. B* 110, 25916–25925.

52. Saha, M.S., Li, R. and Sun, X. 2008. High loading and monodispersed Pt nanoparticles on multiwalled carbon nanotubes for high performance proton exchange membrane fuel cells. *J. Power Sources* 177, 314–322.

53. Hernández-Fernández, P., Montiel, M., Ocón, P., Gómez de la Fuente, J.L., García-Rodríguez, S. et al. 2010. Functionalization of multi-walled carbon nanotubes

and application as supports for electrocatalysts in proton-exchange membrane fuel cell. *Appl. Catal. B: Environ.* 99, 343–352.

54. Yuan, Y., Smith, J.A., Goenaga, G., Liu, D.-J., Luo, Z. and Liu, J. 2011. Platinum decorated aligned carbon nanotubes: Electrocatalyst for improved performance of proton exchange membrane fuel cells. *J. Power Sources* 196, 6160–6167.

55. Saha, M., Chen, Y., Li, R. and Sun, X. 2009. Enhancement of PEMFC performance by using carbon nanotubes supported Pt–Co alloy catalysts. *Asia-Pacific J. Chem. Eng* 4, 12–16.

56. Li, W., Liang, C., Qiu, J., Zhou, W., Han, H. et al. 2002. Carbon nanotubes as support for cathode catalyst of a direct methanol fuel cell. *Carbon* 40, 791–794.

57. Golikand, A.N., Lohrasbi, E. and Asgari, M. 2010. Enhancing the durability of multi-walled carbon nanotube supported by Pt and Pt–Pd nanoparticles in gas diffusion electrodes. *Int. J. Hydrogen Energy* 35, 9233–9240.

58. Lee, K., Zhang, J., Wang, H. and Wilkinson, D.P. 2006. Progress in the synthesis of carbon nanotube- and nanofiber-supported Pt electrocatalysts for PEM fuel cell catalysis. *J. Appl. Electrochem.* 36, 507–522.

59. Jha, N., Leela Mohana Reddy, A., Shaijumon, M.M., Rajalakshmi, N. and Ramaprabhu, S. 2008. Pt-Ru/multi-walled carbon nanotubes as electrocatalysts for direct methanol fuel cell. *Int. J. Hydrogen Energy* 33, 427–433.

60. Thess, A., Lee, R., Nikolaev, P., Dai, H., Petit, P. et al. 1996. Crystalline ropes of metallic carbon nanotubes. *Science* 273, 483–487.

61. Pantea, D., Darmstadt, H., Kaliaguine, S., Smmchen, L. and Roy, C. 2001. Electrical conductivity of thermal carbon blacks: Influence of surface chemistry. *Carbon* 39, 1147–1158.

62. Li, W., Liang, C., Zhou, W., Qiu, J., Zhou et al. 2003. Preparation and characterization of multiwalled carbon nanotube-supported platinum for cathode catalysts of direct methanol fuel cells. *J. Phys. Chem. B* 107, 6292–6299.

63. Li, L. and Xing, Y. 2006. Electrochemical durability of carbon nanotubes in noncatalyzed and catalyzed oxidations. *J. Electrochem. Soc.* 153, A1823–A1828.

64. Rodriguez, N.M., Kim, M.-S. and Baker, R.T.K. 1994. Carbon nanofibers: A unique catalyst support medium. *J. Phys. Chem.* 98, 13108–13111.

65. Park, C. and Baker, R.T.K. 1998. Catalytic behavior of graphite nanofiber supported nickel particles. 2. The influence of the nanofiber structure. *J. Phys. Chem. B* 102, 5168–5177.

66. Park, C., Anderson, P.E., Chambers, A., Tan, C.D., Hidalgo, R. and Rodriguez, N.M. 1999. Further studies of the interaction of hydrogen with graphite nanofibers. *J. Phys. Chem. B* 103, 10572–10581.

67. Salman, F., Park, C. and Baker, R.T.K. 1999. Hydrogenation of crotonaldehyde over graphite nanofiber supported nickel. *Catal. Today* 53, 385–394.

68. Bessel, C.A., Laubernds, K., Rodriguez, N.M. and Baker, R.T.K. 2001. Graphite nanofibers as an electrode for fuel cell applications. *J. Phys. Chem. B* 105, 1115–1118.

69. Yang, G.-W., Gao, G.-Y., Zhao, G.-Y. and Li, H.-L. 2007. Effective adhesion of Pt nanoparticles on thiolated multi-walled carbon nanotubes and their use for fabricating electrocatalysts. *Carbon* 45, 3036–3041.

70. Yin, S., Shen, P.K., Song, S. and Jiang, S.P. 2009. Functionalization of carbon nanotubes by an effective intermittent microwave heating-assisted HF/H2O2 treatment for electrocatalyst support of fuel cells. *Electrochim. Acta* 54, 6954–6958.

71. Okada, M., Konta, Y. and Nakagawa, N. 2008. Carbon nano-fiber interlayer that provides high catalyst utilization in direct methanol fuel cell. *J. Power Sources* 185, 711–716.

72. Joo, S.H., Pak, C., You, D.J., Lee, S.-A., Lee, H.I. et al. 2006. Ordered mesoporous carbons (OMC) as supports of electrocatalysts for direct methanol fuel cells (DMFC): Effect of carbon precursors of OMC on DMFC performances. *Electrochim. Acta* 52, 1618–1626.

73. Arbizzani, C., Beninati, S., Soavi, F., Varzi, A. and Mastragostino, M. 2008. Supported PtRu on mesoporous carbons for direct methanol fuel cells. *J. Power Sources* 185, 615–620.

74. Iijima, S. 1991. Helical microtubules of graphitic carbon. *Nature* 354, 56.

75. Dresselhaus, M.S. and Dai, H. 2004. Special issue on carbon nanotubes. *MRS Bull.* 29, 237–281.

76. Beguin, F. and Ehrburger, P. 2002. Special issue on carbon nanotubes. *Carbon* 40, 1619–1842.

77. Haddon, R.C. 2002. Special issue on carbon nanotubes. *Acc. Chem. Res.* 35, 997–1113.

78. Harris, P.J.F. 1999. *Carbon nanotubes and related structures: New materials for the 21st century.* Cambridge University Press, Cambridge, UK.

79. Dresselhaus, M.S., Dresselhaus, G. and Avouris, P. 2001. *Carbon nanotubes: Synthesis, structure, properties, and applications (Topics in applied physics, Vol. 80).* Springer-Verlag, New York.

80. Saito, R., Dresselhaus, G. and Dresselhaus, M.S. 1998. *Physical properties of carbon nanotubes.* Imperial College Press, London.

81. Meyyappan, M. 2005. *Carbon nanotubes: Science and applications.* CRC Press, Boca Raton, FL.

82. Dai, H.J. 2002. Carbon nanotubes: Synthesis, integration, and properties. *Accounts Chem. Res.* 35, 1035–1044.

83. Durrer, L., Helbling, T., Zenger, C., Jungen, A., Stampfer, C. and Hierold, C. 2008. SWNT growth by CVD on Ferritin-based iron catalyst nanoparticles towards CNT sensors. *Sensors and Actuators B: Chemical* 132, 485–490.

84. Wong, Y., Wei, S., Kanga, W., Davidson, J., Hofmeister, W. and Huang, J. 2004. Carbon nanotubes field emission devices grown by thermal CVD with palladium as catalysts. *Diamond Relat. Mater.* 13, 2105–2112.

85. Makris, T., Giorgi, L., Giorgi, R., Lisi, N. and Salernitano, E. 2005. CNT growth on alumina supported nickel catalyst by thermal CVD. *Diamond Relat. Mater.* 14, 815–819.

86. Liu, H., Zhang, Y., Li, R., Sun, X., Wang, F. et al. 2010. Aligned synthesis of multi-walled carbon nanotubes with high purity by aerosol assisted chemical vapor deposition: Effect of water vapor. *Appl. Surf. Sci.* 256, 4692–4696.

87. Ci, L., Wei, B., Liang, J., Xu, C.D. and Wu, D. 1999. Preparation of carbon nanotubules by the floating catalyst method. *J. Mater. Sci. Lett.* 18, 797–799.

88. Barreiro, A., Kramberger, C., Rümmeli, M., Neis, A., Grimm, D. and Hampel, S. 2007. Control of the single-wall carbon nanotube mean diameter in sulphur promoted aerosol-assisted chemical vapour deposition. *Carbon* 45, 55–61.

89. Pereza, H., Morin, A., Akroura, L., Cremona, C., Baret, B. et al. 2010. Evidence for high performances of low Pt loading electrodes based on capped platinum electrocatalyst and carbon nanotubes in fuel cell devices. *Electrochim. Acta* 55, 2358–2362.

90. Saha, M.S., Li, R., Sun, X. and Ye, S. 2009. 3-D composite electrodes for high performance PEM fuel cells composed of Pt supported on nitrogen-doped carbon nanotubes grown on carbon paper. *Electrochem. Commun.* 11, 438–441.

91. Dubosc, M., Casimirius, S., Besland, M.-P., Cardinaud, C., Granier, A. et al. 2007. Impact of the Cu-based substrates and catalyst deposition techniques on carbon nanotube growth at low temperature by PECVD. *Microelectron. Eng.* 84, 2501–2505.

92. Li, Y., Mann, D., Rolandi, M., Kim, W., Ural, A. et al. 2004. Preferential growth of semiconducting single-walled Carbon nanotubes by a plasma enhanced CVD method. *Nano Lett.* 4, 317–321.

93. Duy, D., Kim, H., Yoon, D., Lee, K., Ha, J. et al. 2009. Growth of carbon nanotubes on stainless steel substrates by DC-PECVD. *Appl. Surf. Sci.* 256, 1065–1068.

94. Hou, X. and Choy, K.L. 2006. Crystal growth of ZnS films by a charged aerosol-assisted vapor deposition process. *Chem. Vap. Deposition* 12, 583–585.

95. Park, C., Engel, E.S., Crowe, A., Gilbert, T.R. and Rodriguez, N.M. 2000. Use of carbon nanofibers in the removal of organic solvents from water. *Langmuir* 16, 8050–8056.

96. Fan, Y.-Y., Liao, B., Liu, M., Wei, Y.-L., Lu, M.-Q. and Cheng, H.-M. 1999. Hydrogen uptake in vapor-grown carbon nanofibers. *Carbon* 37, 1649–1652.

97. Wang, X., Wang, S. and Chung, D.D.L. 1999. Sensing damage in carbon fiber and its polymer-matrix and carbon-matrix composites by electrical resistance measurement. *J. Mater. Sci.* 34, 2703–2713.

98. Rodriguez, N.M. 1993. A review of catalytically grown carbon nanofibers. *J. Mater. Res.* 8, 3233–3250.

99. De Jong, K.P. and Geus, J.W. 2000. Carbon nanofibers: Catalytic synthesis and applications. *Catal. Rev. Sci. Eng.* 42, 481–510.

100. Sebastián, D., Calderón, J.C., González-Expósito, J.A., Pastor, E., Martínez-Huerta, M.V. et al. 2010. Influence of carbon nanofiber properties as electrocatalyst support on the electrochemical performance for PEM fuel cells. *Int. J. Hydrogen Energy* 35, 9934–9942.

101. Vamvakaki, V., Tsagaraki, K. and Chaniotakis, N. 2006. Carbon nanofiber-based glucose biosensor. *Anal. Chem.* 78, 5538–5542.

102. Hao, C., Ding, L., Zhang, X. and Ju, H. 2007. Biocompatible conductive architecture of carbon nanofiber-doped chitosan prepared with controllable electrodeposition for cytosensing. *Anal. Chem.* 79, 4442–4447.

103. Mirabile Gattia, D., Antisari, M.V., Giorgi, L., Marazzi, R., Piscopiello, E. et al. 2009. Study of different nanostructured carbon supports for fuel cell catalysts. *J. Power Sources* 194, 243–251.

104. Ismagilov, Z.R., Kerzhentsev, M.A., Shikina, N.V., Lisitsyn, A.S., Okhlopkova, L.B. et al. 2005. Development of active catalysts for low Pt loading cathodes of PEMFC by surface tailoring of nanocarbon materials. *Catal. Today* 102 (103), 58–66.

105. Zhu, Y.A., Sui, Z.J., Zhao, T.J., Dai, Y.C., Cheng, Z.M. and Yuan, W.K. 2005. Modeling of fishbone-type carbon nanofibers: A theoretical study. *Carbon* 43, 1694–1699.

106. Jong, K.P.D. and Geus, J.W. 2001. Carbon nanofibers: Catalytic synthesis and applications. *Catal. Rev. Sci. Eng.* 42, 481–510.

107. Rodriguez, N.M., Kim, M.S. and Baker, R.T.K. 1993. Promotional effect of carbon monoxide on the decomposition of ethylene over an iron catalyst. *J. Catal.* 144, 93–108.

108. Krishnankutty, N., Rodriguez, N.M. and Baker, R.T.K. 1997. The effect of copper on the structural characteristics of carbon filaments produced from iron catalyzed decomposition of ethylene. *Catal. Today* 37, 295–307.

109. Oberlin, A., Endo, M. and Koyama, T. 1976. Filamentous growth of carbon through benzene decomposition. *J. Crystal Growth* 32, 335–349.

110. Helveg, S., López-Cartes, C., Sehested, J., Hansen, P.L., Clausen, B.S. et al. 2004. Atomic-scale imaging of carbon nanofibre growth. *Nature* 427, 426–429.

111. De Jong, K.P. and Geus, J.W. 2000. Carbon nanofibers: Catalytic synthesis and applications. *Catal. Rev. Sci. Eng.* 42, 481–510.

112. Ermakova, M.A., Ermakov, D.Y. and Kuvshinov, G.G. 2000. Effective catalysts for direct cracking of methane to produce hydrogen and filamentous carbon Part I. Nickel catalysts. *Appl. Catal. A* 201, 61–70.

113. Villacampa, J.I., Royo, C., Romeo, E., Montoya, J.A., Del Angel, P. and Monzón, A. 2003. Catalytic decomposition of methane over Ni-Al2O3 coprecipitated catalysts reaction and regeneration studies. *Appl. Catal. A: General* 252, 557–567.

114. Saha, M.S. and Kundu, A. 2010. Functionalizing carbon nanotubes for proton exchange membrane fuel cells electrode. *J. Power Sources* 195, 6255–6261.

115. Hirsch, A. 2002. Functionalization of single-walled carbon nanotubes. *Angew. Chem. Int. Ed.* 41, 1853–1959.

116. Tasis, D., Tagmatarchis, N., Bianco, A. and Prato, M. 2006. Chemistry of carbon nanotubes. *Chem. Rev.* 106, 1105–1136.

117. Ebbesen, T.W., Hirua, H., Bisher, M.E., Treacy, M.M.J., Shreeve-Keyer, J.L. and Haushalter, R.C. 1996. Decoration of carbon nanotubes. *Adv. Mater.* 8, 155–157.

118. Rocco, A.M., Silva, C.A., Macedo, M.I.F., Maestro, L.F., Herbst, M.H. et al. 2008. Purification of catalytically produced carbon nanotubes for use as support for fuel cell cathode Pt catalyst. *J. Mater. Sci.* 43, 557.

119. Hwang, K. 1995. Efficient cleavage of carbon graphene layers by oxidants. *J. Chem. Soc., Chem. Commun.* 173–174.

120. Xu, C., Chen, J., Cui, Y., Han, Q., Choo, H. et al. 2006. Influence of the surface treatment on the deposition of platinum nanoparticles on the carbon nanotubes. *Adv. Eng. Mater.* 8, 73–76.

121. Tsang, S.C., Chen, Y.K., Harris, P.J.F. and Green, M.L.H. 1994. A simple chemical method of opening and filling carbon nanotubes. *Nature* 372, 159.

122. Lago, R.M., Tsang, S.C., Lu, K.L., Chen, Y.K. and Green, M.L.H. 1995. Filling carbon nanotubes with small palladium metal crystallites: The effect of surface acid groups. *J. Chem. Soc., Chem. Commun.* 1355–1356.

123. Guo, D.-J. and Li, H.-L. 2005. High dispersion and electrocatalytic properties of platinum on functional multi-walled carbon nanotubes. *Electroanalysis* 17, 869–872.

124. He, Z., Chen, J., Liu, D., Tang, H., Deng, W. and Kuang, Y. 2004. Deposition and electrocatalytic properties of platinum nanoparticles on carbon nanotubes for methanol electrooxidation. *Mater. Chem. Phys.* 85, 396–401.

125. Glerup, M., Castignolles, M., Holzinger, M., Hug, G., Loiseau, A. and Bernier, P. 2003. Synthesis of highly nitrogen-doped multi-walled carbon nanotubes. *Chem. Commun.* 2542–2543.

126. Wang, E.G. 2006. Nitrogen-induced carbon nanobells and their properties. *J. Mater. Res.* 21, 2767–2773.

127. Ewels, C.P. and Glerup, M. 2005. Nitrogen doping in carbon nanotubes. *J. Nanosci. Nanotech.* 5, 1345–1363.

128. Sun, C.-L., Chen, L.-C., Su, M.-C., Hong, L.-S., Chyan, O. et al. 2005. Ultrafine platinum nanoparticles uniformly dispersed on arrayed CNx nanotubes with high electrochemical activity. *Chem. Mater.* 17, 3749–3753.
129. Zamudio, A., Elias, A.L., Rodriguez-Manzo, J.A., Lopez-Urias, F., Rodriguez-Gattorno, G. et al. 2006. Efficient anchoring of silver nanoparticles on N-doped carbon nanotubes. *Small* 2, 346–350.
130. Wang, C.-H., Shih, H.-C., Tsai, Y.-T., Du, H.-Y., Chen, L.-C. and Chen, K.-H. 2006. High methanol oxidation activity of electrocatalysts supported by directly grown nitrogen-containing carbon nanotubes on carbon cloth. *Electrochim. Acta* 52, 1612–1617.
131. Maiyalagan, T., Viswanathan, B. and Varadaraju, U.V. 2005. Nitrogen containing carbon nanotubes as supports for Pt–alternate anodes for fuel cell applications. *Electrochem. Comm.* 7, 905–912.
132. Wang, C.-H., Du, H.-Y., Tsai, Y.-T., Chen, C.-P., Huang, C.-J. et al. 2007. High performance of low electrocatalysts loading on CNT directly grown on carbon cloth for DMFC. *J. Power Sources* 171, 55–62.
133. Kuiyang, J., Ami, E., Linda, S.S., Pulickel, M.A., Richard, W.S. et al. 2003. Selective attachment of gold nanoparticles to nitrogen doped carbon nanotubes. *Nano Lett.* 3, 275–277.
134. Ozaki, J.I., Anahara, T., Kimura, N. and Oya, A. 2006. Simultaneous doping of boron and nitrogen into a carbon to enhance its oxygen reduction activity in proton exchange membrane fuel cells. *Carbon* 44, 3358–3361.
135. Li, W., Liang, C., Zhou, W., Qiu, J., Li, H. et al. 2004. Homogeneous and controllable Pt particles deposited on multi-wall carbon nanotubes as cathode catalyst for direct methanol fuel cells. *Carbon* 42, 436–439.
136. Xue, B., Chen, P., Hong, Q., Lin, J. and Tan, K.S. 2001. Growth of Pd, Pt, Ag and Au nanoparticles on carbon nanotubes. *J. Mater. Chem.* 11, 2378–2381.
137. Prabhuram, J., Zhao, T.S., Liang, Z.X. and Chen, R. 2007. A simple method for the synthesis of PtRu nanoparticles on the multi-walled carbon nanotube for the anode of a DMFC. *Electrochim. Acta* 52, 2649–2656.
138. Xu, J., Hua, K., Sun, G., Wang, C., Lv, X. and Wang, Y. 2006. Electrooxidation of methanol on carbon nanotubes supported Pt-Fe alloy electrode. *Electrochem. Commun.* 8, 982–986.
139. Prabhuram, J., Zhao, T.S., Tang, Z.K., Chen, R. and Liang, Z.X. 2006. Multiwalled carbon nanotube supported PtRu for the anode of direct methanol fuel cells. *J. Phys. Chem. B* 110, 5245–5252.
140. Gutierrez, M.C., Hortiguela, M.J., Amarilla, J.M., Jimenez, R., Ferrer, M.L. and Del Monte, F. 2007. Macroporous 3D architectures of self-assembled MWCNT surface decorated with Pt nanoparticles as anodes for a direct methanol fuel cell. *J. Phys. Chem. C* 111, 5557–5560.
141. Zhao, Y., Yang, X., Tian, J., Wang, F. and Zhan, L. 2010. Methanol electrooxidation on Ni@Pd core-shell nanoparticles supported on multi-walled carbon nanotubes in alkaline media. *Inter. J. Hydro. Energy* 35, 3249–3257.
142. Chang, W.-C. and Nguyen, M.T. 2011. Investigations of a platinum-ruthenium/carbon nanotube catalyst formed by a two-step spontaneous deposition method. *J. Power Sources* 196, 5811–5816.
143. Jha, N., Jafri, R.I., Rajalakshmi, N. and Ramaprabhu, S. 2011. Graphene-multi walled carbon nanotube hybrid electrocatalyst support material for direct methanol fuel cell. *Int. J. Hydrogen Energy* 36, 7284–7290.

144. Lo, A.-Y., Yu, N., Huang, S.-J., Hung, C.-T., Liu, S.-H. et al. 2011. Fabrication of CNTs with controlled diameters and their applications as electrocatalyst supports for DMFC. *Diamond Relat. Mater.* 20, 343–350.
145. Zhou, W. and Lee, J.Y. 2007. Highly active core-shell Au@Pd catalyst for formic acid electrooxidation. *Electrochem. Commun.* 9, 1725–1729.
146. Yang, J., Lee, J.Y., Zhang, Q., Zhou, W. and Liu, Z. 2008. Carbon-supported pseudo-core-shell Pd-Pt nanoparticles for ORR with and without methanol. *J. Electrochem. Soc.* 155, B776–B781.
147. Zhao, H., Li, L., Yang, J. and Zhang, Y. 2008. Co@Pt-Ru core-shell nanoparticles supported on multiwalled carbon nanotube for methanol oxidation. *Electrochem. Commun.* 10, 1527–1529.
148. Chen, Y., Yang, F., Dai, Y., Wang, W. and Chen, S. 2008. Ni@Pt core-shell nanoparticles: Synthesis, structural and electrochemical properties. *J. Phys. Chem. C* 112, 1645–1649.
149. Tsuji, M., Hashimoto, M., Nishizawa, Y., Kubokawa, M. and Tsuji, T. 2005. Microwave-assisted synthesis of metallic nanostructures in solution. *Chem. A Eur. J.* 11, 440–452.
150. Liu, Z., Gan, L.M., Hong, L., Chen, W. and Lee, J.Y. 2005. Carbon-supported Pt nanoparticles as catalysts for proton exchange membrane fuel cells. *J. Power Sources* 139, 73–78.
151. Chen, W., Zhao, J., Lee, J.Y. and Liu, Z. 2005. Microwave heated polyol synthesis of carbon nanotubes supported Pt nanoparticles for methanol electrooxidation. *Mater. Chem. Phys.* 91, 124–129.
152. Liu, Z., Lee, J.Y., Chen, W., Han, M. and Gan, L.M. 2004. Physical and electrochemical characterizations of microwave-assisted polyol preparation of carbon-supported PtRu nanoparticles. *Langmuir* 20, 181–187.
153. Li, X., Chen, W.-X., Zhao, J., Xing, W. and Xu, Z.-D. 2005. Microwave polyol synthesis of Pt/CNTs catalysts: Effects of pH on particle size and electrocatalytic activity for methanol electrooxidization. *Carbon* 43, 2168–2174.
154. Hsieh, C.-T., Hung, W.-M., Chen, W.-Y. and Lin, J.-Y. 2011. Microwave-assisted polyol synthesis of Pt-Zn electrocatalysts on carbon nanotube electrodes for methanol oxidation. *Inter. J. Hydro. Energy* 36, 2765–2772.
155. Hsieh, C.-T., Chen, W.-Y., Chen, I.-L. and Roy, A.K. 2012. Deposition and activity stability of Pt-Co catalysts on carbon nanotube-based electrodes prepared by microwave-assisted synthesis. *J. Power Sources* 199, 94–102.
156. Sakthivel, M., Schlange, A., Kunz, U. and Turek, T. 2010. Microwave assisted synthesis of surfactant stabilized platinum/carbon nanotube electrocatalysts for direct methanol fuel cell applications. *J. Power Sources* 195, 7083–7089.
157. Ahmadi, T.S., Wang, Z.L., Green, T.C., Henglein, A. and El-Sayed, M.A. 1996. Shape-controlled synthesis of colloidal platinum nanoparticles. *Science* 272–273, 1924.
158. Bock, C., Paquet, C., Couillard, M., Botton, G.A. and MacDougall, B.R. 2004. Size-selected synthesis of PtRu nano-catalysts: Reaction and size control mechanism. *J. Am. Chem. Soc.* 126, 8028–8037.
159. Li, X. and Hsing, I.-M. 2006. Surfactant-stabilized PtRu colloidal catalysts with good control of composition and size for methanol oxidation. *Electrochim. Acta* 52, 1358–1365.
160. Zhang, X. and Chan, K.-Y. 2002. Water-in-oil microemulsion synthesis of platinumâ ruthenium nanoparticles, their characterization and electrocatalytic properties. *Chem. Mater.* 15, 451–459.

161. Lee, C.-L., Ju, Y.-C., Chou, P.-T., Huang, Y.-C., Kuo, L.-C. and Oung, J.-C. 2005. Preparation of Pt nanoparticles on carbon nanotubes and graphite nanofibers via self-regulated reduction of surfactants and their application as electrochemical catalyst. *Electrochem. Commun.* 7, 453–458.

162. Li, X. and Hsing, I.M. 2006. The effect of the Pt deposition method and the support on Pt dispersion on carbon nanotubes. *Electrochim. Acta* 51, 5250–5258.

163. Kongkanand, A., Vinodgopal, K., Kuwabata, S. and Kamat, P.V. 2006. Highly dispersed Pt catalysts on single-walled carbon nanotubes and their role in methanol oxidation. *J. Phys. Chem. B* 110, 16185–16189.

164. Wang, J., Deng, X., Xi, J., Chen, L., Zhu, W. and Qiu, X. 2007. Promoting the current for methanol electro-oxidation by mixing Pt-based catalysts with CeO2 nanoparticles. *J. Power Sources* 170, 297–302.

165. Cao, J., Du, C., Wang, S.C., Mercier, P., Zhang, X. et al. 2007. The production of a high loading of almost monodispersed Pt nanoparticles on single-walled carbon nanotubes for methanol oxidation. *Electrochem. Commun.* 9, 735–740.

166. Kuo, P.L., Chen, C.C. and Jao, M.W. 2005. Effects of polymer micelles of alkylated polyethylenimines on generation of gold nanoparticles. *J. Phys. Chem. B* 109, 9445–9450.

167. Yoshitake, T., Shimakawa, Y., Kuroshima, S., Kimura, H., Ichihashi, T. et al. 2002. Preparation of fine platinum catalyst supported on single-wall carbon nanohorns for fuel cell application. *Physica B* 323, 124–126.

168. Zhao, Y.E.Y., Fan, L., Qiu, Y. and Yang, S. 2007. A new route for the electrodeposition of platinum-nickel alloy nanoparticles on multi-walled carbon nanotubes. *Electrochim. Acta* 52, 5873–5878.

169. Hussain, S. and Pal, A.K. 2008. Incorporation of nanocrystalline silver on carbon nanotubes by electrodeposition technique. *Mater. Lett.* 62, 1874–1877.

170. Tang, H., Chen, J., Yao, S., Nie, L., Kuang, Y. et al. 2005. Deposition and electrocatalytic properties of platinum on well-aligned carbon nanotube (CNT) arrays for methanol oxidation. *Mater. Chem. Phys.* 92, 548–553.

171. Zhao, Y., Fan, L., Zhong, H. and Li, Y. 2007. Electrodeposition and electrocatalytic properties of platinum nanoparticles on multi-walled carbon nanotubes: Effect of the deposition conditions. *Microchim. Acta.* 158, 327–334.

172. Zhang, X., Shi, X. and Wang, C. 2009. Electrodeposition of Pt nanoparticles on carbon nanotubes-modified polyimide materials for electrocatalytic applications. *Catal. Commun.* 10, 610–613.

173. Quinn, B.M., Dekker, C. and Lemay, S.G. 2005. Electrodeposition of noble metal nanoparticles on carbon nanotubes. *J. Am. Chem. Soc.* 127, 6146–6147.

174. Girishkumar, G., Vinodgopal, K. and Kamat, P.V. 2004. Carbon nanostructures in portable fuel cells: Single-walled carbon nanotube electrodes for methanol oxidation and oxygen reduction. *J. Phys. Chem. B* 108, 19960–19966.

175. Penner, R.M. 2002. Mesoscopic metal particles and wires by electrodeposition. *J. Phys. Chem. B* 106, 3335–3338.

176. Walter, E.C., Zach, M.P., Favier, F., Murray, B.J., Inazu, K. et al. 2003. Metal nanowire arrays by electrodeposition. *Chem. Phys. Chem.* 4, 131–138.

177. Favier, F., Walter, E.C., Zach, M.P., Benter, T. and Penner, R.M. 2001. Hydrogen sensors and switches from electrodeposited palladium mesowire arrays. *Science (New York)* 293, 2227–2231.

178. Zhao, Y., Fan, L., Zhong, H., Li, Y. and Yang, S. 2007. Platinum nanoparticle clusters immobilized on multiwalled carbon nanotubes: Electrodeposition and

enhanced electrocatalytic activity for methanol oxidation. *Adv. Funct. Mater.* 17, 1537–1541.

179. Guo, D.-J. and Li, H.-L. 2004. High dispersion and electrocatalytic properties of Pt nanoparticles on SWNT bundles. *J. Electroanal. Chem.* 573, 197–202.
180. Day, T.M., Unwin, P.R., Wilson, N.R. and Macpherson, J.V. 2005. Electrochemical templating of metal nanoparticles and nanowires on single-walled carbon nanotube networks. *J. Am. Chem. Soc.* 127, 10639–10647.
181. Cha, S.Y. and Lee, W.M. 1999. Performance of proton exchange membrane fuel cell electrodes prepared by direct deposition of ultrathin platinum on the membrane surface. *J. Electrochem. Soc.* 146, 4055–4060.
182. Hirano, S., Kim, J. and Srinivasan, S. 1997. High performance proton exchange membrane fuel cells with sputter-deposited Pt layer electrodes. *Electrochim. Acta* 42, 1587–1593.
183. Chen, C.C., Chen, C.F., Hsu, C.H. and Li, I.H. 2005. Growth and characteristics of carbon nanotubes on carbon cloth as electrodes. *Diamond Relat. Mater.* 14, 770–773.
184. Soin, N., Roy, S.S., Karlsson, L. and McLaughlin, J.A. 2010. Sputter deposition of highly dispersed platinum nanoparticles on carbon nanotube arrays for fuel cell electrode material. *Diamond Relat. Mater.* 19, 595–598.
185. Zacharia, R., Rather, S.-U., Hwang, S.W. and Nahm, K.S. 2007. Spillover of physisorbed hydrogen from sputter-deposited arrays of platinum nanoparticles to multi-walled carbon nanotubes. *Chem. Phys. Lett.* 434, 286–291.
186. Guha, A., Lu, W., Zawodzinski, T.A., Jr. and Schiraldi, D.A. 2007. Surface-modified carbons as platinum catalyst support for PEM fuel cells. *Carbon* 45, 1506–1517.
187. Calvillo, L., Lazaro, M.J., Suelves, I., Echegoyen, Y., Bordeje, E.G. and Moliner, R. 2009. Study of the surface chemistry of modified carbon nanofibers by oxidation treatments in liquid phase. *J. Nanosci. Nanotech.* 9, 4164–4169.
188. Toebes, M.L., van Heeswijk, J.M.P., Bitter, J.H., Jos van Dillen, A. and de Jong, K.P. 2004. The influence of oxidation on the texture and the number of oxygen-containing surface groups of carbon nanofibers. *Carbon* 42, 307–315.
189. Oh, H.-S., Kim, K., Ko, Y.-J. and Kim, H. 2010. Effect of chemical oxidation of CNFs on the electrochemical carbon corrosion in polymer electrolyte membrane fuel cells. *Int. J. Hydrogen Energy* 35, 701–708.
190. Zaragoza-Martin, F., Sopena-Escario, D., Morallon, E. and de Lecea, C.S.-M. 2007. Pt/carbon nanofibers electrocatalysts for fuel cells: Effect of the support oxidizing treatment. *J. Power Sources* 171, 302–309.
191. Calvillo, L., Gangeri, M., Perathoner, S., Centi, G., Moliner, R. and Lázaro, M.J. 2009. Effect of the support properties on the preparation and performance of platinum catalysts supported on carbon nanofibers. *J. Power Sources* 192, 144–150.
192. Guha, A., Zawodzinski, T.A., Jr. and Schiraldi, D.A. 2007. Evaluation of electrochemical performance for surface-modified carbons as catalyst support in polymer electrolyte membrane (PEM) fuel cells. *J. Power Sources* 172, 530–541.
193. Park, I.-S., Park, K.-W., Choi, J.-H., Park, C.R. and Sung, Y.-E. 2007. Electrocatalytic enhancement of methanol oxidation by graphite nanofibers with a high loading of PtRu alloy nanoparticles. *Carbon* 45, 28–33.
194. Tsuji, M., Kubokawa, M., Yano, R., Miyamae, N., Tsuji, T. et al. 2006. Fast preparation of PtRu catalysts supported on carbon nanofibers by the microwave-polyol method and their application to fuel cells. *Langmuir* 23, 387–390.

195. Knupp, S.L., Li, W., Paschos, O., Murray, T.M., Snyder, J. and Haldar, P. 2008. The effect of experimental parameters on the synthesis of carbon nanotube/ nanofiber supported platinum by polyol processing techniques. *Carbon* 46, 1276–1284.

196. Steigerwalt, E.S., Deluga, G.A., Cliffel, D.E. and Lukehart, C.M. 2001. A Pt-Ru/ graphitic carbon nanofiber nanocomposite exhibiting high relative performance as a direct-methanol fuel cell anode catalyst. *J. Phys. Chem. B* 105, 8097–8101.

197. Khosravi, M. and Amini, M.K. 2010. Flame synthesis of carbon nanofibers on carbon paper: Physicochemical characterization and application as catalyst support for methanol oxidation. *Carbon* 48, 3131–3138.

198. Sebastián, D., Lázaro, M.J., Suelves, I., Moliner, R., Baglio, V. et al. 2012. The influence of carbon nanofiber support properties on the oxygen reduction behavior in proton conducting electrolyte-based direct methanol fuel cells. *Int. J. Hydrogen Energy* 37, 6253–6260.

199. Yoshitake, T., Kimura, H., Kuroshima, S., Watanabe, S., Shimakawa, Y. et al. 2002. Small direct methanol fuel cell pack for portable applications. *Electrochem.* 70, 966–968.

200. Sasaki, K., Ujiie, H., Higa, T., Hori, T., Shinya, N. and Uchida, T. 2004. Rabbit aneurysm model mediated by the application of elastase. *Neurol. Med. Chir.* 44, 467–473.

201. Rasheed, A., Howe, J.Y., Dadmun, M.D. and Britt, P.F. 2007. The efficiency of the oxidation of carbon nanofibers with various oxidizing agents. *Carbon* 45, 1072–1080.

202. Tang, H., Chen, J., Nie, L., Liu, D., Deng, W. et al. 2004. High dispersion and electrocatalytic properties of platinum nanoparticles on graphitic carbon nanofibers (GCNFs). *J. Coll. Interface Sci.* 269, 26–31.

203. Li, M.Y., Han, G.Y. and Yang, B.S. 2008. Fabrication of the catalytic electrodes for methanol oxidation on electrospinning-derived carbon fibrous mats. *Electrochem. Commun.* 10, 880–883.

204. Lin, Z., Ji, L. and Zhang, X. 2009. Electrodeposition of platinum nanoparticles onto carbon nanofibers for electrocatalytic oxidation of methanol. *Mater. Lett.* 63, 2115–2118.

205. Maciá, M.D., Campiña, J.M., Herrero, E. and Feliu, J.M. 2004. On the kinetics of oxygen reduction on platinum stepped surfaces in acidic media. *J. Electroanal. Chem.* 564, 141–150.

206. Kuzume, A., Herrero, E. and Feliu, J.M. 2007. Oxygen reduction on stepped platinum surfaces in acidic media. *J. Electroanal. Chem.* 599, 333–343.

207. Stamenkovic, V. and Markovic, N.M., Jr., 2001. Structure-relationships in electrocatalysis: Oxygen reduction and hydrogen oxidation reactions on Pt(111) and Pt(100) in solutions containing chloride ions. *J. Electroanal. Chem.* 500, 44–51.

208. O'Hayre, R., Lee, S.-J., Cha, S.-W. and Prinz, F.B. 2002. A sharp peak in the performance of sputtered platinum fuel cells at ultra-low platinum loading. *J. Power Source* 109, 483–493.

209. Chen, Q., Chenyao, L., Jiao, J., Shuoqi, L., Ji, X. et al. 2011. Characterization of indium tin oxides thin films modified with ion implantation and its application for attachment of platinum nanoparticles. *Electrochim. Acta* 56, 6033–6040.

210. Caillard, A., Charles, C., Boswell, R. and Brault, P. 2007. Integrated plasma synthesis of efficient catalytic nanostructures for fuel cell electrodes. *Nanotechnology* 18, 305603.

211. Kim, H.-T., Lee, J.-K. and Kim, J. 2008. Platinum-sputtered electrode based on blend of carbon nanotubes and carbon black for polymer electrolyte fuel cell. *J. Power Source* 108, 191–194.

212. Zhang, C., Jue, H., Masaaki, N., Xingsheng, S., Hirotaka, T. et al. 2011. Magnetron sputtering of platinum nanoparticles onto vertically aligned carbon nanofibers for electrocatalytic oxidation of methanol. *Electrochim. Acta* 56, 6033–6040.

213. Tian, Z.Q., Jiang, S.P., Liang, Y.M. and Shen, P.K. 2006. Synthesis and characterization of platinum catalysts on muldwalled carbon nanotubes by intermittent microwave irradiation for fuel cell applications. *J. Phys. Chem. B* 110, 5343–5350.

214. Chan, K.Y., Ding, J., Ren, J.W., Cheng, S.A. and Tsang, K.Y. 2004. Supported mixed metal nanoparticles as electrocatalysts in low temperature fuel cells. *J. Mater. Chem.* 14, 505–516.

215. Mehta, V. and Cooper, J.S. 2003. Review and analysis of PEM fuel cell design and manufacturing. *J. Power Sources* 114, 32–53.

216. Jeng, K.-T., Chien, C.-C., Hsub, N.-Y., Yen, S.-C., Chiou, S.-D. et al. 2006. Performance of direct methanol fuel cell using carbon nanotube-supported Pt–Ru anode catalyst with controlled composition. *J. Power Sources* 160, 97–104.

217. Yuan, F. and Ryu, H. 2004. The synthesis, characterization, and performance of carbon nanotubes and carbon nanofibres with controlled size and morphology as a catalyst support material for a polymer electrolyte membrane fuel cell. *Nanotechnology* 15, S596–S602.

218. Biest, O.O.Vd. and Vandeperre, L.J. 1999. Electrophoretic deposition of materials. *Annu. Rev. Mater. Sci.* 29, 327–352.

219. Jeng, K.-T., Huang, W.-M. and Hsu, N.-Y. 2009. Application of low-voltage electrophoreticdeposition to fabrication of direct methanol fuel cell electrode composite catalyst layer. *Mater. Chem. Phys.* 113, 574–578.

220. Zheng, J.S., Wang, M.X., Zhang, X.S., Wu, Y.X., Li, P. et al. 2008. Platinum/carbon nanofiber nanocomposite synthesized by electrophoreticdeposition as electrocatalyst for oxygen reduction. *J. Power Source* 175, 211–216.

221. Hsu, Y.-K., Yang, J.-L., Lin, Y.-G., Chen, S.-Y., Chen, L.-C. and Chen, K.-H. 2009. Electrophoretic deposition of PtRu nanoparticles on carbon nanotubes for methanol oxidation. *Diamond Relat. Mater.* 18, 557–562.

222. Li, W.M., Waje, Z., Chen, P. and Larsen Yan, Y. 2010. Platinum nanopaticles supported on stacked-cup carbon nanofibers as electrocatalysts for proton exchange membrane fuel cell. *Carbon* 48, 995–1003.

223. Bi, Y. and Lu, G. 2009. Control growth of uniform platinum nanotubes and their catalytic properties for methanol electrooxidation. *Electrochem. Commun.* 11, 45–49.

224. Fujigaya, T., Okamoto, M. and Nakashima, N. 2009. Design of an assembly of pyridine-containing polybenzimidazole, carbon nanotubes and Pt nanoparticles for a fuel cell electrocatalyst with a high electrochemically active surface area. *Carbon* 47, 3227–3232.

225. Zhu, Q., Zhou, S., Wang, X. and Dai, S. 2009. Controlled synthesis of mesoporous carbon modified by tungsten carbides as an improved electrocatalyst support for the oxygen reduction reaction. *J. Power Sources* 193, 495–500.

15

Advanced Alkaline Polymer Electrolytes for Fuel Cell Applications

Jing Pan and Lin Zhuang

CONTENTS

15.1 Introduction 481
    15.1.1 How about Replacing the Acidic Polymer Electrolyte with
        an Alkaline One? 482
    15.1.2 Requirements of Alkaline Polymer Electrolytes (APEs) for
        Fuel Cell Applications 482
15.2 Challenges for APE Synthesis and Design 483
    15.2.1 Three Major Approaches for APE Synthesis 483
    15.2.2 Realistic Challenge of APEs 485
    15.2.3 Strategies for Designing High-Performance APEs 486
15.3 Development of High-Performance APEs 488
    15.3.1 Implementations for the Self-Cross-Linking and Self-
        Aggregating Design of APEs 488
    15.3.2 Remarkable Performance of xQAPS and aQAPS 489
    15.3.3 Experimental Proof for the Structural Feature of xQAPS
        and aQAPS 494
    15.3.4 Structural Insights from Molecular Dynamics Simulations 496
15.4 Degradation and Chemical Stability of APEs 497
15.5 Conclusions and Perspectives 499
References 500

15.1 Introduction

Different from the internal combustion engine (ICE), fuel cells convert the chemical energy stored in fuels and oxidants into electricity through electrochemical reactions, which is a quiet, clean, and more efficient fashion of energy conversion; thus they have been considered as a superior power source, particularly suitable for the future hydrogen-based energy system [1]. In the past decade, electric vehicles powered by polymer electrolyte fuel cells (PEFCs) have been demonstrated worldwide. Despite the great success, this promising technology has not yet met the public's expectations, mostly because of its high cost.

In a state-of-the-art PEFC, the most costly components include the graphite bipolar plate, the perfluorosulfonic acid polymer electrolyte, and the platinum catalyst. The costs of the former two components can, in principle, be well reduced by technological improvements, but the dependence on noble-metal catalysts is rather more an issue of resource than just price. The reason that PEFCs have to use Pt catalysts is not only the high catalytic activity, but also the high stability in strongly acid media. Specifically, under the harsh condition of the corrosive acid, only noble metals can remain relatively stable. In fact, even Pt is not stable enough under the fuel-cell operating conditions [2], a worse fact that has been challenging the current PEFC technology.

15.1.1 How about Replacing the Acidic Polymer Electrolyte with an Alkaline One?

To fundamentally get rid of the dependence of Pt catalysts, the fuel cell system needs a revolutionary change, namely, to replace the acidic polymer electrolyte with an alkaline one because under alkaline conditions some transition metals and metal oxides can serve as stable catalysts for the fuel-cell reactions [3–7].

Yet the idea of using alkaline electrolytes seems not new at all as alkaline fuel cells (AFCs) have already been realized and successfully applied in the spaceship a long time before. But things have changed: the desired electrolyte is a polymer version of alkali, rather than the concentrated KOH solution used in AFCs. In an alkaline polymer electrolyte (APE), the cations, typically quaternary ammonia $\left(-NR_3^+\right)$, are attached onto the polymer chain, and OH- is the only dissociative ion in the aqueous phase, serving as the charge carrier. Such a seemingly simple change, however, makes things much different. First, because of the use of solid polymer electrolytes, there is no need to worry about the leakage of liquid electrolytes, a severe issue in AFCs in ground applications, in which the electrode's waterproofing is always threatened by the carbonate precipitates resulting from the carbonation of KOH solution [8]. Some studies demonstrated that the carbonation of APE has little deleterious effect on the performance of APEFCs, which means APEs are tolerant to carbon dioxide when the cathode is operated in air [9,10]. Secondly, by using APEs, there will be "zero gap" between the anode and the cathode because the polymer electrolyte membrane can be made as thin as tens of microns, which will lead to a reduction in the ohmic loss of the cell voltage and thus an improvement in the power density. In summary, the use of APEs is expected to bring a combination of the advantages of PEFCs and AFCs.

15.1.2 Requirements of Alkaline Polymer Electrolytes (APEs) for Fuel Cell Applications

The APE to be applied in fuel cells belongs to the anion exchange resins, traditionally known as the quaternary ammonium polystyrene (Scheme 15.1):

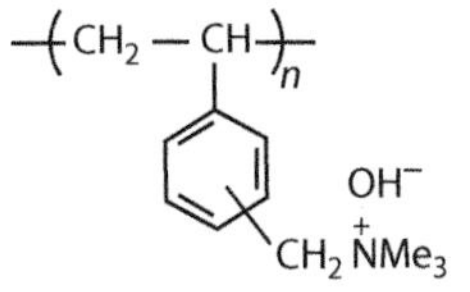

$$-\left(CH_2 - CH\right)_n$$

OH$^-$

CH$_2$ $\overset{+}{N}Me_3$

SCHEME 15.1

which is, however, not suitable for fuel cells because of the low stability of its functional group [11,12] and the disability of forming a flexible membrane. APEs that satisfy the requirement of fuel cell applications should, at least, possess the following properties:

1. High ionic conductivity, no lower than 10^{-2} S/cm, expected to be greater than 10^{-1} S/cm
2. High chemical and thermal stability under fuel cell operating conditions (e.g., 80°C)
3. High mechanical strength and low swelling degree when used in form of a flexible membrane whose thickness is expected to be as small as possible
4. Capability of being used in form of solution (called the ionomer solution), so as to facilitate the preparation of polymer-impregnated electrodes and the membrane-electrode assembly (MEA)

To the best of our knowledge, there has hitherto been no APE completely fulfilling the above requirements. Therefore the most important and urgent mission at present is to develop high-performance APEs that fully suit the fuel cell use.

15.2 Challenges for APE Synthesis and Design

15.2.1 Three Major Approaches for APE Synthesis

In the past decade, there have been literature reports on the synthesis of APEs, which can be categorized into three major approaches:

1. Physical grafting. By radiating an existing membrane, reactive radicals are generated on the polymer chains onto which quaternary ammonia functional groups can then be grafted. Such an approach has been employed by Varcoe and Slade to make APEs based on perfluorocarbon membranes (Figure 15.1) [13,14]. Albeit convenient and clean, this approach seems unable to produce the ionomer solution,

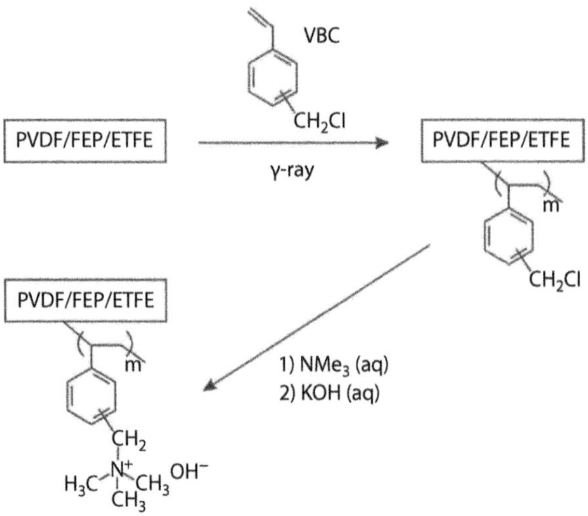

FIGURE 15.1
Synthesis of radiation-grafted PVDF, FEP, and ETFE APEs. (Reprinted with permission from Varcoe, J.R. et al. *Chem. Mater.*, 19, 2686–2693. Copyright 2007 American Chemical Society.) (Danks, T.N. et al. *J. Mater. Chem.*, 2003, 13, 712–721. Reproduced by permission of The Royal Society of Chemistry.)

which then has to be obtained from other approaches when making an MEA [15,16].

2. Chemical grafting. Quaternary ammonia functional groups are grafted fully through chemical reactions onto a linear polymer, which not only can dissolve in certain solvents (to form the ionomer solution) but also are able to form a strong and flexible membrane. This is an efficient and complete approach that has been widely adopted [3,17–19]. Polysulfone (Scheme 15.2) and its analogs are the mostly used polymer backbone, largely because it is a commercially mature product with outstanding stability and capability of forming a flexible thin film with high mechanical strength.

3. Polymerization. The APE is synthesized through polymerization reactions from quaternary ammonia containing monomers [20,21]. The structure of this type of APEs is thought to be more designable; hence this approach is favored by organic chemists like Coates and

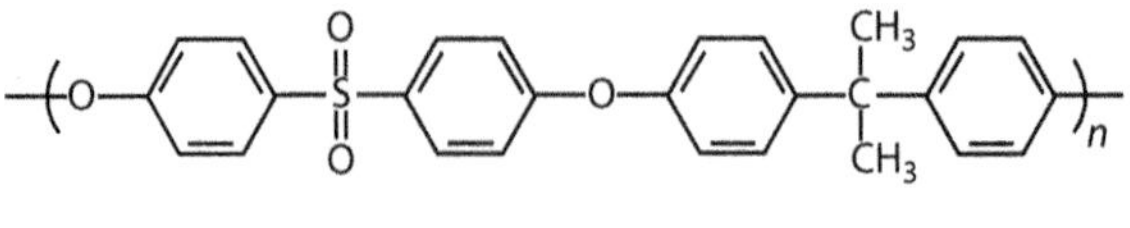

SCHEME 15.2

coworkers [22,23]. Although this method is able to produce APEs with very high ion-exchange capacity (IEC), the overall performance of the membrane, in particular the mechanical strength and the thermal stability, still remains to be further tested.

15.2.2 Realistic Challenge of APEs

In principle, it should not be difficult to find out how to or to design a stable and flexible polymer backbone onto which sufficient cationic groups are then grafted so as to attain high ionic conductivity. However, things are not so simple in reality; the ionic conductivity of polymer electrolytes is usually gained at the expense of the other aspects of performance, in particular the mechanical strength and the swelling degree of the membrane [24]. This problem is especially challenging in the APE development.

In comparison to the acidic polymer electrolytes, such as Nafion, APEs are thought to be intrinsically inferior in terms of ionic conductivity because the conductivity of OH$^-$ ($\sigma_{hydroxide}$) is lower than that of H$^+$ (σ_{proton}) in aqueous phase (for example, in extremely dilute solution, $\sigma_{proton}/\sigma_{hydroxide}$ = 1.76) [25]. In order to remedy this weakness, it is common sense to design APEs with higher IEC so as to enhance the OH$^-$ conduction [17,26–31]. However, upon increasing the IEC, the water uptake of APE will become excessive, leading to significant swelling and a decline in the mechanical strength of the membrane. APEs with high IEC can even dissolve in hot water [17,28].

To restrict the membrane swelling of APEs with high IEC, cross-linking the polymer chains, or casting the APE onto a supporting mesh, is an effective and commonly adopted solution [22,29,30]. Usually, bifunctional reagents were adopted to form cross-links between the polymer chains (see Figure 15.2 for example [19]). For instance, Coates et al. developed a ring-opening metathesis polymerization route to synthesize a cross-linked APE (Figure 15.3), which allows the polymer to possess a very high IEC (IEC = 2.3 ± 0.2 mmol/g) and thus a high ionic conductivity (68.7 ± 0.8 mS/cm at 22°C) [22]. Because the mobility of the cross-linked polymer chains is restricted,

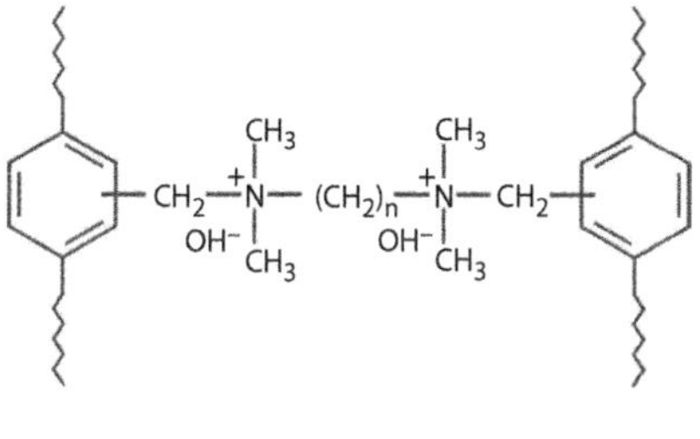

FIGURE 15.2
The structure of cross-linked APEs use dianime with various length of the alkyl chain as the cross-linking reagent. (Reprinted from *J. Power Sources*, 178, Park, J.-S. et al. Performance of solid alkaline fuel cells employing anion-exchange membranes, 620–626, Copyright 2008, with permission from Elsevier.)

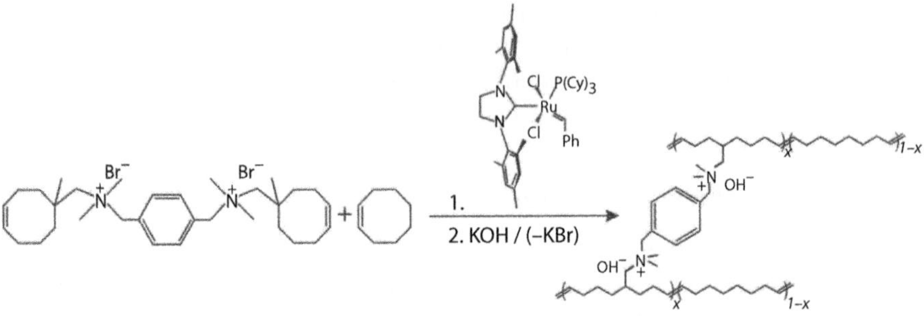

FIGURE 15.3
Synthesis of tetraalkylammonium functionalized cyclooctene based APE membrane. (Reprinted with permission from Robertson, N.J. et al. *J. Am. Chem. Soc.*, 132, 3400–3404. Copyright 2010 American Chemical Society.)

the APE membrane will thus possess higher mechanical strength and lower swelling degree, in comparison to conventional APEs. But the cross-linking has sacrificed the solubility of the polymer; thus no stable ionomer solution will be available for MEA preparations [22,29]. As for the method of using supporting meshes, it can neither be applied to stabilize the APE in electrodes nor prevent the dissolution of APEs with high IEC at elevated temperatures.

Another solution to the dilemma between ionic conductivity and mechanical stability is to keep the APE at moderate IEC and to enhance the OH⁻ conducting efficiency, which is thought to be realizable by designing the Nafion-resembling phase-separation structure in APEs [31,32].

15.2.3 Strategies for Designing High-Performance APEs

To address the above challenging problem and to design APEs specially tailored for fuel cell applications, two strategies have been proven to be particularly effective [33], as illustrated comparatively in Figure 15.4. The performance of APE is evaluated here by two key properties: the ionic conductivity and the membrane swelling degree, represented by the horizontal and vertical axes in Figure 15.4, respectively. Ideal APEs are expected to locate at the right-bottom corner of Figure 15.4, namely, with high ionic conductivity and low swelling degree.

As mentioned above, although increasing the IEC can enhance the ionic conductivity to some degree, the membrane swelling would become unacceptably enormous. This impractical approach corresponds to the change from a to b in Figure 15.4. Two strategies have been used to address this problem, namely self-cross-linking and self-aggregating, respectively. Strategy 1 is to design a short-range cross-linking mechanism for APEs with high IEC (corresponding to the change from b to c in Figure 15.4), such

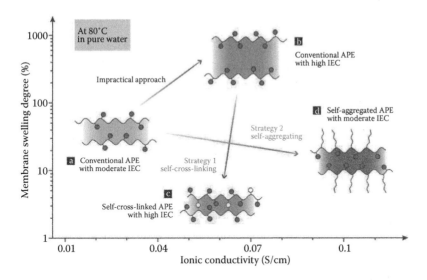

FIGURE 15.4

A road map toward advanced alkaline polymer electrolytes (APEs). To improve the ionic conductivity, increasing the ion-exchange capacity (IEC) is a commonly adopted but impractical approach (from a to b) because of the side effect of severe membrane swelling. Strategy 1 (from b to c) is to develop a short-range cross-linking mechanism, such that the membrane swelling of APEs with high IEC can be greatly reduced due to the thus-formed tight-binding structure while the ionomer solution can remain stable before solidification. Strategy 2 is to design additional hydrophobic side-chains onto the backbone of APEs with moderate IEC (from a to d), so as to drive the microscopic hydrophilic/hydrophobic domain separation and to cause the aggregation of ionic channels, thus enhancing the OH⁻ hopping conduction. (Reprinted with permission from Pan, J. et al. *Acc. Chem. Res.*, 45, 473–481. Copyright 2012 American Chemical Society.)

that the cross-linking process can only take place when the polymer chains are in close proximity, a situation that hardly happens in the solution state but occurs upon solidification. Through such a special cross-linking strategy, both the ionomer solution and the tightly cross-linked membrane can be obtained, thus removing the side effect of conventional cross-linking methods and suppressing significantly the membrane swelling of APEs with high IEC.

Strategy 2 takes a fundamentally different approach: It starts from APEs with moderate IEC so as to prevent excessive membrane swelling and then designs additional hydrophobic side-chains to drive the aggregation of hydrophilic domains (corresponding to the change from a to d in Figure 15.4), such that the local ion concentration will increase and the hopping conduction of OH⁻ can be effectively enhanced. As can be seen, such a self-aggregating strategy can dramatically boost the ionic conductivity of APEs to reach 0.1 S/cm at 80°C.

15.3 Development of High-Performance APEs

15.3.1 Implementations for the Self-Cross-Linking and Self-Aggregating Design of APEs

Both the above strategic designs of advanced APEs can be realized by a structural modification of the quaternary ammonia polysulfone (QAPS, Scheme 15.3), a conventional APE that has been achieved in our previous studies [4,17]. The synthesis of QAPS involves three major steps (Figure 15.5): (1) The polysulfone backbone is grafted with chloromethyl groups to produce a reaction intermediately, denoted as CMPS. (2) The CMPS then reacts with the trimethylamine to produce the chloride form of QAPS. (3) Replacing the Cl⁻ with OH⁻, the final form of QAPS results. The grafting degree of chloromethyl groups in the first step is controllable in our synthetic method and has mostly determined the IEC of the resulting QAPS.

To realize the self-cross-linked QAPS (denoted as *x*QAPS), an additional synthetic step is introduced: Before the CMPS is exposed to trimethylamine, a certain amount of diethylamine is added to produce tertiary amino groups (-NEt$_2$, TA) on the polysulfone backbone, and then the TA-containing CMPS is further functionalized with the quaternary ammonia group (QA), resulting in a TA-containing QAPS (Figure 15.6a). The TA possesses a low reactivity in solution at room temperature, but can react with the residual chloromethyl group on neighboring polysulfone chains in close proximity during membrane casting (Figure 15.6b). Thus such a special form of QAPS can either remain soluble in the form of solution or become cross-linked upon solidification [30].

The implementation of self-aggregated QAPS (denoted as *a*QAPS) is similar to that of *x*QAPS except that the polysulfone backbone is not additionally grafted with TA but with long alkylamine chains. Specifically, before the QA generation step, a certain amount of primary alkylamines with long hydrophobic chains, such as 1-hexanamine, is introduced to react with the CMPS; then the remaining chloromethyl is fully converted to QA (Figure 15.7). The use of primary alkylamines, rather than secondary ones, makes the grafting reaction much easier, as the reactivity of long alkylamines is relatively low. The resulting hydrophobic side-chains can hardly be regarded

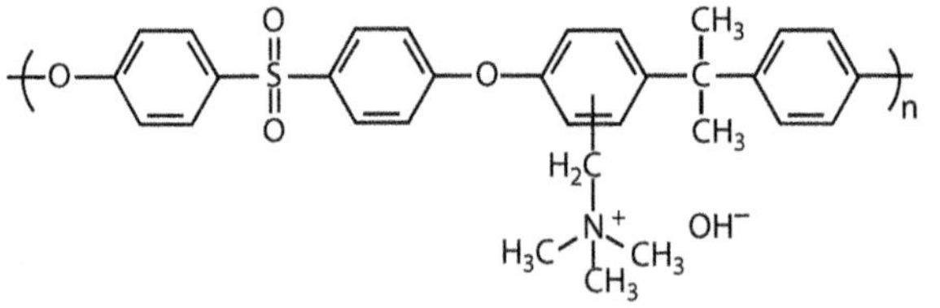

SCHEME 15.3

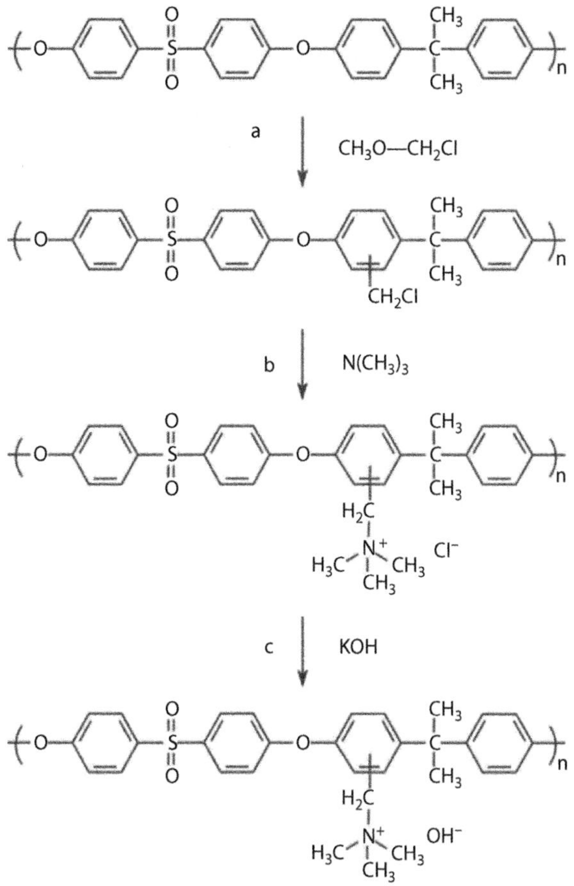

FIGURE 15.5
The synthetic route of quaternary ammonium polysulfone (QAPS). (Pan, J. et al.: High-performance alkaline polymer electrolyte for fuel cell applications. *Adv. Funct. Mater.* 2010. 20. 312–319. Copyright Wiley-VCH Verlag GmbH & Co. KGaA. Reproduced with permission.)

as a functional group in this case, but they actually impose a great impact on the structure and property of the *a*QAPS membrane as to be revealed in the following sections.

15.3.2 Remarkable Performance of *x*QAPS and *a*QAPS

Although only small structural modifications are made in the transformation from QAPS to *x*QAPS and *a*QAPS, the resulting performance improvements turn out to be remarkable. Figure 15.8 demonstrates the distinct change in the temperature-dependent membrane swelling degree. For the conventional QAPS, even its IEC is controlled to be moderate (1.07 mmol/g), the membrane

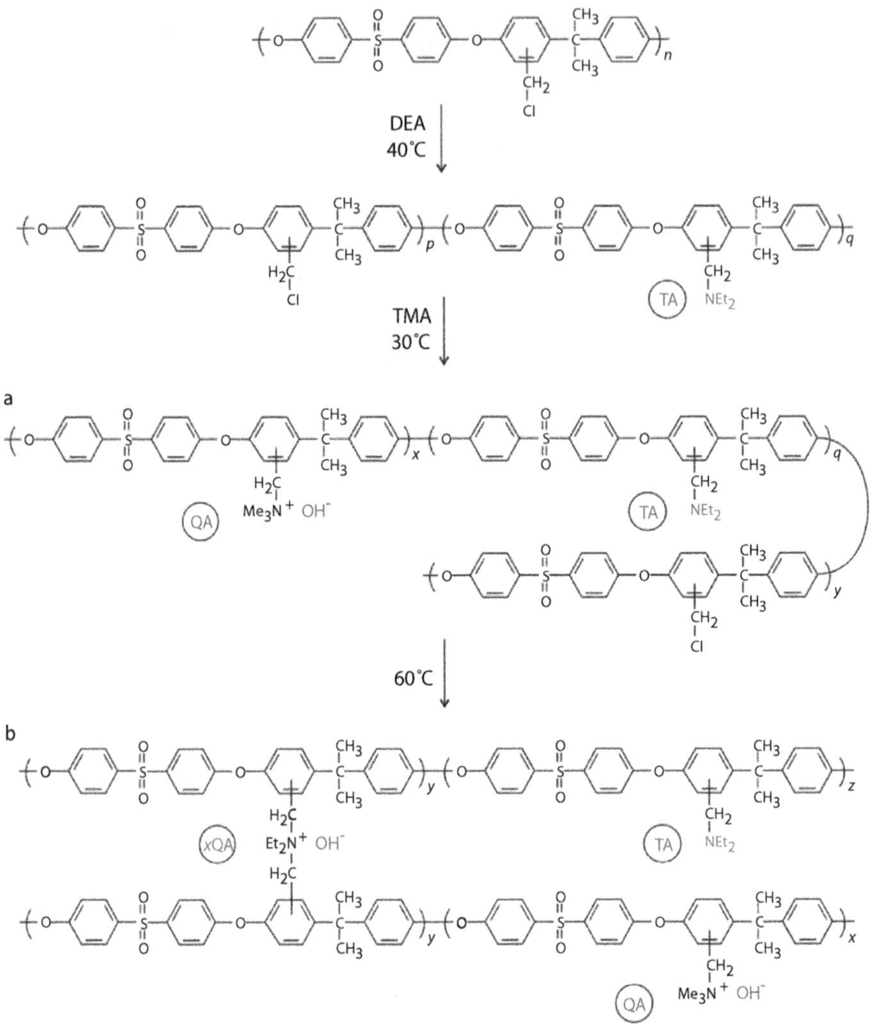

FIGURE 15.6
Synthetic route of self-cross-linked quaternary ammonium polysulfone (*x*QAPS). (Pan, J. et al. *Chem. Commun.*, 2010, *46*, 8597–8599. Reproduced by permission of The Royal Society of Chemistry.)

still swells over 25% at room temperature and expands significantly with increasing temperatures. Such an unacceptable swelling behavior has been suppressed, to a great extent, by the effective structural design of *x*QAPS and *a*QAPS. For the *a*QAPS with moderate IEC (1.02 mmol/g), its membrane swelling degree remains below 10% at room temperature and increases only slightly with temperature. The most surprising effect is caused by *x*QAPS: albeit with a high IEC (1.34 mmol/g), it does not swell at all at temperatures

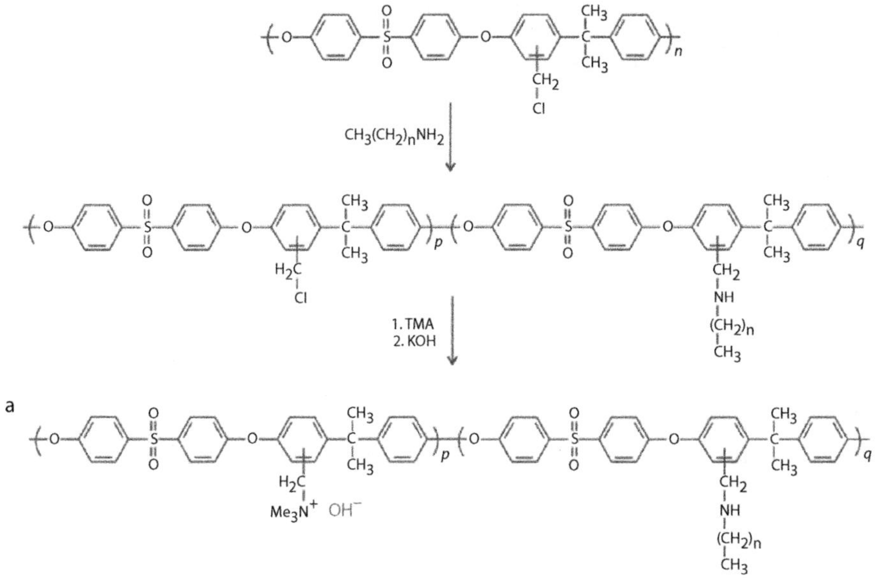

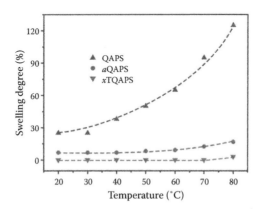

FIGURE 15.7
The synthetic route of self-aggregated quaternary ammonium polysulfone (*a*QAPS). (Reprinted with permission from Pan, J. et al. *Acc. Chem. Res.*, 45, 473–481. Copyright 2012 American Chemical Society.)

FIGURE 15.8
The membrane swelling can be effectively suppressed through either self-cross-linking or self-aggregating; the swelling degree of *x*QAPS membrane is less than 3% even at 80°C. (Reprinted with permission from Pan, J. et al. *Acc. Chem. Res.*, 45, 473–481. Copyright 2012 American Chemical Society.)

below 80°C, and even at 80°C, the membrane swelling degree is less than 3%. Such a highly anti-swelling property can only be ascribed to the tight-binding structure of the *x*QAPS membrane.

In comparison to the QAPS, the *x*QAPS possesses a higher IEC, which should lead to an obvious increase in the ionic conductivity. As revealed in Figure 15.9, however, the ionic conductivity of *x*QAPS is only slightly higher than that of QAPS at room temperature, suggesting that the tight-binding structure has significantly reduced the water uptake and thus restricts, to some extent, the ionic conduction. But upon increasing the temperature, the ionic conductivity of *x*QAPS increases faster than that of QAPS, manifesting the benefit of high IEC. The most intriguing observation in Figure 15.9 is the striking improvement in the hydroxide-conducting efficiency of *a*QAPS. Albeit with the lowest IEC, the ionic conductivity of *a*QAPS is threefold higher than that of QAPS and *x*QAPS at room temperature, clearly indicating that the OH⁻ conduction in *a*QAPS is rather efficient. Furthermore, the ionic conductivity of *a*QAPS increases largely with temperature and becomes comparable to or even slightly higher than that of Nafion at elevated temperatures, for instance, reaching 0.11 S/cm at 80°C. Meanwhile, the ionic conductivity of *a*QAPS is greater than that of the reported perfluorinated APE possessing moderate IEC and the similar structure to that of Nafion (Figure 15.10) [32]. Such an outstanding hydroxide-conducting behavior is due to the superior structural design of *a*QAPS, which also implies that high IEC is not a necessary condition for achieving the high ionic conductivity of APE, the structural factor could be more determining.

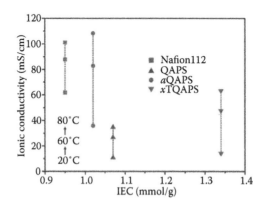

FIGURE 15.9
Ionic conductivity can be improved by increasing IEC (in the case of *x*QAPS) or, more effectively, through aggregating ionic channels (in the case of *a*QAPS). At elevated temperatures, the ionic conductivity of *a*QAPS is comparable to that of Nafion. (Reprinted with permission from Pan, J. et al. *Acc. Chem. Res.*, 45, 473–481. Copyright 2012 American Chemical Society.)

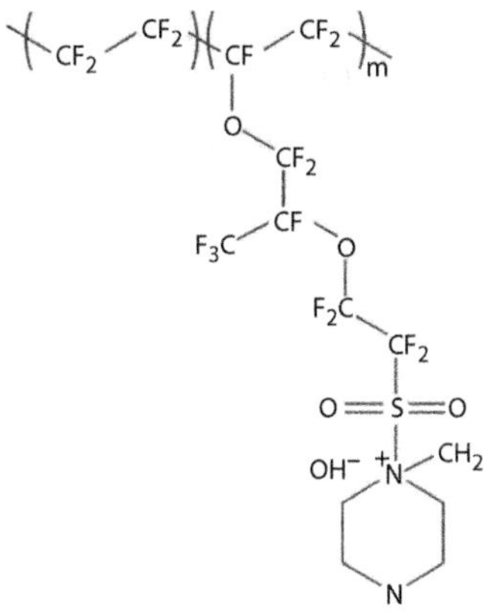

FIGURE 15.10
Illustration of the molecular structure of Nafion-like perfluorinated APE. (Jung, M.J. et al. *J. Mater. Chem.*, 2011, *21*, 6158–6160. Reproduced by permission of The Royal Society of Chemistry.)

Note that the ionic conductivity of polymer electrolytes (including APEs and Nafion) is measured in our works using the two-probe AC impedance method (Figure 15.11) rather than the four probe one [22,23] which is thought to have overestimated the ionic conductivity. Also worth mentioning is that the superior performance gained by *x*QAPS and *a*QAPS is lasting at elevated temperatures. During a test in 90°C water, both the mechanical strength and the ionic conductivity of *x*QAPS and *a*QAPS membranes show no sign of degradation over 1000 h.

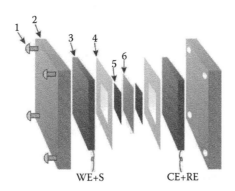

WE+S   CE+RE

1. Securing screw   2. Stainless steel block   3. Carbon block
4. Teflon spacer   5. Carbon cloth   6. Membrane

FIGURE 15.11
Apparatus for ionic conductivity test.

15.3.3 Experimental Proof for the Structural Feature of *x*QAPS and *a*QAPS

It is interesting to see the peculiarities exhibited by *x*QAPS and *a*QAPS: the *x*QAPS is exceptionally anti-swelling even with a high IEC while the *a*QAPS is highly conductive just at a moderate IEC. These unusual behaviors are the consequence of the structural modifications on conventional QAPS (Scheme 15.3) guided by the aforementioned strategies (Figure 15.4). The self-cross-linking feature of *x*QAPS is implemented by deliberately replacing a part of the QA group with a TA group, which can then react with the residual chloromethyl group on neighboring backbones to form a cross-linking QA group (denoted as *x*QA, Figure 15.6b). Thus in an *x*QAPS membrane, there are three types of amino group (QA, TA, and *x*QA), and the formation of *x*QA is a key structural feature. We have employed the x-ray photoelectron spectroscopy (XPS) to ascertain the existence of three different amino groups in *x*QAPS [30]. As revealed in Figure 15.12, the binding energy (BE) of N's 1s-electron differs clearly between the QA in QAPS and the TA in tertiary amino polysulfone (TAPS), a QA-free polymer deliberately prepared for reference. It is understandable that being a cation, the QA has a greater BE (~402 eV) than that of the TA (~400 eV). As for the *x*QAPS, its XPS signal is clearly composed of three peaks with the two low-BE ones corresponding to the QA and the TA, respectively, and the new peak emerging at around 403 eV can only be attributed to the generation of *x*QA. The greatest BE associated with *x*QA is probably because of the enhanced electron-withdrawing effect of the two benzyl groups in proximity [34].

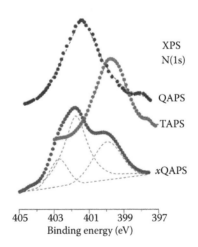

FIGURE 15.12

X-ray photoelectron spectroscopy (XPS) evidence for the existence of three types of amino group in *x*QAPS. XPS spectra of QAPS and tertiary amino polysulfone (TAPS) are also presented for reference. (Pan, J. et al. *Chem. Commun.*, 2010, 46, 8597–8599. Reproduced by permission of The Royal Society of Chemistry.)

For the *a*QAPS, the addition of long alkyl side-chains (Figure 15.7) is expected to enhance the microscopic phase separation between hydrophilic and hydrophobic domains in the membrane, such that the OH⁻ will locally aggregate and the OH⁻ hopping conduction can become more effective. Thus the enhanced hydrophilic/hydrophobic domain separation is a key structural feature of *a*QAPS. To unravel this microscopic characteristic, *a*QAPS and QAPS thin films are cast onto a Cu grid, respectively, followed by exchanging the anions in both films for I⁻, and then subjected to transmission electron microscopy (TEM) observations as manifested in Figure 15.13a

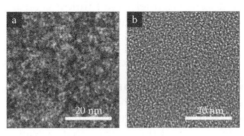

FIGURE 15.13
(a) A transmission electron microscopy (TEM) image of *a*QAPS membrane dyed with I⁻, manifesting the aggregation of the hydrophilic domains (dark spots) and the enhanced microscopic phase separation. (b) A TEM image of QAPS membrane dyed with I⁻, showing the uniform distribution of the hydrophilic/hydrophobic domains with relatively smaller size. (Reprinted with permission from Pan, J. et al. *Acc. Chem. Res.*, 45, 473–481. Copyright 2012 American Chemical Society.)

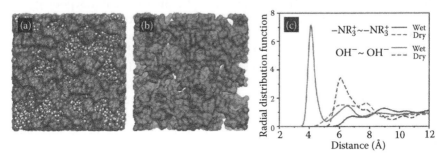

FIGURE 15.14
(See color insert.) Molecular dynamics (MD) simulations for the structure of hydrated *a*QAPS ($\lambda = 30$). (a) A view of the simulated hydrophilic/hydrophobic domain separation: The hydrophilic domains that contain water molecules (represented by red-white balls) and OH⁻ (represented by green-white balls) are intercalated in the hydrophobic network formed by the polysulfone backbone and the alkyl side-chains (represented by the blue continuum). (b) Another view of (a) with the hydrophobic network being hidden and the hydrophilic domains being represented in continuum mode so as to see the interconnections between aqueous clusters. (c) The radial distribution function (RDF) of the $-NR_3^+$ and the OH⁻ in wet and dry *a*QAPS membranes, based on which the aggregation of OH⁻ upon hydration is evident. (Reprinted with permission from Pan, J. et al. *Acc. Chem. Res.*, 45, 473–481. Copyright 2012 American Chemical Society.)

and b whereas the hydrophilic (dark spots) and hydrophobic (light spots) domains in conventional QAPS seem rather small and distribute in a uniformly mixed pattern (Figure 15.13b), the hydrophilic/hydrophobic domains in *a*QAPS are bigger in size and distribute in an aggregated pattern (Figure 15.13a). Furthermore, it is not difficult to find the interconnections among the aggregated hydrophilic domains. In other words, Figure 15.14b provides clear and compelling evidence for the structural peculiarity of *a*QAPS: Upon introducing the hydrophobic side-chains, the OH⁻ containing hydrophilic domains are driven to aggregate to each other, forming larger and more efficiently interconnected ionic channels. We think it is such a superior structure that renders *a*QAPS highly efficient of OH⁻ conduction.

15.3.4 Structural Insights from Molecular Dynamics Simulations

The above-observed aggregation structure of hydrophilic domains in *a*QAPS resembles, to a large extent, the well-known Gierke model proposed for reasoning the highly efficient proton conduction in Nafion [35]. Yet the molecular structure of *a*QAPS is quite different from that of Nafion: For *a*QAPS, the hydrophilic QA groups are closely attached on the relatively rigid polysulfone backbone, which equips separately with hydrophobic alkyl side-chains; while for Nafion, the hydrophilic sulfonic groups are positioned at the end of the long, flexible, hydrophobic side-chains grafted on the polytetrafluoroethylene (PTFE) backbone. It seems that the mechanism of forming the hydrophilic/hydrophobic domain separation should be different for these two distinct polyelectrolytes.

In order to gain molecular insights into the microscopic morphology of *a*QAPS, a series of molecular dynamics (MD) calculations have been performed using methods similar to the structural simulation of Nafion [36–38], and typical results of fully hydrated *a*QAPS are exemplified in Figure 15.14. Figure 15.14a is a representative view of the simulated hydrophilic/hydrophobic separation morphology, showing that the OH⁻ containing water clusters are intercalated in the hydrophobic network formed by the polysulfone backbone and alkyl side-chains. The aqueous clusters in *a*QAPS are smaller in size than that in Nafion [35] but well connected to each other at sufficient hydration level, as illustrated in Figure 15.14b.

In addition to the observation of well-established ionic channels in *a*QAPS, MD simulations have also found important evidence for the aggregation of OH⁻ in hydrated *a*QAPS. According to the radial distribution function (RDF) analysis (Figure 15.14c), the OH⁻ is found to be significantly aggregated upon hydration, indicating that the OH⁻ is strongly solvated and dissociated far away from the cationic group. Such an aggregation effect, however, is not seen for the QA functional groups, which are still closely bound to the rigid backbone. Thus the mechanism of forming hydrophilic/hydrophobic domain separation in *a*QAPS might be fundamentally different from that in Nafion. In the case of Nafion, the separated, broad hydrophilic domains are thought to form through the organization of loose side-chains and the aggregation of sulfonic functional groups [39]. While in the case of *a*QAPS, the formation of ionic channels does not rely on the organization of QA functional groups, but on the dissociation of hydrated OH⁻ into the free space in *a*QAPS. Such free spaces are small and uniformly distributed in conventional QAPS, but enlarged and aggregated in *a*QAPS because of the loose and random stack of the backbones with long side-chains. Nonetheless, the introduction of long, hydrophobic side-chains does cause an enhancement in the microscopic phase separation between the hydrophilic and hydrophobic domains, which is in line with the original design strategy for *a*QAPS.

15.4 Degradation and Chemical Stability of APEs

In addition to the conductive and swelling properties, much attention has been focused on the stability of the bonded organic cations in APEs at high pH. Generally speaking, the organic cations are electrophiles, which are vulnerable to the strong nucleophile hydroxide ion. Among the different cations commonly used in APEs, ammonium groups were considered to be more stable than the phosphonium or sulfonium ones. As a consequence, substituted ammonium cations become the most popular ionic groups in APEs.

Quaternary ammonium (QA) groups are prone to be attacked by OH⁻, especially at elevated temperatures, and suffer from degradation though Stevens or Sommelet-Hauser rearrangement (Figure 15.15) [40], S_N2 nucleophilic substitution (Figure 15.16) [6,12], and E1 or E2 (Hofmann) elimination (Figure 15.17) mechanisms [11,41]. The degradation of QA groups would cause significant IEC decline of APEs and further leads to the decrease in ionic conductivity.

Recent studies combining experimental and computational modeling on the degradation of QA groups arrive at several important findings, which are summarized as follows [11,12,42,43]:

(1) Degradation reactions were slower at higher water contents. The hydroxide ions appear to be much less nucleophilic compared with the OH⁻ and cations are solvated.

(2) In the absence of β-hydrogen atoms, the S_N2 nucleophilic substitution on the α-hydrogen atoms through OH⁻ attack contribute little to the degradation of QA groups.

(3) Benzyltrimethylammonium group is relatively stable under strong alkaline conditions.

(4) Hofmann elimination is the most severe degradation mechanism when β-hydrogen atoms are present in QA groups. However, the carbon chain length and the number of β-hydrogen atoms would affect the efficiency of Hofmann elimination.

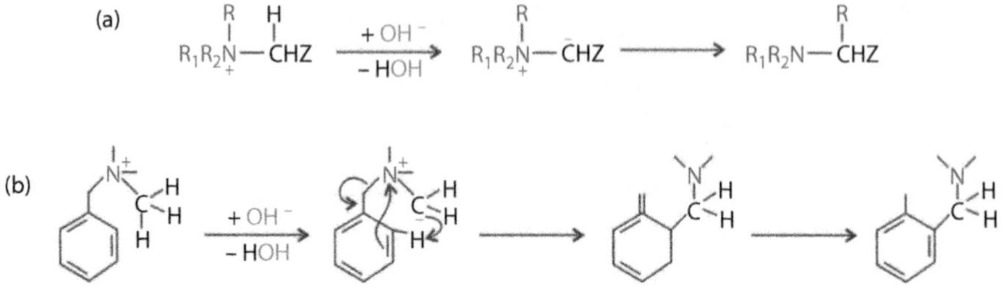

FIGURE 15.15
The degradation mechanisms of quaternary ammonium groups under alkaline conditions by Stevens rearrangement (a) and Sommelet-Hauser rearrangement (b). (Reprinted with permission from Ghigo, G. et al. *J. Org. Chem.*, 75, 3608–3617. Copyright 2010 American Chemical Society.)

In order to improve the chemical stability, other cationic groups, such as tris(trimethoxyphenyl) functionalized quternary phosphonium, guanidinium, methylated melamine, and imidazolium, have been employed in APEs. Most researchers believe that imidazolium cation groups would be more stable than the QA groups because the positive charge is delocalized in the imidazole ring due to the large π-conjugated system. However, Varcoe and colleagues have noted that the benzylmethylimidazolium group (Figure 15.18a) is less stable than benzyltrimethylammonium group (Figure 15.18b) in alkaline conditions and at moderate temperature of 60°C [44].

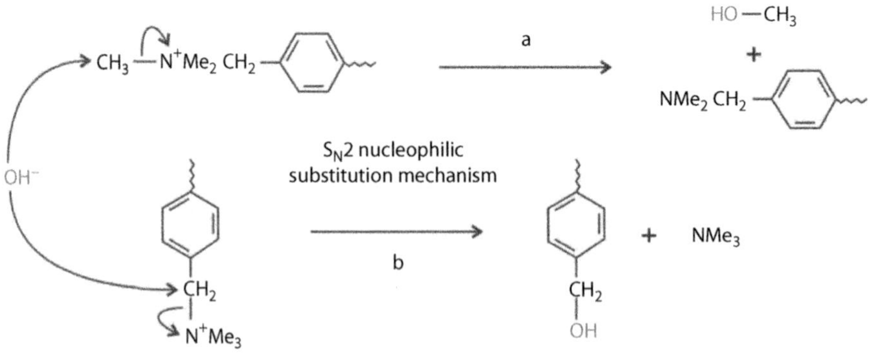

FIGURE 15.16
Degradation mechanism for quaternary ammonium groups under alkaline conditions by S_N2 nucleophillic substitution. (Varcoe, J.R., and Slade, R.C.T.: Prospects for alkaline anion-exchange membranes in low temperature fuel cells, *Fuel Cells*. 2005. 5. 187–200. Copyright Wiley-VCH Verlag GmbH & Co. KGaA. Reproduced with permission.) (Reprinted with permission from Chempath, S. et al. *J. Phys. Chem. C*, 112, 3179–3182. Copyright 2008 American Chemical Society.)

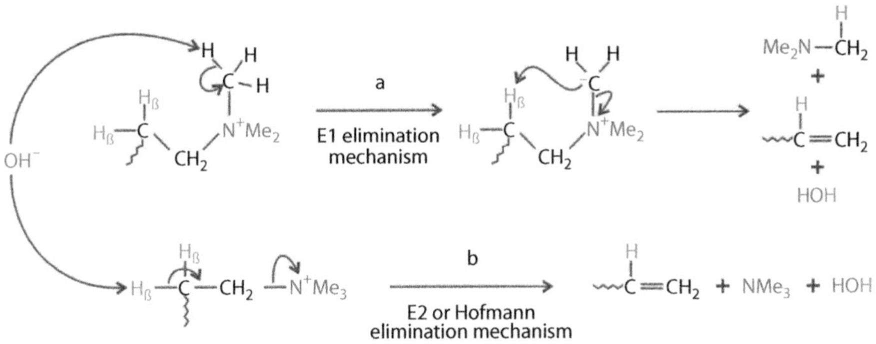

FIGURE 15.17
Degradation mechanism for quaternary ammonium groups under alkaline conditions by E1 elimination (a) and E2 (Hofmann) elimination (b). (Reprinted with permission from Chempath, S. et al. *J. Phys. Chem. C*, 114, 11977–11983. Copyright 2010 American Chemical Society.) (Reprinted with permission from Cope, A.C., and Mehta, A.S., *J. Am. Chem. Soc.*, 85, 1949–1952. Copyright 1963 American Chemical Society.)

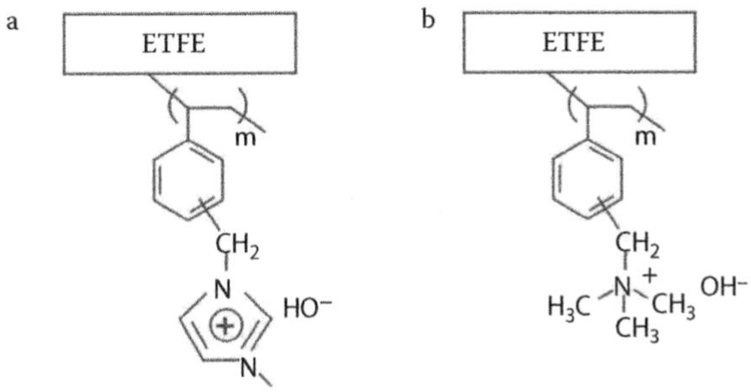

FIGURE 15.18
Chemical structures of the ETFE-based APEs grafted with benzylmethylimidazolium (a) and benzyltrimethylammonium (b). (Deavin, O.I. et al. *Energy Environ. Sci.*, 2012, 5, 8584–8597. Reproduced by permission of The Royal Society of Chemistry.)

15.5 Conclusions and Perspectives

In principle, there are two major approaches to improving the ionic conductivity of alkaline polymer electrolytes (APEs): (1) increasing the ion-exchange capacity (IEC) of the polyelectrolyte and (2) increasing the effective mobility of the charge carrier (OH⁻). While the first approach seems relatively easier to realize, the second one requires superior design of the ionic channels in the polymer. In reality, the first approach turns out not to be as simple as expected; the APE will absorb excessive water when increasing the IEC, leading to severe membrane swelling and mechanical-strength decline. To address the problem, making a cross-linked APE membrane is a usually adopted measure, but it leaves the ionomer solution unavailable for PEFC assembling.

Our strategy for solving the problem of the first approach is to design a "smart" cross-linking mechanism: The cross-linking process will not happen when the APE is in a solution state but occurs upon solidification, such that the ionomer solution and the cross-linked APE membrane can both be obtained. Such a strategy has been realized in our research group by designing the self-cross-linked quaternary ammonia polysulfone (xQAPS), in which the tertiary amino group acts as a short-range cross-linker and can bind with neighboring polymer backbones in close proximity. The resulting tight-binding structure renders the xQAPS exceptionally anti-swelling with its swelling degree being less than 3% at 80°C.

To further enhance the ionic conductivity of APEs, we have focused on designing efficient ionic channels for OH⁻ conduction, namely, taking the aforementioned second approach. Our strategy is to design additional hydrophobic side-chains to drive the microscopic phase separation and the aggregation of ionic channels. Such a strategy has been realized in our research group by designing the self-aggregated quaternary ammonia polysulfone (aQAPS). Both the experimental characterization and the molecular dynamics (MD) simulation have revealed that the ionic channels are enlarged and aggregated in aQAPS. This superior structure makes the aQAPS as conductive as Nafion at fuel-cell operating temperatures; for example, its ionic conductivity is greater than 0.1 S/cm at 80°C.

By now, we think we have successfully passed the first phase of the development of advanced APEs. Through the self-cross-linking and the self-aggregating approaches presented in this chapter, APEs can now be made as good as the acidic polymer electrolyte in terms of the ionic conductivity and the dimensional stability of membrane, which have fulfilled the basic requirements of fuel cell applications. Yet there are still challenges for APEs to become completely suitable for real applications. For instance, the chemical stability of QA groups and the oxidative degradation of APEs under fuel-cell operating conditions have been concerns; how to enhance the chemical stability and oxidation tolerance of APEs remains an open subject. Relevant first-principle calculations are underway in our research group with attempts to identify the possible structural weakness and to find out the resolution.

References

1. Jacobson, M.Z., Colella, W.G. and Golden, D.M. 2005. Cleaning the air and improving health with hydrogen fuel-cell vehicles. *Science* 308, 1901–1905.
2. Zhang, J., Sasaki, K. and Adzic, R.R. 2007. Stabilization of platinum oxygen-reduction electrocatalysts using gold clusters. *Science* 315, 220–222.

3. Gu, S., Cai, R., Luo, T., Chen, Z., Sun, M., Liu, Y., He, G. and Yan, Y. 2009. A soluble and highly conductive ionomer for high-performance hydroxide exchange membrane fuel cells. *Angew. Chem., Int. Ed.* 48, 6499–6502.

4. Lu, S., Pan, J., Huang, A., Zhuang, L. and Lu, J. 2008. Alkaline polymer electrolyte fuel cells completely free from noble metal catalysts. *Proc. Natl. Acad. Sci. U.S.A.* 105, 20611–20614.

5. Asazawa, K., Yamada, K., Tanaka, H., Oka, A., Taniguchi, M. and Kobayashi, T. 2007. A platinum-free zero-carbon-emission easy fuelling direct hydrazine fuel cell for vehicles. *Angew. Chem., Int. Ed.* 46, 8024–8027.

6. Varcoe, J.R. and Slade, R.C.T. 2005. Prospects for alkaline anion-exchange membranes in low temperature fuel cells. *Fuel Cells* 5, 187–200.

7. Wang, Y., Li, L., Hu, L., Zhuang, L., Lu, J. and Xu, B. 2003. A feasibility analysis for alkalinemembrane direct methanol fuel cell: Thermodynamic disadvantages versus kinetic advantages. *Electrochem. Commun.* 5, 662–666.

8. Gülzow, E. and Schulze, M. 2004. Long-term operation of AFC electrodes with CO_2 containing gases. *J. Power Sources* 127, 243–251.

9. Unlu, M., Zhou, J. and Kohl, P.A. 2009. Anion exchangemembrane fuel cells: Experimental comparison of hydroxide and carbonate conductive ions. *Electrochem. Solid-State Lett.* 12, B27–B30.

10. Adams, L.A., Poynton, S.D., Tamain, C., Slade, R.C.T. and Varcoe, J.R. 2008. A carbon dioxide tolerant aqueous-electrolyte-free anion-exchange membrane alkaline fuel cell. *ChemSusChem* 1, 79–81.

11. Chempath, S., Boncella, J.M., Pratt, L.R., Henson, N. and Pivovar, B.S. 2010. Density functional theory study of degradation of tetraalkylammonium hydroxides. *J. Phys. Chem. C* 114, 11977–11983.

12. Chempath, S., Einsla, B.R., Pratt, L.R., Macomber, C.S., Boncella, J.M., Rau, J.A. and Pivovar, B.S. 2008. Mechanism of tetraalkylammonium headgroup degradation in alkaline fuel cell membranes. *J. Phys. Chem. C* 112, 3179–3182.

13. Varcoe, J.R., Slade, R.C.T., Yee, E.L.H., Poynton, S.D., Driscoll, D.J. and Apperley, D.C. 2007. Poly(ethylene-co-tetrafluoroethylene)-derived radiation-grafted anion-exchange membrane with properties specifically tailored for application in metal-cation-free alkaline polymer electrolyte fuel cells. *Chem. Mater.* 19, 2686–2693.

14. Danks, T.N., Slade, R.C.T. and Varcoe, J.R. 2003. Alkaline anion-exchange radiation-grafted membranes for possible electrochemical application in fuel cells. *J. Mater. Chem.* 13, 712–721.

15. Tamain, C., Poynton, S.D., Slade, R.C.T., Carroll, B. and Varcoe, J.R. 2007. Development of cathode architectures customized for H_2/O_2 metal-cation-free alkalinemembrane fuel cells. *J. Phys. Chem. C* 111, 18423–18430.

16. Varcoe, J.R., Slade, R.C.T. and Yee, E.L.H. 2006. An alkaline polymer electrochemical interface: A breakthrough in application of alkaline anion-exchange membranes in fuel cells. *Chem. Commun.* 1428–1429.

17. Pan, J., Lu, S., Li, Y., Huang, A., Zhuang, L. and Lu, J. 2010. High-performance alkaline polymer electrolyte for fuel cell applications. *Adv. Funct. Mater.* 20, 312–319.

18. Wang, J., Li, S. and Zhang, S. 2010. Novel hydroxide-conducting polyelectrolyte composed of an poly(arylene ether sulfone) containing pendant quaternary guanidinium groups for alkaline fuel cell applications. *Macromolecules* 43, 3890–3896.

19. Park, J.-S., Park, S.-H., Yim, S.-D., Yoon, Y.-G., Lee, W.-Y. and Kim, C.-S. 2008. Performance of solid alkaline fuel cells employing anion-exchange membranes. *J. Power Sources* 178, 620–626.

20. Huang, A., Xia, C., Xiao, C. and Zhuang, L. 2006. Composite anion exchange membrane for alkaline direct methanol fuel cell: Structural and electrochemical characterization. *J. Appl. Sci.* 100, 2248–2251.

21. Huang, A., Xiao, C. and Zhuang, L. 2005. Synthesis and characterization of quaternized poly(4-vinylpyridine-co-styrene) membranes. *J. Appl. Polym. Sci.* 96, 2146–2153.

22. Robertson, N.J., Kostalik, H.A., IV, Clark, T.J., Mutolo, P.F., Abruña, H.D. and Coates, G.W. 2010. Tunable high performance cross-linked alkaline anion exchange membranes for fuel cell applications. *J. Am. Chem. Soc.* 132, 3400–3404.

23. Clark, T.J., Robertson, N.J., Kostalik, H.A., IV, Lobkovsky, E.B., Mutolo, P.F., Abruña, H.D. and Coates, G.W. 2009. A ring-opening metathesis polymerization route to alkaline anion exchange membranes: Development of hydroxide-conducting thin films from an ammoniumfunctionalized monomer. *J. Am. Chem. Soc.* 131, 12888–12889.

24. Tang, D., Pan, J., Lu, S., Zhuang, L. and Lu, J. 2010. Alkaline polymer electrolyte fuel cells: principle, challenges, and recent progress. *Sci. China: Chem.* 53, 357–364.

25. Atkins, P.W. 1998. *Atkins' Physical Chemistry*, 6th ed. Oxford University Press: Oxford, Melbourne, Tokyo, Chapter 24, p. 738.

26. Tanaka, M., Fukasawa, K., Nishino, E., Yamaguchi, S., Yamada, K., Tanaka, H., Bae, B., Miyatake, K. and Watanabe, M. 2011. Anion conductive block poly(arylene ether)s: Synthesis, properties, and application in alkaline fuel cells. *J. Am. Chem. Soc.* 133, 10646–10654.

27. Tanaka, M., Koike, M., Miyatake, K. and Watanabe, M. 2010. Anion conductive aromatic ionomers containing fluorenyl groups. *Macromolecules* 43, 2657–2659.

28. Hibbs, M.R., Fujimoto, C.H. and Cornelius, C.J. 2009. Synthesis and characterization of poly(phenylene)-based anion exchange membranes for alkaline fuel cells. *Macromolecules* 42, 8316–8321.

29. Gu, S., Cai, R. and Yan, Y. 2011. Self-crosslinking for dimensionally stable and solvent-resistant quaternary phosphonium based hydroxide exchange membranes. *Chem. Commun.* 47, 2856–2858.

30. Pan, J., Li, Y., Zhuang, L. and Lu, J. 2010. Self-crosslinked alkaline polymer electrolyte exceptionally stable at 90°C. *Chem. Commun.* 46, 8597–8599.

31. Elabd, Y.A. and Hickner, M.A. 2011. Block copolymer for fuel cells. *Macromolecules* 44, 1–11.

32. Jung, M.J., Arges, C.G. and Ramani, V. 2011. A perfluorinated anion exchange membrane with a 1,4-dimethylpiperazinium cation. *J. Mater. Chem.* 21, 6158–6160.

33. Pan, J., Chen, C., Zhuang, L. and Lu, J. 2012. Designing advanced alkaline polymer electrolytes for fuel cell applications. *Acc. Chem. Res.* 45, 473–481.

34. Briggs, D. 1998. *Surface Analysis of Polymer by XPS and Static SIMS*. Cambridge University Press: Cambridge, U.K., Chapter 3.

35. Gierke, T.D., Munn, G.E. and Wilson, F.C. 1981. Themorphology in Nafion perfluorinatedmembrane products, as determined by wide- and small-angle X-ray studies. *J. Polym. Sci., Polym. Phys. Ed.* 19, 1687–1704.

36. Jang, S.S., Molinero, V., Çağın, T. and Goddard, W.A., III 2004. Nanophase-segregation and transport in Nafion 117 from molecular dynamics simulations: Effect of monomeric sequence. *J. Phys. Chem. B* 108, 3149–3157.

37. Mayo, S.L., Olafson, B.D. and Goddard, W.A., III 1990. Dreiding: A generic force field for molecular simulations. *J. Phys. Chem.* 94, 8897–8909.

38. Levitt, M., Hirshberg, M., Sharon, R., Laidig, K.E. and Daggett, V. 1997. Calibration and testing of a water model for simulation of the molecular dynamics of proteins and nucleic acids in solution. *J. Phys. Chem. B* 101, 5051–5061.

39. Mauritz, K.A. and Moore, R.B. 2004. State of understanding Nafion. *Chem. Rev.* 104, 4535–4585.

40. Ghigo, G., Cagnina, S., Maranzana, A. and Tonachini, G. 2010. The mechanism of the Stevens and Sommelet-Hauser rearrangements: A theoretical study. *J. Org. Chem.* 75, 3608–3617.

41. Cope, A.C. and Mehta, A.S. 1963. Mechanism of the Hofmann elimination reaction: An ylide intermediate in the pyrolysis of a highly branched quaternary hydroxide. *J. Am. Chem. Soc.* 85, 1949–1952.

42. Long, H., Kim, K. and Pivovar, B.S. 2012. Hydroxide degradation pathways for substituted trimethylammonium cations: A DFT study. *J. Phys. Chem. C* 116, 9419–9426.

43. Edson, J.B., Macomber, C.S., Pivovar, B.S. and Boncella, J.M. 2012. Hydroxide based decomposition pathways of alkyltrimethylammonium cations. *J. Membr. Sci.* 399–400, 49–59.

44. Deavin, O.I., Murphy, S., Ong, A.L., Poynton, S.D., Zeng, R., Herman, H. and Varcoe, J.R. 2012. Anion-exchange membranes for alkaline polymer electrolyte fuel cells: Comparison of pendent benzyltrimethylammonium- and benzylmethylimidazolium- head-groups. *Energy Environ. Sci.* 5, 8584–8597.

16

High-Temperature Inorganic Proton Conductors for Proton Exchange Membrane Fuel Cells

Haolin Tang, Junrui Li, and Mu Pan

CONTENTS

16.1 Introduction 505
16.2 Inorganic Protonic Acids in Liquid Phase 507
    16.2.1 Acid-Doped Polymer Membranes...507
    16.2.2 Phosphoric Acid–Functionalized Inorganic Solids 508
    16.2.3 Antimonic Acid as Additives in Polymers.................................509
16.3 Inorganic Solid Acid Composites 509
    16.3.1 Family of $A(H,D)_2PO_4$-Type (A = K^+, Rb^+, Cs^+, Tl^+, etc.) Composites...510
    16.3.2 $A(H,D)_2SO_4$-Type (A = K^+, Rb^+, Cs^+, Tl^+, etc.) Family.................510
    16.3.3 Metal Diphosphates...512
16.4 Inorganic Oxides Ceramic Proton Conductors 512
    16.4.1 Yttria-Doped Ceramic Proton Conductors.................................512
    16.4.2 Porous Ceramic Proton Conductors and Periodic Conductive Structures..513
16.5 Heteropoly Acid 515
16.6 Summary and Outlook 517
References..518

16.1 Introduction

Among various fuel cells, proton exchange membrane fuel cells (PEMFCs) are most suitable for applications ranging from portable electronic devices and vehicle transportation to stationary power generation [1]. Compared to other sorts of fuel cells, PEMFCs have promising potential for commercialization, providing a high power density, compact design, and light weight [2,3].

The power density of PEMFCs is closely related to the proton conductivity of proton exchange membranes (PEMs), membrane thickness, and utilization rate or efficiency of catalyst. Other issues that also affect the performance include water management; CO poisoning [4]; and purity of hydrogen, refor-mate, and methanol fuels [5,6]. Increasing the operating temperatures of PEMFCs above 100°C has significant advantages and is one of the increas-ingly important areas in fuel cells [7,8]. The high operation temperature allows for enhanced cathode kinetics, elevated tolerance of the catalyst to contaminants, and easier heat management [9].

The state-of-the-art PEM is a perfluorosulfonic acid (PFSA)–based Nafion membrane due to its good proton conductivity performance in a saturated atmosphere and satisfactory chemical and physical stability [10]. However, the membrane water uptake and the consequent high proton conductivity decrease considerably at low relative humidity (RH), which prevents its uti-lization under high temperatures [11].

There are several approaches to developing new PEMs with adequate per-formance working under high temperatures and low RH. One is to synthe-size inorganic-organic composite membranes, in which the inorganic moiety

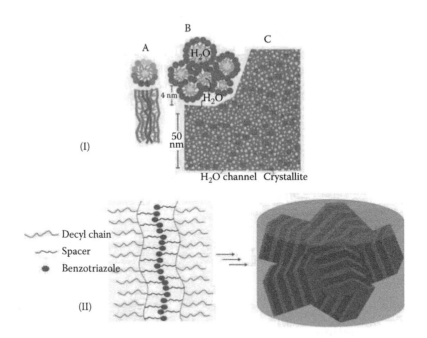

FIGURE 16.1

(I) A. Two views of an inverted-micelle cylinder with the polymer backbones on the outside and the ionic side groups lining the water channel. Shading is used to distinguish chains in front and in the back. B. Schematic diagram of the approximate hexagonal packing of several inverted-micelle cylinders. C. Cross sections through the cylindrical water channels and the Nafion crystallites in the noncrystalline Nafion matrix. (From Schmidt-Rohr, K. and Chen, Q. *Nat. Mater.*, 7, (1), 75–83, 2008.) (II) An illustration of the proposed structure of comb polymers with amphoteric proton-transfer functionalities that can self-assemble into organized supra-molecular structures, and the consequently constructed proton transport channels (benzotri-azole parts). (From Chen, Y.B. et al. *Nat. Chem.*, 2, (6), 503–508, 2010.)

is the proton conductor, and new polymers act as the matrix [12]. The second approach is to incorporate inorganic particles, which are hydrophilic, and once hydrated, can reserve water in a relatively low humidity. Moreover, novel inorganic compounds independent from the water molecules as proton vehicles, which obey a totally different transporting mechanism from Nafion, are also promising candidates for high-temperature PEMFCs. As shown in Figure 16.1, the structure of the proton conductors and the proton transport nanochannels play a dominant role in the proton conductivity, thus the materials with ordered nanostructures and a rigid porous matrix, such as porous silica [13], proton conducting glasses [14,15], mesoporous zirconium phosphate [16] and so on, have been recently studied and employed in PEMFCs operated at elevated temperature.

These inorganic materials have drawn intense attention because they are free from the restriction of glass transition and decomposition temperatures generally associated with polymeric materials and exhibit chemical stability under high temperatures [17,18]. This chapter aims to give a critical review to the development of these promising inorganic proton conductors and provide evidence for researchers in related fields.

16.2 Inorganic Protonic Acids in Liquid Phase

Inorganic acid proton-conducting compounds, such as phosphoric acid, silicic acid, and antimonic acid, have good proton conductivity properties up to 7.9×10^{-2} S cm^{-1} at 200°C and 5% RH [19]. However, due to the soluble nature of these protonic acids, they are often confined in a polymer (PBI, PEO, PAAM, PVP) or inorganic solids (SiO$_2$, zeolite) matrix to enable their use as PEMs in PEMFCs.

16.2.1 Acid-Doped Polymer Membranes

Phosphoric acid–doped polybenzimidazole (PBI) has been investigated as a very promising candidate as a high-temperature proton conductive material. With the unique proton mechanism resulting from the self-ionization and self-dehydration of phosphoric acid molecules, namely the tendency of five phosphoric acid molecules to self-ionize and form into $H_4PO_4^+$, H_3O^+, $H_2PO_4^-$, and $H_2P_2O_7^{2-}$ even in the absence of water, phosphoric acid–doped PBI can operate at low RH and elevated high temperatures as compared to Nafion membranes. Consequently, an increase of the conductivity as a function of the temperatures has been observed above 100°C under an almost dry condition. A dramatic increase has been observed with the increase in the RH, especially at high temperatures. He et al. [19] has also found that the conductivity of phosphoric acid–doped PBI membranes are dependent on the acid-doped capacity. Figure 16.2 shows the conductivities of phosphoric

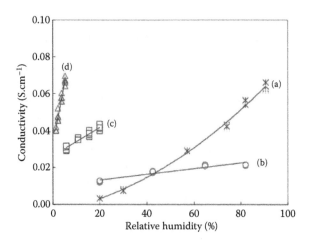

FIGURE 16.2
Conductivity vs. RH for Nafion 117 and acid-doped PBI membranes at a H_3PO_4 doping level of 5.6. (a) Nafion 117, 50°C; (b) PBI, 80°C; (c) PBI, 140°C; (d) PBI, 200°C. (From He, R.H. et al. *J. Membr. Sci.*, 226, (1–2), 169–184, 2003.)

acid and the PBI/phosphoric acid hybrid membranes measured at different temperatures.

Although phosphoric acid–doped PEMFCs can run under temperatures from 120°C to 200°C, leaching of phosphoric acid molecules during operation is still a significant issue. According to Savadogo and Xing [20], after working for 125 min, the current density of the phosphoric acid–polymer electrolyte decreases to about 60% of the original as a result of the flushing function of the generated water. Another problem is the chemical stability of the polymer matrix. Exposed at relatively low temperatures of 68°C for 30 min, structural damage appeared in the PBI matrix due to free radicals attacking with the existence of H_2O_2 and Fe^{2+} [21]. PBI and other PBIs can also be impregnated with different aqueous strong acids, like H_2SO_4, $HClO_4$, HCl, and HBr [22], in which the acids are responsible for the proton transport ability, and PBI plays the role of a matrix to provide sufficient strength and processibility for membranes.

16.2.2 Phosphoric Acid–Functionalized Inorganic Solids

Acid-functionalized proton conductors based on inorganic solids, such as porous silica, are highly accessible by pore filling [23], post grafting [24], and one-pot co-condensing [25]. Jin et al. [26] reported the preparation of a phosphonic-acid functionalized porous silica nanosphere with a conductivity ranging from 3.0×10^{-4} S cm^{-1} at 20% RH to 0.015 S cm^{-1} at 100% RH at 130°C. The most interesting observation is that the proton conductivity at low humidity conditions can be markedly enhanced by the unique morphologies of high specific surface areas as shown in Figure 16.3. In general, proton

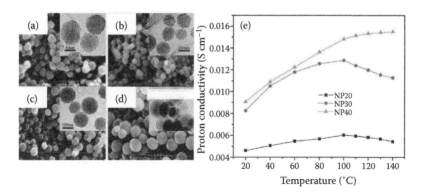

FIGURE 16.3
SEM and TEM (inset) images of (a) NP10, (b) NP20, (c) NP30, and (d) NP40, where NP10, NP20, NP30, and NP40 stands for the nominal P/Si molar ratio of 10%, 20%, 30%, and 40% in the acidified product, respectively, (e) proton conductivity of phosphonic acid functionalized silica nanospheres as a function of temperature at 100% RH. (From Jin, Y.G. et al. *J. Phys. Chem. C*, 113, (8), 3157–3163, 2009.)

conductivities of this type of proton conductors depend largely on the content of phosphoric acid and the porous structures. However, its maximum proton conductivity is 10^{-2} S·cm^{-1}, relatively lower as compared to Nafion membrane (maximized at 10^{-1} S·cm^{-1}) [27]. Thus the challenge is to immobilize phosphoric acid as much as possible without destroying the porous structures.

16.2.3 Antimonic Acid as Additives in Polymers

Antimonic acid with a pyrochlore-like structure [28] is characterized by the three-dimensional network of a corner-sharing $[Sb_2O_6]^{2-}$ octahedral. The interconnected cavities provided by this arrangement can be occupied by acid and water, which give rise to their proton conductive capability. Amarilla et al. [29] employed antimonic acid as an additive in polyvinylidene fluoride (PVDF) and sulfonated polystyrene (SPS) to form a flexible polymer with a thickness of about 200 μm. This type of hybrid membrane possesses a stable structure without significant swelling after hydration, making it quite stable in the alternate circles of hydration and dehydration. However, its application in PEMFCs under elevated temperatures needs further study due to the dependence of the antimonic acid's conductivity on the relative humidity [30].

16.3 Inorganic Solid Acid Composites

Inorganic solid acid composites, such as $CsHSO_4$, CsH_2PO_4 and their derivations, have drawn much considerable attention due to their superprotonic properties under operating temperatures ranging from 100°C to 300°C [31]. These superprotonic properties derive from the phase transition by which the proton conductivity may augment with several orders of magnitude, peaking at 10^{-2} and 10^{-3} S cm^{-1}. What makes these composites unique is that the proton conductivity relies on their mobility but not on water, eliminating the problem of humidifying the electrolyte [32].

16.3.1 Family of $A(H,D)_2PO_4$-Type (A = K$^+$, Rb$^+$, Cs$^+$, Tl$^+$, etc.) Composites

Among these composites, KH_2PO_4 (KDP) is one of the most prominent, being a member of the large family of $A(H,D)_2PO_4$-type (A = K$^+$, Rb$^+$, Cs$^+$, Tl$^+$, etc.). KDP molecules can transform between a ferroelectric phase, paraelectric phase, and superprotonic phase. At high temperatures, KDP exhibits a high proton conductivity up to 10^{-2} S cm^{-1} in the form of superprotonic phase [33]. However, the high dependence of conductivity on the phase transition limit its application in fuel cells; especially in the initial starting stage of a cell when the temperature is quite low, the cell may not be able to start owning to the low conductivity.

CsH_2PO_4 is a typical solid acid compound with a KDP-like structure that shows favorable proton conductivity at high temperatures and low-temperature ferroelectric properties. Thermal stability, conductivity at ambient pressure, and high pressure of CsH_2PO_4 have been studied by Boysen et al. [34]. The results show that the phase transition occurs at 228°C ± 2°C, and the conductivity increases sharply above this temperature. High pressure has also been found to facilitate the stability of the superprotonic state. At a pressure of 1 GPa, the stable high-temperature phase of CsH_2PO_4 over a rather wide temperature regime has been observed [34], thus promoting the application of this type of solid acid composites as PEMs. Zhou et al. [31] studied the conductivities and the thermal properties of $MH(PO_3H)$ phosphites where M is Li$^+$, Na$^+$, K$^+$, Rb$^+$, Cs$^+$, or NH$_4^+$. According to the results, superprotonic properties have been demonstrated to exist in monoclinic phosphites $MH(PO_3H)$ whereas Na$^+$, Rb$^+$, and NH$_4^+$ salt seriously deteriorate in a dry nitrogen atmosphere upon superprotonic transition temperature because of dehydration, which is different from $KH(PO_3H)$ and $CsH(PO_3H)$, which are relatively stable. Although in the same family of periodic table of elements, the superprotonic property of Na$^+$, K$^+$, and Cs$^+$ salt have not been observed in Li$^+$ salt with an orthorhombic form. Superprotonic phase transitions are estimated to only occur in the monoclinic phases.

16.3.2 $A(H,D)_2SO_4$-Type (A = K$^+$, Rb$^+$, Cs$^+$, Tl$^+$, etc.) Family

Solid acid compounds, such as KHSO$_4$, have characteristics between the normal acids (like H_2SO_4) and the normal inorganic salts (such as K_2SO_4) with excellent proton conductive properties but hardly have been selected as members in PEMFCs largely due to their water solubility and the consequent losses under fuel cell operating conditions. In comparison to that of the $A(H,D)_2PO_4$-type family, the superprotonic conductivity of sulfates, although they have no need for humid conditions, are also highly dependent on the phase transition processes. Proton conductivities of some selected solid acid compounds increase with the elevation of temperatures as shown in Figure 16.4a.

Immobilization of these water-soluble compounds into the channels of mesoporous materials provides an efficient approach for the water-soluble issue. Dry-state $CsHSO_4$ has been successfully fixed through the confinement effects of the porous structures of silica without sacrificing its superprotonic conductivity. With the objective to enhance the processability of $CsHSO_4$, Yang mixed $CsHSO_4$ with poly(vinylidene fluoride) to form hybrid membranes [35]. However, $CsHSO_4$ has been observed to experience decomposition to Cs_2SO_4 in the presence of Pt/C catalyst in hydrogen atmosphere, making it difficult to be applied in PEMFCs.

The strong dependence of solid acid compounds on phase transition temperature severely restricts their application in PEMFCs as a result of the difficult start-up of PEMFCs under low temperatures (<100°C). To overcome this problem, Yamane et al. [32] prepared solid solutions of $(CsHSO_4)_{1-x}(CsH_2PO_4)_x$ ($x = 0.25-0.75$) by a mechanical milling method. The composites were shown

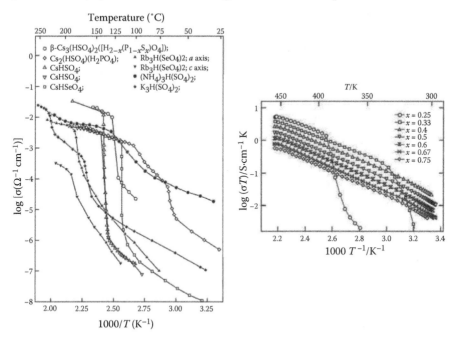

FIGURE 16.4
(a) The conductivities of selected solid acids, shown in Arrhenius form. (From Haile, S.M. et al. *Nature*, 410, (6831), 910–913, 2001.) (b) Temperature dependence of electric conductivity for $(CsHSO_4)_{1-x}(CsH_2PO_4)_x$ ($x = 0.25-0.75$) on the cooling processes. (From Yamane, Y. et al. *Solid State Ionics*, 179, (13–14), 483–488, 2008.)

to have a superprotonic phase between 293 and 420 K for $0.25 \leq x \leq 0.75$ as shown in Figure 16.4b. However, the stability of these composites depends not only on the composition but also on the humid condition.

16.3.3 Metal Diphosphates

In contrast to the oxyanion-contained solid acid compounds, metal diphosphate compounds, one of the derivations of phosphoric acid composites exhibits proton conductivity free from the dependence on phase transition. Heo et al. [36] reported the considerably high proton conductivity of In^{3+}-doped SnP_2O_7 ($Sn_{0.9}In_{0.1}P_2O_7$), 0.1 S cm^{-1} in the temperature range from 150°C to 350°C under unhumidified conditions. As a counterpart of this diphosphate, the anhydrous proton conductor $Sn_{0.95}Al_{0.05}P_2O_7$ has been homogenously distributed in the polystyrene-b-poly(ethylene/propylene)-b-polystyrene matrix to obtain a membrane with a thickness between 65 and 90 μm [37]. Despite the satisfactory tensile strength of the hybrid membrane, its proton conductivity, 3.4×10^{-3} S·cm^{-1} at 200°C, is not high enough for PEMFC applications.

16.4 Inorganic Oxides Ceramic Proton Conductors

Fast proton-conducting ceramics have been investigated to overcome the limitations of the compounds as mentioned above owing to their high chemical and mechanical durability as well as easy formation of films. However, in general, inorganic ceramic composites have relatively low proton conductivity, ranging from 1×10^{-16} to 1×10^{-11} S cm^{-1} under 120°C–370°C. Moreover, the activation energy is also high, 100–140 kJ mol^{-1} [38]. According to Equation 16.1, the activation energy (E and E_0) of the proton conducting process decreases with the increase of humidity (the water concentration, [H_2O]) in proton-conducting ceramics [39]. Moreover, Equation 16.2 demonstrates the dependence of the conductivity of ceramic conductors on temperatures, where E is an activation energy for conduction, T is the temperature, R is the gas constant, and σ_0 is a pre-exponential term called the frequency factor [40].

$$E = E_0 - n\lg\{[H^+][H_2O]\} \tag{16.1}$$

$$\sigma = \sigma_0 \exp(-E/RT) \tag{16.2}$$

16.4.1 Yttria-Doped Ceramic Proton Conductors

Ceramic high-temperature proton conductors are characterized by perovskites, which are appropriately doped to contain oxygen vacancies, for example, BaCe$_{0.9}$Y$_{0.1}$O$_{2.95}$ (BCY10) and BaCe$_{0.8}$Y$_{0.2}$O$_{2.9}$ (BCY20). These compounds react with water according to the chemical Equation 16.3 [41] to give proton conductivity.

$$(H_2O)_{gas} + V_0^{\bullet\bullet} + O_0^x \Leftrightarrow 2OH_0^* \qquad (16.3)$$

Schober [42] studied the conductive abilities of a series of composites, including BCY20-20 wt% Li$_2$CO$_3$–Na$_2$CO$_3$ (2:1). A slope of 0.55 eV in the high-temperature region and the negligible conductive differences between dry and humid conditions indicate its potential application for high-temperature operation. Anhydrous perovskite yttria-doped BaZrO$_3$ (BYZ) has been proven to exhibit high proton conductivity and been successfully fabricated into a three-dimensional crater structural thin film for proton-conducting ceramic fuel cell operated at temperatures from 350°C to 450°C [43]. Despite its high proton conductivity under high temperature, this type of proton conductor has not been reported to be applied in PEMFCs.

16.4.2 Porous Ceramic Proton Conductors and Periodic Conductive Structures

A sol-gel technique [14] was used to prepare a porous P$_2$O$_5$/SiO$_2$ glass, which exhibits a significantly high proton conductivity of 2 × 10^{-2} S cm^{-1} at room temperature. The obtained gels were shown to be porous, containing water and solvent. With humidity-sensitive surfaces, which are terminated by hydroxyl bonds, once they are exposed to the ambient air atmosphere, the porous structural glasses start to absorb water. Nogami et al. [44] concluded that the proton conduction is related to the proton hopping between hydroxyl groups and water molecules, and the conductivity increased as the water content increased. They also found that P$_2$O$_5$/SiO$_2$ glasses show a much higher proton conductivity compared with SiO$_2$ glasses owing to the fact that the protons in POH bonds possess high conductivity, which is about four orders of magnitude greater than that in the SiOH bonds expected from infrared spectra [45]. Proton conductivity of this type of porous glasses is related to the water-maintaining function of the special porous structures, which enable them to work in a relatively low humidity atmosphere. Fan et al. [46] studied the conductive process of a mesoporous silica matrix, discovering the conduction of protons running mainly in an approach of surface-charge-mediated transport, which implies that a highly ordered porous structure may give a higher proton conductivity.

Recently, we developed highly ordered sulfonated mesoporous silica with various nanostructures, including 2-D hexagonal (2D-H), 3-D body-centered cubic (3-D-BC), and 3-D cubic bicontinuous (3-D-CB) functionalized by proton-conducting heteropolyacids, in which sulfonated-2-D-H structural silica exhibit the highest proton conductivity (0.270 S cm^{-1}) under 200°C temperature (displayed in Figure 16.5) [47]. The results show that proton transport conductivities are closely related to morphologies and the magnitudes of the mesoporous silica.

Another use of porous ceramic material is to immobilize inorganic proton conductors, such as zirconium polyphosphates to improve its mechanical properties and permit accurate control over the pore structure of a composite membrane. The porosity of ceramic materials enables the impregnation of a considerable amount of polyphosphates without sacrificing the morphology and the mechanical properties of the original ceramic materials. Proton conductive capability was conferred on ceramic materials through the dipping process, which also resulted in increased hydrogen selectivity when repeating the impregnation process by five to six times [48].

Mesoporous acid-free hematite ceramic membranes have also been studied as proton conductors by Colomer [49]. This α-Fe$_2$–O$_3$ ceramic membrane showed a sigmoidal dependence of the conductivity and the water uptake on the RH at a constant temperature. Despite the unique acid-free property, the highest conductivity of this α-Fe$_2$O$_3$ ceramic membrane is only 2.76×10^{-3} S cm^{-1} at 90°C and 81% RH. Nd$_5$LnWO$_{12}$ and Ln$_6$WO$_{12}$ have been found to have satisfactory proton conductivity at high temperatures in hydrogen-containing

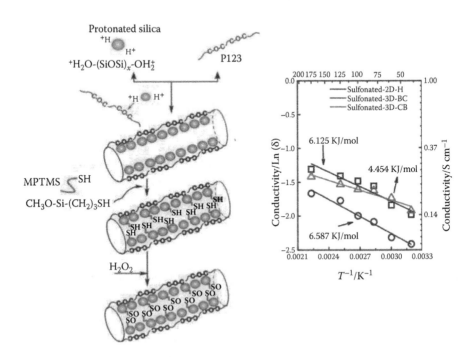

FIGURE 16.5
(a) Schematic diagram of the formation of the periodic ordered sulfonated-silica nanoelectrolytes; (b) temperature-dependent proton conductivity of the die-pressed sulfonated-silica nanoelectrolytes. (From Bangyang, J. et al. *Int. J. Hydrog. Energy*, 37, (5), 4612–4618, 2012.)

atmosphere; however, the coexisting electronic conductivity impedes the use in PEMFCs [50].

16.5 Heteropoly Acid

Nakamura first reported the high proton conductive composite $H_3PM_{12}O_{40}\cdot29H_2O$ (M = Mo,W) in 1979. Heteropoly acids (HPAs) are known to be a Brønsted acid with unique nanosized structures and a very strong acidity. Crystallized HPAs contain two types of protons in their structures: the water-combining proton, which is dissociated and hydrated, the other is the unhydrated proton located on the bridge-oxygen in the HPAs [51].

Most attention has been drawn on the Keggin-type HPAs because of the easy preparation and strong activity. HPAs of 12-phosphotungstic acid (PWA), 12-phosphomolybdic acid (PMA), and 12-silicotungstic acid (SiWA) are commercially available. Among the Keggin-type heteropolyacids, tung-stophosphoric acid ($H_3PW_{12}O_{40}$, abbreviated as HPW) has the strongest

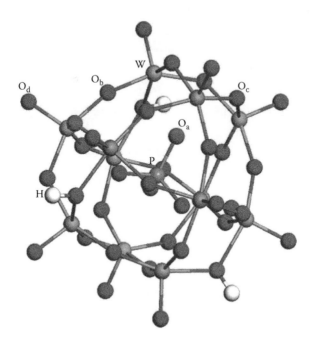

FIGURE 16.6
Keggin structure of $H_3PW_{12}O_{40}$. There are three types of exterior oxygen atoms: O_b, O_c, and O_d in the Keggin unit. O_a is the central oxygen atom. (Reprinted with permission from Yang, J. et al. *J. Am. Chem. Soc.*, 127, (51), 18274–18280, Copyright 2005 American Chemical Society.)

acidity. The Keggin structure of HPW contains three negative charges, which can be neutralized by three protons in acid form (as shown in Figure 16.6) [52]. Many solid high proton conducting HPAs, like $H_5GeW_{10}MoVO_{40}\cdot21H_2O$, $H_6GeW_{10}V_2O_{40}\cdot22H_2O$ and $H_7[In(H_2O)CoW_{11}O_{39}]\cdot14H_2O$ have been reported, all of which are Keggin structures and exhibit conductivities ranging from 10^{-4} to 10^{-3} S cm^{-1} [53]. Recent studies have revealed that the conductive capabilities of HPAs are highly dependent on their structures.

Despite the high conductivity and strong acidity of these solid state HPAs, their acid forms have relatively low surface area (~5 m^2 g^{-1}) and are soluble in water, which limit their application as a PEM in fuel cells. Besides that, proton transportation only occurs on the surface in crystalline HPAs. To overcome these problems, large monovalent ions like Cs$^+$ have been used to partly substitute the protons in HPAs to obtain a salt form, which is insoluble in water.

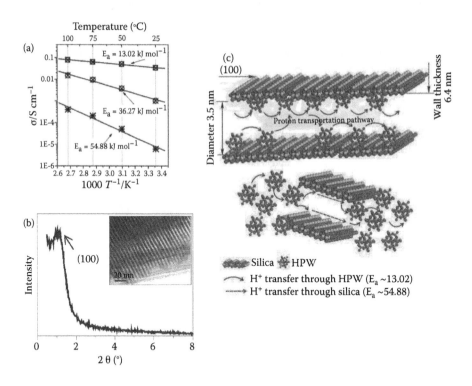

FIGURE 16.7

(See color insert.) (a) Impedance curves of the cell measured at a current density of 100 and 600 mA cm^{-2}. (b) Small angle XRD pattern of the self-assembled 25 wt% HPW/meso-silica electrolyte membrane, (c) the proposed proton transportation pathways of the self-assembled HPW/meso-silica electrolyte membrane and the mixed HPW/meso-silica composite electrolyte membrane, respectively. The insert in (b) is the HRTEM micrograph of the self-assembled 25 wt% HPW/meso-silica electrolyte membrane. (From Tang, H.L. et al. *Chem. Commun.*, 46, (24), 4351–4353, 2010.)

The addition of CsHP inorganics into a Nafion matrix has been confirmed to enhance the water content, reducing the activity of hydrophilic sulfonic group by covering the Nafion clusters. [51] Matsuda et al. [54] used SiO_2 as a support for $H_3PM_{12}O_{40}$ to increase its stability and enhance the membrane formation process. We have reported an innovative one-step synthesizing method to immobilize HPW on a meso-silica matrix. The large specific surface areas of mesostructural silica improve the utilization efficiency of HPW. An extremely high conductivity of 0.06–0.07 S cm^{-1} at 70°C–100°C has been measured in the one-step synthesized 25 wt% HPW/meso-silica as indicated in Figure 16.7 [55].

Lakshminarayana and Nogami [56] prepared a TEOS/PO(OCH$_3$)$_3$/HPA inorganic-organic hybrid nanocomposite membrane by a sol-gel method. This material combines the thermal tolerance of SiO_2 framework and the high proton conductivity of PO(OCH$_3$)$_3$ and HPA. Proton conductivities of 1.59×10^{-2} S cm^{-1} and 1.15×10^{-2} S cm^{-1} have been observed in a membrane based on PWA and PMA, respectively.

16.6 Summary and Outlook

A series of inorganic compounds and their strengths and shortcomings have been discussed in this review. Inorganic liquid acid, such as phosphoric acid, can exhibit rather high proton conductivity as 0.5 S cm^{-1} even under anhydrous conditions at 150°C, thus making it a promising candidate for high-temperature fuel cells. Whereas the losses of acid molecules under humid conditions limit its use. KDP-type conductors need to be combined with other composites due to the high dependence of conductivity on phase transition, which limits their application as PEMs in fuel cells. HPAs have excellent chemical stability and high proton conductivity, and can be immobilized and anchored within a highly ordered porous structure, such as mesoporous silica as promising proton exchange membranes. Inorganic composites holding a promising conductive property under high temperatures may not be limited to these compounds.

Combining different composites and fabricating highly ordered proton transporting structures are becoming a well-focused tendency spreading through the research field [10,55]. Intensive attentions have been paid to the powerful mechanical milling method (MMM) through which complicated inorganic compounds have been successfully prepared [14,40] as well as to the accessible sol-gel [14] and self-assembly [57] approaches. These attractive and prevailing methods will mark the orientation of the research and development of high-temperature inorganic conductors for fuel cells.

References

1. Li, Q.F., He, R.H., Jensen, J.O. and Bjerrum, N.J. 2003. Approaches and recent development of polymer electrolyte membranes for fuel cells operating above 100 degrees C. *Chem Mater* 15 (26), 4896–4915.
2. Gamburzev, S. and Appleby, A.J. 2002. Recent progress in performance improvement of the proton exchange membrane fuel cell (PEMFC). *J Power Sources* 107 (1), 5–12.
3. Curtin, D.E., Lousenberg, R.D., Henry, T.J., Tangeman, P.C. and Tisack, M.E. 2004. Advanced materials for improved PEMFC performance and life. *J Power Sources* 131 (1–2), 41–48.
4. Park, M.J., Downing, K.H., Jackson, A., Gomez, E.D., Minor, A.M., Cookson, D., Weber, A.Z. and Balsara, N.P. 2007. Increased water retention in polymer electrolyte membranes at elevated temperatures assisted by capillary condensation. *Nano Lett* 7 (11), 3547–3552.
5. Velu, S., Suzuki, K. and Osaki, T. 1999. Oxidative steam reforming of methanol over CuZnAl(Zr)-oxide catalysts: A new and efficient method for the production of CO-free hydrogen for fuel cells. *Chem Commun* 23, 2341–2342.
6. Yamanaka, I. and Otsuka, K. 1988. Partial oxidation of methanol using a fuel-cell system at room-temperature. *Chem Lett* 5, 753–756.
7. Garland, N.L. and Kopasz, J.P. 2007. The United States Department of Energy's high temperature, low relative humidity membrane program. *J Power Sources* 172 (1), 94–99.
8. Doyle, M., Choi, S.K. and Proulx, G. 2000. High-temperature proton conducting membranes based on perfluorinated ionomer membrane-ionic liquid composites. *J Electrochem Soc* 147 (1), 34–37.
9. Zhang, J., Xie, Z., Zhang, J., Tang, Y., Song, C., Navessin, T., Shi, Z., Song, D., Wang, H., Wilkinson, D. P., Liu, Z.-S. and Holdcroft, S. 2006. High temperature PEM fuel cells. *J Power Sources* 160 (2), 872–891.
10. Ramani, V., Kunz, H.R. and Fenton, J.M. 2004. Investigation of Nafion (R)/HPA composite membranes for high temperature/low relative humidity PEMFC operation. *J Membr Sci* 232 (1–2), 31–44.
11. Casciola, M., Alberti, G., Sganappa, M. and Narducci, R. 2006. On the decay of Nafion proton conductivity at high temperature and relative humidity. *J Power Sources* 162 (1), 141–145.
12. Piao, J.H., Liao, S.J. and Liang, Z.X. 2009. A novel cesium hydrogen sulfate-zeolite inorganic composite electrolyte membrane for polymer electrolyte membrane fuel cell application. *J Power Sources* 193 (2), 483–487.
13. Tang, H.L., Pan, M., Lu, S.F., Lu, J.L. and Jiang, S.P. 2010. One-step synthesized HPW/meso-silica inorganic proton exchange membranes for fuel cells. *Chem Commun* 46 (24), 4351–4353.
14. Nogami, M., Matsushita, H., Goto, Y. and Kasuga, T. 2000. A sol-gel-derived class as a fuel cell electrolyte. *Adv Mater* 12 (18), 1370–1372.
15. Kasuga, T., Nakano, M. and Nogami, M. 2002. Fast proton conductors derived from calcium phosphate hydrogels. *Adv Mater* 14 (20), 1490–1492.
16. Sahu, A.K., Pitchumani, S., Sridhar, P. and Shukla, A.K. 2009. Co-assembly of a nafion-mesoporous zirconium phosphate composite membrane for PEM fuel cells. *Fuel Cells* 9 (2), 139–147.

17. Schober, T. and Ringel, H. 2004. Proton conducting ceramics: Recent advances. *Ionics* 10 (5–6), 391–395.
18. Ma, G.L., Xu, J., Zhang, M., Wang, X.W., Yin, J.L. and Xu, J.H. 2011. The developments of inorganic proton conductors. *Progr Chem* 23 (2–3), 441–448.
19. He, R.H., Li, Q.F., Xiao, G. and Bjerrum, N.J. 2003. Proton conductivity of phosphoric acid doped polybenzimidazole and its composites with inorganic proton conductors. *J Membr Sci* 226 (1–2), 169–184.
20. Savadogo, O. and Xing, B. 2000. Hydrogen/oxygen polymer electrolyte membrane fuel cell (PEMFC) based on acid-doped polybenzimidazole (PBI). *J New Mater Electrochem Syst* 3 (4), 343–347.
21. Li, Q., Pan, C., Jensen, J.O., Noye, P. and Bjerrum, N.J. 2007. Cross-linked polybenzimidazole membranes for fuel cells. *Chem Mater* 19 (3), 350–352.
22. Jones, D.J. and Roziere, J. 2001. Recent advances in the functionalisation of polybenzimidazole and polyetherketone for fuel cell applications. *J Membr Sci* 185 (1), 41–58.
23. Laberty-Robert, C., Valle, K., Pereira, F. and Sanchez, C. 2011. Design and properties of functional hybrid organic-inorganic membranes for fuel cells. *Chem Soc Rev* 40 (2), 961–1005.
24. Tung, S.-P. and Hwang, B.-J. 2005. Synthesis and characterization of hydrated phosphor-silicate glass membrane prepared by an accelerated sol-gel process with water/vapor management. *J Mater Chem* 15 (34), 3532–3538.
25. Gomes, D., Marschall, R., Nunes, S.P. and Wark, M. 2008. Development of polyoxadiazole nanocomposites for high temperature polymer electrolyte membrane fuel cells. *J Membr Sci* 322 (2), 406–415.
26. Jin, Y.G., Qiao, S.Z., Xu, Z.P., da Costa, J.C.D. and Lu, G.Q. 2009. Porous silica nanospheres functionalized with phosphonic acid as intermediate-temperature proton conductors. *J Phys Chem C* 113 (8), 3157–3163.
27. Zarbin, A.J.G. and Alves, O.L. 1994. Pyrochlore-like compounds derived from antimonic acid. *J Mater Chem* 4 (3), 389–392.
28. Amarilla, J.M., Rojas, R.M., Rojo, J.M., Cubillo, M.J., Linares, A. and Acosta, J.L. 2000. Antimonic acid and sulfonated polystyrene proton-conducting polymeric composites. *Solid State Ionics* 127 (1–2), 133–139.
29. Slade, R.C.T., Hall, G.P. and Skou, E. 1989. Ac and dc conductivity of crystalline pyrochlore antimonic acid, $Sb_2O_5 \cdot nH_2O$. *Solid State Ionics* 35 (1–2), 29–33.
30. Zhou, W., Bondarenko, A.S., Boukamp, B.A. and Bouwmeester, H.J.M. 2008. Superprotonic conductivity in $MH(PO_3H)$ ($M = Li+, Na+, K+, Rb+, Cs+, NH_4+$). *Solid State Ionics* 179 (11–12), 380–384.
31. Yamane, Y., Yamada, K. and Inoue, K. 2008. Superprotonic solid solutions between $CsHSO_4$ and $CsH_2P.O4$. *Solid State Ionics* 179 (13–14), 483–488.
32. In-Hwan Oh, J.J.K., Oh, B.H. and Lee, C.E. 2008. Proton conduction in KH_2PO_4 modied by proton irradiation. *J Korean Phys Soc* 53 (6), 3497–3499.
33. Boysen, D.A., Haile, S.M., Liu, H.J. and Secco, R.A. 2003. High-temperature behavior of CsH_2PO_4 under both ambient and high pressure conditions. *Chem Mater* 15 (3), 727–736.
34. Yang, B., Kannan, A.M. and Manthiram, A. 2003. Stability of the dry proton conductor $CsHSO_4$ in hydrogen atmosphere. *Mater Res Bull* 38 (4), 691–698.
35. Heo, P., Shibata, H., Nagao, M. and Hibino, T. 2008. Pt-free intermediate-temperature fuel cells. *Solid State Ionics* 179 (27–32), 1446–1449.

36. Jin, Y.C. and Hibino, T. 2010. A proton-conducting composite membrane Sn0.95Al0.05P2O7 and polystyrene-b-poly(ethylene/propylene)-b-polystyrene. *Electrochim Acta* 55 (28), 8371–8375.

37. Nogami, Y.A. 1997. Evidence of water-cooperative proton conduction in silica glasses. *Phys Rev B* 55 (18), 12108–12112.

38. Mitsui, A., Miyayama, M. and Yanagida, H. 1987. Evaluation of the activation energy for proton conduction in perovskite-type oxides. *Solid State Ionics* 22 (2–3), 213–217.

39. Nogami, M., Daiko, Y., Akai, T. and Kasuga, T. 2001. Dynamics of proton transfer in the sol-gel-derived P2O5-SiO2 glasses. *J Phys Chem B* 105 (20), 4653–4656.

40. Schober, T. 2005. Composites of ceramic high-temperature proton conductors with inorganic compounds. *Electrochem Solid State Lett* 8 (4), A199–A200.

41. Kim, Y.B., Gur, T.M., Kang, S., Jung, H.J., Sinclair, R. and Prinz, F.B. 2011. Crater patterned 3-D proton conducting ceramic fuel cell architecture with ultra thin Y:BaZrO3 electrolyte. *Electrochem Commun* 13 (5), 403–406.

42. Nogami, R.N., Wong, G., Kasuga, T. and Hayakawa, T. 1999. High proton conductivity in porous P2O5-SiO2 glasses. *J Phys Chem B* 103 (44), 9468–9472.

43. Nogami, M., Miyamura, K. and Abe, Y. 1997. Fast protonic conductors of water-containing P2O5-ZrO2-SiO2 glasses. *J Electrochem Soc* 144 (6), 2175–2178.

44. Fan, R., Huh, S., Yan, R., Arnold, J. and Yang, P. 2008. Gated proton transport in aligned mesoporous silica films. *Nat Mater* 7 (4), 303–307.

45. Bangyang, J., Tang, H. and Pan, M. 2012. Well-ordered sulfonated silica electrolyte with high proton conductivity and enhanced selectivity at elevated temperature for DMFC. *Int J Hydrogen Energy* 37 (5), 4612–4618.

46. Vladimir, L. 2001. New inorganic proton-conductive membranes for hydrogen separation and electro-catalysis. *Membr Technol* 2001 (132), 4–8.

47. M.T., C. 2011. Proton transport, water uptake and hydrogen permeability of nanoporous hematite ceramic membranes. *J Power Sources* 196 (20), 8280–8285.

48. Escolastico, S., Solis, C. and Serra, J.M. 2011. Hydrogen separation and stability study of ceramic membranes based on the system Nd5LnWO12. *Int J Hydrogen Energy* 36 (18), 11946–11954.

49. Amirinejad, M., Madaeni, S.S., Rafiee, E. and Amirinejad, S. 2011. Cesium hydrogen salt of heteropolyacids/Nafion nanocomposite membranes for proton exchange membrane fuel cells. *J Membr Sci* 377 (1–2), 89–98.

50. Yang, J., Janik, M.J., Ma, D., Zheng, A.M., Zhang, M.J., Neurock, M., Davis, R.J., Ye, C.H. and Deng, F. 2005. Location, acid strength, and mobility of the acidic protons in Keggin 12-H3PW12O40: A combined solid-state NMR spectroscopy and DFT quantum chemical calculation study. *J Am Chem Soc* 127 (51), 18274–18280.

51. Kato, C.N., Hara, K., Hatano, A., Goto, K., Kuribayashi, T., Hayashi, K., Shinohara, A., Kataoka, Y., Mori, W. and Nomiya, K. 2008. A dawson-type dirhenium(V)-oxido-bridged polyoxotungstate: X-ray crystal structure and hydrogen evolution from water vapor under visible light irradiation. *Eur J Inorg Chem* 2008 (20), 3134–3141.

52. Matsuda, A., Daiko, Y., Ishida, T., Tadanaga, K. and Tatsumisago, M. 2007. Characterization of proton-conductive SiO2-H3PW12O40 composites prepared by mechanochemical treatment. *Solid State Ionics* 178 (7–10), 709–712.

53. Lakshminarayana, G. and Nogami, M. 2009. Synthesis and characterization of proton conducting inorganic-organic hybrid nanocomposite membranes based on tetraethoxysilane/trimethyl phosphate/3-glycidoxypropyltrimethoxysilane/heteropoly acids. *Electrochimica Acta* 54 (20), 4731–4740.
54. Tang, H., Wan, Z., Pan, M. and Jiang, S.P. 2007. Self-assembled Nafion–silica nanoparticles for elevated-high temperature polymer electrolyte membrane fuel cells. *Electrochem Commun* 9 (8), 2003–2008.
55. Schmidt-Rohr, K. and Chen, Q. 2008. Parallel cylindrical water nanochannels in Nafion fuel-cell membranes. *Nat Mater* 7 (1), 75–83.
56. Chen, Y.B., Thorn, M., Christensen, S., Versek, C., Poe, A., Hayward, R.C., Tuominen, M.T. and Thayumanavan, S. 2010. Enhancement of anhydrous proton transport by supramolecular nanochannels in comb polymers. *Nat Chem* 2 (6), 503–508.
57. Haile, S.M., Boysen, D.A., Chisholm, C.R.I. and Merle, R.B. 2001. Solid acids as fuel cell electrolytes. *Nature* 410 (6831), 910–913.

17

Advances in Microbial Fuel Cells

Chang Ming Li and Yan Qiao

CONTENTS

17.1 Introduction 524
17.2 Principles of MFCs 525
    17.2.1 Classification of MFCs 525
        17.2.1.1 Dual-Chamber MFC 525
        17.2.1.2 Single-Chamber MFC 525
        17.2.1.3 Sediment MFC 527
        17.2.1.4 Tubular Upflow MFC 527
    17.2.2 Catalysis in MFC Anode 527
        17.2.2.1 Biocatalysts 527
        17.2.2.2 Electrode Materials 529
    17.2.3 Catalysis in MFC Cathode 530
        17.2.3.1 Abiotic Cathode 530
        17.2.3.2 Biocathode 531
17.3 Nanostructured Electrode Materials in MFCs 532
    17.3.1 Anode Materials 532
        17.3.1.1 Carbon Nanotube-/Nanoparticle-Based Anode
                Materials 532
        17.3.1.2 Graphene-Based Anode Materials 534
        17.3.1.3 Other Nanomaterials for Anodes 536
    17.3.2 Cathode Materials 537
        17.3.2.1 Pt-Based Cathodes 537
        17.3.2.2 Pt-Free Cathodes 537
17.4 Advances in Biocatalysts 539
    17.4.1 Exoelectrogenic Bacteria 539
    17.4.2 Engineered Microorganisms 540
    17.4.3 Mixed Communities 541
17.5 Other Materials Applied in MFCs 541
    17.5.1 Three-Dimensional Structured Electrode 541
    17.5.2 Membrane Materials 542
17.6 Conclusions and Prospects 543
References 544

17.1 Introduction

Microbial fuel cells (MFCs) have been promising renewable power sources for more than 100 years. In 1911, the occurrence of an electromotive force between electrodes immersed in bacterial or yeast cultures and in a sterile medium with a battery-type setup was reported by Michael C. Potter [1], and it was concluded that electric energy could be liberated from the microbial disintegration of organic compounds. Twenty years later, Cohen [2] confirmed these results and reported a stacked bacterial fuel cell with a voltage of 35 V at a current of 0.2 mA. Although these publications may be considered as the birth of MFCs, the early resonance was doubtful. The rapid advances in other energy technologies like photovoltaic and later also the low prices for mineral oil again led to decreased interest in MFCs. In the 1980s, a slow interest in MFCs came back again [3,4]. Wilkinson [5] proposed the application of MFCs to power self–sustaining (eating) robots. Due to the limitations, including low performance, poor practicability, sustainability, and no application focus, the actual MFC development stagnated at a comparatively low publication level until the turn of the century [6], when the growing awareness of upcoming climate changes and resource depletion fuels environmental concerns to recover energy from organic waste for saving energy and reducing CO_2 emissions, MFCs have come to a new stage of development. The technology is no longer considered as a scientific peculiarity and insignificant issue but a serious and important research subject.

A typical MFC consists of two chambers separated by a proton exchange membrane. In an anodic compartment the microorganisms oxidize the substrates to produce electrons and protons, of which the electrons are transferred to an external electric load and then to the cathode for a complete electric circuit while protons are transported to the cathode compartment through the membrane. At the cathode, oxygen reduction associated with consuming electrons and protons forms water. Over conventional fuel cells, the main difference of the MFC is that the anode uses microbes rather than noble metals as catalysts to convert chemical energy into electricity. The microbes work as combinations of multiple enzymes to catalyze the oxidation of a substrate, such as glucose, and pass electrons to the electrode. As the multiple steps of redox reactions partially happen in microbial cells, the operating conditions of MFCs are milder than most chemical fuel cells. The microbes can self-regenerate, and thus the cost of this kind of catalyst is much lower than platinum. Furthermore, no pollutant, such as heavy metal ions, toxic gases, or organic waste, are generated during the operation of MFCs, thus rendering environmentally friendly energy systems. Many factors may affect the overall performance of MFCs, including the type of the utilized microorganism and substrates to be oxidized, materials of the anode and cathode, the nature of the proton exchange membrane, and the configuration of MFC.

In spite of the various merits of MFCs, the lower power density than conventional H_2-O_2 fuel cells apparently limits its practical applications. Apparently, in a MFC with a comparably sized anode and cathode, the power density is limited by the microbe anode. As a clean energy technology with promising potential applications, MFCs need a breakthrough to overcome its bottleneck of the microbe anode for high performance. In this chapter, after a brief introduction of principles, recent advances in electrode materials of MFCs, especially the application of nanostructured materials to enhance performance, are summarized and discussed profoundly. The challenges and perspectives of MFC development are also presented.

17.2 Principles of MFCs

17.2.1 Classification of MFCs

There is no standard classification criterion for MFCs, but sorting them according their configurations may be the most acceptable method so far. Throughout the development of MFCs, various configurations have been set up by researchers for different applications. Figure 17.1 gives a brief summary of the major MFC configurations, which will be introduced in the following section.

17.2.1.1 Dual-Chamber MFC

The typical "H"-shape dual-chamber model (Figure 17.1a) is the most widely used for basic research in labs. It is very easy to build a dual-chamber model—just connect two bottles with a tube containing a cation-exchange membrane as a separator [7–9]. The tube structure is used for fixing the membrane, but the tube itself is not required. Therefore, the two chambers can be directly pressed onto either side of the membrane and clamped together to decrease the electrode distance, thus lowering the internal resistance. In some models, a salt bridge, which consists of a U-shaped glass tube filled with agar and salt rather than an ion exchange membrane, connects the two chambers [10]. This is an inexpensive way to join bottles, but the power of a salt-bridge MFC is very low due to its high internal resistance. H-shape MFCs are acceptable for fundamental research, such as developing new electrode materials and investigating microbial cell behavior [11,12].

17.2.1.2 Single-Chamber MFC

As the H model produces low power densities due to large inner resistance, a single-chamber MFC lacking a proton exchange membrane was designed,

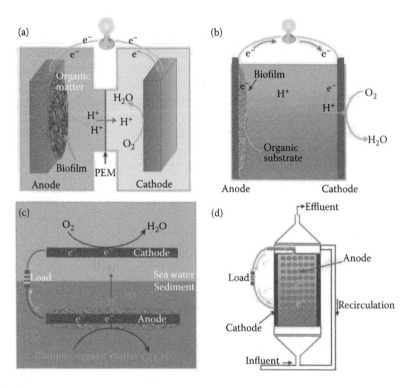

FIGURE 17.1
(See color insert.) Configurations of MFCs. (a) Dual-chamber MFC; (b) single-chamber MFC; (c) sediment MFC; and (d) tubular MFC.

in which an aqueous cathode was replaced by an air cathode (Figure 17.1b). The first single-chamber MFC was developed by Liu et al. [13] in 2004 for continuous wastewater treatment. This single-chamber MFC consisted of a single cylindrical Plexiglas chamber containing eight graphite rods (anode, each 6.15 mm in diameter and 150-mm long) placed in a concentric arrangement around a single air-porous cathode. Although the performance of this MFC is still poor, this reactor design provides a completely new approach to wastewater treatment and brings a breakthrough to MFC development. In the same year, Liu and Logan [14] improved the design of the single-chamber model. The chamber is still made by a Plexiglas cylinder but the size (4-cm long by 3 cm in diameter) is much smaller than the previous one. The graphite rod anode is replaced by carbon paper, and the cathode is carbon cloth; they are placed on opposite sides of the cylindrical chamber. With this design, the power density and Coulombic efficiency of the MFC was increased significantly and was much greater than any reported aqueous-cathode MFCs at that time. After these two reports, single-chamber MFCs were used by lots of research groups to generate electricity from wastewater and degrade or remove specific pollutants such as azo dye [15,16], Congo red [16], and ceftri-axone sodium [17] in the wastewater.

17.2.1.3 Sediment MFC

Another design is for a marine sediment MFC (Figure 17.1c), which consists of a graphite plate anode buried in marine sediment connected by an external circuit to a graphite plate cathode positioned in overlying seawater [18,19]. It utilizes a naturally occurring redox gradient that spans oxidant-rich seawa-ter and reluctant-rich sediment beneath the sediment surface. The chemical composition is different in the seawater with the sediment due to the micro-bial decomposition of the organic matter. The seawater is rich in oxidants, and the sediment is rich in reductants, resulting in a natural redox gradient. In a marine-sediment MFCs, cathodic current is attributed to a reduction in the seawater oxygen while the anodic current comes from the oxidation of sediment under catalysis of the microbes on the anode.

17.2.1.4 Tubular Upflow MFC

The first tubular upflow MFC was reported by He et al. in 2005 [20]. It was designed to generate electricity while simultaneously treating wastewa-ter. This tubular MFC comprised two cylindrical Plexiglas chambers with a 6-cm diameter. The cathode chamber (9-cm height, 250 cm^3 wet volume) was located on the top of the anode chamber (20-cm height, 520 cm^3 wet volume). This tubular upflow MFC continuously generated electricity with a maximum power density of 170 mW/m^2. In same year, Rabaey et al. [21] developed a single-chamber tubular MFC that consisted of a granular anode in the cylindrical reactor surrounded by a cation-exchange membrane and the woven graphite mat cathode tightly matched around the membrane. The cathodic electrolyte was 50 mM $K_3Fe(CN)_6$ solution in a 100 mM KH_2PO_4 buf-fer (pH 7). The maximal power outputs of this tubular MFC was 90 W m^{-3} net anodic compartment for feed streams based on acetate. Two years later, a membraneless tubular MFC with an air cathode was proposed (Figure 17.1d) [22]. The carbon granules in a cylindrical tube were used as the anode and a catalyst-coated carbon cloth cathode was tightly bound around the outside wall of the tube. This tubular air-cathode MFC was capable of continuous electricity generation from glucose-based substrates and produced a maxi-mum volumetric power of 50.2 W m^{-3} at current density of 216 A m^{-3}.

17.2.2 Catalysis in MFC Anode

17.2.2.1 Biocatalysts

In the MFC anode, organic substrates, which can be sugars and organic acids, such as glucose (Equation 17.1) or acetate (Equation 17.2), complex polymers, such as starch [23] and cellulose [24], or even organic wastes are oxidized using microorganisms as biocatalysts.

$$C_6H_{12}O_6 + 6H_2O \rightarrow 6CO_2 + 24H^+ + 24e^- \qquad (17.1)$$

$$C_2H_4O_2 + 2H_2O \rightarrow 2CO_2 + 8H^+ + 8e^- \qquad (17.2)$$

The performance of the anode mainly depends on the biocatalysts—the microbes. Unlike other chemical catalysts, the reaction rate of substrate oxidation is not the criterion to evaluate the microbes. The ability of liberating electrons from cells to the electrode is more important. Generally, because the cell wall and plasma membrane of microbes are nonconductive, the electrons produced in cells need carriers to move them out to reach the electrode surface. Artificial electron mediators are diffusive, small, and biocompatible molecules that can easily access both cells and electrodes, working as external electron acceptors of the electron-transfer chain and rapidly oxidizing mediators on the electrode surface to enhance the kinetics of anodic reactions. Employing mediators could lower the anode overpotential and increase the current density; therefore, they were widely used in early stage MFCs.

Microorganisms that do not require an exogenous redox mediator to transfer electrons to an electrode are currently sparking great research interest. In 1999, Kim et al. reported the first bacterial strain with electrochemical activity, the Fe(III)-reducing bacterium *Shewanella putrefaciens IR-1* [25], with which they developed a mediatorless MFC [26]. Whereas the anaerobically grown cells showed electrochemical activities, the aerobically grown cells did not. To date, this bacterium has been widely used in mediatorless MFCs. Another widely researched MFC bacterium is *Geobacter sulfurreducens*, which was first introduced into MFCs by Bond and Lovley [27]. Finneran et al. [28] reported the novel Fe(III)-reducing bacterium *Rhodoferax ferrireducens*, which had been isolated from anoxic subsurface sediment on Oyster Bay. *Rhodoferax ferrireducens* can oxidize glucose to CO_2 and quantitatively transfer electrons to graphite electrodes without a redox mediator [29]. In the following years, many new strains were identified that possess electrochemical activity, such as *Pseudomonas aeruginosa* [30], *Desulfobulbus propionicus* [31], *Geopsychrobacter electrodiphilus* [32], *Shewanella oneidensis* dsp10 [33], *Shewanella oneidensis* mr-1 [34], and *Klebsiella pneumoniae* [35].

According to the electron transfer pattern between microorganism cells and electrodes, three different types of anodes have been demonstrated so far. Type I, an early-stage MFC, utilizes the metabolic products of microorganisms to generate electricity. In type II, the anode receives electrons through artificial electron mediators, and in type III, electroactive microorganisms directly transfer electrons from electron donors to the anode. The electrocatalytic mechanisms of these three MFCs are quite different. In type I MFCs, the redox reactions of the metabolic products determinate the energy conversion efficiency and power output. In types II and III, the electrocatalytic performance of the anode is mainly dependent on the electron-transfer process between microorganisms and the anode.

17.2.2.2 Electrode Materials

Reduction of the activation energy is the key to improving the reaction kinetics of the anode. The use of unique metal catalysts can lower the activation energy and increase the power output. Early MFCs often utilized electroactive metabolites produced by the bacteria from substrates, such as hydrogen. In this case, Pt is widely used as the anode material due to its superior catalytic activity and stability in extreme environments. Due to its high cost, platinum is not a cost-effective MFC electrocatalyst. Researchers have made significant efforts in searching for replacements for platinum. Rosenbaum et al. [36,37] utilized tungsten carbide (WC), which also demonstrates good electrocatalytic properties toward the oxidation of formate, as an electrocatalyst for hydrogen oxidation in MFCs.

Carbon materials are the most broadly used electrode materials, especially in type II and type III MFCs, due to their high conductivity, good biocompatibility, low cost, and easy handling. For non-modified carbon-based anodes, electrocatalytic capability depends on the oxidation of microorganisms on the anode. Increasing the surface area of the anode is an efficient and widely utilized approach, and various strategies have been used to make carbon materials with a large electrochemically active surface area. Porous carbon materials have been used as the anode in different MFCs, such as dual chamber–mode MFCs [38], continuous-flow single-chamber MFCs [39], and tubular MFCs [40]. The most favorable materials are carbon cloth and carbon paper, both of which are carbon fiber–based porous materials. Alternatively, reticulated vitreous carbon (RVC) can be used due to its open network structure and large surface area, which allow the reaction to occur and provide the microorganisms with easy access to the electrode surface. Although these porous carbon materials can enhance the power and current density, the clogging of the material can be a problem, especially in saturated flow systems. To resolve this problem, Logan et al. [41] developed a new type of brush electrode that bundles large surface-area graphite fibers into a central core. When they examined this brush anode in cube and bottle-air-cathode MFCs inoculated with wastewater, they found that the cube MFCs achieved a maximum power density of 2400 mW m^{-2} or 73 W m^{-3} based on liquid volume. This brush architecture could be scaled up for use in larger reactors.

To date, the most popular anodic materials in MFCs are still porous carbon-based materials, such as carbon cloth, carbon paste, and graphite felt, which are very stable, easy to fabricate, and less costly than most metal-based electrodes. However, because the electrocatalytic activity of carbon is very low, carbon-based materials simply serve as current collectors, accepting the electrons from the biofilm. In addition, they cannot provide a greater number of active reaction centers because their active surface area is too small. Developing nanostructured anode materials could be a promising way to resolve these problems. Because nanostructured materials provide a larger active surface area and superior electrocatalytic activity, they are widely applied in MFC anodes to decrease the activation energy and improve the power output.

17.2.3 Catalysis in MFC Cathode

Cathodic catalysis, including the cathodic electron acceptors and the catalytic activity of the cathodic materials, greatly affects MFC performance. The kinetics of the reduction reaction and the standard redox potential of the oxidants determine the performance of the cathode. In an effort to enhance the electrocatalytic capability of the cathode, different oxidants and various kinds of catalysts were investigated.

17.2.3.1 Abiotic Cathode

In fuel cells, the oxygen-reduction reaction (ORR) at the cathode is the most widely used reaction because oxygen is free to access and its standard redox potential is high (1.229 V vs. NHE for a complete reduction [42]). Two processes could occur during cathodic oxygen reduction: a complete reduction in a four-electron pathway (Equation 17.3) or an incomplete two-electron reduction reaction (Equation 17.4) that needs further reduction to H_2O.

$$O_2 + 4H^+ + 4e^- \rightarrow 2H_2O \ (E'^0 = 1.229 \text{ V}) \tag{17.3}$$

$$2O_2 + 4H^+ + 4e^- \rightarrow 2H_2O_2 \ (E'^0 = 0.695 \text{ V}) \ [42] \tag{17.4}$$

The sluggish kinetics of the ORR is the main limitation of the power output of low-temperature fuel cells. The application of highly active catalysts, such as platinum, and the development of strong pH conditions (acid or alkaline) is the most useful strategy for enhancing ORR efficiency. The mechanism of electrocatalysis for ORR is still not completely understood, but the theory proposed by Damjanovic and Brusic [43], according to which proton transfer to O_2-producing adsorbed HO_2 or HO_2^- occurs prior to cleavage of the O-O bond during the ORR on Pt-group metals, has been widely accepted. The high and still increasing cost of Pt also makes an MFC very expensive if Pt is used as the cathode.

Because the reaction kinetics of the ORR in MFCs is poor, a noble metal catalyst is required. Oxygen can also diffuse across the proton-exchange membrane, reducing the power by taking electrons from the anode [44]. Alternative cathodes to replace oxygen are explored in MFCs, and some have been used for more than 20 years. For example, the $Fe(CN)_6^{3+}$ ion is a good electron acceptor under anaerobic conditions, and has been widely used in dual-chamber MFCs.

$$[Fe(CN)_6]^{3-} + e^- \rightarrow [Fe(CN)_6]^{4-} \ (E'^0 = 0.361 \ V) \ [42] \qquad (17.5)$$

The reduction kinetics of the $Fe(CN)_6^{3+}$ ion are much more rapid than ORR, obviating the need to use catalysts in ferricyanide cathode. A 50%–80% increase in maximum power using a ferricyanide cathode compared to that obtained using an oxygen-dissolved cathode has been reported [45]. The reason for this increase is that the open-circuit potential of a ferricyanide cathode (332 mV vs. Ag/AgCl) is higher and the mass transfer efficiency greater than those of an oxygen-dissolved cathode (268 mV vs. Ag/AgCl). Similar phenomena were also described by Ringeisen et al. [46] and Liu and Li [47]. However, the ferricyanide cathode is not suitable for upscaled MFCs because of the greater cost of the container and $Fe(CN)_6^{3+}$ ion regeneration.

To further increase the MFC cell voltage (normally 0.5 ~ 0.8 V), You et al. applied permanganate as the cathodic electron acceptor [48], which has a higher standard redox potential than oxygen. In acidic conditions, permanganate accepts three electrons and is thus reduced to manganese dioxide, as illustrated in Equation 17.6.

$$MnO_4^- + 4H^+ + 3e^- \rightarrow MnO_2 + 2H_2O \ (E'^0 = 1.70 \ V) \qquad (17.6)$$

The permanganate cathode (pH 3.6) increased the MFC open cell voltage to 1.382 V and delivered a 4.5- and 11.3-fold higher power density than that produced by the ferricyanide cathode and the oxygen cathode, respectively.

17.2.3.2 Biocathode

Microorganisms can also be used as catalysts in the cathode to perform functions similar to those that they perform in the anode. The first research conducted on biocathodes was by Rhoads et al. [49], who employed a cycle of Mn(IV) reduction on a cathode catalyzed by manganese-oxidizing bacteria (MOB). As the oxygen was required for the biocatalyzed reoxidation of Mn(II), manganese dioxide actually served as the mediator for the bacteria-catalyzed oxygen reduction.

$$MnO_{2(s)} + 4H^+ + 2e^- \rightarrow Mn^{2+} + 2H_2O \qquad (17.7)$$

$$Mn^{2+} + O_2 \xrightarrow{\quad MOB \quad} MnO_2 \qquad (17.8)$$

Similarly, iron compounds can be used as mediators in a biocatalyzed oxygen reduction on MFC cathodes. Ter Heijne et al. [50] designed an MFC with a ferric iron–mediated biocathode catalyzed by *Acidithiobacillus ferrooxidans*. In mediatorless biocathodes, bacteria grab electrons directly from the cathode and catalyze the reduction of oxidants. *Geobacter metallireducens* not only exhibited superior electrocatalytic performance in the anode but also catalyzed the reduction of nitrate and fumarate in the cathode [51]. Despite the pure strains, mixed microorganism communities, such as those found in sewage or wastewater, have also been used as biocatalysts in MFC cathodes [52]. The oxidant often used in the cathode of these mixed culture–catalyzed MFCs is nitrate, which has relatively high redox potential, as shown in Equation 17.9. Hence, their denitrification function is the major advantage of using these biocathodes [53,54].

$$2NO_3^- + 10e^- + 12H^+ \rightarrow N_2 + 6H_2O \ (E'^0 = 0.74 \text{ V}) \tag{17.9}$$

17.3 Nanostructured Electrode Materials in MFCs

17.3.1 Anode Materials

As mentioned previously, application of nanostructured electrode materials in anode is an effective way to reduce the activation energy and enhance the current density and power density. Different kinds of nanostructured anode materials are summarized here.

17.3.1.1 Carbon Nanotube-/Nanoparticle-Based Anode Materials

A simple fabrication, just modifying a graphite plate with carbon nanotubes (CNTs) by using epoxy conductive adhesive, can enhance the power density by 148% (267.77 mW m^{-2} vs. 107.51 mW m^{-2}) [55]. By using carbon nanoparticles as a conductive medium to immobilize bacteria on carbon cloth, it was possible to generate appreciable electricity from *Proteus vulgaris* without exogenous mediators [56]. The maximum power density is 269 mW m^{-2} at the cell voltage of ca. 400 mV were obtained using glucose as a substrate. Carbon nanoparticles assisted both the direct electron transfer and self-mediated electron transfer.

Nitrogen-doped CNTs (NCNTs) with a bamboo-like nanostructure were used as anode-modifying materials in MFCs with acetate as a substrate [57]. The bamboo-NCNT anode has increased active surface area and improved biofilm formation and thus possesses lower charge transfer resistance (23 Ω) than a CNT anode (68 Ω) and carbon cloth anode (390 Ω). The bamboo-NCNT MFC delivered a peak current density of 3.63 ± 0.06 A m^{-2}, which is almost 1.6 times greater than that of the CNT MFC and more than four times greater than that of the CC-MFC. The bamboo-NCNT MFC achieved a maximum power density of 1040 mW m^{-2} while the maximum power density of the CNT MFC was 710 mW m^{-2}, and the carbon cloth MFC generated a maximum power density of only 470 mW m^{-2}.

In some reports, conductive polymer modified CNTs were used as anode materials in MFCs [58,59]. The conductive polymers have functions like electron mediators to facilitate the electron transfer between the biofilm and electrode while the CNTs enhance the conductivity of conductive polymers. The conductive polymer@CNT nanocomposites delivered higher power density and current density than carbon cloth.

To further improve the anode performance, some hierarchical-nanostructured materials were used in MFCs. Hierarchical micro/nano structures of electrically conductive carbon composites was constructed by direct CNTs on micro-porous graphite felts at high densities [60]. Using the CNT-modified graphite felt anodes, power outputs from MFCs were increased sevenfold compared to bare graphite-felt anodes. The large increase in double layer capacitance was attributed to the increase in the surface area while the 20-fold decrease in solution resistance likely resulted from an increase in the amount of electron mediators in the diffusion and bulk layers. Xie et al. [61] reported a two-scale porous anode fabricated from a carbon nanotube-textile composite. The macroscale porous structure of the intertwined CNT-textile fibers (Figure 17.2a) creates an open three-dimensional space for efficient substrate transport and internal colonization by a diverse microflora. The conformally coated microscale porous CNT layer displays strong interaction with the microbial biofilm, facilitating electron transfer from exoelectrogens to the CNT-textile anode. An MFC equipped with a CNT-textile anode has a tenfold lower charge-transfer resistance (30 Ω vs. 300 Ω) and achieves considerably better performance than one equipped with a traditional carbon cloth anode: The maximum current density is 157% higher, the maximum power density is 68% higher, and the energy recovery is 141% greater. Two years later, the same group developed another three-dimensional electrode

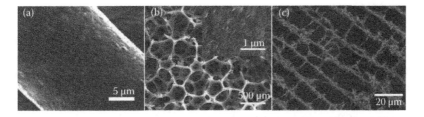

FIGURE 17.2
SEM micrographs of (a) CNT textile fiber. (Reprinted with permission from Zou, Y.J. et al. *Int J Hydrog Energy*, 33: 4856–4862, Copyright 2008 American Chemical Society.) (b) CNT sponge. (Zhao, Y. et al. *Phys Chem Chem Phys*, 13: 15016–21, 2011. Reproduced by permission of The Royal Society of Chemistry.) (c) CNT scaffold. (Xie, X. et al. *Nano Lett*, 2011, 11, 291–296. Reproduced by permission of The Royal Society of Chemistry.)

material, a CNT–sponge composite prepared by coating a sponge with CNTs [62]. The CNT–sponge (Figure 17.2b) composite also provided a three-dimensional scaffold that was favorable for microbial colonization and catalytic decoration. Using a batch-fed H-shaped MFC outfitted with CNT–sponge electrodes, a power density of 1.24 W m^{-2} was achieved when treating domestic wastewater. The maximum volumetric power density of a continuously fed plate-shaped MFC was 182 W m^{-3}. The CNT–sponge achieved a maximum current density of 2.13 mA cm^{-2} or 10.63 mA cm^{-3} in glucose media. This is 48% higher than that obtained from the CNT–textile under the same conditions (1.44 mA cm^{-2} or 7.18 mA cm^{-3}) according to the lower internal resistance and more tunable porous structure of the CNT–sponge.

Besides the coating process, three-dimensional CNT scaffolds can be obtained by ice-segregation induced self-assembly. Three-dimensional microchanneled nanocomposite electrodes fabricated by ice-segregation induced self-assembly of chitosan-dispersed multiwall carbon nanotubes are shown to provide a scaffold (Figure 17.2c) for growth of electroactive bacteria for use as acetate-oxidizing bioanodes in bioelectrochemical systems [63]. The hierarchical structure provides a conductive surface area available for *G. sulfurreducens* colonization with a flow through configuration along the electrode providing a substrate for bacterial colonization and bio-electrochemical processes. This configuration, while resulting in sub-monolayer biofilm coverage over the three-dimensional surface, is capable of providing acetate oxidation current densities of up to 24.5 Am^{-2}. Such bioanodes, when operated in a nonoptimized flow-through MFC configuration, provide a maximum power density of 2.87 W m^{-2}.

17.3.1.2 Graphene-Based Anode Materials

As an emerging miraculous carbon nanomaterial, graphene possesses some kinds of advantages as an electro material and has been applied in various energy systems, such as fuel cells, Li-ion batteries, solar cells, etc. The rapid development of the methods on graphene preparation and fabrication also benefits MFCs. Graphene-modified electrodes possess a high surface area and facilitated bacterial adhesion and exhibit an excellent efficiency of electron transfer. Graphene with high specific surface area (264 m^2 g^{-1}) has been used as an anodic catalyst of MFCs based on *E. coli* with 2-hydroxy-1,4-naphthoquinone as electron mediator [64]. The MFC with a graphene modified stainless steel mesh anode delivers a maximum power density of 2668 mW m^{-2}. The authors attributed the improved performance to the high surface area of the graphene/stainless steel mesh anode and the increase in the number of bacteria attached to anode. In another report, graphene sheets were electrochemically deposited on carbon cloth to fabricate an anode for a *Pseudomonas aeruginosa* mediatorless MFC [65]. The graphene modification improved power density and energy conversion efficiency by 2.7 and 3 times, respectively. It was found that graphene not only possessed high biocompatibility but also promoted the bacteria growth on the electrode surface that resulted in the creation of more direct electron transfer activation centers and stimulated excretion of mediating molecules for a higher electron transfer rate.

To improve the bacteria adhesion on the graphene surface, a one-pot method is exploited by adding graphene oxide and acetate into an MFC in which graphene oxide is microbially reduced, leading to *in situ* construction of a bacteria/graphene network in the anode [66]. The obtained microbially reduced graphene exhibits comparable conductivity and physical characteristics to the chemically reduced graphene. While the number of the actual accessible active biocatalysts was increased and extracellular electron transfer was accelerated, resulting in an apparent enhancement of anodic catalytic activity. As a result, the maximum power density of the MFC was enhanced by 32% (from 1440 to 1905 mW m²) and the Coulombic efficiency was improved by 80% (from 30% to 54%).

Besides two-dimensional graphene, 3-D–structured graphene was also applied in a MFC anode. Crumpled graphene particles (Figure 17.3a) synthesized by capillary compression in rapidly evaporating aerosol droplets were used to modify carbon cloth as electrode materials in a dual-chamber MFC [67]. In comparison to graphene sheets, the graphene particles modified anode has higher current density and lower charge transfer resistance. It is because of its unique three-dimensional open structure over the two-dimensional structure of graphene sheet. Xie et al. who applied CNT sponges in MFCs, also developed a graphene-sponge composite electrode material (Figure 17.3b) for MFC anode [68]. The cost of the graphene sponge is at least one order lower than CNT sponges, but the conductivity of a graphene sponge is about two orders of magnitude less conductive than the CNT sponges. To increase the conductivity, they used stainless steel meshes as current collectors, which had glued a piece of graphene-sponge composite to both sides of it. With the stainless steel mesh current collector, the maximum current

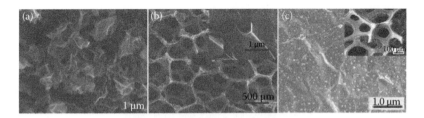

FIGURE 17.3
SEM micrograph of (a) crumpled graphene (Reprinted from *Bioresour Technol*, 114, Liu, J. et al. Graphene/carbon cloth anode for high-performance mediatorless microbial fuel cells, 275–280. Copyright 2012, with permission from Elsevier.) (b) Graphene-sponge (Yuan, Y. et al. *Bioresour Technol*, 2012, 116, 453–458. Reproduced by permission of The Royal Society of Chemistry), and (c) graphene/polyaniline. (Reprinted with permission from Xiao, L. et al. *J Power Sources*, 208, 187–192. Copyright 2012 American Chemical Society.)

density of the anode was increased by seven times and the maximum power density of the MFC was enhanced by 13 times. Thus, the graphene sponge/ stainless steel mesh composites can deliver comparable performance with CNT sponges but the cost is much lower.

Recently, a three-dimensional graphene/conductive polymer composite was reported to be applied in MFCs. A macroporous and monolithic MFC anode based on polyaniline hybridized three-dimensional graphene (Figure 17.3c) has been reported [69]. The charge-transfer resistance of graphene/poly-aniline MFC (~100 Ω) is much smaller than that of carbon cloth MFC (~2800 Ω). The maximum power density obtained from the 3-D graphene/PANI MFC is ~768 mW/m^2, which is about four times higher than that from the carbon cloth MFC (~158 mW/m^2). Furthermore, because graphene/PANI foam is much lighter than carbon cloth (3 g/m^2 vs. 136 g/m^2), it produces about 212 times higher specific power density than carbon cloth (256 mW/g vs. 1.2 mW/g).

17.3.1.3 Other Nanomaterials for Anodes

Except for carbon-based nanomaterials, other nanostructured materials, such as metal oxides and conductive polymers, have also been used for MFC anodes. Qiao et al. reported a mesoporous polyaniline/TiO$_2$ nanocomposite (Figure 17.4) as a MFC anode material [70]. This mesoporous nanocomposite possesses high active surface area, and the uniform pore structure facilitates the diffusion of the electrolyte. With this anode, an *E. coli* MFC fed with glucose achieved maximum power density of 1440 mW m^{-2}. In another report, the activity of Ni/b-Mo$_2$C as anodic electrocatalyst of MFC based on *Klebsiella pneumoniae* was investigated [71]. Ni/b-Mo$_2$C has high electro-catalytic activity toward the oxidation of formate, lactate, ethanol, and 2,6-di-tert-butyl-p-benzoquinon (2,6-DTBBQ), which are the main metabolites of *K. pneumoniae*. The MFC using Ni/b-Mo$_2$C as anodic electrocatalyst delivers a higher power density (4670 mW m^{-3}, 520 mW m^{-2}) than the MFC using b-Mo$_2$C as an anodic electrocatalyst (2390 mW m^{-3}, 270 mW m^{-2}).

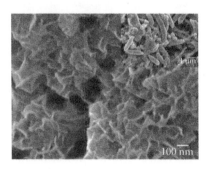

FIGURE 17.4

SEM micrograph of polyaniline/TiO$_2$ nanocomposite. The inset shows the adhesion of *E. coli* cells on the nanocomposite.

17.3.2 Cathode Materials

17.3.2.1 Pt-Based Cathodes

As oxygen is used as an oxide in most MFCs, especially the single-chamber MFC with an air cathode, noble metals are often used as catalysts. The challenge is to enhance the catalytic activity and decrease the loading of noble metals simultaneously. A CNT–textile–Pt cathode was used in a two-chamber MFC with aqueous cathode [72]. The CNT–textile–Pt was prepared by electrochemically depositing Pt nanoparticles on a CNT–textile, which is described in Section 17.3.1.1. Compared with Pt-coated carbon cloth, the surface area utilization efficiency (calculated from the ratio of electrochemically active surface area to the theoretical surface area) of CNT–textile–Pt is two orders higher (53.6% vs. 0.4%). An MFC equipped with a CNT–textile–Pt cathode revealed a 2.14-fold maximum power density with only 19.3% Pt loading, compared to that with a commercial Pt-coated carbon cloth cathode.

17.3.2.2 Pt-Free Cathodes

To save the cost of MFCs, various nanomaterials were used to replace Pt as cathode catalysts. A metal oxide–assisted metal macro cyclic complex was investigated as a catalyst for ORR in an air-cathode MFC [73]. Cobalt oxide (CoO_x) incorporation increased the ORR activity of iron phthalocyanine (FePc). In MFCs, the maximum power density of 654 ± 32 mW m^{-2} was achieved from the C–CoO_x–FePc (Figure 17.5a) cathode, which was 37% higher than the power density of carbon-supported FePc (C–FePc). A reduction peak at -0.015V (vs. SCE) appeared at the C–CoO_x–FePc electrode, which was more positive than those at the C–CoO_x electrode (-0.31 V vs. SCE) and the C–FePc electrode (-0.12 V vs. SCE). The steady state current density for the C–CoO_x–FePc was higher compared to C–CoO_x and C–FePc. It suggests that cobalt oxide is an efficient synergistic component for other ORR catalysts due to its capability of catalyzing the disproportionation reaction of the intermediater in oxygen electroreduction. Liu et al. developed a nanostructured MnOx (Figure 17.5b) cathode material by using an electrochemical deposition method, which achieved the maximum power density of 772.8 mW m^{-3} and removed organics when the MFC was fed with acetate-laden synthetic wastewater [74]. The nanostructured MnOx with controllable size and morphology could be readily obtained with the electrochemical deposition method. Both the morphology and manganese oxidation state of the nanoscale catalyst were largely dependent on the electrochemical preparation process, and they governed its catalytic activity and the cathodic oxygen reduction performance of the MFC accordingly. The MnOx nanorods had an electrochemical activity toward the ORR via a four-electron pathway in a neutral pH solution.

In another work, the catalytic performance of Fe_2O_3 and Mn_2O_3 nanopowders, carbon black powder, and Pt-cathodes were compared in an air-cathode

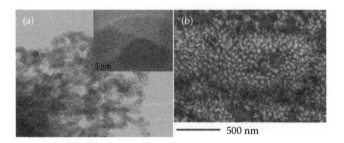

FIGURE 17.5
(a) TEM image of C–CoO$_x$–FePc. (Reprinted from *Int J Hydrog Energy*, 37, Zeng, L.Z. et al. Ni/
beta-Mo2C as noble-metal-free anodic electrocatalyst of microbial fuel cell based on *Klebsiella
pneumoniae*, 4590–4596, Copyright 2012, with permission from Elsevier.) (b) SEM image of
MnOx nanorods. (Reprinted from *Energy Environ Sci*, 4, Xie, X. et al. Nano-structured textiles
as high-performance aqueous cathodes for microbial fuel cells, 1293–1297, Copyright 2011, with
permission from Elsevier.)

MFC [75]. Although the Mn$_2$O$_3$ cathode had the lowest charge transfer resistance, the highest power output in polarization tests was observed for Pt, followed by Mn$_2$O$_3$, Fe$_2$O$_3$, and C. The Mn$_2$O$_3$ exhibited the best ORR potential
whereas Pt exhibited the best volumetric power output. The MFCs based on
these cathodes showed a performance decline with time, most likely due to
the loss of the catalyst, catalyst deactivation, or parasitic reactions. The MFC
based on the carbon cathode showed the most stable behavior. In all tests,
biofilms were, of course, formed at the various cathodes, but a microbially
catalyzed ORR (biocathode) or a biofilm catalytically active for the ORR was
not observed. The Mn$_2$O$_3$ electrode appeared to be the most promising non-
noble electrocatalyst cathode; however, its high overpotential (activation loss)
should be improved in order to increase significantly the power generation.

Besides metal oxides, carbon nanomaterials are also favorable electrode
materials with good catalytic performance. A gram-scale nitrogen-doped
graphene can be obtained at low temperatures using a denotation process
with cyanuric chloride and trinitrophenol as reagents [76]. This nitrogen-
doped graphene exhibited a one-step and four-electron pathway for ORR
at the onset potential of about –0.04 V vs. Ag/AgCl. The voltage and maximum power density produced by MFC with a nitrogen-doped graphene
cathode were comparable with the commercial Pt/C cathode, but the graphene cathode showed better stability than the Pt/C cathode. A construction
of graphene/biofilm composites was conducted by implanting microbially
reduced graphene into cathodic biofilm [77]. The maximum power density of
the MFC with the graphene biocathode generated a maximum power density
of 323.2 mW m^{-2}, which increased by 103% compared with the MFC with carbon cloth–biocathode. Electrochemical analyses showed that the graphene
biocathode had faster electron transfer kinetics and less internal resistance
than the carbon cloth–biocathode.

Modification with metals or metal oxides can enhance the catalytic activity of carbon nanomaterials. Iron- and nitrogen-functionalized graphene (Fe-N-G) as a nonprecious metal catalyst is synthesized via a facile method of thermal treatment of a mixture of Fe salt, graphitic carbon nitride (g-C3N4) and chemically reduced graphene [78]. The Fe-N-G catalyst has more positive onset potential and increased reduction current densities as compared to the pristine graphene catalyst. With the Fe-N-G cathode, the MFC achieves the maximum power density of 1149.8 mW m^{-2}, which is about 2.1 times of that generated with the Pt/C-MFC (561.1 mW m^{-2}) and much higher than that of the pristine graphene MFC (109 mW m^{-2}). Noncovalent functionalized graphene with iron tetrasulfophthalocyanine (FeTsPc) was used as electrocatalyst for ORR in a dual-chamber MFC [79]. The catalytic activity of FeTsPc was enhanced after adsorbing onto a graphene surface. The maximum power density of 817 mW m^{-2} obtained from the MFC with a FeTsPc-graphene cathode is higher than that of 523 mW m^{-2} with a FeTsPc cathode and is close to that of 856 mW m^{-2} with a Pt/C cathode. Surface coating of MnO_2 can enhance the catalytic activity of CNTs to ORR [80]. MnO_2/CNTs electrode shows more positive peak potential (–0.18 V) than either the mechanically mixed MnO_2/CNTs (–0.39 V) or the bare CNTs (–0.41 V). The peak current of the *in situ* MnO_2/CNTs (0.40 mA) was much greater than that of the mechanically mixed MnO_2/CNTs (0.16 mA) and bare CNTs (0.08 mA). A maximum power density of 210 mW m^{-2} produced from the MFC with *in situ* MnO_2/CNTs cathode is 2.3 times of that produced from the MFC using mechanically mixed MnO_2/CNTs (93 mW m^{-2}), and comparable to that of the MFC with a conventional Pt/C cathode (229 mW m^{-2}).

17.4 Advances in Biocatalysts

17.4.1 Exoelectrogenic Bacteria

As discussed in previous sections, some microorganisms can directly transfer electrons to an electrode and are known as exoelectrogenic bacteria. Besides *Shewanella putrefaciens IR-1*, two Fe(III)-reducing bacterial strains are isolated from an MFC and phylogenetically related to *Clostridium butyricum* [81] and *Aeromonas hydrophila* [82], respectively. The direct electron transfer phenomenon has also been discovered in other microorganisms, such as thermophilic Gram-positive bacterium *Thermincola ferriacetica* [83] and yeast cells *Hansenula anomala* [84].

It has been demonstrated [85,86] that a number of genes encoding outer membrane c-type cytochromes are involved in the electron transfer between those exoelectrogenic bacteria cells and solid electron acceptors, but the detail mechanism is still unclear as to how these cytochromes conduct the direct transfer. It was noted that type IV pili were important for *G. sulfurre-ducens* cells to facilitate their extracellular electron transfer to the solid metal oxides [87,88]. A similar phenomenon was observed in *Shewanella* [89] strains that could evolve electronically conducting pili, allowing the organisms to use solid-phase iron and manganese oxides that are not in direct contacts with cells. A recent work by Richter et al. [90] has proposed a new model for electron transfer between the *G. sulfurreducens* biofilm and MFC anodes. They investigated the electrochemical behavior of the biofilm by using cyclic voltammetry, and the results indicate a conductive network of bound (nondif-fusing) electron transfer mediators transferring electrons across the biofilm/electrode interface. They also found that type IV pili played a critical role in the electron transfer process, which agreed with the previous report. The pili-like conductive nanowires connect the membrane-bound cytochromes and metal oxides. The formation of such nanowires may allow the develop-ment of thicker electroactive biofilms for high anode performances [87].

17.4.2 Engineered Microorganisms

In recent years, it has been explored that some non-exoelectrogenic bacteria cells can become electroactive after environmental induction or genetic engineering manipulation. For example, *E. coli* has to be used in combination with artificial electron mediators in MFCs or else the production of electricity is low. Zhang et al. [91] employed *E. coli* cells in a mediator-less single chamber MFC with a carbon/PTFE composite film anode and a conventional air cathode. For the original bacteria, the initial cell voltage was only 0.4 V and decreased to 0.1 V within a few hours. While for the bacteria that pretreated with repeated discharge cycles, the initial cell voltage increased to 0.6 V and formed a discharge plateau at 0.55 V for 2 h and a plateau at 0.25 V for 50 h [92].

In our previous work [93], we found that the electrochemical behavior of *E. coli* cells on carbon cloth electrodes was similar with a quinone redox reaction, and supernatant of cell cultures might contained quinone-like compounds. Other researchers also proposed some possible electron carriers isolated from the metabolites of *E. coli* cells [94]. The production of electroactive mediators from *E. coli* cells might be induced by the electrochemical tension and/or other specific harsh environmental conditions in MFCs that could cause genetic changes of the microbes. As evidence, the surface morphology of *E. coli* cells was changed after electrochemical activation, showing the evolved cells have a much rougher surface than original ones, and there are some pores and clus-ters on the surface of evolved cells. This may increase the permeability of the cell wall and facilitate the diffusion of self-excreted mediators [93]. A transcrip-tome analysis based on microarray chips discovered that 107 genes of *E. coli* were significantly up- or down-regulated during 24 h of MFC discharging [95]. The operon analysis revealed that the outer membrane poring genes and ter-minal oxidase genes were crucial for the direct electrochemistry process. This approach not only could help us to understand the mechanisms of cell direct electrochemistry but also could provide scientific insights to bioengineering new strains for a faster direct electron transfer process.

Recently we constructed a genetically engineered *E. coli* [96] that also possessed direct electrochemical behavior. The engineered *E. coli* was constructed with recombinant plasmid containing overexpressed gldA gene (encoding glycerol dehydrogenase). This enzyme has broad substrate specificity, and it is very likely to result in products containing the endogenous redox mediator. The metabolic pathways that glycerol dehydrogenase participates in and where the mediators come from in more detail still need further investigation.

17.4.3 Mixed Communities

In 2002, Tender et al. [97] reported the development of a fuel cell consisting of an anode embedded in marine sediment and a cathode embedded in overlying seawater. The marine sediment contained various kinds of microbes, some of which served as MFC biocatalysts. The power generation resulted in specific enrichment of electroactive microorganisms on the anode surface [8]. Later, another kind of mixed culture contained in wastewater was used for electricity generation by Liu et al. [13,14]. As bacteria present in wastewater have been demonstrated to possess electroactivity, they have been widely used as mixed biocatalysts in mediatorless MFCs over the past several years [98]. Besides various wastewaters, other organic wastes, such as manure sludge waste [99], heat-treated soil [100], and anaerobic sludge, have also provided mixed bacteria communities for use in MFCs. Although researchers are far from understanding the mechanism behind the electron transfer in this complicated system, the mixed communities apparently deliver much higher power densities than do pure strains [101].

17.5 Other Materials Applied in MFCs

17.5.1 Three-Dimensional Structured Electrode

The first three-dimensional structured electrode for MFCs was developed by Logan et al. in 2007 [41]. They cut carbon fibers to a set length and wound them into a twisted core (consisting of two titanium wires) using an industrial brush manufacturing system to build a brush anode (Figure 17.6a). The brush anode has a high surface area and porous structure that can increase the loading amount of bacteria and thus enhance the catalytic performance. Meanwhile, unlike other high specific surface area materials, such as carbon foams, which could become filled with bacteria and be difficult to unclog, current generation was not adversely affected by biofilm growth on the brush anode surface, and system performance actually improved over

FIGURE 17.6

(a) Graphite brush electrode. (Reprinted with permission from Cheng, S. et al. *Environ Sci Technol*, 40, 2426–2432. Copyright 2006 American Chemical Society.) (b) SEM micrograph of treated brush fiber. (Reprinted from *Electrochem Commun.*, 8, Niessen, J. et al. Heat treated soil as convenient and versatile source of bacterial communities for microbial electricity generation, 869–873, Copyright 2006, with permission from Elsevier.)

time with biofilm development. The cube single-chamber MFC with a brush anode achieved maximum power density of 2400 mW m^{-2}, which is more than double that of the same MFC with carbon cloth anode. The performance of the brush anode can be further improved by combined treatment of acid soaking and heating (Figure 17.6b) [102]. The brush electrode was also used in a biocathode of MFC to improve the power production [103,104].

In another work, Siu et al. reported a microfabricated polydimethylsiloxane (PDMS) MFC with embedded micropillar electrodes [105]. This MFC is characterized by a flexible and biocompatible structure suitable for body implantation as a potential power source for implanted bioMEMS devices. The MFC is biocatalyzed by a microorganism, *Saccharomyces cerevisiae*, which converts chemical energy stored in glucose in the bloodstream to electrical energy. The MFC is a laminate design, consisting of a 0.2-μm-thick gold-evaporated PDMS anode and cathode separated by a Nafion 117 proton exchange membrane. These electrode surfaces feature more than 70,000 8-μm-high micropillar structures in a 1.2-cm Å ~ 1.0-cm geometric area. The MFC is encapsulated by PDMS and has an overall size of 1.7 cm Å~ 1.7 cm Å ~ 0.2 cm and a net weight of less than 0.5 g. Compared with silicon micromachined MFCs working in a phosphate buffer medium, the MFC with micropillar structure showed a 4.9 times increase in average current density and a 40.5 times increase in average power density when operated at identical conditions. Using a 15-μL droplet of human plasma, containing 4.2-mM blood glucose, the PDMS MFC demonstrated a maximum power density of 401.2 nW/cm^2. The Coulombic efficiency of electron conversion from blood glucose was 14.7%.

17.5.2 Membrane Materials

Besides electrodes, the membrane is also a very important component of the MFC that determines the inner resistance and affects the power production. Nafion membranes are often used in MFCs, but the cost is too high especially for large-scale MFCs. To decrease the membrane cost, a custom-made sulfonated poly(ether ether ketone) (SPEEK)/poly(ether sulfone) (PES) membrane was applied in MFCs. The conductivity of a hydrophobic PES membrane increases when a small amount (3%–5%) of hydrophilic SPEEK is added. The PES/SPEEK 5% membrane has the highest performance compared to Nafion 112, Nafion 117, PES, and PES/SPEEK 3% membrane with a maximum power density of 170 mW m^{-2} at the maximum current density of 340 mA m^{-2}. The COD removal efficiency of MFCs with composite membrane PES/SPEEK 5% is nearly 26-fold and twofold higher than that of MFCs with Nafion 112 and Nafion 117 membranes, respectively. In another work, carbon nanofiber (CNF) and activated carbon nanofiber (ACNF) were incorporated with Nafion to develop nanocomposite membranes for MFCs [106]. Nanocomposite membranes had higher production power and Coulombic efficiency (CE) than Nafion 117 and Nafion 112 in MFC systems. The MFC with ACNF/Nafion membrane produce the highest voltage of 57.64 mW m^{-2} while Nafion 112 produces the lowest power density of 13.99 mW m^{-2}.

17.6 Conclusions and Prospects

In comparison to 10 years ago, the power production performance of MFCs has been improved greatly. It might be attributed to the rapid development of material science and technology, especially nanomaterials. Carbon nanomaterials, such as CNTs and graphene, are the most favorable electrode materials for both anodes and cathodes. Three-dimensional carbon nanomaterials could be the best electrode materials for MFCs because they have high active surface area, good conductivity, sound electroactivity, and excellent compatibility to the biocatalysts, etc., to achieve high power density and energy conversion efficiency. For cathodes, the catalytic activity of bare carbon nanomaterials can be improved by modification using metal oxides or metal complexes.

The low power density of MFCs should be mainly ascribed to the microorganisms, the catalysis of which makes organic fuel oxidation possible, but the poor conductivity of microbes significantly limits the power output. It is very encouraging to see the lights of bioengineering new strains of bacteria for MFC performance improvement as discussed above; however, it remains a great distance from what we expected. The main challenge for the research is to motivate more biologists to participate in bioengineering new strains of bacteria in MFC applications.

In the future, with further advances in material science and nanoengineering, nanomaterials with superior chemical and physical properties will be developed and utilized in MFCs for continuous performance improvement. The rapid pace of the development of functional nanomaterials will also lead to the development of novel MFC electrode materials. To be promising candidates for anodic or biocathodic materials in MFCs, biocompatible nanomaterials should allow good biocompatibility for strong cell adhesion, a bacteria-favorable pore structure for cell growth without blocking food (fuel) transport channels, and a large specific surface area for high apparent energy and power density. Electrocatalytic activity can be further improved by tailoring the size distributions and nanostructures. Specific nanostructures, such as nanorods, nanotubes, and nanoneedles, could provide more opportunities to access the redox center inside the bacterium cell membrane through penetration for direct electron transfer. A shape-controlled synthesis of nanostructures is recommended. For abiotic MFC cathodes, the metal catalyst design needs more research on shape, size, and surface property effects on catalytic activity as well as the balance of activity and stability. Further understanding about the atom arrangements of different nanostructures, especially for high aspect ratio structures, will help to increase the catalytic activity of these metal catalysts. Porphyrin- and graphene-based nanocomposites could be promising noble metal–free catalysts for ORR. Catalyst supports with various nanostructures have been well developed, but the interaction between catalyst particles and supporting materials need further investigation.

References

1. Potter, M.C. 1912. Electrical effects accompanying the decomposition of organic compounds. *Proc R Soc London, Ser B* 84, 260–276.
2. Cohen, B. 1931. Thirty-second Annual Meeting of the Society of American Bacteriologists. *J Bacteriol* 21, 1–60.
3. Delaney, G.M., Bennetto, H.P., Mason, J.R., Roller, S.D., Stirling, J.L. and Thurston, C.F. 1984. Electron-transfer coupling in microbial fuel-cells. 2. Performance of fuel-cells containing selected microorganism mediator substrate combinations. *J Chem Technol Biotechnol B—Biotechnol* 34, 13–27.
4. Roller, S.D., Bennetto, H.P., Delaney, G.M., Mason, J.R., Stirling, J.L. and Thurston, C.F. 1984. Electron-transfer coupling in microbial fuel-cells. 1. Comparison of redox-mediator reduction rates and respiratory rates of bacteria. *J Chem Technol Biotechnol B—Biotechnol* 34, 3–12.
5. Wilkinson, S. 2000. "Gastrobots": Benefits and challenges of microbial fuel cells in food powered robot applications. *Auton Robots* 9, 99–111.
6. Schroder, U. 2011. Discover the possibilities: Microbial bioelectrochemical systems and the revival of a 100-year-old discovery. *J Solid State Electrochem* 15, 1481–1486.
7. Park, D.H., Kim, S.K., Shin, I.H. and Jeong, Y.J. 2000. Electricity production in biofuel cell using modified graphite electrode with Neutral Red. *Biotechnol Lett* 22, 1301–1304.

8. Bond, D.R., Holmes, D.E., Tender, L.M. and Lovley, D.R. 2002. Electrode-reducing microorganisms that harvest energy from marine sediments. *Science* 295, 483–485.

9. Logan, B.E., Murano, C., Scott, K., Gray, N.D. and Head, I.M. 2005. Electricity generation from cysteine in a microbial fuel cell. *Water Res* 39, 942–952.

10. Min, B.K., Cheng, S.A. and Logan, B.E. 2005. Electricity generation using membrane and salt bridge microbial fuel cells. *Water Res* 39, 1675–1686.

11. Rabaey, K. and Verstraete, W. 2005. Microbial fuel cells: Novel biotechnology for energy generation. *Trends Biotechnol* 23, 291–298.

12. Logan, B.E., Hamelers, B., Rozendal, R., Schrorder, U., Keller, J., Freguia, S., Aelterman, P., Verstraete, W. and Rabaey, K. 2006. Microbial fuel cells: Methodology and technology. *Environ Sci Technol* 40, 5181–5192.

13. Liu, H., Ramnarayanan, R. and Logan, B.E. 2004. Production of electricity during wastewater treatment using a single chamber microbial fuel cell. *Environ Sci Technol* 38, 2281–2285.

14. Liu, H. and Logan, B.E. 2004. Electricity generation using an air-cathode single chamber microbial fuel cell in the presence and absence of a proton exchange membrane. *Environ Sci Technol* 38, 4040–4046.

15. Sun, J., Hu, Y.Y. and Hou, B. 2011. Electrochemical characteriztion of the bio-anode during simultaneous azo dye decolorization and bioelectricity generation in an air-cathode single chambered microbial fuel cell. *Electrochimica Acta* 56, 6874–6879.

16. Hou, B., Sun, J.A. and Hu, Y.Y. 2011. Simultaneous Congo red decolorization and electricity generation in air-cathode single-chamber microbial fuel cell with different microfiltration, ultrafiltration and proton exchange membranes. *Bioresour Technol* 102, 4433–4438.

17. Wen, Q., Kong, F.Y., Zheng, H.T., Yin, J.L., Cao, D.X., Ren, Y.M. and Wang, G.L. 2011. Simultaneous processes of electricity generation and ceftriaxone sodium degradation in an air-cathode single chamber microbial fuel cell. *J Power Sources* 196, 2567–2572.

18. Reimers, C.E., Tender, L.M., Fertig, S. and Wang, W. 2001. Harvesting energy from the marine sediment-water interface. *Environ Sci Technol* 35, 192–195.

19. Lowy, D.A., Tender, L.M., Zeikus, J.G., Park, D.H. and Lovley, D.R. 2006. Harvesting energy from the marine sediment-water interface II - Kinetic activity of anode materials. *Biosens Bioelectron* 21, 2058–2063.

20. He, Z., Minteer, S.D. and Angenent, L.T. 2005. Electricity generation from artificial wastewater using an upflow microbial fuel cell. *Environ Sci Technol* 39, 5262–5267.

21. Rabaey, K., Clauwaert, P., Aelterman, P. and Verstraete, W. 2005. Tubular microbial fuel cells for efficient electricity generation. *Environ Sci Technol* 39, 8077–8082.

22. You, S.J., Zhao, Q.L., Zhang, J.N., Jiang, J.Q., Wan, C.L., Du, M.A. and Zhao, S.Q. 2007. A graphite-granule membrane-less tubular air-cathode microbial fuel cell for power generation under continuously operational conditions. *J Power Sources* 173, 172–177.

23. Niessen, J., Schroder, U. and Scholz, F. 2004. Exploiting complex carbohydrates for microbial electricity generation - a bacterial fuel cell operating on starch. *Electrochem Commun* 6, 955–958.

24. Rismani-Yazdi, H., Christy, A.D., Dehority, B.A., Morrison, M., Yu, Z. and Tuovinen, O.H. 2007. Electricity generation from cellulose by rumen microorganisms in microbial fuel cells. *Biotechnol Bioeng* 97, 1398–1407.
25. Kim, B.H., Ikeda, T., Park, H.S., Kim, H.J., Hyun, M.S., Kano, K., Takagi, K. and Tatsumi, H. 1999. Electrochemical activity of an Fe(III)-reducing bacterium, Shewanella putrefaciens IR-1, in the presence of alternative electron acceptors. *Biotechnol Tech* 13, 475–478.
26. Kim, H.J., Park, H.S., Hyun, M.S., Chang, I.S., Kim, M. and Kim, B.H. 2002. A mediator-less microbial fuel cell using a metal reducing bacterium, Shewanella putrefaciense. *Enzyme Microb Technol* 30, 145–152.
27. Bond, D.R. and Lovley, D.R. 2003. Electricity production by Geobacter sulfurreducens attached to electrodes. *Appl Environ Microbiol* 69, 1548–1555.
28. Finneran, K.T., Johnsen, C.V. and Lovley, D.R. 2003. Rhodoferax ferrireducens sp. nov., a psychrotolerant, facultatively anaerobic bacterium that oxidizes acetate with the reduction of Fe(III). *Int J Syst Evol Microbiol* 53, 669–673.
29. Chaudhuri, S.K. and Lovley, D.R. 2003. Electricity generation by direct oxidation of glucose in mediatorless microbial fuel cells. *Nat Biotechnol* 21, 1229–1232.
30. Rabaey, K., Boon, N., Siciliano, S.D., Verhaege, M. and Verstraete, W. 2004. Biofuel cells select for microbial consortia that self-mediate electron transfer. *Appl Environ Microbiol* 70, 5373–5382.
31. Holmes, D.E., Bond, D.R. and Lovley, D.R. 2004. Electron transfer by Desulfobulbus propionicus to Fe(III) and graphite electrodes. *Appl Environ Microbiol* 70, 1234–1237.
32. Holmes, D.E., Nicoll, J.S., Bond, D.R. and Lovley, D.R. 2004. Potential role of a novel psychrotolerant member of the family Geobacteraceae, Geopsychrobacter electrodiphilus gen. nov., sp. nov., in electricity production by a marine sediment fuel cell. *Appl Environ Microbiol* 70, 6023–6030.
33. Ringeisen, B.R., Ray, R. and Little, B. 2006. A miniature microbial fuel cell operating with an aerobic anode chamber. In Joint Meeting of the International-Battery-Association/Hawaii Battery Conference (IBA-HBC 2006). Hilo, HI. *J Power Sources* 165, 591–597.
34. Biffinger, J.C., Pietron, J., Ray, R., Little, B. and Ringeisen, B.R. 2007. A biofilm enhanced miniature microbial fuel cell using Shewanella oneidensis DSP10 and oxygen reduction cathodes. *Biosensors and Bioelectronics* 22, 1672–1679.
35. Zhang, L.X., Zhou, S.G., Zhuang, L., Li, W.S., Zhang, J.T., Lu, N. and Deng, L.F. 2008. Microbial fuel cell based on Klebsiella pneumoniae biofilm. *Electrochem Comm* 10, 1641–1643.
36. Rosenbaum, M., Zhao, F., Schroder, U. and Scholz, F. 2006. Interfacing electrocatalysis and biocatalysis with tungsten carbide: A high-performance, noble-metal-free microbial fuel cell. *Angew Chem Int Ed* 45, 6658–6661.
37. Rosenbaum, M., Zhao, F., Quaas, M., Wulff, H., Schroder, U. and Scholz, F. 2007. Evaluation of catalytic properties of tungsten carbide for the anode of microbial fuel cells. *Appl Catal B-Environ* 74, 261–269.
38. Kim, N., Choi, Y., Jung, S. and Kim, S. 2000. Effect of initial carbon sources on the performance of microbial fuel cells containing Proteus vulgaris. *Biotechnol Bioeng* 70, 109–114.
39. Cheng, S., Liu, H. and Logan, B.E. 2006. Increased power generation in a continuous flow MFC with advective flow through the porous anode and reduced electrode spacing. *Environ Sci Technol* 40, 2426–2432.

40. Scott, K., Murano, C. and Rimbu, G. 2007. A tubular microbial fuel cell. *J Appl Electrochem* 37, 1063–1068.

41. Logan, B., Cheng, S., Watson, V. and Estadt, G. 2007. Graphite fiber brush anodes for increased power production in air-cathode microbial fuel cells. *Environ Sci Technol* 41, 3341–3346.

42. Bard, A.J. and Faulkner, L.R. 2001. *Electrochemical Methods: Fundamentals and Applications*, 2nd ed. New York: John Wiley & Sons, Inc. pp. 563–571.

43. Damjanovic, A. and Brusic, V. 1967. Electrode kinetics of oxygen reduction on oxide-free platinum electrodes. *Electrochim Acta* 12, 615–628.

44. Holzman, D.C. 2005. Microbe power! *Environ Health Perspect* 113, A754–A757.

45. Oh, S., Min, B. and Logan, B.E. 2004. Cathode performance as a factor in electricity generation in microbial fuel cells. *Environ Sci Technol* 38, 4900–4904.

46. Ringeisen, B.R., Henderson, E., Wu, P.K., Pietron, J., Ray, R., Little, B., Biffinger, J.C. and Jones-Meehan, J.M. 2006. High power density from a miniature microbial fuel cell using Shewanella oneidensis DSP10. *Environ Sci Technol* 40, 2629–2634.

47. Liu, Z.D. and Li, H.R. 2007. Effects of bio- and abio-factors on electricity production in a mediatorless microbial fuel cell. *Biochem Eng J* 36, 209–214.

48. You, S.J., Zhao, Q.L., Zhang, J.N., Jiang, J.Q. and Zhao, S.Q. 2006. A microbial fuel cell using permanganate as the cathodic electron acceptor. *J Power Sources* 162, 1409–1415.

49. Rhoads, A., Beyenal, H. and Lewandowski, Z. 2005. Microbial fuel cell using anaerobic respiration as an anodic reaction and biomineralized manganese as a cathodic reactant. *Environ Sci Technol* 39, 4666–4671.

50. Ter Heijne, A., Hamelers, H.V.M., De Wilde, V., Rozendal, R.A. and Buisman, C.J.N. 2006. A bipolar membrane combined with ferric iron reduction as an efficient cathode system in microbial fuel cells. *Environ Sci Technol* 40, 5200–5205.

51. Gregory, K.B., Bond, D.R. and Lovley, D.R. 2004. Graphite electrodes as electron donors for anaerobic respiration. *Environ Microbiol* 6, 596–604.

52. Clauwaert, P., Rabaey, K., Aelterman, P., De Schamphelaire, L., Ham, T.H., Boeckx, P., Boon, N. and Verstraete, W. 2007. Biological denitrification in microbial fuel cells. *Environ Sci Technol* 41, 3354–3360.

53. Lefebvre, O., Al-Mamun, A. and Ng, H.Y. 2008. A microbial fuel cell equipped with a biocathode for organic removal and denitrification. *Water Sci Technol* 58, 881–885.

54. Virdis, B., Rabaey, K., Yuan, Z.G., Rozendal, R.A. and Keller, J. 2009. Electron fluxes in a microbial fuel cell performing carbon and nitrogen removal. *Environ Sci Technol* 43, 5144–5149.

55. Mohanakrishna, G., Mohan, S.K. and Mohan, S.V. 2012. Carbon based nanotubes and nanopowder as impregnated electrode structures for enhanced power generation: Evaluation with real field wastewater. *Appl Energy* 95, 31–37.

56. Yuan, Y., Ahmed, J., Zhou, L., Zhao, B. and Kim, S. 2011. Carbon nanoparticles-assisted mediator-less microbial fuel cells using Proteus vulgaris. *Biosensors and Bioelectronics* 27, 106–112.

57. Ci, S.Q., Wen, Z.H., Chen, J.H. and He, Z. 2012. Decorating anode with bamboo-like nitrogen-doped carbon nanotubes for microbial fuel cells. *Electrochem Commun* 14, 71–74.

58. Qiao, Y., Li, C.M., Bao, S.J. and Bao, Q.L. 2007. Carbon nanotube/polyaniline composite as anode material for microbial fuel cells. *J Power Sources* 170, 79–84.

59. Zou, Y.J., Xiang, C.L., Yang, L.N., Sun, L.X., Xu, F. and Cao, Z. 2008. A mediator-less microbial fuel cell using polypyrrole coated carbon nanotubes composite as anode material. *Int J Hydrogen Energy* 33, 4856–4862.

60. Zhao, Y., Watanabe, K. and Hashimoto, K. 2011. Hierarchical micro/nano structures of carbon composites as anodes for microbial fuel cells. *Phys Chem Chem Phys* 13, 15016–15021.

61. Xie, X., Hu, L., Pasta, M., Wells, G.F., Kong, D., Criddle, C.S. and Cui, Y. 2011. Three-dimensional carbon nanotube–textile anode for high-performance microbial fuel cells. *Nano Letters* 11, 291–296.

62. Xie, X., Ye, M., Hu, L.B., Liu, N., McDonough, J.R., Chen, W., Alshareef, H.N., Criddle, C.S. and Cui, Y. Carbon nanotube-coated macroporous sponge for microbial fuel cell electrodes. *Energy Environ Sci* 5, 5265–5270.

63. Katuri, K., Luisa Ferrer, M., Gutierrez, M.C., Jimenez, R., del Monte, F. and Leech, D. 2011. Three-dimensional microchanelled electrodes in flow-through configuration for bioanode formation and current generation. *Energy Environ Sci* 4, 4201–4210.

64. Zhang, Y., Mo, G., Li, X., Zhang, W., Zhang, J., Ye, J., Huang, X. and Yu, C. 2011. A graphene modified anode to improve the performance of microbial fuel cells. *J Power Sources* 196, 5402–5407.

65. Liu, J., Qiao, Y., Guo, C.X., Lim, S., Song, H. and Li, C.M. 2012. Graphene/carbon cloth anode for high-performance mediatorless microbial fuel cells. *Bioresource Technology* 114, 275–280.

66. Yuan, Y., Zhou, S., Zhao, B., Zhuang, L. and Wang, Y. 2012. Microbially-reduced graphene scaffolds to facilitate extracellular electron transfer in microbial fuel cells. *Bioresour Technol* 116, 453–458.

67. Xiao, L., Damien, J., Luo, J.Y., Jang, H.D., Huang, J.X. and He, Z. 2012. Crumpled graphene particles for microbial fuel cell electrodes. *J Power Sources* 208, 187–192.

68. Xie, X., Yu, G.H., Liu, N., Bao, Z.N., Criddle, C.S. and Cui, Y. 2012. Graphene-sponges as high-performance low-cost anodes for microbial fuel cells. *Energy Environ Sci* 5, 6862–6866.

69. Yong, Y.C., Dong, X.C., Chan-Park, M.B., Song, H. and Chen, P. 2012. Macroporous and monolithic anode based on polyaniline hybridized three-dimensional graphene for high-performance microbial fuel cells. *Acs Nano* 6, 2394–2400.

70. Qiao, Y., Bao, S.J., Li, C.M., Cui, X.Q., Lu, Z.S. and Guo, J. 2008. Nanostructured polyanifine/titanium dioxide composite anode for microbial fuel cells. *Acs Nano* 2, 113–119.

71. Zeng, L.Z., Zhao, S.F., Wang, Y.Q., Li, H. and Li, W.S. 2012. Ni/beta-Mo2C as noble-metal-free anodic electrocatalyst of microbial fuel cell based on Klebsiella pneumoniae. *Int J Hydrogen Energy* 37, 4590–4596.

72. Xie, X., Pasta, M., Hu, L.B., Yang, Y.A., McDonough, J., Cha, J., Criddle, C.S. and Cui, Y. 2011. Nano-structured textiles as high-performance aqueous cathodes for microbial fuel cells. *Energy Environ Sci* 4, 1293–1297.

73. Ahmed, J., Yuan, Y., Zhou, L.H. and Kim, S. 2012. Carbon supported cobalt oxide nanoparticles-iron phthalocyanine as alternative cathode catalyst for oxygen reduction in microbial fuel cells. *J Power Sources* 208, 170–175.

74. Liu, X.-W., Sun, X.-F., Huang, Y.-X., Sheng, G.-P., Zhou, K., Zeng, R.J., Dong, F., Wang, S.-G., Xu, A.-W., Tong, Z.-H. and Yu, H.-Q. 2010. Nano-structured manganese oxide as a cathodic catalyst for enhanced oxygen reduction in a microbial fuel cell fed with a synthetic wastewater. *Water Res* 44, 5298–5305.

75. Martin, E., Tartakovsky, B. and Savadogo, O. 2011. Cathode materials evaluation in microbial fuel cells: A comparison of carbon, Mn_2O_3, Fe_2O_3 and platinum materials. *Electrochim Acta* 58, 58–66.

76. Feng, L., Chen, Y. and Chen, L. 2011. Easy-to-operate and low-temperature synthesis of gram-scale nitrogen-doped graphene and its application as cathode catalyst in microbial fuel cells. *Acs Nano* 5, 9611–9618.

77. Zhuang, L., Yuan, Y., Yang, G. and Zhou, S. 2012. In situ formation of graphene/biofilm composites for enhanced oxygen reduction in biocathode microbial fuel cells. *Electrochem Commun* 21, 69–72.

78. Li, S., Hu, Y., Xu, Q., Sun, J., Hou, B. and Zhang, Y. 2012. Iron- and nitrogen-functionalized graphene as a non-precious metal catalyst for enhanced oxygen reduction in an air-cathode microbial fuel cell. *J Power Sources* 213, 265–269.

79. Zhang, Y., Mo, G., Li, X. and Ye, J. 2012. Iron tetrasulfophthalocyanine functionalized graphene as a platinum-free cathodic catalyst for efficient oxygen reduction in microbial fuel cells. *J Power Sources* 197, 93–96.

80. Zhang, Y.P., Hu, Y.Y., Li, S.Z., Sun, J. and Hou, B. 2011. Manganese dioxide-coated carbon nanotubes as an improved cathodic catalyst for oxygen reduction in a microbial fuel cell. *J Power Sources* 196, 9284–9289.

81. Park, H.S., Kim, B.H., Kim, H.S., Kim, H.J., Kim, G.T., Kim, M., Chang, I.S., Park, Y.K. and Chang, H.I. 2001. A novel electrochemically active and Fe(III)-reducing bacterium phylogenetically related to Clostridium butyricum isolated from a microbial fuel cell. *Anaerobe* 7, 297–306.

82. Pham, C.A., Jung, S.J., Phung, N.T., Lee, J., Chang, I.S., Kim, B.H., Yi, H. and Chun, J. 2003. A novel electrochemically active and Fe(III)-reducing bacterium phylogenetically related to Aeromonas hydrophila, isolated from a microbial fuel cell. *FEMS Microbiol Lett* 223, 129–134.

83. Marshall, C.W. and May, H.D. 2009. Electrochemical evidence of direct electrode reduction by a thermophilic Gram-positive bacterium, Thermincola ferriacetica. *Energy Environ Sci* 2, 699–705.

84. Prasad, D., Arun, S., Murugesan, A., Padmanaban, S., Satyanarayanan, R.S., Berchmans, S. and Yegnaraman, V. 2007. Direct electron transfer with yeast cells and construction of a mediatorless microbial fuel cell. *Biosens Bioelectron* 22, 2604–2610.

85. Bretschger, O., Obraztsova, A., Sturm, C.A., Chang, I.S., Gorby, Y.A., Reed, S.B., Culley, D.E., Reardon, C.L., Barua, S., Romine, M.F., Zhou, J., Beliaev, A.S., Bouhenni, R., Saffarini, D., Mansfeld, F., Kim, B.H., Fredrickson, J.K. and Nealson, K.H. 2007. Current production and metal oxide reduction by Shewanella oneidensis MR-1 wild type and mutants. *Appl Environ Microbiol* 73, 7003–7012.

86. Holmes, D.E., Chaudhuri, S.K., Nevin, K.P., Mehta, T., Methe, B.A., Liu, A., Ward, J.E., Woodard, T.L., Webster, J. and Lovley, D.R. 2006. Microarray and genetic analysis of electron transfer to electrodes in Geobacter sulfurreducens. *Environ Microbiol* 8, 1805–1815.

87. Reguera, G., Nevin, K.P., Nicoll, J.S., Covalla, S.F., Woodard, T.L. and Lovley, D.R. 2006. Biofilm and nanowire production leads to increased current in geobacter sulfurreducens fuel cells. *Appl Environ Microbiol* 72, 7345–7348.

88. Reguera, G., McCarthy, K.D., Mehta, T., Nicoll, J.S., Tuominen, M.T. and Lovley, D.R. 2005. Extracellular electron transfer via microbial nanowires. *Nature* 435, 1098–1101.

89. Gorby, Y.A., Yanina, S., McLean, J.S., Rosso, K.M., Moyles, D., Dohnalkova, A., Beveridge, T.J., Chang, I.S., Kim, B.H., Kim, K.S., Culley, D.E., Reed, S.B., Romine, M.F., Saffarini, D.A., Hill, E.A., Shi, L., Elias, D.A., Kennedy, D.W., Pinchuk, G., Watanabe, K., Ishii, S., Logan, B., Nealson, K.H. and Fredrickson, J.K. 2006. Electrically conductive bacterial nanowires produced by Shewanella oneidensis strain MR-1 and other microorganisms. *Proc Natl Acad Sci U S A* 103, 11358–11363.

90. Richter, H., Nevin, K.P., Jia, H.F., Lowy, D.A., Lovley, D.R. and Tender, L.M. 2009. Cyclic voltammetry of biofilms of wild type and mutant Geobacter sulfurreducens on fuel cell anodes indicates possible roles of OmcB, OmcZ, type IV pili, and protons in extracellular electron transfer. *Energy Environ Sci* 2, 506–516.

91. Zhang, T., Cui, C.Z., Chen, S.L., Ai, X.P., Yang, H.X., Ping, S. and Peng, Z.R. 2006. A novel mediatorless microbial fuel cell based on direct biocatalysis of Escherichia coli. *Chem Commun* 21, 2257–2259.

92. Wang, Y.F., Tsujimura, S., Cheng, S.S. and Kano, K. 2007. Self-excreted mediator from Escherichia coli K-12 for electron transfer to carbon electrodes. *Appl Microbiol Biotechnol* 76, 1439–1446.

93. Qiao, Y., Li, C.M., Bao, S.J., Lu, Z.S. and Hong, Y.H. 2008. Direct electrochemistry and electrocatalytic mechanism of evolved Escherichia coli cells in microbial fuel cells. *Chem Commun* 11, 1290–1292.

94. Zhang, T., Cui, C., Chen, S., Yang, H. and Shen, P. 2008. The direct electrocatalysis of Escherichia coli through electroactivated excretion in microbial fuel cell. *Electrochem Commun* 10, 293–297.

95. Qiao, Y., Li, C.M., Lu, Z., Ling, H., Kang, A. and Chang, M.W. 2009. A time-course transcriptome analysis of Escherichia coli with direct electrochemistry behavior in microbial fuel cells. *Chem Commun* 41.

96. Xiang, K., Qiao, Y., Ching, C.B. and Li, C.M. 2009. GldA Overexpressing-engineered E.coli as Superior Electrocatalyst for Microbial Fuel Cells. *Electrochem Commun* 11, 6183–6185.

97. Tender, L.M., Reimers, C.E., Stecher, H.A., Holmes, D.E., Bond, D.R., Lowy, D.A., Pilobello, K., Fertig, S.J. and Lovley, D.R. 2002. Harnessing microbially generated power on the seafloor. *Nat Biotechnol* 20, 821–825.

98. Aelterman, P., Rabaey, K., Clauwaert, P. and Verstraete, W. 2006. Microbial fuel cells for wastewater treatment. *Water Sci Technol* 54, 9–15.

99. Scott, K. and Murano, C. 2007. A study of a microbial fuel cell battery using manure sludge waste. *J Chem Technol Biotechnol* 82, 809–817.

100. Niessen, J., Harnisch, F., Rosenbaum, M., Schroder, U. and Scholz, F. 2006. Heat treated soil as convenient and versatile source of bacterial communities for microbial electricity generation. *Electrochem Commun* 8, 869–873.

101. Logan, B.E. 2009. Exoelectrogenic bacteria that power microbial fuel cells. *Nat Rev Microbiol* 7, 375–381.

102. Feng, Y.J., Yang, Q., Wang, X. and Logan, B.E. 2010. Treatment of carbon fiber brush anodes for improving power generation in air-cathode microbial fuel cells. *J Power Sources* 195, 1841–1844.

103. You, S.J., Ren, N.Q., Zhao, Q.L., Wang, J.Y. and Yang, F.L. 2009. Power generation and electrochemical analysis of biocathode microbial fuel cell using graphite fibre brush as cathode material. *Fuel Cells* 9, 588–596.

104. Zhang, G.D., Zhao, Q.L., Jiao, Y., Zhang, J.N., Jiang, J.Q., Ren, N. and Kim, B.H. 2011. Improved performance of microbial fuel cell using combination biocathode of graphite fiber brush and graphite granules. *J Power Sources* 196, 6036–6041.

105. Siu, C.P.B. and Chiao, M. 2008. A microfabricated PDMS microbial fuel cell. *J Microelectromech Syst* 17, 1329–1341.

106. Ghasemi, M., Shahgaldi, S., Ismail, M., Yaakob, Z. and Daud, W.R.W. 2012. New generation of carbon nanocomposite proton exchange membranes in microbial fuel cell systems. *Chem Eng J* 184, 82–89.

Index

Page numbers followed by f and t indicate figures and tables, respectively.

A

AA–CVD (aerosol-assisted chemical vapor deposition)
  methods for synthesizing CNTs, 448–449
AAO (anodic aluminium oxide), 85, 93–94, 94f
  hard templates, 412, 413, 414t, 420
Abiotic cathode, catalysis in, 530–531
Acid-doped polymer membranes, 507–508, 508f
Acidic polymer electrolyte, with alkaline
  replacement, 482
Acidithiobacillus ferrooxidans, in MFCs, 532
ACR (Advanced Ceramic Reactor) project, 100
Acumentrics Corporation, 99
Adaptive Materials, Inc., 99
Advanced at Industrial Science and Technology (AIST), 110, 111f
Advanced Ceramic Reactor (ACR) project, 100
AE (alkaline earth) metal oxides, 167
AEOR (Alcohol electrooxidation reaction), 376–377, 379, 380, 385, 390, 391, 426
Aeromonas hydrophila, in MFCs, 539
Aerosol-assisted chemical vapor deposition (AA–CVD)
  methods for synthesizing CNTs, 448–449
AES (Auger electron spectroscopy), 22
AFCs (alkaline fuel cells), 6, 8
AFL (anode functional layer), 101, 102
AFRL (Air Force Research Laboratory), 353
A(H,D)$_2$PO$_4$-type (A = K$^+$, Rb$^+$, Cs$^+$, Tl$^+$, etc.) composites, family
  inorganic solid acid composites, 510

A(H,D)$_2$SO$_4$-type (A = K$^+$, Rb$^+$, Cs$^+$, Tl$^+$, etc.) family
  inorganic solid acid composites, 510–512, 511f
Air Force Research Laboratory (AFRL), 353
AIST (Advanced at Industrial Science and Technology), 110, 111f
AITF (Alberta Innovates-Technology Futures), 101, 102, 103, 104, 108, 110, 112, 113, 114
Alberta Innovates-Technology Futures (AITF), 101, 102, 103, 104, 108, 110, 112, 113, 114
Alcohol electrooxidation reaction (AEOR), 376–377, 379, 380, 385, 390, 391, 426
Alcohol oxidation
  on Pd nanocubes, 390–391, 390f
  on Pt alloy nanocubes, 387–388
  on Pt nanocubes, 386, 387, 387f
ALD. *See* Atomic layer deposition (ALD)
ALE (atomic layer epitaxy). *See* Atomic layer deposition (ALD)
Alkaline earth (AE) metal oxides, 167
Alkaline fuel cells (AFCs), 6, 8
Alkaline polymer electrolytes (APEs), for fuel cell applications, 481–499
  acidic polymer electrolyte with alkaline, replacement, 482
  degradation and chemical stability, 497–499, 498f, 499f
  high-performance, development. *See* High-performance APEs, development
  overview, 481–483
  requirements, 482–483
  synthesis and design, challenges, 483–487

cross-linking mechanism,
    485–486, 485f, 486f
high-performance, strategies for
    designing, 486–487, 487f
realistic challenge, 485–486, 485f,
    486f
three major approaches, 483–485,
    484f
Alkalis, 11
Alloy nanowires, 421–424
    application, 423, 424
    cyclic voltammograms, 423f
    cyclic voltammograms of
        electrospun bimetallic
        nanowires, 423f
    morphology of, 421
    synthesis, 421–422
    TEM images, 422f
Anode, MFC
    catalysis in, 527–530
        biocatalysts, 527–529
        electrode materials, 529–530
    materials, 532–536
        CNT-/nanoparticle-based,
            532–534, 533f
        graphene-based, 534–536, 535f
        other nanomaterials, 536, 536f
Anode catalysts, 27f, 28, 29, 31
    for DHFCs, 275, 275f
    for DMFCs, 275, 276–277, 276t, 277f,
        278f
Anode development, in DIR-MCFC,
    179–186
    activity of deactivated catalysts in
        methane steam, 183f
    alkali poisoning effect, 181
    catalytic activity
        deterioration of, 180–183
        liquid-phase status, 180
        vapor phase pollution, 180
    coke formation promoter, 181, 182f
    configuration, 179f
    lanthanum oxycarbonate species,
        formation, 181–182
    lithium aluminates in, 182–183
    mechanism of poisoning, 185–186,
        186f
    methane and ethanol fuel, 179–180
    potassium in, 183

steam reforming of ethanol, 181f
TEM images of bare and alkali-
    doped catalysts, 184f
TPR profiles of electrolyte-added
    catalysts, 185f
transport mechanism of alkali
    species, 180
Anode development, in SOFCs, 26–34
    cermet anode, 27–29, 27f, 28f
    other materials, 33–34
    perovskite oxide anode, 29–33, 30t
    requirements, 26–27
Anode functional layer (AFL), 101, 102
Anode-supported second-generation
    µT–SOFCs, 99–107, 100f, 104f
    AFL, 101, 102
    ASL, 102
    cross-sectional SEM micrograph,
        101f
    EPD technique, 101, 102–103, 102f
        YSZ-electrolyte microtube
        fabricated by, 103f
    polymer hollow fiber, 106–107, 107f
    REM, 104–105, 104f, 105f
    samarium-doped strontium cobaltite
        cathod formation, 105, 105f
    sintered microtubular SOFC without
        cathode, 103f
    thermal cycling test, 106, 106f
    thermal shock resistance, 105–106,
        106f
Anode-supported SOFC, defined, 79
Anode-support layer (ASL), 102
Anodic aluminium oxide (AAO), 85,
    93–94, 94f
    hard templates, 412, 413, 414t, 420
Antimonic acid, as additives in
    polymers, 509
APEs. *See* Alkaline polymer electrolytes
    (APEs)
APUs (auxiliary power units), 6
*a*QAPS. *See* Self-aggregated QAPS
    (*a*QAPS)
Area-selective ALD technique, 271
Area specific resistance (ASR), 72–73
Arylene-based polymers, hydrocarbon
    membranes, 351–353, 352f
ASL (anode-support layer), 102
ASR (area specific resistance), 72–73

Atomic layer deposition (ALD), 81–82,
82f
Atomic layer deposition (ALD), of
metals and metal oxides for
fuel cell applications, 250–279
advantages, 254–255
for catalysts and electrodes, 268, 269,
270, 271f
characteristics, 254–255
defined, 253
for electrolytes, 266–268, 267f, 268f
ALD coating, 267
MEMS processing steps, 266, 268f
YDC, 262, 266–267, 268, 270f
YSZ, 262, 266, 267, 267f, 269f
for fuel cells, 265–277
PEMFCs. *See* Proton exchange
membrane fuel cells (PEMFCs)
SOFCs. *See* Solid oxide fuel cells
(SOFCs), ADL for
fundamentals, 252–257
key deposition factors of, 255–257
precursors, 256, 257
substrate, 255, 256f
temperature, 256, 257f
time, 257
materials fabricated by, 258–265
coatings, 263–265, 264f
metal oxide, 262–263, 263f
metals. *See* Metals for fuel cell
applications
mixed metals, 260–262, 261f
for other components, 271, 272f
overview, 250–252
principle, 253–254, 253f
self-limiting growth mechanism,
254–255
Atomic layer epitaxy (ALE). *See* Atomic
layer deposition (ALD)
Auger electron spectroscopy (AES), 22
Auxiliary power units (APUs), 6

B

Balance of plant (BOP), defined, 71
Ballard advanced materials (BAM)
membrane
in PEMs for DMFCs, 346
chemical structure, 346f

Ballard Power Systems, 346
BAM (Ballard advanced materials)
membrane
in PEMs for DMFCs, 346
chemical structure, 346f
Barium aluminosilicate glass, strain at
fracture of, 54–55
Barium–calcium–aluminosilicate
(BCAS), 56
Barium chromate, formation of, 56–57,
56f
BCAS (barium–calcium–
aluminosilicate), 56
BDD (boron-doped diamond)
supported catalysts, for low-
temperature fuel cells, 236–238,
237f
characteristics of, 240t
Biocatalysts, in MFCs
advances in, 539–541
engineered microorganisms,
540–541
exoelectrogenic bacteria, 539–540
mixed communities, 541
anode, 527–529
Biocathodes, catalysis in, 531–532
Biomass-derived synthetic fuels, in fuel
cells, 9, 11
Bismuth, in DCFCs, 147
BOP (balance of plant), defined, 71
Boride-supported catalysts, for PEMFCs,
302, 302f, 303f
Boron-doped diamond (BDD)
supported catalysts, for low-
temperature fuel cells, 236–238,
237f
characteristics of, 240t
Boron oxide
on CTE, 50
glass volatility and, 57–58, 58t

C

Carbide-supported catalysts, for
PEMFCs, 303–307
silicon carbide (SiC), 306–307, 307t
titanium carbide (TiC), 305–306, 305t,
306f, 306t
tungsten carbide (WC), 303–305, 304t

Carbonate content, influence of
in HDCFC, 152–153, 152f, 153f
Carbon black nanoparticles, in PEMFC
ORR, 204–205
Carbon black particles
CNTs and CNFs–supported catalysts
for DMFCs, 444, 445, 446t
on Pt-based catalysts, 444
Carbon black–supported catalysts, for
low-temperature fuel cells,
224–225
Carbon capture and storage (CCS), 128
Carbon corrosion mechanism, in
PEMFCs, 292–293, 293f, 294f
Carbon gel–supported catalysts, for
low-temperature fuel cells,
227–229
carbon aerogels, 227, 228–229, 228f
carbon cryogels, 228
carbon xerogels, 228
characteristics of, 240t
in DMFC performance test, 228
Carbon materials, catalysis in MFC
anode and, 529
Carbon monoxide (CO)
and CO-like oxidation on Ru@Pt,
396–398, 397f
stripping voltammetry, 396
Carbon nanocoils (CNCs), 224
supported catalysts, for low-
temperature fuel cells, 234–235,
235f
characteristics of, 240t
Carbon nanofibers (CNFs), 224
supported catalysts, for low-
temperature fuel cells, 232–234
characteristics of, 240t
CNT vs., 233
herringbone, 233
for PEMFCs, 233–234
stacked-cup, 233
TEM images, 233f
types, 233
supported catalysts for DMFCs,
449–450. *See also* Direct
methanol fuel cells (DMFCs),
CNTs and CNFs–supported
catalysts for
SC-CNFs, 464–465, 465f

synthesis, 457–462. *See also*
Synthesis, of CNTs and
CNFs–supported catalysts for
DMFCs
Carbon nanohorns (CNHs), 224
supported catalysts, for low-
temperature fuel cells, 234–235
characteristics of, 240t
single-walled, 234
Carbon nanotubes (CNTs), 224
ALD in, 262–263, 263f
characteristics of, 229, 240t
nanoparticle-based anode materials,
in MFCs, 532–534, 533f
nitrogen-doped. *See* Nitrogen-doped
CNTs (NCNTs)
in PEMFC ORR, 205–207, 206f
supported catalysts for DMFCs, 447–
449. *See also* Direct methanol
fuel cells (DMFCs), CNTs and
CNFs–supported catalysts for
CVD method. *See* Chemical vapor
deposition (CVD), methods for
synthesizing CNTs
DWCNTs, 447, 448f
mechanical and electrical
properties, 447
methods for synthesizing, 447, 448
MWCNTs, 447, 448f
SEM images, 449f
SWCNTs, 447, 448f
synthesis, 450–457. *See also*
Synthesis, of CNTs and CNFs–
supported catalysts for DMFCs
supported catalysts for low-
temperature fuel cells, 229–232
carbon black–supported metal
and, 231
chronoamperometry tests, 232
CNF vs., 233
MWCNTs, 229, 230, 230f, 231, 232
Nafion electrodes and, 231–232
SWCNTs, 229, 231, 232
TEM images, 230f
tubular structure, 229
Carbon support materials, CNTs and
CNFs–supported catalysts for
DMFCs, 444, 445–446, 446t
Carnot cycle, 2, 2f

Casting polymer solution methods,
      in hydrocarbon membranes,
      348–349
Catalysis
   in MFC anode, 527–530
      biocatalysts, 527–529
      electrode materials, 529–530
   in MFC cathode, 530–532
      abiotic cathode, 530–531
      biocathodes, 531–532
Catalysts. *See also specific* entries
   ADL for, 268, 269, 270, 271f
   anode, 27f, 28, 29, 31
   nanoparticles, distribution
      challenges to nanocatalysts in
      DMFCs, 466
   noncarbon material–supported,
      for PEMFCs. *See* Noncarbon
      material–supported
      electrocatalysts
Cathode, MFC
   catalysis in, 530–532
      abiotic cathode, 530–531
      biocathodes, 531–532
   materials, 537–539
      Pt-based cathodes, 537
      Pt-free cathodes, 537–539, 538f
Cathode catalysts, for ORR, 273–275,
      274f
Cathode materials, in MCFCs, 164–179
   application, 173
   cerium oxide, 174–175
   CoO–Ni composites, 167–171
   dissolution of composite cathodes,
      178f
   evolution of impedance spectra with
      operation time, 174f
   GSC–NiO, 177, 178
   Li/Na, molten carbonate melts,
      166–167
   lithium, 165–179
   LSC–NiO, 177–178
   neutron diffraction, for structural
      changes of modified cathode
      materials, 172f
   perovskite ABO_3 compounds, 177
   porous Ni plate in, 164–179
   Raman spectrum data, 168f
   SEM observation, 171
      sol-impregnation technique, 173
      stability of materials, 166–168
      TGA analysis of sintered nickel, 169f
      theoretical mechanisms of oxygen
         reduction in carbonate melts,
         177t
      time dependence of Ni solubility,
         175f
      time dependency of ions' mole
         fraction, 170f
      titanium dioxide, 176, 176f
      transport models for, 173
      voltage degradation of, 164f
Cathodes, development, for SOFCs,
      16–26
   lanthanum manganite (LSM)–based
      cathodes, 17–19, 19f
   lanthanum strontium cobalt (LSC)–
      based cathodes, 19–22, 20f
   layered perovskite cathodes, 22–23,
      23f
   materials, modeling approach to,
      23–26, 24f, 25f, 26f
   perovskite oxides, 16–17, 17f
Cathode-supported SOFC, defined, 79
CCS (carbon capture and storage), 128
Ceramic Fuel Cells, Ltd., (CFCL), 11
Ceramic proton conductors, inorganic
      oxides, 512–515
   porous, periodic conductive
      structures and, 513–515, 514f
   yttria-doped, 512–513
Cerium oxide, cathode materials
   in MCFC development, 174–175
Cermet anode, for SOFCs, 27–29, 27f, 28f
CFCL (Ceramic Fuel Cells, Ltd.), 11
CFD (computational fluid dynamic)
      packages, 114
Chemical stability
   APEs, 497–499, 499f
   sealants, in planar SOFCs, 55–57, 56f
Chemical vapor deposition (CVD), 81,
      203
   methods for synthesizing CNTs, 447,
      448–449
      aerosol-assisted CVD (AA–CVD),
         448–449
      floating catalyst CVD (FC–CVD),
         448

Chitosan, 354
Clostridium butyricum, in MFCs, 539
CLPE (cross-linked high-density
　　polyethylene), 360
Cluster-network model, of Nafion
　　membrane, 322–323, 323f
CMK-3 carbons, 226
　　OMC performance in DMFCs, 226
　　TEM images of, 227f
CMK-5 mesoporous carbon, 226
CNCs (carbon nanocoils), 224
　　supported catalysts, for low-
　　　　temperature fuel cells, 234–235,
　　　　235f
　　characteristics of, 240t
CNFs. *See* Carbon nanofibers (CNFs)
CNHs (carbon nanohorns), 224
　　supported catalysts, for low-
　　　　temperature fuel cells, 234–235
　　characteristics of, 240t
　　single-walled, 234
CNTs. *See* Carbon nanotubes (CNTs)
CO (carbon monoxide)
　　and CO-like oxidation on Ru@Pt,
　　　　396–398, 397f
　　stripping voltammetry, 396
Coal, in energy economies, 127–128
Coal-derived synthetic fuels, in fuel
　　cells, 9, 11
Coatings in fuel cell applications, ALD
　　of, 263–265, 264f, 267
Cobalt, cathode materials
　　in MCFC development, 167–179
　　CoO–Ni composites, 167–171
Cobalt oxide (CoO_x), in MFCs, 537
Coefficient of thermal expansion (CTE)
　　matching, 46, 47, 50–52, 50f, 51t
Coke formation promoter, 181, 182f
Colloidal methods
　　in CNFs synthesis, 460–461
　　in CNTs synthesis, 455
Commercial platinum black (CPB), ORR
　　on, 409–410, 410f
Composite membranes, PEMs for
　　DMFCs, 358–368
　　classification, 358
　　electro-osmotic drag, 362–363
　　LDHs, 359
　　Nafion

on conductivity and methanol
　　crossover, 366t
modification of, 363–365
MoPh-a and, 365
PFA nanocomposite, 361
PVDF *vs.*, 367t
silica membranes, 364–365, 365f
and sPEEK + sPSU + PBI,
　　comparison, 362t
sulfonated montmorillonite
　　composite, preparing, 367–368,
　　368f
zirconium membranes, 364
nanocomposite membranes, 367–368
organic-inorganic systems, 362–364
organic-organic system, 360–362
PECVD technique, 359
preparation, blending main-chain
　　polymers, 361, 362f
SPEEK, 359, 360
Compressive sealing, for planar SOFCs,
　　46
Computational fluid dynamic (CFD)
　　packages, 114
COMSOL, 114
Co–N–C catalysts, 198
Conventional SOFCs, 73
Coplanar SC-SOFCs, 75–77, 75f, 77f
Core-shell model, of Nafion membrane,
　　324
Core-shell structure catalysts, 391–405
　　electrochemical performance,
　　　　396–405
　　　　CO and CO-like oxidation on
　　　　　　Ru@Pt, 396–398, 397f
　　　　EG oxidation on Ru@Pt, 401
　　　　ethanol oxidation on, 400–401,
　　　　　　401f
　　　　methanol oxidation on, 398–400,
　　　　　　398f, 399f, 400t
　　　　ORR, 401–405, 402f, 403f, 404f
　　specific activity with test conditions,
　　　　400t
　　structure characteristics, 391–394
　　　　EDS characteristics, 394
　　　　TEM and HRTEM characteristics,
　　　　　　391–392, 392f
　　　　XPS characteristics, 393–394,
　　　　　　393f

XRD characteristics, 392–393, 392f
synthesis methods, 394–396, 395f
Covalently cross-linked ionomer membranes, in PEMs for DMFCs, 354
CPB (commercial platinum black), ORR on, 409–410, 410f
Croferr22 APU, 57
Cross-linked high-density polyethylene (CLPE), 360
Cross-linked polymer membranes, in PEMs for DMFCs, 355
Cross-linking mechanism, for APEs, 485–487, 485f, 486f, 487f
   self-cross-linking design. *See* Self-cross-linked QAPS (*x*QAPS)
Crystallization, planar SOFCs and, 59–61, 60f
CTE (coefficient of thermal expansion) matching, 46, 47, 50–52, 50f, 51t
Current collector supported SOFC, defined, 79
CVD (chemical vapor deposition), 81, 203
   methods for synthesizing CNTs, 447, 448–449
     aerosol-assisted CVD (AA–CVD), 448–449
     floating catalyst CVD (FC–CVD), 448

D

DAFCs. *See* Direct alcohol fuel cells (DAFCs)
Dais membranes, in PEMs for DMFCs, 350–351
DCFCs. *See* Direct carbon fuel cells (DCFCs)
Decorated structure catalysts. *See* Core-shell structure catalysts
DEFCs (direct ethanol fuel cells), 7
Degradation, of APEs, 497–499, 498f, 499f
Density functional theory (DFT) calculations, 402, 403–404, 403f
Deposited carbon, in DCFC, 149
Deposition factors, of ALD precursors, 256, 257

substrate, 255, 256f
temperature, 256, 257f
time, 257
DFAFCs (direct formic acid fuel cells), 7
DFT (density functional theory) calculations, 402, 403–404, 403f
DHFCs (direct hydrogen fuel cells) anode catalysts for, 275, 275f
Digital mobile broadcasting (DMB) systems, 335
Dimethyl ether (DME), in fuel cells, 9, 10t
DIR (direct internal reforming), of methane, 115
Direct alcohol fuel cells (DAFCs), 272, 275
   DMFCs. *See* Direct methanol fuel cells (DMFCs)
   nanostructures, catalysts for, 376–428
     core-shell and decorated structure catalysts. *See* Core-shell structure catalysts
     hierarchical structure catalysts, 424–426, 424f, 425f
     hollow structure catalysts. *See* Hollow structure catalysts
     nanocube catalysts. *See* Nanocube catalysts, for DAFCs
     one-dimensional structure catalysts. *See* One-dimensional structure catalysts
     other shape catalysts, 426–428, 426f, 427f, 428f
     overview, 376–379, 377f, 378f, 379f
     Pt electrocatalysts in, 376–377
     published papers, 378f
     shapes and structures, 379f
   PEMs for. *See* Proton exchange membranes (PEMs), for DAFCs
Direct carbon fuel cells (DCFCs), 7, 127–155
   advantages, 129
   applications, 130
   concept and configuration, 129
   HDCFC, 134, 150–155
     influence of carbonate content, 152–153, 152f, 153f
     optimization, 154–155, 155f
     schematics of, 151f

SOFC geometry, effects, 153–154, 154f
stability test of, 156f
molten salts–based, 130, 134–139
HDCFC. *See* Hybrid DCFC (HDCFC)
molten carbonates. *See* Molten carbonates, in DCFC
molten hydroxide, 134–136, 135f
overview, 127–128, 128f
RDCFCs, 149
research communities, 131t–132t
reviews on, 134
schematics, 133f
solid electrolyte–based, 133
solid oxide, 133, 134, 139–150
deposited carbon, 149
drawback, 140
fluidized bed concept, 142, 143f
HDCFC. *See* Hybrid DCFC (HDCFC)
molten metal, 147–148, 148f, 149f
other contact, 149–150
physical contact, 140–141, 141f, 142t
requirements, 139
in situ (external) gasification, 142, 143, 144f, 145f
solid carbon anode, 145–147, 146f
type, 130
Direct ethanol fuel cells (DEFCs), 7
Direct formic acid fuel cells (DFAFCs), 7
Direct hydrogen fuel cells (DHFCs)
anode catalysts for, 275, 275f
Direct internal reforming (DIR), of methane, 115
Direct internal reforming–molten carbonate fuel cell (DIR–MCFC)
anode development in, 179–186
activity of deactivated catalysts in methane steam, 183f
alkali poisoning effect, 181
coke formation promoter, 181, 182f
configuration, 179f
deterioration of catalytic activity, 180
lanthanum oxycarbonate species, formation, 181–182

liquid-phase status, catalytic activity and, 180
lithium aluminates in, 182–183
mechanism of poisoning, 185–186, 186f
methane and ethanol fuel, 179–180
potassium in, 183
steam reforming of ethanol, 181f
TEM images of bare and alkali-doped catalysts, 184f
TPR profiles of electrolyte-added catalysts, 185f
transport mechanism of alkali species, 180
vapor phase pollution, catalytic activity and, 180
Direct methanol fuel cells (DMFCs), 5, 12, 72, 228
anode catalysts for, 275, 276–277, 276t, 277f, 278f
CMK-3-type OMC performance, 226
CNTs and CNFs–supported catalysts for, 444–467
carbon support materials, 444, 445–446, 446t
catalyst nanoparticles, distribution, 466
challenges to nanocatalysts in, 466–467
CNFs, 449–450
CNTs. *See* Carbon nanotubes (CNTs), supported catalysts for DMFCs
DMFC electrode, fabrication, 466–467
fabrication process of electrodes and corresponding performance, 462–465, 463f, 465f
overview, 444–445
particle size control, 466
significance, 446–447
synthetic methods. *See* Synthesis, of CNTs and CNFs–supported catalysts for DMFCs
operating principle of, 444
papers related to, 377f
Direct methanol fuel cells (DMFCs), PEMs for. *See also* Proton exchange membranes (PEMs)

characteristics, 335
contradictions in, 341, 341f
with fuel supply method
   advantages and disadvantages, 340t
   requirements in materials for, 340t
Li-ion batteries in portable
   electronics, 335
methanol transport in, 345
PEM development for DMFCs,
   337–338
   ionic conductivity, 338, 341
   methanol crossover, 336, 337–338,
     341–342, 345, 346
   polymer matrix, stiffness, 337, 338
performance curve of, 336, 337f
portable, requirements for, 335–336
for portable devices, examples, 339t
requirements for, 342
DIR–MCFC. *See* Direct internal
   reforming–molten carbonate
   fuel cell (DIR–MCFC)
Dissipative particle dynamics (DPD)
   simulation, 318–319
DMB (digital mobile broadcasting)
   systems, 335
DME (dimethyl ether), in fuel cells, 9, 10t
DMFCs. *See* Direct methanol fuel cells
   (DMFCs)
Double-walled carbon nanotubes
   (DWCNTs), 447, 448f
DPD (dissipative particle dynamics)
   simulation, 318–319
Dry- or wet-chemical etching, in thin
   film planar μSOFCs, 91
Dual-chamber MFCs, 525, 526f

E

Ebonex (Altraverde, Ltd.), 295
ECSA (electrochemical surface area)
   Pt active, 293, 295, 297, 302, 305, 307,
     308
EDS (x-ray spectroscopy)
   for DAFCs
     alloy nanocubes, 383
     core-shell structure,
      characteristics, 394
EG (ethylene glycol), oxidation on
   Ru@Pt, 401

Electrical resistance, of sealants
   in planar SOFCs, 59
Electricity production, fossil fuels and, 1
Electrocatalysts, noncarbon
   material–supported
   for PEMFCs. *See* Noncarbon
     material–supported
     electrocatalysts
Electrocatalytic performance, of
   nanowires, 415–419
   Ag nanowires, ORR on, 418, 419
   Pd nanowires
     EOR on, 418, 418f
     propanol and isopropanol
      oxidation on, 417–418, 417f
   Pt nanowires
     EOR on, 416
     ORR on, 416–417, 416f
Electrochemical method
   in CNFs synthesis, 461
   in CNTs synthesis, 455–456, 456f
Electrochemical performance, in
   nanostructures for DAFCs
   of core-shell structure catalysts,
     396–405
     CO and CO-like oxidation on
      Ru@Pt, 396–398, 397f
     EG oxidation on Ru@Pt, 401
     ethanol oxidation on, 400–401, 401f
     methanol oxidation on, 398–400,
      398f, 399f, 400t
     ORR, 401–405, 402f, 403f, 404f
   hollow structure catalysts, 407–410
     ethanol oxidation, 409
     methanol oxidation, 408–409, 408f
     ORR, 409–410, 410f
   of nanocube catalysts, 385–391
     alcohol oxidation on Pd
      nanocubes, 390–391, 390f
     alcohol oxidation on Pt alloy
      nanocubes, 387–388
     alcohol oxidation on Pt
      nanocubes, 386, 387, 387f
     hydrogen adsorption/desorption
      on Pt and Pt alloy nanocubes,
      386, 386f
     oxygen reduction on Pt and Pt
      alloy nanocubes, 388–390, 389f,
      390f

Electrochemical surface area (ECSA)
Pt active, 293, 295, 297, 302, 305, 307,
308
Electrode materials
advanced, for SOFCs. *See* Solid oxide
fuel cells (SOFCs), advanced
electrode materials for
catalysis in MFC anode and, 529–530
development, in MCFCs, 163–187
cathode materials. *See* Cathode
materials, in MCFCs
Electrodes
ADL for, 268, 269, 270, 271f
fabrication, in DMFC, 462–465,
463f, 465f, 466–467. *See also*
Fabrication method
Electrolytes, ADL for, 266–268, 267f, 268f
coating, 267
MEMS processing steps, 266, 268f
YDC, 262, 266–267, 268, 270f
YSZ, 262, 266, 267, 267f, 269f
Electrolyte-supported first-generation
microtubular SOFCs, 98–99, 98f
Electro-osmotic drag, of methanol in
DMFC, 362–363
Electrophoretic deposition (EPD)
techniques, 101, 102–103, 102f,
103f, 464
Engineered microorganisms,
biocatalysts in MFCs and,
540–541
EOR (ethanol oxidation reaction)
on core-shell catalysts, 400–401, 401f
on hollow structure catalysts, 409
on Pd nanowires, 418, 418f
on Pt nanowires, 416
EPD (electrophoretic deposition)
techniques, 101, 102–103, 102f,
103f, 464
Escherichia coli, in MFCs, 534, 536,
540–541
ETFE (ethylene-tetrafluoroethylene), 347
Ethanol oxidation reaction (EOR)
on core-shell catalysts, 400–401, 401f
on hollow structure catalysts, 409
on Pd nanowires, 418, 418f
on Pt nanowires, 416
Ethylene glycol (EG), oxidation on
Ru@Pt, 401

Ethylene-tetrafluoroethylene (ETFE),
347
EXAFS (*ex situ* x-ray absorption fine
structure), 199–200, 200f
Exoelectrogenic bacteria, biocatalysts in
MFCs and, 539–540
Ex situ x-ray absorption fine structure
(EXAFS), 199–200, 200f

F

Fabrication method
of DMFC electrodes and
corresponding DMFC
performance, 462–465
EPD, 464
GDL, 462, 463–464, 465f
MEA, 462–464, 463f, 465, 465f
SC-CNFs based MEA, 464–465,
465f
schematic illustration of, 465f
of forming free-standing electrolyte
on silicon wafer, thin film
planar μSOFCs and, 82–84, 83f
FC-CVD (floating catalyst chemical
vapor deposition)
methods for synthesizing CNTs, 448
Fecralloy, 56
FEM (finite element method), 53
Fe-N-C catalysts, 197, 198
Fe-N-G (iron- and nitrogen-
functionalized graphene)
catalysts, in MFCs, 539
FEP-g-PS (radiation-grafted
poly(tetrafluoro-ethylene-
cohexa-fluoropropylene)-g-
polystyrene) membranes,
346–347
Ferricyanide cathode, catalysts in,
530–531
FeTsPc (iron tetrasulfophthalocyanine)
cathode, in MFCs, 539
Finite element method (FEM), 53
Floating catalyst chemical vapor
deposition (FC–CVD)
methods for synthesizing CNTs, 448
FLUENT, 114
Fluidized bed concept, in DCFC, 142,
143f

Formic acid, for fuel cells, 9, 10t
Fossil fuels, massive use of, 1
Foturan substrate, 91
Free-standing electrolyte on silicon
wafer, fabrication of
thin film planar µSOFCs and, 82–84,
83f
Free-standing membrane, defined, 79
Fuel cells, 128–129
ADL for, 265–277. *See also* Atomic
layer deposition (ALD)
classification, electrolyte types
and, 265
components, materials
fabricated by ALD for use as.
See Materials fabricated by
ALD
SOFCs. *See* Solid oxide fuel cells
(SOFCs), ADL for
applications, 11–13
APEs for. *See* Alkaline polymer
electrolytes (APEs), for fuel cell
applications
DCFCs. *See* Direct carbon fuel cells
(DCFCs)
fuels for, 7–11
hydrogen, methanol, formic acid,
and liquid fuels, 7–9, 10t
natural gas, hydrocarbons, and
coal- and biomass-derived
synthetic fuels, 9, 11
fundamentals, 3–7
classification, 6–7, 6f, 7t
operation principles, 3–6, 5f
micro, types, 72
DMFC, 72
hydrogen-fed, 72
micro-reformed hydrogen, 72
overview, 1–3, 2f
reactions, with OMCs–supported
catalysts, 225–227, 227f, 239t
theoretical efficiency of, 4
Fuel supply method, DMFCs with
advantages and disadvantages for,
340t
requirements in materials for, 340t
Functionalization process
in CNFs synthesis, 457–459, 458f
in CNTs synthesis, 450–451

G

Gadolinium-doped ceria (GDC), 15, 84
electrolytes, ALD in, 262
Gas diffusion layers (GDL), 462,
463–464, 465f
GDC (gadolinium-doped ceria), 15, 84
electrolytes, ALD in, 262
GDL (gas diffusion layers), 462, 463–464,
465f
Gd_x strontium cobalt (GSC)
coated NiO, 177, 178
Gebel's model, of Nafion membrane,
323, 324
Geobacter metallireducens, in MFCs, 532
Geobacter sulfurreducens, in MFCs, 528,
540
Glass and glass-ceramic sealing, for
planar SOFCs, 46, 49
CTE of, 50–52, 50f, 51t
Glass volatility, of sealants
in planar SOFCs, 57–58, 58t
GNSs (graphene nanosheets), 235
GO (graphene oxide counterpart)
supported catalysts, for low-
temperature fuel cells, 235–236,
236f
characteristics of, 240t
Gore-Select membrane, 344
Graft copolymers, in PEMs for DMFCs,
355
Grafting, for APE synthesis
chemical, 484
physical, 483–484, 484f
Graphene, in PEMFC ORR, 207, 208f
Graphene-based anode materials, in
MFCs, 534–536, 535f
Graphene nanosheets (GNSs), 235
Graphene oxide counterpart (GO)
supported catalysts, for low-
temperature fuel cells, 235–236,
236f
characteristics of, 240t
Grotthus mechanisms (structural
diffusion)
for proton transport of Nafion with
increasing water content,
326–327
proton transports, 337

Grove, William, 2
GSC (Gd$_x$ strontium cobalt)
   coated NiO, 177, 178

H

Hansenula anomala, in MFCs, 539
Hard template method, for nanowires
      synthesis, 412, 413–414, 414t,
      420, 422
   AAO, 412, 413, 414t, 420
HCS (hollow carbon sphere)
   supported catalysts, for low-
      temperature fuel cells, 238–239,
      238f
   characteristics of, 240t
HDCFC (hybrid DCFC). *See* Hybrid
   DCFC (HDCFC)
Heating temperature, in PEMFC ORR,
      215–216, 216f
Helium leak test, in hermeticity of seals,
      49
Hermeticity, of seals for planar SOFCs,
      47–49, 48t
Herringbone CNFs, 233
Heteropoly acid (HPA), 515–516, 516f,
      517f
HF (hollow fiber), polymer, 106–107, 107f
Hierarchical structure catalysts,
      424–426, 424f, 425f
High-performance APEs
   development, 488–497
      MD simulations, structural
         insights from, 495f, 496–497
      self-aggregating design, 488–489,
         489f. *See also* Self-aggregated
         QAPS (*a*QAPS)
      self-cross-linking design, 488–489,
         489f. *See also* Self-cross-linked
         QAPS (*x*QAPS)
   strategies for designing, 486–487, 487f
High resolution TEM (HRTEM) image
   for DAFCs
      core-shell structure,
         characteristics, 391–392, 392f
      gold nanowires, 411f
      Pd nanocubes, characteristics,
         382f, 383
      Pt nanocube, 380, 381f

HMPNNs (hollow mesoporous PtNi
      nanospheres), 408–409, 408f
Hofmann elimination, degradation
      mechanism, 497, 499f
Hollow carbon sphere (HCS)
   supported catalysts, for low-
      temperature fuel cells, 238–239,
      238f
   characteristics of, 240t
Hollow fiber (HF), polymer, 106–107,
      107f
Hollow mesoporous PtNi nanospheres
      (HMPNNs), 408–409, 408f
Hollow structure catalysts, 405–410
   electrochemical performance,
      407–410
      ethanol oxidation, 409
      methanol oxidation, 408–409, 408f
      ORR, 409–410, 410f
   structure characteristics, 405–406, 405f
   synthesis methods, 406, 406f
HOR (hydrogen oxidation reaction), 292,
      294, 308
   Pt/WC catalysts for, 304–305
HPA (heteropoly acid), 515–516, 516f,
      517f
HRTEM (high resolution TEM) image
   for DAFCs
      characteristics, core-shell
         structure catalysts, 391–392,
         392f
      gold nanowires, 411f
      Pd nanocubes, characteristics,
         382f, 383
      Pt nanocube, 380, 381f
Hummer's method, 236
Hybrid DCFC (HDCFC), 134, 150–155
   influence of carbonate content,
      152–153, 152f, 153f
   optimization, 154–155, 155f
   schematics of, 151f
   SOFC geometry, effects, 153–154, 154f
   stability test of, 156f
Hydrocarbon membranes
   in PEMs for DMFCs, 348–358
      arylene-based polymers, 351–353,
         352f
      asymmetric acrylic membranes,
         structure, 357f

casting polymer solution methods, 348–349
chitosan-based membranes, 354
covalently cross-linked ionomer membranes, 354
cross-linked polymer membranes, 355
Dais membranes, 350–351
defined, 348
graft copolymers, 355
methanol concentration on methanol crossover, 357t
PBI-based membranes, 356–357, 357f
polymer membranes, 348–357
polyphosphazene-based polymers, 351, 351f
SFRP techniques, 356
S-SIBS membranes, 355
sulfonation methods, 349–350
thickness on methanol crossover for Nafion and 6FCN-35 membranes, 353, 353f
types, 350
Hydrocarbons, for fuel cells, 9, 11
Hydrogen adsorption/desorption, on Pt and Pt alloy nanocubes, 386, 386f
Hydrogen-fed micro fuel cell, 72
Hydrogen gas, for fuel cells, 7–8, 10t
Hydrogen oxidation reaction (HOR), 292, 294, 308
Pt/WC catalysts for, 304–305

I

ICE (internal combustion engine), 2
ICP-AES, 171–172
ICP-MS (inductively coupled plasma mass spectrometry)
for DAFCs, alloy nanocubes, 383
IEC (ion-exchange capacity), APEs with, 485–486
IGFC (Integrated gasification fuel cell), 1
IGR (indirect gradual reforming), of methane, 115
IMH (intermittent microwave heating) technique, 239

Impregnation method
in CNFs synthesis, 459
in CNTs synthesis, 452, 453f
Indirect gradual reforming (IGR), of methane, 115
Inductively coupled plasma mass spectrometry (ICP-MS)
for DAFCs, alloy nanocubes, 383
Infiltration method, SOFC and, 22, 34–35
Inorganic proton conductors for PEMFCs, high-temperature, 505–516
ceramic proton conductors, oxides, 512–515
porous, periodic conductive structures and, 513–515, 514f
yttria-doped, 512–513
heteropoly acid (HPA), 515–516, 516f, 517f
overview, 505–507, 506f
protonic acids in liquid phase, 507–509
acid-doped polymer membranes, 507–508, 508f
antimonic acid as additives in polymers, 509
phosphoric acid–functionalized inorganic solids, 508–509, 509f
solid acid composites, 509–512
$A(H,D)_2PO_4$-type (A = K^+, Rb^+, Cs^+, Tl^+, etc.) composites, family of, 510
$A(H,D)_2SO_4$-type (A = K^+, Rb^+, Cs^+, Tl^+, etc.) family, 510–512, 511f
metal diphosphates, 512
In-plane ionic conductivity, 341
In situ formed graphitized carbon nanostructures
in PEMFC ORR, 209–216, 210f
heating temperature, 215–216, 216f
nitrogen precursors, 210–214, 211f, 212f, 213f
transition metals, 214–215, 215f
In situ (external) gasification, in DCFC, 142, 143, 144f, 145f
Integrated gasification fuel cell (IGFC), 1
Intermittent microwave heating (IMH) technique, 239
Internal combustion engine (ICE), 2

International Union of Pure and
Applied Chemistry (IUPAC),
241
Ion-exchange capacity (IEC), APEs with,
485–486
Ionic conductivity, in PEMs for DMFCs,
338, 341
Ionic conductivity, of APEs, 485–486,
487, 487f. *See also* Self-
aggregated QAPS (*a*QAPS)
Ionomer peak, in Nafion membranes,
322, 325
Ionomers, modifications of
in PEMs for DAFCs, 358–368. *See also*
Composite membranes
IrNi@PtRu catalysts, methanol oxidation
on, 398
Iron- and nitrogen-functionalized
graphene (Fe-N-G) catalysts, in
MFCs, 539
Iron phthalocyanine (FePc), in MFCs,
537
Iron tetrasulfophthalocyanine (FeTsPc)
cathode, in MFCs, 539
Isopropanol oxidation, on Pd
nanowires, 417–418, 417f
IUPAC (International Union of Pure and
Applied Chemistry), 241

K

Key deposition factors, of ALD, 255–257
Klebsiella pneumoniae, in MFCs, 536

L

Lamellar model, of Nafion membrane,
324
Lanthanum manganite (LSM)–based
cathodes, for SOFCs, 17–19, 19f
Lanthanum strontium cobalt (LSC)–
based cathodes
coated NiO, 177–178
for SOFCs, 19–22, 20f
Lawrence Berkeley Laboratory, 84
Layered double hydroxides (LDHs), 359
Layered perovskite cathodes, for SOFCs,
22–23, 23f
LDHs (layered double hydroxides), 359

Leak rates, of seals, 48t, 49
Liquefied petroleum gas (LPG), 110, 112
Liquid fuels, for fuel cells, 8–9, 10t
Liquid-phase status
catalytic activity, in DIR-MCFC, 180
Lithium, in MCFC development
anode materials
lithium aluminates in, 182–183
cathode materials, 165–179, 171f
dissolution mechanism, 172–173
Li/Na, molten carbonate melts,
166–167
Los Alamos National Laboratory, 351
Low-temperature fuel cells
DEFCs. *See* Direct ethanol fuel cells
(DEFCs)
DMFCs. *See* Direct methanol fuel
cells (DMFCs)
nanostructured carbon-
based catalysts for. *See*
Nanostructured carbon-based
catalysts for low-temperature
fuel cells
PEMFCs. *See* Polymer electrolyte
membrane fuel cells
(PEMFCs)
LPG (liquefied petroleum gas), 110, 112
LSC (lanthanum strontium cobalt)–
based cathodes
coated NiO, 177–178
for SOFCs, 19–22, 20f
LSM (lanthanum manganite)–based
cathodes, for SOFCs, 17–19, 19f

M

Magneli phase titania-supported
catalysts, for PEMFCs, 294–295
Mass transfer modeling, of µT–SOFCs,
115–116
Materials fabricated by ALD
for use as fuel cell components,
258–265
coatings, 263–265, 264f
metal oxide, 262–263, 263f
metals. *See* Metals for fuel cell
applications
mixed metals, 260–262, 261f
Matrix peak, defined, 321–322

MCFCs. *See* Molten carbonate fuel cells (MCFCs)
MD (molecular dynamics) simulations
in Nafion membranes, 317–318
for structure of hydrated *a*QAPS and *x*QAPS, 495f, 496–497
MEA (membrane electrode assembly), 6, 336–337, 341
in DMFC, 462–464, 463f, 465, 465f
Mechanical milling method (MMM), 511, 517
Membrane electrode assembly (MEA), 6, 336–337, 341
in DMFC, 462–464, 463f, 465, 465f
Membrane materials, in MFCs, 542–543
MEMS (microelectromechanical systems), 12
processing steps, 266, 268f
Metal diphosphates, inorganic solid acid composites, 512
Metal-doped titania-supported catalysts, for PEMFCs, 295–299, 299t
polarization and power density curves, 298f
single-cell polarization curves, 296f
XANES spectra of PtL edge, 297f
Metal oxide-based supported catalysts, for PEMFCs, 294–301
Magneli phase titania, 294–295
metal-doped titania, 295–299, 299t
polarization and power density curves of PEMFCs, 298f
single-cell polarization curves, 296f
XANES spectra of PtL edge, 297f
NSTO, 299–300
SiO$_2$, 301
sulfated zirconia (S-ZrO$_2$), 300
tin oxide (SnO$_2$)-based material, 300–301
tungsten oxide (WO$_x$), 301
Metal oxide for fuel cell applications, ALD of, 262–263, 263f. *See also* Atomic layer deposition (ALD)
Metals for fuel cell applications, ALD of, 258–260. *See also* Atomic layer deposition (ALD)
palladium (Pd), 259–260, 260f
platinum (Pt), 258–259, 259f

Metal-supported SOFC, defined, 79
Metal–support interaction, on catalyst performance
in PEMFCs, 308
Methanol, in fuel cells, 8–9, 10t
Methanol crossover, in PEMs for DMFCs, 341–342, 345, 346, 358
defined, 337
methanol concentration on, 357t
reduction of, 336, 337–338, 358, 359–360, 362
thickness on, Nafion and 6FCN-35 membrane, 353, 353f
Methanol oxidation reaction (MOR), 463
on core-shell catalysts, 398–400, 398f, 399f, 400t
IrNi@PtRu, 398
Pd@Pt, 398, 398f
Pt@Pd, 399
Ru@Pt, 398–399
on hollow structure catalysts, 408–409, 408f
Methanol transport, in DMFCs, 345
MFCs. *See* Microbial fuel cells (MFCs)
Microbial fuel cells (MFCs), 524–543
anode materials, 532–536
CNT-/nanoparticle-based, 532–534, 533f
graphene-based, 534–536, 535f
other nanomaterials, 536, 536f
biocatalysts, advances in, 539–541
engineered microorganisms, 540–541
exoelectrogenic bacteria, 539–540
mixed communities, 541
cathode materials, 537–539
Pt-based cathodes, 537
Pt-free cathodes, 537–539, 538f
classification, 525–527
dual-chamber, 525, 526f
sediment, 526f, 527
single-chamber, 525–527, 526f
tubular upflow, 526f, 527
configurations, 526f
MFC anode, catalysis in, 527–530
biocatalysts, 527–529
electrode materials, 529–530

MFC cathode, catalysis in, 530–532
abiotic cathode, 530–531
biocathodes, 531–532
nanostructured electrode materials
in, 532–539
other materials applied in, 541–543
membrane materials, 542–543
three-dimensional structured
electrode, 541–542, 542f
overview, 524–525
principles, 525–532
Microelectromechanical systems
(MEMS), 12
processing steps, 266, 268f
Micro-reformed hydrogen fuel cells, 72
Micro SOFCs (μSOFCs), 7, 69–118
microtubular. *See* Microtubular
SOFCs (μT–SOFCs)
overview, 69–73, 70f, 71f, 72f
potential electrical power range, 70f
power-generation system, 71f
single chamber, 73–78, 73f
thin film based planar. *See* Thin film
planar μSOFCs
Microtubular SOFCs (μT–SOFCs), 97–117
anode-supported second-generation.
See Anode-supported second-
generation μT–SOFCs
electrolyte-supported first-
generation, 98–99, 98f
modeling, 114–117
mass transfer, 115–116
mathematical, 115
reliability, 116
PES third-generation, 107–108, 109f,
110f
stack, 110–114, 111f, 112f, 113f
VPD in, 97, 97f
Microwave-heated polyol method
in CNFs synthesis, 459–460, 460f
in CNTs synthesis, 453–455, 454f
Microwave plasma technique, 22
MIECs (mixed ionic and electronic
conductors), 177
Mixed communities, biocatalysts in
MFCs, 541
Mixed ionic and electronic conductors
(MIECs), 177

Mixed metals for fuel cell applications,
ALD of, 260–262, 261f. *See also*
Atomic layer deposition (ALD)
MMM (mechanical milling method),
511, 517
M–N–C catalysts, 196
Mobile devices
with DMFCs, 336
functions, 335
Mobile electronics, thin film planar
μSOFCs for. *See* Thin film
planar μSOFCs
Modeling, of μT–SOFCs, 114–117
mass transfer, 115–116
mathematical, 115
reliability, 116
Modeling approach, to cathode
materials development
for SOFCs, 23–26, 24f, 25f, 26f
Molecular dynamics (MD) simulations
in Nafion membranes, 317–318
for structure of hydrated aQAPS and
xQAPS, 495f, 496–497
Molten carbonate fuel cells (MCFCs), 2,
129
advances in electrode material
development, 163–187
cathode materials. *See* Cathode
materials, in MCFCs
development, 164
DIR. *See* Direct internal
reforming–molten carbonate
fuel cell (DIR–MCFC)
overview, 163–164
schematic diagram, 164f
voltage degradation, 166f
Molten carbonates, in DCFC, 129, 133,
136–138, 139f
carbon oxidation mechanism in, 138f,
139f
HDCFC. *See* Hybrid DCFC (HDCFC)
melting temperatures, 137–138, 137t
schematics, 137f
Molten hydroxide, in DCFC, 134–136,
135f
Molten metal, in DCFC, 147–148, 148f, 149f
Molten salts DCFC, 130, 134–139
molten carbonates, 133, 136–138, 139f

carbon oxidation mechanism in, 138f, 139f

melting temperatures, 137–138, 137t

schematics, 137f

molten hydroxide, 134–136, 135f

Molten Sb anode, in DCFC, 147

Molten tin anode, in DCFC, 148, 149f

Molybdophosphoric acid (MoPh-a), Nafion membranes and, 365

MoPh-a (molybdophosphoric acid), Nafion membranes and, 365

MOR. *See* Methanol oxidation reaction (MOR)

Morphology models, of Nafion membranes, 321–326

    cluster-network model, 322–323, 323f

    core-shell model, 324

    Gebel's model, 323, 324

    ionomer peak, 322, 325

    lamellar model, 324

    local-order model, 324

    matrix peak, 321–322

    parallel inverted-micelle cylinder model, 324–325, 325f

    revised parallel cylindrical water channel model, 325–326, 325f

    sandwich-like model, 324

    SAXS profile, 317, 321, 322, 322f, 324, 325

    three-phase model, 324

MSOFCs. *See* Micro SOFCs (μSOFCs)

Multiscale computational simulations, in Nafion membranes, 317–321, 321f

    DPD simulation, 318–319

    hydration level, 317–318, 318f, 319f

    MD simulations, 317–318

    objectives, 317

    SCMF simulation, 318–319

    SSC PFSA membranes, 316, 319, 320f

Multi-segment nanowires, 419–421

    current density *vs.* potential, 421f

    FE-SEM images of, 419f

    Pt and Ru segments, 419–421, 419f, 420f

    synthesis, 420–421

    XRD profiles, 420f

Multi-walled carbon nanotubes (MWCNTs), 203, 206–207, 211, 229, 230, 230f, 231, 232, 447, 448f

MWCNTs (multi-walled carbon nanotubes), 203, 206–207, 211, 229, 230, 230f, 231, 232, 447, 448f

N

Nafion electrodes, MWCNT and SWCNT, 231–232

Nafion membrane, PEMs for DMFCs, 344, 345, 352, 353

    chemical structure of, 343f

    composite membranes

        on conductivity and methanol crossover, 366t

        modification of, 363–365

        MoPh-a and, 365

        PFA nanocomposite, 361

        PVDF *vs.*, 367t

        silica, 364–365, 365f

        and sPEEK + sPSU + PBI, comparison, 362t

        sulfonated montmorillonite composite, preparing, 367–368, 368f

        zirconium, 364

    PFCA from, 343

    thickness on methanol crossover for, 353f

Nafion membranes, nanostructures in, 315–329

    chemical structures of, 316f

    morphology models of, 321–326

        cluster-network model, 322–323, 323f

        core-shell model, 324

        Gebel's model, 323, 324

        ionomer peak, 322, 325

        lamellar model, 324

        local-order model, 324

        matrix peak, 321–322

        parallel inverted-micelle cylinder model, 324–325, 325f

        revised parallel cylindrical water channel model, 325–326, 325f

        sandwich-like model, 324

SAXS profile, 317, 321, 322, 322f, 324, 325
three-phase model, 324
multiscale computational simulations, 317–321, 321f
DPD simulation, 318–319
hydration level, 317–318, 318f, 319f
MD simulations, 317–318
objectives, 317
SCMF simulation, 318–319
SSC PFSA membranes, 316, 319, 320f
overview, 315–317, 316f
in PEMFCs, 315
proton transport with increasing water content, 326–329, 326t
Grotthus mechanism (structural diffusion), 326–327
vehicle mechanism, 326, 327
PTFE in, 315, 316f
structural evolution, 326, 326t, 327–329, 327f
types, 326
Nanocube catalysts, for DAFCs, 380–391
electrochemical performance, 385–391
alcohol oxidation on Pd nanocubes, 390–391, 390f
alcohol oxidation on Pt alloy nanocubes, 387–388
alcohol oxidation on Pt nanocubes, 386, 387, 387f
hydrogen adsorption/desorption on Pt and Pt alloy nanocubes, 386, 386f
oxygen reduction on Pt and Pt alloy nanocubes, 388–390, 389f, 390f
structure characteristics, 380–383
element analysis, 383
Pd nanocubes, SEM, XRD, SAED, and HRTEM characteristics, 382–383, 382f
Pt and Pt alloy nanocubes, TEM, XRD characteristics, 380–382, 381f
synthesis methods, 383–385
Pd nanocube synthesis, 385

Pt and Pt-M(Co, Fe, Ni, Mn, etc.) nanocube synthesis, 383–385, 384f
Nanoporous proton-conducting membranes (NP-PCM), 366
Nanorods. *See* Nanowires
Nanostructured carbon-based catalysts for low-temperature fuel cells, 223–241
BDD-supported catalysts, 236–238, 237f, 240t
carbon black–supported catalysts, 224–225
carbon gel–supported catalysts, 227–229, 228f, 240t
characteristics of, 239t–240t
CNF-supported catalysts, 232–234, 240t
CNH- and CNC-supported catalysts, 234–235, 235f, 240t
CNT-supported catalysts, 229–232, 230f, 240t
fuel cell reactions with OMCs–supported catalysts, 225–227, 227f, 239t
graphene oxide counterpart (GO)-supported catalysts, 235–236, 236f, 240t
HCS-supported catalysts, 238–239, 238f, 240t
overview, 223–224
Vulcan carbon, 225, 227, 231, 232, 233–234, 236, 239t
Nanostructured nitrogen–carbon– transition metal electrocatalysts for
PEMFCs, ORR in. *See* Polymer electrolyte membrane fuel cells (PEMFCs)
Nanostructured titanium dioxide (NSTO), 299–300
Nanostructures
in catalysts for DAFCs. *See* Direct alcohol fuel cells (DAFCs), nanostructures
in Nafion membranes. *See* Nafion membranes, nanostructures in
Nanowire networks (NWNs), 412

Nanowires
  alloy nanowires. *See* Alloy nanowires
  defined, 410
  gold, HRTEM images, 411f
  multi-segment. *See* Multi-segment
    nanowires
  Pt, SEM images, 411f
  pure metal. *See* Pure metal
    nanowires
Natural gas, for fuel cells, 9, 11
Nature, 74, 77
N-C (nitrogen-doped carbon) catalysts,
  196
NCNTs (nitrogen-doped CNTs), 196,
  203, 265, 451, 457
NEDO (New Energy and Industrial
  Technology Organization), 100
New Energy and Industrial Technology
  Organization (NEDO), 100
Nickel based cermet anode, for SOFCs,
  27–29, 27f, 28f
Nickel oxide
  ALD in, 262–263, 263f
  cathode materials in MCFC
    development, 164–179
    cerium oxide on dissolution of,
      174–175
    CoO–Ni composites, 167–171
    dissolution, 164–177
    GSC–coated, 177, 178
    LSC–coated, 177–178
    perovskite ABO_3 compounds on
      dissolution of, 177
    SEM images, 172f
    TGA analysis of sintered nickel,
      169f
    time dependence of Ni solubility,
      175f
    titanium dioxide on dissolution
      of, 176, 176f
Nicrofer 6045, 56
Nitride-supported catalysts, for
  PEMFCs, 307
Nitrogen, in PEMFCs ORR, 201–203, 204f
  nitrogen-doped carbon structures in
    different forms, 202, 202f
  quaternary nitrogen, 202, 202f
Nitrogen-doped carbon (N-C) catalysts,
  196

Nitrogen-doped CNTs (NCNTs), 196,
  203, 265, 451, 457
  in MFCs, 532–533
Nitrogen precursors, in PEMFC ORR,
  210–214, 211f, 212f, 213f
Noncarbon material–supported
  electrocatalysts, for PEMFCs,
  291–308
  boride-supported catalysts, 302, 302f,
    303f
  carbide-supported catalysts,
    303–307
    silicon carbide (SiC), 306–307, 307t
    titanium carbide (TiC), 305–306,
      305t, 306f, 306t
    tungsten carbide (WC), 303–305,
      304t
  carbon corrosion mechanism,
    292–293, 293f, 294f
  characteristics, 309t
  cost, 292, 292f
  metal oxide-based supported
    catalysts, 294–301
    Magneli phase titania, 294–295
    metal-doped titania, 295–299, 296f,
      297f, 298f, 299t
    NSTO, 299–300
    SiO_2, 301
    sulfated zirconia (S-ZrO_2), 300
    tin oxide (SnO_2)-based material,
      300–301
    tungsten oxide (WO_x), 301
  metal–support interaction on catalyst
    performance, 308
  nitride-supported catalysts, 307
  overview, 291–294, 292f, 293f, 294f
  requirements, 293, 294
Noncarbon supports, in PEMFC ORR,
  207, 208–209, 209f
Nonprecious metal catalysts (NPMCs),
  for ORR, 196–198
  activity and stability of, 197, 197f
  Co–N–C catalyst, 198
  Fe-N-C catalysts, 197
  M–N–C catalysts, 196
  N-C catalysts, 196
  PANI derived catalyst, 197, 197f, 198,
    201f, 208–209, 211, 215–216
  Pt-based catalysts, 196

in situ formed graphitized carbon
nanostructures
heating temperature, 215–216, 216f
nitrogen precursors, 210–214, 211f,
212f, 213f
transition metals, 214–215, 215f
support nanoparticles. *See*
Supporting templates
NPMCs. *See* Nonprecious metal
catalysts (NPMCs)
NP-PCM (nanoporous proton-
conducting membranes), 366
NRVS (nuclear resonance vibrational
spectroscopy), 217
NSTO (nanostructured titanium
dioxide), 299–300
Nuclear resonance vibrational
spectroscopy (NRVS), 217
NWNs (nanowire networks), 412

O

Octadecyltrichlorosilane, 271
OMCs (ordered mesoporous carbons)
supported catalysts, fuel cell
reactions with, 225–227, 227f,
239t
CMK-3 carbon. *See* CMK-3
carbons
Pt electrocatalyst, 127
One-dimensional structure catalysts,
410–424
alloy nanowires. *See* Alloy nanowires
defined, 410
gold, HRTEM images, 411f
multi-segment. *See* Multi-segment
nanowires
Pt, SEM images, 411f
pure metal. *See* Pure metal
nanowires
Ordered mesoporous carbons (OMCs)
supported catalysts, fuel cell
reactions with, 225–227, 227f,
239t
CMK-3 carbon. *See* CMK-3
carbons
Pt electrocatalyst, 127
Organic-inorganic systems, composite
membranes and, 362–364

Organic–organic system, composite
membranes and, 360–362
ORR. *See* Oxygen reduction reaction
(ORR)
Oxide ion–conducting electrolytes, in
coplanar SC-SOFC, 75
Oxygen reduction reaction (ORR), 376,
377, 380
on Ag nanowires, 418, 419
catalysts, core-shell structure and,
401–405, 402f, 403f, 404f
cathode catalysts for, PEMFCs,
273–275, 274f
hollow structure catalysts, 409–410,
410f
in MFCs, 530–531, 537, 538
in PEMFCs, 292
activity and durability, 297, 299,
301, 303, 308
mass activity of catalyst, 295, 298,
301, 302, 304, 306, 307, 307t
in PEMFCs, nanostructured
nitrogen–carbon– transition
metal electrocatalysts for. *See*
Polymer electrolyte membrane
fuel cells (PEMFCs)
on Pt and Pt alloy nanocubes,
388–390, 389f, 390f
on Pt nanowires, 416–417, 416f

P

Pacific Northwest National Laboratory
(PNNL), 52
PAFCs (phosphoric acid fuel cells), 6
Palladium (Pd)
ALD of, fuel cell applications,
259–260, 260f
nanocubes, for DAFCs
alcohol oxidation on, 390–391,
390f
SEM, XRD, SAED, and HRTEM
characteristics, 382–383, 382f
synthesis methods, 385
nanowires
EOR on, 418, 418f
propanol and isopropanol
oxidation on, 417–418, 417f
perovskite oxide anode and, 31

PANI (polyaniline) derived catalyst, 197, 197f, 198, 201f, 208–209, 211, 215–216

PaniNC (polyaniline-Nafion-Silica nanocomposite membranes), 365–366

Parallel inverted-micelle cylinder model, of Nafion membranes, 324–325, 325f

Partially fluorinated membranes, in PEMs for DMFCs, 346–348
  BAM membrane, 346, 346f
  FEP-g-PS membrane, 346–347
  radiation-grafting technique, 346–347

Particle size control, challenges to nanocatalysts in DMFCs, 466

PBI (polybenzimidazole)
  PBI-based membranes, in PEMs for DMFCs, 356–357, 357f
  phosphoric acid-doped, 507–508, 508f

Pd@Pt catalysts
  methanol oxidation on, 398, 398f
  specific activity with test conditions, 400t

PECVD (plasma-enhanced chemical vapor deposition) technique, 358

PEFCs (polymer electrolyte fuel cells), 481, 482

PEMFCs. *See* Polymer electrolyte membrane fuel cells (PEMFCs); Proton exchange membrane fuel cells (PEMFCs)

PEMs. *See* Polymer electrolyte membranes (PEMs); Proton exchange membranes (PEMs)

PEN (positive-electrolyte-negative) design, 75, 79

Perfluorinated carboxylic acid (PFCA), 343

Perfluorinated membranes, in PEMs for DMFCs, 342–346, 343f, 344t
  methanol transport in, 345
  Nafion membrane. *See* Nafion membrane, PEMs for DMFCs
  PFSA, 343, 344, 344t

Perfluorinated sulfonic acid ionomer (PFSI) membrane, 336–337

Perfluorosulfonic acid (PFSA), 343, 344, 348
  membranes, SSC, 316, 319, 320f
  typical functional monomers of commercial, 344t

Periodic conductive structures, porous ceramic proton conductors and, 513–515, 514f

Perovskite ABO₃ compounds, cathode materials
  in MCFC development, 177

Perovskite oxide anode, for SOFCs, 29–33, 30t

Perovskite oxides, for SOFCs, 16–17, 17f

PES (porous electrolyte-supported) third-generation μT–SOFCs, 107–108, 109f, 110f

PFA (polyfurfuryl alcohol), Nafion nanocomposite membranes, 361

PFCA (perfluorinated carboxylic acid), 343

PFSA (perfluorosulfonic acid), 343, 344, 348
  membranes, SSC, 316, 319, 320f
  typical functional monomers of commercial, 344t

PFSI (perfluorinated sulfonic acid ionomer) membrane, 336–337

PGC (porous graphitized carbon), 238

PHCSs (porous hollow carbon spheres), 238

Phosphoric acid–doped PBI, 507–508, 508f

Phosphoric acid fuel cells (PAFCs), 6

Phosphoric acid–functionalized inorganic solids, 508–509, 509f

Photoresist molding with thermosetting polymer (PRM-TP), 77

Physical vapor deposition (PVD), 80

Planar μSOFCs, thin film based. *See* Thin film planar μSOFCs

Planar SOFCs, seals for. *See* Seals, for planar SOFCs

Plasma-enhanced chemical vapor deposition (PECVD) technique, 358

Platinum (Pt)
  ALD of, fuel cell applications, 258–259, 259f

as catalysts and electrodes, 268,
269, 270, 271f
on WC substrates, 273–275, 274f
alloy nanocubes, for DAFCs
alcohol oxidation on, 387–388
hydrogen adsorption/desorption
on, 386, 386f
oxygen reduction on, 388–390,
389f, 390f
TEM, XRD characteristics,
380–382, 381f
in MFCs
Pt-based cathodes, 537
Pt-free cathodes, 537–539, 538f
nanocubes, for DAFCs
alcohol oxidation on, 386, 387,
387f
hydrogen adsorption/desorption
on, 386, 386f
oxygen reduction on, 388–390,
389f, 390f
Pt and Pt-M(Co, Fe, Ni, Mn, etc.)
nanocube synthesis, 383–385,
384f
TEM, XRD characteristics,
380–382, 381f
nanoparticles on reduced GO,
235–236, 236f
nanowires
EOR on, 416
ORR on, 416–417, 416f
SEM images, 411f
Pt–based catalysts
catalytic activity, 444
morphology of, 297
for ORR, 196
in PEMFCs, 292
PLD (pulsed laser deposition), 80, 81,
81f
PMDF (pyrolyzed medium density
fiberboard), 150
PNNL (Pacific Northwest National
Laboratory), 52
Polyaniline (PANI) derived catalyst,
197, 197f, 198, 201f, 208–209, 211,
215–216
Polyaniline-Nafion-Silica
nanocomposite membranes
(PaniNC), 365–366

Polybenzimidazole (PBI)
PBI-based membranes, in PEMs for
DMFCs, 356–357, 357f
phosphoric acid-doped, 507–508, 508f
PolyFuel, Inc., 336, 354–355
Polyfurfuryl alcohol (PFA), Nafion
nanocomposite membranes, 361
Polyhedral and truncated cubic Pt
nanoparticles, 426–427, 426f
Polymer electrolyte fuel cells (PEFCs),
481, 482
Polymer electrolyte membrane (PEM), 72
Polymer electrolyte membrane fuel cells
(PEMFCs), 223, 229
CNFs for, 233–234
Nafion membranes in, 315, 316f
Polymer electrolyte membrane fuel cells
(PEMFCs), ORR in, 195–216
nitrogen, roles, 201–203, 204f
nitrogen-doped carbon structures
in different forms, 202, 202f
quaternary nitrogen, 202, 202f
NPMCs for. *See* Nonprecious metal
catalysts (NPMCs)
overview, 195–198
in situ formed graphitized carbon
nanostructures, 209–216, 210f
heating temperature, 215–216, 216f
nitrogen precursors, 210–214, 211f,
212f, 213f
transition metals, 214–215, 215f
supporting templates, roles, 204–209
carbon black nanoparticles,
204–205
carbon nanotubes, 205–207, 206f
graphene and graphene oxide, 207,
208f
noncarbon supports, 207, 208–209,
209f
transition metals, roles, 198–201
cobalt catalysts, 199
EXAFS spectra, 199–200, 200f
iron catalysts, 199
PANI-derived catalysts and, 201f
polarization curves of oxygen
electrodes, 199f
Polymer hollow fiber (HF), 106–107, 107f
Polymerization, for APE synthesis,
484–485

Polymer matrix, stiffness
  in PEM development for DMFCs, 337,
    338
Polymer membranes,
    hydrocarbon-based
  in PEMs for DMFCs, 348–357
    arylene-based polymers, 351–353,
      352f
    cross-linked, 355
    polyphosphazene-based
      polymers, 351, 351f
    sulfonated phenol-formaldehyde
      polymer, 350
    sulfonation methods of, 349–350
Polyoriented Pt nanoparticles, 427, 427f
Polyphosphazene-based polymers,
    hydrocarbon membranes, 351,
    351f
Polytetrafluoroethylene (PTFE), Nafion
    membranes and, 315, 316f
Polyvinylidene fluoride (PVDF)
    membranes, 360, 361
  modified, 366
  Nafion, 367t
Porod's law, 322
Porous ceramic proton conductors
  periodic conductive structures and,
    513–515, 514f
Porous electrolyte-supported (PES)
    third-generation µT–SOFCs,
    107–108, 109f, 110f
Porous graphitized carbon (PGC), 238
Porous hollow carbon spheres (PHCSs),
    238
Portable DMFCs
  examples, 339t
  requirements for, 335–336
Positive-electrolyte-negative (PEN)
    design, 75, 79
Potassium, deterioration of catalyst in
    DIR-MCFC, 183
Precursors, for ALD, 256, 257
PRM-TP (photoresist molding with
    thermosetting polymer), 77
Propanol oxidation, on Pd nanowires,
    417–418, 417f
Proteus vulgaris, in MFCs, 532
Proton-conducting electrolytes, in
    coplanar SC-SOFC, 75

Proton conductors for PEMFCs, high-
    temperature inorganic. *See*
    Inorganic proton conductors
    for PEMFCs
Proton exchange membrane fuel cells
    (PEMFCs), 2, 6, 6f, 8, 11, 444
  high-temperature inorganic proton
    conductors for. *See* Inorganic
    proton conductors for PEMFCs
  noncarbon material–supported
    electrocatalysts for. *See*
    Noncarbon material–
    supported electrocatalysts
Proton exchange membrane fuel cells
    (PEMFCs), ADL for, 251, 265,
    272–277
  anode catalysts for
    DHFCs, 275, 275f
    DMFCs, 275, 276–277, 276t, 277f,
      278f
  classification, 272
  ORR, cathode catalysts for, 273–275,
    274f
  reaction mechanisms for, 272–273
Proton exchange membranes (PEMs),
    for DAFCs, 335–368
  for DMFCs. *See* Direct methanol fuel
    cells (DMFCs)
  overview, 335–337
  roles and requirements, 337–342
  technical approaches, 342–368
    composite membranes. *See*
      Composite membranes
    hydrocarbon membranes. *See*
      Hydrocarbon membranes
    ionomers, modifications of,
      358–368
    partially fluorinated membranes,
      346–348
    perfluorinated membranes,
      342–346, 343f, 344t
Proton exchange membranes (PEMs)
    electrolyte, 3, 6, 6f
Protonic acids in liquid phase,
    inorganic, 507–509
  acid-doped polymer membranes,
    507–508, 508f
  antimonic acid as additives in
    polymers, 509

phosphoric acid–functionalized inorganic solids, 508–509, 509f
Proton transport of Nafion
with increasing water content, 326–329, 326t
Grotthus mechanism (structural diffusion), 326–327
vehicle mechanism, 326, 327
Proton transports, in PEMs, 337
Pseudomonas aeruginosa, in MFCs, 534
PTFE (polytetrafluoroethylene), Nafion membranes and, 315, 316f
Pt@Pd catalysts, methanol oxidation on, 399
Pulsed laser deposition (PLD), 80, 81, 81f
Pure metal nanowires, 412–419
Ag nanowires, ORR on, 418, 419
durability, 415–416, 415f
electrocatalytic performance of, 415–419
morphology, 412
Pd nanowires
EOR on, 418, 418f
propanol and isopropanol oxidation on, 417–418, 417f
Pt nanowires
EOR on, 416
ORR on, 416–417, 416f
synthesis, 412–414
hard template method, 412, 413–414, 414t
soft template method, 412–413, 413t
PVD (physical vapor deposition), 80
Pyrochlore oxides, anode material, 33–34
Pyrolyzed medium density fiberboard (PMDF), 150

Q

QA (quaternary ammonium) groups
chemical stability, 498, 499f
degradation of, 497, 498f, 499f
QAPS (quaternary ammonia polysulfone)
*a*QAPS. *See* Self-aggregated QAPS (*a*QAPS)
synthesis, 488–489, 489f

*x*QAPS. *See* Self-cross-linked QAPS (*x*QAPS)
Quaternary ammonia polysulfone (QAPS)
*a*QAPS. *See* Self-aggregated QAPS (*a*QAPS)
synthesis, 488–489, 489f
*x*QAPS. *See* Self-cross-linked QAPS (*x*QAPS)
Quaternary ammonium (QA) groups
chemical stability, 498, 499f
degradation of, 497, 498f, 499f
Quaternary ammonium polystyrene, 482–483
Quaternary nitrogen, 202, 202f

R

Radiation-grafted poly(tetrafluoro-ethylene-cohexa-fluoropropylene)-g-polystyrene (FEP-g-PS) membranes, 346–347
Radiation-grafting technique, 346–347
Rare earth (RE) metal oxides, 167
Raspberry-like hierarchical Au/Pt NP assembling hollow sphere (RHAHS), ORR on, 409, 410f
RDCFCs (rechargeable DCFC), 149
Rechargeable DCFC (RDCFCs), 149
Rechargeable lithium ion batteries, in mobile electronics
thin film planar μSOFCs for. *See* Thin film planar μSOFCs
Reduced GO (RGO), Pt nanoparticles on, 235–236, 236f
Reduced graphene oxide (rGO), 207, 208f
REM (reticulated electrolyte matrix), 104–105, 104f, 105f
RE (rare earth) metal oxides, 167
Reticulated electrolyte matrix (REM), 104–105, 104f, 105f
Reverse Boudouard reaction, 150
Revised parallel cylindrical water channel model, of Nafion membranes, 325–326, 325f
RGO (reduced graphene oxide), 207, 208f
Pt nanoparticles on, 235–236, 236f

RHAHS (raspberry-like hierarchical Au/Pt NP assembling hollow sphere), ORR on, 409, 410f

Rhodoferax ferrireducens, in MFCs, 528

Rotating ring-disk electrode (RRDE) tests, 201

RRDE (rotating ring-disk electrode) tests, 201

Ru@Pt catalysts
- CO and CO-like oxidation on, 396–398, 397f
- EG oxidation on, 401
- methanol oxidation on, 398–399
- specific activity with test conditions, 400t

Ruthenium, perovskite oxide anode and, 31

S

Saccharomyces cerevisiae, in MFCs, 542

SAED, for DAFCs
- Pd nanocubes, characteristics, 382f, 383

Samarium-doped strontium cobaltite cathod, formation, 105, 105f

SARA (Scientific Applications and Research Associates, Inc.), 135

SAXS (small-angle x-ray scattering), Nafion membranes from, 317, 321, 322, 322f, 324, 325

Scanning electron microscopy (SEM), for DAFCs
- Pd nanocubes, characteristics, 382–383, 382f

sccm (standard cubic centimeter per minute), 49

SC-CNFs (stacked-cup CNFs), 233, 464–465, 465f

SCFCs (single chamber fuel cells). *See* Single chamber micro SOFCs (SC-SOFCs)

Schott Glass Corp, 91

Science, 74

Scientific Applications and Research Associates, Inc. (SARA), 135

SCMF (self-consistent mean-field) simulation, 318–319

SC-SOFCs (single chamber micro SOFCs), 73–78, 73f

advancement of, 74–78
- cell architecture and experimental setup, 77, 78f
- coplanar, 75–77, 75f, 77f
- electrode configuration, 75–76, 75f

SDBS (sodium dodecylbenzene sulfonate), 455

Sealing temperature, in planar SOFCs, 52, 53f

Seals, for planar SOFCs, 45–62
- critical issues for design, 49–61
  - chemical stability, 55–57, 56f
  - CTE match, 50–52, 50f, 51t
  - electrical resistance, 59
  - glass volatility, 57–58, 58t
  - process, crystallization, 59–61, 60f
  - properties, 49–59
  - sealing temperature, 52, 53f
  - thermomechanical stability, 52, 53–55, 54f
- hermeticity, 47–49, 48t
- overview, 45–46
- techniques, 46–47
  - compressive, 46
  - glass and glass-ceramic, 46

SEBS (styrene–ethylene–butylene–styrene) membrane, 350

SECA (Solid State Energy Conversion Alliance) program, 13

Second-generation μT–SOFCs, anode-supported. *See* Anode-supported second-generation μT–SOFCs

Sediment MFCs, 526f, 527

SEIRAS (surface-enhanced infrared reflection absorption spectroscopy), 398

Self-aggregated QAPS (*a*QAPS), 488–489, 489f
- MD simulations, structural insights, 495f, 496–497
- remarkable performance, 489–493
  - ionic conductivity, 492–493, 492f, 493f
  - membrane swelling degree, 489–492, 491f
  - synthetic route of, 491f
  - structural feature, experimental proof for, 494–496, 495f

Self-aggregating design, of APEs
    *a*QAPS. *See* Self-aggregated QAPS
      (*a*QAPS)
Self-consistent mean-field (SCMF)
    simulation, 318–319
Self-cross-linked QAPS (*x*QAPS),
    488–489, 489f
    MD simulations, structural insights,
      495f, 496–497
    remarkable performance, 489–493
      ionic conductivity, 492–493, 492f,
        493f
      membrane swelling degree,
        489–492, 491f
      synthetic route of, 490f
    structural feature, experimental
      proof for, 494–496, 494f
Self-cross-linking design, of APEs
    *x*QAPS. *See* Self-cross-linked QAPS
      (*x*QAPS)
Self-limiting growth mechanism, in
    ALD, 254–255
SEM (scanning electron microscopy), for
    DAFCs
    Pd nanocubes, characteristics,
      382–383, 382f
SFRP (stable free radical
    polymerization) techniques,
    356
Shewanella putrefaciens IR-1, in MFCs,
    528, 539
Short-side-chain (SSC), PFSA
    membranes, 316, 319, 320f
SiC (silicon carbide)-supported catalysts,
    for PEMFCs, 306–307, 307t
Silicon carbide (SiC)-supported
    catalysts, for PEMFCs, 306–307,
    307t
Silicon wafer, free-standing electrolyte
    on
    fabrication, thin film planar μSOFCs
      and, 82–84, 83f
Silver (Ag) nanowires, ORR on, 418, 419
Single chamber fuel cells (SCFCs). *See*
      Single chamber micro SOFCs
      (SC-SOFCs)
Single-chamber MFCs, 525–527, 526f
Single chamber micro SOFCs
    (SC-SOFCs), 73–78, 73f

advancement of, 74–78
    cell architecture and experimental
      setup, 77, 78f
    coplanar, 75–77, 75f, 77f
    electrode configuration, 75–76, 75f
Single-walled carbon nanotubes
    (SWCNTs), 229, 231, 232, 447,
    448f
Single-walled CNH (SWCNH), 234
SiO_2-supported catalysts, for PEMFCs,
    301
6FCN-35, poly(arylene ether
    benzonitrile) membrane,
    351–353, 352f, 353f
Small-angle x-ray scattering (SAXS),
    Nafion membranes from, 317,
    321, 322, 322f, 324, 325
SMSI (strong metal–support
    interaction), 294
SnO_2 (tin oxide)-based material-
    supported catalysts, for
    PEMFCs, 300–301
Sodium, cathode materials
    in MCFC development
    Li/Na, molten carbonate melts,
      166–167
Sodium dodecylbenzene sulfonate
    (SDBS), 455
SOFCs. *See* Solid oxide fuel cells
    (SOFCs)
Soft template method, for nanowires
    synthesis, 412–413, 413t,
    421–422
Sol-gel method, 366, 513, 515
Solid acid composites, inorganic,
    509–512
    $A(H,D)_2PO_4$-type (A = K^+, Rb^+, Cs^+,
      Tl^+, etc.) composites, family of,
      510
    $A(H,D)_2SO_4$-type (A = K^+, Rb^+, Cs^+,
      Tl^+, etc.) family, 510–512, 511f
    metal diphosphates, 512
Solid carbon anode, 145–147, 146f
Solid electrolyte–based DCFC, 133
Solid Energy Convergence Alliance, 47
Solid oxide DCFC, 129, 133, 134, 139–150
    deposited carbon, 149
    drawback, 140
    fluidized bed concept, 142, 143f

geometry, effects on HDCFC, 153–154, 154f

HDCFC. *See* Hybrid DCFC (HDCFC)

molten metal, 147–148, 148f, 149f

other contact, 149–150

physical contact, 140–141, 141f, 142t

requirements, 139

in situ (external) gasification, 142, 143, 144f, 145f

solid carbon anode, 145–147, 146f

Solid oxide fuel cells (SOFCs), 2, 6, 6f, 11–12, 129

advanced electrode materials for, 15–35

anodes, development. *See* Anode development, in SOFCs

cathodes, development. *See* Cathodes, development

overview, 15–16

conventional, 73

micro, advances in. *See* Micro SOFCs (μSOFCs)

planar, seals for. *See* Seals, for planar SOFCs

tubular designs, 46

Solid oxide fuel cells (SOFCs), ADL for, 251, 265–272, 272f

catalysts and electrodes, 268, 269, 270, 271f

electrolytes, 266–268, 267f, 268f

ALD coating, 267

MEMS processing steps, 266, 268f

YDC, 262, 266–267, 268, 270f

YSZ, 262, 266, 267, 267f, 269f

other components, 271, 272f

Solid State Energy Conversion Alliance (SECA) program, 13

Sol-impregnation technique, cathode polarization in MCFC, 173

SPEEK (sulfonated poly ether ether ketone), hydrocarbon polymer, 359, 360, 363, 364

Sputtering method

in CNFs synthesis, 461–462, 462f

in CNTs synthesis, 456–457

in thin film planar μSOFCs, 80, 81f

radio frequency, 80

SSC (short-side-chain), PFSA membranes, 316, 319, 320f

S-SIBS (sulfonated poly (styrene-isobutylene-styrene)) membranes, 355

Stable free radical polymerization (SFRP) techniques, 356

Stack, μT–SOFCs, 110–114, 111f, 112f, 113f

Stacked-cup CNFs (SC–CNFs), 233, 464–465, 465f

Standard cubic centimeter per minute (sccm), 49

START-CD, 114

Strong metal–support interaction (SMSI), 294

Styrene–ethylene–butylene–styrene (SEBS) membrane, 350

Substrate, deposition factors of ALD, 255, 256f

Sulfated zirconia (S-ZrO$_2$)-supported catalysts, for PEMFCs, 300

Sulfonated poly ether ether ketone (SPEEK), hydrocarbon polymer, 359, 360, 363, 364

Sulfonated poly (styrene-isobutylene-styrene) (S-SIBS) membranes, 355

Sulfonation methods, of polymer membranes, 349–350

Supporting templates, in PEMFC ORR, 204–209

carbon black nanoparticles, 204–205

carbon nanotubes, 205–207, 206f

graphene and graphene oxide, 207, 208f

noncarbon supports, 207, 208–209, 209f

Surface-enhanced infrared reflection absorption spectroscopy (SEIRAS), 398

SWCNH (single-walled CNH), 234

SWCNTs (single-walled carbon nanotubes), 229, 231, 232, 447, 448f

Synthesis, of CNTs and CNFs–supported catalysts for DMFCs, 450–462

CNFs–supported catalysts, 457–462

colloid methods, 460–461

electrochemical method, 461

functionalization process,
457–459, 458f
impregnation method, 459
microwave-heated polyol method,
459–460, 460f
sputtering method, 461–462, 462f
CNTs–supported catalysts, 450–457
colloidal method, 455
electrochemical method, 455–456,
456f
functionalization process, 450–451
impregnation method, 452, 453f
microwave-heated polyol method,
453–455, 454f
sputtering deposition method,
456–457
Synthesis methods, in nanostructures
for DAFCs
core-shell structure, 394–396, 395f
hollow structure catalysts, 406, 406f
nanocubes, 383–385
Pd nanocube synthesis, 385
Pt and Pt-M(Co, Fe, Ni, Mn, etc.)
nanocube synthesis, 383–385,
384f

T

TCPB (three-component acrylic polymer
blend), 357
TEC (thermal expansion coefficient), 21
TEM (transmission electron
microscopy) images
carbon aerogels, 228f
CMK-3 carbons, structure, 227f
CNFs, structure, 233f
CNT supported catalysts, 230f
for DAFCs
alloy nanowires, 422f
core-shell structure,
characteristics, 391–392, 392f
Pt and Pt alloy nanocubes,
characteristics, 380, 381f
Pt/RGO, 236ff
of *a*QAPS membrane, 495, 495f
Temperature, deposition factors of ALD,
256, 257f
Tetrahexahedral (THH) shape platinum
nanocrystals, 427–428, 428f

Thermal cycling test, 106, 106f
Thermal expansion coefficient (TEC), 21
Thermincola ferriacetica, in MFCs, 539
Thermomechanical stability, of sealants
in planar SOFCs, 52, 53–55, 54f
THH (tetrahexahedral) shape platinum
nanocrystals, 427–428, 428f
Thin film planar μSOFCs, 78–96, 79f
advancements of technology, 84–96
AAO in, 85, 93–94, 94f
dense oxide film formation, 93–94,
93f
dry- or wet-chemical etching, 91
fabricated with anodes, 96f
fabrication steps, 91–92, 92f
Foturan substrate, 91
GDC electrolyte, 84
LSCF anodes, 89–90
LSCF cathodes, 89
nanoporous Ni-supported thin
film, 95
Pt-anode|YSZ|Pt-cathode design,
86, 87, 88f
schematic comparison, 88, 89f
structural design, 85
vanadium oxide in, 95–96
wafer with grid-reinforced, 90f
YDC interlayer, 85
YSZ electrolyte, 84–85, 84f, 87f
deposition techniques, 80–82, 81f,
82f
ALD, 81–82, 82f
CVD, 81
PLD, 80, 81, 81f
sputtering, 80, 81f
vacuum deposition technique, 80
fabrication of free- standing
electrolyte on silicon wafer,
82–84, 83f
Third-generation μT–SOFCs, PES,
107–108, 109f, 110f
Three-component acrylic polymer blend
(TCPB), 357
Three-dimensional structured
electrode, for MFCs, 541–542,
542f
Through-plane ionic conductivity, 341
TiB$_2$ (titanium boride) catalysts, for
PEMFCs, 302, 302f, 303f

TiC (titanium carbide)-supported
catalysts, for PEMFCs, 305–306,
305t, 306f, 306t
Time, deposition factors of ALD, 257
Time-programmed reduction (TPR)
profiles of electrolyte-added
catalysts, in DIR-MCFC, 185f
TiN (titanium nitride) catalysts, for
PEMFCs, 307
Tin oxide (SnO₂)-based material-
supported catalysts, for
PEMFCs, 300–301
Titania-supported catalysts, for PEMFCs
Magneli phase, 294–295
metal-doped, 295–299, 296f, 297f, 298f,
299t
Titanium boride (TiB₂) catalysts, for
PEMFCs, 302, 302f, 303f
Titanium carbide (TiC)-supported
catalysts, for PEMFCs, 305–306,
305t, 306f, 306t
Titanium dioxide
cathode materials in MCFC
development, 176
SEM micrographs of, 176f
nanostructured, 299–300
Titanium dioxide nanotubes (TONT),
299–300
Titanium nitride (TiN) catalysts, for
PEMFCs, 307
TONT (titanium dioxide nanotubes),
299–300
TPB (triple phase boundaries), 90, 133
TPR (time-programmed reduction)
profiles of electrolyte-added
catalysts, in DIR-MCFC, 185f
Transition metals, in PEMFCs ORR,
198–201, 214–215, 215f
cobalt catalysts, 199
EXAFS spectra, 199, 200f
iron catalysts, 199
PANI-derived catalysts and, 201f
polarization curves of oxygen
electrodes, 199f
Transmission electron microscopy
(TEM) images
carbon aerogels, 228f
CMK-3 carbons, structure, 227f
CNFs, structure, 233f

CNT supported catalysts, 230f
for DAFCs
alloy nanowires, 422f
core-shell structure,
characteristics, 391–392, 392f
Pt and Pt alloy nanocubes,
characteristics, 380, 381f
Pt/RGO, 236ff
of *a*QAPS membrane, 495, 495f
Triple phase boundaries (TPB), 90, 133
Tubular upflow MFCs, 526f, 527
Tungsten carbide (WC)
platinum on, ALD in fuel cell
applications, 273, 274f
supported catalysts, for PEMFCs,
303–305, 304t
Tungsten oxide (WO$_x$)-supported
catalysts, for PEMFCs, 301

U

United Technologies Corp., 75

V

Vacuum deposition technique, in thin
film planar μSOFCs, 80
Vanadium oxide, in thin film planar
μSOFCs, 95–96
Vapor deposition method (VPD), 300
Vapor phase pollution
catalytic activity, in DIR-MCFC, 180
Vegard's law, 168
Vehicular mechanisms
for proton transport of Nafion with
increasing water content, 326, 327
proton transports, 337
Virginia Polytechnic Institute, 351
Volumetric power density (VPD), 97, 97f
VPD (vapor deposition method), 300
VPD (volumetric power density), 97, 97f
Vulcan carbon, 227, 231, 232, 233–234, 236
XC-72, 225, 234, 239t, 445, 446t, 447

W

WC (tungsten carbide)
platinum on, ALD in fuel cell
applications, 273, 274f

supported catalysts, for PEMFCs, 303–305, 304t
Web of Science, 412
Wet impregnation technique, 28, 30t, 31, 32, 33, 34
W.L. Gore & Associates, 344
WO$_x$ (tungsten oxide)-supported catalysts, for PEMFCs, 301

X

XAFS (x-ray absorption fine structure), 199
XPS (x-ray photoelectron spectroscopy), 22
  for DAFCs
    alloy nanocubes, 383
    core-shell structure, characteristics, 393–394, 393f
  for *x*QAPS, 494, 494f
*x*QAPS. *See* Self-cross-linked QAPS (*x*QAPS)
X-ray absorption fine structure (XAFS), 199
X-ray diffraction (XRD), for DAFCs
  core-shell structure, characteristics, 392–393, 392f
  Pd nanocubes, characteristics, 382–383, 382f
  Pt and Pt alloy nanocubes, characteristics, 380–382, 381f
X-ray photoelectron spectroscopy (XPS), 22
  for DAFCs
    alloy nanocubes, 383
    core-shell structure, characteristics, 393–394, 393f
  for *x*QAPS, 494, 494f

X-ray spectroscopy (EDS), for DAFCs
  alloy nanocubes, 383
  core-shell structure, characteristics, 394
XRD (x-ray diffraction), for DAFCs
  core-shell structure, characteristics, 392–393, 392f
  Pd nanocubes, characteristics, 382–383, 382f
  Pt and Pt alloy nanocubes, characteristics, 380–382, 381f

Y

YDC (yttria-doped ceria), 85
  by ALD, 262, 266–267, 268, 270f
YSZ. *See* Yttria-doped zirconia (YSZ)
Yttria-doped ceramic proton conductors, 512–513
Yttria-doped ceria (YDC), 85
  by ALD, 262, 266–267, 268, 270f
Yttria-doped zirconia (YSZ), in SOFCs, 15, 18–19, 21, 46, 50
  by ALD, 262, 266, 267, 267f, 269f
  based cermet anode, 27–29, 27f, 28f
  electrolyte–based SC-SOFC, 76
  fabrication of free-standing, on silicon wafer, 82–84, 83f
  Pt-anode|YSZ|Pt-cathode design, 86, 87, 88f
  thin film planar μSOFCs, 84–85, 84f, 87f

Z

Zirconium phosphate (ZrP), Nafion membrane and, 364

Printed and bound by CPI Group (UK) Ltd, Croydon, CR0 4YY

18/10/2024

01776262-0019